Magnesium Alloys as Degradable Biomaterials

Magnesium Alloys as Degradable Biomaterials

Yufeng Zheng

CRC Press
Taylor & Francis Group
Boca Raton   London   New York

CRC Press is an imprint of the
Taylor & Francis Group, an **Informa** business

CRC Press
Taylor & Francis Group
6000 Broken Sound Parkway NW, Suite 300
Boca Raton, FL 33487-2742

First issued in paperback 2020

ISBN 13: 978-0-367-57550-2 (pbk)
ISBN 13: 978-1-4665-9804-1 (hbk)

Visit the Taylor & Francis Web site at
http://www.taylorandfrancis.com

and the CRC Press Web site at
http://www.crcpress.com

Contents

xviii Contents
xviii Contents
xviii Contents header

Preface

Biodegradable metals (BMs) are metals expected to corrode gradually in vivo, with an appropriate host response elicited by released corrosion products, then dissolve completely upon fulfilling the mission to assist with tissue healing with no implant residues (Zheng 2013). As the main force of the BM family, Mg and its alloys have been widely studied as potential biomaterials and have attracted the attention of biomaterial scientists and medical device societies. Up to the present, a full chain from material selection and properties optimization of Mg alloys, processing into semiproducts and surface treatment, device design, and manufacturing of the final devices to animal testing and clinical trials of the final Mg alloy implants have been conducted. Various medical device prototypes have been designed, including cardiovascular stents, bone staples, ACL screws, nonvascular stents, and clips, and some of them have been clinically trialed in Europe, China, and South Korea. One kind of Mg alloy device is now available on the European market as an implant for commercial sales. Every year since 2009, an international symposium on biodegradable metals for biomedical applications (http://www.biodegradablemetals.org/) has been held, and, moreover, scientific journals and patent databases have recorded a high increase in publications in this area.

The research on biodegradable Mg alloys designed as degradable metallic biomaterials has been mature, and this book aims to provide a comprehensive description of magnesium and its alloys designed as biodegradable metallic implant materials. The basic concepts of biodegradation mechanisms of Mg and its alloys, the strategy to control their biodegradation mode and rate, microstructure, mechanical property, corrosion resistance to body fluids, and in vitro and in vivo biocompatibility are introduced. Then various types of new biomedical Mg alloy systems will be enumerated with their special properties illustrated. Finally, the medical devices of biomedical Mg alloys will be designed and analyzed with numerical calculation and animal trials. The book provides a wealth of updated research information for graduate students, teachers, and research workers in the fields of materials science, biomedical engineering, and clinical medicine and dentistry and also for professionals involved in implantable medical device design and manufacturing.

I would like to thank colleagues working in the field of biodegradable metals all over the world and especially the majority of the attendants at the international symposiums on biodegradable metals from 2009 to the present. As a co-chair with Prof. Diego Mantovani, Prof. Frank Witte, and Prof. Mark Staiger, I get together with 120–150 scholars during the last week of August every year, sharing the research progress on biomedical Mg alloys with topics on new material fabrication, biodegradation, in vitro and in vivo testing, device design, and clinical trials, with a panel discussion on proposing international standardization on the materials and device evaluations. I get a lot of inspiration through listening and discussing with scholars from all over the world during the conferences.

Special thanks must given to my students at the Laboratory of Biomedical Materials and Devices, Department of Materials Science and Engineering, College of Engineering, Peking University (http://lbmd.coe.pku.edu.cn/) for helping me on the manuscript preparation, typing, and copyright licensing transaction. They are Yang Liu (Chapters 1, 4, and 9), Nan Li (Chapters 2, 12, and 14), Dong Bian (Chapters 5, 10, and 11), Yuanhao Wu (Chapters 6, 7, and 8), Zhen Zhen (Chapter 3), Xiaochen Zhou (Chapter 13), Meng Zhou (Chapter 2), and Weirui Zhou (Chapters 5 and 10). I am very grateful to the funding bodies including the National Basic Research Program of China (973 Program; Grant No. 2012CB619102), the National Science Fund for Distinguished Young Scholars (Grant No. 51225101), the National Natural Science Foundation of China (Grant No. 51431002 and 31170909), the NSFC/RGC Joint Research Scheme under Grant No. 51361165101, the Beijing Municipal Science and Technology Project (Z131100005213002), and the State Key Laboratory for Mechanical Behavior of Materials (Grant No. 20141615).

Author

Dr. Yufeng Zheng earned his PhD in materials science from the Harbin Institute of Technology, China, in 1998. From 1998 to 2004, he was assistant professor (1998–2000), associate professor (2000–2003), and full professor (2003–2004) at the Harbin Institute of Technology, China, and since 2004, he has been a full professor at Peking University in Beijing, China.

Dr. Zheng has authored or co-authored more than 300 scientific peer-reviewed articles, has been cited more than 5800 times, and has an H-index of 37. He served as a member of the editorial board of the *Journal of Biomedical Materials Research-Part B: Applied Biomaterials* (Wiley), *Journal of Biomaterials and Tissue Engineering* (American Scientific Publishers), *Materials Letters* (Elsevier), *Intermetallics* (Elsevier), *Journal of Materials Science & Technology* (Elsevier), *Frontiers of Materials Science* (Springer), *Acta Metallurgica Sinica (English Letters)* (Springer), and *Journal of Orthopaedic Translation* (Elsevier). His areas of special interest include the development of various new biomedical metallic materials (beta-Ti alloys with low elastic modulus, biodegradable Mg alloys and Fe alloys, bulk metallic glass, bulk nanocrystalline materials, etc.). Dr. Zheng has received several awards, including New Century Excellent Talents in University awarded by the Ministry of Education of China (2007) and Distinguished Young Scholars awarded by the Natural Science Foundation of China (2012).

1 Introduction

1.1 PROPERTY AND FUNCTION OF ELEMENT Mg

Magnesium (Mg) is the fourth most abundant mineral in the human body and is also essential for good health. In the human body, approximately 50% of Mg is found in bones, 49% is found inside the cells of body tissues, and organs, and only 1% exists in the blood, which is relatively constant and extremely significant for blood function. Mg is considered to play hundreds of roles in the nervous system, muscle function, and bone strengthening with some important health benefits, including regulating the relaxation and contraction of muscles and benefiting the production of proteins and even the production and transportation of energy throughout the body (Stipanuk and Caudill 2013; Gupta and Gupta 2014). According to previous research, those who drank water deficient in Mg were more likely to contract cardiovascular disease, which revealed that Mg plays an important role in preventing cardiovascular disease (Stipanuk and Caudill 2013). In addition, Mg content is considered to be connected with asthma, kidney disease, metabolic disorders, diabetes, hypertension, and a number of chronic diseases (Champagne 2008; Watson et al. 2013). Moreover, Mg salt content seems to be associated with cancer as hypomagnesemia is one of the adverse side effects that is cancer-caused (Watson et al. 2013).

Magnesium is often taken through food. The recommended daily allowance (RDA) of dietary Mg is 400–420 mg/day for men and 310–320 mg/day for women with higher requirements recommended among older age groups, and more detailed categories can be seen elsewhere (Yates et al. 1998). Although the content of serum magnesium is relatively stable, there is still a chance for patients to suffer from magnesium deficiency or hypermagnesemia. Both significant magnesium deficiency and hypermagnesemia are conditions associated with consequential morbidity and mortality, especially in patients with other comorbidities. Table 1.1 contains some take-home messages on magnesium deficiency and hypermagnesemia summarized by Ismail et al. (2013).

Mg and its alloys, possessing high specific strength, specific stiffness, a modulus similar to human bones, and unique biodegradation, are drawing more and more interest for the application of biodegradable materials.

In this book, the author would like to depict a full image of biodegradable Mg and its alloys, including the biodegradation mechanism of magnesium alloys, novel structure design, surface modification techniques, magnesium alloy systems (i.e., Mg-Ca, Mg-Zn, Mg-Sr, Mg-Ag, Mg-Li, and Mg-RE), common and novel processing methods in biodegradable magnesium alloys, finite element analysis (FEA) structure simulation, and in vivo trial progress. This chapter is a general introduction to biodegradable Mg and its alloys, hoping to provide a brief description of this specific field for better understanding.

TABLE 1.1

Take-Home Messages

Magnesium Deficiency

- Magnesium deficiency is common.
- It is underdiagnosed.
- It has clinical consequences.
- The commonly used serum magnesium is potentially flawed.
- Low serum magnesium indicates deficiency, but normal concentration must not be used to exclude deficiency and negative body store.
- Modus vivendi and medications can help in identifying individuals at risk.
- Magnesium loading test, though laborious, is reliable and informative.
- Treatment is straightforward and clinically beneficial.

Hypermagnesemia

- Clinically difficult to diagnose but biochemically easy to identify.
- Addition of magnesium to a U&E profile marginally increases cost of automated analysis.
- Neurological and CVS are common presentations.
- OTC magnesium-rich medications can be insidious causes of hypermagnesemia even in individuals with normal renal function.
- Reversible with favorable outcome upon restoration to normal magnesium concentration.

Source: Ismail, A. A., Y. Ismail, and A. A. Ismail. In R. R. Watson, V. R. Preedy, and S. Zibadi (eds.) *Magnesium in Human Health and Disease*: Springer, New York, pp. 3–34. 2013. With permission.

1.2 HISTORY OF Mg AND ITS ALLOYS FOR BIOMEDICAL APPLICATIONS

1.2.1 BEFORE 2000

Mg and its alloys have a long history as biodegradable implants. More than 200 years ago, Sir Humphrey Davy used magnesium product as biodegradable magnesium implants for the first time (Witte 2010). Generally speaking, there are two obvious stages at which Mg and its alloys are used for biomedical applications. The first booming period contributed by many surgeons started with Sir Humphrey Davy's effort. Table 1.2 is a comprehensive summary of this period made by Dr. Frank Witte, and more details can be obtained elsewhere (Witte 2010).

However, an overquick degradation rate and hydrogen evolution were two main problems that frustrated surgeons, and then they abandoned the biomedical application of magnesium although there were still reports of investigation of biomedical magnesium off and on.

1.2.2 AFTER 2000

With the improvement of science and technology materials and engineering, around the year 2000, the interest in magnesium and its alloys for biomedical

TABLE 1.2
Historical Overview of Reports on Magnesium and Its Biomedical Application in Historical Order

Author	Year	Magnesium (Alloy)	Application	Human/Animal Model	Reference
Huse	1878	Pure magnesium	Wires as ligature	Humans	Huse 1878
Payr	1892–1905	High-purity Mg	Tubes (intestine, vessel, nerve connector), plates, arrows, wire, sheets, rods	Humans, guinea pigs, rabbits, pigs, dogs	Payr 1901a,b, 1902, 1903, 1905; Payr and Martina 1905
Höpfner	1903	Pure magnesium	Magnesium cylinders as vessel connectors	Dogs	Höpfner 1903
Chlumský	1900–1905	High-purity Mg	Tubes, sheets, and cylinder intestine connector, arthoplastik	Humans, rabbits, dogs	Chlumsky 1900
Lambotte	1906–1932	Pure Mg (99.7%)	Rods, plates, screws	Humans, rabbits, dogs	Lambotte 1932
Lespinasse	1910	Metallic magnesium	Ring-plates for anastomsis	Dogs	Lespinasse et al. 1910
Groves	1913	Pure Mg, mix. of eq. part: Mg/Al, Mg/Cd, Mg/Zn	Intramedullar pegs in bone	Rabbits	Groves 1913
Andrews	1917	Pure Mg (99.99%), distilled in vacuum	Wires, clips as ligature, anastomosis	Dogs	Andrews 1917
Seelig	1924	Pure Mg (99.8–99.9%)	Wires, strips, bands	Rabbits	Seelig 1924
Glass	1925	Pure magnesium	Magnesium arrows	Humans, rats, cats	Glass 1926
Heinzhoff	1928	Dow Metal: Mg–Al6–Zn3–Mn 0.2%-wt. Elektron Mg–A 18%-wt.	Magnesium arrows	Rabbits	
Verbrugge	1933–1937		Plate, band, screws, pegs	Humans, dogs, rats, rabbits	Verbrugge 1933, 1934, 1937
McBride	1938	Mg–Mn 3%-wt., Mg–Al4–Mn 0.3%-wt.	Sheet, plate, band, screw, peg, wire	Humans, dogs	McBride 1938a,b

(Continued)

TABLE 1.2 (CONTINUED)
Historical Overview of Reports on Magnesium and Its Biomedical Application in Historical Order

Author	Year	Magnesium (Alloy)	Application	Human/Animal Model	Reference
Nogara	1939	Elektron (alloy not specified)	Rods	Rabbits	Nogara 1939
Maier	1940	Magnesium	Band, suture from woven	Humans	Maier 1940
Stone	1951	Mg–Al 2%-wt. pure magnesium	Wires for clotting	Dogs	Stone and Lord 1951
Fontenier	1975	Ind.-grade purity: Domal Mg (99.9%), T.L.H. Mg not reported Lab-grade purity: "zone fondue" Mg, R69 Mg MgMn 1.5%-wt., MgAl: GAZ 8%, GAZ 6%, GAZ 3%	Anodes for implantable batteries to feed pacemaker	Dogs	Fontenier et al. 1975
Wexler	1980	Mg–Al 2%-wt.	Wires intravascular	Rats	Wexler 1980
Hussl	1981	Pure Mg (99.8%)	Wires for hemangioma treatment	Rats, rabbits	Hussl et al. 1981
Wilflingseder	1981	Pure Mg (99.8%)	Wires for hemangioma treatment	Humans	Wilflingseder et al. 1981

Source: Witte, F. *Acta Biomaterialia*, 6 (5), 1680–92, 2010. With permission.

application arose again. In 2000, Kuwahara et al. (2000) investigated the surface reaction of AZ91, AZ31, and 3N-Mg in Hank's solution. Considering Mg is an essential element, they demonstrated the potential application of Mg and its alloys as biomaterials. In 2003, Heublein et al. (2003) fabricated a coronary stent with AE21 alloy (containing 2% aluminum and 1% rare earths). After the experiment in pigs, they reported a significant 40% loss of perfused lumen diameter between days 10 and 35 after implantation. In addition, re-enlargement caused by vascular remodeling, inflammation, and neointimal plaque area were also reported. From then on, nearly 200 years after the first trial by Sir Humphrey Davy, great interest and investigations were focused again on Mg and its alloys as biomaterials. First, researchers evaluated the commercial Mg alloys used for aerospace and transportation. Common commercial Mg and its alloys, including aluminum-zinc–containing magnesium alloys (i.e., AZ31, AZ81, AZ91), rare earth element–containing magnesium alloys (i.e., WE42, WE43, LAE), aluminum-manganese–containing alloys (i.e., AM50), and some other commercial Mg alloys, were studied for clinical trials. The reason why they first investigated such alloys is that these Mg alloys have relatively mature processing techniques and, therefore, good mechanical properties. Until now, hundreds of publications covering mechanical tests, in vitro degradation tests, cytotoxicity tests, hemocompatibility evaluation, in vivo animal tests, and human implantation trials have been available for commercial Mg alloys. There is also plenty of access to data on the results of various surface modification methods on commercial Mg and its alloys. When talking about the contribution from commercial Mg alloys as biomaterials, the biodegradable rare earth elements containing WE43 magnesium alloy stents (absorbable metal stent [AMS]) from the company Biotronik cannot be neglected. Di Mario et al. (2004), Erbel et al. (2007), and Haude et al. (2013) successively reported clinical progress in human trials on AMS from generation to generation. Despite the fruitfulness in the investigation of commercial Mg and its alloys, a potential threat still exists. That is, so long as the toxicity of common alloying elements, such as aluminum or rare earth elements, in commercial Mg alloys is confirmed, all these efforts would go in vain.

To serve as biomaterials in vivo, magnesium and its alloys should have good biocompatibility. The uncertain toxicity of commercial Mg alloying elements, such as aluminum and rare earth elements, has a potential threat that may frustrate the application of such alloys in the biomaterial field. In this case, choosing biocompatible elements, such as essential elements, as alloying elements and designing the optimized composition for new biodegradable Mg alloys with good biocompatibility seem to be the most effective ways to eliminate some fundamental problems.

So far, several new systems of biodegradable Mg alloys have been established and widely accepted. Mg-Ca, Mg-Sr, and Mg-Zn are the most investigated binary systems and lay the foundation of ternary and multielement systems. Other new systems include Mg-Ag, Mg-Li, and Mg-RE binary alloy systems. Table 1.3 is a general summary of existing investigations of binary Mg-based alloys. All these novel biodegradable alloy systems will be discussed in detail later, chapter by chapter, in this book.

TABLE 1.3
General Summary of Existing Investigations of Binary Mg-Based Alloys

Alloy System	Year	Application	Human/ Animal Model	Reference
Mg-Ca	2008	Pins	Rabbit	Z. Li et al. 2008
Mg-Zn	2009	Rods	Rabbit	He et al. 2009
Mg-Sr	2012	Rods	Mice	X. N. Gu et al. 2012
Mg-Ag	2013	/	/	Tie et al. 2013
Mg-Li	2013	/	/	W. R. Zhou et al. 2013
Mg-Y	2014	C-rings	Dogs	Luffy et al. 2014

1.3 DESIGN OF Mg AND ITS ALLOYS FOR BIOMEDICAL APPLICATIONS

1.3.1 SELECTION OF ALLOYING ELEMENTS TO MG FOR BETTER PERFORMANCE

The selection of alloying elements in biodegradable Mg alloys should be based not only on the improvement of mechanical properties, but also on the consideration of biocompatibility. Generally speaking, the addition of alloying elements can improve the strength of Mg by means of a grain-refining mechanism and solid solution strengthening. Different phases in the alloys with different solid solubility provide the possibility to control the mechanical properties with different postheat treatment. Table 1.4 shows common alloying elements utilized in biodegradable Mg alloys and their effects on properties (Witte et al. 2008).

The effects of alloying elements on the corrosion properties of binary Mg alloys in different corrosion media are summarized in Figure 1.1 (Hanawalt et al. 1942; Hort et al. 2010; Kirkland et al. 2010; M. Liu et al. 2010; Gu 2011; L. Yang et al. 2011b; Kubasek and Vojtech 2013b). It is obvious that impurity elements, such as Ni, Co, Cu, and Fe, are very detrimental for the corrosion resistance of Mg alloys. Thus the content of such impurities needs strict limits. In addition, except for the individual effect of alloying elements, some joint effect may occur when adding more than one kind of alloying elements into Mg. For example, the existence of Al would lower approximately one order of the content of Fe. However, the addition of Zn and/or Mn would increase the limit of Fe.

1.3.1.1 Calcium

Calcium is the fifth most abundant element in the earth's crust. A certain amount of Ca addition can enhance the metallurgical quality of magnesium alloys, alleviating oxidation both in the melt and casting during heat treatment. The addition of Ca can refine grains and thus improve strength and creep resistance. The reason why Ca is a good alloying element candidate in biodegradable magnesium alloys lies in its benign biological effects. Ca is an essential element in the human body, and almost 99% of its content can be found in bones and teeth. Oral intake of Ca can help develop strong bones and treat some bone-related illnesses, such as osteoporosis (Stipanuk

TABLE 1.4
Common Alloying Elements in Biodegradable Magnesium Alloys and Their Effects

Alloying Element	UTS	Ductility	Creep Resistance	Corrosion Resistance	Biocompatibility/Toxicity
Al	+	+			Neurotoxicity (El-Rahman 2003)
Ca	+		++	–	Essential element
RE			++		Heavy rare earth elements aggregating in brain (Hirano and Suzuki 1996)
Mn	+	+	+	+ (only in combination with Al)	Essential element, toxicity at high content
Zn	+	– (at high content)			Essential element, toxicity at high content
Zr	+	+			Related to liver cancer and lung cancer (Song 2007)
Sr	+	+	+		Trace element with similar effect as Ca
Li	–	+		–	No significant risk (McKnight et al. 2012)
Ag	+		+	–	Uncertain (see Chapter 9)

Source: Witte, F., N. Hort, C. Vogt, S. Cohen, K. U. Kainer, R. Willumeit, and F. Feyerabend. *Current Opinion in Solid State and Materials Science*, 12 (5–6), 63–72, 2008. With permission.
Note: ++, excellent; +, good; –, bad.

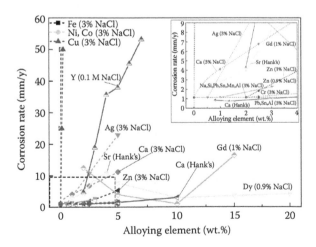

FIGURE 1.1 Effects of alloying elements on corrosion resistance of binary Mg alloys.

and Caudill 2013). There have been reports that inadequate sunlight exposure and Ca intake during rapid growth at puberty can lead to hypocalcemia, hypovitaminosis D, and eventually overt rickets (Dahifar et al. 2007). Detailed information on Ca and Ca-containing biodegradable magnesium alloys will be discussed in Chapter 6.

1.3.1.2   Zinc

Zinc possesses a solid solubility of 6.2 wt.%, providing possibilities for both solid solution strengthening and age hardening. High zinc content in Mg alloys will increase the temperature interval of crystallization, reducing alloy fluidity and thus deteriorating ingot casting. Zinc is an essential trace element in the human body, involved in bone metabolism, maintaining physiological function in the body, and improving the activity of osteoblast phosphatase in bones (Nagata and Lönnerdal 2011). Nevertheless, neurotoxicity is reported when Zn is at higher concentrations (Post et al. 2008). Zn^{2+} is also considered to have antibacterial properties. Detailed information on Zn and Zn-containing biodegradable magnesium alloys will be discussed in Chapter 7.

1.3.1.3   Strontium

Strontium is in the same main group as calcium and, therefore, has similar influence to calcium when added into Mg alloys. Sr addition can enhance the alloy through precipitation strengthening and grain refinement. Sr addition can also improve the creep resistance. Due to the low solid solubility in Mg (0.11 wt.%), more Sr added into Mg alloys would aggregate along the grain boundary and embrittle the alloy. Sr is an essential trace metal element in the human body with 99% stored in bones. Sr is also a natural bone-seeking element that accumulates in the skeleton due to its close chemical and physical properties with Ca (Landi et al. 2008). Detailed information on Sr and Sr-containing biodegradable magnesium alloys will be discussed in Chapter 8.

1.3.1.4   Silver

Silver is a precious metal chemical element. Silver is quite ductile and malleable. The solid solubility of Ag in Mg is 15.14 wt.% (3.83 at.%), which is much larger compared to Ca, Sr, Zn, and many other common alloying elements in magnesium alloys. The addition of silver is considered to improve the strength of the alloys through a strong solid solution effect. Until now, the most significant trait of silver for clinical use has been its antibacterial property. The most important utility of silver currently is as a biocide to prevent infection of long-term problem sites, including burns, traumatic wounds, and diabetic ulcers (Silver et al. 2006). The highly bioactive Ag^+ can penetrate bacterial membranes rapidly, causing cellular distortion and loss of viability due to the interaction with enzymes and other proteins in bacteria. Although silver is considered hazardous for human beings for the risk of argyria and/or argyrosis, Hardes et al. (2007) and Bosetti et al. (2002) found that silver exhibits no bad effects on material biocompatibility. Detailed information on Ag and Ag-containing biodegradable magnesium alloys will be discussed in Chapter 9.

1.3.1.5   Lithium

So far, lithium is the only alloying element in degradable magnesium alloys that can change the crystal structure to body-centered-cubic (BCC) β-phase when Li content reaches to/over 10.3 wt.%, resulting in significant improvement in ductility but reduction of strength and corrosion resistance at the same time. Lithium is also widely used in biomedical application. Many kinds of drug compounds are used for treatment of psychiatric disorders (Loghin et al. 1999). However, possible teratogenic effects (Giles and Bannigan 2006) and nephrological and lung dysfunctions (Sahin et al. 2006) were reported in the literature. In 2012, McKnight et al. (2012) reported that there is no significant risk to lithium usage, hoping to encourage the continuing using of lithium for treatment of psychiatric disorders. Detailed information on Li and Li-containing biodegradable magnesium alloys will be discussed in Chapter 10.

1.3.1.6 Rare Earth Elements

Rare earth elements (REEs) refer to 17 elements, including 15 lanthanides named La, Ce, Pr, Nd, Pm, Sm, Eu, Gd, Tb, Dy, Ho, Er, Tm, Yb, Lu, and other two elements named Sc and Y. REEs have similar chemical properties. They also present a relative high solubility in Mg, which decreases dramatically with temperature lowering. Statistics data revealed that a eutectic reaction occurs between REEs and Mg at the range from 780 K to 890 K (Sato 1999). The solid solution and age-hardening effects increase with the increasing atomic number of REE.

In general, there are two kinds of REEs added into Mg alloys: Cerium mischmetal and mischmetal without Cerium. The former is a natural mixture of La, 50 wt.% Ce, and Nd, and the latter is composed of 85 wt.% Nd and 15 wt.% Pm. REEs possess poor diffusivity, which can improve strength and creep resistance at high temperature. When adding two or more REEs into Mg alloys, the REEs can reduce each other's solubility, resulting in an additional strengthening effect. For Mg alloys used for biomedical applications, the addition of REEs or mischmetal into Mg is supposed to increase the mechanical property, enhance the corrosion resistance, and improve the biocompatibility in the physiological environment. However, it is reported that heavy rare earth elements would aggregate in the brain (Hirano and Suzuki 1996). Therefore, the uncertain toxicity of REEs needs further investigation. Detailed information on RE and RE-containing biodegradable Mg alloys will be discussed in Chapter 11.

1.4 COMPARISON WITH OTHER BIOMATERIALS

1.4.1 Comparison with Bioinert Metallic Materials

Before the introduction of biodegradable Mg alloys, metallic materials used in biomedical applications were mostly bioinert, including biomedical pure metals (i.e., pure Ti, precious metal), biomedical stainless steel (i.e., 304L and 316L), biomedical Co-based alloys (i.e., Co-Cr-Mo alloy and Co-Ni-Cr-Mo alloy), biomedical Ni-Ti alloys, and biomedical Ti alloys (i.e., Ti6Al4V). All of these metallic materials are developed under the idea of being bioinert, staying within the human body. That is, all the materials possess good corrosion resistance and chemical stability. Chemical stability requires the materials to maintain stability and exhibit no change when body fluid changes and calls for good corrosion resistance, antiswelling property, nontoxicity, and durability when implanted into the human body.

Metallic materials have been used as long-term implants for their good mechanical properties. For example, Co-Ni-Cr-Mo alloy is used to fabricate hip joints and knee joints because of its high hardness, good abrasion resistance, and corrosion resistance. Ni-Ti alloys are used to treat cardiovascular diseases for their shape memory effect as well as good strength and ductility. Ti-based alloys and biomedical stainless steels are fabricated to bone screws and plates for bone fixation.

However, traditional biomedical metallic materials (i.e., stainless steel and Ti alloys) have their shortcomings. Common metallic materials for bone implantation possess higher Young's modulus than human bones, which may cause stress shielding, resulting in reduction of bone density. And as for patients with bone fixation, two surgeries, one for implantation and one to remove the implant after it is healed, are required. It increases both the economic burden and pain of patients. Another issue during the application of traditional biomedical metallic materials is the corrosion in vivo. Metallic ions and particles produced by corrosion may cause inflammation and therefore postpone the healing process.

In comparison with traditional biomedical metallic materials, biodegradable Mg alloys have some advantages. Key points of both traditional bioinert metallic materials and biodegradable Mg alloys are summarized in Table 1.5.

Magnesium is essential for good health, that is, so long as the alloying elements are biocompatible. Thus the biocompatibility of Mg alloys is much easy to guarantee. Moreover, biodegradable Mg alloys have the unique feature of biodegradation. That is, when implanted into the human body, magnesium alloy implants can degrade along with the bone/vascular repair process. In addition, the interaction between the tissues and the alloys can stimulate osteogenesis during the degradation when used in orthopedic surgery, which is totally different with traditional bioinert metallic materials. Ideally, there can be a perfect match between the degradation of the alloys and the repair process. Therefore, a second surgery aimed at removing the implants is no longer needed. In addition, Mg has a density (1.74–2.0 g/cm^3) and Young's modulus (41–45 GPa) more

TABLE 1.5
Key Points of Traditional Metallic Materials and Biodegradable Magnesium Alloys

Research Key Points	Traditional Metallic Materials	Biodegradable Magnesium Alloys
Composition design	Alloying elements in Fe alloys Alloying elements in Ti alloys	Alloying elements in Mg alloys
Mechanical property feasibility	Long-term strengthening and toughing, mechanical stability, fatigue resistance, and wear resistance	Strengthening and toughening at first stage Integrity and function degeneration with tissue repair
Degradation feasibility	Bioinert: improvement of corrosion resistance	Controlled degradation: artificial control of degradation rate to fit time and space for tissue repair
Biocompatibility feasibility	Dissolving ions: quantity control for bioinert	Pursuit: nontoxic, bioactive dissolving ions beneficial for tissue repair

close to that of natural bone (1.8–2.1 g/cm^3, 3–20 GPa) compared with stainless steel (7.9–8.1 g/cm^3, 189–205 GPa) and Ti alloys (4.4–4.5 g/cm^3, 110–117 GPa) (Staiger et al. 2006). Thus, Mg-based implants can avoid stress shielding effectively. The fact that Mg alloys have little influence on magnetic resonance imaging (MRI) and computed tomography (CT) also make them a good candidate for biomedical application.

1.4.2 COMPARISON WITH BIODEGRADABLE CERAMICS AND POLYMERS

1.4.2.1 Biodegradable Ceramics

The concept of biodegradable materials has for a long time referred to biodegradable ceramics and biodegradable polymers before the emergence of biodegradable metals. The investigation of resorbable bioceramics can be dated from the 1980s. Typical resorbable bioceramics include calcium sulfate (Plaster of Paris), tricalcium phosphate, calcium phosphate salts, and bioactive glasses (Hench 1998). To achieve good bioactivity and degradable properties, such ceramics usually have a porous microstructure. However, a porous microstructure will reduce the mechanical properties. Therefore, resorbable bioceramics generally possess relatively low strength, low fatigue strength, and high brittleness. These limited properties mean absorbable bioceramics (i.e., β-TCP) are applied in hard tissue replacement only when low mechanical properties are required. The degradation mechanism of absorbable bioceramics is commonly considered with the dissolution in body fluids and the phagocytosis of small particles by macrophages.

1.4.2.2 Biodegradable Polymers

Biodegradable polymers can be classified into two groups. The first is natural-based materials, including polysaccharides (chitosan, starch, alginate), proteins (silk, fibrin gels, collagen), and some biofibers, and the second is synthesized biodegradable polymer, such as PU, PLA, PGA, PLLA, and PMMA (Rezwan et al. 2006). Typically, biodegradable polymers are used for temporary organ replacement, drug-delivery systems, absorbable sutures, and surgical dressing.

As for bulk degradable polymers, the mechanical properties and biodegradation time are summarized in Table 1.6 (Rezwan et al. 2006).

1.4.2.3 Comparison

Compared to biodegradable ceramics and polymers, biodegradable Mg alloys, as a matter of metallic material, possess better mechanical properties, which are critical for bone implantation and stent application. For example, mechanical property deficiency may cause a second fracture. In addition, the degradation products of biodegradable polymers generally possess acidity, which can delay healing and induce inflammation. Key research points for comparison of three kinds of biodegradable materials are shown in Table 1.7.

The degradation mechanisms are different among biodegradable ceramics, polymers, and Mg alloys.

For biodegradable ceramics, taking β-TCP for example, two opinions are generally accepted. One is that ceramics decompose to particles in body fluids, and then the particles are engulfed and transferred by cells. The other theory demonstrates

TABLE 1.6
Biocompatible and Biodegradable Polymers Used as Scaffold Materials

Polymer	Compressive[a] or Tensile Strength (MPa)	Modulus (GPa)	Biodegradation Time (Months)	References
PDLLA	Pellet: 35–150 Film or disk: 29–35	Film or disk: 1.9–2.4	12–16	Lu and Mikos 1999; Middleton and Tipton 2000; Yang et al. 2001
PLLA	Pellet: 40–120[a] Film or disk: 28–50 Fiber: 870–2300	Film or disk: 1.2–3.0 Fiber: 10–16	>24	Middleton and Tipton 2000; Yang et al. 2001
PGA	Fiber: 340–920	Fiber: 7–14	6–12	Kuo and Pu 1999; Yang et al. 2001; Ramakrishna 2004
PLGA	41.4–55.2	1.4–2.8	Adjustable: 1–12	Seal et al. 2001
PPF	2–30[a]		Bulk	Seal et al. 2001; Gunatillake and Adhikari 2003
PHA and blends	20–43		Bulk	Doi et al. 1995

Source: Rezwan, K., Q. Z. Chen, J. J. Blaker, and A. R. Boccaccini. *Biomaterials*, 27 (18), 3413–31, 2006. With permission.

[a] Compressive strength, otherwise tensile strength.

TABLE 1.7
Research Key Points Comparison of Biodegradable Ceramics, Polymers, and Magnesium Alloys

Research Key Points	Biodegradable Ceramics	Biodegradable Polymers	Biodegradable Magnesium Alloys
Composition design	Inorganics	Chemical elements: C, H, O, N, etc.	Alloying elements of magnesium alloys
Mechanical property feasibility	Strength: less than 100 MPa Brittle	Strength: less than 100 MPa High elongation	Strength: more than 200 MPa Elongation: 10–20%
Degradation feasibility		Full degradation in ~2 years Mechanism: swelling	Full degradation in 6–12 months Mechanism: from the surface to center
Biocompatibility feasibility	Degradation products: particles Inflammation	Degradation products: Acidic products Inflammation	Degradation products: alkaline products, metallic ions Side effects of products need further investigation

that ceramics dissolve in body fluids with the dissolving ions transferred into tissue fluids and deposited into a new phase.

As for biodegradable polymers, degradation refers to the chain scission process with polymer chain swelling to form oligomers and monomers. It is generally considered a hydrolysis or enzyme-catalyzed hydrolysis.

The degradation mechanism of biodegradable Mg alloys is the chemical reaction of the Mg matrix with water, and it is detailed in Section 1.4.3 with a comparison of biodegradable Fe and its alloys and Zn and its alloys.

1.4.3 COMPARISON WITH OTHER BIODEGRADABLE METALS

1.4.3.1 General Introduction to Biodegradable Metals

Biodegradable metals are metals expected to corrode gradually in vivo with an appropriate host response elicited by released corrosion products then dissolve completely upon fulfilling the mission to assist with tissue healing with no implant residues (Zheng et al. 2014).

In addition to Mg and its alloys, which will be detailed in this book, Fe and its alloys and Zn and its alloys are also developed as biodegradable metallic materials. Fe and its alloys can provide adequate mechanical support and good biocompatibility (Hermawan et al. 2010). However, pure Fe and Fe-based alloys possess higher Young's modulus than natural bones, which may lead to strength-shielding effects. Therefore, pure Fe and Fe-based alloys are considered only to be used for stent application. Moreover, the degradation rate of pure Fe and Fe-based biodegradable materials is too slow. Figure 1.2 is a summary of some existing Fe and its alloys investigated for biomedical application contributed by H. Li et al. (2014).

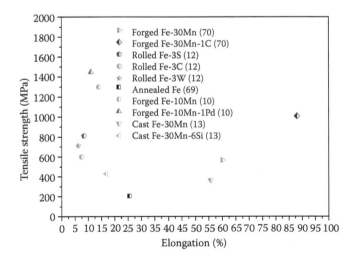

FIGURE 1.2 Mechanical property summary of existing Fe alloys. (Adapted from Li, H., Y. Zheng, and L. Qin. *Progress in Natural Science: Materials International*, 24, 414–22, 2014. With permission.)

Another new type of biodegradable metallic material attracting lots of attention is Zn-based material. Zn is considered to have a moderate corrosion rate between Fe-based materials and Mg-based materials in consideration of standard electrode potential. Moreover, Zn is an essential trace element for humans. As long as the degradation rate is reduced into a certain range, it is considered that Zn-based alloys can perform good biocompatibility. However, because Zn-based alloys are relatively new compared to Mg-based alloys or Fe-based alloys, more investigations and data are needed.

So far, many Mg-based biodegradable alloys still have an overquick degradation rate and relatively insufficient mechanical support. However, compared to Fe-based and Zn-based biodegradable alloys, the biggest advantage of Mg-based alloys is that there are more data and systematic analysis in this field. Lots of work has been done to develop new Mg-based alloy systems, surface modifications, manufacturing techniques, structure design, and optimization of Mg-based stents and degradation mechanisms in vivo. These trials and progress are all helpful to tailor the properties, making a theoretical interpretation and direction for further clinical application. Figure 1.3 is a comprehensive summary of existing research of biodegradable Mg, Fe, and Zn alloys made by H. Li et al. (2014).

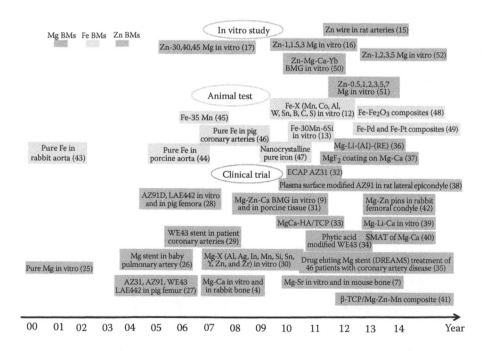

FIGURE 1.3 Summary of existing in vitro and in vivo tests of biodegradable metals. (From Li, H., Y. Zheng, and L. Qin. *Progress in Natural Science: Materials International*, 24, 414–22, 2014. With permission.)

1.4.3.2 Degradation Mechanism

Generally, four reactions (Equations 1.1 through 1.4) are involved in the degradation process in Mg and its alloys and Fe and its alloys. The degradation mechanism of biodegradable metals is shown in Figure 1.4 (Zheng et al. 2014).

$$M \rightarrow M^{n+} + ne^- \text{ (anodic reaction)} \tag{1.1}$$

$$2H_2O + 2e^- \rightarrow H_2 + 2OH^- \text{ (cathodic reaction)} \tag{1.2}$$

$$2H_2O + O_2 + 4e^- \rightarrow 4OH^- \text{ (cathodic reaction)} \tag{1.3}$$

$$M^{n+} + nOH^- \rightarrow M(OH)_n \text{ (production reaction)} \tag{1.4}$$

In a physiological environment, once in contact with the body fluid, the biodegradable metals initiate an electrochemical reaction on the interface. Metal is oxidized

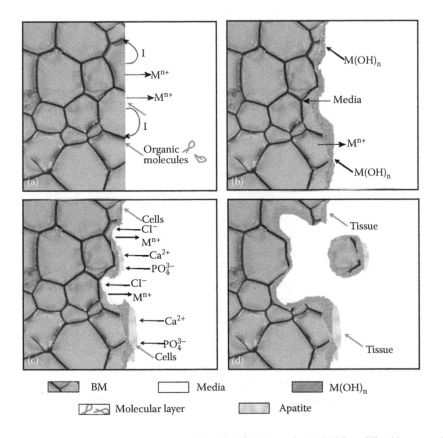

FIGURE 1.4 Degradation schematic of biodegradable metals. (a) BM is oxidized into metal cations, with organic molecules adsorbed, (b) $M(OH)_n$ corrosion products form, (c) chloride adsorption causes the breakdown of the $M(OH)_n$ protective layer and leads to pitting corrosion, and (d) cells will adhere on the surface and eroded BM may disintegrate from the whole BM matrix as irregular particles. (From Zheng, Y. F., X. N. Gu, and F. Witte. *Materials Science and Engineering: R: Reports*, 77, 1–34, 2014. With permission.)

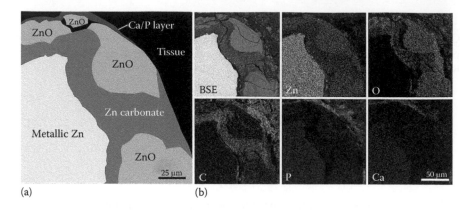

(a)   (b)

FIGURE 1.5   Schematic phase map (a), backscattered electron (BSE) image (center/top), and individual elemental maps (b) of the 4.5-month section presented. The 50-μm scale applies to the BSE image and all elemental maps. (From Bowen, P. K., J. Drelich, and J. Goldman. *Advanced Materials*, 25 (18), 2577–82, 2013. With permission.)

into metallic ions in the body fluid as an anodic reaction (Equation 1.1; Figure 1.4a). Meanwhile, organic molecules, such as proteins and lipids, start to absorb on the implant surface. For the corresponding cathodic reaction, hydrogen evolution (Equation 1.2) and oxygen consumption (Equation 1.3) occurred alternatively based on the galvanic coupling between the metal matrix and intermetallic phase or grain boundary (Figure 1.4b). As corrosion is ongoing, metallic hydroxide deposits on the surface form a protective product layer. From previous literature, physiological fluids provide a Cl-rich environment. Chloride adsorption causes the breakdown of the $M(OH)_n$ protective layer and leads to pitting corrosion, which is the main cause of alloy corrosion in biodegradable metals. Furthermore, calcium/phosphorous compounds from the body fluid attach and deposit on the surface as shown in Figure 1.4c. Cells proliferate and differentiate on this preferred calcium/phosphorous layer to further form tissues adjacent to the corrosion product layer (Figure 1.4d).

As for biodegradable Zn and its alloys, the degradation mechanism remains indistinct. Bowen et al. (2013) investigated the corrosion behavior of zinc through in vivo tests as shown in Figure 1.5.

In their research, due to local alkalization, a thin layer of zinc oxide formed at an early time and protected the matrix from further corrosion. After being implanted for 4.5 months and 6 months, the corrosion layer thickened and the composition of the layer changed. A zinc carbonate and calcium/phosphorous layer formed as shown in Figure 1.5a. However, the precise degradation mechanism needs further investigation.

1.5   GENERAL ANALYTICAL METHODS FOR BIODEGRADABLE Mg AND ITS ALLOYS

Biodegradable Mg and its alloys are newly developed metallic materials for biomedical applications. In general, there are no systematic analytical standards for the evaluations and tests. However, to better understand the mechanism and compare the results all over the world, a comprehensive evaluation is recommended, including chemical definition; microstructural characterization; and mechanical, degradation, and biocompatibility tests. As for chemical definition, the elements and the quantities involved should be declared. In addition, other agents, such as a binder or processing aid, need to be illustrated. As for microstructural characterization, whether the material is cast, rolled, or extruded requires documentation. The same goes for the presence of heat treatment and other special methods. In addition, in this part, the confirmation of grain size and the distribution, number, and size of the second phase is needed. As for property evaluation, such as mechanical tests, corrosion tests, cytotoxicity tests, hemocompatibility tests, and animal tests, recommended routines and matters that need attention will be given in this section.

1.5.1 CHEMICAL AND MICROSTRUCTURAL CHARACTERIZATION

Before the analysis of the material, the chemical composition of the resulting alloys needs to be rationed according to ASTM-B954 and/or ISO 3116 and is recommended to be measured by inductively coupled plasma atomic emission spectrometry (ICP-AES) or other similar equipment.

1.5.1.1 Metallographic Observation and Scanning Electron Microscope (SEM) Observation

Metallographic observation and SEM observation are the most common observation methods in material investigation because the grain distribution, grain size, and grain boundary area all have significant influence on the property of alloys. The general analytical method is as follows. The samples are ground with SiC emery papers of 400 grit, 800 grit, 1200 grit, and 2000 grit, successively. A chemical polishing solution (composition: 20 ml glycerinum, 2 ml hydrochloric acid, 3 ml nitric acid, and 5 ml acetic acid) can also be used for a mirror-like surface. As Mg-based alloys easily corrode, we recommend 3% nitric acid as an etchant to obtain a clear metallographic structure. Then the samples are washed with deionized water and dried. This method is a general sample preparation for all the experiments.

The prepared sample observation is via metalloscope or scanning electron microscope (SEM). Energy dispersive spectrometer (EDS) analysis is also needed for element analysis.

1.5.1.2 X-Ray Diffraction (XRD) Characterization

Phase composition can be confirmed through XRD patterns. The recommended scan range and scan rate are 10–90° and 4°/min, respectively. Because the alloying elements added to Mg alloys are usually in small amounts, with the scan rate of 4°/min, some phases in small amounts can be detected, which may be neglected under the scan rate of 8°/min.

1.5.2 MECHANICAL PROPERTY

First of all, all of the mechanical property experiments require at least three duplicates for statistical analysis to assure reliability.

1.5.2.1 Tension

The size of the tensile sample is processed according to ASTM E8-09, ASTM B557M, which is derived from ASTM E8 or ISO 6892-1. We recommend that the tensile samples are ground with SiC emery papers of 400 grit, 800 grit, and 1200 grit, successively, before the tensile experiment. To our knowledge, a different speed of testing will result in different tensile strength. According to ASTM E8-09, because most biodegradable Mg alloys can reach an elongation greater than 5%, when determining only the tensile strength or after the yield behavior has been recorded, the speed of the testing machine shall be set between 0.05 and 0.5 mm/mm (or in./in.) of the length of the reduced section (or distance between the grips for specimens not having a reduced section) per minute. Alternatively, an extensometer and strain rate indicator may be used to set the strain rate between 0.05 and 0.5 mm/mm/min (or in./in./min). Moreover, the impact of temperature should also be considered; 37°C is the best temperature utilized to monitor body temperature. Otherwise, room temperature, 25°C, is also acceptable.

1.5.2.2 Three-Point Bending

Not only for stent application, but also for orthopedic application, the bending property is important. ISO 7438-2005 and ASTM E290 both provide detailed testing guidelines for the bending test. Three-point bending is a generally used method to determine the bending strength of materials. For a typical bending test via a bending device with supports and a former, unless otherwise specified, the distance between the supports, l, shall satisfy the relationship in Equation 1.5:

$$l = (D + 3a) \pm \frac{a}{2} \qquad (1.5)$$

where l is distance between supports, a is the thickness or diameter of the test piece (or diameter of the inscribed circle for pieces of polygonal cross-section), and D is the diameter of the former.

As is the same with a tensile test sample, the sample for the bending test also needs the same pretreatment as polishing to gain a smooth surface.

Currently, a three-point bending test can be conducted via a universal mechanical testing machine. Thus the sample preparation may be more important.

1.5.2.3 Wear

Wear debris is a real issue when the implant is implanted into the human body. Despite that Mg alloys are not intended for load-bearing bone implants, wear debris from Mg-alloy implants can also induce inflammation. However, there is little literature concentrating on the wear properties of biodegradable Mg alloys. X. P. Zhang et al. (2007) reported an improvement of 1.5 times on a MAO-treated AZ91D sample compared to an untreated sample. Another reason we consider wear property important is its role in tribocorrosion or fretting corrosion, which is really important in traditional bioinert metallic materials. Although Mg alloys are not intended for load-bearing bone implants, considering the overquick corrosion rate, better wear property may help improve corrosion resistance.

1.5.2.4 Fatigue

Table 1.8 has some specific fatigue testing of surgical implant materials and devices according to ASTM (Teoh 2000).

Compared to higher amounts of literature on the corrosion investigation of bio-degradable Mg-based alloys, corrosion fatigue attracts little attention. In fact, corrosion fatigue strength of AZ31 and WE43 alloys is less than that of conventional metallic biomaterials, such as 316L stainless steel and Ti alloys (Gu et al. 2010c). Moreover, the fatigue strength of the Mg alloy is significantly reduced in humid environments, and the fatigue limit drops drastically in NaCl solution (Antunes and de Oliveira 2012). Therefore, if the Mg alloys are intended for plates, screws, and pins for repairing bone fracture or vascular devices, fatigue testing is critical. For orthopedic implants, the ultimate implant failure is usually associated with corrosion fatigue, which is the synergetic effect of electrochemical corrosion and cyclic mechanical loading (Azevedo 2003; Magnissalis et al. 2003). For vascular devices, the failure of a stent due to fatigue may result in loss of radial support of the stented vessel or in perforation of the vessel by the stent struts (Gu et al. 2010c).

In general, especially for commercial material testing, fatigue testing can be conducted either in air or in corrosion medium in consideration of different applications. To better mimic the in vivo environment, we recommend fatigue testing conducted in simulated body fluids (the selection of the corrosion medium will be discussed in Section 1.5.3).

TABLE 1.8
Specific Fatigue Testing of Surgical Implant Materials and Devices

Fatigue Tests	ASTM Reference
Practice for cyclic fatigue testing of metallic stemmed hip arthroplasty femoral components without torsion	F 1440-92
Test method for bending and shear fatigue testing of calcium phosphate coatings on solid metallic substrates	F 1659-95
Test method for constant amplitude bending fatigue tests of metallic bone staples	F 1539-95
Test methods for static and fatigue for spinal implant constructs in a corpectomy model	F 1717-96
Guide for evaluating the static and fatigue properties of interconnection mechanisms and subassemblies used in spinal arthrodesis implants	F 1798-97
Practice for corrosion fatigue testing of metallic implant materials	F 1801-97
Test method for cyclic fatigue testing of metal tibial tray components of total knee joint replacements	F 1800-97
Practice for constant stress amplitude fatigue testing of porous metal-coated metallic materials	F 1160-98

Source: Teoh, S. *International Journal of Fatigue,* 22 (10), 825–37, 2000. With permission.

1.5.3 BIODEGRADATION BEHAVIOR AND ION RELEASE

In vitro biodegradation is the most discussed and investigated property in biodegradable Mg alloys. The in vitro biodegradation property, and also different corrosion medium selection, different measurement methods, and the advantages and disadvantages of different measurements are summarized in Table 1.9 (Waizy et al. 2012).

1.5.3.1 Corrosion Medium Selection

To mimic the in vivo environment in vitro and gain more convincing data from in vitro corrosion tests, corrosion medium selection is the premier factor. Various solution systems have been used in the last 10 years to mimic the body fluid, including 0.9% NaCl aqueous solution, SBF (simulated body fluid), Hank's, PBS (phosphate-buffered saline), and DMEM.

Because different corrosion medium has different composition and mimics in vivo environments to varying degrees, materials would perform different corrosion behavior in these medium. The electrochemical data of AZ91 alloys in various simulated body solutions are listed in Table 1.10.

According to the literature, both inorganic ions and organic components have significant influence on the degradation procedure of metal alloys. A buffer electrolyte solution containing similar components to human blood plasma should be used for the in vitro biodegradation test (Z. Zhen et al. 2013).

1.5.3.2 Electrochemical Measurement

Electrochemical measurement is the most common corrosion test for both industrial and biomedical alloys for its convenience. Electrochemical measurement is considered to gain instantaneous data on the corrosion rate of the materials while also elucidating some of the mechanisms resulting in this corrosion behavior (Song 2005). Typically, the test includes open circuit potential (OCP), electrochemical impedance spectroscopy (EIS), and potentiodynamic polarization curves via a three-electrode system. EIS is used to establish a simulating circuit for the understanding of the corrosion mechanism. Besides, for the various examinations of surface coatings, EIS is an efficient method. A potentiodynamic polarization curve can provide quick evaluation of biodegradable Mg alloys at a specific stage. However, as the surface morphology alters with the corrosion process, several potentiodynamic polarization tests at different time points are needed to give a full image of corrosion behavior over time.

For these electrochemical experiments, the scanning rate of polarization curves and frequency of EIS are two major parameters affecting results. Another influential parameter is testing temperature, for which we recommend 37°C as the best. It is necessary to note that preventing the electrode from oxidation before the test is critical. In this case, a nitrogen atmosphere is preferred before and throughout the test process.

1.5.3.2.1 Scanning Rate of Potentiodynamic Polarization Curves

With the increase in scanning rate, J_{corr} rises, and zero current potential shifts to a more negative value. The electrode system stays approximately steady when the scanning rate is low. However, as scanning rate increases, the steady state of the

TABLE 1.9
Summary of In Vitro Corrosion Tests

Magnesium Alloy	Corrosive Media	Immersion Time (h)	Temperature (°C)
AZ31	NaCl, PBS	2, 24, 48, 72, 96, 144	37 (Alvarez-Lopez et al. 2010)
AZ31	Hank's solution	Up to 480	37 ± 0.5 (Gu et al. 2011b)
AZ31, AZ61, AZ91D, Mg	m-SBF	24, 48, 120, 192, 384, 480, 576	36.5 ± 0.5 (Z. Wen et al. 2009)
AZ31, AZ91, MgCa	Hank's solution, DMEM, and DMEM + FBS	0, 24, 48, 72, 96, 120, 144, 168	37 (Gu et al. 2009d)
AZ31, AZ91, MgCa	Hank's solution, MEM, MEM + FBS	–	37 ± 1 (Kirkland et al. 2010)
AZ31, LAE442	NaCl, PBS, PBS + albumin	–	Not mentioned
AZ61, AZ61Ca, AZ91, AZ91Ca	m-SBF	–	36.5 ± 0.5 (Kannan and Raman 2008)
AZ91	Hank's solution	–	20 ± 2 and 36.5 ± 0.5
AZ91, Mg	SBF	–	36.5 ± 0.5 (Kannan 2010)
AZ91	Four solutions with NaCl, $NaHCO_3$, K_2HPO_4, and Na_2SO_4	Up to 168	37 ± 0.5 (Xin et al. 2008)
AZ91	C-SFB, Hank's solution, PBS, NaCl, DMEM	Up to 96	37 ± 0.5 (Xin et al. 2010)
AZ91	SBF, SBF + BSA	168	37 ± 1 (Liu et al. 2007c)
AZ91D, LAE442	NaCl, PBS, PBS + albumin	–	Not mentioned
AZ91, Mg, WE54	SBF	–	36.5 ± 0.5 (Walter and Kannan 2011)
M1A	SBF, SBF + albumin	1–24	37 (Y. Wang et al. 2011)
Mg	SBF	24	37 ± 0.5 (Xin et al. 2011c)
Mg	SBF	72, 120, 168, 336, 504	37
Mg	NaCl, NaCl + HEPES, NaCl + $NaHCO_3$, Earle(+), Eagle's minimal essential medium (EMEM), E-MEM + FBS	Up to 336	37 (Yamamoto and Hiromoto 2009)
Mg, Mg–Al, Mg–Ag, Mg–In, Mg–Mn, Mg–Si, Mg–Y, Mg–Zn, Mg–Zr	SBF, Hank's solution	Up to 250	37 (Gu et al. 2009b)

(Continued)

TABLE 1.9 (CONTINUED)
Summary of In Vitro Corrosion Tests

Magnesium Alloy	Corrosive Media	Immersion Time (h)	Temperature (°C)
Mg, AZ31, MgCa0.8, Mg-Zn, Mg–Mn, Mg–1.34Ca–3Zn	EBSS, MEM, MEM + bovine serum albumin	168, 336, 504	37 (Walker et al. 2012)
Mg, AZ91, ZE41, Mg2Zn0.2Mn	Hank's solution	Up to 312	37 ± 2 (Zainal Abidin et al. 2011)
Mg, MgCa	SBF	–	Not mentioned (Wan et al. 2008b)
Mg, Mg–Zn	SBF	72, 720	37 (S. Zhang et al. 2009)
Mg, Mg–Zn	SBF	Up to 72	37 ± 0.5 (Zhang et al. 2010b)
Mg, WE43, E11	Hank's solution, DMEM, DMEM + FBS	Up to 180	37 (Feyerabend et al. 2012)
MgCa	SBF	12, 24, 36, 48, 72, 84, 96	37 (Harandi et al. 2011)
MgCa	SBF	Up to 250	37 (Z. Li et al. 2008)
MgCa	Distilled water + bovine serum albumin, NaCl, NaCl + Ab	Up to 120	37 ± 1 (C. Liu et al. 2010)
MgCa, Mg–Ca–Y	SBF, modified-MEM + FBS	0, 20, 65, 120	37 (Li et al. 2011b)
Mg–Mn, Mg–Mn– Zn, WE43	SBF	24, 48, 96, 216	37 ± 1
Mg–Mn–Zn	Hank's solution, simulated blood plasma (SBP)	Up to 288	Not mentioned (Yang and Zhang 2009)
Mg–Y–Zn, WE43	SBF (H), SBF (T), MEM, PBS	Up to 170	37 ± 2 (Hänzi et al. 2010)
Mg–Zn–Ca, Mg–Zn–Mn, Mg–Zn–Si	Ringer's solution	–	37
Mg–Zn–Ca–Mn	Hank's solution	–	37 ± 1 (Zhang and Yang 2008)
Mg–Zn–Zr	Hank's solution, DMEM, DMEM + FBS	Up to 480	37 ± 0.5 (Gu et al. 2011c)
Mg–Zn–Zr, WE43	Hank's solution	Up to 3528 (21 weeks)	37 (Huan et al. 2010a)
WE43	SBF	Up to 654	37 ± 2
WE43	NaCl, NaCl + CaCl$_2$, NaCl + K$_2$HPO$_4$, m-SFB, m-SFB + albumin	Up to 120	37 (Rettig and Virtanen 2008, 2009)

Source: Waizy, H., J.-M. Seitz, J. Reifenrath, A. Weizbauer, F.-W. Bach, A. Meyer-Lindenberg, B. Denkena, and H. Windhagen. *Journal of Materials Science*, 48 (1), 39–50, 2012. With permission.

TABLE 1.10

Electrochemical Data of AZ91 Alloys in Various Simulated Body Solutions

Alloy	Roughness	Solution	φ_{corr}/V	J_{corr}/(μA·cm^{-2})	Reference
AZ91	4000	Blood plasma	−1.53		Xin et al. 2008
		NaCl	−1.518		
		NaCl + K$_2$HPO$_4$	−1.774		
		NaCl + K$_2$HPO$_4$ + NaHCO$_3$	−1.789		
AZ91	4000	SBF	−1.836	3.75	Xin et al. 2009a
AZ91	2400	m-SBF	−1.713	65.7	Kannan and Raman 2008
AZ91D	1000	Hank's	−1.36	297	Song et al. 2008
AZ91D		0.9% NaCl	−1.528	22.56	Yao et al. 2009
AZ91E	0.5 μm diamond	Hank's	−1.593	4.927	E. L. Zhang et al. 2009

Source: Zhen, Z., T.-F. Xi, and Y.-F. Zheng. *Transactions of Nonferrous Metals Society of China*, 23 (8), 2283–93, 2013. With permission.

system is disturbed, and the speed of electron transfer is larger than that of electron consumption in the cathode reaction, which leads to the accumulation of electrons on the surface of the electrode and causes the negative shift of zero current potential (Z. Zhen et al. 2013). Therefore, the scanning rate of the potentiodynamic polarization measurement should be slow enough, and the testing time should be short. In this case, 0.5 mV/s or 1 mV/s is suggested.

1.5.3.2.2   Frequency of EIS

The frequency range has a strong influence on EIS spectra. The frequency should be low enough to make the measurement time longer if the electrode system is stable enough, or the frequency should be higher to ensure a short test time if the electrode system is active and the surface character has changed during the long test time, making the data nonsense. The range of 100 kHz to 10 mHz is generally adopted (Kannan and Raman 2008; Song et al. 2008, 2009; Xin et al. 2008, 2009a,b; S. X. Zhang et al. 2009). The lowest frequency was set at 10 mHz in order to reduce the time and potential noise interference. The upper frequency limit was set at 100 kHz to reduce the effects caused by phase shifts from the potentiostat in the high-frequency region (Xin et al. 2008).

1.5.3.2.3   Corrosion Rate Calculation via Electrochemical Test

Gravimetric measurements and electrochemical measurements are the most common methods to determine the corrosion rate in vitro for its convenience. Either the corrosion current density i_{corr} obtained from potentiodynamic polarization curves or the corrosion rate calculated from the potentiodynamic polarization curves can indicate the corrosion performance. Corrosion rate is calculated via Equation 1.6.

$$R = K \frac{J_{corr}}{\rho} m_e \qquad (1.6)$$

where R is the corrosion rate, K is 3.273×10^{-3} mm·g/(μA·cm·a), J_{corr} is the current density, and m_e is equivalent mass.

1.5.3.3 Immersion Test

The immersion test is also a typical corrosion test in biodegradable Mg alloys. There are two standards for a standard immersion test, namely ISO 10993-15 (ISO 2000; Biological evaluation of medical devices—Part 15: Identification and quantification of degradation products from metals and alloys) and ASTM G31-72 (ASTM 2004; Standard practice for laboratory immersion corrosion testing of metals). According to ISO 10993-15, the test cell should be tightly closed to prevent evaporation and maintained at 37°C for 7 days and then analyzed. Factors affecting the results include the ratio of solution volume to sample surface area (SV/SA), flow rate, and immersion time. These factors will be discussed in detail in Chapter 2. Ion release detection and calculation via weight loss/gain are common evaluating methods in the immersion test (Z. Zhen et al. 2013) and will be discussed here.

1.5.3.3.1 Ion Release Detection

Ion release is not as common a test as weight loss measuring Mg corrosion in the immersion test. The general method is to immerse Mg or Mg alloy samples into a solution for extended periods of time. Samples of the immersion solution are then collected at regular intervals, and the ionic concentrations of Mg are measured either by colorimetric techniques (xylidyl blue assay; Yamamoto and Hiromoto 2009), flame atomic absorption spectroscopy (Lu et al. 2011b), or inductively coupled plasma atomic emission spectrometry (ICP-AES; Walker et al. 2014). Compared to the former two techniques, ICP-AES can detect not only the amount of Mg^{2+}, but also other ions existing in the corroding solution. Therefore, data on the different ion amounts in the corroding solution is thought to provide a possibility for interpretation of Mg alloy corrosion mechanism (Walker et al. 2014). Meanwhile, as a specific detection technique in the immersion test, the results of ion release detection also have a connection to the environmental parameters, such as pH value and temperature. Considering its potential to provide data for a corrosion mechanism study, ICP-AES-based ion release detection is suggested.

1.5.3.3.2 Corrosion Rate Calculation of Immersion Test (Weight Loss/Gain)

Weight loss is a common calculation method to determine the corrosion rate. It is based on the measurement of weight change of samples before and after the immersion test using the following equation:

$$R = \frac{W}{At\rho} \qquad (1.7)$$

where R is the corrosion rate, W is the change of mass, A is the original surface area exposed to the corrosive media, t is the exposure time, and ρ is the standard density of the sample.

Weight loss/gain is a convenient method, but it cannot be used to analyze the corrosion mechanism because it is a macroscopic process (Champagne 2008). Many factors have influence on the results, and thus adequate duplicates are needed. Moreover, corrosion product removal is thought to be critical to obtain convincing results because ineffective removal of corrosion products will lead to a reduced indication of the extent of corrosion, and the overzealous removal of corrosion products has the potential to degrade the remaining substrate, providing inaccurately high corrosion results (Champagne 2008). So far, chromic acid (i.e., 180 g/L chromic acid) processing is a generally accepted method to remove the corrosion product because it minimally affects Mg-based biomaterials and, therefore, does not corrode the remaining substrate (Makar and Kruger 1993; Champagne 2008). Thus, the most important part aimed at obtaining convincing data is the acid pickling process. In addition, more duplicates are really needed to eliminate personal error.

1.5.3.4 Hydrogen Evolution Test

The hydrogen evolution test is a unique measurement utilized in the investigation of biodegradable Mg alloys. The unique biodegradation property comes from Equation 1.8:

$$Mg\ (s) + 2H_2O\ (aq) \rightarrow Mg(OH)_2\ (s) + H_2\ (g) \tag{1.8}$$

The hydrogen evolution test can be seen as an immersion test with a special setup. Therefore, some basic discipline, such as the ratio of solution volume to the specimen surface and sensitivity to the environment, are all the same as the immersion test. Until now, hydrogen gas is still ineluctable during the corrosion process of Mg alloys. In this case, considering the relationship of the amount of substance between the dissolving Mg and the hydrogen gas produced, the monitor of hydrogen evolution can imply the corrosion rate of biodegradable Mg alloys. Figure 1.6 is a schematic of a hydrogen evolution test setup. However, some shortcomings also exist in this test. First of all, it is hard to calculate the amount of dissolving hydrogen, which makes the results biased. Second, the corrosion mechanism is still uncertain, and it makes the calculation via the molar ratio of Mg and H_2 in Equation 1.8 undefined (Witte et al. 2008). Last but not least, some researchers have reported the possibility of H_2 escape from the apparatus especially when the burette is plastic instead of glass (Kirkland et al. 2012). Therefore, when applying the hydrogen evolution test, researchers must pay great attention and make several duplicates to make sure the results are robust.

1.5.4 Cytotoxicity (Extract Preparation)

In cytotoxicity evaluation of biodegradable Mg alloys, two advised guidelines are ISO 10993-5 and ISO 10993-12. The former is a guidance standard for in vitro

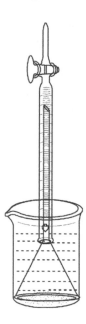

FIGURE 1.6   Schematic of hydrogen evolution test setup.

cytotoxicity tests, including the extract cytotoxicity test and the direct contact test, and the latter is the guidance standard for sample preparation and reference materials. To be frank, ISO 10993 is not established for biodegradable materials but for bioinert materials, such as stainless steel and Ti alloys. However, for a long time, the guidance standard has been used in the biocompatibility evaluation of biodegradable Mg alloys. Typical cell lines used for orthopedic applications include human cells of noncancerous origin (i.e., bone marrow-derived mesenchymal stem cells), human cells of cancerous origin (i.e., MG63), animal cells of noncancerous origin (i.e., MC3T3-E1), and animal cells of cancerous origin (i.e., L-929). As for stent application, cytotoxicity of VSMC and ECV304 cell lines are necessary. According to ISO 10993-5, common cytotoxicity tests include the neutral red uptake (NRU) cytotoxicity test, the colony formation cytotoxicity test, the MTT cytotoxicity test, and the XTT cytotoxicity test. Table 1.11 is a summary of cytotoxicity experiments for evaluation of Mg alloys for orthopedic applications made by Waizy et al. (2012).

In this section, the extract preparation and some common factors affecting the experiment results, including extract concentration, cell line selection, and the influence of protein, are discussed.

1.5.4.1   Extract Preparation

A common method to prepare the extract is as follows. First, the polished samples, as mentioned previously, are washed in acetone, ethanol, and deionized water for 10 min, successively. Second, the sample is dried at room temperature and then sterilized. Third, the sample is placed in a sterilized Petri dish, appropriate culture medium

TABLE 1.11

Survey of Cytotoxicity Experiments for Evaluation of Magnesium Alloys for Orthopedic Applications Excluding Coated Samples

Test Method	Subject of Test	Cell Lines			
		Human Cells of Noncancerous Origin	Human Cells of Cancerous Origin	Animal Cells of Noncancerous Origin	Animal Cells of Cancerous Origin
MTT assay	Coloring agent MTT (3-(4,5-dimethylthiazol-2-yl)-2,5-diphenyl tetrazolium bromide) is metabolically transformed into formazan	Human umbilical cord perivascular (HUCPV) (Feyerabend et al. 2010); Isolation of primary human osteoblasts	MG63 (Feyerabend et al. 2010; Gu et al. 2011b; Y. Wang et al. 2011)	NIH3T3 (Gu et al. 2009b); MC3T3-E1 (Gu et al. 2009b; S. Zhang et al. 2009; Datta et al. 2011); RAW 264.7 (Feyerabend et al. 2010); rabbit bone marrow stroma cells (rBMSC) 9	L-929 (Z. Li et al. 2008; Gu et al. 2009b; Zhang et al. 2010b)
XTT assay	XTT (sodium 3,3'-(1-((phenylamino)carbonyl)-3,4-tetrazolium)-bis(4-methoxy-6-nitro)benzene sulfonic acid hydrate) is transformed by living cells to a formazan product				L-929
Cell count and cell attachment	Adherent cells were stained with DAPI (4',6-diamidino-2-phenylindole dilactate) and counted via fluorescence microscope, tryptophan blue exclusion method, or other cell count methods	Bone marrow–derived mesenchymal stem cells (Johnson et al. 2012)	SaOS2 (Li et al. 2011b)	MC3T3-E1 (S. Zhang et al. 2009)	
Cell protein	Determination of the alkaline phosphatase (ALP) activity		MG63 (Gu et al. 2011c)		L-929 (Gu et al. 2011c)

Source: Waizy, H., J.-M. Seitz, J. Reifenrath, A. Weizbauer, F.-W. Bach, A. Meyer-Lindenberg, B. Denkena, and H. Windhagen. *Journal of Materials Science*, 48 (1), 39–50, 2012. With permission.

is added at a solution/surface area ratio of 1.25 cm^2/ml (different ratio according to the sample thickness can be seen elsewhere; ISO 2008), and then it is placed in an incubator for 72 h at 37°C, 95% relative humidity, 5% CO_2. The third step otherwise can only be sustained for 24 h if the culture medium contains 10% FBS. Finally, the extract is centrifuged to remove impurities to become extract stoste, which can be diluted to different concentrations for further investigation. More detail can be seen in ISO 10993-12.

1.5.4.2 Extract Concentration

As mentioned previously, ISO 10993 is established for bioinert materials not specifically designed for biodegradable materials. Therefore, the degradation process of Mg materials led to a highly concentrated extract with a very high osmolality and Mg concentration (Fischer et al. 2011). More problems, such as pH value increase and corrosion product formation, occur during the extract preparation due to the more corroding process compared to bioinert materials. In this case, a dynamic system, such as a bioreactor, can provide a better simulation of the in vivo environment. However, the extract concentration needs to be well concerned if the flow system is not available. Fischer et al. (2011) reported a tenfold diluted extract in as-cast Mg–xCa (x = 0.6, 1.0 wt.%) to still have a measurable effect on the extract, which is not caused by osmotic pressure via experience. Therefore, because more data is needed for the diluting multiple, the extract concentration simulation needs to be as close as the specific tissue planned for application.

1.5.4.3 Cell Lines or Primary Cells

To simulate the in vivo environment the most, the optimal cell line is the cell line in the tissue that is exactly prepared for the in vivo test. However, this situation is hard to achieve in most cases. Nowadays, the most focused rethink on cell selection is the utility of primary cells. Cell lines are transformed cells. If a transformed cell type is used, the "healthy" pathways may already have been altered, and any observed effects may be attributable in large part to the disease pathophysiology of the original tumor of the cell line (Staiger et al. 2011). In addition, primary isolated human osteoblasts or mesenchymal stem cells are able to produce a bone matrix, which is especially significant for the evaluation of suitable orthopedic implant materials (Fischer et al. 2011). Therefore, primary cells are recommended for the evaluation of orthopedic implant materials. Moreover, the utilization of a bioreactor to better simulate in vivo environment is also a trend and more convincing.

In addition, the influence of Mg ions on vascular endothelial cells, embryonic stem cells, adult stem cells, osteoblasts, and osteoclasts are all reported. Maier et al. (2004) reported that a supplement of Mg ions ranging from 2.0 mM to 10.0 mM has a positive effect on the proliferation of human umbilical vein endothelial cells. What's more, the expression of nitric oxide synthase is enhanced at a high Mg ion content. Nguyen et al. (2013) illustrated that embryonic stem cells could tolerate a Mg content of 10 mM, under which the proliferation of embryonic stem cells would not be affected. They also reported a pH limit of 8.1 for culture medium. Shimaya et al. (2010) found that Mg ions with a concentration of 10 mM could add an extracellular matrix of SMSCs and promote differentiation to cartilage cells through activation of some integrin-related signal pathways. Recently, Dai's group reported a strong inhibition of RANKL-induced osteoclast differentiation caused by Mg degradation products (Zhai et al. 2014). They also confirmed that the inhibition process is achieved by inhibition of NF-kappaB and NFATc1 signal pathways.

1.5.4.4 Protein Effect

In fact, the best culture medium is human blood, which is complex and contains both inorganic and organic components. However, considering the ethical and economical issues, human blood cannot be utilized on a large scale. Scheideler et al. (2013) evaluated the difference between common cell culture medium and 100% serum. The result revealed that, among most of the investigated alloys, metabolic activity was significantly reduced when extraction of Mg alloys was performed in serum. Therefore, to avoid ethical and economical problems, cell culture medium with a more similar content of serum is needed and recommended in consideration of the mismatch of in vivo and in vitro cytotoxicity test results.

Apart from the factors above, other common parameters, such as pH value and osmotic pressure, all affect in vitro cytotoxicity test results. It needs to be noted that the relatively high activity of Mg^{2+} also needs to be concerned. For example, it has been found that Mg ions interfere with the assay by promoting the conversion to formazan, which affects formazan-related experiments, such as XTT and MTT. Therefore, BrdU is a more appropriate test for the cytotoxicity of Mg materials (Fischer et al. 2010). Thus, the experiments need to be carefully designed.

As for in vitro cytotoxicity tests, the tests can simulate different magnesium ion concentrations, alkaline environments, and other biological effects through careful design. However, the human body is a complex environment. The best simulation of the in vivo environment with quick results obtained at the same time seems more important.

1.5.5 HEMOCOMPATIBILITY

Mg-based alloys are used for stent application and bone implantation. They both are involved in an environment of blood. Previous cytotoxicity tests can only reflect the cytotoxicity and cellular response. Thus, the evaluation of hemocompatibility is required to avoid hemolysis after implantation. ISO 10993-4 (ISO 2002) provides a guideline for the evaluation of hemocompatibility on biomedical materials. Common test methods include a hemolysis test and a platelet adhesion test.

For the hemolysis test, healthy human blood from a volunteer containing sodium citrate (3.8 wt.%) at a ratio of 9:1 was taken and diluted with normal saline (4:5 ratio by volume). The pretreated sample is then dipped in a standard tube containing 10 ml of normal saline, which has been previously incubated at 37°C for 30 min. Next, 0.2 ml of diluted blood is added to the tube and incubated at 37°C ± 0.5°C for 60 min. A negative control of 10 ml of normal saline and a positive control of deionized water are also needed. After 5 min centrifuge at 3000 rpm, spectroscopic analysis of supernatant at 545 nm is analyzed to calculate the hemolysis based on Equation 1.9. At least three duplicates are needed for statistical analysis.

$$\text{hemolysis} = \frac{\text{OD(test)} - \text{OD(negative control)}}{\text{OD(positive control)} - \text{OD(negative control)}} \times 100\% \qquad (1.9)$$

where OD is optical density.

Another test for hemocompatibility evaluation is the platelet adhesion test. Platelet-rich plasma (PRP) is prepared by centrifuging whole blood for 10 min at 1000 rpm. Then, samples with PRP are incubated at 37°C ± 0.5°C for 60 min. The adhered platelets are fixed in 2.5% glutaraldehyde solution for 1 h at room temperature followed by dehydration in a gradient ethanol/distilled water mixture (50%, 60%, 70%, 80%, 90%, and 100%) for 10 min each and dried in a hexamethyldisilazane (HMDS) solution. The quantity and morphology of the platelets adhered are then observed via SEM. At least five fields of view are needed for each sample for statistical analysis.

1.5.6 ANIMAL TESTING

Because there is always a deviation on corrosion performance and toxicity reported in the literature, in vitro tests can only be used to preliminarily screen material for further in vivo testing, which is critical for more accurate analysis of toxicity and corrosion performance. However, considering the time period and the economic factor, in vivo tests are not well applied as in vitro experiments. Taking the deviation between in vitro and in vivo results into consideration, in vivo tests are strongly recommended because they are more convincing. Animal tests for biomedical application can be divided more specifically as animal tests for bone repair, animal tests for stent application, and some tests for other applications.

1.5.6.1 Bone Repair

As for bone repair, here we discuss some techniques and parameters involved in animal tests, including the selection of animal model, test period, implantation site, and some common measurements, such as micro-CT, histological analysis serum Mg analysis, and explant analysis.

1.5.6.1.1 Animal Model

According to ISO 10993-6-2009, animals used for tests are divided into two categories. One is for short-term tests, including mice, rats, hamsters, and rabbits. The other one is for long-term tests, including rodents, rabbits, dogs, sheep, goats, and pigs, and other animals with a relatively long life expectancy are suitable. As for bone repair, both short-term and long-term tests are needed based on the different implant sites and recovery rates.

1.5.6.1.2   Testing Period

As for degradable Mg alloys, the testing period shall be in coincidence with the degradation property of the material. In this case, a pre-experiment is needed for the determination of the degradation period and thus the determination of time points after the conducting of the in vivo test. During the test period, because biodegradable Mg alloys degrade with time, a burst release of ions and inflammation may occur, which needs to be well observed.

1.5.6.1.3   Implantation Site

Biodegradable Mg alloys are often in either intramedullary or intraosseous locations, and they are relatively rare in a subcutaneous or intramuscular environment as Mg alloys are often intended for orthopedic application (Walker et al. 2014). However, in Walker's view, the soft tissue environment has several benefits specific to the use of Mg, such as easy removal of the implant and better tolerance of the inevitable production (Walker et al. 2014). In general, according to ISO 10993-6-2009, the test samples shall be implanted into the tissues most relevant to the intended clinical use of the material.

1.5.6.1.4   Micro-CT

As for bone implantation tests, the healing process will undergo an inflammation period (a couple of days), blood vessel formation, proliferation, and remodeling. Micro-CT provides an effective way to analyze the region of interest without executing the animal.

Micro-CT is one of the most important techniques used for in vivo analysis. Different from clinically used CT, it has a higher resolution ratio to micron order because of which it is also called μ-CT. The biggest advantage of micro-CT is that the analysis doesn't need to execute a testing animal at each time point, making the observations more consecutive. In addition, micro-CT can not only identify the structure of the bone tissue surrounding intraosseously implanted materials, but can also quantify bone–implant contact and new bone growth at the perimeter of the implantation site (Walker et al. 2014). With the help of software, 3D reconstruction and thus corrosion rate calculation from the 3D model can be obtained.

1.5.6.1.5   Histological Analysis

Histological analysis is the most common analysis method for tissue response and bone response for orthopedic application. However, as mentioned previously, Mg and its alloys possess relatively higher activities than traditionally bioinert materials. In this case, when preparing tissue sections, the selection of dye is important because dye with relatively high acidity will damage the surface of the tissue sections. A different purpose calls for different dye. For example, toluidine blue staining can be used to distinguish mineralized bone or collagenous fiber. As for histological analysis, SEM and TEM are both used to determine element distribution and contents. Bone histomorphometry is also adopted to quantitatively

TABLE 1.12

Common Terminology and Computational Formula Used in Bone Histomorphometry

Terminology	Abbreviation	Unit	Computational Formula
Tissue area	T.Ar	mm²	
Trabecular bone area	Tb.Ar	mm²	
Trabecular perimeter	Tb.Pm	mm	
Single label perimeter	SL.Pm	mm	
Double label perimeter	DL.Pm	mm	
Interlabeled width	IL.Wi	μm	
Osteoclast number	N.Oc	#	
Percent trabecular area	BV/TV	%	Tb.Ar/T.Ar × 100
Trabecular thickness	Tb.Th	μm	(2000/1.199)(Tb.Ar/Tb.Pm)
Trabecular number	Tb.N	#/mm	(1.199/2)(Tb.Pm/T.Ar)
Trabecular separation	Tb.Sp	μm	(2000/1.199)(T.Ar − Tb.Ar)/Tb.Pm
Percent labeled perimeter	%L.Pm	%	(DL.Pm + SL.Pm/2)Tb.Pm × 100
Mineral apposition rate	MAR	μm/d	IL.Wi/Intervel
Bone formation rate	BFR/BS	μm/d*100	(DL.Pm + SL.Pm/2) × MAR/Tb.Pm × 100
	BFR/BV	%/year	(DL.Pm + SL.Pm/2) × (MAR/100 × 365)/ Tb.Pm × 100
Osteoclast number	OC.N	#/mm²	N.Oc/Tb.Ar

examine the effects of biodegradable Mg implants on tissue regeneration. Some terminology and equations used in bone histomorphometry are summarized in Table 1.12.

1.5.6.1.6   Serum Mg

Serum Mg content is an important factor during implantation because the Mg content in serum must remain in a certain range to maintain physical health as mentioned previously. In addition, the content change of Mg and other ions in serum during implantation can provide the possibility for mechanism interpretation. In general, two milliliters of blood is collected from the helix vein of the implantation site to measure the change of serum Mg before and at every time point after implantation.

1.5.6.2   Stent Application

For stent application, in vivo tests can be a lot different from the aforesaid in vivo tests for bone repair application. A stent, no matter whether it is in a vascular or other nonvascular orifice, such as a bile duct or the esophagus, is intended to treat stenosis phantom, and basically the stent provides mechanical open support in the orifice and prevents early recoil. To observe the performance of the stent in vivo, the reaction between the stent and tube walls, and considering the most common problem of restenosis during stent application, different methods are adopted. In this section, techniques and parameters existing in in vivo stent tests are discussed. For convenience, all the discussion here is based on cardiovascular stent application. Generally, cardiovascular absorbable implants can follow the guidance of IOS/TS 17137.

1.5.6.2.1 Objects: Animal Models and Patients

As for stent application in vivo tests, larger animals are needed, considering the size of the stent and vasculature and to mimic the human body environment. Until now, pigs (Di Mario et al. 2004; Waksman et al. 2006, 2007; Loos et al. 2007) have been commonly used for in vivo experiments; rabbits (Chen et al. 2014b) and dogs (Luffy et al. 2014) are also adopted.

With the solid data from animal tests and such preliminary experiments, further clinical trials can be conducted. Until now, several clinical experiments have been conducted with AMS from generation to generation (Peeters et al. 2005; Bosiers et al. 2006; Barlis et al. 2007; Erbel et al. 2007; McMahon et al. 2007; Waksman et al. 2009; Haude et al. 2013).

1.5.6.2.2 Testing Period

The ideal degradation period is shown in Figure 1.7 (Zheng et al. 2014). It is necessary for stents to retain their integrity for at least 4 months to serve their function. However, until now, most of the stents implanted into animal bodies revealed a faster degradation rate as shown in Figure 1.7. It also needs to be noted that the loss of mechanical integrity is faster than the degradation (Zheng et al. 2014). The overquick degradation rate limited the selection of an end point during in vivo tests. Therefore, for now, short testing periods such as 1, 2, and 3 months are reasonable.

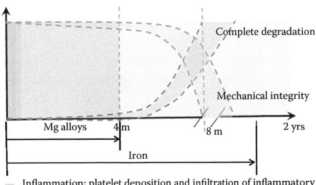

Inflammation; platelet deposition and infiltration of inflammatory cells, which lasts for several days

Granulation; endothelial cells migrate to cover the injured surface, and smooth muscle cells modulate and proliferate lasting for 1–2 weeks

Remodeling; extracellular matrix deposition and remodeling continues for months. This phase is variable in duration but largely complete at 90–120 days

FIGURE 1.7 Ideal degradation curve for biodegradable cardiovascular stent materials. (From Zheng, Y. F., X. N. Gu, and F. Witte. *Materials Science and Engineering: R: Reports*, 77, 1–34, 2014. With permission.)

Nevertheless, on the other hand, once the material possesses an improved corrosion resistance, longer periods of in vivo tests are recommended to examine if it matches the ideal curve.

1.5.6.2.3 Angiography

Angiography is a basic technique adopted during stent implantation and postoperative observation. With the help of angiography, the stent can be precisely transported and implanted to the target. Before and after implantation, angiographic images can help quantitatively determine minimum lumen diameter, reference vessel diameter, late lumen loss, percentage diameter stenosis, binary restenosis rate, acute recoil, acute gain, and lesion angulation. Figure 1.8 shows a general adoption of angiography to testify to the vessel angulation that remained at the inserted vessel site after 12 months implantation (Haude et al. 2013).

1.5.6.2.4 Intravascular Ultrasound

Before the application of intravascular ultrasound (IVUS), angiography was considered best to diagnose a coronary artery lesion. However, angiography can only present the situation of the lumen regardless of the lumen wall and atheromatous plaque. Therefore, sometimes coronary artery stenosis is underestimated if one can only apply angiography, and IVUS is of help to patients. IVUS places an ultrasonic probe in the vasculature and provides the organizational structure and geometrical morphology of the cardiovasculature. It can help confirm atheromatous plaque, especially during remodeling. Figure 1.9 shows an IVUS image used by Serruys et al. (2009) to sequentially examine the changes in stress (polymer stents). Figure 1.9a showed a long area of high strain, which is color-coded for strain, ranging from

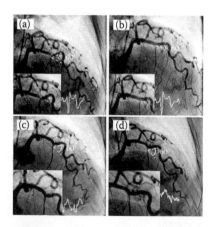

FIGURE 1.8 Angiographic appearance of the scaffolded vessel segment (a) before the procedure, (b) after scaffold implantation, and at (c) 6-month and (d) 12-month follow-up. (Adapted from Haude, M., R. Erbel, P. Erne, S. Verheye, H. Degen, D. Böse, P. Vermeersch, I. Wijnbergen, N. Weissman, F. Prati, R. Waksman, and J. Koolen. *The Lancet*, 381, 9869, 2013. With permission.)

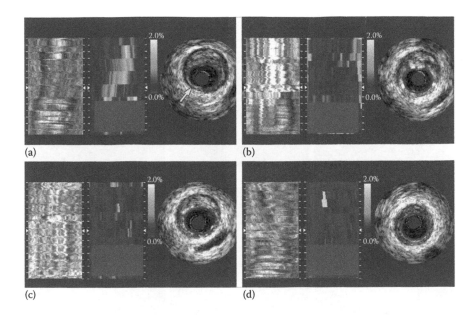

FIGURE 1.9 Sequential changes in strain assessed by intravascular ultrasound palpography. (a) Before stenting, (b) abolition of the high strain spot after stenting, (c) absence high strain at 6 months, and (d) absence high strain 2 years after stenting. (Adapted from Serruys, P. W., J. A. Ormiston, Y. Onuma, E. Regar, N. Gonzalo, H. M. Garcia-Garcia, K. Nieman, N. Bruining, C. Dorange, and K. Miquel-Hébert. *The Lancet*, 373 (9667), 897–910, 2009. With permission.)

1% to 2% with the high strain region marked by an arrow. High strain continuously disappears over time (Figure 1.9b and c).

1.5.6.2.5   Optical Coherence Tomography

Optical coherence tomography (OCT) is a high-resolution imaging technique based on a weak coherent optical interferometer. OCT is similar to IVUS except that it uses infrared light waves rather than acoustic waves. Due to this, OCT provides a higher axial resolution of 10 μm than the resolution of 100 to 150 μm provided by IVUS. Figure 1.10 is a comparison of OCT and IVUS images used by Slottow et al. (2008). Mean stent areas (mm²) based on IVUS for postimplantation 28 days and 3 months are 6.25 ± 0.81, 4.20 ± 1.51, and 4.76, respectively. The counterpart mean stent areas (mm²) based on OCT are 8.02 ± 0.49, 2.48, and 5.90 ± 0.09, respectively. Because OCT revealed a precise resolution, it is recommended as a method for in vivo degradation rate calculation.

1.5.6.2.6   Histological Analysis

Histological analysis can also be adopted for in vivo stent application experiments. It can be used to examine the vascular integrity after stent implantation. However,

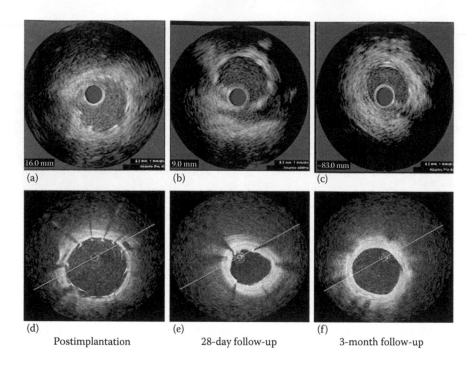

<table>
<tr><td>(a)</td><td>(b)</td><td>(c)</td></tr>
<tr><td>(d)</td><td>(e)</td><td>(f)</td></tr>
<tr><td>Postimplantation</td><td>28-day follow-up</td><td>3-month follow-up</td></tr>
</table>

FIGURE 1.10   (a–c) IVUS and (d–f) OCT images of porcine coronary arteries implanted with AMS. (a, d) Following stent implantation; (b, e) 28 days after deployment; (c, f) 3 months following implant. (Adapted from Slottow, T. L., R. Pakala, T. Okabe, D. Hellinga, R. J. Lovec, F. O. Tio, A. B. Bui, and R. Waksman. *Cardiovascular Revascularization Medicine*, 9 (4), 248–54, 2008. With permission.)

as illustrated previously, histological analysis requires a pretreatment of euthanasia of the animals, which increases economic and ethical issues. With histological analysis, injury score can be determined. Mural injury, inflammation, vascularization, intimal fibrin, intimal smooth muscle, adventitial fibrosis, and endothelialization can also be available.

2 Biodegradation Mechanism and Influencing Factors of Mg and Its Alloys

2.1 BIODEGRADATION MECHANISM OF PURE Mg IN THE PHYSIOLOGICAL ENVIRONMENT

Magnesium has the lowest standard potential of all the engineering metals, which is −2.37 V at 25°C (Song and Atrens 1999). Due to the formation of corrosion product film, the corrosion potential of Mg is slightly more negative than −1.5 V in dilute chloride solution or a neutral solution with respect to the standard H electrode (Ghali et al. 2013). Magnesium dissolution in an aqueous environment generally proceeds by an electrochemical reaction with water to produce magnesium hydroxide and hydrogen gas, a mechanism that is relatively insensitive to oxygen concentration as shown in Figure 2.1. Immediately after contact with the body fluid, the anodic dissolution of magnesium occurred as is expressed by the partial reaction presented in Equation 2.1; at the same time, the reduction of protons occurs at the cathode as is expressed in Equation 2.2.

$$\text{Anodic reaction: } Mg \rightarrow Mg^{2+} + 2e^- \tag{2.1}$$

$$\text{Cathodic reaction: } 2H_2O + 2e^- \rightarrow 2OH^- + H_2 \tag{2.2}$$

Therefore, at the corrosion potential, the overall reaction of the corrosion process can be expressed as Equation 2.3:

$$Mg\ (s) + 2H_2O\ (l) \rightarrow Mg(OH)_2\ (s) + H_2\ (g) \tag{2.3}$$

These reactions occur arbitrarily over the entire surface on which a galvanic coupling forms due to different potentials between the metal matrix and intermetallic phase or grain boundary. It is reported that the anodic partial reaction (Equation 2.1) possibly involves intermediate steps that produce the monovalent magnesium ion (Mg^+), which exists for a short time (Hoey and Cohen 1958; Makar and Kruger 1990).

It can be seen from the electrochemical reactions that the migration of ions from the metal surface into solution results in the formation of an oxide layer that adheres to the metal surface. When the oxide layer fully covers and seals the metal surface, it forms a kinetic barrier or passive layer, which physically limits or prevents further

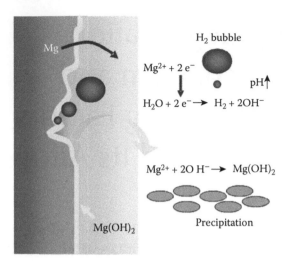

FIGURE 2.1 Schematic representation of Mg degradation. Mg dissolves as Mg^{2+} which reacts with water, generates hydrogen bubbles, creates hydroxyl groups, and increases pH. Microbubbles that get larger and float in the air are generated from pitting or active corrosion sites. Dissolved divalent Mg^{2+} ions react with hydroxyl groups (OH^-) and precipitate as Mg hydroxide. Also, a passive interlayer of Mg hydroxide or Mg oxide is formed on the degradation surface. (From Yun, Y. H., Dong, Z. Y., Lee, N. et al., *Materials Today*, 12, 22, 2009. With permission.)

migration of ionic species across the metal oxide/solution interface. In reality, there is always a natural oxidation layer (MgO) already formed on the Mg surface in air. For the pure Mg exposed to air, the thickness of the oxidation layer is 2.65 nm after 1 min exposure and is 5–6 nm independent of the exposure time between 7 days and 7 years (Yao et al. 2000). Once immersed into an aqueous solution, the original MgO film will react with water, and the outer layer of the surface film will become mainly $Mg(OH)_2$. The $Mg(OH)_2$ film is normally quite loose and cannot offer significant protection for the Mg substrate. Simultaneously, the hydrogen bubbles being generated from the corroding areas can stir the deposited $Mg(OH)_2$ and prevent the corroding areas from being fully covered by the deposited $Mg(OH)_2$; thus, the corrosion cannot be easily self-inhibited.

Although the native $Mg(OH)_2$ surface layers are loose in nature and cannot provide sufficient protection to resist corrosion encountered in the physiological environment, which contains a large amount of chloride ions (~140 mmol·l⁻) (Hassel et al. 2005), chloride ions start to convert the surface $Mg(OH)_2$ into more soluble $MgCl_2$, and dissolution of $Mg(OH)_2$ makes the surface more active, decreasing the protected area and promoting further dissolution of Mg. The reaction can be summarized as follows:

$$Mg + 2Cl^- \rightarrow MgCl_2 \qquad (2.4)$$

$$Mg(OH)_2 + 2Cl^- \rightarrow MgCl_2 \qquad (2.5)$$

In addition to magnesium oxide or hydroxide, it has been proven that the corrosion product layer on the sample surface also contains magnesium/calcium carbonates and phosphates. The reactions can be described as follows:

$$Mg^{2+}(\text{or } Ca^{2+}) + OH^- + HCO_3^- + (n-1)H_2O \rightarrow Mg(\text{or } Ca)CO_3 \cdot nH_2O \quad (2.6)$$

$$3Mg^{2+}(\text{or } Ca^{2+}) + 2OH^- + 2HPO_4^{2-} + (n-2)H_2O \rightarrow Mg_3(\text{or } Ca)(PO_4)_2 \cdot nH_2O$$
$$(2.7)$$

Willumeit et al. (2011) summarized some of the initial steps of corrosion under cell culture conditions (DMEM, 37°C, 20% O_2, 5% CO_2, 95% rH) as Figure 2.2. (Calcium carbonate and magnesium phosphates are not presented in this figure where they should be.) The fast precipitation of insoluble carbonates and phosphates at these corrosion sites is also suspected to be the main reason for observed passivation behavior. It is reported that magnesium phosphates present uniformly whereas calcium phosphates appear preferentially in many isolated regions and evolve gradually (Xin et al. 2011a). This is because surface-absorbed Mg ions can inhibit nucleation of calcium phosphates in an aqueous solution (Golubev et al. 1999). The local concentration of magnesium ions at the corrosion sites can be very high, and hence, magnesium phosphates are much easier to precipitate than calcium phosphates. After the whole surface has undergone corrosion, magnesium phosphates will cover the entire surface to serve as considered protection. It is easier for calcium phosphates to nucleate and grow at passive regions than corrosive regions. Finally, nonuniform distributions of calcium phosphates in the corrosion product layer formed (Xin et al. 2011a).

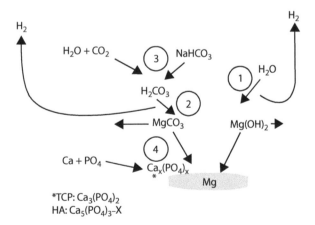

FIGURE 2.2 Summary of some of the initial steps of corrosion under cell culture conditions. (1) $Mg(OH)_2$ formation due to contact with water, (2) $MgCO_3$ formation due to the presence of carbon dioxide, (3) increase in the content of hydrogen carbonate due to the buffering system, and (4) biomineralization due to the introduction of calcium. The asterisk denotes possible products (TCP = tricalcium phosphate, HA = hydroxyapatite, X = OH, Cl,...). (From Willumeit, R., Fischer, J., Feyerabend, F. et al., *Acta Biomaterialia*, 7, 2704, 2011. With permission.)

In general, on account of the large area of fresh surface and inferior protection offered by the corrosion products, intense corrosion takes place during early exposure, and the corrosion rate is high. Subsequently, passivation of the active surface and accumulation of corrosion products such as $Mg(OH)_2$, magnesium/calcium phosphates, or carbonates may decelerate the corrosion behavior. After immersion for a sufficiently long time, equilibrium between the formation and dissolution of the corrosion products is established, leading to stable degradation rates.

While the corrosion proceeds, the organic molecules, such as proteins, amino acids, and lipids, will adsorb over the metal surface, thus influencing the dissolution of Mg. Cells are also observed to adhere on the surface. With progressing implantation time, the adhered cells proliferate to form tissues adjacent to the corrosion product layer. Meanwhile, eroded magnesium may disintegrate from the matrix as irregular particles and fall into the surrounding media. These particles might be enclosed by the fibrous tissue or macrophages, depending on the particle size, and further degraded until the metallic phase is completely exhausted.

2.2 INFLUENCE OF ALLOYING ELEMENTS AND IMPURITIES ON THE BIODEGRADATION OF Mg AND ITS ALLOYS

Due to the high chemical activity of Mg, any of the alloying elements or impurities in its pure form or intermetallic phase will increase the galvanic corrosion of Mg and Mg alloys. Figure 2.3 schematically summarizes the microgalvanic processes involved in the corrosion of a Mg alloy. The Mg matrix with a lower content of

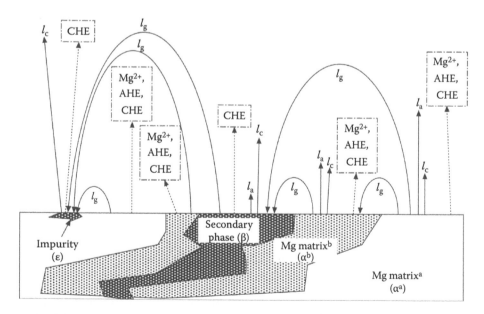

FIGURE 2.3 Corrosion model of Mg alloy with various microgalvanic cells. (From Song, G. L., Corrosion electrochemistry of magnesium (Mg) and its alloys. In *Corrosion of Magnesium Alloys*, Woodhead Publishing, 2011. With permission.)

alloying elements (α^a) is always a microanode and is preferentially corroded, and the impurity-containing particles (ε), secondary phases (β), and the matrix phase with a higher concentration of solid solutes (α^b) can act as microcathodes.

For pure Mg, purification remarkably slows down the corrosion rate. The corrosion resistance of pure Mg is relevant to the tolerance limits of impurities. When the impurity concentration exceeds the tolerance limit, the corrosion rate is greatly accelerated. The most harmful impurities to pure Mg are Fe, Cu, and Ni with tolerance limits of 170 ppm, 1000 ppm, and 5 ppm, respectively (Makar and Kruger 1993). The tolerance limits are influenced by the method of manufacture as well as the presence of a third element. J.Y. Lee et al. (2009) claimed that the corrosion behavior of pure Mg depends on the content ratio of impurities, such as the Fe/Mn ratio, rather than the content value of them. Similarly, the addition of Zr can lead to a higher purity and hence a more corrosion-resistant Mg alloy. Zinc inhibits the harmful effects of Fe and Ni impurities on the corrosion (Ding et al. 2014).

Besides combining the impurity elements, alloying elements influence the corrosion resistance of Mg alloy in several ways. First of all, the solid solution of alloying elements in the Mg matrix changes the corrosion potential. For example, Al containing Mg matrix phase can become more passive as the Al content increases. The distribution of Zr focuses in the central area of a grain. Thus, the alloy is less corrosion resistant along the grain boundaries (Song 2011). Second, when the concentration of the alloying element exceeds solid solubility, intermetallic phases are formed, almost all of which are more noble and also more stable than the Mg matrix. The secondary phases play a dual role in the corrosion of a Mg alloy as shown in Figure 2.4. They can act as microgalvanic cathodes to accelerate the corrosion of the matrix as well as

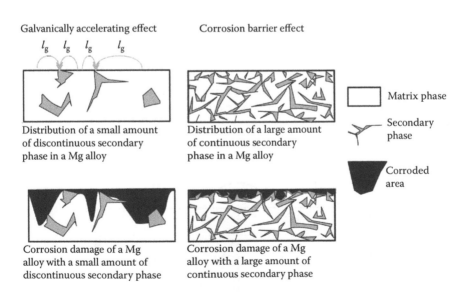

FIGURE 2.4 Schematic illustration of the dual-role model of the secondary phase in a Mg alloy in its corrosion. (From Song, G. L., Corrosion electrochemistry of magnesium (Mg) and its alloys. In *Corrosion of Magnesium Alloys*, Woodhead Publishing, 2011. With permission.)

acting as a corrosion barrier to retard or confine the corrosion development in a Mg alloy. A finely and continuously distributed β-phase is more effective in stopping the development of corrosion in an alloy whereas the presence of a small amount of discontinuous β-phase accelerates the corrosion. Third, the alloying elements, especially some rare earth elements, can change the property of the corrosion product layer. La and Nd oxide combined with $Mg(OH)_2$ enhance corrosion resistance, and Er enhances the stability of the $Mg(OH)_2$ layer, thus reducing corrosion resistance. Fourth, elements such as Zr, Sr, and Ca can improve the corrosion resistance of the Mg alloy through grain refinement.

2.3 DEGRADATION MODE AND RATE OF Mg AND ITS ALLOYS IN THE PHYSIOLOGICAL ENVIRONMENT

2.3.1 MODES OF CORROSION

2.3.1.1 Galvanic Corrosion

The active nature of Mg means that galvanic effect is always an issue. As Mg is more active than most other engineering alloys, consequently, Mg is the anode and corrodes preferentially in any galvanic. In Section 2.2, we discussed that the cathodes can be internal as second or impurity phases. In addition, cathodes can be external as other metals in contact with Mg are illustrated in Figure 2.5. It has been reported that the galvanic corrosion rate is increased by the following factors: high conductivity of the medium, large potential difference between anode and cathode, low polarizability of anode and cathode, large area ratio of cathode to anode, and small distance from anode to cathode (Bosiers et al. 2009). In order to avoid external galvanic corrosion, it would be good design practice to use metals with similar electrochemical properties when designing implant devices. Considerable attention needs to be paid to protecting against galvanic corrosion in any application using Mg alloys.

2.3.1.2 Localized Corrosion

Mg alloys always suffer from nonuniform corrosion due to the existence of internal galvanic coupling or electrochemical nonuniformity. So localized or nonuniform

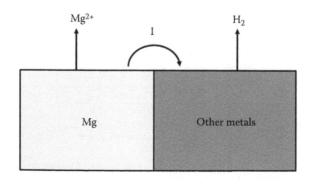

FIGURE 2.5   Galvanic corrosion between dissimilar metals.

damage is the biggest concern about corrosion of Mg alloys in applications, which may reduce the load-bearing capability. Mg is a naturally passive metal, so it undergoes pitting corrosion at its free corrosion potential, E_{corr}, when exposed to chloride ions in a nonoxidizing medium (Tunold et al. 1977). As a result, the corrosion of Mg alloys in a physiological environment typically takes the form of pitting. However, in contrast to typical pitting corrosion, the localized corrosion of Mg or a Mg alloy does not penetrate deeply. This is a result of the alkalization effect at the tips of the corroding pits, which prevent the solution at the tips from acidifying or forming occlusive autocatalytic cells (Song 2011). The pits spread laterally although the spreading degree of corrosion can vary markedly from alloy to alloy, depending on the alloy chemistry.

Filiform or "wormtrack" corrosion is another form of localized corrosion that can be observed in Mg. It is caused by an active corrosion cell, which moves across a metal surface. The head is the anode and the tail the cathode. For other metals, filiform corrosion usually appears on coated surfaces and is driven by the oxygen reduction reaction (Ruggeri and Beck 1983). However, the filiform corrosion has already been observed on the bare Mg, and it alloys in various solutions. The protective surface film covering the Mg substrate acts as a coating, and the cathodic reaction of filiform corrosion is driven by hydrogen evolution (Lunder et al. 1994; Schmutz et al. 2003). Figure 2.6 shows the filiform corrosion and pitting corrosion morphology of the ZEK100 and MgCa0.8 alloys in HBSS (Weizbauer et al. 2014).

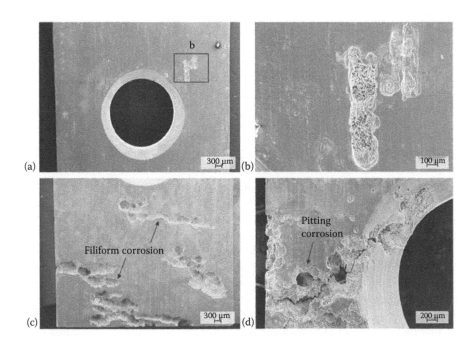

FIGURE 2.6 Filiform corrosion and pitting corrosion morphology after 96 h immersion and removal of the corrosion layer. (a, b): ZEK100 and (c, d): MgCa0.8. (From Weizbauer, A., Modrejewski, C., Behrens, S. et al., *Journals of Biomaterials Applications*, 28, 1264, 2014. With permission.)

2.3.1.3 Fretting Corrosion

Fretting is a particular form of degradation of materials that occurs when two surfaces in contact but in "small" relative motion with respect to one another result in "damage" to the surface in the contact region (Hoeppner and Chandrasekaran 1994). Fretting can pose a serious threat because it may occur at any location where two surfaces are in contact and are subject to vibration. It mainly results in two kinds of damage: fretting wear and fretting fatigue (Huang et al. 2006). Fretting wear results in the generation of debris and a loss of fit between contacting surfaces. Rapid crack nucleation and propagation resulting from fretting fatigue leads to failure of the component. For orthopedic implants, fretting has been identified at the stem/neck and neck/head contacts of modular implants, at the stem/bone and stem cement interfaces of cemented and uncemented implants, and at the screw/plate junction of fixation plates due to the daily activity of the patients (Hoeppner and Chandrasekaran 1994; Tritschler et al. 1999; Hiromoto and Mischler 2006). Fretting results in aseptic loosening, early loss of mechanical integrity, and eventual failure of the implant, thus causing suffering to the patients and additional surgery (Geringer et al. 2006; Kumar et al. 2010). Fretting also accelerates corrosion of implants that are exposed to the physiological medium due to the mechanical disruption of the passive surface oxide layer. The release of corrosion products and particulate debris could cause adverse biological reactions. In the case of Mg and its alloys, metallic ions released during fretting can be considered physiologically beneficial because these ions can be consumed or absorbed by the surrounding tissues or be dissolved and readily excreted through the kidneys. Previous studies showed fretting of AZ91D and AM60 Mg alloys in a dry condition is influenced by normal load and slip amplitude (Chen et al. 2006; Huang et al. 2006), and the fatigue strength of the AZ61 alloy is reduced due to fretting (Saengsai et al. 2009). Yet as a new type of potential biomedical metallic material, there is hardly any research on the fretting behavior of Mg alloys in a physiological environment.

2.3.1.4 Stress Corrosion Cracking

Stress corrosion cracking (SCC) is the cracking of a susceptible alloy induced from the combined influence of tensile stress (directly applied stress or residual stress) and a corrosive environment. SCC is also related to hydrogen embrittlement (HE), and hydrogen is often postulated to be involved in the mechanism by which SCC initiates and propagates (Al-Abdullat et al. 2001). This stress-initiated corrosion mechanism effectively increases the corrosion rate by two to three times (Poinern et al. 2012).

Pure Mg (Meletis and Hochman 1984; Stampella et al. 1984; Lynch and Trevena 1988) and all Mg alloys (Busk 1987) have been reported to be susceptible to SCC to some extent. Alloying additions, such as Al and Zn, promote stress corrosion cracking (Chakrapani and Pugh 1975; Busk 1987; Avedesian and Baker 1999). Reports are varied on the effect of other elements (Winzer et al. 2005). Wrought Mg alloys are more susceptible to SCC than cast alloys due to the influence of residual stress or microstructural defects resulting from the extrusion or rolling processes (Winzer et al. 2005).

SCC of Mg alloys is generally attributed to one of two groups of mechanisms: anodic dissolution or cleavage (Winzer et al. 2005). For the dissolution mechanism, in its simplest form, dissolution at a film-free crack tip causes crack advance. The cleavage type mechanism is (a) an embrittled region forms ahead of the crack tip, (b) a crack propagates through the embrittled region, and (c) the crack is stopped as it enters the ductile parent material.

SCC is particularly dangerous because it can lead to fast fracture at mechanical loads considered safe in the absence of the environment, and it often occurs without any visible deformation and is difficult to detect in its early stages. Thus, the issue of SCC should not be neglected in the design of Mg-based stents due to the proven material susceptibility, the in vivo condition uncertainty, and the high mechanical property requirements for the device structure (Gastaldi et al. 2011). Usually, stress corrosion cracks are considered to be macroscopically brittle, which means that they occur at stresses below general yield and propagate in an essentially elastic body even though local plasticity may be necessary. It is widely believed that there is a threshold, σ_{SCC}, below which SCC does not occur, and it has been suggested that σ_{SCC} is related to the yield stress (Wearmouth et al. 1973). Thus, in the prevention of SCC for Mg alloy, the general principle is to avoid loading a susceptible alloy above a critical stress during exposure to the corrosive environment.

2.3.1.5 Corrosion Fatigue

Corrosion fatigue may be defined as the combined action of an aggressive environment and cyclic stress leading to premature failure of metals by cracking (Gastaldi et al. 2011). For orthopedic implants, the ultimate implant failure is usually associated with corrosion fatigue, which is the synergetic effect of electrochemical corrosion and cyclic mechanical loading (Azevedo 2003; Magnissalis et al. 2003). For vascular devices, the failure of a stent due to fatigue may result in loss of radial support of the stented vessel or in perforation of the vessel by the stent struts. It was estimated that approximately two thirds of the failures of vascular devices would result in patient death (James and Sire 2010).

The corrosion fatigue behavior of Mg alloys examined by the standard metallic biomaterial testing method in a physiological environment was studied (Gu et al. 2010b). The die-cast AZ91D alloy indicated a fatigue limit of 50 MPa at 10^7 cycles in air compared to 20 MPa at 10^6 cycles tested in SBF at 37°C. A fatigue limit of 110 MPa at 10^7 cycles in air was observed for extruded WE43 alloy compared to 40 MPa at 10^7 cycles tested in SBF at 37°C.

Figure 2.7a illustrates the fatigue strength of the die-cast AZ91D, extruded WE43 and other Mg alloys for biomedical application, including AZ31, AZ61, and AM50 at 10^6 cycles tested in air and corrosion medium. The corrosive environment reduces 30–67% of the fatigue strength, and the die-cast AZ91D seems to be more sensitive to the corrosion medium.

Figure 2.7b compares the fatigue strength of Mg alloys with some clinically used biomaterials tested in a physiological environment. It can be seen that Mg alloy shows a relatively wide fatigue strength range, which is much lower than that of alumina, Ti alloy, and ISO5832-9 stainless steel, but it indicates a higher fatigue strength than that of calcium phosphate bone cement. Moreover, the high range of fatigue strength for Mg alloy is comparable with that of 316L stainless steel and also much greater than that of polymer.

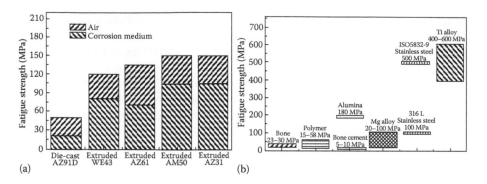

FIGURE 2.7   (a) Comparison of the fatigue strength of die-cast AZ91D, extruded WE43, and other Mg alloys for biomedical application at 106 cycles tested in air and in corrosion medium. Note that uniaxial tension–compression fatigue was conducted for extruded AZ61 alloy with a frequency of 20 Hz and a stress ratio of –1 in 5 wt.% NaCl solution, whereas a rotating beam-type fatigue was conducted for extruded AM50 and AZ31 alloy with a frequency of 30 Hz and a stress ratio of –1 in 3.5 wt.% NaCl solution. (b) Fatigue strength of Mg alloys and some clinically used biomaterials tested in a physiological environment. Note that the fatigue strength of the polymer was tested in air. (From Gu, X. N., Zhou, W. R., Zheng, Y. F. et al., *Acta Biomaterialia*, 6, 4605, 2010. With permission.)

2.3.2 In Vitro and In Vivo Degradation Rate

Corrosion rate is a critical parameter for biodegradable implants. The measurement of corrosion rate is an essential step in evaluating the corrosion performance of the Mg alloy implants. To measure the in vitro degradation rates of Mg alloys, two methods are usually employed, namely the immersion test and electrochemical test.

In the immersion test, the corrosion can be calculated based on the weight loss. Before weighing, the corrosion products over the samples must be removed with chromate acid (200 gL^{-1} CrO$_3$ + 10 gL^{-1} AgNO$_3$) (Xin et al. 2011b). The weight loss method is perhaps the simplest method to measure the corrosion rate of Mg alloy as it typically requires only a sample, corrosion medium, and an accurate microbalance. Although weight loss experiments reveal how much corrosion has occurred, they do not divulge the mechanisms involved in the corrosion process (Kirkland et al. 2012).

Corrosion rate also can be calculated from the volume of hydrogen generated during the immersion test. According to Equation 2.3, the evolution of one mole hydrogen gas (22.4 L) directly corresponds to the dissolution of one mole of Mg (24.31 g). This method is reliable, easy to implement, and not prone to errors that are inherent with the weight loss method. In addition, the hydrogen evolution method allows the study of the variation in degradation rates with exposure time. Experimental data have shown that the corrosion products do not influence the relationship between hydrogen emission and Mg dissolution (Song and Atrens 2003). However, Z. Abidin et al. (2011b) found the corrosion rates calculated from hydrogen evolution were less than expected from the corresponding corrosion rate calculated from weight loss, which may be explained by the fact that some hydrogen dissolves in the Mg metal rather than being evolved as a gas.

Potentiodynamic polarization is the most commonly used electrochemical technique for studying in vitro corrosion of Mg alloys. The advantage of potentiodynamic polarization over the weight loss and hydrogen evolution methods is that it can not only provide the corrosion rate, but it also provides information on the kinetic and thermodynamic differences between various alloys and solutions. It allows quantification of the relative rates of the anodic and cathodic reaction over a range of potentials. Nevertheless, the corrosion rates obtained from potentiodynamic polarization usually are not in good agreement with those obtained from weight loss and hydrogen evolution. Typical deviations have been ~50–90% (Shi et al. 2010). On one hand, potentiodynamic polarization is an instantaneous test and, as such, represents only a snapshot of the corrosion at the time it is performed. On the other hand, the results are influenced by various parameters, such as dwell time prior to the potentiodynamic polarization, scan rate, and potential range (Kirkland et al. 2012). Even the analysis of the same data can result in different conclusions because small changes in the determined Tafel slopes can result in large variations in I_{corr}. Furthermore, Song and Atrens (2003) indicated that, for Mg alloys, Tafel extrapolation had not estimated the corrosion rate reliably because there are multiple reactions for both anodic and cathodic polarization.

For the measurement of in vivo corrosion rate of Mg implants, the weight loss method is also commonly used, and the results can be compared directly with weight loss data gathered from in vitro studies. The limitation of this method is that the removal of the implant from the surrounding tissue eliminates the opportunity to investigate the implant–tissue interface. Another method is to analyze the remaining cross-sectional area of explanted Mg samples after embedding in methyl methacrylate. However, this method only provides local information about the sample because the implant is usually not uniformly corroded. Microcomputed tomography (μ-CT) can analyze the remaining volume of the implanted material, new bone growth, bone-implant contact, and osseo-integration. Typical in vivo μ-CT scanners have resolutions ranging from 100 to 30 μm, and ex vivo scanners have resolutions from 30 to 1 μm. In vivo μ-CT applies to small animals, such as rats and rabbits. It allows continuous monitoring of one animal over time without interfering with the ongoing experiment. The 3-D micro-CT reconstruction can represent both the new bone formation and 3-D degradation of the implants, and 2-D cross-sectional images provide information about new bone formation, the bone–implant contact, and cavity formation (Zagorchev et al. 2010). Ex vivo μ-CT is generally performed on explants of tissue containing an implant that have been fixed and embedded in resin. After μ-CT imaging, the samples can be used for further histological analysis (Dziuba et al. 2013).

No single technique discussed here provides all of the information required to fully understand the corrosion behavior of Mg and its alloys in an SBF. So it is recommended that these techniques are used complementarily to each other.

By far, the in vitro tests did not well predict the in vivo degradation rate of Mg implants. Sanchez et al. (2015) did a systematic literature survey and attempted to find a correlation between in vitro and in vivo corrosion rates of Mg and Mg alloys. The average in vitro and in vivo corrosion rates of 19 Mg alloys and pure Mg are presented in Figure 2.8. For all the Mg and Mg alloys represented in Figure 2.8 (except Mg6Zn), the in vivo corrosion rates are lower than are the in vitro ones. In addition, the corrosion rate in vivo is in a smaller range of values than is the in vitro corrosion

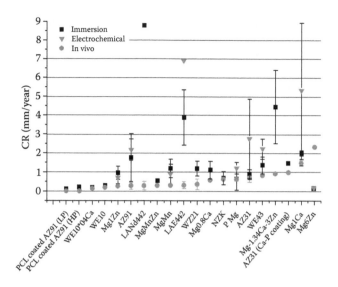

FIGURE 2.8 Comparison of the averaged in vitro and in vivo corrosion rates of 20 different materials. (From Sanchez, A. H. M., Luthringer, B. J., Feyerabend, F., and Willumeit, R., *Acta Biomaterialia*, 13, 16, 2015. With permission.)

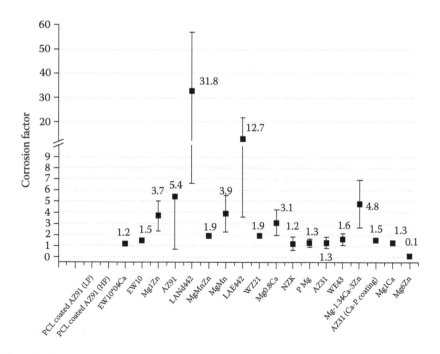

FIGURE 2.9 Corrosion factors in vitro versus in vivo in the same order as in Figure 2.7. (From Sanchez, A. H. M., Luthringer, B. J., Feyerabend, F., and Willumeit, R., *Acta Biomaterialia*, 13, 16, 2015. With permission.)

rate. In an attempt to quantify the correlation between the in vitro and in vivo corrosion rates, the corrosion factor (cf) was calculated by dividing the overall corrosion rate in vitro by the overall corrosion rate in vivo (Equation 2.8).

$$Cf = (X \pm dX)/(Y \pm dY) = (X/Y) \pm dZ \qquad (2.8)$$

where X is the average corrosion rate in vitro, dX is the standard error of the corrosion rate in vitro, Y is the average corrosion rate in vivo, dY is the standard error of the corrosion rate in vivo, and dZ is the error propagation. The calculated corrosion factors vary mainly from 1 to 4.9 as shown in Figure 2.9, which indicates that the possible corrosion rate in vivo is, on average, 1–5 times lower than the corrosion rate obtained in vitro. They also found EBSS, MEM, and SBF result in relatively good correlations between in vitro and in vivo results. The difference between the in vitro and in vivo degradation rates is resulted from the different environmental variables in vitro and in vivo, which we will discuss in Section 2.4.

2.4 INFLUENCE OF ENVIRONMENTAL VARIABLES ON THE BIODEGRADATION OF Mg AND ITS ALLOYS

Biodegradable implants serve in the particular environment of the human body. Figure 2.10 presents a schematic illustration of a Mg implant in a physiological environment. After implantation, Mg alloy implants are exposed to an environment that consists of blood, protein, and other constituents of the body fluid, such as chloride, phosphate, bicarbonate ions, and cations (Na^+, K^+, Ca^{2+}, Mg^{2+}, etc.). This physiological environment makes an extremely complex corrosive medium. The corrosion rate and the underlying corrosion process of Mg and its alloys depend on a variety of environmental factors, including but not limited to temperature, composition of the media, and flow. Only if we fully understand the factors that influence in vivo degradation of Mg implants, can we improve the accuracy of in vitro prediction and design Mg-based biodegradable devices with an appropriate degradation rate.

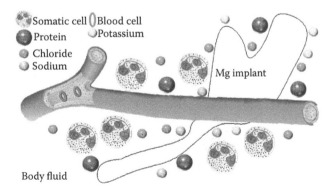

FIGURE 2.10 Schematic illustration of a Mg implant in a physiological environment. (From Ding, Y., Wen, C., Hodgson, P., and Li, Y., *Journal of Materials Chemistry B*, 2, 1912, 2014. With permission.)

2.4.1 Temperature

Temperature can affect the degradation of Mg and its alloys in several ways. The corrosion rate increases exponentially with increasing temperature, following an Arrhenius-type relationship (Jones 1992). Temperature also affects the Gibbs free energy of adsorption for proteins and thus affects the affinity and number of proteins that adsorb to a surface (Sasha and Roscoe 1999). Merino et al. (2010) reported that an increase in temperature from 20°C to 35°C caused an increase in the corrosion attack of Mg, AZ31, AZ80, and AZ91D Mg materials exposed to salt fog. Kirkland and Birbilis (2013) found that the temperature change from 20°C to 37°C seemed to at least double the mass loss of pure Mg, Mg-0.8Ca, and Mg-1Zn after 7 days of exposure to the Hank's balanced salt solution (HBSS). They also found pure Mg and 12 Mg alloys displayed increased i_{corr} at 37°C compared to that at 20°C; the changes ranged from 64% to 840% in HBSS and 14–437% in minimum essential medium containing 10% fetal bovine serum (MEM_{FBS}). The extent of change in degradation rate with temperature is different for each alloy. The rate of the anodic reaction increases with increasing temperature. The influence of temperature on anodic reaction is little because the cathodic reaction is an electron transfer reaction. This study suggests that measurements of corrosion rate that are not executed at physiological temperature (37°C) are inaccurate, and assertions from data not at physiological temperature will not necessarily translate correctly to physiological conditions.

2.4.2 Chemical Composition of Corrosion Media

Currently, pseudophysiological solutions that mimic the composition of body fluids have been used in in vitro experiments and include 0.9 wt.% NaCl solution, c-SBF, r-SBF, Hank's solution, PBS, artificial plasma, DMEM, and so on. The compositions and ion concentrations of human plasma and eight common solutions are listed in Table 2.1.

2.4.2.1 Inorganic Ions

As discussed in Section 2.1, chloride ions are detrimental to the corrosion resistance of Mg. SBF and Hank's solution are more aggressive than artificial plasma (Quach et al. 2008; Yang and Zhang 2009) mainly due to the higher chloride concentration. Hank's solution is more aggressive than DMEM (Gu et al. 2009d), which may also be explained by the higher chloride concentration and the lower hydrocarbonate concentration. Sulfate ions also attack Mg, and phosphate ions can retard the corrosion rate effectively and delay the emergence of pitting corrosion (Xin et al. 2008). The degradation behavior of pure Mg in 0.125 M NaCl, Earle solution containing calcium and Mg salts [Earle(+)], Eagle's minimum essential medium (E-MEM), and E-MEM supplemented with fetal bovine serum (E-MEM + FBS) have been studied (Yamamoto and Hiromoto 2009). All the solutions have the same chloride ion concentration. The total Mg^{2+} release during 14 days of immersion into six kinds solutions are shown in Figure 2.11. The Mg^{2+} dissolutions in Earle(+), E-MEM, and EMEM + FBS were much lower than those of the three kinds of NaCl solutions due

TABLE 2.1

Ion Concentrations of Human Plasma and Eight Common Solutions

	Human Plasma	0.9% NaCl	PBS	Hank's	c-SBF	Artificial Plasma	Earle	E-MEM	DMEM (L)
Na^+ (mmol^{-1})	142	153	157	140.1	142	144.5	143.5	143.5	153.45
K^+ (mmol^{-1})	5		4.1	6.2	5.0	5.4	5.37	5.37	5.37
Ca^{2+} (mmol^{-1})	2.5			1.9	2.5	1.8	1.80	1.80	1.80
Mg^{2+} (mmol^{-1})	1.5			0.8	1.5	0.8	0.81	0.81	0.81
HCO_3^- (mmol^{-1})	27			3.2	4.2	26.2	26.2	26.2	44.05
Cl^- (mmol^{-1})	103	153	140	145.8	147	125.3	125	125	118.37
HPO_4^{2-} (mmol^{-1})	1		11.5	0.4	1	3			
SO_4^{2-} (mmol^{-1})	0.5			0.8	0.5	0.8	0.81	0.81	0.81
$H_2PO_4^-$ (mmol^{-1})				0.7			1.09	1.09	1.04
Tris (g L^{-1})					6.069				
Protein (g L^{-1})	63–80								
Amino acids (g L^{-1})	Unknown							0.860	1.6
Glucose (mmol^{-1})	Unknown							1	4.5
Hepes (g L^{-1})									5.97
Ref.	Kokubo and Takadama 2006; Xin et al. 2010	Xin et al. 2010	Xin et al. 2010	Gu et al. 2009	Kokubo and Takadama 2006	Quach et al. 2008	Yamamoto and Hiromoto 2009	Yamamoto and Hiromoto 2009	Gu et al. 2009d

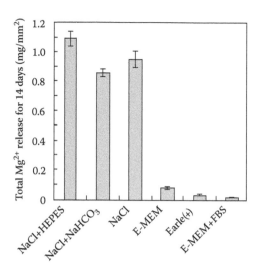

FIGURE 2.11 Total Mg^{2+} release during 14 days of immersion of pure Mg into six kinds of solutions. (From Yamamoto, A. and Hiromoto, S., *Materials Science & Engineering C*, 29, 1559, 2009. With permission.)

to the precipitation of insoluble phosphate salts. It is important to realize that the chloride concentration of all the above solutions is higher than that in human plasma.

2.4.2.2 Buffer

The corrosion of Mg and its alloys is also markedly affected by the presence of different buffers. The commonly used buffering agents in simulated physiological fluids include HEPES, Tris–HCl, and HCO_3^-. HCO_3^- is the most important buffering agent in body plasma (~27 mmol L^{-1} in body fluids). As shown in Figure 2.6, buffering the NaCl solution with HEPES increased the total Mg^{2+} release during 14 days of immersion whereas buffering with NaHCO$_3$ under the atmosphere of 5 vol.% of CO$_2$ decreased the Mg^{2+} release (Yamamoto and Hiromoto 2009). Other studies also indicated HEPES (Walker et al. 2012) and Tris-HCl (Xin and Chu 2010) promoted corrosion of Mg. Hydrocarbonate ions promote the dissolution of Mg during early immersion but induce rapid surface passivation due to precipitation of magnesium carbonate, which totally suppresses pitting corrosion (Xin et al. 2008).

2.4.2.3 Amino Acid and Proteins

The Earle(+), E-MEM, and EMEM + FBS contain the same concentration of inorganic salts, and E-MEM contains amino acid, and EMEM + FBS contains both amino acid and proteins. As shown in Figure 2.11, the lowest degradation rate was obtained in E-MEM + FBS, followed by Earle(+) and E-MEM in that order. Yamamoto and Hiromoto (2009) suggested amino acids and some organic chelating compounds can form a complex with Mg^{2+}, which inhibits the formation of insoluble salts and thus accelerates the corrosion of Mg. Meanwhile, proteins can form an absorbed layer on the surface of Mg alloy, which make the insoluble salt layer dense and more effective as a barrier against corrosion. Liu et al. (2007c) also reported that proteins may serve a "blocking effect" on Mg corrosion. However, the influence of proteins on the corrosion of Mg may depend on the composition of the alloy. Gu et al. (2009d) found the Mg–Ca alloy indicated a faster corrosion rate with the addition of FBS in DMEM whereas the AZ91 alloy showed the opposite tendency.

From the above discussion, a NaCl-only solution is not suitable for the in vitro test of Mg biodegradation because of the complete lack of other inorganic salts. Hank's solution and SBF consist of similar concentrations of inorganic ions to those in body plasma, and consequently, they are suitable media for in vitro investigations of degradation of Mg alloys. However, Hank's solution and SBF do not contain proteins and amino acids, which prevents, to some extent, its having as accurate a degradation performance as that in a physiological environment. Xin et al. (2011b) suggested that DMEM is most desirable for in vitro degradation studies of biomedical Mg-based alloys. Nevertheless, the influence of proteins and amino acids on the corrosion mechanism of Mg alloys needs further understanding.

2.4.3 SOLUTION VOLUME AND FLOW

According to ASTM G31-72, the minimum values of V/S are 0.20 ml/mm^2 (20 ml/cm^2) and 0.40 ml/mm^2 (40 ml/cm^2) to ensure the volume of the test solution is large enough to avoid any appreciable change in corrosion during the test, either from the using up of corrosive constituents or aggregation of corrosion products, which might affect further corrosion (Z. Zhen et al. 2013). Also, a sample with larger surface area in unit mass and smaller area ratio of lateral to total area is recommended. However, according to ISO10993-15, the V/S should be less than 1 ml/cm^2. In this case, various ratios of V/S, from 0.33 ml/cm^2 (Q. Wang et al. 2011) to 500 ml/cm^2, have been adopted (Xu et al. 2008). Yang and Zhang (2009) found a low solution volume/surface area (SV/SA) ratio resulted in a high pH, which resisted the corrosion. But when the ratio was high enough, 6.7, for example, the influence was negligible. They suggest that Hank's solution with a high SV/SA ratio, such as 6.7, and Hank's solution with a low SV/SA ratio, such as 0.67, should be selected to simulate the in vivo degradation behavior of a Mg bone screw in a bone marrow cavity and a Mg plant and screw in cortical bone or muscle tissue, respectively. SBP solution with a high SV/SA ratio, for example, 6.7, should be chosen to simulate the degradation of a Mg stent in an artery.

To understand the corrosion behavior of Mg stents in a blood vessel, it is necessary to take the influence of flow into consideration. Some researchers developed devices driven by a pump as illustrated in Figure 2.12 (Levesque et al. 2008; Y. Chen et al. 2010; Liu and Zheng 2011). Hiromoto et al. (2008b) found that the existence of flow prevented the accumulation of the corrosion product and promoted uniform corrosion, leading to an increase of the anodic current density and a decrease of the impedance. Levesque et al. (2008) reported that when the stress applied by the flow is low, it protects the surface from localized corrosion, and when it is very high, in addition to high uniform corrosion, some localized corrosion also occurs.

We have studied the corrosion behavior of as-cast pure Mg and as-extruded WE43 alloy in static, stirring, and flowing Hank's solution. Both pure Mg and the WE43 alloy exhibited the highest corrosion rate in flowing solution and the lowest corrosion

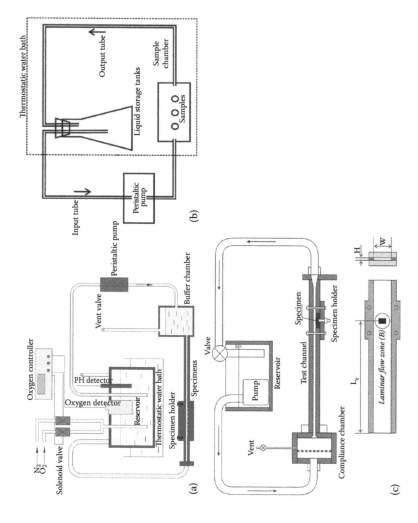

FIGURE 2.12 Diagram of dynamic corrosion test devices. (From Liu, B. and Zheng, Y., *Acta Biomaterialia*, 7, 1407, 2011; Zhu, S., Huang, N., Xu, L. et al., *Materials Science & Engineering C*, 29, 1589, 2009; Levesque, J., Hermawan, H., Dube, D., and Mantovani, D., *Acta Biomaterialia*, 4, 284, 2008. With permissions.)

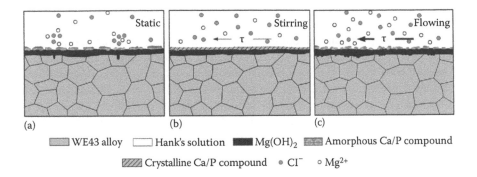

FIGURE 2.13 Illustration of the corrosion mechanism of WE43 alloy in (a) static, (b) stirring, (c) flowing Hank's solution. (From Li, N., Guo, C., Wu, Y. H., Zheng, Y. F., and Ruan, L. Q., *Corrosion Engineering Science Technology*, 47, 346, 2012. With permission.)

rate in stirring solution and were more susceptible to corrosion in the flowing solution. A schematic illustration of how the flow condition of solution influenced the corrosion behavior of materials is shown in Figure 2.13. Samples in the stirring test exhibited the best corrosion resistance, and a crystalline Ca/P compound layer formed on the surface. In the static state (Figure 2.13a), deleterious chloride ions may congregate near the surface and promote pitting corrosion. In the flowing condition (Figure 2.13c), the existence of the flow may influence corrosion behavior of Mg in two aspects, that is, applying shear stress on the surface and promoting diffusion, both of which accelerate the corrosion. The shear stress applied by the stirring solution is low, and it protects the surface from localized corrosion. Consequently, lower Mg^{2+} concentration near the surface allowed Ca/P crystal growth, forming a more compact and protective layer than an amorphous Ca/P compound layer (Figure 2.13b).

From the discussion, we can indicate that with different temperature and corrosive medium the degradation behavior of Mg alloys in different sites of the human body will be quite distinct. In general, when designing biomedical materials based on Mg alloys, the designers should take the specific environment (temperature, fluids, and stress) into consideration. In the future, it is critical to formulate a standard test protocol for more reliable and accurate evaluation of the corrosion rates.

2.5 BIOLOGICAL EFFECTS OF DEGRADATION PRODUCTS

Corrosion products of Mg-based implants will need to be accommodated (by some means) in the body. Based on the current experimental results, it is suggested that corrosion is the inherent response of Mg alloys to body fluid with the following corrosion products: released Mg^{2+} ions, hydrogen bubbles, alkalization of solution by the OH^- ions, released alloying element metal ions, and peeled-off particles from the implant. The corrosion products should be biocompatible and biosafe to the host tissue at the minimum requirements. Subsequently, the next logical question is how to fully exploit the biofunction of the degradation products.

2.5.1 Biological Effects of Mg and Its Alloying Elements

Mg is essential to human metabolism and is found naturally in bone tissue. Mg^{2+} is the fourth most abundant cation in the human body and can be considered physiologically beneficial with the adult body storing around 30 g of Mg in both muscle and bone tissue (Saris et al. 2000). The homeostasis of Mg^{2+} in the human body fluid is controlled by the excretion and reabsorption of Mg^{2+} in the kidneys and by uptake of Mg^{2+} in the intestinal tract as Figure 2.14 shows (Yamamoto and Hiromoto 2009). In a healthy adult human, 2.4–3.5 g of Mg is filtered from blood plasma through the kidneys, and 95% of filtered Mg is reabsorbed in 1 day, which is about 10 and 33 times higher than the amounts of Mg in extracellular fluid and in blood plasma. If the uptake of Mg^{2+} at the intestinal tract increases, the reabsorption of Mg^{2+} at the kidney decreases, which means the excess amount of Mg^{2+} is rapidly excreted into the urine to maintain a similar concentration of Mg^{2+} in the body fluid (Yamamoto and Hiromoto 2009).

The importance of Mg to the body stems from the fact that it is a bivalent ion, which is used to form apatite in the bone matrix and is also used in a number of metabolic processes within the body (Kim et al. 2003). It is also a cofactor in many enzymes and a key component of the ribosomal machinery that translates the genetic information encoded by mRNA into polypeptide structures (Hartwig 2001; Maguire and Cowan 2002; Vormann 2003; Staiger et al. 2006). An increased Mg supply might contribute to a risk reduction toward various diseases, such as coronary artery disease or osteoporosis (Vormann 2003). Additionally, Slutsky et al. (2010) studied the effect of increasing brain Mg using Mg-L-threonate on learning and memory in rat models. Their results suggested that elevating brain Mg enhanced both short-term

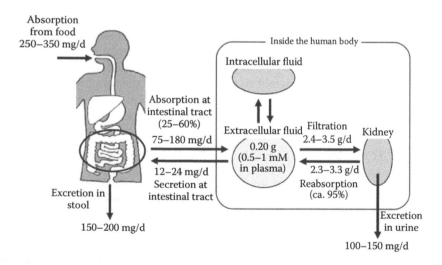

FIGURE 2.14 Schematic explanation of homeostatic control of magnesium in the human body. (From Yamamoto, A. and Hiromoto, S., *Materials Science & Engineering C-Materials for Biological Applications*, 29, 1559, 2009. With permission.)

synaptic facilitation and long-term potentiation as well as improved learning and memory functions.

It is believed that the release of Mg ions from corroding Mg alloys causes no toxicity (local or systemic) and may even have beneficial effects on some of the structures, including the cells, in the relevant local tissue (Williams 2006). Increased Mg levels induced by degradation of Mg implants could stimulate new bone growth. Numerous studies on various Mg alloys have reported enhanced bone growth around the corroding Mg implant (Witte et al. 2005, 2007; E.L. Zhang et al. 2009a; Witte 2010; Zou et al. 2011). Park et al. (2012) compared the bone-healing capacity of Mg ion-incorporated porcine bone, prepared with hydrothermal treatment in an alkaline Mg-containing solution, in rabbit calvarial defects with that of untreated porcine cancellous bone and deproteinized bovine bone. The Mg-incorporated porcine bone with surface nanostructures achieved rapid new bone formation in the osseous defects of rabbit calvaria compared with untreated xenografts of porcine and bovine origin. Recently, pure Mg wires were implanted into femora in STZ-induced diabetic rats (W. Yang et al. 2011). The serum Mg level and bone mineral contents increased significantly, and as a result, the bone mineral density in the Mg treatment group was higher than that in the diabetic group. Besides, blood biochemical analysis indicated that Mg implants had no toxic effect on the liver and kidney function. All these indicated that the implantation of Mg could stimulate new bone growth and has potentially antiosteoporotic activity (W. Yang et al. 2011).

Alloying elements play an important role in Mg alloys, and the mechanical properties are usually the primary consideration when introducing alloying elements into the materials. Moreover, the biological effects, such as biocompatibility and toxicity are crucial. The short- and long-term effects of introducing foreign metals (most likely as ions) into the human body present an important alloy design consideration. The common alloying elements in Mg alloys include Al, Zn, Ca, Zr, Mn, Li, and some rare earths (RE). Table 2.2 summarizes in three categories the pathophysiology and toxicology of Mg as commonly used alloying elements: essential nutrient, potential essential metals, and the other common alloying elements.

Aluminum: Of all the alloying elements, Al is the most common addition to Mg alloys to improve both mechanical properties and corrosion resistance. However, perhaps Al is also the most controversial element. Aluminum is well known as a neurotoxicant. The accumulation of Al has been suggested to be associated with various neurological disorders, inducing dementia and potentially leading to Alzheimer's disease (although data are not conclusive for humans) (Flaten 2001; El-Rahman 2003). Hence, the amount of Al released from the Mg alloys must be carefully controlled. However, the long-term effects of exposure to Al is still unclear, and it needs to be further studied.

Calcium: It is known to be the essential nutrient element and the most abundant mineral in the body, about 1000–1300 g for a healthy adult. It can regulate the normal physiological function of the organs, tissues, and systems. Calcium is the major component of bone and teeth, playing a crucial role in the formation of bone. Ion calcium is a medium between nerves and muscles, always controlling the relaxation and contraction of muscles, thus maintaining the normal activity of nerve–muscle. Calcium can also promote the activity of enzymes, and the activity of many enzymes

TABLE 2.2

Summary of the Pathophysiology and Toxicology of Mg, Fe, and the Commonly Used Alloying Elements

Element	Human Amount	Blood Serum Level	Pathophysiology	Toxicology	Daily Allowance	Bone Cell[a]	Vascular Cell[a]
			Essential Nutrients				
Mg	25 g	0.73–1.06 mM	Activator of many enzymes; coregulator of protein synthesis and muscle contraction; stabilizer of DNA and RNA	Excessive Mg leads to nausea	0.7 g	+	+
Fe	4–5 g	5.0–17.6 g/l	Component of several metalloproteins; crucial in vital biochemical activities, i.e., oxygen sensing and transport	Iron toxicity gives rise to lesions in the gastrointestinal tract, shock, and liver damage	10–20 mg	+–	+–
Ca	1100 g	0.919–0.993 mM	More than 99% has a structure function in the skeleton; the solution Ca has a signal function, including muscle contraction, blood clotting, cell function, etc.	Inhibits the intestinal absorption of other essential minerals	0.8 g	+	+
Zn	2 g	12.4–17.4 μM	Trace element; appears in all enzyme classes; most Zn appears in muscle	Neurotoxic and hinders bone development at higher concentrations	15 mg	–	–

(Continued)

TABLE 2.2 (CONTINUED)
Summary of the Pathophysiology and Toxicology of Mg, Fe, and the Commonly Used Alloying Elements

Element	Human Amount	Blood Serum Level	Pathophysiology	Toxicology	Daily Allowance	Bone Cell[d]	Vascular Cell[d]
Mn	12 mg	<0.8 µg/l	Trace element; activator of enzyme; Mn deficiency is related to osteoporosis, diabetes mellitus, atherosclerosis	Excessive Mn results in neurotoxicity	4 mg	–	–
Potential Essential Metal							
Sr	0.3 g	0.17 mg[a]	99% is located in bone; shows dose dependent metabolic effect on bone; low doses stimulated new bone formation	High doses induce skeletal abnormalities	2 mg	+	+
Si	–	–	Cross-linking agent of connective tissue; necessary for growth and bone calcification	Silica and silicate caused lung diseases	–	–	–
Sn	30 mg	–	Tin-deficient diets in rat studies resulted in poor growth, reduced feeding efficiency, hearing loss, and bilateral (male) hair loss	Some organic compounds are poison, i.e., methyl and ethyl compounds	–	+–	+–
Other Element							
Li	–	2–4 ng/g	Used in the treatment of manic-depressive psychoses	Plasma concentration of 2 mM is associated with reduced kidney function and neurotoxicity, 4 mM maybe fatal	0.1 g[b]	+	+

(Continued)

TABLE 2.2 (CONTINUED)
Summary of the Pathophysiology and Toxicology of Mg, Fe, and the Commonly Used Alloying Elements

Element	Human Amount	Blood Serum Level	Pathophysiology	Toxicology	Daily Allowance	Bone Cell[d]	Vascular Cell[d]
Other Element							
Al	<300 mg	2.1–4.8 µg/l	–	Primarily accumulated in bone and nervous system; implicated Al in the pathogenesis of Alzheimer's disease	–	+	+
Zr	<250 mg		Probably excreted in feces; low systematic toxicity to animals	High concentration in liver and gall bladder	3.5 mg	+	+
Y and Lanthanides		<47 µg[c]	Substituted for Ca^{2+} and matters when the metal ion at the active site; compound of drugs for treatment of cancer	Basic lanthanides deposited in liver; more acidic and smaller cations deposited in bone	–	+–	+–

Source: Zheng, Y. F., Gu, X. N., and Witte, F., *Mater. Sci. Eng., R,* 77, 1–34, 2014. With permission.

a Sr concentration in total blood (Seiler and Sigel 1988).

b The therapeutic dose for lithium carbonate is up to about 0.1 g/d in divided doses (Seiler and Sigel 1988).

c The concentrations for Y and Lanthanides (La, Ce, Nd, Sm, Eu, Gd, Tb, Dy, Ho, Er, Tm, and Yb) are below 0.1, 0.44, 0.03, 0.07, 0.2, 0.1, 0.09, 0.1, 0.2, 0.03, 0.1, and 0.1 mg/l, respectively (Seiler and Sigel 1988).

d The toxicity levels for bone- and vascular-related cells are according to the cytotoxicity test of the metal salts (Yamamoto et al. 1998; Hallab et al. 2002; Feyerabend et al. 2010), + stands for mild toxicity, +– stands for moderate toxicity, and – stands for severe toxicity.

involved in cell metabolism requires the activation of ion calcium, such as lipase and amylase.

Zinc: Zinc is one of the most abundant nutritionally essential elements in the human body, and about 90% is found in the muscle and bone tissue. Zinc has been shown to be essential to the structure and function of a large number of macromolecules and for more than 300 enzymatic reactions (Tapiero and Tew 2003). Optimal nucleic acid and protein metabolism as well as cell growth, division, and function require sufficient availability of zinc. Zinc can also enhance the immunity of the human body and promote wound healing and trauma recovery (Wellinghausen 1998) although it would be neurotoxic in high concentrations and may hinder bone development (Poinern et al. 2012).

Lithium: Although Li has been used in medicine for almost 150 years (Bruckner et al. 1998; Stippich et al. 1998), it has not been used widely in implanted biomaterials, with which continual exposure may occur on a milligram per day level. It is known that lithium can be used in drugs to treat psychiatric disorders although overdose would cause central nervous center disorders, lung dysfunctions, and impaired kidney functions (Poinern et al. 2012).

Manganese: As the essential trace element (≤ 0.8 µg l^{-1}) in blood serum, it has been proven that excessive amounts of Mn can produce neurological disorder. Cell culture studies on Mg-1Mn indicate that the extracted media with a Mn concentration of 1.8 ± 0.5 µmol l^{-1} induces serious cytotoxicity. The poisonous effect of Mn from magnesium alloys on cell viability and proliferation has also been observed (Loos et al. 2007).

Rare earths: Rare earth (RE) elements can be used to improve the mechanical characteristics, corrosion properties, and creep resistance of Mg alloys. In the use of RE elements in Mg alloys for biomedical purposes should also be considered the perspective of their potential cytotoxicity. It has been proven that many rare earth elements have anticancer properties and are used in the treatment of cancer (Poinern et al. 2012), but some studies have also found cytotoxicity for an intraperitoneal LD_{50} dose of $GdCl_3$ of 550 mg kg^{-1} in mice. Excess yttrium can change the expression of some rat genes and impose adverse effects on DNA transcription factors (Dumas et al. 2001; Yang et al. 2006). This creates the potential that the significant body of work that has been and will be performed using alloys containing Al and/or RE may, in the end, be unexploited if these materials cannot be proven to be nontoxic. In spite of these scattered studies, systematic in vitro studies of RE elements dissolved from Mg alloys are still rare. Hence, it is critical to investigate systematically the potential cytotoxicity of dissolved RE elements from biomedical Mg implants in the future.

Zirconium: Corrosion performance of Zr-containing alloys, such as the ZK and ZE series, is relatively satisfactory although not as good as that of the AZ class of alloys. Results to date suggest that Zr at low levels in multielement alloys is innocuous; however, excess Zr will cause significant corrosion, and Mg-Zr binary alloys also suffer excessive corrosion.

Silicon: As one of the trace elements necessary for the human body, the average intake of silicon ranges from about 20 to 50 mg/day with the lower values for animal-based diets and higher for plant-based diets. A trace amount of Si has been proven to be essential in mammals (Song 2007) and to be important for the growth and development of bone, cartilage, and connective tissue (Sripanyakorn et al. 2007), which can link the mucopolysaccharides to each other and bind them to the protein to form the fibrous structure, thus increasing the flexibility and strength of connective tissue to maintain structure integrity. Meanwhile, as a component of collagen, silicon is quite safe, not reacting with the immune system or breeding bacteria.

Strontium: Strontium, along with Ca and Mg, belongs to group IIA of the periodic table and shares similar chemical, biological, and metallurgical properties. There is about 140 mg Sr in the human body, and 99% of the body content of Sr salts is used in the treatment of osteoporotic patients to increase bone mass and reduce the incidence of fractures (Boivin et al. 1996; Christoffersen et al. 1997).

Tin: The main function of Sn is its performance as an antitumor as the compounds containing Sn produced in the human thymus can inhibit the formation of cancer cells. Sn can promote the synthesis of proteins and nucleic acids, which are beneficial for the body's growth and are also involved in biological reactions, enhancing the stability of the internal environment.

As bioabsorbable materials, Mg alloys are expected to be totally degraded in the body, and their corrosion products not deleterious to the surrounding tissues. Considering the security of biodegradable Mg as implants in the human body, in the design of materials, elements with potential toxicological problems should be ideally avoided if possible, and these elements should only be used in minimal, acceptable amounts if they cannot be excluded from the design.

2.5.2 BIOLOGICAL EFFECT OF HYDROGEN

Molecular hydrogen (H_2) is the smallest gas molecule and the most abundant gas in the universe. It is not easily dissolved in water, and 100% saturated hydrogen water contains 1.6 ppm or 0.8 mM hydrogen at room temperature. For a long time, hydrogen has been regarded as a physiologically inert gas by biologists, and high-pressure hydrogen can be used as a breathing medium in the diving medical field. However, recent studies have strongly suggested that hydrogen is an excellent antioxidant. A major breakthrough in hydrogen research occurred after Ohsawa et al. (2007) reported a prominent effect of molecular hydrogen on a rat model of cerebral infarction. It has been proven that inhaled H_2 has antioxidant and antiapoptotic properties that protect the brain against ischemia-reperfusion injury and stroke by selectively reducing hydroxyl radical ($\cdot OH$) and peroxynitrite ($ONOO^-$) in a cell-free system (Ohsawa et al. 2007). Since then, the effects of molecular hydrogen on various diseases have been documented for more than 60 disease models and human diseases as shown in Table 2.3. Hydrogen has been administered to animals and humans in the forms of hydrogen gas, hydrogen-rich water, hydrogen-rich saline, instillation, and dialysis solution. Most studies have been performed on rodents. Protective effects are observed especially in oxidative stress-mediated diseases, including neonatal cerebral hypoxia; Parkinson's disease; ischemia/reperfusion of spinal cord, heart, lung, liver, kidney, and intestine as well as transplantation of lung, heart, and kidney (Ohno et al. 2012). For example, Buchholz et al. (2008) found hydrogen inhalation ameliorates oxidative stress in transplantation-induced intestinal graft injury in a rat model. And for the first time, they found the transplantation-induced upregulation

TABLE 2.3
Sixty-Three Disease Models and Human Diseases for Which Beneficial Effects of Hydrogen Have Been Documented

Diseases	Species	Administration
Brain		
Cerebral infarction (Ohsawa et al. 2007; Ji et al. 2011; Y. Liu et al. 2011; Ono et al. 2011)	Rodent, human	Gas, saline
Cerebral superoxide production (Sato et al. 2008)	Rodent	Water
Restraint-induced dementia (Nagata et al. 2008)	Rodent	Water
Alzheimer's disease (Li et al. 2010d; C. Wang et al. 2011)	Rodent	Saline
Senile dementia in senescence-accelerated mice (Y. Gu et al. 2010)	Rodent	Water
Parkinson's disease (Fu et al. 2009; Fujita et al. 2009)	Rodent	Water
Hemorrhagic infarction (Chen et al. 2010a)	Rodent	Gas
Brain trauma (Ji et al. 2010)	Rodent	Gas
Carbon monoxide intoxication (Q. Sun et al. 2011a)	Rodent	Saline
Transient global cerebral ischemia (Hugyecz et al. 2011)	Rodent	Gas
Deep hypothermic circulatory arrest-induced brain damage (Shen et al. 2011)	Rodent	Saline
Surgically induced brain injury (Eckermann et al. 2011)	Rodent	Gas
Spinal Cord		
Spinal cord injury (Chen et al. 2010b)	Rodent	Saline
Spinal cord ischemia/reperfusion (Y. Huang et al. 2011)	Rabbit	Gas
Eye		
Glaucoma (Oharazawa et al. 2010)	Rodent	Instillation
Corneal alkali-burn (Kubota et al. 2011)	Rodent	Instillation
Ear		
Hearing loss (Kikkawa et al. 2009; Taura et al. 2010; Lin et al. 2011)	Tissue, rodent	Medium, water
Lung		
Oxygen-induced lung injury (Huang et al. 2010; Zheng et al. 2010c; C.-S. Huang et al. 2011; Q. Sun et al. 2011b)	Rodent	Saline
Lung transplantation (Kawamura et al. 2010)	Rodent	Gas
Paraquat-induced lung injury (S. Liu et al. 2011)	Rodent	Saline
Radiation-induced lung injury (Qian et al. 2010a; Chuai et al. 2011; Terasaki et al. 2011)	Rodent	Water
Burn-induced lung injury (Fang et al. 2011)	Rodent	Saline
Intestinal ischemia/reperfusion-induced lung injury (Mao et al. 2009)	Rodent	Saline

(Continued)

TABLE 2.3 (CONTINUED)

Sixty-Three Disease Models and Human Diseases for Which Beneficial Effects of Hydrogen Have Been Documented

Diseases	Species	Administration
Heart		
Acute myocardial infarction (Hayashida et al. 2008; Sun et al. 2009; Y. Zhang et al. 2011)	Rodent	Gas, saline
Cardiac transplantation (Nakao et al. 2010a)	Rodent	Gas
Sleep apnea-induced cardiac hypoxia (Hayashi et al. 2011)	Rodent	Gas
Liver		
Schistosomiasis-associated chronic liver inflammation (Gharib et al. 2001)	Rodent	Gas
Liver ischemia/reperfusion (Fukuda et al. 2007)	Rodent	Gas
Hepatitis (Kajiya et al. 2009a)	Rodent	Intestinal gas
Obstructive jaundice (Q. Liu et al. 2010)	Rodent	Saline
Carbon tetrachloride-induced hepatopathy (H. Sun et al. 2011a)	Rodent	Saline
Radiation-induced adverse effects for liver tumors (K.-M. Kang et al. 2011)	Human	Water
Kidney		
Cisplatin-induced nephropathy (Nakashima-Kamimura et al. 2009; Kitamura et al. 2010; Matsushita et al. 2011)	Rodent	Gas, water
Hemodialysis (Nakayama et al. 2009; 2010)	Human	Dialysis solution
Kidney transplantation (Cardinal et al. 2009)	Rodent	Water
Renal ischemia/reperfusion (F. Wang et al. 2011)	Rodent	Saline
Melamine-induced urinary stone (Yoon et al. 2011)	Rodent	Water
Chronic kidney disease (W.-J. Zhu et al. 2011)	Rodent	Water
Pancreas		
Acute pancreatitis (H. Chen et al. 2010)	Rodent	Saline
Intestine		
Intestinal transplantation (Buchholz et al. 2008, 2011; Zheng et al. 2009)	Rodent	Gas, medium, saline
Ulcerative colitis (Kajiya et al. 2009b)	Rodent	Gas
Intestinal ischemia/reperfusion (H. Chen et al. 2011)	Rodent	Saline
Blood Vessel		
Atherosclerosis (Ohsawa et al. 2008)	Rodent	Water
Muscle		
Inflammatory and mitochondrial myopathies (Ito et al. 2011)	Human	Water

(Continued)

TABLE 2.3 (CONTINUED)
Sixty-Three Disease Models and Human Diseases for Which Beneficial Effects of Hydrogen Have Been Documented

Diseases	Species	Administration
Cartilage		
NO-induced cartilage toxicity (Hanaoka et al. 2011)	Cells	Medium
Metabolism		
Diabetes mellitus type I (Li et al. 2011a)	Rodent	Water
Diabetes mellitus type II (Kajiyama et al. 2008)	Human	Water
Metabolic syndrome (Nakao et al. 2010b; Hashimoto et al. 2011)	Human, rodent	Water
Diabetes/obesity (Kamimura et al. 2011)	Rodent	Water
Perinatal Disorders		
Neonatal cerebral hypoxia (Cai et al. 2008, 2009; Domoki et al. 2010)	Rodent, pig	Gas, saline
Preeclampsia (X. Yang et al. 2011)	Rodent	Saline
Inflammation/Allergy		
Type I allergy (Itoh et al. 2009)	Rodent	Water
Sepsis (Xie et al. 2010a)	Rodent	Gas
Zymosan-induced inflammation (Xie et al. 2010b)	Rodent	Gas
LPS/IFNγ-induced NO production (Itoh et al. 2011)	Cells	Gas
Cancer		
Growth of tongue carcinoma cells (Saitoh et al. 2008)	Cells	Medium
Lung cancer cells (Ye et al. 2008)	Cells	Medium
Radiation-induced thymic lymphoma (Zhao et al. 2011)	Rodent	Saline
Others		
UVB-induced skin injury (Yoon et al. 2010)	Rodent	Bathing
Decompression sickness (Ni et al. 2011)	Rodent	Saline
Viability of pluripotent stromal cells (Kawasaki et al. 2010)	Cells	Gas
Radiation-induced cell damage (Qian et al. 2010b,c)	Cells	Medium
Oxidized low-density lipoprotein-induced cell toxicity (Song et al. 2011)	Cells	Medium
High glucose-induced oxidative stress (Yu et al. 2011)	Cells	Medium

Source: Ohno, K., Ito, M., Ichihara, M., and Ito, M., *Oxidative Medicine and Cellular Longevity*, 2012, 11, 2012. With permission.

in the inflammatory mediators CCL2, IL-1b, IL-6, and TNF-a were mitigated by hydrogen. Ohsawa et al. (2008) found consumption of H_2-dissolved water has the potential to prevent arteriosclerosis.

Cai et al. (2009) investigated the neuroprotective effect of peritoneal administration of saturated H_2 saline in neonatal HI rats, and found H_2 saline reduces brain injury and improves long-term neurological and neurobehavioral functions. Some intestinal bacteria can produce a remarkable amount of molecular hydrogen. Kajiya et al. (2009a) reported that H_2 released from intestinal bacteria can suppress inflammation induced in the liver by Concanavalin A (ConA).

Several human diseases have been studied to date: diabetes mellitus type 2 (Kajiyama et al. 2008), metabolic syndrome (Nakao et al. 2010b), hemodialysis (Nakayama et al. 2009, 2010), inflammatory (Ito et al. 2011) and mitochondrial myopathies (Ono et al. 2011), brain stem infarction, and radiation-induced adverse effects. No adverse effect of hydrogen has been documented in the six human diseases described above. Among the six diseases, the most prominent effect was observed in subjects with metabolic syndrome, who consumed 1.5–2.0 l of hydrogen water per day (Nakao et al. 2010b).

The therapeutic effects of hydrogen on various diseases have been mainly attributed to its ability to reduce reactive oxygen species (ROS). ROS or free radicals, such as the superoxide anion $\left(O_2^-\right)$, are generated through the leak of electrons to molecular oxygen in the mitochondria during the production of adenosine triphosphate by oxidative phosphorylation (Figure 2.15). In the normal state, mitochondria generate 2–3 nmol of superoxide anion per minute per milligram of protein, resulting in the generation of the superoxide radical, hydrogen peroxide, and the highly toxic hydroxyl radical. These ROS can be safely reduced to water by superoxide dismutase, catalase, and glutathione peroxidase (Figure 2.15). During ischemia-reperfusion injury,

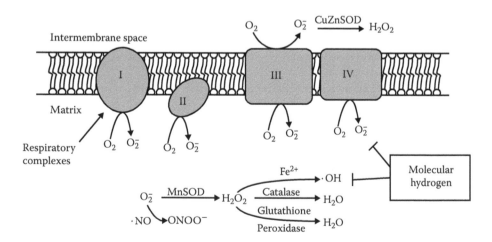

FIGURE 2.15 Schematic of reactive oxygen species generation in the mitochondria. Oxygen radicals are O_2^-, $\cdot OH$, and $\cdot NO$. The potential sites for the neutralizing antioxidant effects of molecular hydrogen are shown. SOD: super oxide dismutase. (From George, J. F. and Agarwal, A., *Kidney International*, 77, 85, 2010. With permission.)

the balance between the generation of ROS and the mechanisms to detoxify them can be upset, resulting in the accumulation of ROS in the tissues, where they quickly react with lipids, proteins, and nucleic acids, causing disordered cellular functions (George and Agarwal 2010). The administration of molecular H_2 can effectively reduce the ROS. In addition, H_2 can regulate various gene expressions and signal-modulating activities. However, it remains unclear whether such regulations are the cause or consequence of the reduction of ROS (Ohta 2011).

The findings on the biological effects of hydrogen are quite inspiring. However, there is no research that connects the biological effects of hydrogen to Mg corrosion up to now. Mg alloy can be regarded as a source that continuously produces hydrogen. The role that hydrogen plays while the Mg device degrades in vivo needs exploring.

2.5.3 BIOLOGICAL EFFECTS OF OH⁻

The increase of OH⁻ induced by Mg corrosion leads to an antibacterial effect. Robinson et al. (2010) found (a) when Mg corrosion products were added to growth media, the growth of *Escherichia coli*, *Pseudomonas aeruginosa*, and *Staphylococcus aureus* was inhibited; (b) the addition of Mg^{2+} alone will not inhibit bacterial growth; but (c) increasing the OH⁻ ions (i.e., higher pH) will inhibit bacterial growth as shown in Figure 2.16. Y. Li et al. (2014) also demonstrated that pure Mg (99.9%) reduced bacterial adhesion and prevented biofilm formation in vitro, and implantation of an Mg intramedullary nail into the bone cavity of a rat protected the implant from bacterial contamination and improved new peri-implant bone formation, which was most likely due to the increased local alkalinity caused by degradation of the metal.

Moreover, OH⁻ ion might also influence bone growth in vivo. Experiments in rabbits with implanted $Mg(OH)_2$ cylinders have demonstrated enhanced bone remodeling adjacent to the dissolving cylinder (Janning et al. 2010). This indicates that either local alkalinity and/or enhanced Mg ion concentration might stimulate bone growth. Zhai et al. (2014) found Mg leach liquor (MLL) has antiosteoclast activity in vitro and prevents wear particle-induced osteolysis in vivo. Both the Mg ions and pH values in the solution contribute to the inhibitory effects of MLL on osteoclastogenesis.

2.6 CONCLUDING REMARKS

Based on the above discussion, we have understood the corrosion mechanism and influence factors of Mg alloys as biodegradable materials, which are quite useful for our further investigation into improving the corrosion resistance of Mg. The modern research on biodegradable Mg alloys for biomedical applications rises in the beginning of this century. However, the development of biodegradable Mg has been hampered by a lack of coordination along with a lack of appropriate standards and insufficient cross-disciplinary collaboration. To date, there is no widely accepted criterion to evaluate the corrosion resistance of Mg alloy. It is hard to compare results achieved by different researchers. In some in vitro tests, important variables (such as pH, temperature) are not correctly controlled to mimic physiological conditions. So generally it is very difficult to relate the in vitro results to in vivo performance. Thus,

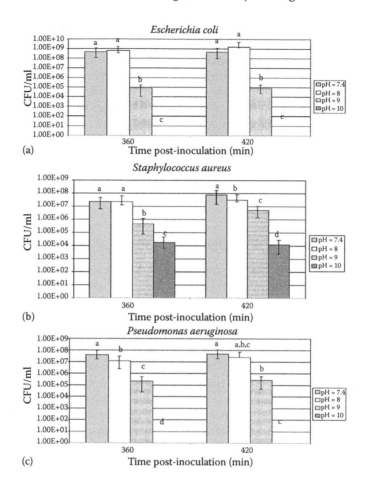

(a)

(b)

(c)

FIGURE 2.16 Culture plate counts for (a) *E. coli*, (b) *S. aureus*, and (c) *P. aeruginosa* with pH 7.4 (control), 8, 9, and 10 treatment groups. Data are presented as median CFU ml^{-1} with the error bars representing the 25th and 75th percentiles. Columns labeled with the same letter were not significantly different ($P > 0.05$) at the given time point. (From Robinson, D. A., Griffith, R. W., Shechtman, D., Evans, R. B., and Conzemius, M. G., *Acta Biomaterialia*, 6, 1869, 2010. With permission.)

a reliable in vitro test bench and a database are needed to be established to ensure the comparisons between reported literature as well as to facilitate the creation of suitable models to predict performance of future alloys. For the in vivo test, a sensor to record the pH value in situ is needed, which is quite important for the study of corrosion behavior of Mg. We are also looking forward to figuring out the exact metabolic pathways and distribution of the corrosion product (hydrogen gas, Mg ions, and alloying element ions). A more comprehensive study of the effect of organic compounds on Mg corrosion is needed, which requires a multidisciplinary approach based on biomolecular chemistry and electrochemistry.

3 Novel Structure Design for Biodegradable Mg and Its Alloys

3.1 INTRODUCTION

Despite numerous series of metallic biomaterials being developed or underdeveloped as bulk and microcrystalline structures, some novel structures of biomaterials have been designed, such as porous structures, composites with other biomaterials, nanostructures with severe plastic deformation, and so on.

A porous structured scaffold is one of the most important structures of biomaterials for bone remodeling. The scaffolds serve as osteoconductive moieties for new bone deposition by creeping substitution from adjacent living bone (Karageorgiou and Kaplan 2005), and they have been maturely applied in bone tissue engineering. Porous ceramics, such as hydroxyapatite (Jin et al. 2000; Dong et al. 2001; Woodard et al. 2007; Bose and Tarafder 2012), calcium metaphosphate (Lee et al. 2001), and bioglasses (Gong et al. 2001; Sepulveda et al. 2002; Jones et al. 2006); polymers (Liu and Ma 2004), such as collagen (Rocha et al. 2002), silk fibroin (Nazarov et al. 2004), PLA (Taboas et al. 2003), and propylene glycol-co-fumaric acid (Trantolo et al. 2003); and titanium alloys (Wen et al. 2002; Sikavitsas et al. 2003; van den Dolder et al. 2003; Spoerke et al. 2005; Willie et al. 2010; Kato et al. 2012) have been developed by salt-leaching, gas-foaming, phase-separation, freeze-drying, and sintering techniques. The application of porous structure in Mg and its alloys is a breakthrough for Mg bone implants, which are degradable with superior mechanical strength to polymers and lower elastic modulus than ceramics.

Composite structures are used to combine the advantages of different materials. For metallic materials, the weak bioactivity is the defect for biomedical uses; therefore, materials of more bioactivity have been chosen for addition to metallic materials to improve their biocompatibility. For example, spark plasma sintered or powder metallurgy method Ti alloy composites: Ti/TCP (Kumar et al. 2013), Ti/HA (Ning and Zhou 2002; Kumar et al. 2013), Ti/titanium boride (Makau et al. 2013), Ti/HA/bioactive glass (Ning and Zhou 2004). For Mg and its alloys, the bioactivity is one aspect to improve, and another aspect, the mechanical strength, should also be ameliorated. Consequently, some bioactive materials or reinforcements are added into a Mg-based matrix in order to form biocompatible composites with appropriate strength.

Nanofine/ultrafine-grained metallic materials have been shown to be greatly interesting for biomaterials in recent years. Due to the increasing grain boundaries, nanofine/ultrafine-grained metallic materials obtain higher strength than coarse-grained materials. In addition, the grain size effect and stronger hydrophilicity of the novel materials have a positive influence on cell attachment, spread, and proliferation, which has been proven by experiments on preosteoblasts, stem cells, osteoblasts cultured with nanofine/ultrafine-grained stainless steel (Misra et al. 2009a,b), Ti and its alloys (Faghihi et al. 2007; Estrin et al. 2011; Yao et al. 2012; Lee et al. 2013), or Zr (Saldaña et al. 2007).

Contrary to nanofine/ultrafine-grained structures, the glassy structure of metallic biomaterials is trying to avoid grain boundaries and crystallinity. With the tremendous critical cooling rates, alloys could form a homogeneous single-phased structure (Park and Kim 2005). Therefore, metallic glasses display superior mechanical properties with high corrosion and wear resistance over common metallic materials (Schroers et al. 2009) and become promising biomaterials. Many bulk metallic glasses, such as Ti-based (Oak et al. 2007; Sugiyama et al. 2009), Zr-based (Nagase et al. 2008; Wada et al. 2009), and Fe-based (Pang et al. 2002; L. Liu et al. 2008) glasses have been investigated in vitro and in vivo. Results indicate that the bulk metallic glasses are biocompatible to cells and tissue function.

Inspired from the structures of other biomaterials, especially the metallic biomaterials, bio-magnesium scientists have developed many novel structured Mg alloys in order to improve the biocompatibility and reduce the corrosion rate. We illustrate them in the following sections.

3.2   POROUS STRUCTURE

The density and size of biomaterial scaffolds are important factors in bone rebuilding. In early studies, the minimum pore size of 100 μm is considered beneficial for the migration and transport of osteoblasts. However, in practice, pore sizes >300 μm are preferred, because larger pores benefit the vascularization, which leads to direct osteogenesis (Karageorgiou and Kaplan 2005).

Several types of porous Mg have been developed by various methods.

a. Gu et al. (2010e) produced a lotus-type porous pure Mg by the metal/gas eutectic unidirectional solidification method (GASAR process). Compared to the compressive yield strength of bulk pure Mg, that of the lotus-type porous pure Mg degrades more slowly with the extension of the immersion period in simulated body fluid (SBF), and the cytotoxicity of the porous pure Mg is within Grade I RGR (no cytotoxicity).

b. The powder metallurgical process is a common method for making foams, and Wen et al. (2001) fabricated Mg foams for bone substitution applications by this method using carbamide ($CO(NH_2)_2$) particles as space holder particles. They fabricated a series of porous Mg with porosity of 35–55% and pore size of about 70–400 μm. Results indicated that the Young's modulus and peak stress increased with decreasing porosity and pore size (Wen et al. 2004). However, the reactant of carbamide and Mg during sintering remained on the wall of holes, which may influence the biocompatibility of the porous Mg. Others applied the same fabrication procedure to make Mg foams and immersed the foams in physiological saline solution (PSS, 0.9% NaCl solution) to evaluate their in vitro degradation. Because more pores lead to more reaction-specific surface area, the foams with larger porosity degraded faster (Zhuang et al. 2008).

c. Topologically ordered porous Mg (TOPM) was developed by Staiger et al. (2010) using computer-aided design (CAD) and 3-D printing with several following steps. They aimed for the possible use of TOPM in biomedical implant applications; however, the foams were fragile if strut thicknesses were less than 0.8 mm.

d. Aghion's group also made Mg foams by powder metallurgy technology as degradable drug delivery platforms using ammonium hydrogen carbonate (NH_4HCO_3) with 50- to 200-mm-diameter space holders (Aghion et al. 2010). They tested the corrosion property and released profiles of gentamicin in PBS. The results showed that the release profile of gentamicin from Mg foam with 10% and 25% spacer in PBS solution was in accord with the common dissolution kinetics of an active ingredient from polymeric drug delivery systems.

e. Three-dimensional honeycomb-structured Mg scaffolds were fabricated by the laser perforation technique using a programmable multifunctional laser processing machine (Tan et al. 2009). Finite element methods were applied to analyze the influence of pore arrangement, size, and porosity on the compressive property of porous Mg, and the results showed the pore arrangement is the most significant factor. According to the Taguchi method, the optimum combination of pore parameters is porosity = 70%, pore size = 300 μm, pore arrangements angle = 90°, and the calculated result agreed well with the experimental result. After that, they coated β-TCP on the surface of the porous Mg to improve the biocompatibility and corrosion properties (Geng et al. 2009b).

f. The relationship between gas pressure and the porosity in lotus-type porous Mg fabricated by a metal/gas eutectic unidirectional solidification (GASAR process) in a mixture of hydrogen and argon with high pressure was evaluated (Yuan et al. 2005). Results found that the porosity decreased with the increase of total gas pressure and decrease of the partial pressure of hydrogen.

g. To compare the mechanical property of Mg foam with other metal foams, open cellular SG91A Al and AZ91 Mg were fabricated (Yamada et al. 1999) using a polyurethane form and plaster mold. Results indicated that although the ductility of the two solid materials are much different, the stress–strain relationship of the cellular AZ91 Mg was in agreement with that of the cellular SG91A Al, which meant the ductility of solid materials did not have much effect on the mechanical properties of the cellular alloys.

h. Witte et al. (2007b,c) fabricated AZ91D scaffolds using the negative salt pattern molding process and inserted them into the right knee of adult New Zealand white rabbits with autologous bone cylinders in the left knee as control. After 3- and 6-month implantation, the bone volume per tissue volume index (BV/TV) were 45.7% and 43.4%, which were higher than that of the control group, about 37.2% and 39.6%, indicating that significantly more regenerated bone was formed on the AZ91 side.

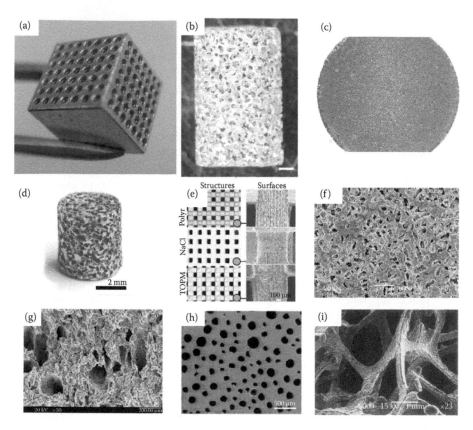

FIGURE 3.1   Various types of Mg foams. (a) Three-dimensional honeycomb-structured Mg, (b) porous Mg structure, (c) Mg foams, (d) AZ91D scaffolds, (e) topologically ordered porous Mg, (f and g) Mg foams, (h) lotus-type porous pure Mg, and (i) open-cellular AZ91. ([a] From Tan, L., Gong, M., Zheng, F., Zhang, B., and Yang, K., *Biomedical Materials*, 4, 015016, 2009. With permission; [b] From Lietaert, K., Weber, L., Van Humbeeck, J., Mortensen, A., Luyten, J., and Schrooten, J., *Journal of Magnesium and Alloys*, 2013. With permission; [c] From Aghion, E., Yered, T., Perez, Y., and Gueta, Y., *Advanced Engineering Materials*, 12, B374, 2010. With permission; [d] From Witte, F., Ulrich, H., Palm, C., and Willbold, E., *Journal of Biomedical Materials Research Part A*, 81A, 757, 2007. With permission; [e] From Staiger, M. P., Kolbeinsson, I., Kirkland, N. T., Nguyen, T., Dias, G., and Woodfield, T. B., *Materials Letters*, 64, 2572, 2010. With permission; [f] From Wen, C., Yamada, Y., Shimojima, K., Chino, Y., Hosokawa, H., and Mabuchi, M., *Materials Letters*, 58, 357, 2004. With permission; [g] From Zhuang, H., Han, Y., and Feng, A., *Materials Science and Engineering: C*, 28, 1462, 2008. With permission; [h] From Gu, X., Zhou, W., Zheng, Y., Liu, Y., and Li, Y., *Materials Letters*, 64, 1871, 2010. With permission; [i] From Yamada, Y., Shimojima, K., Sakaguchi, Y., Mabuchi, M., Nakamura, M., Asahina, T., Mukai, T., Kanahashi, H., and Higashi, K., *Materials Science and Engineering: A*, 272, 455, 1999. With permission.)

i. Four open cellular structured Mg alloys (AZ63, M2, ZM21, and MZX211) were processed with a pore size near 500 mm and a porosity of 75% (Lietaert et al. 2013) using NaCl as space holder. Among these alloys, AZ63 displayed super corrosion resistance and high tensile strength. Only 3.2% fraction weight of AZ63 foam was lost after being immersed in PBS for 24 h with 0.2 ml/(d·cm^2) H$_2$ release, followed by M2, ZM21, and MZX211. Compared to other alloy foams, MZX211 corroded the fastest due to Ca additions, of which the fraction weight loss and H$_2$ release were 67% and 6.2 ml/(d·cm^2). After the immersion test, the 0.2% yield tensile strength of AZ63 decreased from 3.3 to 2.1 MPa. But MZX211 and ZM21 completely failed in mechanical integrity. Figure 3.1 displays the various structures of Mg foams, and their mechanical properties are shown in Table 3.1.

Although various types of Mg foams are fabricated for bone implantation applications, none of them are successful in the in vitro cell culture of osteoblasts or in vivo implant experiments. The main reason may be the fast corrosion rate. Because the foams are built up by holes, they degrade much faster than the solid metal alloys, which is the main problem for further applications for Mg foams.

TABLE 3.1
Mechanical Properties of Mg Foams

Metal and Alloy	E (GPa)	Porosity and Pore Size (µm)	CS (MPa)	Ref.
Mg	–	28% ± 1.3%, 170 ± 19	23.9	Gu et al. 2010e
Mg	1.8	35%, 250	18	Wen et al. 2004
	1.2	40%, 250	15	
	0.8	45%, 250	13	
	1	50%, 250	12	
Mg	1.3	45%, 100	16	Zhuang et al. 2008
	0.9	45%, 200	14	
	0.9	45%, 300	13	
	0.6	45%, 400	12	
Mg	–	800	–	Staiger et al. 2010
Mg	–	10%, 120	–	Aghion et al. 2010
		25%, 50		
Mg	–	70%, 300	15.16	Tan et al. 2009
Mg	18	36%, 200–400	25	Zhuang et al. 2008
	8	43%	20	
	3.6	55%	15	
AZ91D	–	97.2%	0.07–0.13	Yamada et al. 1999
AZ63	0.8	75%, 578	3.3 (YS)	Lietaert et al. 2013
M2	0.6		1.4 (YS)	
ZM21	0.7		1.6 (YS)	
MZX211	0.6		0.6 (YS)	

3.3 COMPOSITES

Calcium phosphate salts, fluorapatite, bioactive glass, and metallic materials were added into Mg alloys as reinforcements to form Mg metal matrix composites (MMCs) and to improve the corrosion and mechanical properties shown in Table 3.2. Meanwhile, most of the selected reinforcement materials are bioactive or biocompatible; therefore, the MMCs exhibit better compatibility compared to the magnesium matrix. The microscopic images of different Mg-based composites are illustrated in Figure 3.2.

3.3.1 CALCIUM PHOSPHATE SALTS

The most common additives used in MMCs are calcium phosphate salts (Van der Stok et al. 2011; Wagoner Johnson and Herschler 2011; Bose and Tarafder 2012), which are the most widely used substitutes in hard tissue engineering. Calcium phosphates possess similar composition to bone mineral and exhibit excellent biocompatibility in vivo.

As a natural bone composition, hydroxyapatite (HA) (Oguchi et al. 1995; Grandfield et al. 2010; Zhou and Lee 2011; Bose and Tarafder 2012) can bind directly to bone without intervening fibrous tissue and has been added into various types of Mg alloys in order to develop novel HA-MMC for hard tissue reparation.

HA was first chosen as a reinforcement to add into AZ91D matrix in 2007 (Witte et al. 2007a). The MMC-HA was investigated by mixing it with 20 wt.% HA powder with AZ91D in the extruded rod. After immersion in artificial sea water for 24 h and 72 h, $CaCO_3$ and $Mg(OH)_2$ were formed on the surfaces of MMC-HA samples. But as for the samples immersed in DMEM with or without proteins, no $CaCO_3$ was found on the surfaces. The fast corrosion rate was observed for the specimens immersed in sea water, followed by immersion in DMEM with protein addition and by pure DMEM. Less corrosion attack and lower corrosion rates were found on MMC-HA compared to the AZ91D sample in all kinds of immersion media. Cell viability in MTT tests of direct and preincubation (the Mg samples were incubated for 2 days in cell culture medium prior to cell seeding) assay of the AZ91D and MMC-HA samples illustrate that preincubation treatment led to an increase in cell viability, and more human bone-derived cells (HBDCs), MG-63 (human osteosarcoma cell line), and RAW 264.7 (mouse tumor-derived macrophage cell line) were found on MMC-HA samples compared to AZ91D.

Mg-20wt.% HA, Mg-4wt.%Zn-20wt.% HA, Mg-40wt.% HA, and Mg-4wt.%Zn-40wt% HA (D.-B. Liu et al. 2010) were produced and immersed in SBF for 100 h. With the increase of HA, the immersed samples corroded more slowly and kept more integral surface by inducing the deposition of Ca-P compounds on the surfaces of composites. Furthermore, the addition of Zn was favorable to improve the corrosion resistance of HA/Mg composites due to the densification of composites.

Ye et al. (2010) fabricated MMC using Mg-2.9Zn-0.7Zr alloy as the matrix and 1 wt.% nano-hydroxyapatite (n-HA) particles as reinforcements. MMC and Mg-Zn-Zr alloy samples immersed in the SBF for 3, 10, and 20 days. After 10 days of immersion, the surface of the Mg-Zn-Zr alloy suffered a severe corrosion attack, and the

TABLE 3.2

Mechanical and Chemical Properties of Mg Composites

Components	CS (MPa)	UTS (MPa)	YS (MPa)	E (GPa)	I_{corr} (μm/cm^2)	E_{corr} (V)	CR	Ref.
ZK60/20%CPP	495	228	212.77	39	0.088	–	–	Feng and Han 2010
ZK60/5%CPP	–	~330	~300	~46	–	–	–	Feng and Han 2011
AZ91/20%FA	–	~130	123.2	37.5	2.3	–1.43	0.8 mg/(h·cm^2) (SBF)	Razavi et al. 2010; Fathi et al. 2011
AZ91/20%HA	–	–	264.3	40	–	–	2.0–3.2 mm/yr (DMEM)	Witte et al. 2007a
Mg/1Ca	–	217.28	147.78	–	15.82	–1.65	–	Y.-F. Zheng et al. 2010
Mg/MgO/MgZn	–	131.8	–	–	95.02	–1.475	0.25 mg/(h·cm^2) (SBF)	Lei et al. 2012
Mg/10HA	–	171.6	117.3	–	60.02	–1.604	1.43 mm/y (SBF)	Gu et al. 2010d
As-extruded pure Mg	–	197.1	107.2	–	51.34	–1.8	–	Gu et al. 2010d
MgCa-HA-TCP	128.7	–	–	–	15.23	–1.49	0.029 ml/(h·cm^2) (Hank's)	Gu et al. 2011d
ZK30-10BG	~273	–	–	–	–	–	–	Huan et al. 2011
Mg/5.6 Ti	–	248	163	–	–	–	–	Hassan and Gupta 2002

Note: CR, corrosion rate; CS, compressive strength; UTS, ultimate tensile stress; YS, yield strength.

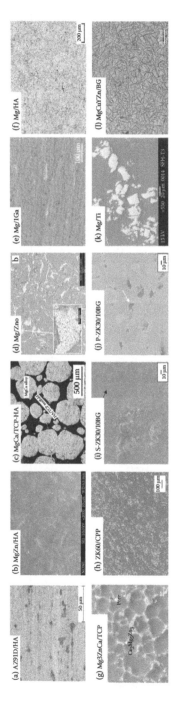

FIGURE 3.2 Various types of Mg composites. (a) AZ91D/HA, (b) MgZn/HA, (c) MgCa/TCP-HA, (d) Mg/ZnO, (e) Mg/1Ca, (f) Mg/HA, (g) Mg3ZnCa/TCP, (h) ZK60/CPP, (i) S ZK30/10BG, (j) P ZK30/10BG, (k) Mg/Ti, and (l) MgCaYZn/BG. ([a] From Witte, F., Feyerabend, F., Maier, P., Fischer, J., Störmer, M., Blawert, C., Dietzel, W., and Hort, N., *Biomaterials*, 28, 2163, 2007. With permission; [b] From Liu, D.-B., Chen, M.-F., and Ye, X.-Y., *Frontiers of Materials Science in China*, 4, 139, 2010. With permission; [c] From Gu, X., Wang, X., Li, N., Li, L., Zheng, Y., and Miao, X., *Journal of Biomedical Materials Research Part B: Applied Biomaterials*, 99, 127, 2011. With permission; [d] From Lei, T., Tang, W., Cai, S.-H., Feng, F.-F., and Li, N.-F., *Corrosion Science*, 54, 270, 2012. With permission; [e] From Zheng, Y., Gu, X., Xi, Y., and Chai, D., *Acta Biomaterialia*, 6, 1783, 2010. With permission; [f] From Khanra, A. K., Jung, H. C., Hong, K. S., and Shin, K. S., *Materials Science and Engineering: A*, 527, 6283, 2010. With permission; [g] From Liu, D., Zuo, Y., Meng, W., Chen, M., and Fan, Z., *Materials Science and Engineering: C*, 32, 1253, 2012. With permission; [h] From Feng, A. and Han, Y., *Journal of Alloys and Compounds*, 504, 585, 2010. With permission; [i] From Huan, Z., Zhou, J., and Duszczyk, J., *Journal of Materials Science: Materials in Medicine*, 21, 3163, 2010. With permission; [j] From Huan, Z., Zhou, J., and Duszczyk, J., *Journal of Materials Science: Materials in Medicine*, 21, 3163, 2010. With permission; [k] From Hassan, S., and Gupta, M., *Journal of Alloys and Compounds*, 345, 246, 2002. With permission; [l] From Zhang, X., Chen, G., and Bauer, T., *Intermetallics*, 29, 56, 2012. With permission.)

integrity of the surface was destroyed after 20 days of immersion. On the contrary, Mg-Zn-Zr MMC maintained its integrity with some white precipitates containing C, O, Mg, P, and Ca on both surfaces after 20 days of immersion. The average corrosion rate calculated by the immersion test of MMC was 2.5×10^{-3} mm/d, which was only half of that of the Mg-Zn-Zr alloy although there was no significant difference in their electrochemical properties. During 5 days of coculture, the osteoblasts near Mg-Zn-Zr alloy samples were killed while cells growing near MMC increased and attached onto the MMC surface.

Magnesium-hydroxyapatite (Mg/HA) and ZM61/HA composites were made by the melting and extrusion route (Khanra et al. 2010); 0, 5, 10, and 15 wt.% HA powders were mixed with metal matrix. The presence of HA particles could refine the matrix grain during the extrusion procedure and improve the strength of the matrix. However, the agglomeration of HA in high concentrations caused brittle fractures in the MMC. ZM61-HA showed higher strength than Mg-HA.

Mg/HA (10, 20, and 30 wt.%) was prepared by the powder metallurgy method (Gu et al. 2010d). The main phases of Mg/HA were α-Mg and HA. The yield tensile strength of the Mg-10HA composites increased compared with that of as-extruded Mg, but with the increase of HA, the tensile strength and ductility of the composites decreased. In the electrochemical test in SBF, Mg/HA composites showed weaker corrosion resistance than the as-extruded Mg attributed to more galvanic coupling between the Mg matrix and HA particles. The results were different from the result of the references (Witte et al. 2007a; D.-B. Liu et al. 2010; Ye et al. 2010) because of the fabrication method, and this MMC was compared to as-extruded Mg, which has a different status. The extract of Mg/10HA and Mg/20HA had less effect on the cell viability of L929 cells, but the extract of Mg/30HA could lead cytotoxicity to L929 cells.

Although HA is widely used in bone reparation, it will stay for a long time in host because of its poor bioabsorbability. In contrast, tricalcium phosphate (TCP) could degrade during bone regeneration and be completely substituted by bone formation (Kamitakahara et al. 2008).

A nanosized 1 wt.% β-tricalcium phosphate (β-TCP)/Mg–3Zn–Ca composite (D. Liu et al. 2012) was fabricated using a novel melt shearing technology combined with a high-pressure die casting (HPDC) process, reducing the agglomerate phenomenon of β-TCP particles in the matrix. The tensile strength of this kind of MMC was only 125.4 MPa, and the elongation was 2.85%, which could not match the performance requirement. Therefore, it was suggested to use hot extrusion and heat treatment processes in order to improve the mechanical properties.

Gu et al. (2011d) fabricated a MgCa/HA-TCP (hydroxyapatite-tricalcium phosphate) composite by infiltrating molten Mg-1Ca into a HA-TCP porous scaffold. The compressive strength of this MMC was 128.7 MPa, which was about 200-fold higher than that of the original porous HA-TCP scaffold and half of that of the MgCa alloy. The electrochemical test and immersion test carried out in Hank's solution illustrated that the corrosion resistance of MgCa/HA-TCP MMC was improved by 68% compared to the MgCa alloy. After a long immersion, infiltrated MgCa degraded completely, and the HA-TCP scaffold remained with a lower corrosion rate. The 100% extract of this MCC could indicate Grade II cytotoxicity to L929 and MG63 cells, but after 50% and 10% degrees dilution, the cell compatibilities were improved to Grade I.

Feng and Han (2010) mixed 0, 10, 20, and 30 wt.% calcium polyphosphate powder (CPP) with ZK60A magnesium powder to fabricate a matrix composite by powder metallurgy. With the addition of CPP, the corrosion resistance of the composites was improved, of which the I_{corr} decreased from 0.116 to 0.077 μA/cm^2. After immersion in SBF for 240 h, the composites accelerated hydroxyapatite precipitation from the simulated body fluid, which could be a good point for bone implant application. On the contrary, the mechanical property decreased with the increase amount of CCP; the ultimate tensile strength and yield strength were weakened from 280 to 210 MPa and 245 to 205 MPa, separately, with the increase of E modulus due to the detachment and CPP particle fracture. Therefore, they redesigned the components of the composite with 2.5, 5, 7.5, and 10 wt.% calcium polyphosphate particles (Feng and Han 2011). Higher concentrations of CPP lead to voids and agglomerations for the composites. The mechanical properties of the composites raised to the top as the content of CPP increased from 0 to 5 wt.% and then decreased with the further increase of CPP from 5 to 10 wt.%. Samples were immersed in physiological saline for 1 to 12 days to evaluate their chemical properties by weight losses, pH value changes, and Mg ion concentrations of the solutions. Similar to the results of the former studies, more CPP resulted in a lower degradation rate.

3.3.2 BIOGLASSES

Bioglass is another core component of biomaterials used to repair, restore, and regenerate bone and tissues in the human body (Tilocca 2010; Kaur et al. 2014). The influence of bioglass on the biocompatibility and mechanical property of Mg alloys has been studied.

Bioactive glass (BG, 45S5) particles at 5, 10, and 20 wt.% (3.4%, 6.9%, and 14.3% by volume) were added into ZK30 (3 wt.% Zn, 0.6 wt.% Zr) Mg alloys by a semisolid high-pressure casting (SSC) process to form ZK30-BG MMC, respectively (Huan et al. 2010b). The porosity of ZK30-BG MMC increased with bioactive glass particles, which led to the decrease of the compressive strength of MMC from 330 to 240 MPa as the BG particles increased from 1 to 20 wt.%. Immersed in cell culture medium-MEM, there was a bone-like apatite layer with Ca/P ratios of 1.2–1.35 forming on the MMC surface. And then the author tried another method to fabricate denser MMC using the powder metallurgy (P/M) method (Huan et al. 2011). By comparing the two methods, it was found that SSC composites exhibited higher porosity and a faster degradation rate than both ZK30 alloy and P/M composites. P/M composites showed the highest corrosion resistance by the dynamic immersion test in Earle's balanced salts (E-MEM). The extraction of P/M composites exhibited higher biocompatibility to rat bone marrow stromal cells (rBMSC) than that of the ZK30 matrix. All of these good characteristics were attributed to the protective Ca/P layer induced by the dissolution of BG particles. Apart from that, MMC significantly enhanced the ALP activity and osteoblastic differentiation of rBMSC (Huan et al. 2012).

X. Zhang et al. (2012a) fabricated a Mg75Cu13.33Y6.67Zn5/metallic glass MMC by Bridgman solidification. In the composite, the flake-shaped, microsized, solid solution phase of α-Mg dispersed homogeneously in the glassy matrix. The compressive strength (1040 MPa) of the composite was two times more than that (350 MPa) of AZ31, and the elongation reached 19%, which was greatly improved compared to the 2% elongation of the BMG. In addition, although the corrosion resistace of BMG composite in Hank's solution was reduced a little compared to BMG, it was incomparable for crystalline alloys, such as AZ31.

3.3.3 METALS

Y.-F. Zheng et al. (2010) fabricated Mg/Ca composites by mixing 1 wt.%, 5 wt.%, and 10 wt.% Ca powders into Mg powders using the powder metallurgy method without forming Mg_2Ca. With the increase of Ca, the mechanical properties of the composites decreased because of increased amount of local damage caused by Ca particles, of which the UTS tensile strength dropped from 217.28 MPa to 200.25 MPa, and the elongation shortened from 14.36% to 7.74%. However, the corrosion resistance of the composites in DMEM were improved by the addition of Ca, which was attributed to the precipitation of $CaCO_3$ and $Ca(OH)_2$ on the surface of the samples against further corrosion. Cell viabilities of the L929 cells incubated with extracts of the Mg/1Ca and Mg/5Ca composites were kept at the same level as the control group whereas the Mg/10Ca composite indicates round-shaped cells, suggesting Mg/10Ca composite is cytotoxic to L929 cells, which might be caused by the high concentration ions (Mg 479.6 μg/ml, Ca 91.4 μg/ml) and pH value (9.23) of the Mg/10Ca composite.

Ti particulate-reinforced Mg materials were synthesized using the disintegrated melt deposition technique followed by hot extrusion (Hassan and Gupta 2002). The existence of Ti in Mg/5.6Ti MMC improved the elongation of Mg from 7.7% to 11.1%, and the 0.2% yield strength by 50% from 100 to 150 MPa.

3.3.4 OTHERS

A magnesium/silicon carbide composite (Nunez-Lopez et al. 1996) was prepared by magnetron sputtering Mg onto a substrate surface impregnated with silicon carbide particles. The MMC of ZC71(Mg-Zn-Cu)/SiC revealed a low corrosion current density of about 10 μA/cm^2 in the electrochemical test in 3.5 wt.% NaCl solution, and after immersion in 3.5 wt.% NaCl solution of pH 6.5 for 1 and 36 h, the surface topography was relatively unchanged; therefore, silicon carbide caused no microgalvanic corrosion.

Fluorapatite (FA) nanocomposites at 10 wt.%, 20 wt.%, and 30 wt.% (Razavi et al. 2010; Fathi et al. 2011) were added into AZ91 alloy using powder metallurgy (PM), respectively. When added to 20 wt.% FA, the component got to the highest tensile stress of 123.2 MPa. Electrochemical tests carried out in the Ringer solution and the immersion test carried out in the SBF solution show that the addition of FA could strengthen the corrosion resistance of the AZ91 alloy by forming a calcium magnesium phosphate layer on the surface of the Mg alloys.

Nanosize ZnO particles were mixed with pure Mg (Lei et al. 2012), and reactions took place when the mixture was heated to 550°C as the following: Mg + ZnO → MgO + Zn, Mg + Zn → Mg_xZn_y. As reviewed by XRD, the composited was made of

Mg, MgO, and Mg_xZn_y. The ZnO particles induced the refinement of the grains and formation of intermetallic, which enhanced the overall average hardness to 66.2 HB, increased by 30% as compared to pure Mg. The tensile strength of MMC increased to 131.8 MPa, but the elongation of MMC was significantly reduced to 6.0%. In addition, the corrosion rate of MMC was four times lower than pure Mg in SBF.

3.4 ULTRAFINE-GRAINED STRUCTURE

One of the reasons for the fast corrosion rate is the galvanic corrosion between secondary phases and the Mg matrix. Therefore, ultrafine-grained Mg alloys with uniformly distributed second phases are in desire. Rapidly solidified (RS) (Guo and Shechtman 2007; Gu et al. 2010a; Z. Xu et al. 2011), high-pressure torsion treatment (HPT) (Gao et al. 2011), equal channel angular pressing (ECAP) (Kang et al. 2010; Gu et al. 2011b), and cyclic extrusion and compression (CEC) (X. Zhang et al. 2012b) processing were applied to obtain ultrafine-grained Mg alloys. Theses ultrafine-grained Mg alloys are displayed in Figure 3.3.

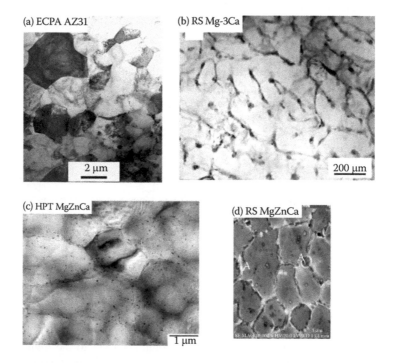

FIGURE 3.3 Ultrafine-grained Mg alloys. (a) ECPA AZ31, (b) RS Mg-3Ca, (c) HPT MgZnCa, and (d) RS MgZnCa. ([a] From Zuberova, Z., Kunz, L., Lamark, T., Estrin, Y., and Janeček, M., *Metallurgical and Materials Transactions A*, 38, 1934, 2007. With permission; [b] From Gu, X., Li, X., Zhou, W., Cheng, Y., and Zheng, Y., *Biomedical Materials*, 5, 035013, 2010. With permission; [c] From Gao, J., Guan, S., Ren, Z., Sun, Y., Zhu, S., and Wang, B., *Materials Letters*, 65, 691, 2011. With permission; [d] From Xu, Z., Smith, C., Chen, S., and Sankar, J., *Materials Science and Engineering: B*, 176, 1660, 2011. With permission.)

Reciprocating extrusion (RE) of rapidly solidified (RS) Mg-6Zn-1Y-0.6Ce-0.6Zr alloy was fabricated (Guo and Shechtman 2007). The increased RE passes led to a more homogeneous microstructure of the material and a smaller size of the strengthening particles. After two passes RE, the average grain size of the matrix was reduced to ~1.2 μm with nanoscaled precipitates (<50 nm) homogeneously dispersed inside. This unique microstructure contributed to two distinct yield regions in the tensile test and a high yield strength of about 332 MPa with high ductility (27%) of RE-4 alloy.

Rapidly solidified (RS) Mg-3Ca alloy ribbons (Gu et al. 2010a) were fabricated at different wheel-rotating speeds (15, 30, and 45 m/s) by a melt-spinning technique. The decreasing surface area ratio of the second phase Mg_2Ca to the Mg matrix caused by the RS process enhanced the microstructural and electrochemical homogeneities. The higher the rotating speed and the thinner the ribbons, the stronger the corrosion resistance that was obtained. In the cytotoxity test, as-cast Mg-3Ca alloy extract showed toxicity to the L929 cells due to its poor corrosion resistance; on the contrary, the RS-Mg-3Ca alloy ribbons exhibited good cytocompatibility.

MgZnCa (Z. Xu et al. 2011) alloys were melted from powder and solidified by different cooling methods, of which the cooling rates were ~40°C/min (cooling in molds with thermal insulation), ~100°C/min (cooling in molds open to air), and ~3000°C/min (cooling with liquid nitrogen), respectively. The SEM images showed clearly that the high cooling rates reduced the grain size from 100 to 3 μm smaller grain size, caused secondary phase uniformly precipitating and forming a continuous 3-D network around the very small grains, and resulted in high Zn and Ca contents in matrix Mg. The supersaturating of alloying elements and more homogeneous microstructures enhanced the corrosion resistance of the alloy measured by electrochemical tests, of which the corrosion potential and current for as-cast Mg-20Zn-1Ca and ~3000°C/min RS Mg-20Zn-1Ca were −1.54 V, 2.07×10^{-4} A/cm^2 and −1.58 V, 1.88×10^{-4} A/cm^2, respectively.

Mg-2 wt.% Zn-0.24 wt.% Ca was prepared using an electric resistance heating furnace following by high-pressure torsion (HPT) treatment (Gao et al. 2011). After HPT treatment, the averaged large number of the average grain size of α-Mg was reduced from 100 μm to only 1.2 μm with nanosized second phase particles precipitated in the grain interiors instead of in the grain boundaries. Immersed for 2 days in SBF, the as-cast alloy suffered from pitting corrosion around the second phase (grain boundaries), and the conventional extrusion alloy was peeled off partly on the surface; however, the HPT-treated sample exhibited uniform corrosion without pitting corrosion, peeling off due to the uniform distribution of the second phase, the decrease of the volume fraction of the second phase, or the relieving of internal stress. In the electrochemical test, the corrosion current density of HPT-treated alloy decreased from 5.3×10^{-4} A/cm^2 to 3.3×10^{-6} A/cm^2 compared with the as-cast Mg-Zn-Ca alloy.

As-extruded AZ31 alloy bars were treated by equal channel angular pressing (ECAP) with or without 125 MPa back pressure (BP) (Kang et al. 2010). Results found that with an increase in the number of passes of ECAP, the grain size of the sample was refined, and a homogeneous structure of the sample was formed. In addition, the grain-refining effect of BP-ECAP was far superior to that without back pressure. The average grain size was refined to 8.5 μm after 4-ECAP and further

TABLE 3.3
Mechanical and Chemical Properties of Ultrafine-Grained Mg

Alloy	Grain Size (μm)	YS (MPa)	UTS (MPa)	Elongation (%)	E_{corr} (V)	I_{corr} (μA/cm²)	CR	Ref.
RS45 Mg-3Ca	0.2–0.5	–	–	–	−1.516	17.1	0.39 mm/yr (SBF)	Gu et al. 2010a
RS Mg-20Zn-1Ca	3	–	–	–	−1.54	188	–	X. Xu et al. 2011
HPT MgZnCa	1.2	–	–	–	–	3.3	–	Gao et al. 2011
BP-ECAP AZ31	1.7	290	430	31	–	25.37	–	Kang et al. 2010
SC ZA31	450	50	170	10	–	–	0.15 mg/(d-cm²) (Hank's)	Zuberova et al. 2007
HR ZA31	20	175	277	21	–	–	0.13 mg/(d-cm²) (Hank's)	Zuberova et al. 2007
ECAP AZ31	2.5	115	251	27	–	–	0.153 mg/(d-cm²) (Hank's)	Zuberova et al. 2007
RE-2-RS Mg-6Zn-1Y-0.6Ce-0.6Zr	1.2	117	340	20	–	–	–	Guo and Shechtman 2007
CEC Mg-2.73Nd-0.16Zn-0.45Zr alloy	1	270	300	30	–	3.31	1.25 mm/y (Hank's)	Z. Zhang et al. 2012b

Note: CR, corrosion rate; CS, compressive strength; UTS, ultimate tensile stress; YS, yield strength.

refined to 1.78 μm after 4-BP-ECAP. Meanwhile, splitting texture peaks were developed. These reasons contributed to the significant improvement of both strength and ductility of the samples, especially that of samples treated by ECAP with back pressure. After 4-BP-ECAP, the YS and UTS of the samples were 290 and 430 MPa, which were significantly higher than 107 and 290 MPa after 4-ECAP with the same elongation of 31%. The corrosion resistance of the AZ31 alloy was also improved by 4-ECAP and 4-BP-ECAP revealed by 20 days of immersion in Hank's solution (Gu et al. 2011b). The current densities of 4-ECAP and 4-BP-ECAP AZ31 samples are 16.52 μA/cm^2 and 25.37 μA/cm^2 in comparison with that of as-extruded AZ31 alloy 91.52 μA/cm^2. The extract of these samples exhibits acceptable toxicity to MG63 cells with grade I toxicity.

Squeeze cast (SC), hot rolled (HR) and four passes of ECAP AZ31 were compared to study the effect of grain structure on the characteristics of biomaterials (Wang et al. 2007; Zuberova et al. 2007). The average grain size of SC, HR, and 4-ECAP AZ31 samples were 450, 20, and 2.5 μm, separately. As shown in Table 3.3, the mechanical property was improved after hot rolling because of the refined grains, but it was not further improved after 4-ECAP. The same phenomenon was observed for the fatigue properties in the fatigue tests and the corrosion resistance in the immersion test in Hank's solution. The grain size and the boundary effect should be further studied.

RS-RE-RS66 (Willbold et al. 2013) with 1-μm grain size was obtained by rapid solidification processing and reciprocal extrusion. In vitro, the pure extract medium or a 1:2 dilution of RS66 extract could induce severe cytotoxicity to primary human osteoblasts. However, in a 1:11 dilution of which the ion concentration was more like physiological situations, cells grew in good condition. In vivo, RS66 cylinders corroded fastest when implanted under the skin, followed by that implanted intramuscularly and bonily, which were related with the local blood flow prior to implantation. No inflammation or sterile sinuses eliminating debris were observed, and only small gas cavities were formed at the subcutaneous and intramuscular implantation, which were completely absorbed later. Unfortunately, no enhanced bone had been formed around the RS66 implantation in the condyles compared to the blank control.

After treatment by cyclic extrusion and compression (CEC) processing, as-extruded Mg-2.73Nd-0.16Zn-0.45Zr alloy (X. Zhang et al. 2012b) had a refined grain size of ~1 μm, and due to the homogenous microstructure, the mechanical property and corrosion property were significantly improved, of which the yield strength, ultimate tensile strength, and elongation of the alloy were improved by ~71%, ~28%, and ~154%, respectively, with ~20% reduction of the corrosion rate.

3.5   GLASSY STRUCTURE

Mg-based metallic glassy alloys (MGA) were fabricated for biomedical applications to improve the corrosion properties of crystalline Mg alloys. A MgZnCa glassy alloy system was the focus of Mg-MGA, which has been developed various types of metallic glassy alloys for biomedical use, for example, $Mg_{66}Zn_{30}Ca_4$ (Y.-Y. Zhao et al. 2008), $Mg_{67}Zn_{28}Ca_5$ (Zberg et al. 2009b), $Mg_{72}Zn_{23}Ca_5$ (González et al. 2012),

$Mg_{66}Zn_{23}Ca_5Pd_6$ (González et al. 2012), by atomic concentration. Due to homogeneous single-phased structures, the compressive strengths of the MGA increased to over 800 MPa, and the corrosion resistance was significantly enhanced because the galvanic corrosion effect was minimized. However, the poor ductility of MGA is the fatal obstacle for clinical application.

Y.-Y. Zhao et al. (2008) examined the composition effects on the failure reliability by comparing two MGAs, $Mg_{66}Zn_{30}Ca_4$ and $Mg_{71}Zn_{25}Ca_4$. Results indicated that higher Mg content led to a lower shear flow barrier, which reduced the yield in metallic glasses.

Zn-rich $Mg_{60+x}Zn_{35-x}Ca_5$ (0 < x < 7) glassy alloys were fabricated by Zberg et al. (2009b), which possessed super corrosion resistance and high strength but poor ductility. They found the increasing Zn content decreased significantly the hydrogen evolution of the glassy alloys immersed in SBF; in particular, there was a distinct drop in the gas volume when Zn content reached around 28 at.%. Glassy $Mg_{60}Zn_{35}Ca_5$ discs, together with crystalline reference WZ21 alloy bars were implanted into domestic pigs. After 27 and 91 days of implantation, amorphous and crystalline samples both induced a typical fibrous capsule foreign-body reaction, but only the crystalline samples showed hydrogen evolution.

The cytocompatibility of $Mg_{66}Zn_{30}Ca_4$ and $Mg_{70}Zn_{25}Ca_5$ MGA samples were evaluated by indirect and direct tests (Gu et al. 2010c), showing higher cell viability than that of as-rolled pure Mg. Moreover, L929 and MG63 cells were found to be well adhered and proliferated on the surface of the $Mg_{66}Zn_{30}Ca_4$ samples.

The influence of Pd on MgZnCa MGA was studied (González et al. 2012) by adding Pd into Mg–Zn–Ca metallic glasses to form $Mg_{72-x}Zn_{23}Ca_5Pd_x$ (x = 0, 2, and 6 at.%). By adding Pd, the glassy structure turned into a fully crystalline structure, containing $CaZn_5$, $MgZn$, Mg_6Pd, and $Mg_6Zn_3Ca_2$ phases. The OCP shifted toward more positive values, but the corrosion current was increased from 1.7 to 2.7 mA/cm² in Hank's solution. Cytotoxicity tests confirmed that these alloys were not toxic, and preosteoblasts cultured with $Mg_{66}Zn_{23}Ca_5Pd_6$ alloy grew in a good condition.

The effects of Mn substitution for Mg on $Mg_{69-x}Zn_{27}Ca_4Mn_x$ MGA (x = 0, 0.5, and 1 at.%) were investigated (Hollister 2005). Crystalline peaks appeared and became stronger on the broad diffraction hump as Mn increased. Due to the change of crystallinity, the strength of the alloys decreased from 545 MPa to 364 MPa; however, the corrosion resistance had been improved by adding 0.5% Mn. The cell viability of Mn-doped Mg–Zn–Ca bulk metallic glass extracts showed the same level as the control group cell viability rather than the control extraction medium.

The rare-earth element ytterbium (Yb) (Yu et al. 2013) at an atomic concentration of 10% was added to the brittle MgZnCa MGA in order to improve the ductility of the MGA. $Mg_{66}Zn_{30}Ca_{4-x}Yb_x$ (x = 0, 2, 4, 6, 10) metallic glassy ribbons were fabricated. The plastic strain of Yb2 and Yb4 MGA achieved about 0.5%, and the Yb0 was too brittle and broke below 100 MPa; because of that, Yb could improve the transformation behavior of MgZnCa MGA from inhomogeneous to homogeneous deformation behavior. Low Yb concentration alloys were at their as-cast amorphous state. On the contrary, high Yb concentration up to 10 at.% weakened the glass-forming ability of MGA and induced the formation of crystalline phases. In the indirect cytotoxicity test, the $Mg_{66}Zn_{30}Ca_{4-x}Yb_x$ extract concentration of 60% displayed good cell compatibility to both fibroblasts and osteoblasts. Meanwhile, in the direct cell adhesion test, there were comparable numbers of cells growing on the surface of Yb-MGA ribbons to the control comparable, which was 80% lower on the surface of Yb0.

Production of 100-m-long, 100-μm-diameter amorphous $Mg_{67}Zn_{28}Ca_5$ wires was fabricated by Zberg et al. (2009a) with flawless surface quality bending plasticity. Because the glassy wires were below a critical size, they displayed significant bending plasticity with three times higher tensile strength (675–894 MPa) than that of crystalline samples. This unique property was interesting for biomedical application, for example, self-expandable stents.

3.6 CONCLUDING REMARKS

This chapter reviews novel structured Mg and its alloys fabricated for biomaterial application. These different structures bring new characteristics to typical Mg bulk alloys: porous structured Mg with controlled porosity and pore size that promote osteogenesis and bone rebuilding; composites of Mg that are more biocompatible and even improve the corrosion resistance; ultrafine-grained and glassy Mg that possesses superior strength with high corrosion resistance.

Based on the former studies, we propose some possible work that could be done in future research on novel structured Mg-based biodegradable metals.

a. Precise porous structure with 3-D printing technology: Computer-aided design with 3-D printing technique has developed quickly in the recent decade and has been introduced into tissue engineering to print porous scaffolds (Lam et al. 2002; Cooke et al. 2003; Pfister et al. 2004; Leukers et al. 2005; Silva et al. 2008; Rengier et al. 2010). As referenced, this technique has already been used in porous Mg (Staiger et al. 2010), and with the rapid development of 3-D printing, this layered fabrication process could print more meticulous framework.

b. Composite with biodegradable polymers: Polymers, especially natural polymers, have excellent biocompatibility, but how to make Mg/polymer composites rather than only coatings should be studied.

c. Enhanced ductility of nanostructured/glassy structure Mg-based biodegradable materials: Nanostructured/glassy structure Mg-based materials possess super compressive strength; however, the elongation is limited within 1% (Zberg et al. 2009b; Yu et al. 2013). Therefore, to enhance the ductility is the breakthrough for further applications, especially for stent use. Nonbiomedically used MGA composites by adding Mo or Ti porous particles (Jang et al. 2010, 2011, 2012; Oka et al. 2013) were fabricated for better plastic. The mechanical property has been improved; in particular, the plasticity of the material was improved by 17% due to dispersions of Mo, and no deterioration of the glassy structure of the matrix was observed. We can get ideas from these works to apply in Mg biomaterial.

d. Novel surface structured Mg-based biodegradable materials: For example, surface pattern structures have great influence on the adhesion and immigration of cells (Tziampazis et al. 2000; Segura et al. 2005; D. H. Kim et al. 2009; Jeon et al. 2010). We can fabricate various biofunctional patterns or roughness on the surface of the current novel structured Mg-based biodegradable metals to regulate the growth of cells and also control the degradation rate.

e. Single crystalline Mg or its alloys: In 2013, at the TMS Annual Meeting & Exhibition, Pravahan et al. (2013) proposed the use of Mg single crystal for biodegradable implant applications, which could reduce the corrosion and increase mechanical strength without grain boundaries. However, in a personal view, to produce large size single crystal Mg or its alloys is difficult, and the directional dependence of various properties should be taken into consideration.

f. Combined novel structures: These novel structures of Mg and its alloys referenced here could be combined together to avoid the weaknesses and obtain other superior properties from the partnerships. Many possible combined structures may be fabricated, such as porous Mg-based composites, porous ultrafine-grained/glassy Mg-based biodegradable metals, composited (the combination of microcrystalline/ultrafine-grained/glassy structures, depending on the annealing temperature and time for the original ultrafine-grained/glassy raw materials) Mg-based biodegradable metals. In this way, millions of new types of structured Mg alloys could be manufactured. For example, a two-layer structured Mg constituted of an interconnected porous structure and an inner compact Mg/salt structure was developed (Zhang et al. 2013e). The outside porous structure Mg works as scaffold material, and the interior composite structure can enhance mechanical compatibility.

To conclude, many more novel structured biodegradable Mg and Mg alloys will be developed in the future. The research direction still will be focused on the following points: to obtain better biocompatibility, suitable mechanical properties, and regulate the corrosion rate.

4 Surface Modification Techniques for Biodegradable Mg and Its Alloys

4.1 INTRODUCTION

Traditional biomedical materials include stainless steels, NiTi alloys, Ti-based alloys, and Co-based alloys. They are all bioinert materials. Therefore, traditional surface treatments on bioinert materials focus on the enhancement of surface bioactivity or the prevention of the toxic metallic ion release from the matrix, which would benefit subsequent cell adhesion, cell proliferation, and cell differentiation. Moreover, inflammation and infection caused by corrosion products in these materials also need to be taken into consideration. Thus, the surface treatments on traditional biomedical materials often contain a process to improve the bioactivity of the surface.

Different from traditional biomedical materials, biodegradable Mg and its alloys are novel bioactive materials, which implies benign biocompatibility and cellular response during degradation. However, biodegradable Mg alloys have encountered a big problem, which is that they usually have an overquick corrosion rate, which leads to the loss of mechanical property integrity at an early stage. One more issue is its pitting corrosion mode, which, in general, results in nonuniform corrosion on the surface and needs to be improved. Faced with such problems, surface treatments on biodegradable Mg alloys often concentrate more on the retardation of degradation to sustain the mechanical property integrity of the materials and guarantee good biocompatibility at the same time; moreover, a homogeneous corrosion mode is another expected target.

Common surface treatments on biodegradable Mg alloys can be summarized with three specific aspects, including mechanical surface treatment, physical surface treatment, and chemical surface treatment. Mechanical surface treatment often serves as pretreatment for subsequent treatment, but it can also be an independent treatment of the material industry for its effectiveness and cheapness. Physical surface treatment consists of plasma surface treatment and laser surface treatment. Chemical surface treatment introduced here covers chemical conversion coating, electrochemical treatment, biomimetic deposition, the sol-gel method, polymer and organic coating, and molecular self-assembled coating.

4.2 MECHANICAL SURFACE TREATMENT

Mechanical surface treatment is generally utilized to achieve desired surface morphology and, thus, desired surface properties. It is often used as a pretreatment in surface modification, including shot peening (or shot blasting) and burnishing (or brightening).

As for biodegradable Mg alloys, mechanical surface treatment can be an effective way to improve the properties of the alloys. Surface roughness is a critical property for both bone implantation and stent application. As for bone implantation, a rougher surface is beneficial for cell adhesion, cell proliferation, and cell differentiation. In addition, a rough surface provides a larger area contacting the bone tissue after implantation. However, inflammation, worse mechanical property, and corrosion resistance occur when surface roughness is beyond a certain value. Nguyen et al. (2012) found that increasing the surface roughness of Mg-based implants can increase the corrosion rate although there was no pitting corrosion observed. Walter et al. (2013) investigated different surface roughness caused by 120-grit SiC paper grinding and 120- to 2500-grit SiC paper grinding. The results revealed that despite the surface roughness of AZ91D Mg alloy not showing a significant effect on general degradation resistance in SBF, it played a critical role in the localized degradation behavior of the alloy. Walter and Kannan (2011) also reported that the corrosion current and the pitting tendency of the alloy also increased with the increase of surface roughness. Moreover, the influence of surface roughness on hemocompatibility also needs to be noted. Generally speaking, a smoother surface achieves better hemocompatibility owing to less adsorption of plasma protein and blood platelets. In that case, cellular morphology and structure are less affected, and the blood flow will not be disrupted. Therefore, the surface roughness needs to be controlled within a certain range. Mechanical surface treatment can be an effective way to alter the surface roughness and thus affect properties of alloy matrix.

Residual stress is another important factor affecting alloys' properties when applying a mechanical surface treatment such as shot peening. The high residual compressive stress generated in the subsurface via a deep rolling process was claimed to reduce the corrosion rate of a biphasic magnesium-calcium alloy by nearly two orders (Denkena and Lucas 2007). Mhaede et al. (2014) utilized shot peening followed by a dicalcium phosphate dihydrate (DCPD)-coating treatment to improve the mechanical properties and corrosion resistance of AZ31 Mg alloys. The results indicated that shot peening pretreatment altered the surface morphology before coating deposition. As surface morphology can affect the size of dicalcium phosphate dihidrate, the shot peening pretreatment would induce different coating morphology and control the mechanical properties and corrosion resistance. Figure 4.1 shows the different surface morphology of DCPD coating formed by electrodeposition after various surface conditions. It is obvious that shot peening at higher Almen intensity corresponded to smaller DCPD crystal sizes. However, shot peening treatment negatively influenced the corrosion resistance of AZ31 Mg alloys compared to a grinded-only surface.

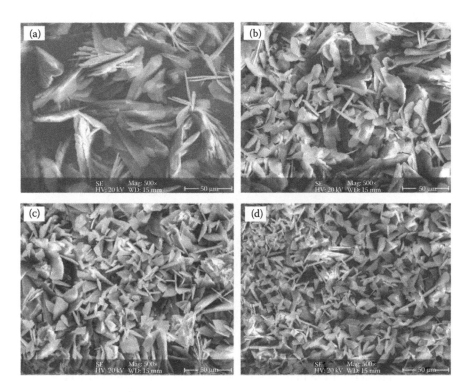

FIGURE 4.1 Surface morphology of DCPD coating formed by electrodeposition after various surface conditions. (a) grinded + DCPD, (b) SP at 0.042 mmN + DCPD, (c) SP at 0.140 mmN + DCPD, and (d) SP at 0.262 mmN + DCPD. (Adapted from Mhaede, M., F. Pastorek, and B. Hadzima. *Materials Science and Engineering: C*, 39, 330–35, 2014. With permission.)

Other than shot peening, Pu et al. (2011c, 2012b) reported a grain-refined and basal-textured surface developed by a severe plastic deformation (SPD) process, severe plasticity burnishing (SPB), as illustrated schematically in Figure 4.2. Burnishing, with or without liquid nitrogen, is conducted with a fixed roller. The AZ31B disk was fixed in the lathe chuck and rotated during processing. A machining clearance cut using an uncoated carbide insert was conducted to reduce the diameter from 130 to 128 mm in order to standardize initial burnishing conditions. The circumferential residual stress mode turned out to be less compressive or even tensile after burnishing. After 200 h immersion in 5 wt.% NaCl solution, the surface of the dry burnishing and cryogenic burnishing multistep-treated samples were much more even, which indicated better corrosion resistance.

In previous similar experiments, Pu et al. (2011a, 2012a) investigated the influence of different SPB conditions, such as different edge radius and with or without liquid nitrogen, on the performance of the surface integrity. The results showed that application of liquid nitrogen reduced the surface roughness by about 20% compared to dry machining using both 30-μm and 70-μm edge radium

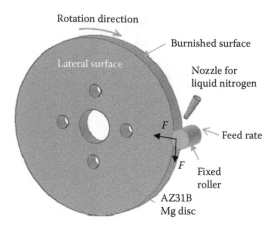

FIGURE 4.2 Schematic of the severe plasticity burnishing (SPB) process. (From Pu, Z., G. L. Song, S. Yang, J. C. Outeiro, O. W. Dillon, D. A. Puleo, and I. S. Jawahir. *Corrosion Science*, 57, 192–201. 2012b. With permission.)

tools. In addition, the combination of cryogenic cooling and a large cutting edge radius achieved a depth of compressive residual stresses of about 200 µm below the surface, increasing the compressive areas by a factor of 10 times compared to the sample before machining experiments. They also reported a "white layer" with such a method for the first time in biodegradable Mg alloys (Pu et al. 2010, 2011b) as Figure 4.3 shows. The difference was significant in the surface layer on differently machined samples. Although the slowest corrosion rate occurred in the machined sample with the thickest white layer, the sequence of corrosion rates were not coincident with the sequence of white layer thickness. S. Yang et al. (2011) also utilized a similar machine to investigate the surface integrity and product sustainability and found that the cryogenic burnishing process led to a more than 2-mm-thick surface layer with remarkably refined microstructures formed on the burnished surface. About a 95% increase in hardness was obtained on the burnished surface.

Salahshoor and Guo (2011a,b,c, 2013; Guo and Salahshoor 2010) investigated the influence of high-speed dry face milling and deep rolling on the surface performance of Mg-Ca alloys. High compressive residual strength was obtained by these methods, which enhanced the corrosion resistance of the orthopedic Mg-Ca alloys. Kannan et al. (2011) investigated the effects of friction stir processing (FSP) on the corrosion resistance of AZ31 Mg alloy, and the results revealed that samples with FSP possessed a better corrosion resistance in SBF.

As important as the relationship between surface roughness and some properties of the materials, such as corrosion behavior and biocompatibility, is, surface quality needs more attention. Moreover, mechanical surface treatment is effective and also easy to achieve. Thus mechanical surface treatment can be combined with other surface treatments to get the desired corrosion behavior as well as good biocompatibility on degradable biomedical Mg alloys.

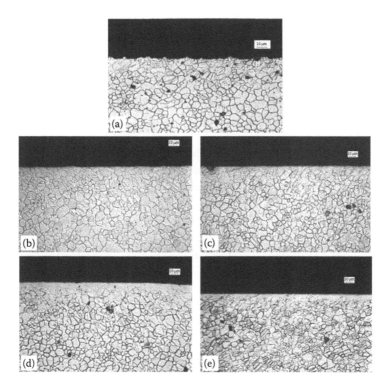

FIGURE 4.3 Microstructure of magnesium alloy discs before machining experiments (a), after machining: (b) dry machining, edge radius = 30 μm, (c) cryogenic machining, edge radius = 30 μm, (d) cryogenic machining, edge radius = 68 μm, and (e) cryogenic machining, edge radius = 74 μm. (From Pu, Z., D. A. Puleo, O. W. Dillon, and I. S. Jawahir. In *Magnesium Technology 2011*, edited by W. H. Sillekens, S. R. Agnew, N. R. Neelameggham and S. N. Mathaudhu. 2011b. With permission.)

4.3 PHYSICAL SURFACE TREATMENT

The current physical surface treatment of biodegradable Mg alloys includes plasma surface treatment and laser surface treatment. As for plasma surface treatment, it can be classified by the temperature of surface treatment. Plasma sputtering and ion implantation can be achieved near room temperature, and they belong to the category of low-temperature/pressure plasma surface treatment. Another option is plasma spray, which can induce temperature elevation over thousands centigrade during a short treatment. Some research on ion beam-assisted deposition (IBAD) are also summarized here. As for laser surface treatment, laser surface polishing and roughening, laser surface peening (LSP), laser surface melting, and laser surface cladding are discussed. It must be noted that some common plasma surface treatment in traditional NiTi alloys and biomedical stainless steels, such as plasma cladding and plasma surface quenching, are seldom used for biomedical Mg alloys. Similar situations are found for laser surface treatments such as laser surface alloying and pulsed laser deposition (PLD) for the surface modification on biomedical Mg alloys.

4.3.1 PLASMA SURFACE TREATMENT

Plasma is considered the fourth fundamental state of matter besides solid, liquid, and gas. When gas is ionized to become a system of charged particles and neutral particles, plasma comes into being.

In the field of degradable biomedical Mg alloys, plasma surface treatment is an effective and relatively cheap way to modify the surface and to obtain the desired corrosion resistance as well as good biocompatibility.

4.3.1.1   Low Temperature/Pressure Plasma Surface Treatment

4.3.1.1.1   Plasma Sputtering

With gas ionization caused by gas discharge, positive ions bombard the material on the cathode at high speed to modify the surface of the materials, achieving ideal hardness, abrasive resistance, corrosion resistance, and good biocompatibility.

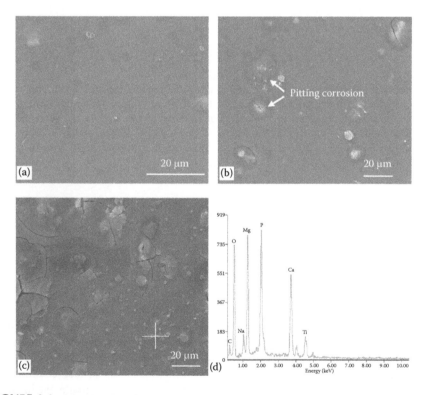

FIGURE 4.4   Morphology of the TiO_2 films after immersion for 1 (a), 3 (b), and 5 (c) days; (d) the EDS result of the particles on the film. (Adapted from Chen, S., S. Guan, B. Chen, W. Li, J. Wang, L. Wang, S. Zhu, and J. Hu. *Applied Surface Science*, 257 (9), 4464–7, 2011. With permission.)

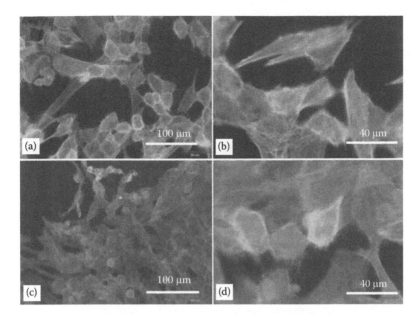

FIGURE 4.5 Cell morphologies after 1 day and 3 days of culture on as-coated AZ91 magnesium: (a) 1 day, low magnification; (b) 1 day, high magnification; (c) 3 days, low magnification; (d) 3 days, high magnification. (Adapted from Xin, Y., J. Jiang, K. Huo, G. Tang, X. Tian, and P. K. Chu. *Journal of Biomedical Materials Research. Part A*, 89 (3), 717–26, 2009a. With permission.)

Chen et al. (2011a) deposited a TiO_2 film on a Mg-Zn alloy by direct current reactive magnetron sputtering and investigated the degradation behavior in SBF. As Figure 4.4 shows, the failure of the film was attributed to pitting corrosion after a period of immersion. The results revealed that TiO_2 film improved the corrosion resistance significantly. The corrosion rate was reduced from 4.13 mm/y of the substrate sample to 1.95 mm/y of the sample with TiO_2 film.

Xin et al. (2009a) reported a-Si:H coatings deposited on as-cast AZ91 Mg alloys by direct current magnetron sputtering deposition. Besides the enhanced corrosion resistance, hFOB 1.19 cells were observed to attach well on the coating; as Figure 4.5 shows, cells attached well and proliferated normally, and no cell attachment was observed on the untreated sample surface.

G.S. Wu et al. (2008) deposited Cr on the surface of as-extruded AZ31 Mg alloys with a hybrid ion beam deposition system combining a linear ion source and a magnetron sputtering ion source. In addition to the improved corrosion resistance, the surface hardness was also enhanced by the Cr coating.

4.3.1.1.2 Ion Implantation

Common surface treatments, such as kinds of coatings, introduce an interface between the coatings and the substrates. Often the interface will introduce new issues, such as binding force and an unmatched expansion coefficient between the coating and substrate. However, ion implantation can avoid such problems because it accelerated and implanted ions into the surface so that no abrupt interface and layer delamination would be introduced (G. Wu et al. 2013). In addition, ion implantation provides the possibility of introducing different species into a substrate independent of thermodynamic limitations, such as solubility. By altering the species and amount of the implanted ions, researchers can control the properties of the materials in an effective way (G. Wu et al. 2013). Widely used ion implantation techniques in biodegradable Mg alloys include plasma immersion ion implantation (PIII) and metal vapor vacuum arc (MEVVA) source techniques. Both techniques can achieve metallic ion implantation. The difference lies in that PIII can implant nonmetallic ions and achieve implantation on pieces with a complex shape. Table 4.1 is a summary of plasma ion implantation utilized in biomedical pure Mg and Mg alloys.

R. Xu et al. (2011, 2012a) implanted chromium ions and oxygen ions into pure Mg plates. Although chromium ion implantation induced worse corrosion behavior in SBF due to a galvanic effect, chromium and oxygen ion dual implantation produced a thicker surface layer of chromium oxide, which slowed down the corrosion rate, especially at the beginning stage. In addition, it implied that pitting corrosion was the common behavior due to the defects existed on the chromium oxide layer.

Liu et al. (2007a,b) investigated the corrosion behavior of AZ91D after aluminum, zirconium, and titanium plasma immersion ion implantation and deposition (PIII and D), respectively. All the implantations improved the corrosion resistance of AZ91D in artificial physiological fluids because of an oxidized layer that was formed, and the specimen after aluminum ion implantation and deposition showed the best corrosion resistance of the three in electrochemical tests. In addition to a single type of metallic ion implantation and metallic ion and oxygen dual implantation, multiple metallic ion implantation is also examined as Table 4.1 summarizes.

Xu et al. (2012b) investigated the influence of Zn and Al dual implantation on the corrosion behavior of pure Mg, AZ31 Mg alloy, and AZ91 Mg alloy. A three-layer surface film composed of an outer layer of magnesium oxide/aluminum oxide; a middle layer alloyed with metallic Zn, Al, and Mg; and an inner layer rich in metallic Mg was formed. The corrosion resistance was also enhanced with such a three-layer structure in SBF.

Xin et al. (2009b) deposited a ZrN/Zr coating of about 1.5 μm in thickness on AZ91 with a filtered cathodic arc deposition system. Zirconium played the role of an interlayer to diminish the mismatch of zirconium nitride and the substrate, and the zirconium nitride served as a biocompatible buffer to retard the corrosion rates. With the help of such a bilayer, the corrosion potential of the coated alloy was much more positive compared to the uncoated alloy, shifting from −1830 mV to −1420 mV. A much lower corrosion current density was also obtained, which meant a much better corrosion resistance.

Besides metal ions, nonmetallic ions can also be implanted to metal and alloys through plasma immersion ion implantation (PIII). Similar to metal ion implantation, the implantation of nonmetallic ions also results in an oxide on the surface of the substrate.

4.3.1.1.2.1 Si Implantation   Jamesh et al. (2013) investigated the effects of silicon implantation on a WE43 alloy. With the help of a silicon oxide layer on the

TABLE 4.1

Summary of Ion Implantation on Biomedical Pure Mg and Magnesium Alloys

Substrate	Implanted Ion	Implant Condition	Corrosion Resistance	Cytocompatibility	References
Pure Mg	Cr and O	Cr: 20 kV, 1.5×10^{-3} Pa, 1 h O: 30 kV, 3 h	+ (SBF)		R. Xu et al. 2011, 2012a
As-extruded AZ91D	Al, Zr, Ti, respectively	10 kV	+ (SBF) Al best		Liu et al. 2007a,b
As-extruded AZ91	ZrN/Zr	Filtered cathodic arc deposition system	+ (SBF)		Xin et al. 2009b
As-cast AZ31	CO_2	20 kV, 200 Hz, 1 h	+ (SBF) + (DMEM)		R. Xu et al. 2013
As-cast WE43	Si	35 kV, 4.2×10^{-4} Pa, 20 min	+ (SBF)		Jamesh et al. 2013
As-extruded Mg-3.0Nd-0.2Zn-0.4Zr	O	20 kV	No obvious improvement (SBF)		Wu et al. 2012a
As-extruded Mg-1Ca	Zr and O	25 kV, 1.5×10^{-3} Pa, 0.5 h	+ (SBF)	+ (MC3T3-E1)	Zhao et al. 2014a
As-cast Mg-0.5Sr	Zr and O	25 kV, 1.5×10^{-3} Pa, 0.5 h	+ (SBF)	+ (MC3T3-E1)	Zhao et al. 2014a
As-cast WE43	Al and O	Al: 35 kV, 1.5×10^{-3} Pa, 1 h O: 20 kV, 100 Hz, 3 h	+ (SBF)		Zhao et al. 2012
As-cast pure Mg	Zn and Al	Zn: 15 kV, 30 min Al: 15 kV, 120 min	+ (SBF)		R. Xu et al. 2012b
AZ31	Zn and Al	Zn: 15 kV, 30 min Al: 15 kV, 120 min	+ (SBF)		R. Xu et al. 2012b
AZ91	Zn and Al	Zn: 15 kV, 30 min Al: 15 kV, 120 min	+ (SBF)		R. Xu et al. 2012b

(Continued)

TABLE 4.1 (CONTINUED)
Summary of Ion Implantation on Biomedical Pure Mg and Magnesium Alloys

Substrate	Implanted Ion	Implant Condition	Corrosion Resistance	Cytocompatibility	References
As-cast WE43	Ti and O	Ti: 20 kV, 1.5×10^{-3} Pa, 0.5 h; O: 25 kV, 120 Hz, 3 h	+ (SBF)		Y. Zhao et al. 2013
As-cast Mg-Ca	Zn	60 kV, 1×10^{-3} Pa	− (SBF)		Wan et al. 2008a
As-extruded Mg-3.0Nd-0.2Zn-0.4Zr	C_2H_2 to form diamond-like carbon film (DLC)	20 kV, 3×10^{-3} Pa, 3 h	+ (0.9 wt% NaCl solution)		G. Wu et al. 2014
As-cast pure Mg	Zn	35 kV, 1.8×10^{-4} Pa, 1 h	− (SBF)		G. Wu et al. 2011
As-cast pure Mg	Al	35 kV, 2×10^{-4} Pa, 1 h	+ (SBF)		Wu et al. 2012b
As-cast AZ31	Al	35 kV, 2×10^{-4} Pa, 1 h	+ (SBF)		Wu et al. 2012b
As-cast AZ91	Al	35 kV, 2×10^{-4} Pa, 1 h	+ (SBF)		Wu et al. 2012b
As-cast AZ91	Al and O	Al: 15 kV, 10 Hz; O: 12.6 kV, 10 Hz	+ (SBF)	+ (MC3T3-E1)	Wong et al. 2013

surface, the corrosion current density was significantly reduced from 642 ± 125 to 27 ± 32 μA/cm², and the corrosion potential shifted from −1972 ± 7 to −1895 ± 12 mV versus SCE in SBF.

4.3.1.1.2.2 C-O Implantation   On the basis of the good biocompatibility of carbon-based materials, such as pyrolytic carbon, R. Xu et al. (2013) implanted carbon dioxide onto an AZ31 alloy and examined the properties. Figure 4.6 is the surface morphology of the AZ31 alloy before and after carbon dioxide implantation and before and after immersion in DMEM and SBF. It was obvious that carbon dioxide did not change the surface much. In addition, the electrochemical test results revealed that pitting corrosion was replaced by homogeneous corrosion after carbon dioxide implantation.

4.3.1.1.2.3 O Implantation   Wu et al. (2012a) examined the effect of oxygen implantation on the corrosion behavior of the Mg–3Nd–0.2Zn–0.4Zr alloy. Although oxygen implantation thickened the surface oxide layer, no obvious improvement of corrosion resistance was found by them. Zhao et al. (2014a) did systematic research on the corrosion behavior, cytotoxicity, and antimicrobial property of plasma ion-implanted Mg-Ca and Mg-Sr alloys. An obvious surface roughness value increase was obtained from 2.16 nm (untreated sample) to 7.29 nm (Zr-O implanted sample) in the Mg-Ca alloy, and similar numbers were also obtained in the Mg-Sr alloy. Therefore, ion implantation is a suitable technique to produce a rough but uniform surface on Mg alloys. They also found that the corrosion current densities measured from the untreated, Zr-implanted,

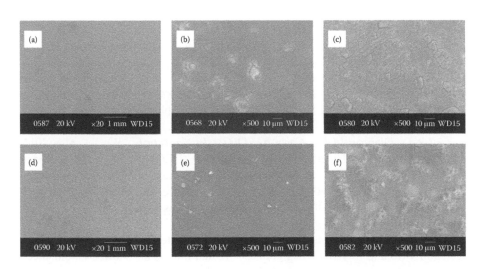

FIGURE 4.6 Surface morphology: (a) unimplanted sample, (b) unimplanted sample after electrochemical test in DMEM, (c) unimplanted sample after electrochemical test in SBF, (d) CO₂ PIII sample, (e) CO₂ PIII sample after electrochemical test in DMEM, and (f) CO₂ PIII after electrochemical test in SBF. (Adapted from Xu, R., X. Yang, X. Zhang, M. Wang, P. Li, Y. Zhao, G. Wu, and P. K. Chu. *Applied Surface Science*, 286, 257–60, 2013. With permission.)

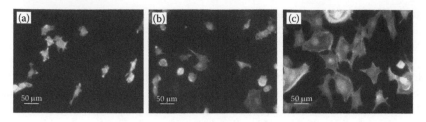

FIGURE 4.7 Fluorescent images of MC3T3-E1 pre-osteoblasts after culturing for 5 h on (a) unimplanted Mg–Ca, (b) Zr-implanted Mg–Ca, and (c) Zr–O-implanted Mg–Ca. (Adapted from Zhao, Y., M. I. Jamesh, W. K. Li, G. Wu, C. Wang, Y. Zheng, K. W. Yeung, and P. K. Chu. *Acta Biomaterialia*, 10 (1), 544–56, 2014. With permission.)

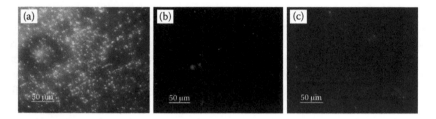

FIGURE 4.8 Fluorescent microscopic views of magnesium samples after 30 min of bacteria culture: (a) unimplanted Mg–Ca, (b) Zr-implanted Mg–Ca, and (c) Zr–O-implanted Mg–Ca. (Adapted from Zhao, Y., M. I. Jamesh, W. K. Li, G. Wu, C. Wang, Y. Zheng, K. W. Yeung, and P. K. Chu. *Acta Biomaterialia*, 10 (1), 544–56, 2014. With permission.)

and Zr-O-implanted Mg–Ca and Mg–Sr exhibit descent trends not only in the SBF but also in tryptic soy broth (TSB) and cell culture medium. Cytotoxicity test and antibacterial test results of the Mg-Ca alloy are shown in Figures 4.7 and 4.8 with similar results in the Mg-Sr alloy. From the fluorescent images, it could be seen that the Zr-O-implanted Mg-Ca sample possessed the best biocompatibility and antibacterial property.

4.3.1.1.2.4 Diamond-Like Carbon (DLC) Implantation Wu et al. (2014) introduced a chemical inert diamond-like carbon (DLC) coating on as-extruded Mg-3.0Nd-0.2Zn-0.4Zr by ion implantation. Surface morphology of the untreated sample and the sample with DLC coating is as Figure 4.9 illustrates. With such a chemical inert coating, the corrosion resistance was significantly increased in 0.9 wt.% NaCl solution.

What needs to be noted is that Al and O ions implanted AZ91 Mg alloy has been tested in vitro and in vivo (Wong et al. 2013), and the results showed an enhanced corrosion resistance and biological performance both in vitro and in vivo. Figure 4.10 is the typical histological images of untreated and Al_2O_3-treated implants.

So far, the examination of ion implantation on biomedical Mg alloys focuses on the improvement of corrosion resistance. The results are inspiring that ion implantation can be an effective way for enhancement of corrosion resistance. Dual ion implantation with subsequent treatment is the trend in this certain field. However, not all of these varied

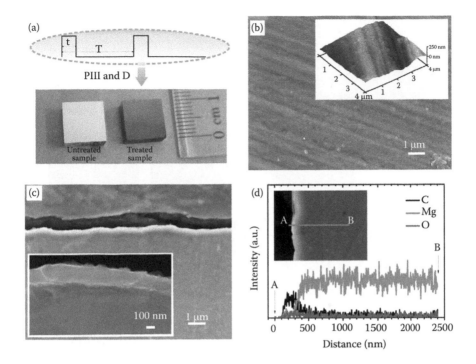

FIGURE 4.9   (a) Surface appearance of the untreated and treated samples. (b) SEM picture of surface morphology of the film with the inset showing the magnified surface obtained by AFM. (c) SEM view of the cross-section of the plasma-modified sample with the inset showing a magnified picture of the film. (d) EDS line scan of the film. (From Wu, G., X. Zhang, Y. Zhao, J. M. Ibrahim, G. Yuan, and P. K. Chu. *Corrosion Science*, 78, 121–9, 2014. With permission.)

ion implanted materials have been tested for their cytocompatibility in vitro or even in vivo. Thus, the biocompatibility of the ion-implanted Mg alloys needs further convincing investigation. In addition, the unique degradation property of biomedical Mg alloys may be affected by some chemical inert layer produced by ion implantation because of the nonbiodegradable properties and uncertain biological effects of these particulates in vivo, which also needs consideration in further research and applications.

4.3.1.2 Plasma Spray

Plasma spray is a common surface modification of biomedical materials. Typically, coating material is heated and melted by plasma and then sprayed on the substrate material. Plasma spray can provide a thicker protective coating compared to other processes, such as chemical vapor deposition and physical vapor deposition.

Zeng et al. (2008a) produced a TiO_2 coating on an AM60 Mg alloy by plasma spray and investigated its corrosion behavior in Hank's solution. The results revealed that if the TiO_2 coating was not sealed, the corrosion rate became faster than the bare specimen because of more severe galvanic corrosion caused by the porosity of TiO_2. However, the corrosion rate would be effectively reduced when the holes were sealed with Na_2SiO_3.

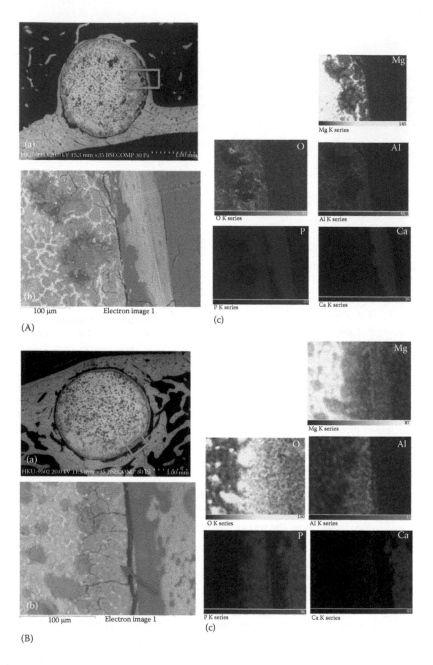

FIGURE 4.10 Histology of untreated (A) and Al_2O_3-treated (B) implants viewed under scanning electron spectroscopy. Element mapping was conducted in the rectangular box in (a) as magnified in (b) and the examined elements are as shown in (c). Aluminum and oxygen can be detected on the Al_2O_3-treated implant surface as indicated by arrow. (From Wong, H. M., Y. Zhao, V. Tam, S. Wu, P. K. Chu, Y. Zheng, M. K. To, F. K. Leung, K. D. Luk, K. M. Cheung, and K. W. Yeung. *Biomaterials*, 34 (38), 9863–76, 2013. With permission.)

Naked

Coated

0 days   7 days   11 days   15 days

FIGURE 4.11 Comparison of degradation behavior of naked and post-treated samples after different immersion time of 0, 7, 11, and 15 days during degradation process. (From Yang, J. X., Y. P. Jiao, F. Z. Cui, I.-S. Lee, Q. S. Yin, and Y. Zhang. *Surface and Coatings Technology*, 202 (22–23), 5733–6. 2008. With permission.)

Plasma spray introduces high temperatures when utilized for surface modification. Grain growth occurs at a relatively lower temperature in Mg alloys compared to other biomedical material systems, such as Ti alloys, and coarse grains would cause worse corrosion behavior and mechanical properties. This may explain why plasma spray is not commonly used in biomedical Mg alloys as biomedical Ti alloys.

In the field of plasma spray surface treatments, ion-beam assisted deposition (IBAD) is a representative method used to alter the surface of biodegradable Mg alloys. Yang et al. deposited CaP coating (Yang et al. 2008c) and C-N coating (Yang et al. 2008b) on AZ31 Mg alloys. Figure 4.11 shows the morphology of a naked sample and the sample coated with CaP; an obvious improvement of corrosion resistance by the CaP coating can be obtained. As for the C-N coating experiment, the C-N coating is amorphous C_3N_4. However, the corrosion resistance and biocompatibility of such a coating require further investigation.

4.3.2 LASER SURFACE TREATMENT

When a laser beam is focused, the temperature can reach into the thousands centigrade near the focal point. Almost all materials can be melted with this kind of high temperature, and thus, the surface property can be altered. Common laser surface treatment consists of laser polishing and roughening, laser surface melting, laser surface cladding, laser surface alloying, and pulsed laser deposition (PLD).

4.3.2.1 Laser Polishing and Roughening

Russo investigated the effects of laser shock peening (LSP) on an AZ91D Mg alloy systematically (Russo 2012). The results showed that significant compressive residual stresses were produced by double-sided LSP, and the compressive residual stresses were not so apparent through single-sided LSP. In addition, DC polarization tests proved that LSP can decrease corrosion rates from 2.5 to 8 times that of the untreated sample. Guo et al. (2012) also examined the effects of LSP with different overlap ratios, as Figure 4.12 shows, on the corrosion behavior of the Mg-0.8Ca alloy. The corrosion behavior of the LSP-treated samples with different parameters is as Figure 4.13 illustrates. All

Magnesium Alloys as Degradable Biomaterials

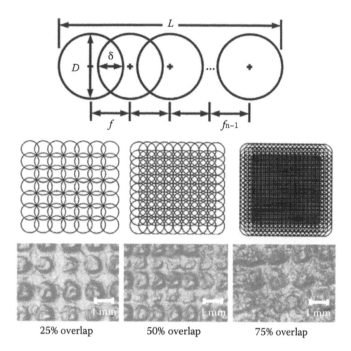

25% overlap   50% overlap   75% overlap

FIGURE 4.12 Theoretical and experimental LSP patterns with different peening overlap ratios. (From Guo, Y., M. P. Sealy, and C. Guo. *CIRP Annals—Manufacturing Technology*, 61 (1), 583–6, 2012. With permission.)

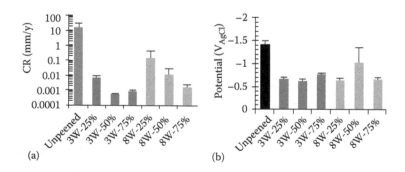

FIGURE 4.13 (a) Corrosion rate (CR) and (b) potential of peened MgCa surfaces. (From Guo, Y., M. P. Sealy, and C. Guo. *CIRP Annals—Manufacturing Technology*, 61 (1), 583–6, 2012. With permission.)

the LSP-treated samples showed an enhanced corrosion resistance, and especially the 3W-50% overlap-treated sample possessed a more than hundredfold improvement calculated from the electrochemical test in Hank's solution. Sealy and Guo (2010) also simulated the LSP procedure on Mg-0.8Ca with 3-D semi-infinite simulation.

4.3.2.2 Laser Surface Melting

Ng et al. (2011) evaluated the microstructure of selective laser melting (SLM)-treated pure Mg. The molten zone consisted of fine equiaxed grains of the α-Mg phase after the SLM process. Guan et al. (2009) investigated the corrosion behavior of SLM-treated AZ91D with the technology parameters listed in Table 4.2. All four specimens exhibited better corrosion resistance. In addition, lower scanning speed (i.e., A and B) could induce a refined continuous network of β-$Mg_{17}Al_{12}$ phases and the increased average Al concentration. A higher laser scanning rate (i.e., C and D) produced smaller cracks in the molten zones, which accelerated the corrosion rates. Thus, lower scanning speeds seem more suitable for SLM treatment on Mg alloys.

4.3.2.3 Laser Surface Cladding

Santhanakrishnan et al. (2012) produced a hydroxyapatite coating on AZ31B by laser cladding. Among all treated specimens, only the specimen treated with a laser energy density of 7×10^6 showed a lower corrosion rate compared to untreated AZ31B. Thus, the protection of a hydroxyapatite coating remains uncertain. Huang et al. (2013b) produced a Zr-based coating, which consists of Zr, zirconium oxides, and Zr aluminides, on AZ91D by the laser cladding method, and the corrosion resistance was enhanced compared to a bare AZ91 Mg alloy. Yue and Huang (2011) fabricated a Zr-based coating by Zr and Mg cladding on AZ91D. Similar results were obtained as mentioned (Huang et al. 2013b). Y. Gao et al. (2009b) also found improved biocompatibility by hydroxyapatite laser cladding for the AZ91D Mg alloy.

As a summary of Section 4.3, the reports and investigations on plasma surface treatment are major/dominant with both in vitro and in vivo biocompatibility being tested whereas only limited reports about the laser surface treatment of Mg alloys can be found without deep study on the in vitro and in vivo biocompatibility.

TABLE 4.2

Main Processing Parameters Used for Laser Surface Melting of AZ91D Mg Alloy

Specimen ID	Power Density (W/cm²)	Scanning Speed (mm/s)	Frequency (Hz)	Overlap (%)	Pulse Duration (ms)
A	3.82×10^4	5	100	50	1.0
B	3.82×10^4	10	100	50	1.0
C	3.82×10^4	20	100	50	1.0
D	3.82×10^4	30	100	50	1.0

Source: Guan, Y. C., W. Zhou, and H. Y. Zheng. *Journal of Applied Electrochemistry*, 39 (9), 1457–64, 2009. With permission.

4.4 CHEMICAL SURFACE TREATMENT

Various chemical surface treatment methods have been used for the degradable biomedical Mg alloy, which includes chemical conversion coating, electrochemical treatment, sol-gel treatment, biomimetic deposition, organic and polymer coating, and molecular self-assembled coating (Zheng et al. 2014). As for chemical conversion coating, fluoride conversion coating is a common method of treatment as is alkali treatment. As for electrochemical treatment, anodic oxidation, micro-arc oxidation and electrodeposition are widely explored.

4.4.1 CHEMICAL CONVERSION COATING

Chemical conversion coatings are produced by chemically bonding a superficial metal oxide or metal salt layer to the Mg alloy surface. This method can be used individually or as a pretreatment before polymer coating (Zheng et al. 2014).

4.4.1.1 Fluoride Conversion Coating

Two main methods have been reported to produce a MgF_2 protective coating on pure Mg and Mg alloy substrates so far (Zheng et al. 2014). The conventional method is to use hydrofluoric acid to fabricate the MgF_2 coating directly. The other method is to use a sodium hydroxide solution to establish a $Mg(OH)_2$ layer and then convert it to a MgF_2 coating, which is thicker compared to the coating by hydrofluoric acid. In addition, the MgF_2 coating is known to benefit bone formation (N. Li et al. 2013), which may be a good candidate for orthopedic application. Table 4.3 presents a summary of existing MgF_2 coatings produced in degradable biomedical Mg and its alloys.

As shown in Table 4.3, the HF solution method is the most widely adopted method. With the help of the resulting MgF_2 coating, the corrosion resistance of the materials is enhanced, especially at the early stage of the corrosion process. Yet the MgF_2 coating formed by the HF method also has its weaknesses. As it is a relatively thin coating (Chiu et al. 2007), the protection may only last for a short time. Thus, consequent research moved to promoting the thickness of the coating by altering the concentration of the HF solution (da Conceicao et al. 2010; Bakhsheshi-Rad et al. 2013b), prolonging immersion time (Witte et al. 2010; Yan et al. 2010), or changing methods. However, the higher concentration of HF can't always provide a thicker coating (Assadian et al. 2012; Bakhsheshi-Rad et al. 2013b), so utilization of KF solution was tried (Pereda et al. 2010, 2011). The good thing is that the thickness of the MgF_2 can be improved by prolonged immersion time (da Conceicao et al. 2010; Yan et al. 2010). Therefore, an appropriate F concentration and prolonged immersion time should be combined to obtain the MgF_2 coating with good quality. Before HF treatment, pretreatment with NaCl (Drynda et al. 2010) and NaOH (Lalk et al. 2013) were tried as an initial step. In this way, a $Mg(OH)_2$ layer will form first and then convert to a MgF_2 coating when immersed in HF solution. The MgF_2 coating produced in this way is thicker compared to the method of direct immersion in HF (Zheng et al. 2014).

L. Wu et al. (2013) utilized KF solution instead of HF solution to fabricate the MgF_2 coating on the surface of the AZ31 alloy. With such a method, the conversion

TABLE 4.3
Summary of Current Investigation on MgF$_2$ Coating of Biomedical Mg and Its Alloys

Method	Immersion Condition, Solution Concentration and Time	Substrate	Corrosive Medium	Corrosion Rate (mm/y)	Thickness of Coating (μm)	References
HF	5–48%, 6–24 h	Mg-0.5Ca	Kokubo solution	0.734 (40% HF, 24 h)	12.6 (40% HF, 24 h)	Bakhsheshi-Rad et al. 2013b
	7–45%, 24 h	Mg-1Ca	Kokubo solution	1.01 (40% HF)		Assadian et al. 2012
	7–28 M, 1–24 h	AZ31	3.5 wt.% NaCl	0.46 ± 0.08 (20 M HF, 24 h)		da Conceicao et al. 2010
	40%, 24 h	Mg-Ca-Zn	SBF	0.11 (HA/MgF$_2$ coating)		Bakhsheshi-Rad et al. 2013a
	40%, 96 h	As-cast LAE442	In vivo		150–200	Witte et al. 2010
	47–51 M, 72 h	MgYZrRE, MgZnYZrRE	Hank's	2.26 and 1.5, respectively		N. Zhao et al. 2014
	48%, 6–24 h	Pure Mg	Hank's		1.5 (24 h)	Chiu et al. 2007
	48%, 24 h	Pure Mg, As-cast Mg, and AZ31	DMEM + 10% FBS			Carboneras et al. 2011a
		AZ31	SBF			Ren et al. 2014
		MAO-coated ZK60	Hank's	– (Compared to MAO-treated)		Lin et al. 2013a
		Pure Mg	SBF	1.3 (As an interlayer)		Jo et al. 2011
		Powder metallurgy and extruded Mg, as-cast pure Mg	DMEM + 10% FBS			Carboneras et al. 2011b
		As-cast AZ91	SBF			Rojaee et al. 2013b

(Continued)

TABLE 4.3 (CONTINUED)
Summary of Current Investigation on MgF₂ Coating of Biomedical Mg and Its Alloys

Method	Immersion Condition, Solution Concentration and Time	Substrate	Corrosive Medium	Corrosion Rate (mm/y)	Thickness of Coating (µm)	References
	50%, 3–168 h	AZ31B	SBF	Increase with treating time		Yan et al. 2010
	50%, 48 h	AZ31B	Simulated blood plasma			Yan et al. 2014
	48% HF tailored with galvanic coupling and K_2CO_3, 1 h	Pure Mg	Hank's		0.4–1.6	Sankara Narayanan et al. 2014b
NaCl + HF	NaCl: 200 g/l, 3 h HF: 40%, 96 h	Mg-Ca	0.9% NaCl		15–20	Drynda et al. 2010
NaOH + HF	NaOH: 200 g/l, 2 h HF: 40%, 96 h	AX30	In vivo			Lalk et al. 2013
KF	0.01–0.3 M, 1 h	Powder metallurgy Mg	8 g/l NaCl			Pereda et al. 2011
	0.1 M, 1 M, 1–168 h	Powder metallurgy and extruded Mg	8 g/l NaCl and PBS	0.1 M better		Pereda et al. 2010
MgF_2	0.1 M, 1 h	As-extruded Mg-1Ca	8 g/l NaCl and PBS Hank's	0.14 ± 0.02		L. Wu et al. 2013 N. Li et al. 2013
	MgF_2 powder evaporation, 3 min					
Saturated NH_4HF_2	Micro arc fluorination, 3 min, 120–200 V	Pure Mg	SBF	0.03 (200 V)	5.5 (200 V)	Jiang et al. 2014

coating was produced through $Mg(OH)_2$, MgF_2, and $KMgF_3$ to provide a prolonged protection of the layer.

A double-layer or multilayer structure with multisurface treatment has been explored recently. For example, Lin et al. (2013a) developed an MgF_2 coating on the surface of MAO-treated ZK60. The corrosion resistance is somewhat decreased with the MgF_2 coating compared to the only MAO-treated sample. Rojaee et al. (2013b) produced a nanostructured hydroxyapatite (n-HAp) coating on a MgF_2 pre-coated AZ31 alloy and a MAO pretreated AZ31 alloy. The results revealed a good in vitro bioactive response both in the n-HAp/MgF_2-treated sample and n-HAp/MAO-treated sample. However, compared to the n-HAp/MAO-treated sample, the n-HAp/MgF_2-treated sample had a relatively faster corrosion rate. Hou et al. (2012) used a fluoride coating to be an interlayer to deposit Ti-O film on a Mg-Zn-Y-Nd alloy and achieved a better corrosion resistance.

Regarding the cytotoxicity evaluation of the HF-treated Mg alloy, N. Zhao et al. (2014) studied the endothelialization of the HF-treated MgYZrRE and MgZnYZrRE alloys and reported that apparently better results were obtained on the HF-treated MgZnYZrRE alloys. A similar result was also seen on MgYZrRE.

Ren et al. (2011) compared the antibacterial property of pure Mg and AZ31 by different surface treatments, including MAO treatment and fluoride conversion coating. The results revealed that the Si-F-containing surface treated by fluoride conversion coating possessed little antibacterial property whereas the MAO-treated Si-containing surface showed a strong antibacterial property.

The in vivo tests on MgF_2-coated Mg alloys have been reported in references (Witte et al. 2010; Lalk et al. 2013). Figure 4.14 shows the results of ex vivo μ-computed tomographies (μCT80) after 6-, 12-, and 24-week implantation periods of a MgF_2-coated AX30 alloy and a CaP-coated AX30 alloy obtained by Lalk et al. (2013). After 6 weeks, a few thin trabeculae extended into the outer implant area, indicating good bone sponge contact of the MgF_2 sponges (Figure 4.14a). Moreover, deeper trabeculae ingrowth and isles of bone-like materials were found after 12 weeks and 24 weeks, indicating a good bone response to the MgF_2 sponge.

4.4.1.2 Alkali Treatment

Alkali treatment and alkali heat treatment are typical surface modification techniques because they are simple and easy to realize and the biocompatibility can be reserved after these processes.

For biodegradable Mg and its alloys, alkali treatment can produce a dense MgO or $Mg(OH)_2$ layer on the surface. Gu et al. (2009a) investigated the influence of three alkali heat treatments, Na_2HPO_4, Na_2CO_3, and $NaHCO_3$, on the corrosion behavior and cytotoxicity of a Mg-Ca alloy. The corrosion rates of alkaline heat-treated samples decreased by an order of magnitude in comparison to the bare Mg-Ca alloys and followed the ranking order $NaHCO_3$ heated $<$ Na_2HPO_4 heated $<$ Na_2CO_3 heated. In addition, all the treated samples show no cytotoxicity to L929 after 7 days of culture. Al-Abdullat et al. (2001) evaluated the influence of treating parameters, such as pH value and the concentration of $NaHCO_3$, on the corrosion behavior of pure Mg in Hank's solution for at least 75 days. The results revealed that the concentration of carbonate ions $\left(HCO_3^-\right)$ was an important factor affecting the coating. In

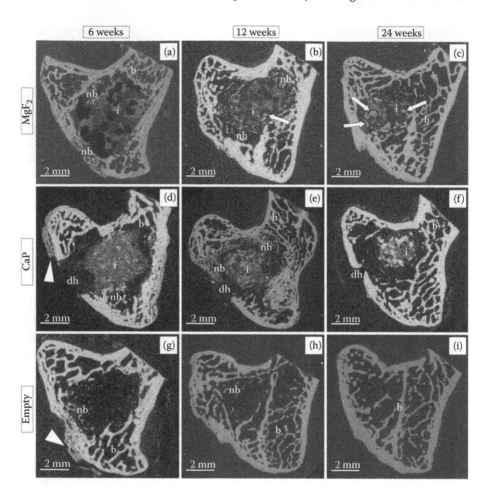

FIGURE 4.14 Ex vivo μ-computed tomographies (μCT80) after 6, 12, and 24 weeks implantation period. (a–c) Bone with MgF$_2$ sponge (proceeding implant degradation and bone-like material in the remaining implant can be seen); (d–f) bone with CaP sponge (merely at the beginning bone trabeculae approach the sponge and the drill hole closure is proceeding, but even after 24 weeks incomplete); (g–i) bone with empty drill hole (increasing infiltration of the defect with trabeculae getting thicker from g to i); arrow: bone-like material inside the sponge pores; b: Bone; dh: drill hole; i: implant; nb: new bone; triangle: periosteal bone formations. (Adapted from Lalk, M., J. Reifenrath, N. Angrisani, A. Bondarenko, J. M. Seitz, P. P. Mueller, and A. Meyer-Lindenberg. *Journal of Materials Science. Materials in Medicine*, 24 (2), 417–36, 2013. With permission.)

addition, when the pH value was around 8.3, the thickest coating could be obtained. Y. Song et al. (2010) fabricated a HA coating transferred from brushite (DCPD, CaHPO$_4$·2H$_2$O) through alkali heat treatment. However, it was fragile and less stable compared to Ca-P coating synthesized by electrodeposition in their research and showed worse corrosion resistance. Gao et al. (2004) reported good corrosion resistance on pure Mg treated by alkali heat treatment. In addition, Li et al. (2004)

reported the results of noncytotoxicity to marrow cells from the femora of mice blood in alkali heat-treated pure Mg. Wei et al. (2011) and Du et al. (2011b) fabricated a calcium silicate and calcium phosphate ($CaSiO_3$/$CaHPO_4 \cdot 2H_2O$) composite coating on Mg-Zn-Ca and Mg-Zn-Mn-Ca alloys and reported improved corrosion resistance and biocompatibility, respectively.

Alkali heat treatment is often followed with a subsequent treatment, such as electrodeposition and other methods. Guan et al. (2012) utilized alkali heat treatment as a pretreatment to obtain a dense MgO layer on the surface of as-rolled Mg-4.0Zn-1.0Ca-0.6Zr. Such treatment helped to enhance the corrosion resistance with a subsequent electrodeposition process of HA. Gray-Munro et al. (2009) used a multistep pretreatment involving alkali heat treatment for the subsequent deposition of HA. Y. Chen et al. (2013) used alkali treatment as a pretreatment to form the hydroxyl group on the pure Mg surface and subsequently fabricate a phytic acid coating. Figure 4.15 depicts the surface morphology of pure Mg, pure Mg after alkali pretreatment (Mg@OH), pure Mg after phytic acid treatment (Mg@PA), and Mg after alkali pretreatment and phytic acid treatment (Mg@OH@PA). A more homogeneous and smoother surface was obtained with NaOH pretreatment. In addition, the hemolysis ratio decreased dramatically from 60.5% (bare Mg) to 2.35% (Mg@OH[24 h]@PA). Geng et al. (2009a) deposited a β-TCP coating on the surface of pure Mg by an alkali heat pretreatment of immersion in supersaturated Na_2HPO_4 solution at room temperature for 3 h and a heat treatment at 400°C for 10 h. By using such pretreatment, the thickness of the β-TCP coating was 85 μm. As Figure 4.16 shows, the thickness of the β-TCP coating decreased after immersion in Hank's solution, and a number of tiny needle-like precipitates simultaneously formed on the surface of the β-TCP coating.

In vivo animal testing has been done for the Si-containing coated AZ31B alloys by alkali heat treatment (Tan et al. 2014). As Figure 4.17b shows, many cracks were found in the new bone layer whereas Figure 4.17a the new bone layer and degradation layer of the noncoated AZ31B were not dense. The chemical composition results also revealed that the new bone layer and degradation layer of both noncoated and coated AZ31B were mainly composed of Mg, Ca, P, and O.

4.4.1.3 Hydrothermal Treatment

In addition to common chemical conversion coatings, including fluoride coating and alkali treatment, the hydrothermal method is also being adopted to produce a cerium conversion coating on AZ31 (Rocca et al. 2010; Cui et al. 2011), a hydroxyapatite coating on pure Mg (Hiromoto and Tomozawa 2010; Tomozawa and Hiromoto 2011) and AZ31 (Onoki and Yamamoto 2010), a calcium phosphate conversion coating on AZ91D (Chen et al. 2012a), a $Mg(OH)_2$ coating on AZ31 (Y. Zhu et al. 2009, 2011, 2012b), and a hydroxyapatite-$Mg(OH)_2$ coating on pure Mg (X.B. Chen et al. 2011).

Chen et al. (2012a) systematically evaluated the effects of Ca^{2+} and PO_4^{3-} concentration on the corrosion behavior of the AZ91D magnesium alloy. The nine different concentrations they utilized are as illustrated by the diamonds (◆) in Figure 4.18. The blue line is the stoichiometric ratio of hydroxyapatite, and the boundary between direct hydroxyapatite and $CaHPO_4 \cdot 2H_2O$ precipitation is illustrated in red. The results

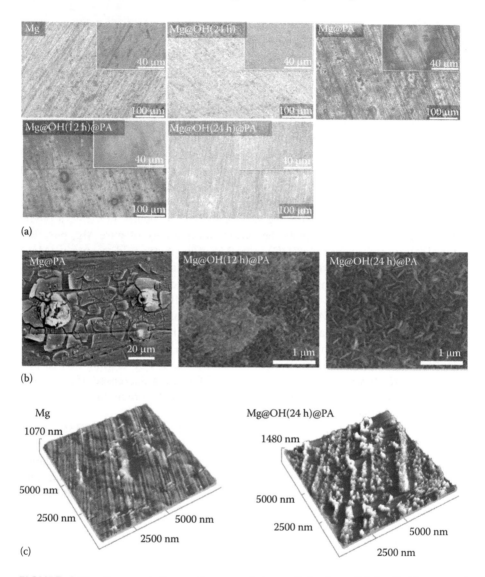

FIGURE 4.15 Representative surface morphology of the bare (Mg), alkaline-treated (Mg@OH; 24 h), direct phytic acid deposited (Mg@PA), and alkaline pretreated phytic acid coated magnesium (Mg@OH[12 h]@PA and Mg@OH[24 h]@PA): (a) optical images; (b) SEM photos; and (c) AFM images. (From Chen, Y., G. Wan, J. Wang, S. Zhao, Y. Zhao, and N. Huang. *Corrosion Science*, 75, 280–6, 2013. With permission.)

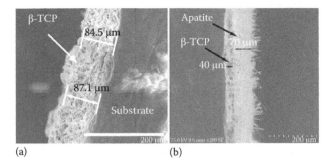

(a)   (b)

FIGURE 4.16 Cross-section morphology of β-TCP coated Mg (a) before and (b) after immersion in Hank's solution for 2 months. (From Geng, F., L. L. Tan, X. X. Jin, J. Y. Yang, and K. Yang. *Journal of Materials Science. Materials in Medicine*, 20 (5), 1149–57, 2009. With permission.)

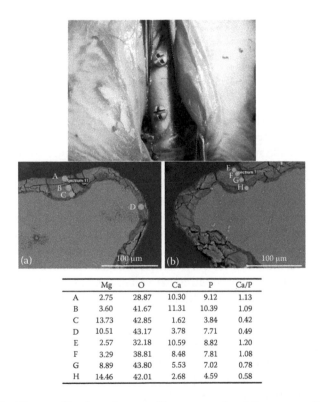

	Mg	O	Ca	P	Ca/P
A	2.75	28.87	10.30	9.12	1.13
B	3.60	41.67	11.31	10.39	1.09
C	13.73	42.85	1.62	3.84	0.42
D	10.51	43.17	3.78	7.71	0.49
E	2.57	32.18	10.59	8.82	1.20
F	3.29	38.81	8.48	7.81	1.08
G	8.89	43.80	5.53	7.02	0.78
H	14.46	42.01	2.68	4.59	0.58

FIGURE 4.17 Picture of implanted screws (the upper one) and the cross-section morphologies and EDS analysis of the bone–implant interface after 21 weeks of implantation: (a) noncoated AZ31B screw; (b) coated AZ31B screw. (From Tan, L., Q. Wang, X. Lin, P. Wan, G. Zhang, Q. Zhang, and K. Yang. *Acta Biomaterialia*, 10 (5), 2333–40, 2014. With permission.)

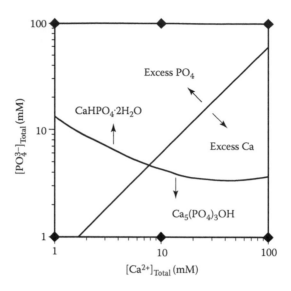

FIGURE 4.18   The nine conversion coating compositions evaluated are shown by the diamonds (◆). (From Chen, X. B., N. Birbilis, and T. B. Abbott. *Corrosion Science*, 55, 226–32, 2012. With permission.)

revealed that the Ca^{2+}-rich and PO_4^{3-}-poor coating baths generated coatings with the highest amount of protection, which they thought was attributed to the reduction of anodic reaction on the surface. Also, better corrosion resistance corresponded to the more diluted bath solution.

The pH value of the solution used for hydrothermal treatments is a critical parameter. Hiromoto and Tomozawa (2010a) investigated the influence of pH value (6.3, 7.3, and 11.3) and treatment time (8 h and 24 h) on the fabrication of HAp on pure Mg. According to their report, as for the 8 h-treated samples, samples treated at the pH value of 7.3, which was more equivalent to the m-Hanks they used, possessed the best corrosion resistance. Tomozawa and Hiromoto (2011) evaluated the effect of treating pH value, 5.9, 8.9, and 11.9, on the formation of HAp coating on pure Mg. XRD patterns revealed that not only HAp, but also octacalcium phosphate (OCP) formed only at the pH value of 5.9. In addition, both the HAp coating and the HAp/OCP coating consisted of an inner dense layer and a coarse outer layer. Different morphologies formed at a pH value of 5.9 and 8.9 after the anodic polarization test shown in Figure 4.19. More rod-like HAp was found on the pH 8.9-treated sample after the anodic polarization test in 3.5% NaCl solution, indicating a better corrosion protection compared to the HAp/OCP coating formed at pH 5.9.

Yet no in vitro biocompatibility tests or in vivo animal tests are reported. Therefore, considering the importance of biocompatibility of biodegradable Mg and its alloys, study in this specific area needs more attention.

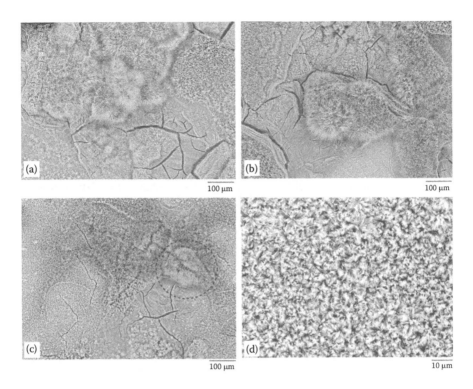

FIGURE 4.19 SEM images of specimens after the anodic polarization test: (a) uncoated magnesium; (b) specimen treated at pH 5.9; and (c and d) specimen treated at pH 8.9. (From Tomozawa, M., and S. Hiromoto. *Acta Materialia*, 59 (1), 355–63, 2011. With permission.)

4.4.2 ELECTROCHEMICAL TREATMENT

4.4.2.1 Anodic Oxidation and Micro-Arc Oxidation

Micro-arc oxidation, also called plasma electrolytic deposition (PED) or plasma electrolytic oxidation (PEO), is one kind of anodic oxidation under high voltage and heavy current.

By utilizing anodic oxidation and micro-arc oxidation, there will be a porous film on the surface of Mg alloy substrates. The porous film produced by anodic oxidation and micro-arc oxidization is generally considered beneficial for cell adhesion and osteogenesis but damaging to corrosion resistance.

Hiromoto and Yamamoto (2010b) studied the effects of different anodization voltage on the corrosion behavior of as-extruded pure Mg. The results indicated that a nonporous films were fabricated at 2 V and 20 V, and porous films were produced at 7 V. The nonporous film can retard local corrosion due to its dense and smooth surface. Lei et al. (2010a,b) produced a compact MgO coating on an as-cast Mg-Zn-Ca alloy specimen through anodic oxidation. A short-time immersion test in SBF indicated that the dense MgO coating can suppress the corrosion process. Hiromoto et al.

(2008a) also improved the corrosion resistance of pure Mg with an anodic oxidized calcium phosphate coating. In addition, Song (2007) reported an anodic oxidized coating improved the corrosion resistance of pure Mg in Hank's solution.

So far, micro-arc oxidation on the commercial magnesium alloys AM50 (Srinivasan et al. 2010), AM60 (Da Forno et al. 2013), AZ31 (Ren et al. 2011; Chu et al. 2012, 2013; Gu et al. 2012a,b; Seyfoori et al. 2012, 2013a,b; Sreekanth and Rameshbabu 2012; L. Chang et al. 2013; M.-A. Chen et al. 2013; Da Forno et al. 2013; Tang et al. 2013b; Tang and Wang 2013; Q. Zhao et al. 2013), AZ80 (Shi et al. 2011; W. Yang et al. 2013), AZ81 (Lu et al. 2011a), AZ91 (X. P. Zhang et al. 2007; J. Chen et al. 2008; Yao et al. 2009; C. Wu et al. 2010; J. Zhang et al. 2012a; Razavi et al. 2013, 2014; Rojaee et al. 2013b; Wang et al. 2013a,e; Zhang et al. 2013a; S. Wang et al. 2014), WE42 (Guo et al. 2011), WE43 (Liu et al. 2012b), ZK60 (Pan et al. 2012, 2013a,b; Lin et al. 2013a,b,c, 2014), ZX50 (Fischerauer et al. 2013), and MB26 (X. Fan et al. 2013) have been investigated. The study of micro-arc oxidation-modified pure Mg (Shi et al. 2009; Zhao et al. 2010; Durdu et al. 2011; Liu et al. 2011b; Ren et al. 2011; L.-H. Li et al. 2014a), Mg-Ca (Gu et al. 2011a), Mg-Zn-Ca (Lei et al. 2010a; Gao et al. 2011a; W. Li et al. 2011; Zhang et al. 2011b; K. Bai et al. 2012; S. Chen et al. 2012; Y. Zhang et al. 2012a; Xia et al. 2013; Ma et al. 2014), and Mg-Y-Zn (T.-F. Lu et al. 2012) are also reported.

As mentioned, moderate surface roughness possessed a better property of cell adhesion. The porous microstructure produced by micro-arc oxidation always exhibits good biocompatibility (Gu et al. 2011a; Pan et al. 2012, 2013b; Lin et al. 2013c, 2014; Seyfoori et al. 2013a) and hemocompatibility (Zhang et al. 2011b).

Typical morphology of micro-arc oxidation is a porous film as shown in Figure 4.20 (Wang et al. 2013e). It is easy to know that the applied voltage, frequency and time, and alkali solution concentration have strong effects on the porous film produced through micro-arc oxidation.

Gu et al. (2011a) investigated the corrosion resistance and biocompatibility of Mg-Ca treated by MAO at 300–400 V. The results indicated that 360 V MAO-treated Mg-Ca displayed the best corrosion resistance after 50 days of immersion in Hank's solution. Various studies on the effects of different voltage on the microstructure, corrosion behavior, and biocompatibility can be found in the literature (Yao et al. 2009; Gao et al. 2011a,c; Liu et al. 2011a; K. Bai et al. 2012; T.-F. Lu et al. 2012; Y. Zhang et al. 2012a; L. Chang et al. 2013; Lin et al. 2013c; Q. Zhao et al. 2013).

Su et al. (2011) investigated the effects of frequency on the structure and corrosion resistance of PEO-coated ZK60 Mg alloy. With the increase of frequency from 100 Hz to 1000 Hz, a less porous MgO film was produced, and thus, the corrosion resistance was improved.

Gu et al. (2012a) examined the effects of oxidation time on the coating characteristic and corrosion behavior of micro-arc oxidized AZ31 Mg alloy. The microstructure analysis revealed that the thickness of the coating increased from 1 to 5 min (20 μm). The sample coated at 5 min exhibited a relatively smooth and uniform microstructure. However, when the oxidation time increased to 8 min, the coating thickness decreased, and the surface became rough. The corrosion resistance showed a tendency in agreement with the coating thickness and surface finish. That is, the sample coated at 5 min possessed the best corrosion resistance in SBF.

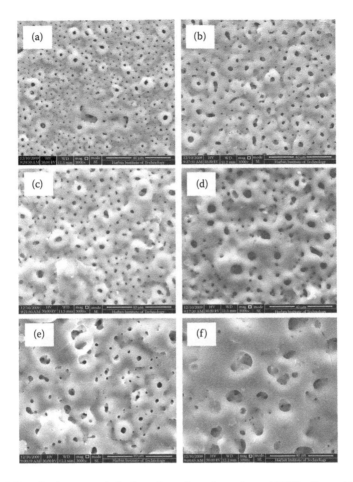

FIGURE 4.20 Surface morphologies of coatings formed on AZ91D alloy with different $Ca(H_2PO_4)_2$ concentration and various applied voltages: (a) Si-Ca5 at 450 V; (b) Si-Ca5 at 500 V; (c) Si-Ca10 at 450 V; (d) Si-Ca10 at 500 V; (e) Si-Ca15 at 450 V; and (f) Si-Ca15 at 500 V. (Adapted from Wang, Y. M., J. W. Guo, Z. K. Shao, J. P. Zhuang, M. S. Jin, C. J. Wu, D. Q. Wei, and Y. Zhou. *Surface and Coatings Technology*, 219, 8–14, 2013. With permission.)

Zhao et al. (2010) evaluated the microstructure and corrosion resistance of pure Mg treated with MAO under different treating times. The microstructure of the MAO coating under different times is shown in Figure 4.21. The results revealed that the thickness of the MAO coating increased with an increasing time from 1 min to 10 min, and the porosity is relatively low. Further prolonging the treating time to 17 min, the thickness of the MAO coating reached its peak, but the pore size was also increased. When treating time reached 22 min, the thickness decreased, the pore size became larger, and cracks also appeared. Other groups also reported the influence of different treating time on the microstructure and corrosion resistance of biomedical Mg and its alloys (Ren et al. 2011; K. Bai et al. 2012; T.-F. Lu et al. 2012; Ma et al. 2014).

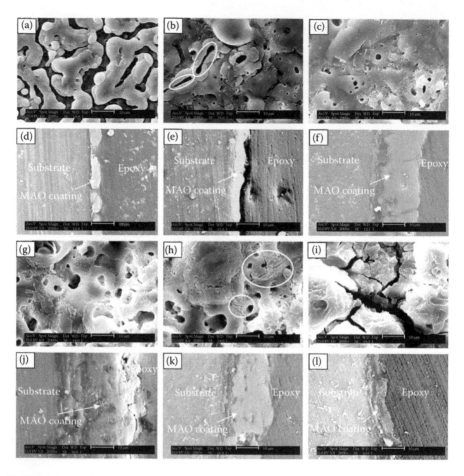

FIGURE 4.21 Surface and cross-section micrographs of the samples anodized for different times: (a) and (d) 1 min, (b) and (e) 3 min, (c) and (f) 10 min, (g) and (j) 17 min, (h) and (k) 19 min, (i) and (l) 22 min. (From Zhao, L., C. Cui, Q. Wang, and S. Bu. *Corrosion Science*, 52 (7), 2228–34, 2010. With permission.)

The resulting MAO coatings produced on the surface of degradable Mg alloys are porous, which is considered detrimental and changeable for further enhancement of corrosion resistance. Despite different electrolytes being used as a one-step method to deposit bioactive ceramics on the surface through MAO treatment, the surface remained porous (Yao et al. 2009; Liu et al. 2011a; Wang et al. 2013e; Ma et al. 2014). Thus, sealing treatment on MAO-treated samples or further-deposited coating on the surface of MAO-treated samples seems promising to further improve the corrosion resistance of MAO-treated biomedical Mg alloys. Table 4.4 presents a summary of some typical sealing agents and methods and their corrosion resistance compared to

TABLE 4.4

Summary of Sealing Agent/Method Used in MAO-Treated Biomedical Magnesium Alloys

Substrate	Sealing Agent/Method	Corrosion Resistance	References
As-cast WE42	PLLA	+ (Hank's)	Guo et al. 2011
As-extruded ZK60	HF treatment	− (Hank's)	Lin et al. 2013a
AZ91	Nano-diopside (CaMgSi$_2$O$_6$)	+ (SBF)	Razavi et al. 2014
As-cast Mg-Zn-Ca	TiO$_2$	+ (SBF)	Y. Zhang et al. 2012a
AZ81	PLLA	No data	Lu et al. 2011
AZ31	Ni-P containing epoxy resin	− (3.5% NaCl)	M.-A. Chen et al. 2013
AZ91D	Calcium phosphate + chitosan	+ (m-SBF)	Zhang et al. 2013a
AZ31B	ZrO$_2$ + gelatin–HA	+ (SBF)	Chu et al. 2013
AZ91	n-Hap	+ (SBF)	Rojaee et al. 2013b
Mg-Zn-Ca	Propolis	+ (SBF)	Gao et al. 2011b
Mg-Zn-Ca	Rod-like nano-hydroxyapatite	+ (SBF)	Gao et al. 2011c
AZ91D	Chitosan + n-HA	PBS	C. Wu et al. 2010
As-cast Mg-Zn-Ca	Chitosan	+ (SBF)	K. Bai et al. 2012
CP Mg	HA + DCPD	No data	Liu et al. 2011b
AZ31	DCPD + β-TCP	+ (Hank's)	L. Chang et al. 2013
WE43	Chitosan + poly(styrene sulfonate)	+ (SBF)	Liu et al. 2012b
CP Mg	TiO$_2$	No data	Shi et al. 2009
AZ80	DCPD	+ (SBF)	Shi et al. 2011
AZ80	DLC-doped TiN	+ (3.5% NaCl)	W. Yang et al. 2013
AZ31	Solution treatment	+ (SBF)	Tang et al. 2013b
MB26	Ni-P composite	− (3.5% NaCl)	X. Fan et al. 2013
As-cast Mg-2Zn-0.2Ca	Electrochemical deposition	+ (SBF)	W. Li et al. 2011
AZ31B	HA	+ (SBF)	Q. Zhao et al. 2013
AZ91D	Chitosan + n-HA	+ (m-SBF)	J. Zhang et al. 2012a
AZ91	Nano-bredigite	+ (SBF)	Razavi et al. 2013
AZ31	TCP + n-HAp	+ (r-SBF)	Seyfoori et al. 2013b

counter-specimens by MAO treatment only. Basically, it is critical if the outer coating on the MAO-treated sample is smooth because if microcracks are introduced by the sealing procedure, a worse corrosion resistance would occur.

Chu et al. (2013) investigated the influence of three sealing treatments, including boiling water, zirconia sol-gel, and organic gelatin-hydroxyapatite (HA) coatings, on the corrosion resistance of MAO-treated AZ31B wires with the surface morphology shown in Figure 4.22. The surface is covered by a smooth organic layer (Figure 4.22d) as the gelatin solution has good fluidity, and with the other three post-treatments a porous morphology still remained on the surface. Razavi et al. (2013) fabricated a nanobredigite coating on the surface of MAO-treated AZ91 Mg alloy. The coating performed good protection during 672 h immersion in SBF. The pH value increased to 8.75 at 72 h and then went back and stabilized around 7.5.

The in vivo animal experiments of MAO-treated Mg alloys are also reported (S. Chen et al. 2012; Fischerauer et al. 2013; Lin et al. 2013b). Taking Fischerauer's research (Fischerauer et al. 2013) as an example, the MAO-treated ZX50 pins were implanted into rat femoral bone. Figure 4.23 is the μ-CT images (3-D reconstruction)

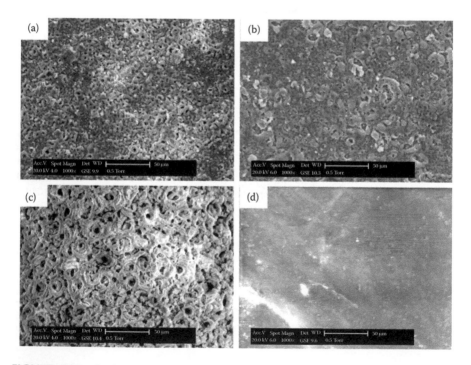

FIGURE 4.22 Surface morphologies of the micro-arc oxidized magnesium alloy wires after different surface treatment: (a) control (untreated); (b) boiling water; (c) zirconia sol-gel coating; (d) gelatin–HA coating. (Adapted from Chu, C. L., X. Han, F. Xue, J. Bai, and P. K. Chu. *Applied Surface Science*, 271, 271–5, 2013. With permission.)

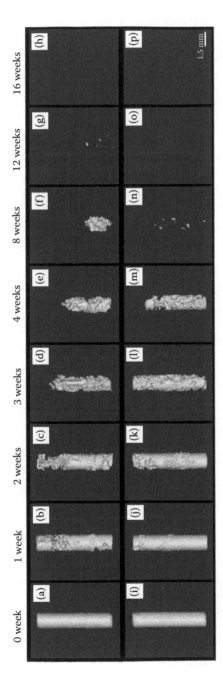

FIGURE 4.23 μ-CT images (3-D reconstruction) of implanted ZX50 (a–h) and MAO (i–p) pins. (Adapted from Fischerauer, S. F., T. Kraus, X. Wu, S. Tangl, E. Sorantin, A. C. Hanzi, J. F. Loffler, P. J. Uggowitzer, and A. M. Weinberg. *Acta Biomaterialia*, 9 (2), 5411–20, 2013. With permission.)

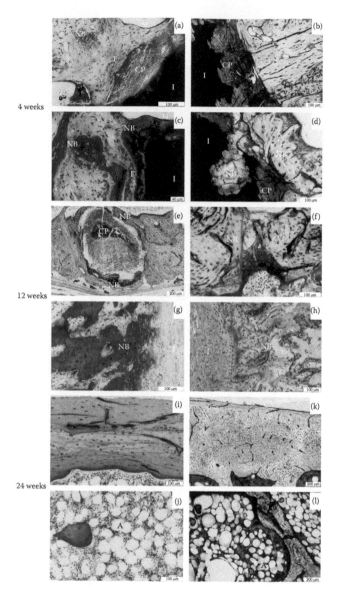

FIGURE 4.24 Histological thin slides of ZX50 and MAO pins in Levai–Laczko staining for ZX50 samples (left) and Toluidine Blue O for the MAO implants (right). Situation after (a–d) 4 weeks, (e–h) 12 weeks, and (i–l) 24 weeks. A = adipocytes, CP = corrosion products, F = fibroblast band, G = hydrogen gas bubble, I = implant/initial implant site, NB = new bone formation. (Adapted from Fischerauer, S. F., T. Kraus, X. Wu, S. Tangl, E. Sorantin, A. C. Hanzi, J. F. Loffler, P. J. Uggowitzer, and A. M. Weinberg. *Acta Biomaterialia*, 9 (2), 5411–20, 2013. With permission.)

of implanted bare ZX50 and MAO-treated ZX50 pins. The pins degrade over time and vanish completely after 12–16 weeks. The MAO-treated ZX50 pin performed better corrosion resistance at the early stage (first 4 weeks) but a worse corrosion resistance after an 8-week implantation. Figure 4.24 is the histological thin slides of pins in Levai–Laczko staining for bare ZX50 samples (left) and Toluidine Blue O for the MAO ZX50 implants (right). The inhibition of corrosion by MAO coating suppressed the corrosion products, such as hydrogen, and thus stimulated earlier new bone formation.

4.4.2.2   Electrodeposition

Electrodeposition has a long history in surface modification in material science and engineering. Therefore, it is a relatively mature technique used in degradable biomedical Mg and its alloys to deposit a certain coating on the surface to retard corrosion but also sustain good biocompatibility.

So far, the reports about electrodeposition on biomedical Mg and its alloys include AZ91 (Song et al. 2008; Zhang et al. 2010a; Kannan and Orr 2011; H. Yang et al. 2011; Kannan 2012; X. Zhang et al. 2012j; Grubač et al. 2013a; Kannan and Liyanaarachchi 2013; Kannan and Wallipa 2013), AZ31 (C. Wen et al. 2009; Salman et al. 2013), AZ80 (Shi et al. 2011), ZK60 (Lu et al. 2013), WE43 (Liu et al. 2012a), pure Mg (Ng et al. 2010a; Jamesh et al. 2011b; Luo and Cui 2011), Mg-Ca (Bakhsheshi-Rad et al. 2014a), Mg-Zn (Li et al. 2010a; Y. Song et al. 2010; J. Li et al. 2011; Song et al. 2013), Mg-Zn-Ca (H. X. Wang et al. 2010b; Liu et al. 2011b; Meng et al. 2011; H. Wang et al. 2011; Bakhsheshi-Rad et al. 2013a), and Mg-Gd-Y-Zr (Yi et al. 2009). The most common application of electrodeposition in degradable biomedical Mg and its alloys is the fabrication of hydroxyapatite (HA)/Ca-P/dicalcium phosphate dehydrate (DCPD) (Song et al. 2008, 2010, 2013; C. Wen et al. 2009; Jamesh et al. 2011b; Kannan and Orr 2011; W. Li et al. 2011; Shi et al. 2011; Kannan 2012; Bakhsheshi-Rad et al. 2013a; Grubač et al. 2013a; Kannan and Wallipa 2013; Lu et al. 2013; Salman et al. 2013). Typical electrolyte solutions used in electrodeposition are $Ca(NO_3)_2$ and $NH_4H_2PO_4$ (Kannan and Orr 2011; Grubač et al. 2013a; Kannan and Liyanaarachchi 2013; Kannan and Wallipa 2013; Song et al. 2013). In addition, H_2O_2 (Y. Song et al. 2008, 2010; Jamesh et al. 2011b; Meng et al. 2011; Shi et al. 2011; Zhang et al. 2012j; Bakhsheshi-Rad et al. 2013a; Lu et al. 2013), and $NaNO_3$ (C. Wen et al. 2009; Li et al. 2010a; Y. Song et al. 2010; H. X. Wang et al. 2010b; J. Li et al. 2011; W. Li et al. 2011; H. Wang et al. 2011, 2014; Bakhsheshi-Rad et al. 2013a) are also introduced to achieve better Ca-P coating on the surface. Bakhsheshi-Rad et al. (2013a) reported that the addition of $NaNO_3$ led to an enhancement of the ionic strength in the electrolyte. Typical morphology of Ca-P fabricated on the surface of Mg alloys is a regular flake-like structure with a certain number of pores and cracks as shown in Figure 4.25 (Jamesh et al. 2011b). These pores and cracks are beneficial for cell adhesion but induce more local corrosion. More specifically, Ca-deficient hydroxyapatite coatings were deposited (H. X. Wang et al. 2010b; Liu et al. 2011b; H. Wang et al. 2011), and the corrosion resistance was improved. It must be noted that many reports indicated that the coating after deposition is DCPD; thus a post-heat treatment is needed after electrodeposition to convert DCPD into HA.

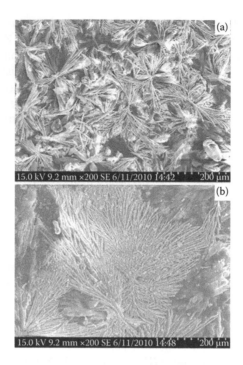

FIGURE 4.25 Surface morphology of (a) as-deposited coating, and (b) after the coating is immersed in 1 M NaOH at 80°C for 2 h. (From Jamesh, M., S. Kumar, and T. S. N. Sankara Narayanan. *Journal of Coatings Technology and Research*, 9 (4), 495–502, 2011. With permission.)

As a multiparameter controlled technique, the change of parameters used during electrodeposition will produce different films. Meng et al. (2011) utilized a pulse reverse current (PRC) mode to fabricate fluoride-doped HA on the surface of Mg-Zn-Ca alloy. The results exhibited that PRC-fabricated coating was even and smooth without pores and cracks that traditional electrodeposition induced. Lu et al. (2013) investigated the influence of electrodeposition at different temperatures on the corrosion resistance of ZK60 alloy. The results revealed that the particle size and surface roughness in the coatings increased with increasing electrolyte temperatures, and the 60°C electrodeposited ZK60 sample possessed the best corrosion resistance among the experimental temperature range of 20–60°C.

Apart from Ca-P coating deposited through electrodeposition, other films were also obtained by altering the type and concentration of the electrolyte. Ng et al. (2010a) fabricated a Ce-based coating on pure Mg through electrodeposition followed by a hydrothermal treatment to enhance the corrosion resistance. The corrosion current density results indicated a significant two-order improvement of the corrosion resistance in Hank's solution. Yi et al. (2009) also reported an enhanced corrosion resistance of Mg-Gd-Y-Zr with Ce-based coating electrodeposited. Zhang et al. (2010a) deposited a Ni-TiO$_2$ composite coating on the surface of AZ91D to improve the corrosion resistance. Electroless Ni coating was an interlayer to enhance the adhesion between TiO_2 and the AZ91D substrate. Compared to the specimen only with Ni coating, the sample with Ni-TiO_2 composite coating exhibited a prolonged corrosion resistance. H. Yang et al. (2011) deposited a dense FCC structured Al-containing coating followed by heat treatment on the surface of AZ91. A marked increase of adhesion strength was obtained by post-coating heat treatment. The results also revealed that the sample after 200°C heat treatment exhibited a better corrosion resistance compared to the sample heat-treated at 420°C. Luo and Cui (2011) reported an electrodeposited poly(3,4-ethylenedioxythiophene) (PEDOT) coating, which is uniform and can not only improve the corrosion resistance, but also load the anti-inflammatory drug dexamethasone during the electrodeposition (PEDOT) coatings on Mg.

Regarding the in vitro biocompatibility, J. Li et al. (2010a, 2011) investigated the cytocompatibility of fluoride hydroxyapatite-coated Mg-6Zn on human bone marrow stromal cells (hBMSC). As Figure 4.26 shows, all cells spread on the surface of both treated and untreated samples after 24 h culture. LSCM fluorescence micrographs (Figure 4.26c, d) showed long red bundles of stress fibers composed of actin filaments, displaying normal cell cytoskeleton morphology. In addition, through molecular biological tests, it was concluded that the FHA coatings were beneficial to osteogenic genes after 21 days of culture.

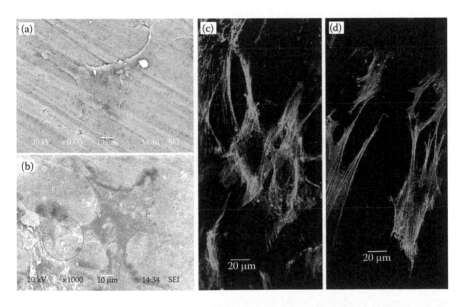

FIGURE 4.26 The typical hBMSC morphologies of all the materials seeded with a density of 1×10^4/disc and cultured for 24 h. (a) Typical SEM image of Mg-6Zn alloy. Scale bar = 10 μm. (b) Typical SEM image of FHA-coated Mg-6Zn alloy. Scale bar = 10 μm. (c) The typical LSCM image of Mg-6Zn alloy. Scale bar = 20 μm. (d) The typical LSCM image of FHA-coated Mg-6Zn alloy. Scale bar = 20 μm. (Adapted from Li, J., Y. Song, S. Zhang, C. Zhao, F. Zhang, X. Zhang, L. Cao, Q. Fan, and T. Tang. *Biomaterials*, 31 (22), 5782–8, 2010. With permission.)

In vivo animal testing of electrodeposition-treated biomedical Mg alloys have been carried out (J. Li et al. 2011; H. Wang et al. 2011). H. Wang et al. (2011) deposited a Ca-deficient hydroxyapatite on the surface of Mg-Zn-Ca and implanted the treated sample into New Zealand white rabbit left femurs with the untreated counterpart sample implanted into the right femurs. Representative micro-CT 2-D reconstruction images of the femora containing Mg implants are illustrated in Figure 4.27. The reconstruction images and data calculated after 24 weeks of implantation indicated that the Ca-deficient HA coating provided a better protection compared to the untreated sample. However, after 8 weeks of implantation, the degradation of the Ca-deficient HA-coated sample exhibited an increasing corrosion rate. Figure 4.28 is the pathological photographs of bone tissues containing Mg alloy implants after 8, 12, 18, and 24 weeks postimplantation.

J. Li et al. (2011) carried out an in vivo test of FHA-coated Mg-6Zn with in vivo implant morphology as shown in Figure 4.29. After implantation into the femoral condyle of adult rabbits for 1 month, the FHA-coated sample performed more direct contacts to the tissue compared to the untreated sample, which can be regarded as the consequence of the protection of FHA coating, which controlled hydrogen evolution and ion release, according to their research.

4.4.2.3 Electrophoretic Deposition

Electrophoretic deposition is also utilized in degradable biomedical Mg alloys. Suspended particles are transferred and deposited on the surface of materials under the effect of an electric field. Because it is a process at relatively low temperatures, phase change and embrittlement caused by high-temperature coating can be effectively avoided. C. Wu et al. (2010) deposited a calcium phosphate/chitosan coating on AZ91D Mg alloy through electrophoretic deposition followed by a conversion process. The deposited coating after electrophoretic deposition in different electrolytes revealed different morphology and a different ratio of HA/Ca(OH)$_2$. A further immersion in Ca(OH)$_2$ can convert DCPD into HA. As for adhesion strength, the results indicated that better adhesion strength was achieved with the electrolyte at a volume concentration of the n-HA/chitosan–acetic acid aqueous solution of 60% and 80%. Rojaee et al. (2013b) fabricated a nanostructured hydroxyapatite coating on the surface of AZ91 Mg alloy with a different electrophoretic deposition method. The results revealed that the MAO-n-Hap-coated sample achieved better corrosion resistance compared to the MgF$_2$-n-Hap-coated sample although they were all produced through electrophoretic deposition.

4.4.3 Biomimetic Deposition

The biomimetic process is based on simple chemical immersion techniques to produce a Ca-P coating on the substrate wherein different kinds of solutions have been adopted (Zheng et al. 2014).

As for biomimetic deposition, the Ca/P ratio is usually sustained at 1.67 or in the range of 1.33–1.67, which is known as the Ca/P ratio of hydroxyapatite. Thus, by stabilizing this ratio in the solution, researchers could fabricate a Ca-P coating with a similar ratio to benefit osteogenesis and cell adhesion.

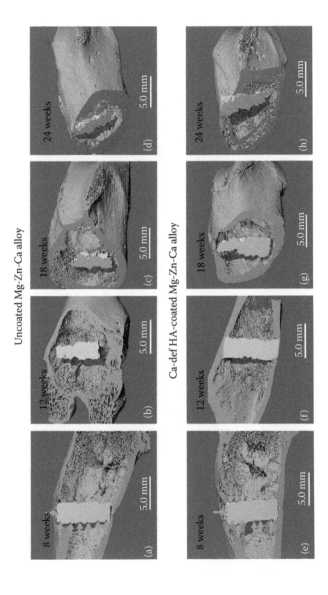

FIGURE 4.27 Representative micro-CT 2-D reconstruction images of the femora containing magnesium implants. (From Wang, H., S. Guan, Y. Wang, H. Liu, H. Wang, L. Wang, C. Ren, S. Zhu, and K. Chen. *Colloids Surf B Biointerfaces*, 88 (1), 254–9, 2011a. With permission.)

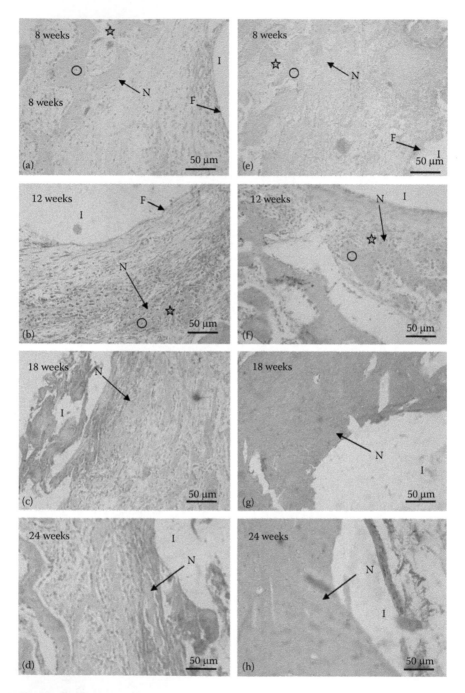

FIGURE 4.28 Pathological photographs of bone tissues containing magnesium alloy implants after 8 (a, e), 12 (b, f), 18 (c, g), and 24 (d, h) weeks post-implantation (HE stained). (From Wang, H., S. Guan, Y. Wang, H. Liu, H. Wang, L. Wang, C. Ren, S. Zhu, and K. Chen. *Colloids Surf B Biointerfaces*, 88 (1), 254–9, 2011a. With permission.)

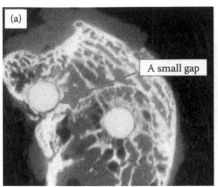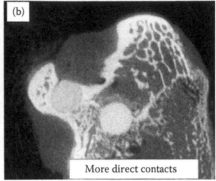

FIGURE 4.29 The in vivo implant morphologies of (a) the Mg–Zn alloy and (b) the FHA-coated Mg–Zn alloy detected by micro-CT scans. (Adapted from Li, J., P. Han, W. Ji, Y. Song, S. Zhang, Y. Chen, C. Zhao, F. Zhang, X. Zhang, and Y. Jiang. *Materials Science and Engineering: B*, 176 (20), 1785–8, 2011. With permission.)

Common solutions used to produce biomimetic Ca-P coating include modified SBF (Y. Zhang et al. 2009; Keim et al. 2011; Waterman et al. 2011; W. Lu et al. 2012; Ou et al. 2013) and $CaCl_2/Ca(NO_3)_2 + NaH_2PO_4$ (Yang et al. 2008a, 2009; Gray-Munro et al. 2009; Gray-Munro and Strong 2009; Singh et al. 2011; Q. Wang et al. 2011; Cui et al. 2013; Shadanbaz et al. 2013, 2014). In addition, two typical pretreatments are also used prior to biomimetic deposition. One typical pretreatment is alkali treatment followed by heat treatment (Gray-Munro and Strong 2009; Singh et al. 2011; Cui et al. 2013; Ou et al. 2013; Shadanbaz et al. 2014). The other is acid etching followed by neutralization in NaOH solution (Gray-Munro et al. 2009; Xu et al. 2009b; Du et al. 2011b; W. Lu et al. 2012). Alkali treatment followed by heat treatment is considered to increase the surface concentration of hydroxyl groups. Acid etching is utilized for surface activation.

Singh et al. (2011) reported a Ca-P-Si coating produced on the surface of as-rolled AZ31 and Mg-4Y alloys through biomimetic deposition. Although such a coating did not improve the corrosion resistance obviously, it significantly lowered the concentration of Mg and Zn released. Moreover, phosphating bath coating revealed a more effective improvement on the bioactivity of AZ31 than Mg-4Y. Keim et al. (2011) produced Ca-P coating on the surface of CP Mg by immersing the sample in SBF 5 solution and subsequent treatments. A significant improvement of HeLa cell density was obtained compared to the bare sample after 24 h culture.

Yanovska et al. (2012) reported a simple method to deposit Ca-P coating by applying a permanent magnetic field set by Sm/Co at 0.3 T. The DCPD phase in the coating performed a certain preferred orientation, and HA crystals showed no preferred orientation at c-axes under the magnetic field. Cui et al. (2013) deposited a novel 3-DSS peptide and Ca-P coating for the first time. A better corrosion resistance was obtained for that after 72 h immersion, noncoated AZ31B still generated 1.30 times more hydrogen than 3-DSS-coated AZ31B. Y. Zhang et al. (2009) adjusted the biomimetic solution to be 3CaP-SBF + 4-(2-hydroxyethyl)-1-piperazineethanesulfonic acid (HEPES). By immersing pure Mg into such a solution at 42°C for 24 h and 48 h, the results revealed a thick coating was obtained after 48 h immersion. The corrosion resistance was also improved with such apatite coating. These results implied that altering the composition of biomimetic solution can get rid of post-heat treatment. W. Lu et al. (2012) prepared a Ca-deficient HA coating on the surface of ZK60 Mg alloy in a revised SBF at pH values of 5, 6, and 7 for 7 days. The results revealed an improved corrosion resistance of all coated samples under different pH values, and the sample with HAp coating deposited from the solution with a pH of 6 had the best corrosion resistance. Besides the improved corrosion resistance obtained, the cytotoxicity test on L-929 was also executed. The results indicated that the coated samples had good biocompatibility, and the sample coated from the solution with a pH of 6 had the best biocompatibility.

Q. Wang et al. (2011) produced a Ca-P coating by immersing an as-extruded AZ31B sample in biomimetic solution consisting of $Ca(NO_3)_2$ and $NH_4H_2PO_4$ at a pH value of 4 for 24 h. A much lower pH value compared to the uncoated sample was obtained during 26 days of immersion in PBS. Moreover, the hemolytic rate was under 5% during 90 days of immersion in PBS, and the uncoated sample possessed

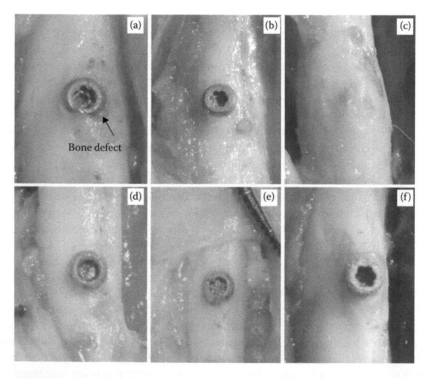

FIGURE 4.30 Top view on AZ31B alloy/bone system after implantation, the uncoated alloy was implanted for 30 days (a), 60 days (b), and 90 days (c); the coated alloy was implanted for 30 days (d), 60 days (e), and 90 days (f). (Adapted from Wang, Q., L. Tan, W. Xu, B. Zhang, and K. Yang. *Materials Science and Engineering: B*, 176 (20), 1718–26, 2011. With permission.)

a hemolytic rate more than 10% at the beginning, which declined to lower than 5% after 10 days immersion. They also implanted the coated and uncoated samples into the femora of adult white rabbits with a top view on the alloy/bone system as shown in Figure 4.30. An obvious bone defect could be seen after 30 days implantation of the uncoated sample, and a better contact to the tissue could be seen on the coated sample. However, with prolonged implantation, the bone defect seen in the uncoated group disappeared. In addition, more new tissue was generated in the uncoated group than the coated group.

4.4.4 Sol-Gel Treatment

Sol-gel coating is prepared using a colloidal solution as the precursor to synthesize an integrated network, which will be further crystallized at high temperatures to generate a porous, nanocrystalline coating with a controlled microstructure (Zheng et al. 2014).

So far, the sol-gel method successfully synthesized many coatings on degradable biomedical Mg alloys, including Ca-P coating (Roy et al. 2011; Ren et al. 2013; Rojaee et al. 2013a; Tang et al. 2013a; X. Wang et al. 2014), TiO_2 coating (Shi et al. 2009; W. Chen et al. 2010; Hu et al. 2011; Amaravathy et al. 2012), silica coating (López et al. 2010), glycidoxypropyltrimethoxysilane (GPTMS) coating (Zomorodian et al. 2012), bioactive glass-ceramic coating (Ye et al. 2012; Dou et al. 2013; Huang et al. 2013a, 2014; Wang and Wen 2014), and composite coating (Lamaka et al. 2008, 2009; Karavai et al. 2010; H. Wang et al. 2010a; Gao et al. 2011c; Y. Zhang et al. 2012a; X. Wang et al. 2013a; Ren et al. 2014).

Rojaee et al. (2013a) produced a nanostructured hydroxyapatite coating on AZ91 through the sol-gel method followed with calcination and sintering processes. The surface morphology during immersion in SBF is shown in Figure 4.31. Rice-shaped precipitations nucleated at the early stages of immersion, and then crystal nuclei grew to a certain size, got stabilized, and became clusters. The clusters grew to spherical particles in the presence of sufficient Ca- and P-derived ions within further immersion time.

Dou et al. (2013) produced a 45S5 bioactive glass-ceramic coating on the surface of AZ31 Mg alloy by a synthesized sol-gel method, according to the molar ratios of SiO_2, Na_2O, CaO, and P_2O_5 in 45S5. The sol was prepared by mixing two separate solutions (solution 1: tetraethyl orthosilicate [TEOS] dissolved in HNO_3; solution 2: $NaNO_3$ and $Ca[NO_3]_2$ dissolved in distilled water). Through dip-coating, the coatings were deposited on AZ31 Mg alloy for one to five cycles. The results revealed that, with such coatings consisting of amorphous phase and $Na_2Ca_2Si_3O_9$, the mass loss was significantly decreased from 78.04% to 2.31% after 7 days immersion in m-SBF, indicating an improvement of corrosion resistance. Ye et al. (2012) also produced a 45S5 bioactive glass-ceramic coating and obtained an improved corrosion resistance during 7 days immersion in SBF and, however, a declined corrosion resistance after 14 days immersion in SBF. Zomorodian et al. (2012) produced a modified glycidoxypropyltrimethoxysilane (GPTMS) coating with a different pretreatment on the surface of AZ31 Mg alloy through the sol-gel method. The modified silane sol was synthesized by mixing 3-glycidoxypropyltrimethoxysilane (GPTMS),

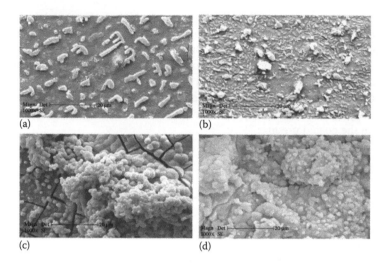

FIGURE 4.31   SEM photomicrographs of sol-gel-derived n-HAp coating on AZ91 substrate after soaking in c-SBF2 solution at 37°C ± 1°C within (a) 1, (b) 3, (c) 14, and (d) 28 days. (From Rojaee, R., M. Fathi, and K. Raeissi. *Materials Science & Engineering. C, Materials for Biological Applications*, 33 (7), 3817–25, 2013. With permission.)

methyltriethoxysilane (MTES), and diethylenetriamine (DETA). In addition, the pretreatment played an important role in the protection performance of the GPTMS coating. Samples with hydrofluoric acid etching pretreatment possessed an improved corrosion resistance, and samples with dc polarization treatment in NaOH showed little difference in corrosion resistance compared to the uncoated samples.

Usually, the coatings synthesized by the sol-gel method are not dense, smooth, and uniform. Thus, some methods need to be adopted to improve the coating quality. Zhong et al. (2010) introduced zinc nitrate as a healing agent to repair the sol-gel organic silane coatings in NaCl solution. The silane sol was synthesized by mixing 3-glycidoxypropyltrimethoxysilane (GPTMS), tetraethylorthosilicate (TEOS), distilled water, and ethanol in 3:1:13:40 molar ratios. Then, the coating, deposited on as-cast AZ91D Mg alloy by spin-coating, could provide better corrosion protection according to their electrochemical results with 0.005 M NaCl as a corrosion electrolyte.

Moreover, the sol-gel method is a generally accepted method to seal the pores induced by a fluoride conversion coating treatment, anodic oxidation, and MAO to develop a coating with good morphology for further improvement of corrosion resistance. Lamaka et al. (2009) synthesized a polymer composite of titania nanoparticles and (3-glycidoxypropyl)-trimethoxysilane (GPTMS) to seal the pores on the anodic oxidized ZK30 alloy. This composite coating enhanced the corrosion resistance in NaCl for its fine morphology. What needs to be noted is that the doping of Ce+ in the anodic oxidized layer can improve the corrosion resistance as well as the post-treated composite coating did. Y. Zhang et al. (2012a) coated MAO-treated Mg-Zn-Ca with TiO_2 coating through the dip-coating method, followed with an annealing treatment.

The electrochemical test results revealed an enhancement of corrosion resistance in SBF because the corrosion current density decreased nearly one order of magnitude. Ren et al. (2014) reported using a sol-gel method to deposit Ca-P glass on MgF_2-coated AZ31 to fabricate a Ca-P/MgF_2 composite coating through the sol-gel method. The MgF_2 coating was fabricated by the fluoride conversion coating method. An alkyl phosphate, prepared by dissolving P_2O_5 in anhydrous ethanol, was used as the phosphorus precursor for the sol-gel Ca-P coating. The results indicated that such composite coating exhibited uniform, dense, and smooth coating morphology, which significantly decreased corrosion current density by two orders of magnitude, according to electrochemical test results in SBF.

Regarding the in vitro biocompatibility of Mg alloys coated with the sol-gel method, Amaravathy et al. (2012) evaluated the cytocompatibility of TiO_2-coated AZ31 Mg alloy with MG63 cell line. As shown in Figure 4.32, more spread and larger area occupied cells were found for the coated group than the uncoated group. In addition, the relative cell number adhered to the coated sample surface after 7 days

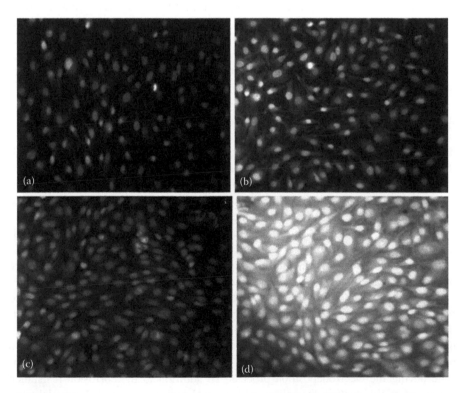

FIGURE 4.32 Fluorescence images of MG-63 oesteoblast-like cells growing on each substrate after culturing for 7 days. (a) Uncoated AZ31 first day, (b) TiO_2-coated AZ31 first day, (c) uncoated AZ31 seventh day, and (d) TiO_2-coated AZ31 seventh day. (From Amaravathy, P., C. Rose, S. Sathiyanarayanan, and N. Rajendran. *Journal of Sol-Gel Science and Technology*, 64 (3), 694–703, 2012. With permission.)

culture was nearly 8000, and around 3000 cells were calculated on the uncoated counterpart after the same period.

4.4.5 ORGANIC AND POLYMER COATING

The organic and polymer coatings fabricated on the surface of biomedical Mg alloys not only perform efficiently, that is, with corrosion protection and good biocompatibility as with other kinds of coatings, but they also bring new functions, such as drug delivery, which makes it a good candidate for surface modification of degradable biomedical Mg alloys.

To date, several polymers/organisms and composites of polymers have been successfully fabricated on Mg alloys for biomedical application, including poly(ether imide) (PEI) (da Conceicao et al. 2011; Kim et al. 2013), vanadate (Y. Ma et al. 2012), chitosan (Gu et al. 2009c; K. Bai et al. 2012; Liu et al. 2012b; Zhang et al. 2012a, 2013a), poly(lactic acid) (PLA, PLLA) (Gray-Munro et al. 2009; Lu et al. 2011a; Xu and Yamamoto 2012b; Abdal-hay et al. 2014), poly(lactic-co-glycolic acid) (PLGA) (Li et al. 2010b), polypyrrole (Turhan et al. 2011), stearic acid (SA) (Ng et al. 2010b), oxalate (Jiang et al. 2009), phytic acid (Gupta et al. 2013), silane (Zucchi et al. 2006, 2008), poly(3,4-ethylenedioxythiophene) (PEDOT) (Luo and Cui 2011), and complex polymer-based composite (Moulton et al. 2008; Wong et al. 2010; Xu et al. 2010; Correa et al. 2011; Z. Kang et al. 2011; H. Wang et al. 2012; Xu and Yamamoto 2012a; Zomorodian et al. 2012; Abdal-hay et al. 2013; Kartsonakis et al. 2013; Rosemann et al. 2013) as summarized in Table 4.5.

Kim et al. (2013) produced both porous and dense poly(ether imide) (PEI) coatings on pure Mg through spin-coating. The pH changes during immersion in SBF for 5 days exhibited better control of the alkalization on porous coating rather than dense coating, which is attributed to the formation of strong chemical bonding with the substrate on the porous sample. da Conceicao et al. (2011) produced a PEI coating on AZ31 by the dip-coating method using N'N'-dimethyl acetamide (DMAc) and N'-methyl pyrrolidone solutions and evaluated the influence of solvent and substrate pretreatment. The results showed that the best coating (about 13 μm in thickness), by hydrofluoric acid-treated substrates coated using DMAc solution, showed impedance on the order of 10^7 Ω cm^2 even after 3312 h of exposure to a 3.5 wt.% NaCl solution. In addition, only when the coating can cover the defects induced by surface irregularities can the increase in coating thickness show improved protection.

In addition to the improvement in corrosion resistance, polymer coating can also realize drug delivery according to existing experiments. A conducting polymer coating (CPC) makes it possible to electrically control drug release, which would be very useful for on-demand drug delivery. Luo and Cui (2011) successfully produced a poly(3,4-ethylenedioxythiophene) (PEDOT) coating on Mg and loaded dexamethasone during the electrodeposition process. In addition to the corrosion resistance possessed by this coating, they also reported an electrically controlled drug delivery. Upon an applied potential of −2 V for 20 s, an average of about 16.3 μg Dex was released from the PEDOT coatings with Dex (PEDOT/IL/Dex). Moreover, every time PEDOT coating loaded with Dex was applied potential of −2 V for 20 s, drug release was found. Lu et al. (2011a) reported a MAO/PLLA/drug delivery composite

TABLE 4.5

Summary of Organic and Polymer Coatings Used in Biodegradable Mg and Its Alloys

Substrate	Coating Components	Method	Corrosion Resistance	Biocompatibility	References
AZ31	Polycaprolactone (PCL) and dichloromethane (DCM)	Layer by layer deposition	+ (SBF)	+ (SaOS-2 human cell)	Wong et al. 2010
AZ31	3-glycidoxypropyltrimethoxysilane (GPTMS)	Dip-coating	+ (Hank's, with HF pretreatment)	No data	Zomorodian et al. 2012
As-extruded pure Mg	Amorphous PLLA or semicrystalline PCL	Spin-coating	No data	+ (SaOS-2 human cell)	Xu and Yamamoto 2012a
AZ81	MAO/PLLA-PLGA drug delivery coating	Chemical method	No data	Good hemocompatibility	Lu et al. 2011a
AZ91	Methyltriethoxysilane (MTES) doped with cerium ions	Chemical method	+ (0.1 M Na$_2$SO$_4$)	No data	Correa et al. 2011
AZ31	Poly(ether imide) (PEI)	Dip-coating	+ (3.5% NaCl)	No data	da Conceicao et al. 2011
MAO-treated WE42	Cross-linked gelatin and PLGA nanoparticles composite	Chemical method	+ (Hank's)	No data	Xu et al. 2010
Pure Mg	poly(3,4-ethylenedioxythiophene) (PEDOT)	Electrochemical deposition	+ (PBS)	No data	Luo and Cui 2011
AZ91D	Polypyrrole (PPy)	Electrodeposition	+ (0.1 M Na$_2$SO$_4$)	No data	Turhan et al. 2011
Mg-6Zn	Dicalcium phosphate dihydrate (DCPD) and polycaprolactone (PCL)	Electrodeposition	No data	No data	H. Wang et al. 2012
AM50	Ti-O, PLA top layer	Dip-coating	+(SBF)	No data	Abdal-hay et al. 2014
HP Mg	Phytic acid conversion coating	Chemical method	+ (PBS)	No data	Gupta et al. 2013
Mg-Mn-Ce	DHN + ATP nanofilm	Polymer plating	+ (0.1 M NaCl)	No data	Z. Kang et al. 2011
AZ21	PPy and Dex polymer film	Galvanic coupling	No data	No data	Moulton et al. 2008

(Continued)

TABLE 4.5 (CONTINUED)
Summary of Organic and Polymer Coatings Used in Biodegradable Mg and Its Alloys

Substrate	Coating Components	Method	Corrosion Resistance	Biocompatibility	References
ZK30	Hybrid organic-inorganic coating with CeMo nanocontainer and corrosion inhibitor MBT	Chemical method	+ (0.5 M NaCl)	No data	Kartsonakis et al. 2013
As-cast AZ31	Hydroxyapatite-doped poly(lactic acid)	Chemical method	+ (SBF)	+ (MC3T3 cells)	Abdal-hay et al. 2013
As-extruded Mg-6Zn	PLGA	Dip-coating	+ (0.9% NaCl)	+ (MC3T3-E1 cells)	Li et al. 2010b
As-extruded Mg	Poly(L-lactic acid) (PLLA) or poly(ε-caprolactone) (PCL)	Spin-coating	+ (DMEM + 10% FBS)	No data	Xu and Yamamoto 2012b
WE43	Silanic-based coating	Dip-coating	+ (0.1 M Na$_2$SO$_4$)	No data	Zucchi et al. 2006
AZ31	PLA or poly(DTH carbonate)	Spray-coating	+ (SBF)	No data	Gray-Munro et al. 2009
AZ91D	Oxalate conversion coating	Chemical method	+ (5% NaCl)	No data	Jiang et al. 2009
PCO-treated Mg-1Ca	Poly(L-lactid-co-caprolacton) (PLLC)	Dip-coating	+ (SBF)	No data	Rosemann et al. 2013
Pure Mg	Stearic acid (SA)	Chemical method	+ (Hank's)	No data	Ng et al. 2010b
As-extruded Mg-1Ca	Chitosan	Dip-coating	+ (SBF)	No data	Gu et al. 2009c
Pure Mg	Poly(ether imide) (PEI)	Spin-coating	+ (SBF)	+ (MC3T3-E1)	Kim et al. 2013

coating on AZ81. The base PLGA/PTX drug releasing film was fabricated on the MAO/PLLA-coated sample by dropping the solution on the coating. A top blank PLGA drug-controlled releasing film was finally fabricated to evaluate the effects of plasticizer on drug delivery. The results revealed that a top blank PLGA film effectively controlled the delivery of PTX during 50 days of evaluation, and a higher ratio of LA:GA in the blank PLGA film led to a more obvious control effect. This regular pattern could also avoid burst release of the drug.

As a polymer coating, the molecular weight may be an impact factor that does not exist in other coatings. Gu et al. (2009c) produced a chitosan coating on an as-extruded Mg-1Ca alloy with four types of chitosan (the difference lies in the degree of deacetylation, intrinsic viscosity, and molecular weight). The results revealed that both the chitosan molecular weight and layer numbers of coating affect the corrosion resistance. The coatings, produced by 1.5×10^5 and 2.7×10^5 molecular weight chitosan, presented a smooth surface whereas microholes could be seen in coatings produced by 1.0×10^4 and 6.0×10^5 molecular weight chitosan. Moreover, the coating with a molecular weight of 2.7×10^5 for six layers possessed the best corrosion protection. R. Xu et al. (2012b) investigated the influence of molecular weight on the properties of poly(L-lactic acid) (PLLA)-coated Mg and poly(ε-caprolactone) (PCL)-coated Mg. The results suggested that PLLA coating had better corrosion protection compared to PCL coating in DMEM with 10% FBS. In addition, the polymer with high molecular weight had better corrosion resistance both in PLLA coating and PCL coating.

The in vitro biocompatibility of the polymer/organic-coated Mg alloy has been investigated. Wong et al. (2010) evaluated the cytotoxicity of PCL-coated AZ91 at different PCL porosities, with LPM indicating low-porosity membrane and HPM high-porosity membrane. Both the cytotoxicity tests on SaOS-2 human cell viability and GFP mouse osteoblast viability showed better viability on the coated sample. Moreover, the HPM-coated sample had better cell viability on both tests, and the HPM-coated sample possessed a slightly higher cell viability than the LPM-coated sample with microscopic photos of GFP mouse osteoblast culture as shown in Figure 4.33. R. Xu et al. (2012a) reported a significant improvement of cytocompatibility on SaOS-2 cells in both PLLA- and PCL-coated samples. Abdal-hay et al. (2013) reported good cytocompatibility on MC3T3-E1 cells in a HAp-doped PLA-coated AZ31 sample. Other polymer/organic coating compatibility, such as PLGA (Li et al. 2010b) and PEI (Kim et al. 2013), were reported to be good.

Moreover, the in vivo biocompatibility of the polymer/organic-coated Mg alloy has been studied. Wong et al. (2010) fabricated a composite polymer coating consisting of polycaprolactone and dichloromethane on AZ91 and implanted both coated and uncoated samples into New Zealand white rabbits' greater trochanter. Figure 4.34 is the micro-CT reconstruction images of the greater trochanter containing coated and uncoated samples with LPM indicating low-porosity membrane and HPM high-porosity membrane. PCL membrane and corrosion could be recognized. Newly formed bone can be seen in both the uncoated and coated groups with a 3-D model of the newly formed bone on both the coated and uncoated implants as Figure 4.35 shows. The LMP sample had the greatest amount of new bone formation of 10.79 mm^3 and did not have any implant volume reduction whereas the HPM sample

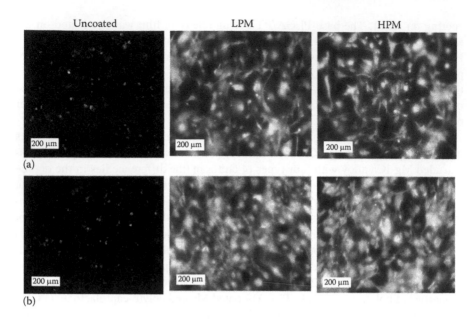

FIGURE 4.33 Microscopic views of GFP mouse osteoblasts cultured on PCL-coated and uncoated AZ91 magnesium alloy after 1 and 3 days. (a) 1 day; (b) 3 days. 5000 GFP osteoblasts were cultured on the coated and uncoated samples for 1 and 3 days so as to evaluate the cytocompatibility of the polymer-coated magnesium alloys. (Adapted from Wong, H. M., K. W. Yeung, K. O. Lam, V. Tam, P. K. Chu, K. D. Luk, and K. M. Cheung. *Biomaterials*, 31 (8), 2084–96, 2010. With permission.)

had 5.17 mm^3 new bone formation and 0.05% implant reduction; the untreated sample had 1.36 mm^3 new bone formation and 0.33% implant reduction after 2 months of implantation.

4.4.6 MOLECULAR SELF-ASSEMBLED COATING

For the surface modification of degradable biomedical Mg and its alloys, as-reported molecular self-assembled coating includes a self-assembling monolayer (SAM) and multilayer coating produced by a layer-by-layer technique. The formation of the SAM using oxy-acid species can be either an acid–base reaction to give a carboxylate or phosphonate complex of a surface metal cation or simple hydrogen bonded adducts (Gouzman et al. 2006). Moreover, the number and sequence of the layer can be controlled when applying the layer-by-layer technique, and as a consequence, the thickness of the coating is controlled. In general, a molecular self-assembled coating can get rid of expensive equipment, and it can also achieve the goal of improving corrosion resistance and biocompatibility in an effective way.

Guo et al. (2013a) fabricated self-assembled nanoparticles (SANP) on the surface of PEO-treated Mg-Gd-Y followed with or without a top layer of the fluorocarbon (FC) paint. Two SANP solutions were used to fabricate SANP; one contained phytic acid (SANP-P), and the other SANP solution contained acetic acid (SANP-A). The

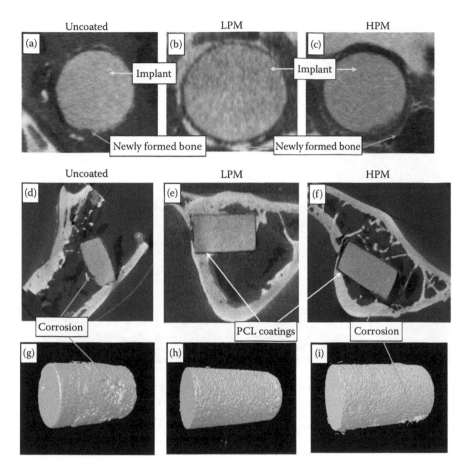

FIGURE 4.34 Micro-CT reconstruction images of the greater trochanter containing coated and uncoated sample. (a–c) transverse view; (d–f) coronal view, and (g–i) 3-D view of the uncoated, LPM, and HPM samples. The corrosion condition of the samples can be scanned and viewed in a micro-computed tomography device. (Adapted from Wong, H. M., K. W. Yeung, K. O. Lam, V. Tam, P. K. Chu, K. D. Luk, and K. M. Cheung. *Biomaterials*, 31 (8), 2084–96, 2010. With permission.)

results revealed that the corrosion current density decreased four orders of magnitude when coated with PSPF (PEO/SANP-P/FC) film and three orders of magnitude when coated with PSAF(PEO/SANP-A/FC) film, compared to the bare Mg-Gd-Y sample. Guo et al. (2013b) also developed a composite film on PEO-treated Mg-Gd-Y through a self-assembly process and gained improved corrosion resistance.

In addition, the SANP-P layer possessed a better adhesion strength to the substrate compared to the SANP-A layer. Grubač et al. (2013b) evaluated the influence of chain length on the corrosion behavior of self-assembled film-coated AZ91D. The SAM film included dodecylphosphonic ($CH_3[CH_2]_{11}PO[OH]_2$, DDP) and octadecyl-phosphonic acid ($CH_3[CH_2]_{17}PO[OH]_2$, ODP). There was also a heat treatment after

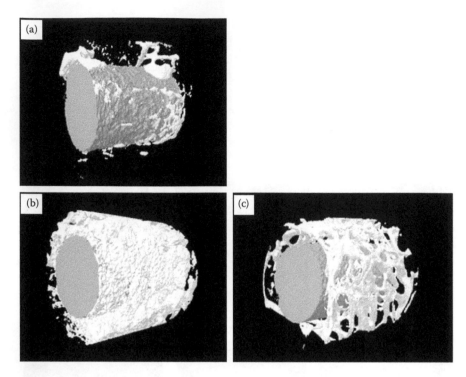

FIGURE 4.35 Micro-CT 3-D reconstruction models of newly formed bone (white in color) on both coated and uncoated implants. (a) Uncoated, (b) LPM, and (c) HPM. (Adapted from Wong, H. M., K. W. Yeung, K. O. Lam, V. Tam, P. K. Chu, K. D. Luk, and K. M. Cheung. *Biomaterials*, 31 (8), 2084–96, 2010. With permission.)

the process of self-assembly. The results showed that the post-heat treatment significantly improved the corrosion resistance of both DDP-coated and ODP-coated samples. Moreover, the ODP-coated sample had slightly better corrosion resistance than the DDP-coated sample.

Ishizaki et al. (2011) produced several kinds of monolayers derived from alkyl- and perfluoro-phosphonic acid on AZ31, including OP: $CH_3(CH_2)_7PO(OH)_2$, DP: $CH_3(CH_2)_{11}PO(OH)_2$, ODP: $CH_3(CH_2)_{17}PO(OH)_2$ dissolved in phosphonic acid, and PFEP: $CF_3(CF_2)_5CH_2CH_2PO(OH)_2$ dissolved in phosphonic acid. The results revealed the highest chemical stability and thickest coating in ODP-modified AZ31. The order of corrosion resistance performance in 5 wt.% NaCl was ODP > DP > OP.

Zeng et al. (2013a) produced Ag nanoparticles (NPs) on 3-aminopropyltrime-thoxysilane (APTMS)-modified Mg alloy AZ31 and investigated the antibacterial property. The schematic of the APTMS modification process and AgNPs attachment onto the APTMS/Mg surface is illustrated in Figure 4.36. The antibacterial results are as shown in Figure 4.37. An obvious inhibition to *Escherichia coli* suspension of Ag NPs could be obtained.

Cai et al. (2011) produced a 25-cycle multilayer consisting of multilayers that were composed of poly(ethylene imine) (PEI), poly(styrene sulfonate) (PSS), and

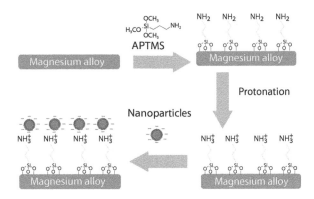

FIGURE 4.36 Schematic of the APTMS modification process and AgNPs attachment onto the APTMS/Mg surface. (From Zeng, R., L. Liu, S. Li, Y. Zou, F. Zhang, Y. Yang, H. Cui, and E.-H. Han. *Acta Metallurgica Sinica (English Letters)*, 26 (6), 681–6, 2013. With permission.)

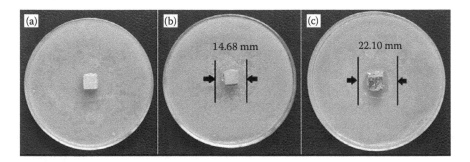

FIGURE 4.37 Photographic images of the zone of inhibition of Mg (a), APTMS/Mg (b), and AgNPs/APTMS/Mg (c) samples. (Adapted from Zeng, R., L. Liu, S. Li, Y. Zou, F. Zhang, Y. Yang, H. Cui, and E.-H. Han. *Acta Metallurgica Sinica (English Letters)*, 26 (6), 681–6, 2013. With permission.)

8-hydroxyquinoline (8HQ) with or without GA crosslinking on AZ91D. Although the coating possessed enhanced corrosion resistance and biocompatibility, the GA crosslinking performed an adverse effect on biocompatibility due to the ALP test results. Liu et al. (2012b) reported an improvement of corrosion resistance when combining layer-by-layer self-assembly with MAO treatment.

Cytocompatibility tests and in vivo tests are done for Mg alloys after molecular self-assembly surface modification. Ostrowski et al. (2013) utilized three cationic polymer solutions (marked as A, B, and C) to fabricate a multilayer film following a certain sequence of ABCBCBCBCB on the surface of AZ31. In addition, the composition and concentration of solution B also covered a range to develop film with differentiation. The results revealed an improved corrosion resistance in all the treated samples. MC3T3-E1 and hMSCs cell culture tests are as Figure 4.38 shows. Improved

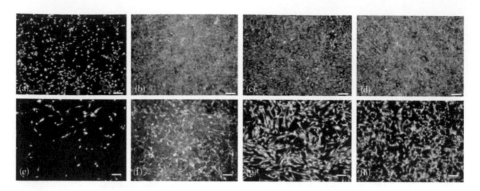

FIGURE 4.38 Live/dead staining of MC3T3 osteoblasts (a–d) and hMSCs (e–h) on day 3 of culture post seeding on uncoated AZ31 (a and e), PCL (b and f), PLGA 50:50 (c and g), and PLGA 75:25 (d and h), indicating improved biocompatibility of coated substrates over uncoated substrates. Scale bar = 200 μm. (From Ostrowski, N., B. Lee, N. Enick, B. Carlson, S. Kunjukunju, A. Roy, and P. N. Kumta. *Acta Biomaterialia*, 9 (10), 8704–13, 2013. With permission.)

cell ability and proliferation were obtained in all three coated substrates. However, no discernible difference can be seen among the three types of polymer coatings.

Gao et al. (2010b) reported a heat-self-assembled monolayer (HSAM) on as-cast pure Mg. The pure Mg samples were subjected to heat treatment at 773 K for 10 h, and then a self-assembled monolayer was formed on the surface in stearic acid alcohol solution. The immersion test results revealed a reduced corrosion rate of 0.5 mm/y compared to 0.11 mm/y of the bare sample in SBF. Figure 4.39 shows the morphology of the interface between implants and bone after 12 weeks of implantation in New Zealand flap-eared white rabbits' thigh bones. New bone directly grew on the surface of the Mg implants and osteoclast cells appeared on the Mg. Moreover, more new bone and closer bonds were found on the modified Mg implants compared to the TC4 sample surface.

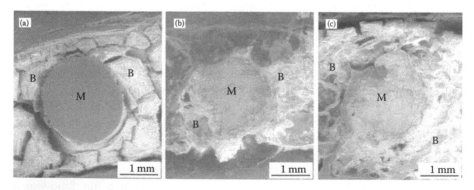

FIGURE 4.39 Morphologies of the histological interface of implants after 12 weeks' implantation (M: material, B: bone): (a) Ti6Al4V as control; (b) 4N-Mg; (c) HSAM-Mg. (From Gao, J.-C., L.-Y. Qiao, and R.-L. Xin. *Frontiers of Materials Science in China*, 4 (2), 120–5, 2010. With permission.)

4.5 CONCLUDING REMARKS

In this chapter, a comprehensive demonstration of the surface modification of biodegradable Mg and its alloys is given, including mechanical surface treatment, physical surface treatment, and chemical surface treatment.

Different from traditional surface modification on bioinert metallic materials, the surface modification on biomedical Mg and its alloys focuses on retarding corrosion and sustaining good biocompatibility at the same time. Therefore, various attempts have been made. When applying surface treatments to biodegradable Mg and its alloys, some issues may need to be considered in the future.

4.5.1 COATING ROUGHNESS

As mentioned, the surface roughness has a strong influence on the corrosion resistance, cytocompatibility, and bone response. Thus, the ideal coating would have a suitable surface roughness. For the surface treatments on biodegradable Mg alloys, mechanical surface treatments and physical surface treatments can control the surface roughness more easily compared to chemical surface treatments. However, chemical surface treatments can provide many more possibilities for surface modification due to endless chemical substances and numerous treatment processes. Therefore, surface roughness control need to be considered in more detail.

4.5.2 ADHESION

It is preferred that coatings provide 3–6 months of protection after implantation (Zheng et al. 2014), which calls for good adhesion strength between the coating and the substrate to remain stable after implantation. Nevertheless, a lot of research works involving surface coating on biodegradable Mg alloys do not mention the adhesion strength between the substrate and the coating. In addition, a relatively short in vitro immersion test cannot provide useful data on the adhesion situation with a macroscopic view. Too much attention has been concentrated on the protection of the coating itself while whether the coating and substrate can sustain as a whole for sufficient time is more or less ignored. The adhesion strength tests should be considered more in a future study.

4.5.3 LONG-TERM CORROSION BEHAVIOR

In this chapter, almost all the reported coatings applied on biomedical Mg alloys have retarded the corrosion rate. However, most of the research works did not report long-term corrosion behavior; some research reported an enhanced corrosion rate after the integrity loss of the coating compared to bare material, let alone the different corrosion behavior in vivo. Figure 4.40 is a typical change of corrosion rate with applied coating on biodegradable Mg alloys (Li et al. 2010b).

In addition, the long-term corrosion mechanism after coating treatment needs further investigation. It is widely accepted that pitting corrosion is the typical corrosion form in biomedical Mg alloys, which deprives the mechanical integrity at an early stage. According to the research mentioned in this chapter, some coatings can

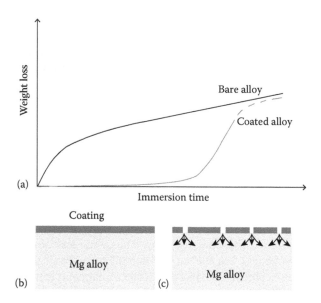

FIGURE 4.40 Degradation model for PLGA coatings reported by Li et al. (a) Schematic showing the degradation performance of the uncoated (bare) and the coated magnesium alloy. (b) Initial surface condition of the coated magnesium alloy. (c) Penetrated polymeric coating upon degradation. After the onset of coating breakdown, the reactive area exposed increases. (From Li, J. N., P. Cao, X. N. Zhang, S. X. Zhang, and Y. H. He. *Journal of Materials Science*, 45 (22), 6038–45, 2010. With permission.)

alter the corrosion mechanism at the early stage of corrosion, but the long-term corrosion behavior data is still unsatisfied.

4.5.4 BIOCOMPATIBILITY

Good biocompatibility is one of the most important properties if biodegradable Mg alloys are utilized for bone implantation or stent application. As mentioned, the surface treatment research on biodegradable Mg alloys often only focus on one or two specific properties, and cytotoxicity tests and hemocompatibility tests are usually not covered, let alone in vivo tests. Therefore, to assure good biocompatibility, systematical biocompatibility and hemocompatibility tests are needed.

Moreover, only a single surface treatment can hardly solve all the problems that exist; for example, micro-arc oxidation would produce a porous surface, Ca-P coating produced by the chemical method usually had bad adhesion strength. Therefore, multiple/combined surface treatments, including one step or multistep pretreatments and post treatments, which can provide better results, may be the future development trend in the field. In addition, because different clinical applications require different properties, specific target-defined surface treatments need to be further developed.

5 Mg with High Purity for Biomedical Applications

5.1 INTRODUCTION

Because magnesium (Mg) is an essential element in the human body and it has well-known biocompatibility (Zreiqat et al. 2002; Brar et al. 2009), Mg and its alloys have been regarded as a potential biodegradable material in biomedical applications (Yun et al. 2009a; Witte 2010; Zheng et al. 2014). Recently, a large number of researchers have been focused on the development of novel Mg alloys with high mechanical properties, excellent biocompatibility, and biodegradability that can satisfy clinical requirements (Z. Li et al. 2008; Hort et al. 2010; Zhang et al. 2010b; Li and Zheng 2013). Simultaneously, some researchers are also interested in the development of pure Mg as a biomedical material, and pure Mg is regarded as the best choice for the fundamental learning about interactions between Mg and the physiological environment (Lorenz et al. 2009; Yamamoto and Hiromoto 2009; Gu et al. 2010e; Cheng et al. 2013; Hofstetter et al. 2014).

Despite the many desirable properties of Mg for biomedical implants and devices, overquick degradation of Mg and its alloys in vitro and in vivo remains a critical challenge (Yun et al. 2009a; Chen et al. 2014a; Farraro et al. 2014; Gu et al. 2014; Sankara Narayanan et al. 2014a). Rapid degradation of Mg implants in vivo results in high local pH value, accumulated hydrogen (gas pocket), premature mechanical loss, and if not well solved, all of them may lead to implant failures (Staiger et al. 2006). Typically, two ways are adopted to improve and adjust the corrosion behavior of biomedical Mg: alloying and surface treatment (Song and Atrens 2003; Song 2005; Zeng et al. 2008b; Chen et al. 2014a). As clarified in the literature, impurities, such as Fe, Ni, Cu, and Co, in Mg have adverse effects on the corrosion of Mg, so purification is another effective way to improve the corrosion resistance, and this method has been adapted for decades in engineering Mg applications (Eliezer et al. 1998).

In this chapter, impurities and their effects on the corrosion of Mg will be introduced at first, and then previous reports on pure Mg with different degrees of purity as biomaterials will be comprehensively reviewed. At the end, new advances in property improvement on pure Mg will be proposed.

5.1.1 COMMERCIAL PURE MG (CP-MG)

Mg was first extracted by Humphrey Davey in 1808, and commercial production commenced in Germany in 1886 (Polmear 1994; Witte 2010). The purity required for commercial Mg according to ASTM B92 (2007) for 9980A grade is a minimum of 99.80 wt.% Mg with impurities such as Ca, Al, Si, and Fe below 0.05 wt.% for each (Wulandari et al. 2010). Ordinary commercial-grade pure Mg and Mg alloys have a Fe content of 0.01 to 0.03 wt.% (Avedesian and Baker 1999).

Mg is chemically active and has great power as a reducing agent. It can be used in the extraction of reactive metals, such as titanium, zirconium, and uranium (Kroll et al. 1950; Lee and Kim 2003). Mg has cathodic-free corrosion potential and consequently can be used as a sacrificial anode (Aurbach et al. 2007; Du et al. 2014). Presently, Mg alloys are very attractive in applications for aerospace, automotive, and other transport industries due to their low density, high specific strength (strength/weight ratio), and high damping capacity (Polmear 1994; Eliezer et al. 1998; Froes et al. 1998; Kainer and Mordike 2000; Anilchandra and Surappa 2012). However, the low strength of commercial pure Mg (CP-Mg) restricts its practical applications to a great extent (X. S. Hu et al. 2005). In addition, poor galvanic corrosion resistance resulting from impurities and a higher cost than commercial aluminum alloys also hinder pure Mg from being used in structural components from an engineering perspective (Song and Atrens 1999). Therefore, CP-Mg is rarely used for engineering applications without being alloyed with other metals (Polmear 1994).

5.1.2   HIGH-PURITY MG (HP-MG)

The poor corrosion resistance of Mg is a result of the high intrinsic corrosion tendency of Mg (standard electrode potential −2.37 V), and as for Mg alloys with second phases or pure Mg with impurities, microgalvanic corrosion can accelerate the corrosion. Normally, Mg alloys have corrosion rates greater than that of high-purity Mg (HP-Mg) or ultra-high-purity Mg (XHP-Mg) (Qiao et al. 2012). Chemical applications account for most of the consumption of higher purity Mg.

For pure Mg, impurities can significantly affect its corrosion rate. Noble metal impurities (Fe, Ni, Cu, and Co) are detrimental to Mg corrosion resistance because of their low solid-solubility in α-Mg and due to their tendency to form cathodic sites for microgalvanic corrosion in corrosive media (Makar and Kruger 1993; Song and Atrens 1999; Zeng et al. 2008b). It is widely accepted that there is a tolerance limit for each of the impurities (Fe, Ni, Cu, and Co) in Mg (McNulty and Hanawalt 1942; Makar and Kruger 1993). Under this tolerance limit, influences of impurities on corrosion resistance are usually insignificant. Otherwise, if the contents of impurities exceed this limit, corrosion rates will substantially increase. A schematic diagram is drawn in Figure 5.1.

The detailed mechanism that can explain the existence of a tolerance limit is not fully clear yet. But basically, the most widely accepted theory is that when the content of the specific impurity exceeds its tolerance limit, the impurity particles precipitate and act as active sites to form galvanic corrosion and, in this way, significantly accelerate the corrosion (M. Liu et al. 2009). The actual tolerance limit of each impurity (Fe, Ni, Cu, and Co) will be discussed in detail in Section 5.2.1: Impurities. ASTM standard B-93 specified an upper limit for Fe concentration of 0.004 wt.% for most common HP-Mg alloy ingots and 0.005 wt.% for die-cast parts (Tathgar et al. 2006). From the aspect of impurities, HP-Mg means that the impurity elements (Fe, Ni, Cu, and Co) are each below their (alloy-dependent) tolerance limit (Song and Atrens 2003; Atrens et al. 2011a; Z. Abidin et al. 2013). Similarly,

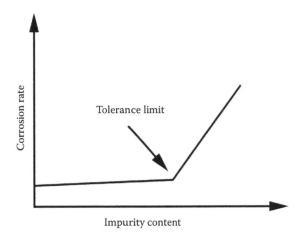

FIGURE 5.1 Generalized curve showing the influence of impurity content (Fe, Ni, Cu, Co) on the corrosion rate of Mg.

in XHP-Mg, the impurity contents should be much less. Typical impurity levels in CP-Mg, HP-Mg, and XHP-Mg are listed in Table 5.1.

5.1.3 Previous Studies on Pure Mg for Industrial/ Engineering Applications

Owing to its low density and high damping capacity, Mg has been attractive in lightweight engineering applications, for instance, the aerospace, electronic, and automobile industries (Stalmann et al. 2001). However, its high intrinsic reactivity and low strength restrict its use. Current studies on pure Mg for engineering purposes are concentrated on new application area exploitation and also on how to solve the remaining disadvantages of pure Mg, such as high corrosion rates, low mechanical properties, and insufficient deformability.

Polycrystalline pure Mg exhibits extraordinarily high damping capacity at engineering stress levels, and it could be a promising material in the control of vibration and noise (James 1969). Recently, some research work has focused on the excellent damping properties of pure Mg (X. S. Hu et al. 2005, 2012; Watanabe et al. 2015). Two damping peaks (P_1, P_2) were found at 80–100°C and 230°C, respectively, according to X. S. Hu et al. (2005, 2012). Arc-bending followed by annealing was reported to improve the damping capacity of pure Mg (99.996 wt.%) (J. Wang et al. 2014).

Mg has many advantages as a hydrogen storage material in terms of hydrogen storage capacity and cost. The remaining problem is the slow hydrogen absorption and desorption rates (Jain et al. 2010; Lang and Huot 2011; Lal and Jain 2012). Some fundamental studies are also carried out on the hydrogen absorption behavior of pure Mg (Song et al. 2014). Pure Mg-based porous foams have been fabricated, trying to improve the hydrogen storage capacity (Gil Posada and Hall 2014). In another

TABLE 5.1

Typical Compositions (Impurity Levels in ppm) of CP-Mg, HP-Mg, and XHP-Mg

Mg Quality	Material Condition	Impurity Level (in ppm)									References
		Fe	Si	Mn	Cu	Ni	Al	Zn	Pb	Sb	
CP-Mg	Not mentioned	500	/	80	40	10	200	100	100	600	Jamesh et al. 2011a
HP-Mg	As-cast, annealed	37	11	9	<1	<1	17	/	/	/	Hofstetter et al. 2014
XHP-Mg	As-cast, annealed	2.2	0.3	2.7	<1	<1	1	/	/	/	Hofstetter et al. 2014

work, hydrogen absorption was improved by high-pressure torsion (HPT), and a total hydrogen absorption of 6.9 wt.% could be achieved (Edalati et al. 2011).

Without alloying elements, the only strengthening method for pure Mg is based on the structure design, such as grain refinement, texture modification, and hybridization with other reinforcing materials. The processing of Mg is relatively difficult at ambient temperature due to its limited ductility, which is essentially attributed to limited slip systems owing to its hexagonal close-packed structure (HCP). Diverse severe plastic deformation processes (SPD) have been adapted to improve the mechanical properties and deformability of pure Mg, such as equal channel angular pressing (extrusion) (ECAP) (Gan et al. 2009; D. Song et al. 2010; Li et al. 2011b; Fan et al. 2012, 2013; Poggiali et al. 2012, 2013; Kitahara et al. 2014), HPT (Bonarski et al. 2008, 2010; Edalati et al. 2011; Qiao et al. 2014), high ratio extrusion (H.-F. Sun et al. 2012), and accumulative roll bonding (ARB) (Chang et al. 2010).

According to the results from Sun (2012), the yield strength (YS), ultimate tensile strength (UTS), and elongation of as-cast pure Mg (99.95 wt.%) were 24 MPa, 86 MPa, and 4.8% and could be improved to 124 MPa, 199 MPa, and 10.7% after high ratio extrusion, respectively. Biswas et al. (2010) reported that ultrafine grains (−250 nm) could be obtained after equal channel angular extrusion, and the hardness could be significantly improved. The elongation of ECAPed pure Mg (99.88 wt.%) was enhanced to 27% without sacrificing strength, resulting from the effective refinement of grain size and the weakening of the basal plane texture (Fan et al. 2012).

Besides the research progress on pure Mg for industrial purposes mentioned here, there are still some other studies on pure Mg, including pure Mg-based composites (Habibnejad-Korayem et al. 2010; Sun et al. 2013; Rashad et al. 2014; Sankara Narayanan et al. 2014b), various ways to improve the corrosion resistance of pure Mg (Tsubakino et al. 2003; Brunelli et al. 2005; Chino et al. 2006; Shi et al. 2006; X. M. Wang et al. 2006; Hwang et al. 2011), recycling of pure Mg (Chino et al. 2003, 2006), and fundamental studies on factors that affect Mg deformation and strengthening (Beausir et al. 2009; Beyerlein et al. 2011; Sandlöbes et al. 2011, 2012; Sankara Narayanan et al. 2011; Qiao et al. 2013b; T. Fan et al. 2014).

5.2 DEVELOPMENT OF PURE Mg AS DEGRADABLE METALLIC BIOMATERIAL

No matter what methods are adopted (alloying, surface treatment, or purification), the final goal is to improve the biocompatibility and biodegradation behaviors of biomedical Mg. According to previous studies, no element has been identified that can produce a Mg alloy with a lower corrosion rate compared to HP-Mg in a technically relevant testing solution, such as a 3.5% NaCl solution (Song and Atrens 1999, 2003, 2007; Atrens et al. 2011b). Compared with alloying and coating, purification of Mg improves its corrosion resistance without introducing another element or any other substances. So the possible adverse effects on the alloying elements, possible second phases resulting from alloying and coatings, can be avoided. Moreover, from the biological point of view, the use of pure Mg is more advisable than Mg alloys because it prevents the release of toxic ions (Pereda et al. 2011).

5.2.1 Impurities

Commercially, two principal extraction processes of Mg are electrolysis of molten Mg chloride and thermal reduction of Mg oxides (Froes et al. 1998; Kainer and Mordike 2000). Impurities are inevitably in pure Mg. Even if refinement can further reduce the impurity content, there are still remnants (Persaud and McGoran 2011). Impurities can be derived from raw materials and also can be introduced in the production process (Atchison and Beamer 1952; Witte 2010). For example, Fe impurity can be introduced during the melting process in a steel crucible or from the steel mold or die during casting.

In principle, any other element in pure Mg should be recognized as an impurity. Meanwhile, some impurities in quite limited amounts do not have any obvious influences on the properties of Mg, and that is why we don't pay much attention to these elements. It has been reported that up to 5 wt.% Pb, Sn, or Al has little or no effect on the corrosion rate of pure Mg. Na, Si, and Mn, likewise, showed no adverse effect even when present in excess of their solid solubility limit. Subsequent measurements also indicated that Th, Zr, Ce, Pr, and Nd each showed no adverse effects (Ferrando 1989). On the contrary, impurities such as Be, Fe, Cu, Ni, and Co have apparently adverse effects on the corrosion properties of pure Mg even at a low content, and on these occasions, they should be well controlled (Hanawalt et al. 1942; Inoue et al. 1998), especially in pure Mg for biomedical applications.

Different impurities have different influences or different severities on the corrosion resistance of Mg, and some elements (Be, Fe, Ni, Cu, and Co) have been found to be extremely detrimental to the corrosion performance of Mg (Song and Atrens 2003). Fe can be picked up from the steel melting pot or the steel casting mold, and those tools are commonly used in the extraction of Mg. Therefore, among these impurities, Fe is the most common impurity, and its effect on corrosion should be carefully concerned (Yang et al. 2015).

Here, common impurities in Mg and their effects on corrosion will be discussed separately, and finally, the tolerance limits of those impurities will be concluded.

5.2.1.1 Fe

The solubility of Fe in Mg is 0.018 wt.% at 650°C, and this value increases to approximate 0.04 wt.% at 750°C according to the ASM phase diagram (Nayeb-Hashemi 1988). Recently, the tolerance of Fe in Mg was explained by M. Liu et al. (2008, 2009) using the calculated phase diagram. It was proposed that Fe particle formation could be inhibited when Fe content is <180 ppm due to the extremely narrow solidification interval, and Fe particles could formed only when the Fe content was >180 ppm during nonequilibrium solidification. However, under quasiequilibrium solidification conditions or heat-treated conditions, the tolerance limit could be significantly reduced to 5–10 ppm. Above this limit, Fe particles could form, and accelerated corrosion could happen. The calculated Fe tolerance limit of 180 ppm is close to the reported limit in as-cast pure Mg, 170 ppm (Song and Atrens 1999; Makar and Kruger 1993) and 150 ppm (Froats et al. 1987).

Although the tolerance limit of Fe is acknowledged widely as 170 ppm (Makar and Kruger 1993; M. Liu et al. 2009), and it was cited in the ASM handbook

(Cramer and Covino 2003), there are still observations showing high corrosion rates of pure Mg even with Fe content <50 ppm (Pardo et al. 2008; Qiao et al. 2012; Birbilis et al. 2013; King et al. 2014). Qiao et al. (2012) found a high corrosion rate of as-cast HP-Mg with 26–48 ppm Fe, and they predicted that Fe particles had already formed in the HP-Mg. Birbilis et al. (2013) attributed the high corrosion rate of pure Mg (99.97 wt.%) with 40 ppm Fe to the formation of Fe-rich particles even if the material condition was not mentioned. Wrought pure Mg (99 wt.%) with 40 ppm Fe also resulted in a high corrosion rate in a 3.5 wt.% NaCl solution according to Pardo's results (2008). Yang et al. (2015) found that Fe-rich particles were formed in as-cast CP-Mg, which contained less than 25 ppm Fe (far below 170 ppm). They also concluded that trace silicon (Si at low ppm level), which promoted the Fe-rich particle formation, was one of the main reasons for the high corrosion rate of pure Mg. In the work of Matsubara et al. (2013), they found that corrosion rates of AM series alloys depended on the impurity concentration and increased with the Fe/Mn ratio.

Based on this analysis, we might conclude that the tolerance limit of Fe in pure Mg is closely related to the material conditions (processing details: heat treatments, rolling, forging, etc.) and interference of other impurities, Si for instance. As-cast pure Mg has a relatively high tolerance limit, but heat-treated and deformation-processed Mg exhibits a low tolerance limit. Tolerance limits in pure Mg and in Mg alloys are totally different. Basically, the actual tolerance limit of Fe in different pure Mg should be judged by whether there are any Fe (Fe-rich) particles formed.

5.2.1.2 Ni

Nickel (Ni) is more harmful than Fe, both in pure Mg and Mg alloys due to a lower tolerance limit (Hawke 1975). At 650°C, the liquid Mg dissolves approximately 32 wt.% Ni (Nayeb-Hashemi 1988). Alloying elements, such as Al, reduce the solubility of Ni considerably (Tathgar 2001). Ni may be reduced from about 0.2% to as low as 0.001% by adding zirconium and aluminum to the Mg melt (Tathgar et al. 2006). To ensure acceptable corrosion resistance, ASTM B-93 standards specified a Ni concentration below 0.001 wt.% for Mg alloy ingots, and for some reasons, the upper limit in die-cast parts is 0.002 wt.% (Tathgar et al. 2006). The tolerance limit of Ni is also dependent on the casting form. Sand and permanent mold casting have a significant lower limit of 10 ppm Ni compared to 50 ppm for high-pressure die castings (Song and Atrens 1999).

5.2.1.3 Cu

The solubility of copper (Cu) in molten Mg is about 70 wt.% at 650°C (Nayeb-Hashemi 1988). According to ASTM B-93 standards, an upper limit for the Cu concentration is 0.008 wt.% for high-purity AM alloys and 0.025 wt.% for die-cast parts. The adverse effects of Cu on the AZ and AM alloys are considerably less than Ni and Fe (Tathgar et al. 2006). A small amount of Cu has a beneficial effect on the creep strength of Mg die castings but strongly accelerates the saltwater corrosion (Hawke 1975). A limit of 300 ppm has been set as the tolerance of Cu (Song and Atrens 1999).

TABLE 5.2
Recommended Tolerance Limits (in ppm) of Common Impurities in Magnesium

Impurity	Be	Fe	Ni	Cu
Tolerance limit	2–4	30–50	20–50	100–300
Appendix	Widely accepted tolerance limit of as-cast pure Mg is 170 ppm, and this value can be decreased to 5–10 in heat-treated conditions (M. Liu et al. 2008, 2009). The exact tolerance limit of impurities differs with different pure Mg or Mg alloys, and also influenced by their thermal histories.			

Source: Chen, Y., Z. Xu, C. Smith, and J. Sankar, *Acta Biomaterialia*, 10, 11, 4561–73, 2014; Song, G. L., and A. Atrens, *Advanced Engineering Materials*, 1, 1, 11–33, 1999; Witte, F., N. Hort, C. Vogt, S. Cohen, K. U. Kainer, R. Willumeit, and F. Feyerabend, *Current Opinion in Solid State and Materials Science*, 12, 5–6, 63–72, 2008; Poinern, G. E. J., S. Brundavanam, and D. Fawcett, *American Journal of Biomedical Engineering*, 2, 6, 218–40, 2012.

5.2.1.4 Co

Cobalt (Co) has strongly adverse effects on the corrosion of Mg alloys, but it is not a common impurity in Mg, and its tolerance limit has not been well clarified or documented (Hillis 1983). Co content in samples from different producers is reported to be less than 10 ppm. Tathgar et al. (2006) recommended Co should be added into the list of elements that is specified in international standards, and tentatively, maximum allowed Co content should be the same as Ni.

At the same concentration, the detrimental effect of these impurity elements decreases as follows: Ni > Fe > Cu (Song and Atrens 1999, 2003). Recommended tolerance limits of common impurities are listed in Table 5.2 based on previous reports. Synergisms and antagonisms between different impurities are still under investigation and should be further understood.

5.2.2 FABRICATION, HOT AND COLD WORKING, AND HEAT TREATMENT

As mentioned, two dominating methods for producing raw pure Mg are electrolysis of molten Mg chloride and thermal reduction of Mg oxides. Among the two traditional methods, the Pidgeon process, the silico-thermic reduction of Mg using ferrosilicon (FeSi) as a reduction agent, is still a popular route to produce Mg (see Figure 5.2). In China, it is also the only Mg production process at present (Ehrenberger et al. 2008). The Mg vapor condenses in a water-cooled condenser, and high-purity Mg can be obtained because the vapor pressure of potential impurities (Ca, Fe, Cu, etc.) is low under these conditions (Wulandari et al. 2010). Because the Pidgeon process suffers from low productivity, a high labor requirement, and high energy consumption (not a green route), significant work has been done on trying to develop more sustainable routes for raw Mg production around the world (Wulandari et al. 2010).

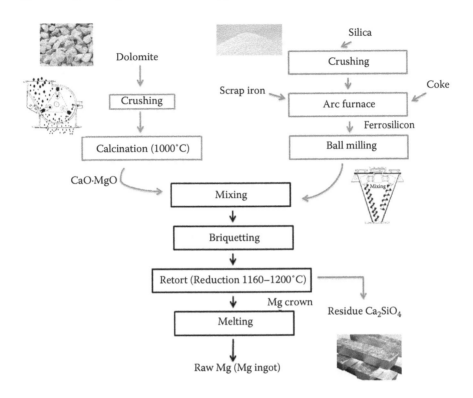

FIGURE 5.2 Schematic flow sheet of the Pidgeon process.

The above method is applied in the industrial field, and the purity of Mg is restricted. On account of HP-Mg or even XHP-Mg considered for biomedical applications, advanced methods have been adapted to further purify the raw pure Mg. The first important application of zone melting was to purify germanium, and this method is also applicable to Mg purification (Pfann 1957). The zone melting method can satisfactorily remove a number of impurities, including Fe, Cu, and Zn, but is not so effective in Mn elimination, but this is not a problem when the raw pure Mg is electrolyzed because electrolytically produced raw Mg is almost Mn free. Vacuum distillation is another effective way to remove the preponderant impurities. It eliminates most of the impurities still remaining in industrial Mg. Revel et al. (1978) reported, among about 40 elements that were determined by neutron activation analysis, only zinc content was found to be more than 1 ppm in the usable fraction of the distillate.

Casting is still the basic and predominant process to manufacture pure Mg parts even for biomedical implants. Inhomogeneous grain size and grain size distribution derived from casting can be solved by hot or cold working and various heat treatments. A static magnetic field and an alternating electric field were simultaneously imposed during solidification of pure Mg (99.9 wt.%) in Mizutani's et al. work (2005). The solidified crystal grains of pure Mg were refined from 1 mm to 200 μm by the generated electromagnetic vibrations. Similarly, ultrasonic irradiation during

Mg solidification has emerged as an effective method for microstructure refinement (Qian et al. 2009).

By studying the effects of rolling temperature on microstructures, mechanical properties, texture, and dynamic recrystallization in a HP-Mg (99.99 wt.%), Qiao et al. (2013a) found that the suitable temperature of pure Mg rolling with a rolling reduction of 40% is 230°C. Lu et al. (2012) found that a series of twins turned up during annealing. Furthermore, the micro-hardness of the annealed Mg (99.8 wt.%) was reduced, which could be explained by the high annealing temperature promoting recrystallization and grain growth while shortening the recovery and recrystallization duration.

5.2.3 MICROSTRUCTURE AND MECHANICAL PROPERTIES

The typical microstructures of pure Mg in different conditions are illustrated in Figure 5.3. The phase composition of pure Mg is a single-phase α (Mg). Grain size of the as-cast pure Mg ranges from several microns to several millimeters. The as-cast

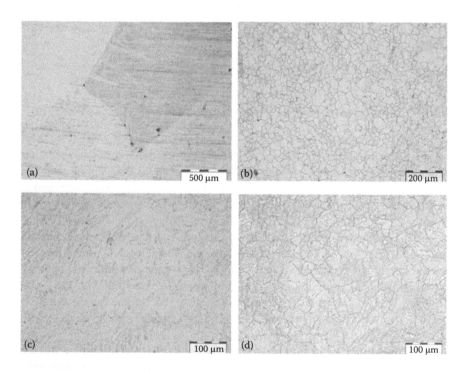

FIGURE 5.3 Typical optical microscope images of Mg (99.8 wt.%) in (a) the as-cast condition, (b) the extruded condition, (c) the as-cast Mg processed by one half turn of HPT, and (d) the extruded Mg processed by one turn of HPT. (From Qiao, X. G., Y. W. Zhao, W. M. Gan, Y. Chen, M. Y. Zheng, K. Wu, N. Gao, and M. J. Starink. *Materials Science and Engineering: A*, 619, 95–106, 2014. With permission.)

pure Mg has coarse equiaxed grains, and finer grains can be achieved by deformation processes, such as extrusion and HPT.

Pure Mg typically has a yield strength of 55 MPa, lower than natural bone (e.g., femur ≈ 110 MPa) (Pereda et al. 2011). Mechanical properties of as-cast pure Mg are quite low, and the ultimate tensile strength is around 100 MPa or even lower. Table 5.3 lists some typical mechanical property data of pure Mg with different purities and under different treatment conditions.

Because only limited impurities exist in pure Mg, there is no solid solution strengthening and precipitation strengthening in pure Mg. Then, grain refinement becomes the main strengthening method in pure Mg. Mechanical properties of pure Mg are closely related to its microstructure, especially in grain size, and various methods have been adapted to obtain fine and homogeneous grains, including extrusion, SPD methods, and powder metallurgy methods. The tensile strength of extruded CP-Mg can reach 200 MPa after appropriate extrusion and rolling with grain size reducing to 2 μm (see Table 5.3). According to the results of Pereda et al. (2010, 2011), pure Mg (99.8 wt.%) fabricated from a powder metallurgy method has high TYS and UTS with the values of 280 MPa and 320 MPa, respectively.

5.2.4 Degradation in Simulated Body Fluids

Pure Mg has the potential to be used as a biodegradable metallic material, and to be a biodegradable implant material, the degradation rate should be consistent with the healing rate of affected tissues. Some previous studies have reported that conventional pure Mg degraded rapidly, and this is undesirable (Inoue et al. 2002; Song 2007; Xu et al. 2008; Gu et al. 2009b; Mueller et al. 2009; Wang and Shi 2011). Recently, with the development of refinement technology, some Mg with high purity can be acquired, and there have been some studies focusing on the in vitro and in vivo degradation of pure Mg with high purity. Effects of purity levels on the degradation of pure Mg have attracted extensive attention.

Table 5.4 gives some typical data on the pure Mg impurity levels, treatment conditions, and their corresponding electrochemical corrosion parameters in different simulated body fluids.

Quite different corrosion current densities (corrosion rates) were found for different pure Mg with different impurity levels. Even in the same simulated body fluids and at the same purity levels, corrosion data of pure Mg still differ a lot in different treatment conditions or in different authors' work.

There are too many factors that can affect the corrosion of pure Mg. From the corrosion environment aspect, the presence or absence of physiological salt ions in the immersion solution influences the degradation rate and mode. In the immersion test of Johnson and Liu (2013), the as-rolled commercial pure Mg foil (99.9 wt.%) with or without a thermal oxide layer degraded faster in phosphate-buffered saline (PBS) than in distilled water (DI), and no big differences in the pH change and mass loss were found between the CP-Mg with or without a thermal oxide layer. Xin et al. (2011c) found that corrosion behavior of a CP-Mg in SBF was significantly influenced by the HCO_3^- concentration. EIS results confirmed that a higher concentration of HCO_3^- induced a more effective protection layer. In another work, an overdose of

TABLE 5.3
Typical Mechanical Properties of Pure Mg with Different Purities in Different Treatment Conditions

Purity of Mg (wt.%)	Condition	Grain Size (µm)	TYS (MPa)	UTS (MPa)	Elongation (%)	References
99.8 wt.%	325 mesh Mg powder, cold-pressed + extruded at 420°C (powder metallurgy)	/	280	320	2	Pereda et al. 2010, 2011
99.9 wt.% (CP-Mg)	Extruded	43	55	167	18	Diez et al. 2014
99.9 wt.% (CP-Mg)	Extruded + screw rolling	2–15	109–146	177–205	7–15	Diez et al. 2014
99.9 wt.% (CP-Mg)	Extruded + screw rolling (11 passes)	2–3	146	205	7	Diez et al. 2014
>99.9 wt.%	Mg turnings, disintegrated melt deposition (DMD) + extruded at 350°C	Several microns to Several millimeters	125 ± 9	169 ± 11	6.2 ± 0.7	Sankara Narayanan et al. 2011
99.95 wt.% (CP-Mg)	As-cast	/	24	86	4.8	H.-F. Sun et al. 2012
99.95 wt.% (CP-Mg)	High ratio extrusion with different passes + (annealing)	9–35	84–124	162–199	7–12	H.-F. Sun et al. 2012
99.99 wt.% (CP-Mg)	As-cast	>100	/	58 ± 1.8	7.2 ± 0.8	Pan et al. 2015

Note: TYS, tensile yield strength; UTS, ultimate tensile strength.

TABLE 5.4

Collection of Corrosion Data from Pure Mg with Different Purity Levels in Different Simulated Body Fluids

Purity of Mg	Description of Impurities	Condition	Electrolyte	E_{corr}	i_{corr} ($\times$ μA/cm^2)	Corrosion Rate	References
99.8 wt.% (CP-Mg)	/	As-cast	0.9 wt.% NaCl	-1.43 ± 0.042	36.1 ± 11	/	Zakiyuddin et al. 2014
99.8 wt.% (CP-Mg)	/	Hot-rolled		-1.62 ± 0.041	20.4 ± 11	/	Zakiyuddin et al. 2014
99.92 wt.%	/	As-cast		-1.638	145	/	Zhang et al. 2005
99.9 wt.%	70 ppm Al, 20 ppm Cu, 280 ppm Fe, 170 ppm Mn, <10 ppm Ni, 50 ppm Si, 20 ppm Zn	As-drawn	0.11 mol l^{-1} NaCl	/	/	1.1 ± 0.5 mg/ (d·cm^2) (calculated from mass loss)	Doepke et al. 2013
99.9 wt.% (CP-Mg)	/	Ribbons	SBF	-1.805	101.7	/	Zhang et al. 2013f
99.9 wt.%	/	Not mentioned		-1.8 ± 0.02	23.5 ± 3.6	/	Alabbasi et al. 2014a
99.95 wt.% (CP-Mg)	30 ppm Fe, 100 ppm Si, 10 ppm Ni, 20 ppm Cu, 100 ppm Al, 100 ppm Mn	As-cast		1.85	13.5	/	Zhao et al. 2010
99.95 wt.%	/	As-cast		-1.976	359	/	R. Xu et al. 2012a
99.95 wt.%	/	As-cast plates		-1.998	173.8	/	G. Wu et al. 2011
99.99 wt.%	/	Not mentioned		-1.98	6831	/	S.-M. Kim et al. 2014
99.99 wt.%	/	As-cast		-1.941	1082	/	Pan et al. 2015

(Continued)

TABLE 5.4 (CONTINUED)
Collection of Corrosion Data from Pure Mg with Different Purity Levels in Different Simulated Body Fluids

Purity of Mg	Description of Impurities	Condition	Electrolyte	E_{corr}	i_{corr} ($\times \mu A/cm^2$)	Corrosion Rate	References
99.99 wt.% (HP)	37 ppm Fe, 11 ppm Si, 9 ppm Mn, <1 ppm Cu, 1 ppm Ni, 17 ppm Al	As-cast		/	/	28 ± 2 μm/y (calculated from the hydrogen evolution)	Hofstetter et al. 2014
99.99 wt.% (HP)	37 ppm Fe, 11 ppm Si, 9 ppm Mn, <1 ppm Cu, 1 ppm Ni, 17 ppm Al	As-annealed		/	/	39 ± 3 μm/y (calculated from the hydrogen evolution)	Hofstetter et al. 2014
99.999 wt.% (XHP)	0.2–2.2 ppm Fe, 0.1–0.9 ppm Si, 1.2–2.7 ppm Mn, <1 ppm Cu, <1 ppm Ni, <1 ppm Al	As-cast/ as-extruded/ as-annealed		/	/	Average corrosion rate 10 ± 3 μm/y (calculated from the hydrogen evolution)	Hofstetter et al. 2014
99.8 wt.% (CP-Mg)	/	As-cast	Tas-SBF	−1.58 ± 0.039	5.8 ± 0.13	/	Zakiyuddin et al. 2014
99.8 wt.% (CP-Mg)	/	Hot-rolled		−1.66 ± 0.039	13.6 ± 4.2	/	Zakiyuddin et al. 2014
99.9 wt.% (CP-Mg)	/	Not mentioned	Hank's	−1.886	86.06	1.94 mm/y	Cheng et al. 2013
99.96 wt.% (CP-Mg)	<30 ppm Fe, <10 ppm Cu, <2 ppm Ni	Not mentioned		−1.85	430	/	Ng et al. 2010a

(Continued)

TABLE 5.4 (CONTINUED)

Collection of Corrosion Data from Pure Mg with Different Purity Levels in Different Simulated Body Fluids

Purity of Mg	Description of Impurities	Condition	Electrolyte	E_{corr}	i_{corr} ($\times$ μA/cm^2)	Corrosion Rate	References
99.9 wt.% (CP-Mg)	<280 ppm Fe, <170 ppm Mn, <70 ppm Al, <50 ppm Si, <20 ppm Zn, <10 ppm Ni	Not mentioned		-1.7--1.8	13-33	100 mg/(cm^2·d) in the first 19-24 h (calculated from weight loss)	Cabrini et al. 2014
99.99 wt.%	14 ppm Fe, 1 ppm Ni, 2 ppm Cu, 16 ppm Si, 4 ppm Al, 4 ppm Mn	As-cast plate		-1.70	23.4	/	Lu et al. 2014a
>99.99 wt.% (HP-Mg)	<10 ppm Fe, <10 ppm Ni, <20 ppm Cu, <50 ppm Al, <80 ppm Zn, <50 ppm Mn, <20 ppm Sn, <20 ppm Pb	Not mentioned		-1.54	2.2	0.25 mm/y (static)/ 0.8 mm/y (dynamic)	Wang and Shi 2011
99.9 wt.% (CP-Mg)	/	As-rolled	PBS	/	/	2.08 ± 0.39 mg/ (d·cm^2) (calculated from mass loss)	Johnson and Liu 2013
99.99 wt.%	14 ppm Fe, 10 ppm Ni, 20 ppm Cu, 16 ppm Si, 4 ppm Mn	As-cast		-1.795	136	/	Lu et al. 2014b
99.8 wt.% (CP-Mg)	/	As-cast	Ringer's	-1.5 ± 0.039	96 ± 8.1	/	Zakiyuddin et al. 2014
99.8 wt.% (CP-Mg)	/	Hot-rolled		-1.65 ± 0.049	1.21 ± 0.8	/	Zakiyuddin et al. 2014

bicarbonate addition was found to activate the Mg surface in the SBF (Z. Li et al. 2014). Yamamoto and Hiromoto (2009) also found that pure Mg (3N) degraded fastest in NaCl, followed by E-MEM, Earle's solution, and E-MEM + FBS. Their results showed that protein adsorption and insoluble salt formation retarded Mg degradation whereas organic compounds, such as amino acids, encouraged the dissolution of Mg. Buffering the solution also influenced the degradation rate. They also recommended that appropriate solution, such as E-MEM + FBS, is important for in vitro evaluation. Effects of Tris in SBF and albumin on the corrosion of pure Mg have also been clarified by other researchers (Xin and Chu 2010; C.-L. Liu et al. 2014). Studies on these factors not only improve our understanding of Mg degradation in the physiological environment, but also present the key factors that we should be concerned with when developing biodegradable Mg implants.

From the aspect of the material itself, impurity contents, microstructures (can be modified by deformation or thermal treatments), and surface conditions all affect the corrosion of pure Mg. As mentioned, corrosion of pure Mg is sensitive to impurities, especially for Ni, Fe, and Cu. Zakiyuddin et al. (2014) observed the corrosion behavior of CP-Mg (99.8 wt.%) under different treatment conditions in different corrosion media. The as-cast specimens in 0.9% NaCl and Ringer's solutions had lower corrosion potentials and corrosion rates than the hot-rolled specimens. However, the as-cast specimen in the Tas-SBF solution had a higher corrosion potential and a lower corrosion rate than the hot-rolled specimen. Only the specimens in the Tas-SBF solution showed passivation behavior due to the presence of bicarbonate ions. The hot-rolled specimen could form apatite more easily on its surface than the as-cast specimen in Tas-SBF.

Wang and Shi (2011) found HP-Mg (>99.99 wt.%; Ni, Fe, Cu < tolerance limit) had degradation rates of 0.25 mm/year and 0.8 mm/year in static and in dynamic Hank's solution, respectively. Their in vitro results showed that the corrosion products accumulated on the material surface as a protective layer in a static situation, which resulted in a lower degradation rate than the dynamic condition. Moreover, they concluded purification is an effective way to improve Mg corrosion resistance. Z. Abidin et al. (2011a,b) studied the corrosion behavior of HP-Mg (Be, Cu, Ni < tolerance limit, Fe: 70 ppm) in Hank's solution compared with some common Mg alloys at room temperature and at 37°C. Corrosion rates at steady state gave the following ranking: HP-Mg < Mg2Zn0.2Mn < ZE41 < AZ91. HP-Mg showed better corrosion resistance than the common Mg alloys.

XHP-Mg was fabricated via a vacuum distillation apparatus, and impurities were well controlled in Löffler's group (0.2–2.2 ppm Fe, <1 ppm Cu, <1 ppm Ni) (Hofstetter et al. 2014). The in vitro degradation rate, which was evaluated in a self-made testing setup, was about 10 ± 3 μm/year. The results of XHP-Mg were also compared with HP-Mg (37 ppm Fe, <1 ppm Cu, <1 ppm Ni) in the as-cast and as-annealed states. The less pure specimens exhibited significantly higher degradation rates due to the formation of Fe-containing precipitates during casting and annealing (see Figure 5.4).

Generally, there is a trend that the corrosion rates of CP-Mg > HP-Mg > XHP-Mg. The actual tolerance limits of impurities in different pure Mg are critical to pure Mg corrosion. For each impurity below this limit, pure Mg can have high corrosion

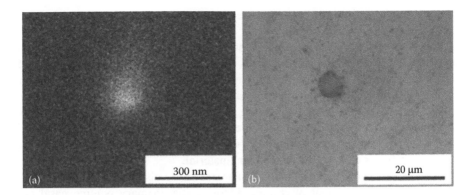

FIGURE 5.4 EDX Fe Kα mapping of (a) annealed HP-Mg, a Fe–Mn–Si particle with a size of ≈150 nm could be detected. (b) Micropore derived from distillation (not Fe precipitant) in XHP-Mg. (From Hofstetter, J., E. Martinelli, A. M. Weinberg, M. Becker, B. Mingler, P. J. Uggowitzer, and J. F. Löffler. *Corrosion Science*, 91, 29–36, 2014. With permission.)

resistance. Because the tolerance limit of a specific impurity can be largely influenced by the thermal history or hot and cold working, the impurity content control should be based on the processing history in the biodegradable Mg fabrication.

5.2.5 CYTOTOXICITY

Mg corrosion takes place under H_2 gas formation and leads to alkalization of the surroundings. On the one hand, both factors could strongly affect the biocompatibility of Mg. On the other hand, they all can be well solved if the corrosion of Mg is precisely controlled as expected.

Even if the Mg element is biocompatible, fast corrosion may also impair the cell functionality. According to our previous work (Gu et al. 2009b), CP-Mg extracts (pH 9.18 ± 0.14) significantly reduced the cell viabilities of five cell lines: murine fibroblast cells (L-929 and NIH3T3), murine calvarial preosteoblasts (MC3T3-E1), human umbilical vein endothelial cells (ECV304), and rodent vascular smooth muscle cells (VSMC). Lorenz et al. (2009) found that CP-Mg (99.9 wt.%) showed high reactivity in the cell culture medium, leading to a pH shift in the alkaline direction, and CP-Mg did not enable adhesion and survival of HeLa cells on the surface. HP-Mg (99.99 wt.%) was adapted to assess the cell response of MC3T3-E1 cells by S.-M. Kim et al. (2014). Only a few cells with a spherical shape were attached to the surface of the HP-Mg specimen after 1 day of culture. Cell proliferation evaluated by NDA quantification after 5 days of culturing found the DNA content of the specimen was close to zero.

In the work of R. W. Li et al. (2014), the HP-Mg (99.99 wt.%) had a superior effect on the proliferation and differentiation of human mesenchymal stem cells (hMSCs), and this was also observed from gene expression data. Zhai et al. (2014) demonstrated that pure Mg leach liquor suppressed osteoclast formation, polarization, and osteoclast bone resorption in vitro. The author considered Mg had significant

potential for the treatment of osteolysis-related diseases caused by excessive osteoclast formation and function.

Yun et al. (2009b) have cocultured the osteoblast cells U2OS with a pure Mg
specimen (99.95 wt.%) for up to a week. The proliferation and viability of U2OS
cells were proven to be not significantly affected via MTT assay and visual observation; meanwhile, the bone tissue formation study using von Kossa and alkaline
phosphatase (ALP) staining showed that expression and enzymatic activity of the
phosphatase and mineralization in the osteoblast cells were not significantly altered.
ALP staining results are presented in Figure 5.5. All those results indicate that Mg
may be suitable as a biodegradable implant material.

Besides orthopedic implants and vascular stents, Mg also offers a new therapeutic option for the treatment of chronic rhinosinusitis. Primary porcine nasal epithelial cells were used to test the biocompatibility of degrading pure Mg (extruded
99.92 wt.%) and investigate whether the degradation products may also affect cellular metabolism (Schumacher et al. 2014). The results showed pure Mg did not
induce apoptosis, and no major influence on enzyme activities or protein synthesis

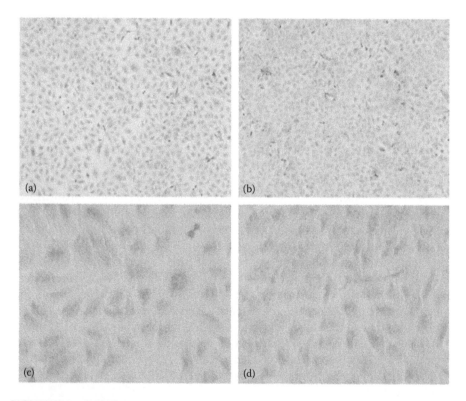

FIGURE 5.5 ALP staining of an adherent U2OS cell after 5 days incubation. (a and c) cultured as controls and (b), (d) cocultured with Mg. Magnification (a, b): 10× and (c, d): 40×.
(From Yun, Y., Z. Dong, D. Yang, M. J. Schulz, V. N. Shanov, S. Yarmolenko, Z. Xu, P.
Kumta, and C. Sfeir. *Materials Science and Engineering: C*, 29 (6), 1814–21, 2009. With
permission.)

was found, but cell viability was reduced, and elevated interleukin-8 secretion indicated proinflammatory reactions. Whether this may pose a problem for the in vivo application of Mg-based biomaterials needs further research. Y. Zhang et al. (2012b) found a great increase in pH value, caused by the degradation of pure Mg (99.9 wt.%) in the culture medium, which showed strong cytotoxic effects on U2OS cells. They supposed this might provide an alternative way to cure bone cancers through creating a high alkalinity surrounding the cancer cells.

5.2.6 HEMOCOMPATIBILITY

Our previous studies found the hemolysis percentage of pure Mg (99.9 wt.%) reached 37% (Cheng et al. 2013), which exceeded the recommended value of 5% considerably and meant that Mg would lead to severe hemolysis according to ISO10993-4: 2002. Even after extrusion, pure Mg still had a high hemolysis ratio, exceeding 20% (Gu et al. 2009b). This might be attributed to the high corrosion rate of pure Mg and large pH variation after 1 h incubation in saline solution. The morphologies of adhered human platelets were characterized, and the results showed that platelets present nearly round with only one or two short pseudopodia spreading on the surface of as-cast and as-rolled pure Mg (see Figure 5.6).

Y. Chen et al. (2011) investigated the hemolytic ratios of pure Mg (99.99%) and the normal saline extracts of pure Mg with various pH. It was found that pure Mg and the extract with a pH of 12.01 or 2.48 possessed hemolysis ratios much higher than 5%, but the extract with a pH of 7.35 or 4.93 did not induce any hemolysis, which implied that the high hemolysis ratio of Mg was probably due to the high pH instead of released Mg^{2+}.

5.2.7 ANTIBACTERIAL PROPERTIES

Robinson et al. (2010) assessed the in vitro antibacterial properties of Mg metal against *Escherichia coli*, *Pseudomonas aeruginosa*, and *Staphylococcus aureus*. The experimental results demonstrated that the addition of Mg and increased pH resulted in an antibacterial effect on three common aerobic bacterial organisms;

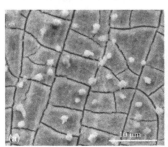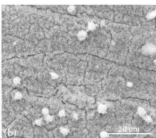

FIGURE 5.6 SEM images of platelets adhering to (a) as-cast and (b) as-rolled pure Mg (99.95 wt.%) samples. (From Gu, X., Y. Zheng, Y. Cheng, S. Zhong, and T. Xi. *Biomaterials*, 30 (4), 484–98, 2009. With permission.)

however, a simple increase in Mg^{2+} concentration alone had no effect. Thus, it can be concluded that the mechanism of the antibacterial effect of Mg appears to be an alkaline pH.

5.2.8 IN VIVO PERFORMANCE IN ANIMAL MODELS

An in vitro test is a powerful way to predict the degradation behavior in vivo in consideration of time and cost. It is vital to use appropriate in vitro tests to prescreen Mg to determine its suitability for subsequent in vivo studies, and eventually the development of a clinically relevant biomedical Mg implant would require thorough in vivo testing, initially using animal models and eventually humans. Using in vitro tests to predict in vivo behavior is based on a correlation between them (Kirkland et al. 2012).

Xue et al. (2012b) investigated the in vivo degradation behavior of pure Mg (99.9 wt.%) by implanting Mg samples at the subcutis of mice. X-ray imaging showed that the implanted Mg samples did not exhibit severe corrosion or dissolution during 2 months implantation, and no abnormality or toxicity to the tissues was observed from the implantations of pure Mg in the liver, kidney, heart, skin, and lung.

Bowen et al. (2013a) studied the physiological corrosion of pure Mg wires (>99.9%) from the mechanical behavior, and a new in vitro–in vivo correlation was proposed. After being corroded both in the murine artery (in vivo) and in static cell culture media (in vitro), the Mg wires were subjected to mechanical analysis by tensile testing. The experimental results were presented by four selected metrics: the effective tensile strength with a multiplier of 2.2; strain to failure (3.1); the time to functional degradation (2.3); and a combined metric (3.1), which indicated that in vitro corrosion in static media was more rapid than in vivo corrosion.

W. Yang et al. (2011) have assessed the effect of pure Mg wire (99.9%, 0.4 mm in diameter) on bone growth after 6 weeks implantation into the femur of streptozotocin (STZ)-induced diabetic rats. The results showed that, along with the degradation of Mg implants in the STZ-induced diabetic rats, the serum Mg level and bone mineral contents of Ca, P, Mg, K, Sr, Zn, and S were increased significantly. Therefore, the femoral bone mineral density (BMD) in the diabetic group was restored by Mg implantation. In addition, blood biochemical analysis indicated that Mg implants had no toxicity in STZ-induced diabetic rats. So it was concluded that the implantation of Mg could stimulate new bone growth and may act as a potential agent to treat osteoporosis.

In order to create a quantitative in vitro–in vivo correlation based on an accepted measure of corrosion, Bowen et al. (2015) implanted CP-Mg wires (99.9%, 0.25 mm in diameter) into the abdominal aortas of rats for 5–32 days; simultaneously, an in vitro test using Dulbecco's modified Eagle medium (DMEM) was carried out for 14 days. The in vitro penetration rates were consistently higher than comparable in vivo rates by a factor of 1.2–1.9 × (±0.02×), analyzed from empirical modeling. For a sample <20% corroded, an approximate in vitro–in vivo multiplier of 1.3 ± 0.2× was applied whereas a multiplier of 1.8 ± 0.2× became appropriate when the Mg specimen was 25–35% degraded. Extruded XHP-Mg with Fe impurities of 0.2–2.2 ppm was employed to observe its in vitro and in vivo degradation behaviors and their correlation (Hofstetter et al. 2014). In both cases, very low and homogeneous degradation rates were found, 10 ± 3 µm/y in vitro and 13 ± 3 µm/y from the 3-month in vivo test, which showed very good agreement. This very slow and homogeneous biodegradation could be explained by very low amounts of impurities, such as Fe, Si, Mn, Cu, Ni, and Al. The less pure specimens exhibited significantly higher degradation rates due to the formation of Fe-containing precipitates during casting and annealing.

Many publications are available on the in vivo physiological degradation of Mg and its alloys for biodegradable implant applications. However, few focused on the characterization of explanted materials. Bowen et al. (2014) implanted CP-Mg wire (99.9 wt.%, 300 ppm Fe, 250 ± 25 µm in diameter) in the artery walls of rats for up to 1 month and then removed and characterized both the bulk and surface products. Surface characterization using infrared spectroscopy revealed a duplex structure comprising heavily Mg-substituted hydroxyapatite that later transformed into an A-type (carbonate-substituted) hydroxyapatite. Elemental mapping of the bulk products of biocorrosion revealed the elemental distribution of Ca, P, Mg, and O in the outer and Mg, O, and P in the inner layers.

In addition to the CP-Mg, HP-Mg, or XHP-Mg bulk materials, Salunke et al. (2011) made efforts on the development of HP-Mg coatings and their biological assessments. HP-Mg coating was acquired through a physical vapor deposition process (PVD). The coating consisted of Mg grains with different sizes, which could be controlled by the total reactor pressure. The foreign body response to Mg coating grains deposited on the silicon substrate was studied by implantation subcutaneously in mice. The results confirmed that the Mg crystallites were stable in vivo for the duration of 48 h implantation. The foreign body reaction showed only a thin fibrous capsule comparable to the fibrous capsule formed around the titanium control samples. The obtained results demonstrated promising properties of the developed Mg coating that could be applied on different metals and used for designing and manufacturing biodegradable implants.

5.3 NEW ADVANCES IN PROPERTY IMPROVEMENTS ON PURE Mg FOR MEDICAL PURPOSES

As-cast pure Mg has poor mechanical properties, and to be orthopedic implants, the strength of pure Mg needs to be improved. Some newly developed processing methods are introduced in this section. In addition, bioactive coatings on pure Mg to retard corrosion and to enhance biocompatibility will be discussed.

5.3.1 New Techniques for Enhancing Mechanical Properties

One of the limitations of pure Mg with regard to its use in medical implants is its low mechanical strength and poor ductility (Staiger et al. 2006). If pure Mg is chosen to be that material, the only path to strengthening it is through grain refinement and texture control (Diez et al. 2014). Severe plastic deformation (SPD) techniques have been widely used to refine the grain structure and improve the mechanical properties of metallic materials. Here, two SPD methods are shown to introduce their effects on grain refinement and mechanical property improvement. Schematic illustrations of equal channel angular pressing (ECAP) and three-roll planetary milling (screw rolling) are presented in Figure 5.7.

According to the results of Fan et al. (2012), as-cast CP-Mg (99.88 wt.%) was hot quadratic extruded into a rectangular bar at a reduction ratio of 12:1, then followed by ECAP. Mechanical test results showed that the elongation to failure of the ECAPed pure Mg was enhanced to 27% without sacrificing the strength after extrusion. In addition, fine grains smaller than 5 µm were obtained. Yamashita et al. (2001) examined the microstructure and mechanical property of pure Mg (99.9%) subjected to ECAP processing at 400°C. Reduced grain sizes of Mg were observed

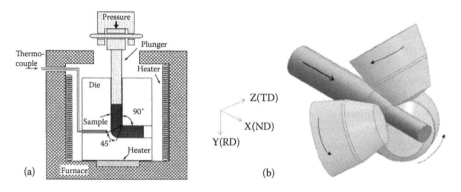

FIGURE 5.7   Schematic illustrations of the facility used for: (a) ECAP and (b) screw rolling. ([a] From Yamashita, A., Z. Horita, and T. G. Langdon. *Materials Science and Engineering: A*, 300 (1), 142–7, 2001. With permission; [b] From Diez, M., H.-E. Kim, V. Serebryany, S. Dobatkin, and Y. Estrin. *Materials Science and Engineering: A*, 612, 287–92, 2014. With permission.)

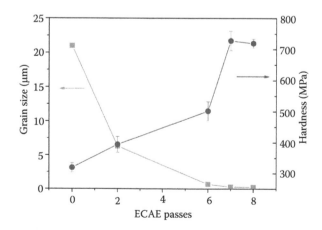

FIGURE 5.8   Correlations between the grain size, hardness, and ECAE passes. (Data were collected from Biswas, S., S. S. Dhinwal, and S. Suwas, *Acta Materialia*, 58, 9, 3247–61, 2010.)

under all pressing conditions attributed to the occurrence of recrystallization during pressing. Meanwhile, the tensile testing indicated a significant improvement in both strength (improvement > 160%) and ductility. Equal channel angular extrusion (ECAE) is one of the most effective ways for the grain refinement of pure Mg. Biswas et al. (2010) demonstrated a way to impart severe plastic deformation to hot-rolled CP-Mg (99.93%) at room temperature to produce an ultrafine grain size of ~250 nm with significantly improved strength through ECAE as shown in Figure 5.8.

Besides SPD, previous studies have shown that pure Mg can be significant reinforced by powder metallurgy (PM). The tensile yield strength and ultimate tensile strength could be improved to 280 MPa and 320 MPa, respectively. However, the degradation of PMed pure Mg should be further improved (Pereda et al. 2010, 2011).

5.3.2 New Methods for Enhancing Corrosion Resistance

Apart from further purification, surface treatments or coatings are another effective way to improve the corrosion resistance of Mg, especially in the early implantation period, and various coatings have been extensively researched in recent years. Table 5.5 lists the corrosion parameters of various coatings on pure Mg. Almost every coating in the table more or less improved the corrosion resistance of pure Mg. The minimum corrosion current density was found in MAO-coated CP-Mg, which was carried out by Gan et al. (2013). PEO coating and PLLA-sealed PEO coating also significantly reduced the corrosion current densities in SBF (Alabbasi et al. 2014a). In this chapter, we care more about the effects of purity on the in vitro and in vivo degradation behaviors of Mg. Therefore, we have chosen several typical examples; for more information about different coatings on the corrosion behavior of pure Mg, refer to Chapter 4.

MAO treatment has received considerable attention in recent years due to its advantages of simplicity, cost-effectiveness, and environmental characteristics in many fields. A thick, hard, well-adhered oxide layer is formed on surfaces of Mg by this method, thus greatly improving both the corrosion and wear resistance. Durdu et al. (2011) prepared MAO coatings on pure Mg (99.96 wt.%) samples in different aqueous solutions of Na_2SiO_3 and Na_3PO_4. The coating was observed to have a two-layer structure, a porous outer layer and a dense inner layer. A super corrosion resistance was obtained with MAO coatings, and the coating consisting of Mg_2SiO_4 exhibited more resistance to corrosion than that containing $Mg_3(PO_4)_2$.

Chemical conversion treatment is an effective and simple approach and utilizes chromate, permanganate, phosphate, and phosphate/permanganate treatments. Chromate conversion treatment is the traditional chemical conversion method, which has been used extensively in engineering fields for many years. However, these chromium compounds have been shown to be highly toxic carcinogens. As a consequence, there is an urgent need to develop novel environmentally friendly conversion coatings. Liu et al. (2012) formed a conversion coating by immersing the Mg samples (as-cast CP-Mg, >99.95 wt.%) in a phytic acid solution. This conversion coating had a multideck structure with netlike morphology with the thickness varying from 1.0 μm to 15 μm according to the processing parameters. The OCP curve exhibited a relatively swift increase during the conversion treatment processing, indicating that this

TABLE 5.5

Electrochemical Corrosion Data on Various Surface-Treated Pure Mg

Substrate	Coating	E_{corr} (V)	i_{corr} (μA/cm^2)	Coating on the Degradation Corrosion Rate	Test Solution	References
99.9 wt.% (CP-Mg)	/ (means without coating)	−1.732	0.962	/	Hank's	Gan et al. 2013
	MAO coating	−1.768	0.0061			
99.9 wt.% (CP-Mg)	/	−1.7~−1.8	13–33	100 mg/(cm^2 · d)		Cabrini et al. 2014
	Hydrofluoric acid treatment (MgF$_2$)	−1.55~−1.51	1.8	14.5 mg/(cm^2 · d)		
	DLC coating	Around −1.7~−1.8	14	35 mg/(cm^2 · d)		
99.96 wt.% (CP-Mg)	/	−1.85	430	/		Ng et al. 2010
	Cerium-based coating	−1.92	343	/		
	Cerium-based coating + hydrothermal treatment	−1.85	4.0	/		
99.99 wt.%	/	−1.70	23.4	/		Lu et al. 2014a
	Pre-CaP coating	−1.65	29.5	/		
	Sr containing Ca–P MAO coatings	−1.66	7.34	/		
99.9 wt.%	/	−1.8 ± 0.02	23.5 ± 3.6	/	SBF	Alabbasi et al. 2014a
	PEO coating	−1.92 ± 0.02	8.3 ± 3	/		
	PLLA-sealed PEO coating	−1.54 ± 0.01	0.03 ± 0.2	/		
99.9 wt.%	/	−1.8 ± 0.02	23.5 ± 3.6			Alabbasi et al. 2014b
	PEO coating	−1.92 ± 0.02	8.3 ± 3			
	Double-layer coating (PEO + CaP)	−1.66	0.85 ± 0.07			

(Continued)

TABLE 5.5 (CONTINUED)

Electrochemical Corrosion Data on Various Surface-Treated Pure Mg

Substrate	Coating	E_{corr} (V)	i_{corr} (µA/cm²)	Coating on the Degradation Corrosion Rate	Test Solution	References
99.9 wt.%	/	−1.71 ± 0.11	31.6 ± 2.6			Khakbaz et al. 2014
	Magnesium phosphate cement (MPC) coating	−1.64 ± 0.03	5.5 ± 1.5			
99.95 wt.% (as-cast)	/	−1.7998	477			R. Xu et al. 2012b
	Cr + O plasma-modified	−1.5983	134			Zhao et al. 2010
99.95 wt.% CP-Mg	MAO coating	−1.85	13.5			
	/	−1.69	0.173			
99.95 wt.% (as-cast)	/	−1.976	359			R. Xu et al. 2012a
	Zn and Al coimplanted	−1.8125	22.7			S.-M. Kim et al. 2014
99.99 wt.% (HP-Mg)	/	−1.98	6831			
	HA coating	−1.56	3.062			
99.9 wt.%	/	−1.65	38.3		HBSS	Li et al. 2014a
	MAO coating	−1.47	0.264			
99.9 wt.%	/	−1.57	38.25			Li et al. 2014b
	MAO coating	−1.43	2.45			
99.92 wt.% (as-cast)	/	−1.638	145		0.9 wt.% NaCl	Zhang et al. 2005
	Ion-implanted Ti coating	−1.352	21.5			
99.99 wt.% (as-cast)	/	−1.795	136		PBS	Lu et al. 2014b
	Ca–Sr–P coating	−1.395	3.75			

phytic acid conversion coating could improve the electrochemical properties and provide effective protection of Mg. In addition, Rudd et al. (2000) reported cerium, lanthanum, and praseodymium conversion coatings formed on pure Mg (99.9 wt.%) with demonstrated improved corrosion resistance by the DC polarization and AC impedance techniques.

Ion implantation affords the unique possibility of introducing a controlled concentration of an element to a thin surface layer. X. M. Wang et al. (2006) performed a yttrium ion implantation on Mg (as-cast pure Mg, 99.9 wt.%), and the corrosion resistance was assessed by electrochemical polarization. Based on the results, an improved corrosion resistance of implanted Mg was observed compared with that of pure Mg.

Zuleta et al. (2011) carried out a comparative study, in which three different chromium-free methods of surface modification on pure Mg (99.9 wt.%) were studied; namely, a cerium conversion coating, a carbonated coating, and an anodic film. The resultant polarization resistances for bare Mg, conversion-coated Mg, anodized Mg, and carbonate-coated Mg were 1.02 kΩ cm^2, 2.46 kΩ cm^2, 3.51 kΩ cm^2, and 7.16 kΩ cm^2, respectively, indicating that calcium carbonated coating provided higher corrosion protection of pure Mg. Furthermore, the corrosion protection effectiveness on powder metallurgy Mg of three environmentally friendly coatings, a silane coating, an anticorrosive paint formulated with ion exchangeable pigments (IEPs), and a chemical conversion treatment to form a MgF$_2$ layer, have been compared by electrochemical measurements (Carboneras et al. 2010). The experimental results indicated that both the silane coating and the anticorrosive paint with Ca/Si IEPs exhibited an initial protective character, which arose from the resistive nature of the layer. For longer exposure times, the fluoride conversion coating may be considered the only viable and effective barrier to protect PM Mg from degradation because the MgF$_2$ layer prevents the metallic substrate from getting into contact with the electrolyte due to its high compactness and excellent adherence to the base metal.

5.3.3 NEW WAYS FOR ENHANCING BIOCOMPATIBILITY

Various surface modification methods have been developed to form a protective coating on the surface of Mg in recent years as discussed in Section 5.3.2. These coatings can enhance the corrosion resistance and retard the process of hydrogen evolution and alkalization and thereby improve the biocompatibility of a Mg substrate to different degrees on the condition that the coating itself is biocompatible.

Jo et al. (2012) have prepared a MgO coating layer on a pure Mg (99.5%) substrate by anodization and MAO, respectively. The DNA levels of the surface-treated Mg samples were about 6–10 times higher than the bare Mg, and the ALP activities were also more than double after the anodization or MAO followed by post-treatments, indicating significantly improved biocompatibility of Mg. Lorenz et al. (2009) have passivated the Mg samples (99.9%) by soaking in 1 M NaOH or M-SBF. The cell culture experiment showed that passivation of the Mg surface in 1 M NaOH significantly increased the cell survival rate, and M-SBF passivated Mg initially enhanced cell adhesion on the surface (Figure 5.9). It can be concluded that

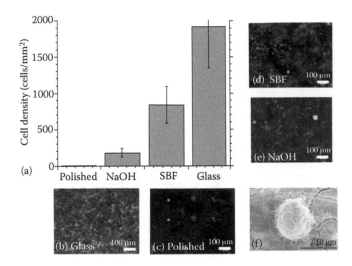

FIGURE 5.9 Behavior of Hela cells on differently pretreated Mg surfaces as well as on glass: (a) cell density after 24 h in cell culture; (b–e) fluorescence imaging of stained cells; (f) SEM image of a cell attached to the surface of Mg soaked in M-SBF. (From Lorenz, C., J. G. Brunner, P. Kollmannsberger, L. Jaafar, B. Fabry, and S. Virtanen. *Acta Biomaterialia*, 5, 7, 2783–9, 2009. With permission.)

the surface pretreatments provided a viable strategy to facilitate cell survival on otherwise nonbiocompatible Mg surfaces. By soaking in cell culture medium for 24 h, a biomimetic Ca-, P-, and C-containing layer was formed on the surface of Mg (99.9%) (Keim et al. 2011). HeLa cells were cultured for 24 h on the bare and treated Mg surfaces, and the results showed that cell density on treated sample surfaces was significantly increased compared to the polished Mg surface, where almost no cells survived. Li et al. (2004) investigated the in vitro biocompatibility of an alkali heat-treated Mg (99.9%) by a screening test on in vitro cell cultures. No signs of morphological changes on cells or inhibitory effects on cell growth were detected. Moreover, biodegradable polymer films produced by spin-coating (Xu and Yamamoto 2012), siloxane-containing vaterite/poly(L-lactic acid) composite coating (Yamada et al. 2013), hydroxyapatite/poly(e-caprolactone) double coating (Jo et al. 2013), and β-tricalcium phosphate coating (Geng et al. 2009a) have also been reported to modify the surface of Mg with excellent cytotoxicities proven.

Gao et al. (2006) investigated the hemolysis effect of heat organic film-treated Mg (99.99 wt.%). Compared to the hemolytic ratio of untreated Mg of 59.24%, the heat organic film-treated Mg exhibited excellent hemocompatibility with a hemolytic ratio of 0. However, Ren et al. (2011) indicated that with fluorine-containing and Si coatings, pure Mg (99.9 wt.%) lost its antibacterial abilities, owing to the very dense coatings on the surface. S.-M. Kim et al. (2014) coated Mg (99.99%) with hydroxyapatite (HA) in an aqueous solution containing calcium and phosphate sources. The in vitro cell tests indicated that the biological response, including cell attachment, proliferation, and differentiation of the HA-coated samples was enhanced considerably compared to bare Mg, and the in vivo tests indicated promoted bone growth,

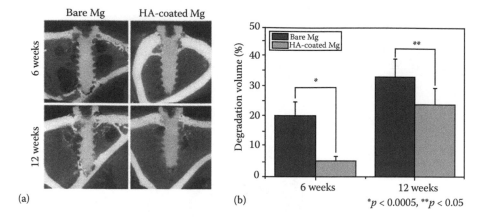

FIGURE 5.10 (a) Histological images of the stained sections and (b) bone-to-implant contact (BIC) ratios of the bare Mg and HA-coated after 6 and 12 weeks of implantation. (From Kim, S.-M., J.-H. Jo, S.-M. Lee, M.-H. Kang, H.-E. Kim, Y. Estrin, J.-H. Lee, J.-W. Lee, and Y.-H. Koh. *Journal of Biomedical Materials Research Part A*, 102 (2), 429–41, 2014. With permission.)

and a higher bone-to-implant contact was observed for the HA-coated Mg screws after 6 and 12 weeks of implantation as illustrated in Figure 5.10. These findings suggest that providing Mg with a protective coating by a surface treatment is a promising avenue for affording Mg with required biocompatibility as a biomedical implant material.

5.4   CONCLUDING REMARKS

In this chapter, we elaborated on the impurities in Mg and their effects on the corrosion of Mg. Previous studies (in vitro and in vivo) on pure Mg with different degrees of purity as biomaterials have been reviewed. Some new advances in property improvement on pure Mg have also been discussed.

Plenty of research has been carried out on pure Mg and Mg alloys for biomedical applications with only limited studies focused on HP-Mg. Studies on biomedical XHP-Mg are quite limited. Traditional CP-Mg has a high impurities level, and detrimental impurities, such as Ni, Fe, and Cu, may exceed their tolerance limits, and this is fatal to Mg degradation. Poor corrosion resistance of commercial-level pure Mg has been widely reported in the literature, and the accompanying bad biocompatibility has also been reported.

The reason why people investigate pure Mg with high purities to be biomaterials is based on their excellent corrosion resistance and biocompatibility. If corrosion of HP-Mg/XHP-Mg is well monitored, biocompatibility of these materials is not a concern because the Mg element itself is biocompatible under limited dosage. Currently, the slowest degradation rate of XHP-Mg is about 10 μm/year, and this value can be adjusted by impurity level control. The corrosion behaviors of 99.999 wt.% and 99.9999 wt.% Mg are very interesting and need to be explored in future research.

The development of single crystalline pure Mg with high purity may be another research topic because no grain boundaries exist in single crystals.

Apart from the corrosion and biocompatibility, the strength of pure Mg remains a problem when pure Mg is aimed at some clinical load-bearing conditions. To meet the specific requirements of different applications, the mechanical properties of Mg can be improved and adjusted via hot and cold working, such as extrusion or various severe plastic deformation processes. Powder metallurgy is another choice to improve mechanical properties. Because alloying is an effective way to strengthen Mg, anticorrosive high pure Mg alloys doped with some biocompatible or biofunctional elements could be another development trend.

6 Mg-Ca-Based Alloy Systems for Biomedical Applications

6.1 INTRODUCTION

6.1.1 CALCIUM

Calcium (Ca), with a density of 1.55 g/cm^3, is similar to low-density Mg (1.74 g/cm^3), which will maintain its specific properties. In terms of physiological functions, Ca is an essential metal element for human beings. It is the major component of human bone, and the majority of element Ca is stored in bones (Yin et al. 2013). It is recommended that an adult should take 1000 mg Ca per day (Drynda et al. 2010). In addition, the Ca ion has been associated with many bioreactions in the human body. For orthopedic applications, the release of Ca ions help improve the bone healing procedure (Zhang and Yang 2008; Harandi et al. 2011). As an alloying element to Mg, Ca can refine the grain size and improve the strength and creep properties under elevated temperatures due to the formation of thermally stable intermetallic phases (Hirai et al. 2005). It was also reported that a low amount of Ca can increase the mechanical properties and corrosion resistance of Mg alloys (Erdmann et al. 2011).

6.1.2 PREVIOUS STUDIES ON MG-CA ALLOY SYSTEMS FOR ENGINEERING APPLICATIONS

The addition of Ca to Mg alloys can retard the oxidation rate during the melting process by the formation of a thin and dense CaO film on the surface of the molten alloy and improve the oxidation resistance of the Mg in elevated temperatures (You et al. 2000). It was reported that Mg-Ca alloys possess good age-hardening response (Nie and Muddle 1997; Ortega and del Rio 2005). Moreover, Mg-Ca alloys were used as high damping materials to reduce noise pollution (D. Q. Wan et al. 2008).

Mg-Ca binary alloys are commercially available as a master alloy to fabricate other ternary or quaternary alloys. The contents of Ca in the Mg-Ca master alloys used for metallurgy are far from the equilibrium and range from 13.3% to 40% (L. Geng et al. 2009; Oh-ishi et al. 2009; Park et al. 2009; Jayaraj et al. 2010; Bakhsheshi-Rad et al. 2012b) because the solubility of Ca in Mg is 1.34% at room temperature.

6.2 DEVELOPMENT OF Mg-Ca BINARY ALLOYS AS DEGRADABLE METALLIC BIOMATERIALS

6.2.1 ALLOY COMPOSITION DESIGN

Mg-Ca alloys with a large range of Ca addition were fabricated during their development as biomaterials. Z. Li et al. (2008) fabricated a series of Mg-Ca alloys with the Ca content ranging from 1 wt.% to 20 wt.% among which they found that the Mg-5Ca, Mg-10Ca, and Mg-20Ca alloys were very brittle at room temperature, and the alloy plates could be easily broken by bare hands. And their results suggested that low Ca contents (such as 1 wt.%, 2 wt.%, and 3 wt.%) might be suitable for designing Mg-Ca system biomaterials, and the Mg-1Ca alloys exhibited the best biocompatibility property, mechanical property and corrosion resistance property. Kirkland et al. (2010) also developed a large series of Mg-Ca alloys with the Ca content ranging from 0.4 wt.% to 28 wt.% to investigate the biocorrosion properties in different corrosion media. Jung et al. (2012) developed a Mg-10Ca alloy and studied the corrosion mechanism by elemental diffusion in vivo. W. C. Kim et al. (2008) fabricated Mg-0.8Ca and Mg-5Ca alloys and found that the Mg-0.8Ca alloy exhibited less hydrogen evolution. Gu et al. (2010a) developed rapid solidified Mg-3Ca ribbons by the melt-spinning technique. The ribbons with the wheel rotating at a speed of 45 m/s showed more uniform corrosion morphology and did not induce toxicity of L-929 cells. Wan et al. (2008b) fabricated Mg-(0.6, 1.2, 1.6, 2.0)Ca and reported that Mg-0.6Ca showed good corrosion and mechanical properties. In spite of the development of Mg-Ca bulk alloys, Mg/Ca composites were also studied. Y.-F. Zheng et al. (2010) fabricated Mg/1Ca, Mg/5Ca, and Mg/10Ca composites by utilizing Mg powder (99.9%, particle size < 150 μm) and Ca powder (99.5%, particle size < 80 μm) via conventional powder metallurgy methods and found that the mechanical property of Mg/1Ca composite was comparable with as-extruded Mg-1Ca alloy; moreover, the Mg/1Ca composite exhibited no significant toxicity to L-929 cells.

6.2.2 FABRICATION, HOT AND COLD WORKING, AND HEAT TREATMENT

6.2.2.1 Metallurgy Method

Metallurgy is a low-cost and effective way to produce as-cast alloys, and the Ca in Mg-Ca alloys can be introduced with Mg-Ca master alloys (13.3%, 26%, 33%, and 40% Ca master alloys are commercially available). Recommended addition temperatures for these master alloys depend on the Ca content and range from 670–800°C. For the melting process, a dissolution or mixing time of 15–30 min after addition is recommended (Friedrich and Mordike 2006b). To ensure the homogeneity of the molten metal, mechanical stirring should be applied during the whole melting procedure. Mg has very low burning and melting points of approximately 480–510°C and 648.8°C, respectively, and thus during the whole fabrication process, a protective atmosphere (such as high pure argon or mixture of CO_2 and SF_6) is required. After melting, the molten metal can be poured into a mold to get the as-cast alloys. Generally speaking, the molds were usually made of mild steel preheated at a specific temperature.

6.2.2.2 Powder Metallurgy Method

Conventional powder metallurgy (PM) is also an effective way to develop composites. Y.-F. Zheng et al. (2010) utilized Mg powder (99.9%, particle size < 150 μm) and calcium powder (99.5%, particle size < 80 μm) as raw materials. Before PM, the mixed powder was dried in a vacuum drying oven at 200°C for about 12 h and mixed by ball milling with agate balls for about 1 h in an argon atmosphere. After the mixing procedure, the powder mixture was cold pressed into a cylindrical compact at the pressure 400 MPa. Then the green compact was hot pressed at 320°C at 350 MPa for 20 min.

Due to the inherent *hcp* crystal structure of Mg, which lacks an active slip system at room temperature, cold working for Mg would be difficult (Harandi et al. 2011). Nevertheless, the critical resolved shear stress can be reduced with the increasing temperatures and improved the formability of Mg and its alloys.

6.2.2.3 Hot and Cold Working

Forging has been considered to be an effective way to improve the mechanical properties of the cast parts. Harandi et al. (2011) developed the forged Mg-1Ca alloy using a stamping machine with a nominal force of 1080 kN and different forging speeds ranging from 40 to 65 stock per minutes and found that after the forging process equiaxed grains were obtained due to the dynamic recrystallization. In other words, the forging process can refine the grain size of the Mg-1Ca alloy. With the increasing forging temperature, more twinning planes were formed as well as the precipitated Mg_2Ca at grain boundaries. But the XRD data indicated that the diffraction intensity of Mg_2Ca decreased with the increasing forging temperature.

Koleini et al. (2012) studied the as-rolled Mg-1Ca alloy and found that the rolling process can refine the grain size of the Mg-1Ca alloy due to the continuous dynamic recrystallization effects. The higher percentage of thickness reducing, the more positive the effect on the refining of the grain. Furthermore, the secondary phase of Mg_2Ca reduced in the matrix and relocated along the grain boundaries after the rolling process.

6.2.3 Microstructure and Mechanical Properties

6.2.3.1 Microstructure

The typical Mg-Ca phase diagram is illustrated in Figure 6.1 (Kozov et al. 2008). The solubility of Ca is relatively limited to only about 1.34 wt.% under equilibrium condition (Bakhsheshi-Rad et al. 2012b). From the phase diagram, it can be seen that, in the ambient environment, the main phase in Mg-Ca alloys would be α-Mg matrix and Mg_2Ca phase when the Ca content is less than 33.3 wt.%. When the Ca content is reduced to 10.6 wt.%, the eutectic reaction occurred at the temperature of 516.5°C.

A typical optical microstructure of Mg-1Ca alloys is depicted in Figure 6.2. It can be found that the second phase Mg_2Ca particles distributed mainly along the grain boundary of the α-Mg matrix, and a few Mg_2Ca particles can also be seen within the α-Mg grain. With the increasing Ca content, the second phase formed a continuous network structure (Z. Li et al. 2008). It should be noted that the addition of Ca can

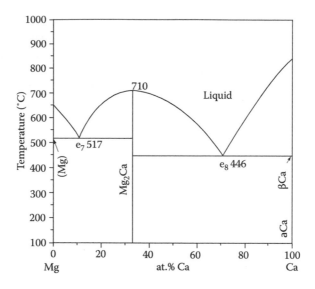

FIGURE 6.1   Mg-Ca binary phase diagram. (Modified from Kozov, A., Ohno, M., Arroyave, R., Liu, Z. K., and Schmid-Fetzer, R., *Intermetallics*, 16, 299, 2008. With permission.)

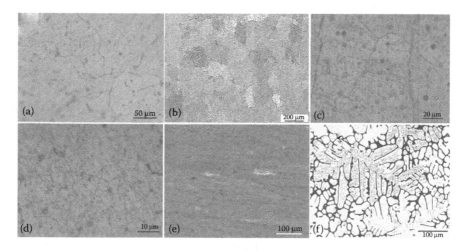

FIGURE 6.2   A typical optical microstructure of Mg-1Ca alloys: (a) as-cast, (b) as-forged, (c) as-rolled, (d) as-extruded, (e) Mg/1Ca composite, and (f) Mg-5Ca alloy. ([a] Modified from Li, Z. J., Gu, X. N., Lou, S. Q., and Zheng, Y. F., *Biomaterials*, 29, 1329, 2008. With permission; [b] Modified from Harandi, S. E., Idris, M. H., and Jafari, H., *Materials and Design*, 32, 2596, 2011. With permission; [c] Modified from Li, Z. J., Gu, X. N., Lou, S. Q., and Zheng, Y. F., *Biomaterials*, 29, 1329, 2008. With permission; [d] Modified from Li, Z. J., Gu, X. N., Lou, S. Q., and Zheng, Y. F., *Biomaterials*, 29, 1329, 2008. With permission; [e] Modified from Zheng, Y. F., Gu, X. N., Xi, Y. L., and Chai, D. L., *Acta Biomater*, 6, 1783, 2010. With permission; [f] Modified from Bakhsheshi-Rad, H. R., Idris, M. H., Kadir, M. R. A., and Farahany, S., *Materials and Design*, 33, 88, 2012b. With permission.)

reduce the grain size of the Mg-Ca alloys (Kannan and Raman 2008). Subsequent working processes can also refine the grain size (Gu et al. 2010a; Harandi et al. 2011).

X-ray diffraction results also indicated that the second phase in the Mg-Ca alloys was mainly the Mg_2Ca phase. The diffraction intensity that arose from the Mg_2Ca phase increased with the increasing Ca contents in the alloys. Hagihara (2013) investigated the corrosion and mechanical properties of Mg_2Ca as an intermetallic compound. Mg_2Ca exhibited a faster corrosion rate and higher mechanical properties when compared with Mg_2Si compounds.

6.2.3.2 Mechanical Properties

It has been reported that, due to the grain refinement mechanism, the addition of Ca to Mg can both increase the strength and the elongation rate (Hirai et al. 2005). Moreover, the addition of Ca not only improves the creep resistance at elevated temperatures (Luo and Pekguleryuz 1994), but it also promotes hot tearing (Hirai et al. 2005). The Ca addition to the Mg can refine the grain size, disperse the particle, and break down the dendritic morphology of the particle into round and well-distributed small particles (Hirai et al. 2005).

Z. Li et al. (2008) found that the yield strength, ultimate tensile strength, and elongation of as-cast Mg-xCa (x = 1, 2, 3) alloy samples decreased with increasing Ca content. The as-cast Mg-1Ca alloy presented the best mechanical properties. The elastic modulus of the Mg-xCa alloys increased from 15.0 GPa for the Mg-0.5Ca to 34.8 GPa for the Mg-20Ca, and the Vickers hardness increased from 18 to 110 HV for the lowest Ca content and the highest Ca content (Y. C. Li et al. 2011). Nevertheless, the elongation decreases with the increasing Ca content in the Mg-xCa alloy, which is consistent with other reports (Li et al. 2008; J. H. Kim et al. 2009). A summary of the mechanical properties of the binary Mg-Ca alloys is shown in Table 6.1. From the table, it can be concluded that both the yield strength and ultimate strength decrease with the increasing Ca contents in the Mg-Ca binary alloys except for the alloys in Li et al. (2010a). The elongation decreases with the increasing Ca contents, indicating that increasing Ca in the Mg would be detrimental for the alloys. Under the same casting conditions, subsequent working, such as extrusion and rolling, can significantly improve the mechanical behavior of the Mg-Ca alloys. The grain size decreases with the increasing Ca content, and the hardness of the alloys exhibited a reverse tendency. A summary of the grain size and mechanical properties of binary Mg-Ca alloys is shown in Table 6.1.

6.2.4 DEGRADATION IN SIMULATED BODY FLUIDS

The degradation of Mg-Ca alloys results in hydrogen release and the precipitation of corrosion products. Because the Mg-Ca alloys in an equilibrium state only consist of α-Mg matrix and Mg_2Ca second phase, the contents and distribution of the Mg_2Ca phase play an important role in the degradation of Mg-Ca alloys. It was reported (W. C. Kim et al. 2008) that galvanic corrosion might occur between the Mg_2Ca phase and α-Mg matrix, which resulted in the release of hydrogen. The amount of Mg_2Ca increased with the increasing of Ca in the Mg-Ca alloys, which accelerated the galvanic corrosion. Therefore, the corrosion rate can be controlled by adjusting the content and distribution of Mg_2Ca phase in the Mg-Ca alloy system.

TABLE 6.1

Summary of the Grain Size and Mechanical Property of Binary Mg-Ca Alloys

Sample	Condition	Mechanical Property						Ref.
		Elastic Modulus (GPa)	Yield Strength (MPa)	Ultimate Strength (MPa)	Elongation (%)	Vickers Hardness (HV)	Grain Size	
Mg-0.5Ca	Cast	15	70.1[a]	166.2	14.5	51.7	N/A	Li et al. 2010a
Mg-1Ca	Cast	16.2	72[a]	179.5	11.5	51.5	N/A	
Mg-2Ca	Cast	16.7	77.2[a]	184.6	11.2	52	N/A	
Mg-5Ca	Cast	18	94.1	188.4	9.4	66	N/A	
Mg-10Ca	Cast	21.7	109.4	190	9.2	71.9	N/A	
Mg-15Ca	Cast	26.8	172.3	208.1	3.2	87.3	N/A	
Mg-20Ca	Cast	34.8	234.9	291.3	1.7	108.4	N/A	
Mg-0.7Ca	Cast	N/A	N/A	N/A	N/A	~34	0.51 mm	Harandi et al. 2013
Mg-1Ca	Cast	N/A	N/A	N/A	N/A	~39	0.44 mm	
Mg-2Ca	Cast	N/A	N/A	N/A	N/A	~42	0.31 mm	
Mg-3Ca	Cast	N/A	N/A	N/A	N/A	~46	0.17 mm	
Mg-4Ca	Cast	N/A	N/A	N/A	N/A	~48	0.12 mm	
Mg-3Ca	Cast	N/A	110 ± 5	118 ± 5	0.26 ± 0.04	N/A	N/A	Du et al. 2011a
Mg/1Ca	Composites	N/A	147.78	217.28	14.36	N/A	N/A	Y.-F. Zheng et al. 2010
Mg/5Ca	Composites	N/A	183.32	202.72	9.03	N/A	N/A	
Mg/10Ca	Composites	N/A	119.63	200.25	7.74	N/A	N/A	

(Continued)

TABLE 6.1 (CONTINUED)
Summary of the Grain Size and Mechanical Property of Binary Mg-Ca Alloys

		Mechanical Property						
Sample	Condition	Elastic Modulus (GPa)	Yield Strength (MPa)	Ultimate Strength (MPa)	Elongation (%)	Vickers Hardness (HV)	Grain Size	Ref.
Mg-2Ca	Cast	N/A	47.3	115.2	3.05	43.2	135	Bakhsheshi-Rad et al. 2014b
Mg-4Ca	Cast	N/A	34.5	77.4	2.1	53.3	92	
Mg-1Ca	Cast	45.5	39	105 ± 4	4.1 ± 0.5	N/A	N/A	Zhang et al. 2011a
Mg-0.6Ca	Cast	46.5 ± 0.5	114.4 ± 0.8[a]	273.2 ± 6.1[a]	N/A	N/A	N/A	Wan et al. 2008b
Mg-1.2Ca	Cast	49.6 ± 0.9	96.5 ± 6.6[a]	254.1 ± 7.9[a]	N/A	N/A	N/A	
Mg-1.6Ca	Cast	54.7 ± 2.4	93.7 ± 7.8[a]	252.5 ± 3.3[a]	N/A	N/A	N/A	
Mg-2Ca	Cast	58.8 ± 1.2	73.1 ± 3.4[a]	232.9 ± 3.7[a]	N/A	N/A	N/A	
Mg-1Ca	Cast	N/A	N/A	71.38	1.87	N/A	N/A	Z. Li et al. 2008
Mg-1Ca	As-extruded	N/A	N/A	239.63	10.63	N/A	N/A	
Mg-1Ca	As-rolled	N/A	N/A	166.7	3	N/A	N/A	

[a] Compressive strength.

Li evaluated the degradation behavior of the binary Mg-Ca alloys with the immersion method (Z. Li et al. 2008). After immersion in SBF for 5 h, the Mg-1Ca alloy showed an integrity corrosion film. But cracks and peeled off feature morphology were observed in the Mg-2Ca alloy. With further increasing Ca content, the Mg-3Ca alloy significantly suffered severe local corrosion, and even deep pits were left on the sample surface. With an increased immersion time of 250 h, entire samples were found to be covered with white deposits, which were rich in carbon, oxygen, Mg, phosphorus, and chloride. The immersed morphology and the chemical composition of the surface corrosion layer are shown in Figure 6.3. All the immersion samples

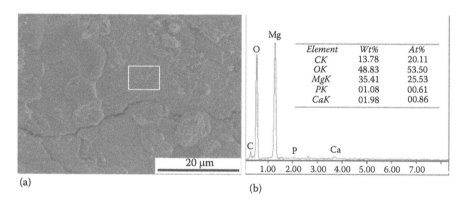

Element	Wt%	At%
CK	13.78	20.11
OK	48.83	53.50
MgK	35.41	25.53
PK	01.08	00.61
CaK	01.98	00.86

(a)

(b)

FIGURE 6.3 Illustration of (a) corrosion morphology of Mg-1Ca alloy after immersion in SBF for 250 h, (b) chemical composition of corrosion products in the framed area in (a). (Modified from Li, Z. J., Gu, X. N., Lou, S. Q., and Zheng, Y. F., *Biomaterials*, 29, 1329, 2008. With permission.)

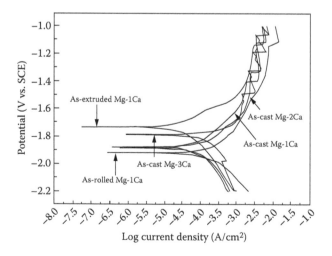

FIGURE 6.4 Potentiodynamic polarization curves of the binary Mg-Ca alloys in SBF. (Modified from Li, Z. J., Gu, X. N., Lou, S. Q., and Zheng, Y. F., *Biomaterials*, 29, 1329, 2008. With permission.)

were severely destroyed to the naked eye. The Mg-3Ca alloys disintegrated into fragments after immersion for 24 h, indicating that the corrosion resistance decreased with the Ca content in the Mg.

Typical potentiodynamic polarization curves of binary Mg-Ca alloys are depicted in Figure 6.4. From Figure 6.4, the current plateaus with different breakdown potentials that were observed in the anode parts of the curves in electrochemical tests indicated that a protective film was formed on all the sample surfaces. The higher potential range of the current plateau and higher breakdown potentials of Mg-1Ca alloy indicated a better corrosion resistance. The higher breakdown potentials observed from the as-extruded and as-rolled Mg-1Ca alloy also implied that the surface protective films were more protective than the as-cast alloy (Z. Li et al. 2008).

A summary of corrosion rates of binary Mg-Ca alloys in different corrosive mediums is exhibited in Table 6.2.

6.2.5 CYTOTOXICITY AND HEMOCOMPATIBILITY

Z. Li et al. (2008) evaluated the cytotoxicity of Mg-1Ca alloy to L-929 cells by both indirect and direct assays. It was reported that even after 7 days of culture in extraction medium the viability of L-929 cells can reach 150% when compared with a control group. The morphologies of the cells exhibited a healthy morphology with a flattened spindle shape. The typical morphology of the cells cultured on Mg-1Ca alloys is shown in Figure 6.5b.

Gu et al. (2009a) reported the L-929 cells' response to the alkaline-heated and untreated Mg-1.4Ca alloy by MTT test. The results showed that the cells cultured in the untreated Mg-Ca alloy sample extraction medium showed a relatively higher cell viability than that of the negative control. The cell morphology also indicated the alkaline-heated treatment of the Mg-1.4Ca alloy showed no obvious toxicity to L-929 cells.

Gu et al. (2010) evaluated the biocompatibility of as-spun Mg-3Ca alloy ribbon by indirect assay. All the as-spun alloys exhibited a higher cell viability when compared with as-cast alloys. Furthermore, a higher wheel rotating speed showed a higher cell viability. The cell attachment experiment showed that all the cells on the experimental samples exhibited a spindle-like morphology. The typical cell morphology that grew on the sample surface or in extract medium is shown in Figure 6.5a.

6.2.6 IN VIVO PERFORMANCE IN ANIMAL MODELS

Jung et al. (2012) evaluated the elemental interdiffusion of Mg10Ca alloy in vivo. The elemental diffusion is shown in Figure 6.6. The interface of implant and bone tissue included a reaction layer. Inside the interface, the observation of P, O, and Ca indicated the formation of calcium phosphate. On the outside of the interface, which is the healed reaction layer, a hydroxyapatite phase with a thickness of about 2 μm was formed. Inside the implants, the decrease of Mg particle size followed power-law dependency with time. The findings exhibited that the lamellar phase corroded faster, and the three-dimensionally connected lamellar structure led to rapid bulk corrosion. After 3 days implantation, the alloy had already corroded because the O

TABLE 6.2

Summary of Corrosion Rates of Binary Mg-Ca Alloys in Different Corrosive Media

Sample	Condition	Corrosion Rate in Different Corrosive Media				I_{corr}	Ref.
		SBF	Kokubo Solution	MMEM	Hank's Solution		
Mg-0.5Ca	As-cast	0.2 mg/day	N/A	0.1 mg/day	N/A	N/A	Li et al. 2010a
Mg-1Ca	As-cast	0.5 mg/day	N/A	0.87 mg/day	N/A	N/A	
Mg-2Ca	As-cast	2.03 mg/day	N/A	2.03 mg/day	N/A	N/A	
Mg-0.7Ca	As-cast	N/A	1479 mpy	N/A	N/A	$1.97 \text{ A/cm}^2 \times 1000$	Harandi et al. 2013
Mg-1Ca	As-cast	N/A	1876 mpy	N/A	N/A	$2.24 \text{ A/cm}^2 \times 1000$	
Mg-2Ca	As-cast	N/A	2017 mpy	N/A	N/A	$3.12 \text{ A/cm}^2 \times 1000$	
Mg-3Ca	As-cast	N/A	2881 mpy	N/A	N/A	$3.95 \text{ A/cm}^2 \times 1000$	
Mg-4Ca	As-cast	N/A	3604 mpy	N/A	N/A	$4.7 \text{ A/cm}^2 \times 1000$	
Mg-1Ca	As-cast	1.47×10^{-3} m/yr	N/A	N/A	N/A	$1.97 \times 10^{-3} \text{ µA/cm}^2$	Harandi et al. 2011
Mg-1Ca	Forged	1.90×10^{-3} m/yr	N/A	N/A	N/A	$3.71 \times 10^{-3} \text{ µA/cm}^2$	
Mg-2Ca	As-cast	N/A	6.89 mm/yr	N/A	N/A	301.9 µA/cm^2	Bakhsheshi-Rad et al. 2014b
Mg-4Ca	As-cast	N/A	9.04 mm/yr	N/A	N/A	395.7 µA/cm^2	
Mg-1Ca	As-cast	N/A	N/A	N/A	3.16 ± 0.5 mm/yr	N/A	Zhang et al. 2011a
Mg-3Ca	As-cast	21 ± 1.4 mm/yr	N/A	N/A	N/A	$929.3 \pm 60.1 \text{ µA/cm}^2$	Gu et al. 2010a
RS15 Mg-3Ca	As-rolled	1.68 ± 0.2 mm/yr	N/A	N/A	N/A	$74.2 \pm 9.9 \text{ µA/cm}^2$	
RS30 Mg-3Ca	As-rolled	1.25 ± 0.3 mm/yr	N/A	N/A	N/A	$55.6 \pm 14.3 \text{ µA/cm}^2$	
RS45 Mg-3Ca	As-rolled	0.39 ± 0.1 mm/yr	N/A	N/A	N/A	$17.1 \pm 4.9 \text{ µA/cm}^2$	
Mg-1Ca	As-cast	12.56 mm/yr	N/A	N/A	N/A	N/A	Z. Li et al. 2008
Mg-1Ca	As-extruded	1.74 mm/yr	N/A	N/A	N/A	N/A	
Mg-1Ca	As-rolled	1.63 mm/yr	N/A	N/A	N/A	N/A	

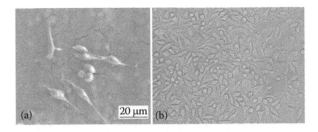

FIGURE 6.5 Typical L-929 cell morphology cultured with (a) RS45-Mg-3Ca alloy for 2 days and (b) Mg-1Ca 100% extracts for 2 days. ([a] Modified from Gu, X. N., Li, X. L., Zhou, W. R., Cheng, Y., and Zheng, Y. F., *Biomedical Materials*, 5, 035013, 2010. With permission; [b] Modified from Li, Z. J., Gu, X. N., Lou, S. Q., and Zheng, Y. F., *Biomaterials*, 29, 1329, 2008. With permission.)

already occupied the entire implant except for the Mg phase. At 8 weeks, the surface was covered with a uniform layer of MgO with no detectable Ca inside the implant, but coarse calcium phosphates still existed near the implant surface.

Mg-1Ca alloy pins, with the same geometric size as a cortical bone Ti pin (AO company), were implanted into the femoral shaft of New Zealand rabbits (Z. Li et al. 2008). The radiographs of rabbit femora showed that the Mg-1Ca pins degraded gradually and totally absorbed in 3 months. Moreover, the new bone was formed around the Mg-1Ca pins whereas there was no new bone formed around the Ti pins. Gas bubbles were observed around the Mg-1Ca pins in the first month and vanished in the second month. Although gas bubbles were formed, no adverse effects were detected. The histological examination (Figure 6.7d–f) showed that a large number of osteoblasts and osteocytes were found around the Mg-1Ca pins, meaning the formation of new bone. But in the control cortical bone Ti pin group, no evidence of newly formed bone was seen in the 3-month implantation. These findings suggested that Mg-1Ca alloy possesses the potential as a biodegradable orthopedic material.

Erdmann et al. (2011) evaluated the in vivo biocompatibility of Mg0.8Ca alloy with New Zealand white rabbits as an animal model. All the implant materials were well tolerated, and mild reddening and swelling were observed near the wound but resolved completely no later than 14 days after surgery. The mild tissue formation occurred in the first week after surgery. With increasing implantation time, the growing tissue not only increased in size but also in compactness. After 2–3 weeks, newly formed tissue similar to natural bone could be palpated in both Mg-Ca and S316L groups, and no difference was detected. Accumulation of gas derivate from the degradation of Mg0.8Ca alloy was clinically and radiographically detected from the first week after surgery throughout the remaining observation period. The pull-out testing results exhibited that the pull-out force of S316L was significantly higher than that of Mg0.8Ca alloy. The pull-out force for Mg0.8Ca alloy remained approximately equal after 4 weeks and decreased with increasing implantation time from 6 to 8 weeks. The Mg0.8Ca alloy suffered a slight weight loss during the first 2 weeks after implantation. After 8 weeks, the implantation reduced more than than 90% of its initial volume, and all the samples were covered with a thin corrosion layer (Figure 6.7a–c).

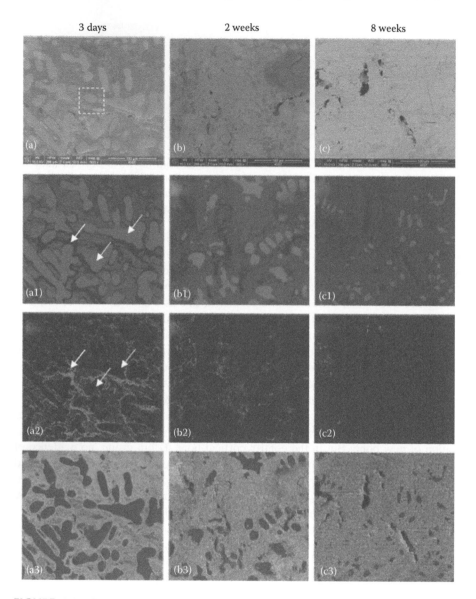

FIGURE 6.6 Comparison of elemental distributions at (a) 3 days, (b) 2 weeks, and (c) 8 weeks. (a–c) Cross-sectional SEM images acquired by a BSE detector. EDX elemental maps showing elemental distributions of Mg (a1–c1), Ca (a2–c2), and O (a3–c3) at 500× magnification. The white dotted arrows indicate Mg-depletion traces (a1) and Ca-rich lines (a2) between primary Mg particles and the lamellar structures. (Modified from Jung, J. Y., Kwon, S. J., Han, H. S., Lee, J. Y., Ahn, J. P., Yang, S. J., Cho, S. Y., Cha, P. R., Kim, Y. C., and Seok, H. K., *Journal of Biomedical Materials Research Part B-Applied Biomaterials*, 100B, 2251, 2012. With permission.)

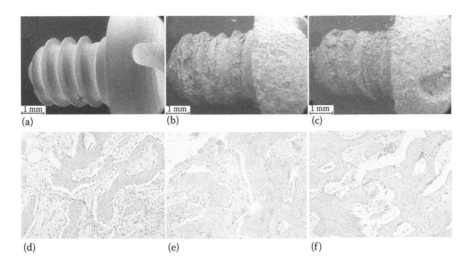

FIGURE 6.7 SEM images of Mg-0.8Ca screws, (a) before implantation, (b) after 2 weeks, (c) after 8 weeks, histological graphs of hematoxylin and eosin stained sections of tissues around, (d) Mg-1Ca alloy pins 1 month after implantation, (e) after 2 months, and (f) after 3 months. ([a–c] Modified from Erdmann, N., Angrisani, N., Reifenrath, J., Lucas, A., Thorey, F., Bormann, D., and Meyer-Lindenberg, A., *Acta Biomaterialia*, 7, 1421, 2011. With permission; (d–f) Modified from Li, Z. J., Gu, X. N., Lou, S. Q., and Zheng, Y. F., *Biomaterials*, 29, 1329, 2008. With permission.)

6.3 DEVELOPMENT OF Mg–Ca–X TERNARY ALLOYS AS DEGRADABLE METALLIC BIOMATERIALS

6.3.1 ALLOY COMPOSITION DESIGN

The Mg-Ca binary alloys exhibited good biocompatibility and comparable mechanical property with natural bone in both in vitro and in vivo evaluations. But the fast corrosion rate might hinder its wide usage in clinic applications. In order to improve the corrosion resistance of Mg-Ca alloy, more alloying elements were tried to be added into Mg-Ca alloys.

1. Zr can refine the crystal grain and improve both the strength and corrosion resistance of Mg alloys (Huaiying et al. 2011), and the Ca can increase the solubility of Zr in Mg. W. J. Zhang et al. (2012) developed the Mg-1Ca-5Zr alloy and evaluated the microstructure and in vitro corrosion property. Y. L. Zhou et al. (2013) fabricated four kinds of Mg-Ca-Zr alloys, including Mg-1Ca-0.5Zr, Mg-2Ca-0.5Zr, Mg-1Ca-1Zr, and Mg-2Ca-1Zr.
2. Zn is an important alloy element for Mg alloys. For binary Mg-Ca alloys, the addition of Zn can increase the tensile strength, hardness, and ductility of the ternary alloys (Datta et al. 2011; Farahany et al. 2012; Y. Sun et al. 2012; Cho et al. 2013). Mg-Ca-Zn alloys with large ranges of Ca and Zn content are presented in Table 6.1.

3. It has been reported that oral administration of strontium (Sr) was found to stimulate bone formation in rats. Other clinical reports also indicated that strontium ranelate can increase bone mineral density and bone strength; thus it was very effective for treating osteoporosis (Berglund et al. 2012). Mg-xCa-ySr alloy system (where x = 0.5–7.0 wt.%, y = 0.5–3.5 wt.%) was fabricated by Berglund et al., with the microstructure and mechanical property as well as biocompatibility being evaluated.

4. Lithium (Li) has been reported to have no influence on cell viability or the inflammatory response (Zeng et al. 2014). The dual phase structure depends on the Li content in Mg alloys and can improve the superplasticity and ductility of the alloys. For biomedical application consideration, Zeng et al. (2014) developed Mg-9.29Li-0.8Ca alloy, and Song and Kral (2005) fabricated a series of Mg-Li-Ca alloys with Ca content ranges from 0 to 15 wt.%. By utilizing electrochemical deposition of Mg, Li, and Ca from salts melts, Yan developed Mg-Li-Ca alloys on inert electrodes (Yan et al. 2009a).

5. Some other elements, such as Mn (Yamauchi and Asakura 2003; Stanford 2010), Y (Li et al. 2010a; Y. C. Li et al. 2011), Sn (Abu Leil et al. 2009), Bi (Remennik et al. 2011), Ce (Hampl et al. 2007), and Si (E. L. Zhang et al. 2010b) were also added as an alloy element in Mg-Ca alloys.

6.3.2 FABRICATION, HOT AND COLD WORKING, AND HEAT TREATMENT

Metallurgy is the most common method to fabricate the Mg-Ca-X ternary alloys. The high-purity Mg, Ca (>99.8%), or Mg-Ca master alloys were often used as raw materials whereas other alloying elements, which are mentioned here, were introduced as either high-purity bulk material or Mg-X master alloys, depending on their chemical and physical properties.

6.3.2.1 Mg-Ca-Zn Alloys

High-purity Mg, Ca, and Zn granules or Mg-Ca master alloys with different Ca contents (Oh-ishi et al. 2009; Bakhsheshi-Rad et al. 2012a; S. Chen et al. 2012; Chen et al. 2012b; Yin et al. 2013) were used as raw materials to fabricate the Mg-Ca-Zn ternary alloys. The melting temperature of raw materials ranged from 700°C to 810°C (Du et al. 2011a; Bakhsheshi-Rad et al. 2012a; S. Chen et al. 2012; Chen et al. 2012b; Farahany et al. 2012; Y. Sun et al. 2012; Yin et al. 2013). During the whole melting process, applying mechanical stir was suggested to get a homogeneous melt. To prevent oxidation of the melt, a protective atmosphere provided by high-purity argon gas or a mixture of CO_2 and SF_6 was applied. After the melts were kept at about 700°C for 10–30 min to ensure all the added alloying elements were dissolved in the melt alloy, the melts were then poured into a mold, which was preheated to 200°C (Y. Sun et al. 2012; Xia et al. 2012; Yin et al. 2013) or 300°C (Bakhsheshi-Rad et al. 2012a) to get the cast ingots. The Mg-Ca-Zn ingots were then heated, solution treated, or aging treated. Heat treatments for Mg-Ca-Zn alloys were performed at 315°C to 470°C—the heated time ranged from 5 to 24 h to remove any nonequilibrium eutectic mixture—and finally quenched in water. After the heat treatment, the solutionizing heat treatment was performed at 400°C for 5 h (Y. Sun et al. 2012) or 480°C for 24 h (Gao et al.

2005; Langelier et al. 2012) and then quenched in water. Other researchers had reported the extrusion procedure of Mg-Ca-Zn alloys. Gao et al. (2011b) extruded the Mg-2Zn-0.24Ca alloy under the temperature of 320°C at an extrusion rate of 17.4:1. Xia et al. (2012) developed the extruded Mg-4Zn-0.2Ca alloy. The extrusion was carried out at the temperature of 270°C with an extrusion ratio of 16:1 and an extrusion speed of 2 mm/s.

6.3.2.2 Mg-Ca-Zr Alloys

In order to introduce Zr as an alloying element into the Mg-Ca alloys, both high-purity Zr granules and Mg-Zr master alloys were applied, including commercial master alloys, such as Mg-10Zr (W. J. Zhang et al. 2012), Mg-33Zr (Y. L. Zhou et al. 2013), Mg-33.3Zr (Chou et al. 2013), and Mg-65Zr (Chang et al. 1997). Chang et al. (1997) fabricated a large series of Mg-Ca-Zr alloys (with Zr contents ranging from 0% to 1.6% and Ca contents ranging from 0% to 1.9%) at the temperature of 800°C in a mixed gas atmosphere of SF_6 and CO_2. Then the melts were squeeze-cast under the pressure of 100 MPa with a plunger speed of 23 mm/s and a mold temperature of 300°C. For further processing, Zhou et al. (2011) and Y. L. Zhou et al. (2013) prepared the hot-rolled Mg-Ca-Zr alloys after a homogeneous process at 450°C for 0.5 h, and the reduction rate was 62%.

6.3.2.3 Mg-Ca-Li Alloys

Due to its very high reactivity, Li is extremely difficult to add as an alloying element. For the melting process, LiCl-LiF flux was used to float on the top of the melts, providing surface protection. Li was usually added by means of an inverted cup plunged into the melts, and the possible temperature was 680–700°C. During the whole melting process, a thorough mechanical stir was required. Song and Kral (2005) developed a rolling process for Mg-12Li-5Ca alloy. The alloy was first homogenized at 350°C for 2 h and then cold rolled to sheets with a reduction of 87.5%. Subsequently, an annealing heat treatment of the sheets was conducted at 200°C for 15 h. Zeng et al. (2014) fabricated the as-extruded Mg-9.29Li-0.88Ca alloy with a extrusion ratio of 20.4:1 and an extrusion rate of 1 m/min. Yan et al. (2009a) fabricated the Mg-Li-Ca alloy by electrochemical codeposition from $LiCl-KCl-MgCl_2-CaCl_2$ melts.

6.3.2.4 Mg-Ca-Sr Alloys

As an alloying element, strontium was usually added for creep resistance in die-casting and gravity-casting alloys to reduce microshrinkage porosity in gravity-cast alloys (Pekguleryuz 2000; Labelle et al. 2001; Luo and Powell 2001). The Sr can be introduced as pure Sr metal, and the Sr amount may be less than 3%. Due to the high reactivity, the Sr element should be added at the very end to avoid excessive contact with flux if the fluxes were applied. The melting time and temperature should be optimized to avoid contaminants. In order to get high-purity alloys that are free of oxide and chloride contaminants, mechanical agitation and equilibration periods were required to assure effective alloying (Friedrich and Mordike 2006b). Berglund et al. (2012) developed a series of Mg-Ca-Sr alloys. The high-purity Mg chips, Ca granules, and Sr granules were used as raw materials. The melting temperature was carried out between 725°C and 825°C. Each melt was held for about 40 min and mechanically stirred prior to casting. During the melting and casting processes, a protective atmosphere of high purity Ar was applied to avoid oxidation.

6.3.2.5   Other Ternary Mg-Ca-X Alloys

Mg-1Mn-0.5Ca alloy was cast by melting the Mg-1Mn master alloy and Ca in a resistance heating furnace at about 720°C under the protection of high-purity Ar gas (Yamauchi and Asakura 2003). The as-cast Mg-Ca-Mn alloy was then solution treated at 500°C for 6 to 24 h. After the solution treatment, the alloys were extruded at 450°C with an extrusion rate of 1 mm/s and an extrusion ratio of 30.

Mg-Ca-Sn alloys were developed by melting the high-purity Mg, Ca, and Sn at approximately 720°C under the atmosphere of a mixture of argon and SF_6. Then the cast alloys were heat treated at 500°C for 6 h followed by extrusion at 350°C with an extrusion ratio of 30 and extrusion rate of 5.5 mm/s (Abu Leil et al. 2009).

Additionally, Kozlov et al. (2008) developed 12 kinds of Mg-Ca-Sn alloys. E. L. Zhang et al. (2010) reported the fabrication of Mg-Ca-Si alloys, and Remennik et al. (2011) reported the fabrication of Mg-Ca-Bi alloys. Mg-1Ca-1Y alloy was also reported in Li et al. (2010a). The material preparation details can be found accordingly.

6.3.3   MICROSTRUCTURE

6.3.3.1   Mg-Ca-Zn Alloys

The ternary phase diagram of the as-cast Mg-Ca-Zn alloys is illustrated in Figure 6.8. The presence of Zn had a dramatic influence on the microstructure of the Mg-Ca alloy. From the phase diagram (Brubaker and Liu 2004; Du et al. 2011a), it can be seen that when the ternary Mg-Ca-Zn alloy melts cool down, the possible eutectic reaction happened:

$$L \rightarrow \alpha - Mg + Mg_2Ca + Ca_2Mg_6Zn_3 \tag{6.1}$$

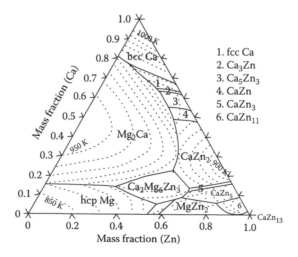

FIGURE 6.8   Ternary phase diagram of the as-cast Mg-Ca-Zn alloys. (Modified from Du, H., Wei, Z. J., Liu, X. W., and Zhang, E. L., *Materials Chemistry Physics*, 125, 568, 2011. With permission.)

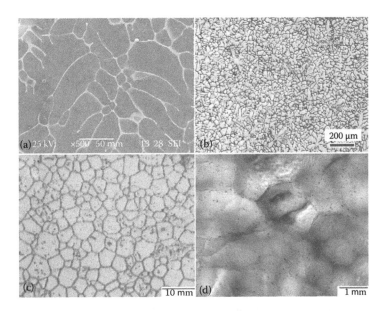

FIGURE 6.9 Microstructure of different Mg-Ca-Zn alloys with different working conditions: (a) as-cast Mg-1Ca-5Zn, (b) as-cast Mg-3Ca-2Zn, (c) as-extruded, and (d) high pressure torsion Mg-0.24Ca-2Zn. ([a] Modified from Yin, P., Li, N. F., Lei, T., Liu, L., and Ouyang, C., *Journal of Materials Science-Materials in Medicine*, 24, 1365, 2013. With permission; [b] Modified from Du, H., Wei, Z. J., Liu, X. W., and Zhang, E. L., *Materials Chemistry and Physics*, 125, 568, 2011. With permission; [c, d] Modified from Gao, J. H., Guan, S. K., Ren, Z. W., Sun, Y. F., Zhu, S. J., and Wang, B., *Materials Letters*, 65, 691, 2011. With permission.)

Larionova et al. (2001) reported that the eutectic (α-Mg + $Ca_2Mg_6Zn_3$) phase formed when the Zn/Ca atomic ratio is more than 1.2, and the eutectic (α-Mg + Mg_2Ca + $Ca_2Mg_6Zn_3$) phase precipitates when the Zn/Ca ratio is less than 1.2. The addition of Zn in the Mg-Ca-Zn alloys can lead to a grain refinement, and further addition of Zn up to 9% can lead to a more significant grain refinement (Bakhsheshi-Rad et al. 2012a). When the ternary Mg-Ca-Zn alloy contains small amount of Zn (such as less than 4%), the secondary phases mainly distributed along the grain boundary; however, with the increasing of Zn, small-sized second phase particles were found in grain interiors. The typical microstructure of Mg-Ca-Zn alloys is shown in Figure 6.9.

6.3.3.2 Mg-Ca-Zr Alloys

The binary Mg-Zr phase diagram is illustrated in Figure 6.10 (Bhan and Lal 1993). Zr has a maximum solubility of 0.6% in Mg and precipitates with the form of single Zr during the cooling process according to W. J. Zhang et al. (2012). The typical microstructure of Mg-Ca-Zr ternary alloys is shown in Figure 6.11. The microstructure of the Mg-Ca-Zr alloys exhibited large grains with secondary phase distributed along the grain boundaries. Y. L. Zhou et al. (2012) fabricated four kinds of

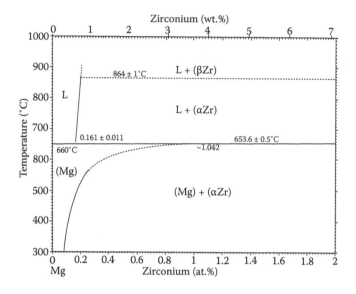

FIGURE 6.10 The binary phase diagram of Mg-Zr alloy. (Modified from Bhan, S. and Lal, A., *Journal of Phase Equilibria*, 14, 634, 1993. With permission.)

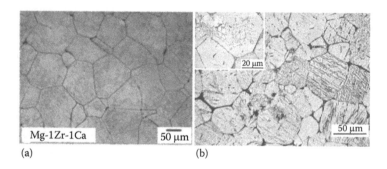

FIGURE 6.11 Typical microstructure of Mg-Ca-Zr alloys: (a) as-cast Mg-1Ca-1Zr and (b) as-cast Mg-1Ca-5Zr, the insert image was in higher magnification. ([a] Modified from Zhou, Y. L., Li, Y. C., Luo, D. M., Wen, C. E., and Hodgson, P., *Journal of Materials Science*, 48, 1632, 2013. With permission; [b] Modified from Zhang, W. J., Li, M. H., Chen, Q., Hu, W. Y., Zhang, W. M., and Xin, W., *Materials and Design*, 39, 379, 2012. With permission.)

Mg-Ca-Zr alloys (Mg-0.5Zr-1Ca, Mg-0.5Zr-2Ca, Mg-1Zr-1Ca, Mg-1Zr-2Ca), but the presence of Zr can hardly be detected by the XRD measurement. The absence of Zr in the XRD results may suggest that Ca can improve the solubility of Zr in Mg. And the addition of Zr can also refine the grain size of Mg alloys (Y. L. Zhou et al. 2012, 2013).

6.3.3.3 Mg-Ca-Li Alloys

The binary phase diagram of Mg-Li is illustrated in Figure 10.1. Li has a body-centered cubic structure, and the addition of Li as an alloying element can change

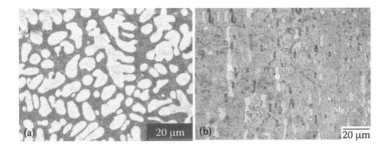

FIGURE 6.12 Typical microstructure of Mg-Li alloys: (a) as-cast Mg-12Li-10Ca and (b) as-extruded Mg-9.29Li-0.88Ca. ([a] Modified from Song, G. S. and Kral, M. V., *Materials Characterization*, 54, 279, 2005. With permission; [b] Modified from Zeng, R.-C., Sun, L., Zheng, Y.-F., Cui, H.-Z., and Han, E.-H., *Corrosion Science*, 79, 69, 2014. With permission.)

the structure of Mg. Mg-Li alloys exhibit a dual-phase structure consisting of α-Mg and β-Li phase when the Li content is between 5.5% and 10.2%. The single β-phase structure only exists when Li content is greater than 10.2% (Yan et al. 2009a). The typical microstructure of Mg-Li-Ca alloys is shown in Figure 6.12. From the microstructure image, it can be concluded that the as-cast Mg-Li-Ca alloys mainly consist of primary dendrites and interdendritic eutectic regions. Song and Kral (2005) also reported the increasing content of Ca may lead to the increasing amount of eutectic reaction products as the volume fraction of the interdendritic eutectic increased with the increasing Ca contents. Zeng et al. (2014) found the Mg_2Ca phase in the Mg-Li-Ca alloys rather than the Li_2Ca phase due to Ca having a larger negative enthalpy of formation with Mg than with Li. And the Mg_2Ca particles were mainly distributed in the grain boundaries and the phase interiors.

6.3.3.4 Mg-Ca-Sr Alloys

The typical Mg-Sr binary phase diagram is illustrated in Figure 8.1, from which we can see that for Mg-Sr binary alloys that contains less than 10% Sr, α-Mg matrix and secondary phases of $Mg_{17}Sr_2$ coexist at room temperature. Regarding the Mg-Ca-Sr ternary alloy, the addition of Sr into Mg-Ca alloys has a significant influence on the grain size, the formation of precipitates, and the distribution of the secondary phase. Berglund et al. (2012) reported that large, irregular, ellipsoidal-shaped α-Mg dendrites and intermetallic compounds were found in the Mg-Ca-Sr alloys. Except for the Mg-0.5Ca-0.5Sr alloy, all the other alloys had a continuous precipitate and eutectic network along the dendrites. Furthermore, the Mg-0.5Ca-0.5Sr exhibited the largest dendrite spacing. The XRD results indicated that the intermetallic compounds were Mg_2Ca and $Mg_{17}Sr_2$ phase. With the increasing Ca and Sr contents, an increase in the amount of intermetallic compounds were found along the dendrite boundaries.

6.3.3.5 Mg-Ca-Mn Alloys

Mn can be precipitated in the form of stable fine particles due to its low solubility, and it diffuses slowly in the Mg matrix. According to the microstructure

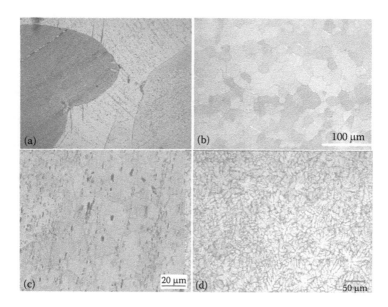

FIGURE 6.13 Microstructure of some typical Mg-Ca ternary alloys: (a) as-cast Mg-1Ca-1Y, (b) as-cast Mg-1.17Mn-0.1Ca, (c) as-cast Mg-1Ca-3Sn, and (d) as-cast Mg-0.2Ca-0.6Si. ([a] Modified from Li, Y. C., Hodgson, P. D., and Wen, C. E., *Journal of Materials Science*, 46, 365, 2011. With permission; [b] Modified from Stanford, N., *Materials Science and Engineering a-Structural Materials Properties Microstructure and Processing*, 528, 314, 2010. With permission; [c] Modified from Prasad, Y. V. R. K., Rao, K. P., Hort, N., and Kainer, K. U., *Materials Science and Engineering a-Structural Materials Properties Microstructure and Processing*, 502, 25, 2009. With permission; [d] Modified from Zhang, E. L., Yang, L., Xu, J. W., and Chen, H. Y., *Acta Biomaterialia*, 6, 1756, 2010. With permission.)

exhibited in Figure 6.13, the as-extruded Mg-0.1Ca-1.17Mn mainly consists of a fully equiaxed and recrystallized microstructure. There were three phases observed in the Mg-Ca-Mn alloys according to Stanford's work (Stanford 2010). The matrix of the alloy was primarily Mg. The second phase was enriched in Ca and confirmed to be the Mg_2Ca eutectic. The third phase particle was found to be particles of Mn. It should be noted that the increasing Ca content can decrease the grain size of Mg-Mn-Ca alloys. Other researchers reported that the presence of fine, stable particles can lead to grain refinement (Robson et al. 2011).

6.3.3.6 Mg-Ca-Sn Alloys

Sn has a high solid solubility over a wide temperature range due to the formation of Mg_2Sn intermetallic precipitates. It has been reported that the existing Ca and Sn in the Mg-Ca-Sn alloys were apt to form CaMgSn particles, and the remaining Ca and Sn eventually form the Mg_2Sn and Mg_2Ca phases (Yang et al. 2010). Hort et al. (2006) reported that the volume fraction of the CaMgSn phase is in proportional to the content of Ca and found that the Mg_2Sn phase decreased with the formation of

CaMgSn, indicating that the Ca suppressed the formation of the Mg_2Sn phase in the Mg-Ca-Sn alloy system.

6.3.3.7 Mg-Ca-Bi Alloys

The RS and extruded Mg-1Ca-5Bi alloys consist of very fine grain sizes less than 2 μm. The Mg-1Ca-5Bi alloys mainly consist of Mg_2Bi_2Ca, which precipitated along the grain boundary, and nanosized precipitation of Mg_2Bi_3 was found within the grain (Remennik et al. 2011).

6.3.3.8 Mg-Ca-Si Alloys

Mg-0.2Ca-0.6Si alloys mainly consist of the needle-like Mg_2Si phase and CaMgSi phase. The needle-like Mg_2Si phase tended to distribute along the grain boundary. The addition of Ca has an effect on the refinement of the Mg_2Si and the grain size (E. L. Zhang et al. 2010).

6.3.4 MECHANICAL PROPERTIES

A summary of the yield strength, ultimate strength, elongation, elastic modulus, hardness, and grain size of the ternary Mg-Ca-X alloys are displayed in Table 6.3. From the table, we can conclude that the selection of alloying elements significantly influences the mechanical behavior of the Mg-Ca-X ternary alloys. With further working, processing would benefit both the mechanical property and the ductility of the alloys. The influences of the alloying elements and the amounts of alloying elements on the mechanical behavior of the alloys are introduced in detail as follows.

6.3.4.1 Mg-Ca-Zn Alloys

It has been reported that both Zn and Ca alloying elements play an important role in refining the grain size and improving the tensile strength of Mg alloys (Kannan and Raman 2008; Xu et al. 2008; Yin et al. 2013). The lamellar Mg_2Ca phase was harmful for the ductility of the binary Mg-Ca alloys (Chino et al. 2002). Additional Zn in Mg-Ca alloys can refine the grain size, and the grain boundary became thinner, which can benefit the ductility of the Mg-Ca alloys. When the Zn contents are higher than the maximum solubility of Mg, the melt Zn atom would be rejected by the growing of α-Mg and enriched in the residual liquid, which is prone to form a eutectic structure and micropores and is harmful to the tensile strength. Zhang fabricated Mg-1Ca-xZn (x = 1 – 6) alloys (Zhang et al. 2011a). When the Zn content was less than 4%, the tearing edges and big dimples could be seen on the fracture surface, indicating the fracture type was ductile. However, the fracture type became brittle when the Zn concentration was up to 6% (Zhang et al. 2011a). Yin et al. (2013) reported that the increasing Ca content in the Mg-Ca-Zn alloys could also improve the tensile strength. This may due to the distribution of the $Ca_2Mg_6Zn_3$ and Mg_2Ca phases along the grain boundaries, which promote the strength by dispersion strengthening (Yin et al. 2013). Hot working can refine the grain size and increase the yield strength and the tensile strength of Mg. Sun et al. (2012) reported that the yield strength, ultimate tensile strength, and elongation of a Mg-4.0Zn-0.2Ca alloy were significantly increased after hot extrusion.

TABLE 6.3

Summary of the Grain Size and Mechanical Properties of Ternary Mg-Ca Alloys

Mg-Ca-X Ternary Alloys	Condition	Mechanical Property					Grain Size (μm)	Ref.
		Elastic Modulus (GPa)	Yield Strength (MPa)	Ultimate Strength (MPa)	Elongation (%)			
Mg-1Ca-5Zn	As-cast	N/A	~65	~86	N/A		N/A	Yin et al. 2013
Mg-2Ca-5Zn	As-cast	N/A	~71	~94	N/A		N/A	
Mg-3Ca-5Zn	As-cast	N/A	~75	~84	N/A		N/A	
Mg-3Ca-2Zn	As-cast	N/A	117 ± 5	145 ± 5	0.57 ± 0.04		N/A	Du et al. 2011a
Mg-2Zn-0.24Ca	As-cast	N/A	N/A	N/A	N/A		97	Gao et al. 2011b
Mg-2Zn-0.24Ca	As-extruded	N/A	N/A	N/A	N/A		5.4	
Mg-2Zn-0.24Ca	High-pressure torsion treated	N/A	N/A	N/A	N/A		1.2	
Mg-1Ca-1Zn	As-cast	43.9	45	125 ± 5	5.7 ± 1.0		N/A	Zhang et al. 2011a
Mg-1Ca-2Zn	As-cast	44.7	52	143 ± 5	7.3 ± 1.5		N/A	
Mg-1Ca-3Zn	As-cast	45.3	57	160 ± 10	8.3 ± 1.0		N/A	
Mg-1Ca-4Zn	As-cast	45.9	63	182 ± 5	9.1 ± 2.5		N/A	
Mg-1Ca-5Zn	As-cast	45	65	173 ± 5	8.2 ± 0.5		N/A	
Mg-1Ca-6Zn	As-cast	45.3	67	145 ± 5	4.5 ± 0.5		N/A	
Mg-(1,2)Ca-(0.5,1)Zr	As-cast	N/A	140–170[a]	180–275[a]	5–7[a]		N/A	Y. L. Zhou et al. 2013
Mg-(1,2)Ca-(0.5,1)Zr	Hot-rolled	N/A	250–300[a]	300–330[a]	7–8.2[a]		N/A	
Mg-4Zn-0.2Ca	As-cast	N/A	60	185	12.5		100–130	Y. Sun et al. 2012
Mg-4Zn-0.2Ca	As-extruded	N/A	240	297	21.3		3–7	

(Continued)

TABLE 6.3 (CONTINUED)

Summary of the Grain Size and Mechanical Properties of Ternary Mg-Ca Alloys

Mg-Ca-X Ternary Alloys	Condition	Mechanical Property					Grain Size (μm)	Ref.
		Elastic Modulus (GPa)	Yield Strength (MPa)	Ultimate Strength (MPa)	Elongation (%)			
Mg-0.5Ca-0.5Sr	As-cast	N/A	274.3 ± 7.2	N/A	N/A		N/A	Berglund et al. 2012
Mg-1Ca-0.5Sr	As-cast	N/A	274.2 ± 4.0	N/A	N/A		N/A	
Mg-1Ca-1Sr	As-cast	N/A	214.5 ± 3.5	N/A	N/A		N/A	
Mg-9.29Li-0.88Ca	As-cast	N/A	74	98	4.7		279	Zeng et al. 2014
Mg-9.29Li-0.88Ca	As-extruded	N/A	111	118	53		59	
Mg-0.3Ca-1.8Zn	As-extruded	43.5	291	329	N/A		N/A	Somekawa and Mukai 2007
Mg-1Ca-3Sn	As-cast	N/A	N/A	N/A	N/A		520	Abu Leil et al. 2009
Mg-1Ca-3Sn	As-extruded	N/A	N/A	N/A	N/A		34	
Mg-2Ca-3Sn	As-cast	N/A	N/A	N/A	N/A		455	
Mg-2Ca-3Sn	As-extruded	N/A	N/A	N/A	N/A		18	
Mg-0.2Ca-0.6Si	As-cast	N/A	50.05 ± 1.12	154.4 ± 5.3	6.62 ± 0.59		N/A	E. L. Zhang et al. 2010
Mg-0.4Ca-0.6Si	As-cast	N/A	56.85 ± 0.99	156.8 ± 4.8	6.22 ± 0.23		N/A	
Mg-1Ca-5Bi	As-cast	N/A	205	240	40		N/A	Remennik et al. 2011

[a] Compressive strength.

6.3.4.2 Mg-Ca-Zr Alloys

Zr was reported to be a very effective alloying element for improving strength in Al-free Mg alloys (Gu et al. 2009b). A low amount of Ca in the Mg-Ca-Zr alloys can improve the strength due to the refinement of grain size and solid solution strengthening. Nevertheless, the addition of Ca above 0.5% results in the formation of a brittle eutectic compound at the grain boundaries and decreases the tensile strength and elongation (Chang et al. 1997). Chang et al. also reported that the tensile strength and yield strength of Mg-1.2Ca alloy increased with Zr content. But the tensile strength and elongation of Mg-0.5Ca alloy slightly decreased because of the brittle eutectic compounds (Chang et al. 1997). Zhou reported that the as-cast Mg-Ca-Zr alloys exhibited low strength and poor ductility because of the coarse microstructure. An increase in Zr content from 0.5% to 1% improves the strength of Mg-Ca-Zr alloys with 1% Ca concentration, but an increase in Ca content decreased the strength of alloys with the same Zr concentration because of the formation of Mg_2Ca phase on the grain boundaries. According to their studies, Mg-1Ca-1Zr alloys exhibited the highest strength and best ductility (Y. L. Zhou et al. 2012).

6.3.4.3 Mg-Ca-Li Alloys

The alloying element Li and Ca in Mg alloys can refine the microstructure and improve the strength by forming the intermetallic compound Mg_2Ca (Haferkamp et al. 2000). A low amount of precipitation in Mg-Li alloys can significantly improve the ductility (Song and Kral 2005). Zeng et al. reported that the yield strength and ultimate strength of Mg-9.29Li-0.88Ca alloy were significantly increased by hot extrusion. And the elongation of the extruded alloy was 12.7 times that of the cast alloys. They concluded that the improvement in mechanical property was ascribed to the decrease in grain size and the increase of Mg_2Ca phase (Zeng et al. 2014).

6.3.4.4 Mg-Ca-Si Alloys

The Mg_2Si phase formed during the solidification by eutectic reaction has a coarse lamellar structure. It breaks the α-Mg matrix under tensile or compressive conditions. Thus, the Mg_2Si phase needs to be modified in order to improve its mechanical properties. E. L. Zhang et al. (2010) reported that the addition of Ca into the Mg-Si alloys not only refines the grain size but also modifies the Mg_2Si phase to a short bar shape. The Ca consumed Si and reduced the fraction of Mg_2Si to form the CaMgSi phase. However, both the CaMgSi and Mg_2Si phases were brittle; thus, there was no improvement in elongation and tensile strength with the addition of Ca from 0.2% to 0.4%.

6.3.4.5 Mg-Ca-Sr Alloys

Small amounts of Ca and Sr can refine the grain size, thereby improving the mechanical properties of Mg (Lee et al. 2000; Brar et al. 2012). The mechanical properties of the Mg-Ca-Sr alloys depend on the amounts and distribution of the secondary phases along the grain boundaries (Berglund et al. 2012). The $Mg_{17}Sr_2$ phase was shown to have low ductility; thus, increasing the Sr content in the Mg alloys may be harmful to the mechanical property of the resulting alloys (Brar et al. 2012). Berglund et al. (2012)

reported that the Mg-0.5Ca-0.5Sr and Mg-1Ca-0.5Sr alloys have similar compressive strength, but with the increase in Sr content, the compressive strength decreases.

6.3.4.6 Mg-Ca-Sn Alloys

A low amount of Sn, less than 5%, can improve the tensile strength and ductility (Hort et al. 2006). And stable intermetallic CaMgSn particles in the Mg alloys can improve creep strength at elevated temperatures (Prasad et al. 2009). A ratio of 3:1 would consume the entire Ca in forming CaMgSn particles while the remaining Sn stays in solid solution (H. M. Liu et al. 2007; Prasad et al. 2009). Yang et al. (2010) fabricated Mg-1Ca-3Sn and Mg-2Ca-3Sn alloys. The former alloy exhibited a relatively higher ultimate tensile strength and elongation at room temperature. They concluded that the coarse, needle-like primary CaMgSn phase in the Mg-2Ca-3Sn alloy had a detrimental effect on the mechanical property. The tensile fracture of the Mg-2Ca-3Sn alloy exhibited larger cleavage-type facets and small cracks. A summary of the grain size and mechanical properties of ternary Mg-Ca alloys are shown in Table 6.3.

6.3.5 Degradation in Simulated Body Fluids

The in vitro degradation tests of ternary Mg-Ca-X alloys included immersion tests, hydrogen evaluation tests, and electrochemical measurement in various simulated body fluids. Table 6.4 summarizes the corrosion behavior of the ternary Mg-Ca-X alloys in different corrosive media by both electrochemical tests and immersion tests. As can be seen from the table, the increasing Ca content in the Mg-Ca-Zn alloys would increase the corrosion rate in the same corrosion medium. The working process, on the contrary, improved the corrosion resistance.

Yin et al. (2013) evaluated the hydrogen evaluation of three kinds of Mg-Ca-Zn alloys with different Ca contents, and the average hydrogen evaluation rate increased with the increasing addition of Ca. The electrochemical test showed similar results. The corrosion potential decreased with the increasing of Ca, and the Mg-5Zn-1Ca alloy exhibited the lowest corrosion current density. In other words, the increasing Ca content may decrease the corrosion resistance of the Mg-Ca-Zn alloy. After being immersed in the SBF for 5 days, a dense layer of corrosion products was observed on the Mg-5Zn-1Ca alloy (Figure 6.14a). Deep pits superimposed on the whole sample surface were observed after removing the corrosion layer, indicating that a substantial localized corrosion occurred during the immersion period (Figure 6.14b). Du et al. (2011a) studied the influence of Zn on the corrosion behavior of the Mg-3Ca alloy by electrochemical test, and found that the addition of Zn can shift the corrosion potential toward the noble direction and significantly increase the corrosion resistance and decrease the corrosion current density. More positive pitting potential of Mg-Ca-Zn alloy also indicated that localized corrosion may be less likely to occur when compared with Mg-3Ca alloy. The morphology of the Mg-3Ca-2Zn alloy immersed in Hank's solution for 8 h is shown in Figure 6.15, and pitting corrosion can still be seen on the sample surface. Chen et al. (2012b) drew a similar conclusion that for Mg-3Zn-xCa alloy, the mass loss rate increased with the rising Ca content, and Zn addition decreased the corrosion rate. Zhang et al. (2011) suggested that less than 3% Zn can enhance the corrosion resistance of the Mg-Zn-1Ca alloys.

TABLE 6.4

Summary of Corrosion Rates of Ternary Mg-Ca Alloys in Different Corrosive Media

Sample	Condition	SBF	Kokubo Solution	Hank's Solution	5% NaCl	I_{corr}	Ref.
		Corrosion Rates in Different Corrosive Media					
Mg-1Ca-1Y	As-cast	27.4 mg/day	N/A	N/A	N/A	N/A	Li et al. 2010a
Mg-1Ca-5Zn	As-cast	0.28 mm/y	N/A	N/A	N/A	12.16 μA/cm²	Yin et al. 2013
Mg-2Ca-5Zn	As-cast	0.34 mm/y	N/A	N/A	N/A	14.85 μA/cm²	
Mg-3Ca-5Zn	As-cast	0.44 mm/y	N/A	N/A	N/A	19.39 μA/cm²	
Mg-3Ca-2Zn	As-cast	N/A	N/A	N/A	N/A	3.86 μA/cm²	Du et al. 2011a
Mg-1Ca-1Zn	As-cast	N/A	N/A	2.13 ± 0.2 mm/y	N/A	N/A	Zhang et al. 2011a
Mg-1Ca-2Zn	As-cast	N/A	N/A	2.38 ± 0.3 mm/y	N/A	N/A	
Mg-1Ca-3Zn	As-cast	N/A	N/A	2.92 ± 0.5 mm/y	N/A	N/A	
Mg-1Ca-4Zn	As-cast	N/A	N/A	4.42 ± 1.0 mm/y	N/A	N/A	
Mg-1Ca-5Zn	As-cast	N/A	N/A	6.15 ± 1.5 mm/y	N/A	N/A	
Mg-1Ca-6Zn	As-cast	N/A	N/A	9.21 ± 1.5 mm/y	N/A	N/A	
Mg-(1,2)Ca-(0.5,1)Zr	As-cast	6–15 mg/cm²/day	N/A	N/A	N/A	N/A	Y. L. Zhou et al. 2013
Mg-(1,2)Ca-(0.5,1)Zr	Hot-rolled	<3 mg/cm²/day	N/A	N/A	N/A	N/A	
Mg-4Zn-0.2Ca	As-cast	2.05 mm/yr	N/A	N/A	N/A	2.67×10^{-4} A/cm²	Y. Sun et al. 2012
Mg-4Zn-0.2Ca	As-extruded	1.98 mm/yr	N/A	N/A	N/A	2.43×10^{-4} A/cm²	

(Continued)

TABLE 6.4 (CONTINUED)

Summary of Corrosion Rates of Ternary Mg-Ca Alloys in Different Corrosive Media

Sample	Condition	Corrosion Rates in Different Corrosive Media					I_{corr}	Ref.
		SBF	Kokubo Solution	Hank's Solution	5% NaCl			
Mg-0.5Ca-(0.5–9)Zn	As-cast	N/A	N/A	N/A	N/A		178–465 μA/cm^2	Bakhsheshi-Rad et al. 2012a
Mg-9.29Li-0.88Ca	As-cast	N/A	N/A	1.268 mg/cm^2/day	N/A		8.36×10^{-5} A/cm^2	Zeng et al. 2014
Mg-9.29Li-0.88Ca	As-extruded	N/A	N/A	0.248 mg/cm^2/day	N/A		6.74×10^{-5} A/cm^2	
Mg-1Ca-3Sn	As-cast	N/A	N/A	N/A	2.34 ± 0.21 mm		N/A	Abu Leil et al. 2009
Mg-1Ca-3Sn	As-extruded	N/A	N/A	N/A	1.91 ± 0.40 mm		N/A	
Mg-2Ca-3Sn	As-cast	N/A	N/A	N/A	5.99 ± 0.44 mm		N/A	
Mg-2Ca-3Sn	As-extruded	N/A	N/A	N/A	2.28 ± 0.16 mm		N/A	
Mg-0.2Ca-0.6Si	As-cast	N/A	N/A	0.39 mg/cm^2/day	N/A		36.3 μA/cm^2	E. L. Zhang et al. 2010
Mg-0.4Ca-0.6Si	As-cast	N/A	N/A	0.15 mg/cm^2/day	N/A		14.3 μA/cm^2	

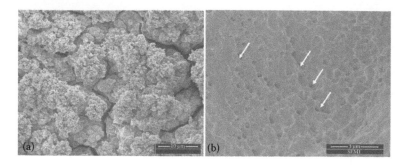

FIGURE 6.14 (a) As-cast Mg-1Ca-5Zn alloy immersed in SBF for 5 days, (b) after removal of corrosion products of as-cast Mg-1Ca-5Zn alloy immersed in SBF for 5 days, arrows indicate the corrosion hole on the alloy surface. (Modified from Yin, P., Li, N. F., Lei, T., Liu, L., and Ouyang, C., *Journal of Materials Science-Materials in Medicine*, 24, 1365, 2013. With permission.)

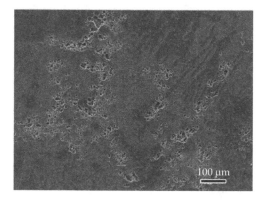

FIGURE 6.15 Morphology of the as-cast Mg-3Ca-2Zn alloy immersed in Hank's solution for 1 day. (Modified from Du, H., Wei, Z. J., Liu, X. W., and Zhang, E. L., *Materials Chemistry and Physics*, 125, 568, 2011. With permission.)

Li et al. (2010a) reported that the addition of 1% Y accelerated the corrosion rate of Mg-1Ca-Y alloy both in MMEM and SBF corrosive media. Y. L. Zhou et al. (2013) evaluated the influence of Zr in Mg-Ca-Zr alloy. Increasing Ca content from 1% to 5% may result in a higher corrosion rate, but the entire Mg-Ca-Zr alloys exhibited lower corrosion rates than pure Mg, and Mg-1Ca-1Zr alloy exhibited the best corrosion resistance property.

The degradation rate of Mg-Ca-Sr alloys was evaluated by the hydrogen evolution method (Berglund et al. 2012). The alloys with higher Sr and Ca content (Mg-1Ca-2Sr, Mg-7Ca-3.5Sr) suffered a very rapid corrosion procedure and totally dissolved within 24 h. The alloy with 1% Ca and 0.5% Sr exhibited the lowest degradation rate.

Zeng et al. (2014) studied the mass loss, hydrogen evolution, pH value changes, and electrochemical property of the Mg-9.29%Li-0.88%Ca alloy. The as-extruded alloy showed a higher corrosion potential and lower corrosion current density, which was in accordance with the mass loss results. The hydrogen evolution procedure consists of two distinct stages. A dense corrosion products layer was rapidly formed on the as-cast alloy, which led to the decrease of hydrogen evolution rate. In the subsequent immersion period, the corrosion protective layer broke down the exposed fresh surface to the medium. A summary of the corrosion rate of ternary Mg-Ca alloys in different corrosive media is shown in Table 6.5.

6.3.6 CYTOTOXICITY AND HEMOCOMPATIBILITY

The content of alloying elements, the corrosion behavior, and even the mechanical property may have an influence on the cytotoxicity and hemocompatibility of the ternary Mg-Ca-X alloys.

The hemolysis rate of the Mg-5Zn-1Ca alloy was 4.07%, which was below the tolerance level according to the ISO 10993-4 standard, implying that the degradation in vitro of the alloy had no destructive effect on erythrocytes. The L-929 cells exhibited more than 80% cell viability in 100% concentrate extracts even after 7 days of incubation. The cell toxicity scale is 0–1 according to ISO 10993-5:1999, indicating that Mg-5Zn-1Ca alloy does not induce toxicity in L-929 cells and is suitable for biomedical applications (Yin et al. 2013).

For the Mg-4Zn-0.2Ca alloy, MC3T3-E1 cells exhibited nearly 100% viability in 100% concentration extract after 7 days of culture when compared with the control group, indicating that the alloy was safe (Xia et al. 2012).

Zhang et al. (2011) reported that the pH values of the extract for Mg-1Zn-1Ca, Mg-2Zn-1Ca, and Mg-3Zn-1Ca were below 8.0. All the extracts exhibited a significantly high viability, and the toxicities of the extracts were Grade 0–1.

The cytotoxicity of other rare earth alloying elements were also investigated (Y. C. Li et al. 2011). When compared with Mg-Ca binary alloys, the Mg-1Ca-1Y alloy exhibited comparable cell viability, which was 89.1%. Furthermore, the amount of live and dead cells was similar to the control group. The cells cultured in Mg-1Ca-1Y extracts grew and spread well according to the optical microscopy, implying that the Y addition in the Mg-Ca alloy did not deteriorate the biocompatibility.

6.3.7 IN VIVO PERFORMANCE IN ANIMAL MODELS

Cho et al. (2013) studied the in vivo performance of Mg-Zn-Ca alloy in New Zealand white rabbits. During the whole implantation period, no or very little presence of polymorphonuclear cells, lymphocytes, and plasma cells were found adjacent to the Mg-Ca-Zn screws. After 52 weeks of implantation, the mean score of macrophage and the multinucleated giant cell inflammatory score were slightly lower around the Mg-Ca-Zn screw than the SR-PLLA screw. The typical histological microscopy is shown in Figure 6.16e. All the elemental concentration was shown to decrease over time except for Mg content at 26 weeks in soft tissue. After 8 weeks, the implanted screw retained its 70.33% strength and totally degraded after 52 weeks of implantation. The morphology of the screws is shown in Figure 6.16a–d.

Xia et al. (2012) evaluated the in vivo biocompatibility of a Mg-4Zn-2Ca alloy in adult New Zealand rabbits. After 3 months of implantation, the Mg alloy implant

TABLE 6.5

Summary of the Grain Size and Mechanical Properties of Quaternary Mg-Ca Alloys

Sample	Condition	Mechanical Property					Grain Size (μm)	Ref.
		Elastic Modulus (GPA)	Yield Strength (MPa)	Ultimate Strength (MPa)	Elongation (%)	Vickers Hardness (HV)		
Mg-1Y-0.6Ca-0.4Zr	As-cast	~60	~60	~125	~3	N/A	79	Chou et al. 2013
Mg-4Y-0.6Ca-0.4Zr	As-cast	~50	~80	~160	~5.4	N/A	98	
Mg-2Ca-0.5Mn-2Zn	As-cast	N/A	78.3	168.5	7.84	64.5	78	Bakhsheshi-Rad et al. 2014b
Mg-2Ca-0.5Mn-4Zn	As-cast	N/A	83.1	189.2	8.71	69.1	59	
Mg-2Ca-0.5Mn-7Zn	As-cast	N/A	45.4	140.7	4.15	82.2	N/A	
Mg-0.3Ca-1.8Zn-1.1Mn	As-cast	N/A	~60	~160	~8	N/A	175 ± 15	Zhang and Yang 2008
Mg-0.5Ca-2.0Zn-1.2Mn	As-cast	N/A	~70	~190	~9	N/A	63 ± 7	
Mg-1Ca-1.5Zn-1.1Mn	As-cast	N/A	~80	~140	~2.5	N/A	51 ± 5	
Mg-5.3Zn-0.2Ca-0.5Ce	As-cast	N/A	77	168	~6	N/A	N/A	Du et al. 2013

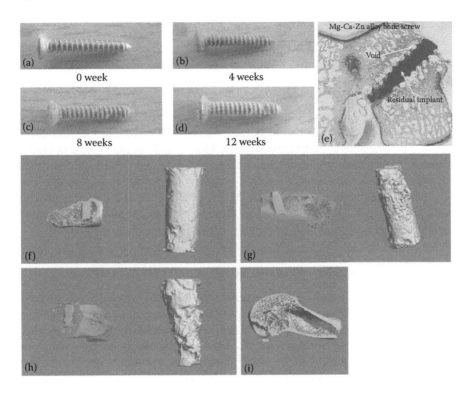

FIGURE 6.16 (a–d) Morphology of Mg-Ca-Zn bone screws following immersion in Hank's solution, (e) histology image of Mg-Ca-Zn screws implanted for 12 weeks, two-dimensional and 3-D images of the MAO-coated Mg-2Zn-0.2Ca at (f) 8 weeks, (g) 12 weeks, (h) 18 weeks, and (i) 50 weeks, postoperatively. ([a–e] Modified from Cho, S. Y., Chae, S. W., Choi, K. W., Seok, H. K., Kim, Y. C., Jung, J. Y., Yang, S. J., Kwon, G. J., Kim, J. T., and Assad, M., *Journal of Biomedical Materials Research Part B-Applied Biomaterials*, 101B, 201, 2013. With permission; [f–i] Modified from Chen, S., Guan, S., Li, W., Wang, H., Chen, J., Wang, Y., and Wang, H., *Journal of Biomedical Materials Research Part B-Applied Biomaterials*, 100, 533, 2012. With permission.)

corroded to an irregular shape, and a degradation layer, newly formed bone, can be clearly seen. The lymphocytes were observed 1 month after operation. A large number of disorganized trabeculars was observed after 2 months of implantation, implying the formation of an active bone. New bone tissue was formed around the Mg implant after 3 months.

S. Chen et al. (2012) investigated the in vivo degradation and bone response of a MAO and electrochemical deposition coated Mg-2Zn-0.2Ca alloy. The newly formed bone tissue was observed around the coated sample, and no obvious degradation occurred at 8 weeks. After 12 weeks, more bone cells were found between the trabeculars. At 18 weeks, the bone trabeculars align compactly and osteoid tissue can be seen. The micro-CT images of the implants are shown in Figure 6.16f–i.

6.4   DEVELOPMENT OF Mg-Ca QUATERNARY
AND MULTIELEMENTARY ALLOYS
AS DEGRADABLE METALLIC BIOMATERIALS

6.4.1   Alloy Composition Design

In addition to the binary and ternary Mg-Ca alloys, quaternary and multielementary Mg-Ca alloys were also studied as biodegradable metals. For the consideration of bio-compatibility, the alloying elements should be biosafe and not cause acute or chronic problems. Furthermore, the mechanical properties, biocompatibility, and corrosion resistance properties of the alloys should benefit from the alloying elements. Until now, the essential elements, such as Zn, Mn, Sn, Sr, and Si, and the rare earth elements, such as Zr, Y, and the noble metal element Pd, were reported to fabricate the degradable quaternary and multielementary alloys for biomedical applications.

6.4.2   Fabrication, Hot and Cold Working, and Heat Treatment

A conventional metallurgy method was utilized to fabricate the quaternary and multielementary alloys. Chou et al. (2013) reported a two-step method to fabricate the Mg-Y-Ca-Zr alloy. High purity of Mg, Ca, and Y were melted in an induction furnace under an atmosphere of ultrapure Ar gas. Subsequently, the impurities and oxides were thoroughly removed. And the Mg-Zr master alloy was added into the remelted metal at the temperature of 780°C. During the remelting process, the melts were stirred for 10 s at intervals of 1–5 min. The as-cast alloys were then solution treated at 525°C for about 6 h in a tubular furnace, which was covered with UHP Ar gas.

Other multielementary alloys, such as AZ31-Ca (J. H. Kim et al. 2009), Mg-Ca-Zr-Sn-Sr (W. J. Zhang et al. 2012), Mg-Ca-Mn-Zn (Bakhsheshi-Rad et al. 2014b), Mg-Zn-Ca-Ce (Du et al. 2013), Mg-Zn-Ca-(Zr, Co) (Park et al. 2002), Mg-Ca-Si-Zr (Ai et al. 2007), Mg-Ca-Zn-Si (E. L. Zhang et al. 2010), and Mg-Ca-Sn-Si (Kozlov et al. 2011) alloys, were fabricated by the same metallurgy methods under an atmosphere of argon gas or a mixture of CO_2 and SF_6, and the melting temperature ranged from 650°C to 810°C, depending on the alloying element and the alloy composition.

For solid solution treatment, Mg-Ca-Zn-Zr (Shepelev et al. 2009) alloys were heated for 96 h at 300°C, followed by slow heating and ended with holding for an additional 10 h or 96 h, followed by rapid quenching in water. After the solid solution treatment, the samples were then aged at 175°C for up to 24 h. Mg-Ca-Y-Zr alloy was solution treated at 525°C for 6 h (Chou et al. 2013). Park et al. (2002) reported that the Mg-5Zn-4Ca-0.5Zr exhibited the highest hardness value after aging at 400°C for 1 h. AZ31-Ca alloys were extruded at the temperature of 380°C with a ram speed of 33 mm/s and a ratio of 32:1. The extrusion ingots were finally annealed at 300°C for 24 h.

6.4.3   Microstructure and Mechanical Properties

The addition of Ca in the AZ31 alloy leads to a homogeneous microstructure and a finer grain size when compared with the AZ31 alloy. Due to the limited solubility of Ca in the alpha-Mg matrix, the nucleation sites may be saturated and the constitutional undercooling effect maximized after adding 0.7% Ca. Kim et al. (2009) suggested 0.7% Ca would be sufficient for grain refinement. The Sr and Sn addition into the Mg-Zr-Ca alloys led to an increase of grain size. The grain boundaries were thinner, and some were fuzzy. New blob-shaped phase Mg_2Sn was formed within the grain interior (W. J. Zhang et al. 2012). The Y and Zr can greatly decrease the grain size of the Mg-Ca alloys. The alloying elements tended to precipitate along the grain boundary (Chou et al. 2013).

The addition of Mn and Zn into Mg-Ca alloys led to the formation of a diffusion layer ahead of the solid–liquid interface and resulted in the rejection of alloying elements at the front of the grain growth, which limits the growth stage. Thus, the grain size decreased (Fu and Yang 2011; Bakhsheshi-Rad et al. 2014b). The Mg-Ca-Mn-Zn alloys contain alpha-Mg + $Ca_2Mg_6Zn_3$ + Mg_2Ca lamellar eutectic along the grain boundary when the Zn:Ca ratio is less than 1.25. With the Zn:Ca ratio above 1.25, the Mg_2Ca phase was replaced by the $Mg_{12}Zn_{13}$ phase (Bakhsheshi-Rad et al. 2014b). A typical microstructure of the alloys mentioned is shown in Figure 6.17.

A summary of the yield strength, ultimate strength, elongation, elastic modulus, hardness, and grain size of the ternary Mg-Ca-based quaternary or multielementary

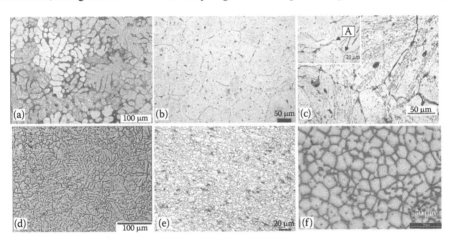

FIGURE 6.17 Microstructure of (a) as-cst AZ31 + 2.0Ca, (b) as-cast Mg-4Y-0.6Ca-0.4Zr, (c) as-cast Mg-1Ca-5Zr-8(Sn + Sr), (d) as-cast Mg-0.5Ca-0.5Mn-2Zn, (e) AZ31 + 2.0Ca compressed at 400°C, and (f) as-cast Mg-1.49Ca-4.2Zn-1.06Zr. ([a] Modified from Kim, J. H., Kang, N. E., Yim, C. D., and Kim, B. K., *Materials Science and Engineering a-Structural Materials Properties Microstructure and Processing*, 525, 18, 2009. With permission; [b] Modified from Chou, D. T., Hong, D., Saha, P., Ferrero, J., Lee, B., Tan, Z. Q., Dong, Z. Y., and Kumta, P. N., *Acta Biomaterialia*, 9, 8518, 2013. With permission; [c] Modified from Zhang, W. J., Li, M. H., Chen, Q., Hu, W. Y., Zhang, W. M., and Xin, W., *Materials and Design*, 39, 379, 2012. With permission; [d] Modified from Bakhsheshi-Rad, H. R., Idris, M. H., Abdul-Kadir, M. R., Ourdjini, A., Medraj, M., Daroonparvar, M., and Hamzah, E., *Materials and Design*, 53, 283, 2014b. With permission; [e] Modified from Kim, J. H., Kang, N. E., Yim, C. D., and Kim, B. K., *Materials Science and Engineering a-Structural Materials Properties Microstructure and Processing*, 525, 18, 2009. With permission; [f] Modified from Shepelev, D., Bamberger, M., and Katsman, A., *Journal of Materials Science*, 44, 5627, 2009. With permission.)

alloys is displayed in Table 6.5. From the table, it can be found that the selection of Y, Zr, Mn, and Zn alloying elements significantly increased the mechanical property when compared with the binary and ternary Mg-Ca-based alloys at the same state.

Regarding the mechanical properties of Mg-1Y-0.6Ca-0.4Zr and Mg-4Y-0.6Ca-0.4Zr alloys, an increase in compressive yield strength was observed with increasing Y content, and the Mg-4Y-0.6Ca-0.4Zr alloy exhibited higher ultimate tensile strength than the Mg-1Y-0.6Ca-0.4Zr alloy (Chou et al. 2013). The T4 solution treatment to the two alloys resulted in dissolution of second phase precipitates from the grain boundaries as well as grain coarsening; thus a reduction in both compressive strength and strain was observed. The higher Y content in the alloys resulted in a higher presence of secondary phases in the Mg-Y-Ca-Zr alloys. This may contribute to precipitation strengthening by acting as impediments for dislocation during plastic deformation and thus improving the tensile strength.

For the quaternary Mg-Ca-Zn-Mn alloy, the grain size decreased with increasing Ca content, which may contribute to the increasing of yield strength based on the Hall-Petch law. Nevertheless, the fraction of second phase also increased with the increasing Ca content. When further increasing Ca content from 0.5% to 1%, the lamellar Mg_2Ca phase was observed, which was harmful to the ductility of the Mg-Ca alloy (Chino et al. 2002). Zhang and Yang (2008) reported that the yield strength slightly increased with the increasing Ca concentration in the Mg-Ca-Mn-Zn alloys, and the ultimate tensile strength and the elongation increased when the Ca content increased from 0.3% to 0.5% and then decreased sharply with increasing Ca content to 1%.

Du et al. (2013) developed the Mg-5.3Zn-0.2Ca-0.5Ce alloy and found that the ultimate tensile strength decreased with the increasing temperature of the alloy, and it may result from the activation of the nonbasal slip systems. The stable $Ca_2Mg_6Zn_3$ and T1 phase (Mg-Zn-Ce) with a high melt point and thermal stability, which distributed at the grain boundaries, effectively restricted the grain boundary sliding, leading to the slow decrease of strength. A summary of the grain size and mechanical property of quaternary Mg-Ca alloys is shown in Table 6.5.

6.4.4 DEGRADATION IN SIMULATED BODY FLUIDS

The alloying elements can tailor the corrosion behavior of the Mg-Ca based alloys in corrosive medium. In Table 6.6, the corrosion rate in different corrosive media of the Mg-Ca–based quaternary alloys can be seen. It suggests that lower Y contents and Zn contents may improve the corrosion resistance of the quaternary Mg-Ca-based alloys.

The formation of Mg_2Sn and Sr phases due to the addition of Sr and Sn into the Mg-Ca-Zr alloys protected the Mg matrix from corrosion and deferred the corrosion rate (W. J. Zhang et al. 2012). Thus, the Mg-Ca-Zr-Sn-Sr alloys exhibited higher open circuit potential and lower corrosion current density. The mass loss of the Mg-Ca-Zr-Sn-Sr alloy was stable with the increasing immersion time in SBF.

The alloying element Y also plays an important role in the corrosion resistance of the Mg-Ca-Zr-Y alloys by improving the corrosion potential. With the Y content increased from 1% to 4%, the corrosion rate decreased according to the electrochemical tests. The immersion results also indicated a similar result, the alloy with higher Y content exhibited a lower corrosion rate and lower Mg ion release (Chou et al. 2013).

TABLE 6.6
Summary of Corrosion Rates of Quaternary Mg-Ca Alloys in Different Corrosive Media

| Sample | Condition | Corrosion Rates in Different Corrosive Media | | I_{corr} | Ref. |
		Kokubo Solution	DMEM + 10% FBS		
Mg-1Y-0.6Ca-0.4Zr	As-cast	N/A	0.34 mm/yr	N/A	Chou et al. 2013
Mg-4Y-0.6Ca-0.4Zr	As-cast	N/A	0.11 mm/yr	N/A	
Mg-2Ca-0.5Mn-2Zn	As-cast	1.78 mm/yr	N/A	78.3 µA/cm²	Bakhsheshi-Rad
Mg-2Ca-0.5Mn-4Zn	As-cast	2.27 mm/yr	N/A	99.6 µA/cm²	et al. 2014b
Mg-2Ca-0.5Mn-7Zn	As-cast	3.98 mm/yr	N/A	174.1 µA/cm²	

The quaternary Mg-2Ca-0.5Mn-2Zn showed a much higher corrosion potential, and the corrosion current density was two orders of magnitude lower than Mg-Ca binary alloys (Bakhsheshi-Rad et al. 2014b). The plateau area in the polarization curves indicated the formation of a passive film on the surface of the Mg-Ca-Mn-Zn alloy when immersed in the SBF. The increased break potential also revealed that localized corrosion was less likely to occur. But increasing the Zn content from 2% to 7% may decrease the corrosion resistance. The morphology of the Mg-0.5Ca-0.5Mn-2Zn alloys after being immersed in Kokubo solution for 6 days is illustrated in Figure 6.18. It exhibited a typical morphology with cracks, and corrosion products can be seen on the immersion sample surface.

Mg-0.5Ca-0.5Mn-2Zn   40 µm

FIGURE 6.18 Morphology of Mg-0.5Ca-0.5Mn-2Zn after immersion in Kokubo solution for 6 days. (Modified from Bakhsheshi-Rad, H. R., Idris, M. H., Abdul-Kadir, M. R., Ourdjini, A., Medraj, M., Daroonparvar, M., and Hamzah, E., *Materials and Design*, 53, 283, 2014a. With permission.)

Zhang evaluated the influence of Ca content on the biocorrosion properties in the Mg-Zn-Mn-Ca alloys. The corrosion resistance increased when the Ca content increased from 0.3% to 1%. However, no substantial change was observed in the corrosion potential with the increasing of Ca content from 0.3 wt.% to 0.5 wt.%, and the corrosion resistance increased, and the corrosion current density decreased slightly. When the Ca content increased from 0.5 wt.% to 1.0 wt.%, the corrosion resistance increased significantly, and the corrosion current density decreased significantly (Zhang and Yang 2008).

A summary of corrosion rates of quaternary Mg-Ca alloys in different corrosives is shown in Table 6.6.

6.4.5 CYTOTOXICITY AND HEMOCOMPATIBILITY

The MC3T3-E1 cells were used to evaluate the in vitro cytotoxicity of Mg-Y-Ca-Zr alloys (Chou et al. 2013). The cell viability reduced rapidly with 100% extract medium mainly because the high ion concentration led to osmotic shock, and no cytotoxicity was observed for 25% and 10% diluted extract media. The Mg-Y-Ca-Zr alloy with a small amount of Y exhibited significantly high cell viability at 50% extract concentration. After 1 day of culture with extract medium, the Mg-Y-Ca-Zr alloys showed significantly higher cell viability compared with pure Mg at 25% extract concentration. The MC3T3-E1 cell attachment after being directly cultured on the Mg-1Y-0.6Ca-0.4Zr and Mg-4Y-0.6Ca-0.4Zr alloys with live and dead staining is shown in Figure 6.19. From the fluorescent images, it can be concluded that the

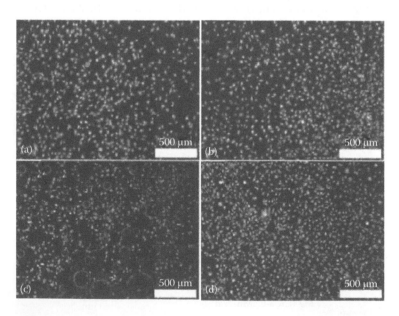

FIGURE 6.19 Fluorescent images of live and dead MC3T3-E1 cells attached after 1 day: (a) Mg-1Y-0.6Ca-0.4Zr, (c) Mg-4Y-0.6Ca-0.4Zr and 3 days: (b) Mg-1Y-0.6Ca-0.4Zr, (d) Mg-4Y-0.6Ca-0.4Zr cultured on sample surface. (Modified from Chou, D. T., Hong, D., Saha, P., Ferrero, J., Lee, B., Tan, Z. Q., Dong, Z. Y., and Kumta, P. N., *Acta Biomaterialia*, 9, 8518, 2013. With permission.)

7 days   40 days   70 days

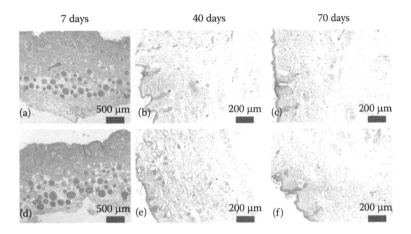

FIGURE 6.20 Histology images (H&E staining) of the skin above the implants: (a–c) Mg-1Y-0.6Ca-0.4Zr, (d–f) Mg-4Y-0.6Ca-0.4Zr. (Modified from Chou, D. T., Hong, D., Saha, P., Ferrero, J., Lee, B., Tan, Z. Q., Dong, Z. Y., and Kumta, P. N., *Acta Biomaterialia*, 9, 8518, 2013. With permission.)

Mg-Y-Ca-Zr with lower Zr contents exhibited higher cell attachment on the first day. All the samples exhibited good cell attachment after 3 days of incubation.

6.4.6 IN VIVO PERFORMANCE IN ANIMAL MODELS

Healthy nude mice were selected as model animals to access the in vivo biocompatibility of the Mg-Y-Ca-Zr alloys (Chou et al. 2013). A plate with a diameter of 5 mm and thickness of 1.4 mm was embedded in a subcutaneous pocket on the back of the mouse. The H&E staining results indicated that the surrounding tissue appeared to be undergoing normal tissue repair. After 7 days of implantation, no significant accumulation of inflammatory cells was observed, and collagen fibers and reactive fibroblasts were seen. A high density of fibroblasts adjacent to the as-cast Mg-Y-Ca-Zr alloys indicated that their implantation did not inhibit the normal healing response in the implantation site (Figure 6.20a,d). After 40 and 70 days of implantation, dense collagen connective tissue and normal adiposities can be seen around the implants without the presence of a high density of chronic inflammatory cells. Furthermore, normal adipocytes could be faintly observed past the dermis (Figure 6.20b,c,e,f).

6.5 CONCLUDING REMARKS

Due to the good biocompatibility and biodegradability of Mg and Ca, the Mg-Ca alloys have been widely studied during the past decades. The degradation products Mg^{2+}, Ca^{2+}, and calcium phosphate of Mg-Ca alloy can promote the formation of new bone according to the in vivo experiments. Nevertheless, the fast corrosion rate and pit corrosion in vitro and in vivo may impede its application in clinic. Thus, further investigation may mainly focus on the following aspects.

6.5.1 ALLOY COMPOSITION DESIGN

Because the Mg-Ca-based alloys possess good biocompatibility and comparable mechanical property with natural cortical bones, further development of novel Mg-Ca-based alloys should pay special attention to the selection of the alloying elements. Biofunctional elements, such as Sr; biocompatible elements, such as Zr; essential trace elements, such as Mn and Zn, would be good choices for the development of novel Mg-Ca-based alloys, but the release of alloying elements in the degradation periods should be carefully controlled because an overdose of the alloying elements would be harmful to the metabolism of the human body. With the development of computing science, computer-aided simulation can also be applied to predict the mechanical and corrosive behavior of the novel alloys before the in vitro and in vivo evaluations.

6.5.2 ENHANCING MECHANICAL PROPERTIES

As we can see from Tables 6.2 through 6.4, further working processing, such as hot extrusion and hot rolling can obviously improve the mechanical property of the alloys. Somekawa et al. reported that the as-extruded Mg-0.3Ca-1.8Zn alloy after proper solution treatment exhibited higher strength and fracture toughness when compared with conventional wrought Mg alloys due to the control of the microstructure and dispersion of precipitates (Somekawa and Mukai 2007). The ultimate tensile strength of the extruded Mg-4Zn-0.2Ca alloy exhibited 60.54% higher than the as-cast alloy (Y. Sun et al. 2012). Harandi et al. (2011) reported that the forging process could increase the hardness of the Mg-1Ca alloy. Other severe deformation processing, such as equal channel angular pressing (ECAP) and high-pressure torsion (HPT) also play an important role in improving the mechanical behavior of the Mg-Ca-based alloys. Hradilová reported that the ECAP process could significantly increase the hardness as well as the compressive yield strength of the Mg-4Zn-0.4Ca alloy (Hradilova et al. 2013). Other researchers reported similar results to those of Hradilová (Tong et al. 2010).

6.5.3 ENHANCING CORROSION RESISTANCE

The fast corrosion rate of the Mg and Mg alloys in vivo hindered its wild application in the clinic. Thus, further improving the corrosion resistance property is still needed. Coating seems to be a very effective way to improve the corrosion resistance. B. Wang et al. (2012), Gu et al. (2011a), S. Chen et al. (2012), W. Li et al. (2011), and Jia et al. (2014) fabricated the MAO coating on the Mg alloy, and the immersion tests, electrochemical tests, and hydrogen evolution tests all showed a lower corrosion rate in the early stages, indicating the MAO coating plays an important role in the corrosion resistance of Mg alloys. Drynda et al. (2010) formed a homogeneous corrodible fluoride coating and improved the degradation kinetics. Bakhsheshi-Rad et al. (2013b) also formed the MgF_2 film and suggested that a higher HF concentration exhibited higher corrosion resistance. C. Y. Zhang et al. (2010), Chen et al. (2012b), and Zeng et al. (2013b) developed a calcium phosphate coating on the Mg-1Ca alloy and AZ31 alloy, and a higher corrosion potential and lower corrosion current density of the coated samples were observed by electrochemical tests. Other coating methods, such as alkaline heat treatment (Gu et al. 2009a), chitosan coating (Gu et al. 2009c), titanium coating (M. H. Li et al. 2011), and pulsed laser deposition (Chung et al. 2011) were also reported. Except for the surface coating methods, mechanical processing was also a useful tool to improve the corrosion resistance of Mg and Mg alloys. Salahshoor and Guo (2011d) reported that the burnishing process provided a smoother surface than the as-machined surface, a higher subsurface hardness and residual compressive stresses, which will all improve the corrosion behavior of the alloys. Sequential laser peening (Sealy and Guo 2011), high-speed milling (Guo and Salahshoor 2010), and laser shock peening (Sealy and Guo 2010) processing methods were also reported to improving the surface integrity. Mao et al. (2010) reported the Ag-ion implantation in Mg-Ca alloys to improve the corrosion resistance.

6.5.4 Enhancing Biocompatibility

Further improving the biocompatibility of the Mg-Ca-based alloys can broaden their usage in clinical applications. Surface coating with bioactive films can improve the cell interactions with the implants, which thus may benefit the cell attachment and cell migration, which were essential for the initial healing procedure of the tissue. The selection of bioactive compounds to fabricate composites also may be helpful to improving the biocompatibility. Gu developed a MgCa-HA/TCp metal-ceramic composite with liquid metal infiltration. The 50% and 10% extraction exhibited Grade I toxicity to L-929 and MG63 cells. They concluded that the dissolving Mg ion would accelerate the bone restoration process, and the remaining HA/TCP scaffold structure would exhibit good osteoconductivity and permit the invasion of cells (Gu et al. 2011d).

7 Mg-Zn-Based Alloy Systems for Biomedical Applications

7.1 INTRODUCTION

7.1.1 ZINC

Zinc, with a density of 7.14 g/cm^3, is one of the most abundant essential metal elements in the human body, and it is also an essential component of more than 300 various enzymes (Brar et al. 2012). For biological consideration, it is critical for cellular metabolism and supports protein, DNA synthesis, and the sense of taste and smell (Zhang et al. 2010b; Peng et al. 2012). For adults, the recommended dietary allowance of Zn is about 10–15 mg/day, and higher amounts than these values are often regarded as relativity nontoxic, and amounts near 100 mg/day can be tolerated for some time (Vojtech et al. 2011; Kubasek and Vojtech 2013b). It has been reported that the alloying element Zn for Mg alloys can enhance the ductility and deformability (X. B. Zhang et al. 2012e). Furthermore, introducing Zn into Mg alloys can increase the tensile strength and the hardness through a solid solution hardening mechanism and aging strengthen effect (Jiang et al. 2013). The element Zn can also decrease impurities, such as iron and nickel, in Mg alloys, thus enhancing the corrosion resistance properties (E. L. Zhang et al. 2009b; Vojtech et al. 2011; Kubasek and Vojtech 2013b).

7.1.2 PREVIOUS STUDIES ON MG-ZN ALLOY SYSTEMS FOR ENGINEERING APPLICATIONS

Zn is one of the most common alloying elements of Mg, but it seldom serves as the major alloying element. In industrial and engineering applications, it is usually used in conjunction with elemental aluminum (Al) or with elemental zirconium (Zr) or rare earth elements (RE). Commercial Mg-Zn alloys, such as Mg-Zn-Al alloys, Mg-Zn-Mn alloys, and Mg-Zn-RE alloys, are widely used in the automotive industry (Kulekci 2008) and aerospace industry due to their low density and good mechanical properties in an elevated temperature atmosphere. Conventional commercial Mg-Zn alloys are listed in Table 7.1.

TABLE 7.1

Normal Composition of Commercial Mg-Zn System Alloys

Designation (ASTM)	Nominal Composition (wt.%)							References
	Al	Zn	Mn	Si	Zr	Rare Earth	Y	
AZ91	9	1	0.5	N/A	N/A	N/A	N/A	Zafari et al. 2014
AZ31	2.5–3.5	0.6–1.4	0.20–1.0	N/A	N/A	N/A	N/A	Olguín-González et al. 2014
AZ61	5.8–7.2	0.40–1.5	0.15–0.50	N/A	N/A	N/A	N/A	Olguín-González et al. 2014
AM50	5	N/A	0.3	N/A	N/A	N/A	N/A	Ding et al. 2010
AM20	2	0.2	0.5	N/A	N/A	N/A	N/A	Yi et al. 2004
AS21	2	N/A	0.5	0.1	N/A	N/A	N/A	Yi et al. 2004
ZK60	N/A	6	N/A	N/A	0.5	N/A	N/A	Lin et al. 2013c
WE43	N/A	N/A	N/A	N/A	0.5	3.25	4	Polmear 1994

7.2 DEVELOPMENT OF Mg-Zn BINARY ALLOYS AS DEGRADABLE METALLIC BIOMATERIALS

7.2.1 Alloy Composition Design

The binary phase diagram is illustrated in Figure 7.1 (Wei et al. 1995). According to the diagram, it can be seen that the maximum solid solubility of Zn in Mg is about 6.2 wt.% at the eutectic temperature of 341°C (Cai et al. 2012). During the past decades, three groups of Mg-Zn alloys for different applications were reported (Kubasek et al. 2012). The first group includes alloys containing less than 6% Zn, which is below the solubility limit of Zn in Mg. Lots of binary Mg-Zn alloys with different Zn contents, such as Mg-0.5Zn (Peng et al. 2012), Mg-1Zn (Cai et al. 2012; Kubasek et al. 2012; Peng et al. 2012; Kubasek and Vojtech 2013b), Mg-1.5Zn (Peng et al. 2012), Mg-2Zn (Peng et al. 2012), Mg-3Zn (Kubasek and Vojtech 2013b), Mg-4Zn (Kubasek et al. 2012; Wei et al. 2014), Mg-5Zn (Cai et al. 2012), Mg-6Zn (Zhang et al. 2010b; Kubasek et al. 2012), have been reported. The microstructure of these alloys mainly consist of α-Mg solid solution, which improves the formability and machining property at elevated temperatures. Thus, the low Zn content alloys can be formed into their final shapes by hot working processing. In order to investigate the influence of Zn contents on the microstructure, corrosion behavior, or the microstructure, other Mg-Zn alloys that contain higher Zn contents than the solubility of Zn in Mg were also fabricated (Wei et al. 1995;

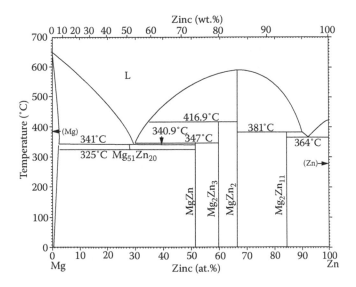

FIGURE 7.1 Binary phase diagram of Mg-Zn alloys. (Modified from Wei, L. Y., Dunlop, G. L., and Westengen, H., *Metallurgical and Materials Transactions A-Physical Metallurgy and Materials Science* 26 (8):1947–55, 1995. With permission.)

Gao and Nie 2007; Cai et al. 2012). The second group of Mg-Zn-based alloys has a chemical composition that is close to the deep eutectic point in the Mg-Zn system with a high glass-forming ability. Alloys of this group can easily form an amorphous phase when rapidly cooled and will be discussed in Section 7.3.2. The third group is Zn-based alloys, of which the Mg concentration is limited. More details can be found in Vojtech et al. (2011).

7.2.2 FABRICATION, HOT AND COLD WORKING, AND HEAT TREATMENT

The conventional fabrication methods of Mg-Zn binary alloys refer to gravity casting and die casting. A typical gravity casting procedure is as follows. High-purity Mg (>99.5%) and Zn (>99.9%) are selected as the raw materials. The raw materials are then melted at the temperature in the range of 700°C to 800°C according to the Zn content in the resistance furnace with a steel crucible or graphite crucible. The melts were usually kept at the melting temperature with constant mechanical stirring for about 15 to 60 min in order to homogenize the molten alloy prior to pouring it into the mold. Subsequently, the melt was poured into the preheated steel mold. During the whole melting procedure, high-purity argon or a mixture of CO_2 and SF_6 was applied to protect the melts from oxidation. For further improving the mechanical behavior of the binary Mg-Zn alloy, solution treatment or age-hardening treatment was applied. For the Mg-4Zn alloy, the as-cast samples were homogenized at 330°C for 90 h and then quenched by cold water (Wei et al. 2014). To determine the effect of Zn distribution on the mechanical and corrosion property of the Mg-Zn alloys, Mg-1Zn, Mg-4Zn, and Mg-6Zn cast alloys were annealed at 300°C for 150 h, followed by water quenching (Kubasek et al. 2012). A Mg-8Zn alloy was first homogenized for 120 h at 325°C, then solution treated at the temperature of 335°C for 24 h, and finally aged in an oil bath at 200°C (Gao and Nie 2007). Mandal et al. (2014) applied the homogenized treatment at 340°C for 48 h to Mg-2.4Zn alloy. Zhang et al. (2010b) developed the as-extruded Mg-6Zn alloy with the extrusion temperature being 250°C and the extrusion ratio being 8:1. Peng et al. (2012) developed a series of backward extruded Mg-Zn alloys with an extrusion ratio and speed of 12.25 and 50 mm/min, respectively.

7.2.3 MICROSTRUCTURE AND MECHANICAL PROPERTIES

7.2.3.1 Microstructure

There are significant differences between the microstructure of the Mg-Zn alloys with different Zn contents. In the alloys containing less than 2% Zn, the as-cast alloys are mainly composed of Mg matrix and polygon petal-shaped secondary dendrites, and all the grain sizes are similar to each other. But the average secondary

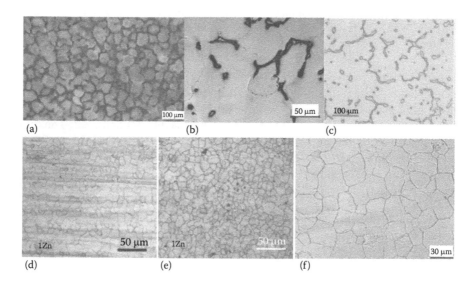

FIGURE 7.2 Microstructure of the binary Mg-Zn alloy: (a) as-cast Mg-5Zn, (b) as-cast Mg-6Zn, (c) as-cast Mg-9Zn, (d) as-extruded Mg-1Zn (parallel to the extruded direction), (e) as-extruded Mg-1Zn (perpendicular to the extruded direction), and (f) as-extruded Mg-6Zn. ([a] Modified from Cai, S. H., Lei, T., Li, N. F., and Feng, F. F., *Materials Science and Engineering C-Materials for Biological Applications* 32 (8):2570–7, 2012. With permission; [b, f] Modified from Zhang, S. X., Zhang, X. N., Zhao, C. L., Li, J. A., Song, Y., Xie, C. Y., Tao, H. R., Zhang, Y., He, Y. H., Jiang, Y., and Bian, Y. J., *Acta Biomaterialia* 6 (2):626–40, 2010. With permission; [c] Modified from Wei, L. Y., Dunlop, G. L., and Westengen, H., *Metallurgical and Materials Transactions A-Physical Metallurgy and Materials Science* 26 (8):1947–55, 1995. With permission; [d, e] Modified from Peng, Q. M., Li, X. J., Ma, N., Liu, R. P., and Zhang, H. J., *Journal of the Mechanical Behavior of Biomedical Materials* 10:128–37, 2012. With permission.)

dendrite arm spacing decreased with the increasing Zn content. The EDX spectrum indicated that the precipitates were MgZn$_2$ phase (Peng et al. 2012). With increasing Zn content from 3% to 7%, the grain size decreased with the increasing Zn content, but the volume fraction of the second phase increased with the increasing Zn content. And the addition of 5% Zn can significantly refine the grain size; the refinement efficiency decreased with further addition of Zn over 7%. Eutectic (MgZn + α-Mg) phase that distributed at the grain boundary was detected in Mg-Zn alloys containing Zn higher than 5% (Cai et al. 2012). The solution treatment has an important effect on the distribution of the secondary phase. Both Zhang et al. (2010b) and Wei et al. (2014) reported that the dendrite structure of the as-cast Mg-Zn alloys disappears, and the microstructure becomes more homogeneous after solid solution treatment, indicating that the secondary phases have been dissolved into the α-Mg matrix. X. B. Liu et al. (2010) reported a similar result. The T4 treatment significantly reduced the MgZn particles and increased the grain size of Mg-3Zn alloy. The single phase microstructure of the Mg-Zn alloys made it easier to apply hot working, which will surely improve the mechanical property. Kubasek et al. (2012) suggested that an annealing temperature of 300°C for Mg-Zn alloys was not sufficient to accelerate solid state diffusion and to drive complete dissolution of the MgZn phase in Mg matrix. The typical microstructure of the Mg-Zn binary alloys are illustrated in Figure 7.2.

7.2.3.2 Mechanical Properties

Zn was an effective alloying element to improve mechanical properties of Mg alloys (Zhang et al. 2010b; Peng et al. 2012). Grain size and the dispersion of the secondary phase may have an influence on the mechanical property of the Mg-Zn binary alloys. In the binary Mg-Zn alloys in which the Zn content is less than 2%, the yield strength and ultimate strength increased with the increasing Zn concentration (Peng et al. 2012). Peng et al. (2012) suggested that the improved mechanical strength was mostly related to two aspects. The first one is that the Zn element was homogeneously distributed in the Mg matrix, thus increasing critical resolved shear stress of dislocation and greatly enhancing the mechanical strength. The second one is that the increased fraction of boundaries resulted from the refined microstructure preventing the dislocation movement effectively. The mechanical property of the as-cast Mg-Zn binary alloy can benefit from hot-working processing, such as hot extrusion (Zhang et al. 2010b). The as-extruded Mg-Zn binary alloy exhibited higher hardness of the MgZn$_2$ precipitate; the fine particle effectively blocked the deformation of the matrix under the tensile condition. Furthermore, the well-distributed grain after extrusion reduced the fluctuation of hardness; thus local strain segregation can be effectively eliminated during deformation, resulting in the enhanced elongation (Peng et al. 2012). Boehlert et al. (2006) reported that the strength values of the Mg-Zn binary alloy were similar for Zn concentrations between 2.9% and 4.1%. Further increasing Zn content to 4.4% led to a significantly lower strength value and higher elongation. Cai et al. (2012) reported that the hardness, ultimate strength in yield, tension, and compression of Mg-Zn binary alloys increased with the increase of Zn concentration up to 5%. Further increasing Zn concentration to 7% may deteriorate mechanical properties. According to the Mg-Zn binary phase diagram, the

TABLE 7.2
Summary of the Mechanical Properties of the Binary Mg-Zn Alloys

Sample	Condition	Elastic Modulus (GPa)	Yield Strength (MPa)	Ultimate Strength (MPa)	Elongation (%)	Vickers Hardness (HV)	Grain Size (µm)	References
Mg-2Zn	As-cast	N/A	65 ± 3	121 ± 11	5.3 ± 1.9	~51	N/A	Peng et al. 2012
Mg-2Zn	As-extruded	N/A	111 ± 1	198 ± 6	15.7 ± 1.6	~79	N/A	
Mg-1Zn	As-cast	24.23	60.62	187.73	13.77	47.33 (HB)	100	Cai et al. 2012
Mg-5Zn	As-cast	36.47	75.6	194.59	8.5	53.8 (HB)	55	
Mg-7Zn	As-cast	39.6	67.28	135.53	6	56.26 (HB)	56	
Mg-6Zn	As-extruded	42.3 ± 0.1	169.5 ± 3.6	279.5 ± 2.3	18.8 ± 0.8			Zhang et al. 2010b
Mg-1Zn	As-cast	N/A	N/A	N/A	N/A	36.4 (HB)	15–20	Kubasek and Vojtech 2013b
Mg-3Zn	As-cast	N/A	N/A	N/A	N/A	39.8 (HB)	15–20	
Mg-2.9Zn	As-cast	N/A	84	219	4.7	N/A	106.5	Boehlert and Knittel 2006
Mg-3.3Zn	As-cast	N/A	90	210	4.6	N/A	155.6	
Mg-4Zn	As-cast	N/A	95	216	4.1	N/A	70.2	
Mg-4.4Zn	As-cast	N/A	68	155	8.4	N/A	112.9	

maximum solubility of Zn in Mg is 1.6% at room temperature under the equilibrium conditions. When the Zn content is less than 5%, the Zn mainly dissolves into the Mg matrix, which generates solid-solution strengthening. What is more, the MgZn phase precipitate from the Mg matrix along the grain boundaries can promote the strength by dispersion strengthening. However, the increasing precipitated secondary phase may increase the dislocation density and act as a new crack source, which can eventually result in brittle breakage failure. When the Zn content increased up to 7%, a network structure with dendritic segregation was formed along the grain boundaries and resulted in residual defects and sharply decreased the strength and elongation of the Mg-Zn binary alloy (Cai et al. 2012). A summary of the mechanical properties of various binary Mg-Zn alloys in the literature is shown in Table 7.2.

7.2.4 DEGRADATION IN SIMULATED BODY FLUIDS

The corrosion behavior of the Mg-Zn alloys strongly depends on the Zn contents and the distribution of the MgZn phase. The standard potential of Zn is -0.762 V, which is significantly nobler than Mg (-2.37 V). Furthermore, Zn is characterized by a relatively high hydrogen over potential (Fontana and Stactile 1970; Kubasek et al. 2012), and the Zn addition into Mg alloys can increase the over potential of hydrogen evolution of Mg alloys. Thus, the addition of Zn may improve the corrosion resistance of the Mg alloys. On the other hand, the distribution of MgZn precipitate results in microgalvanic corrosion and causes an increase in the corrosion rate of the alloys. Peng et al. (2012) reported that the increasing Zn content reduced the corrosion resistance property for both as-cast and as-extruded Mg-Zn binary alloys due to the microgalvanic cells between the Zn and Mg matrix. Furthermore, the corrosion products of the Mg-Zn binary alloys on the surface were too loose to prevent further corrosion (Peng et al. 2012). Peng et al. also pointed out that the corrosion modes of Mg-Zn binary alloys were different from other systems, such as stainless steel. The corrosion pits can spread laterally by covering the entire surface instead of growing in depth and then provoke the undermining of grain. And the particles would fall away when enough mass had been lost around them (Peng et al. 2012). Kubasek et al. (2012) reported a similar result. Due to the low concentration of Zn, the MgZn secondary phase was not detected in the Mg-1.4Zn alloy. Thus, the as-cast Mg-1.4Zn alloy exhibited the lowest corrosion rate whereas the interdendritic MgZn phase in the Mg-4Zn and Mg-6Zn alloys significantly accelerated their corrosion rate.

In order to improve the corrosion resistance of the Mg-Zn alloys, heat treatment would be an effective method. The heat treatment procedure led to a partial dissolution of the coarse MgZn phase, and the heterogeneous continuous MgZn network was replaced by a more homogeneous distribution of Zn and noble MgZn phase. Thus, the effect of microgalvanic corrosion diminished (Kubasek et al. 2012). X. B. Liu et al. (2010) reported that T4 treatment can also improve the corrosion property of Mg-3Zn alloy. Cai et al. (2012) reported that the as-cast Mg-7Zn alloy, which contained a higher volume fraction of second phase than the Mg-5Zn alloy, suffered a rapid corrosion. A summary of the corrosion rates of the Mg-Zn binary alloys in different corrosive media is shown in Table 7.3. A typical corrosion surface morphology of Mg-6Zn binary alloy is shown in Figure 7.3.

TABLE 7.3
Summary of the Corrosion Rates of the Binary Mg-Zn Alloys in Different Corrosive Media

Sample	Condition	Corrosion Rates in Different Corrosive Media		I_{corr}	References
		SBF	NaCl Solution		
Mg-2Zn	As-cast	a	N/A	$(4.23 \pm 0.23) \times 10^{-1}$ mA/cm^2	Peng et al. 2012
Mg-2Zn	As-extruded	a	N/A	$(5.33 \pm 0.11) \times 10^{-2}$ mA/cm^2	
Mg-3Zn	T6-treated	N/A	0.1 mol/L	10 μA/cm^2	X. B. Liu et al. 2010
Mg-1Zn	As-cast	2.01 mm/y	N/A	23.4 μA/cm^2	Cai et al. 2012
Mg-5Zn	As-cast	1.26 mm/y	N/A	11.72 μA/cm^2	
Mg-7Zn	As-cast	3.18 mm/y	N/A	51.79 μA/cm^2	
Mg-6Zn	As-extruded	0.16 mm/y	N/A	N/A	Zhang et al. 2010b
Mg-1Zn	As-cast	0.08 mm/y	N/A	N/A	Kubasek et al. 2012
Mg-4Zn	As-cast	0.36 mm/y	N/A	N/A	
Mg-6Zn	As-cast	3 mm/y	N/A	N/A	
Mg-1Zn	As-cast	N/A	1.27 mm/y[b]	N/A	Kubasek and Vojtech
Mg-3Zn	As-cast	N/A	2.51 mm/y[b]	N/A	2013b

[a] Corrosive medium was SBF.
[b] Concentration of the NaCl solution was 9 g/l.

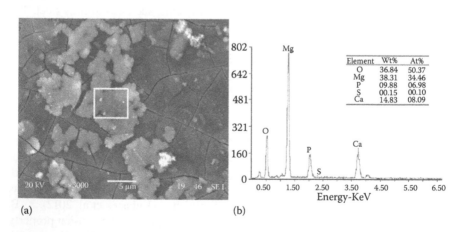

(a)   (b)

FIGURE 7.3 Typical corrosion morphology of binary Mg-Zn alloys in corrosive media: (a) morphology of as-extruded Mg-6Zn alloy after immersion in SBF for 30 days, (b) composition of the corrosion products in the white frame of (a). (Modified from Zhang, S. X., Zhang, X. N., Zhao, C. L., Li, J. A., Song, Y., Xie, C. Y., Tao, H. R., Zhang, Y., He, Y. H., Jiang, Y., and Bian, Y. J., *Acta Biomaterialia*, 6 (2): 626–40, 2010. With permission.)

7.2.5 Cytotoxicity and Hemocompatibility

As an essential component of some enzymes and proteins in the human body, Zn is regarded as absolutely necessary and is biosafe except at extreme exposures. Mg plays an important role in cell adhesion mechanisms (Hynes 1992) and calcium incorporation (Serre et al. 1998; Paul and Sharma 2006). It has been reported that Mg ion resulted in an increase in osteoblast adhesion (Zreiqat et al. 2002). Zn ions are also known to be an important component in controlling the function of osteoblasts, osteoblast adhesion, and upgrading the level of ALP (Yamaguchi et al. 1987).

Zhang et al. (2010b) evaluated the cytotoxicity of Mg-6Zn alloy using the L-929 cell line by indirect assay with different concentration extracts. After 7 days of incubation, the cells exhibited normal and healthy morphologies in different extracts, which was similar to that of the negative control group. The relative growth rate of the cells in the extracts with different concentrations showed no significant different from the negative control. The cytotoxicity of these extracts was Grade 0–1, according to ISO 10993-5:1999, indicating that the Mg-6Zn alloy was biosafe for biomaterial applications (Zhang et al. 2010b).

S. X. Zhang et al. (2009) studied the early cell adhesion of MC3T3-E1 cells on the Mg-6Zn alloy by direct assay. In the early stage, a loose and rough corrosion layer was found on the alloy surface. The pH value of DMEM increased rapidly to about 8.90 during the first 3 h incubation with the Mg-6Zn alloy sample. The microscopy results released that the Mg-Zn alloy was able to support the earlier adhesion of pre-osteoblasts. But the relatively higher pH value of the incubation medium resulted in some elliptical cells being spread on the sample surface.

For hemolysis tests, the hemolysis rate of the Mg-6Zn alloy was 3.4%, lower than 5%, meaning that the Mg-6Zn alloy would not cause damage to red blood cells. It was suggested that the rough protective corrosion layer would benefit the adhesion of the cells, and surface modification of the alloy would be necessary to avoid a sharp pH change of the culture medium (S. X. Zhang et al. 2009).

D. Y. Chen et al. (2011) studied the influence of the Mg-5.6Zn alloy on cell attachment and osteogenic mineralization of MC3T3-E1 cells. Their findings suggested that more cells were attached to the surface of the Mg-5.6Zn alloy. The area of mineralized nodules formed on the Mg-5.6Zn alloy surface was significantly greater than that formed on the control. The qualitative real-time PCR results released that the expression level of COL1alpha1, ALP, and OC transcript were also significantly higher than the control (D. Y. Chen et al. 2011). Morphology of the cells cultured with both Mg-5.6Zn alloy and its extract are shown in Figure 7.4.

7.2.6 In Vivo Performance in Animal Models

Zhang et al. (2010b) implanted Mg-6Zn alloy rods into the femoral shaft of New Zealand rabbits, and the same hole drilled in the femoral shaft without any implantation was used as control. The biochemical tests revealed that the degradation of Mg-6Zn alloy implants did not raise serum Mg and Zn ion levels. Moreover, the H&E stained slices of heart, liver, and spleen showed the tissues were in a normal state. According to the radiographic evaluation results, the implant started to

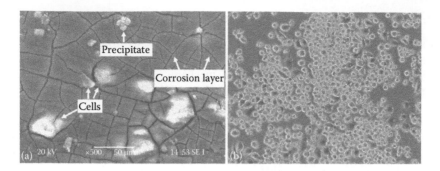

FIGURE 7.4 Typical cell morphology cultured with both alloys and alloy extracts: (a) morphology of MC3T3-E1 cells cultured with Mg-6Zn alloy for 2 h, (b) L929 cell cultured with 100% extracts of Mg-6Zn alloy for 7 days. ([a] Modified from Zhang, S. X., Li, J. A., Song, Y., Zhao, C. L., Zhang, X. N., Xie, C. Y., Zhang, Y., Tao, H. R., He, Y. H., Jiang, Y., and Bian, Y. J., *Materials Science and Engineering C-Materials for Biological Applications* 29 (6):1907–12, 2009. With permission; [b] Modified from Zhang, S. X., Zhang, X. N., Zhao, C. L., Li, J. A., Song, Y., Xie, C. Y., Tao, H. R., Zhang, Y., He, Y. H., Jiang, Y., and Bian, Y. J., *Acta Biomaterialia* 6 (2):626–40, 2010. With permission.)

degrade in the first 3 weeks whereas the implant was too blurry to be recognized after 12 weeks. Newly formed trabecular and osteoblasts were observed in the slices of the bone tissue histology (Figure 7.5h, i), and the corrosion rate of the implants according to the weight loss was determined to be 2.32 mm/year. They concluded that the Mg-6Zn alloy did not cause any side effects during the implantation period, but the corrosion rate in vivo needs to be further reduced (Zhang et al. 2010b).

Yan et al. (2013) implanted Mg-6Zn alloy in the cecum incision 1 cm from the distal end of the Sprague–Dawley rat cecum. The edge of the implants became fuzzy in week 3. In week 4, the implants disintegrated. The degradation of the implants during the implantation did not raise serum Mg ion levels and did not bring organic damage to vital organs. According to the immunohistochemical evaluation, the TGF-β1 in the Mg-6Zn alloy group was at the highest level, which indicated that it promoted the healing procedure. The inhibited TNF-α expression in the cecum indicated that the Mg-6Zn alloy may reduce stimulation and inflammation and shorten the inflammatory period in the healing process. And they concluded that the Mg-6Zn alloy would be a promising candidate for gastrointestinal reconstruction (Yan et al. 2013).

He et al. (2009) fully investigated the biocompatibility of the Mg-Zn alloy. Subcutaneous gas bubbles were observed after 3 weeks implantation and disappeared within 6 weeks. Bone absorption occurred at 12 weeks, and the implants totally degraded at 24 weeks. An X-ray graph of the implantation site is shown in Figure 7.5a, b. Bone resorption was observed around the implant 6 weeks after surgery (Figure 7.5a). New bone formation outside of the femur cortex was observed 12 weeks after surgery (Figure 7.5b). Histological examination of the viscera showed that the heart, liver, kidney, and spleen did not change morphologically, and no

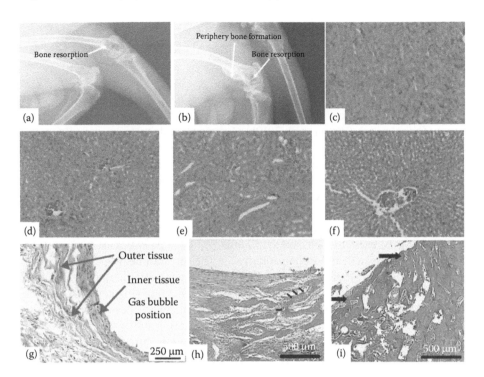

FIGURE 7.5 In vivo evaluation of the binary Mg-Zn alloys: bone absorption around the Mg-6Zn implants (a) 6 weeks, (b) 12 weeks after surgery; pathological section of (c) heart cells, (d) liver tissue, (e) kidney tissue, (f) spleen tissue; (g) histological section of Mg-6Zn alloy 12 weeks after surgery; H&E stained bone tissue surrounding Mg-6Zn rods (h) 6 weeks, and (i) 18 weeks postimplantation, the black arrow indicated the newly formed bones. ([a–f] Modified from He, Y. H., Tao, H. R., Zhang, Y., Jiang, Y., Zhang, S. X., Zhao, C. L., Li, J. N., Zhang, B. L., Song, Y., and Zhang, X. N., *Chinese Science Bulletin* 54 (3): 484–91, 2009. With permission; [g–i] Modified from Zhang, S. X., Zhang, X. N., Zhao, C. L., Li, J. A., Song, Y., Xie, C. Y., Tao, H. R., Zhang, Y., He, Y. H., Jiang, Y., and Bian, Y. J., *Acta Biomaterialia* 6 (2):626–40, 2010. With permission.)

infiltration was observed (Figure 7.5c–f). Furthermore, the gas bubbles can be seen in the H&E slices as exhibited in Figure 7.5g. Chen (2011) implanted Mg-6Zn alloy rods in the femoral marrow cavity of the rabbits. After 12 weeks implantation, all the implants were tightly fixed although the gradual degradation of the implant changed from a rod to an irregular shape. And newly formed bone can be seen. When compared with the control group, no fibrous membrane was present at the implant–bone interface. Furthermore, newly formed bone can be observed surrounding the degradation layer 12 weeks after implantation. They concluded that the Mg-Zn alloy has excellent biocompatibility as a degradable metallic material for orthopedic application.

A typical histological microstructure and X-ray image of in vivo tests of Mg-Zn binary alloys are shown in Figure 7.5.

7.3 DEVELOPMENT OF Mg-Zn TERNARY ALLOYS
AS DEGRADABLE METALLIC BIOMATERIALS

7.3.1 ALLOY COMPOSITION DESIGN

Despite the fact that Mg-Zn binary alloys possess good biocompatibility both in vitro and in vivo, the corrosion rate in vivo needs to be further improved according to Zhang's work (Zhang et al. 2010b). Among the efforts to improve the corrosion resistance of the binary Mg-Zn alloys, further alloying seems to be an effective method. Until recently, studies of biodegradable Mg-Zn alloys have been focused on AZ(Mg-Al-Zn), ZE(Mg-Zn-RE), MZ(Mg-Zn-Mn), WZ(Mg-Y-Zn), ZK(Mg-Zn-Zr), and Mg-Zn-Ca alloys. However, the majority of these ternary Mg-Zn-X alloys were mainly designed for engineering materials used in the automotive and aerospace industries. For biomedical applications, alloy elements and the amount of alloy elements should be carefully selected to minimize or avoid potential toxic effects.

Due to the excellent biocompatibility of Ca, crystalline Mg-Zn-Ca ternary alloys were widely studied among the Mg-Zn-X ternary alloys, and this information has been summarized in Sections 6.3.3 through 6.3.5 of Chapter 6. In addition, the Mg-Zn-Ca alloy has a chemical composition that approaches the deep eutectic point in the Mg-Zn system and shows a high glass-forming ability. These amorphous Mg-Zn-Ca alloys were reported to have excellent strength, corrosion resistance, and biocompatibilities (Zberg et al. 2009b; Gu et al. 2010c; Cao et al. 2012).

Zr plays an important role in grain refinement when used at less than the solubility limit of 0.6% in a Mg matrix, which seems to be a suitable alloying addition (Hong et al. 2013). It has been reported that Zr is an effective alloying element in improving corrosion resistance when its concentration is less than 0.48% without the formation of Zr-containing precipitates (Song 2005). Recent studies also revealed that Mg-Zn-Zr alloys possess suitable corrosion resistance and mechanical properties as well as in vitro biocompatibility as orthopedic implants (Huan et al. 2010a; Gu et al. 2011c). A daily intake of 50 µg Zr is tolerated for an adult (Emsley 2011; Hong et al. 2013).

It has been reported that the addition of Y element could increase the strength and ductility (Bae et al. 2002a; Kim and Bae 2004). The addition of Y can also provide excellent corrosion resistance (Hänzi et al. 2010). For the consideration of biocompatibility, Y is also biocompatible with a tolerant daily intake of 4.2 mg/d (Zhang et al. 2000; Zhao et al. 2013a). However, Y is not an essential metal element for the human body; thus the concentration of Y should be minimized in the alloys for biomedical applications (Zhao et al. 2013a).

Mn has made an effort to remove iron and other heavy metal elements into relatively harmless intermetallic compounds, which may be separated out during the melting procedure. Thus, it can improve the saltwater resistance of Mg alloys (Xu et al. 2008). However, Mn does not have much effect on tensile strength, but it does increase yield strength slightly (E. L. Zhang et al. 2009b). In sense of biocompatibility, Mn has no toxic effect unless after extreme occupational exposure. It is also associated with the activation of multiple enzyme systems. And a daily intake of 2–5 mg was recommended for people between the ages of 11 and 51 (Xu et al. 2008; E. L. Zhang et al. 2009b).

Other alloying elements, such as Sn (Sasaki et al. 2009, 2011), Ag (Ben-Hamu et al. 2006a), Dy (Bi et al. 2011a), Si (Rosalbino et al. 2010), Gd (R. Zhen et al. 2013), Sr (Cipriano et al. 2013), and Li (Yan et al. 2009b; X. H. Liu et al. 2012), are also reported in the literature.

7.3.2 Fabrication, Hot and Cold Working, and Heat Treatment

7.3.2.1 Mg-Zn-Ca Alloys

The fabrication and working of a crystalline Mg-Zn-Ca alloy can be found in Chapter 6. Mg-Zn-Ca bulk metallic glass has a very poor glass-forming ability (GFA); only a small critical casting thickness size of about 5 mm was reported (Gu et al. 2005). Furthermore, the thermoplastic forming (TPF) window of these BMG is only 20°C. GFA can only be achieved in the temperature above the glass transition and below first crystallization. The small TPF window and small critical thickness would be the limitation of the fabrication of Mg-based BMG (Cao et al. 2012). A typical fabricating procedure is as follows: The raw materials were melted in the furnace at a temperature of about 700°C. The melts were then cooled to the determined injection temperature followed by injection casting (Laws et al. 2008a,b; Gu et al. 2010c; Wessels et al. 2012). Moreover, Zhao et al. (2014b) reported a hot pressing sintering method by using Mg, Zn, and Ca powder with diameters of 48–75 μm.

7.3.2.2 Mg-Zn-Zr Alloys

Commercial Mg-Zr master alloys with different compositions, such as Mg-20%Zr (Gu et al. 2011c), Mg-30%Zr (Huan et al. 2010a), Mg-33.3%Zr (Hong et al. 2013), and Mg-34.6%Zr (Bhattacharjee et al. 2012), were available for introducing the Zr element. The melting temperature was in the range of 700°C to 800°C, and a holding time of 30 min was recommended. After the addition of Zr, mechanical stir should be applied to dissolve it, and it will be helpful for dispersing the Zr particles uniformly into the melts. Subsequently, a further hold for another 30 min is necessary to get the homogenous melts. After the melting procedure, the melts were then poured into the mold, and the mold was preheated at 200–500°C. During the whole procedure, high-purity Ar or a mixture of CO_2 and SF_6 was applied to protect the melts from oxidation.

7.3.2.3 Mg-Zn-Y Alloys

For conventional metallurgy methods, both high-purity Y (Bae et al. 2002b; Muller et al. 2007; J. S. Zhang et al. 2012b; Rosalie et al. 2013) and Mg-Y master (E. Zhang et al. 2008; Zhao et al. 2013a) alloys were utilized to introduce element Y. The melting temperature ranged from 700°C to 810°C for the raw materials that were reported in different literature (E. Zhang et al. 2008; Zhao et al. 2013a). The holding time of the melts was usually kept at 30 min. Instead of the high-purity Ar, the mixture of N_2 and CO_2 (Zhao et al. 2013a) or CH_2FCF_3 and N_2 (J. S. Zhang et al. 2012b) were also applied as protective gases. Powder metallurgy was also reported to produce Mg-Zn-Y alloys. These powders were fabricated from the Mg-Zn-Y cast alloys either by the electrode induction–melting gas atomization method (Mora et al. 2009) or the argon atomization method (Asgharzadeh et al. 2014). Subsequently, these powders were compressed and heated at a determined temperature, and the green compacts were hot-extruded to get the extruded Mg-Zn-Y alloys (Mora et al. 2009; Asgharzadeh et al. 2014).

7.3.2.4 Mg-Zn-Mn Alloys

High-purity Mn element (Rosalbino et al. 2013) or commercial Mg-Mn master alloys (D. F. Zhang et al. 2011; Wang et al. 2013b; H. J. Zhang et al. 2013b) were available as the raw materials to introduce Mn into the Mg-Zn-Mn alloys. The melting temperature was about 750°C. Subsequently, the melts were cast in a metal mold, which was preheated to 200°C (Yin et al. 2008; E. L. Zhang et al. 2009b). Moreover, E. L. Zhang et al. (2009b) developed the Mg-Zn-Mn alloys by high-purity Mg, Zn, and analytically pure $MnCl_2$ whereas Yin et al. (2008) fabricated the Mg-Zn-Mn alloys by high-purity Mg, Zn, Al, and analytical-grade $MnCl_2$.

7.3.2.5 Other Mg-Zn-X Alloys

Bi et al. (2011a) developed an Mg-2Dy-0.5Zn alloy by the metallurgy method. Pure Mg, Zn, and Mg-20Dy master alloys were used as raw materials. The melts were homogenized at 750°C for 0.5 h followed by being poured into a water-cooling mold at 720°C. The as-cast ingots were then homogenized at 525°C for 10 h. The extrusion procedure was carried out by an extrusion ratio of 17 at 360°C. The extruded alloy was further aged at 180°C for 99 h.

Leng et al. (2013c) prepared the Mg-8.3RY-3.8Zn alloy with commercial Mg, Zn, and Mg-20RY master alloy. After homogenizing at 500°C for 12 h, the as-cast ingots were extruded at 420°C with an extrusion ratio of 17.

R. Zhen et al. (2013) fabricated the Mg-11Gd-1Zn alloy and homogenized it at 510°C for 12 h followed by extrusion at 430°C into 20-mm rods. The extrusion rods were solution treated at 500°C for 8 h and quenched into cold water. Aging was applied at the temperature between 200°C and 250°C.

H. Li et al. (2012) cast a series of Mg-Zn-Er alloys with different Zn/Er ratios with pure Mg, Zn, and Mg-20Er master alloys. The raw materials were melted at 740°C under antioxidizing flux. The melts were poured into the mold preheated at 400°C after holding for 30 min, followed by cooling in the protective atmosphere.

Ben-Hamu et al. (2006a) studied the microstructure and corrosion behavior of Mg-Zn-Ag alloy. The Mg-6Zn-(1,2,3)Ag ingots were homogenized at 400°C and water cooled, followed by extruding into 10-mm-diameter rods at an extrusion ratio of 25:1 and a temperature between 275°C and 300°C. The extrusion rods were finally double aged.

Both the metallurgy method (X. H. Liu et al. 2012) and the electrochemical codeposition method (Yan et al. 2009b) were reported to develop Mg-Li-Zn ternary alloys. The Mg-8Li-2Zn alloys were first homogenized at 300°C for 24 h followed by a two pass extrusion (X. H. Liu et al. 2012).

7.3.3 Microstructure and Mechanical Properties

7.3.3.1 Mg-Zn-Ca Alloys

For the constitutional phases of crystalline Mg-Zn-Ca alloys, refer to Section 6.3.3 in Chapter 6. Mg-Zn-Ca bulk metal glasses possess an amorphous structure. Thus, only a halo peak appeared in the XRD patterns, and no crystalline diffraction peaks were detected (Gu et al. 2005, 2010c; Q. F. Li et al. 2008; Cao et al. 2012; Qin et al. 2013). The broad scattering maximum peak of the MgZnCa bulk glasses ranges from 30° to 60°, which depended on the composition of the metal glass. Cao et al. (2012) reported that the higher Zn concentration resulted in the shape of the major halo moving slightly toward higher scattering angles.

7.3.3.2 Mg-Zn-Zr Alloys

The binary Mg-Zr phase diagram is shown in Figure 6.8. Zr has been reported to be an effective grain refiner for Mg. The presence of Zr in the MgZn solute can result in reduction to an effective barrier for nucleation; thus a large number of Zr nuclei would serve as active sites for nucleation and, hence, result in effective refinement (Hong et al. 2013). The typical microstructure of Mg-Zn-Zr alloys contains equiaxed grains with eutectic constituents distributed at the grain boundaries (Shahzad and Wagner 2009a,b; Hong et al. 2013). The uniform and equiaxed grain throughout the microstructure indicated the excellent grain refinement ability of Zr. Combined with the XRD analysis, EDS analysis, and the Mg-Zn binary phase diagram, these Zn/Zr-rich secondary phases were identified as (α-Mg + MgZn) intermetallic in as-cast ZK40 alloy (Hong et al. 2013) (Figure 7.6a). After T4 treatment, the volume fraction

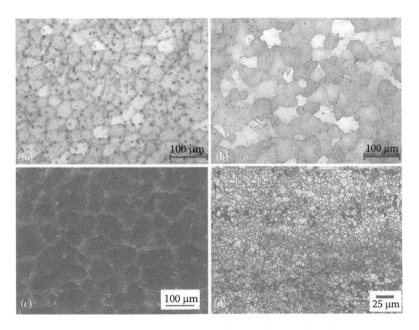

FIGURE 7.6 Typical microstructure of Mg-Zn-Zr alloys: (a) as-cast ZK40, (b) heat-treated ZK40, (c) as-cast ZK60, and (d) as-extruded ZK60. ([a, b] Modified from Hong, D., Saha, P., Chou, D. T., Lee, B., Collins, B. E., Tan, Z. Q., Dong, Z. Y., and Kumta, P. N., *Acta Biomaterialia* 9 (10):8534–47, 2013. With permission; [c] Modified from Gu, X. N., Li, N., Zheng, Y. F., and Ruan, L. Q., *Materials Science and Engineering B-Advanced Functional Solid-State Materials* 176 (20):1778–84, 2011. With permission; [d] Modified from Somekawa, H., Singh, A., and Mukai, T., *Journal of Materials Research* 22 (4):965–73, 2007. With permission.)

of MgZn phase decreased due to the rapid dissolution of MgZn intermetallic into α-Mg grain during solution treatment (Hong et al. 2013) (Figure 7.6b). Gu et al. (2011c) reported that Mg_2Zn phase appeared in the as-extruded ZK60 alloy (Figure 7.6c). Unidirectional deformation during extrusion can lead to the diffusion of Zn from Zn-rich regions and the decomposition of the eutectic constituents (Shahzad and Wagner 2009a). Typical microstructures of Mg-Zn-Zr ternary alloys are shown in Figure 7.6.

7.3.3.3 Mg-Zn-Y Alloys

The Mg-Y binary phase diagram is depicted in Figure 7.7 (C. P. Guo et al. 2007). The addition Y in the Mg-Zn alloys results in the formation of heat stable intermetallic phases, of which composition depended on the Zn/Y ratio (Tsai et al. 2000; Garces et al. 2008). There were four different Mg-Zn-Y ternary phases being reported (Luo and Zhang 2000; Tsai et al. 2000; Singh et al. 2004; Lee et al. 2005): I-phase Mg_3YZn_6 (icosahedral), W-phase $Mg_3Y_3Zn_2$ (fcc), H-phase $MgYZn_3$ (hexagonal), and Z-phase $Mg_{12}YZn$ (hexagonal).

Zhang et al. reported that the $Mg_{12}YZn$ phase is an 18R long period stacking ordered (LPSO) structure, and the phase volume fraction increased with increasing Zn and Y content (J. S. Zhang et al. 2012b). Various novel LPSO phases with 6H, 10H, 14H, 18R, and 24R have also been discovered (Itoi et al. 2004; Nishida et al. 2004; Matsuda et al. 2005). The LPSO structure has a similar a-axis with hcp Mg; nevertheless, the stacking periodicity is lengthened along the c-axis (Luo and Zhang 2000). It played an important role in the mechanical strength. And the presence of

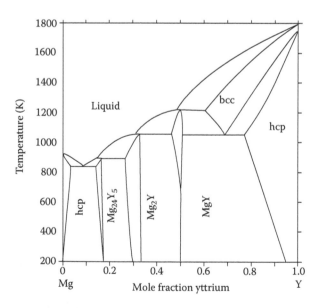

FIGURE 7.7 Typical Mg-Y binary phase diagram. (Modified from Guo, C. P., Du, Z. M., and Li, C. R., *Calphad-Computer Coupling of Phase Diagrams and Thermochemistry* 31 (1):75–88, 2007. With permission.)

the LPSO phase in the α-Mg matrix can increase the critical resolved shear stress of basal slip and activate nonbasal slip; the kinking bands of the LPSO phase formed in the deformation processes also act as an additional contributor for strengthening (Matsuda et al. 2005; Hagihara et al. 2010b).

Garces et al. (2008) reported that the quasicrystalline I-phase possessing low interfacial energy exhibited a strong bonding of the I-phase/matrix; thus, it was very promising for the strengthening of the mechanical property of the Mg-Zn-Y alloys. The typical microstructures of Mg-Zn-Y alloys are shown in Figure 7.8.

7.3.3.4   Mg-Zn-Mn Alloys

The typical binary Mg-Mn phase diagram is illustrated in Figure 7.9 (Okamoto 2012). From the phase diagram, we can see that there was no Mg-Mn phase formed during the cooling procedure, and Mn was mainly precipitated as Mn particles. The typical microstructure of Mg-Zn-Mn alloys varies with the concentration of the Zn and Mn. With the increasing of Zn content, the grain size significantly reduced both in as-cast and as-extruded alloys (Yin et al. 2008; E. L. Zhang et al. 2009b). Zhang et al. (2011) reported that the as-cast Mg-(4-9)Zn-1Mn alloys exhibited a dendritic crystal microstructure with a second phase distributed along the grain boundaries and found that high Zn content had an effect on refining the grain and increasing interdendritic compounds. Moreover, the as-cast ZM61 alloy cooled in a salt bath exhibited a finer dendritic morphology when compared with alloys cooled in air. And the rapid solidification ZM61 alloy exhibited extremely small dendritic microstructure with about 25 μm (H. J. Zhang et al. 2013b).

Al-Mn phase particulates were found at the grain boundary and the interdendrite matrix of the Mg-1Zn-1Mn alloy whereas the Mg-Zn phase with a Mg:Zn atom ratio of 7:3 was observed in the Mg-3Zn-1Mn alloy. When the Zn content increased to 2%, Mg-Zn phase particles can also be found in addition to the Al-Mn phase (E. L. Zhang et al. 2009b). Rosalbino et al. (2013) found that Mn can totally dissolved in the Mg matrix with less than 1% content in the Mg-1.5Zn-1Mn alloy. The typical microstructure of Mg-Zn-Mn alloys is shown in Figure 7.10.

7.3.3.5   Other Mg-Zn-Based Ternary Alloys

Both the Mg-Dy-Zn alloy and Mg-RY-Zn alloy contained the LPSO structure, which is similar to the Mg-Zn-Y alloy system. With heat treatment, the 18R LPSO structure in the Mg-Dy-Zn alloy transformed to the lamellar 14H LPSO phase. Precipitation particles that contained the $(Mg,Zn)_xDy$ phase were also observed at the grain boundaries and in the grain interior (Bi et al. 2011a). Long strip phases and fine lamellae in the grains were found in Mg-Ry-Zn alloy, of which both had a 14H LPSO structure (Leng et al. 2013c).

Ag seems to be an effective grain refiner in the Mg-Zn-Ag alloys because the grain size decreased with increasing Ag content in the alloy. Furthermore, both the precipitate density in the grain boundary and within the grains increased considerably with increased Ag content. Higher temperatures for working conditions would facilitate the dynamic recrystallization, thus resulting in a larger grain size for the Mg-Zn-Ag alloy (Ben-Hamu et al. 2006a).

Figure 7.11 shows the typical microstructure of other Mg-Zn ternary alloys.

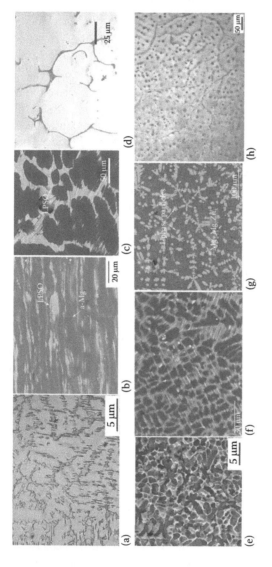

FIGURE 7.8 Typical microstructure of Mg-Zn-Y alloys: (a) as-cast Mg-1Zn-2Y, (b) as-extruded Mg-1Zn-2Y, (c) LPSO structure in the Mg97Zn1Y2 alloy, (d) as-cast Mg-4Zn-0.8Y, (e) as-cast Mg-4.3Zn-0.7Y by powder metallurgy, (f) as-cast Mg-2Zn-0.36Y. ([a] Modified from Zhang, J. S., Xu, J. D., Cheng, W. L., and Kang, J. J., *Journal of Materials Science and Technology* 28 (12):1157–62, 2012. With permission; [b] Modified from Mine, Y., Yoshimura, H., Matsuda, M., Takashima, K., and Kawamura, Y., *Materials Science and Engineering A-Structural Materials Properties Microstructure and Processing*, 570, 63, 2013. With permission; [c] Modified from Shao, X. H., Yang, H. J., De Hosson, J. T. M., and Ma, X. L., *Microscopy and Microanalysis*, 19, 1575, 2013. With permission; [d] Modified from Lee, J. Y., Do, H. K., Lim, H. K., and Kim, D. H., *Materials Letters* 59 (29–30):3801–5, 2005. With permission; [e] Modified from Asgharzadeh, H., Yoon, E. Y., Chae, H. J., Kim, T. S., Lee, J. W., and Kim, H. S., *Journal of Alloys and Compounds* 586:S95–100, 2014. With permission; [f] Modified from Itoi, T., Inazawa, T., Yamasaki, M., Kawamura, Y., and Hirohashi, M., *Materials Science and Engineering A-Structural Materials Properties Microstructure and Processing*, 560, 216, 2013. With permission; [g] Modified from Geng, J. W., Teng, X. Y., Zhou, G. R., and Zhao, D. G., *Journal of Alloys and Compound*, 577, 498, 2013. With permission; [h] Modified from Zhang, E., He, W. W., Du, H., and Yang, K., *Materials Science and Engineering A-Structural Materials Properties Microstructure and Processing* 488 (1–2):102–11, 2008. With permission.)

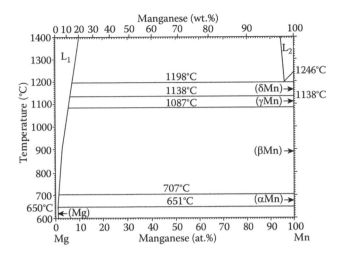

FIGURE 7.9 Typical binary Mg-Mn phase diagram. (Modified from Okamoto, H., *Journal of Phase Equilibria and Diffusion* 33 (6):496, 2012. With permission.)

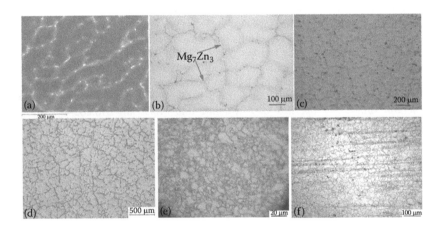

FIGURE 7.10 Typical microstructure of Mg-Zn-Mn alloys: (a) as-cast Mg-2Zn-0.2Mn, (b) as-cast Mg-6Zn-1Mn, (c) as-cast Mg-1Zn-1Mn, (d) as-cast Mg-5Zn-1Mn, (e) as-extruded Mg-1Zn-1Mn, and (f) as-extruded Mg-6Zn-1Mn. ([a] Modified from Rosalbino, F., De Negri, S., Saccone, A., Angelini, E., and Delfino, S., *Journal of Biomedical Materials Research Part A* 101A (3):704–11, 2010. With permission; [b, c] Modified from Zhang, E. L., Yin, D. S., Xu, L. P., Yang, L., and Yang, K., *Materials Science and Engineering C-Biomimetic and Supramolecular Systems* 29 (3):987–93, 2009. With permission; [d] Modified from Yuan, J. W., Zhang, K., Li, T., Li, X. G., Li, Y. J., Maa, M. L., Luo, P., Luo, G. Q., and Hao, Y. H., *Materials and Design* 40:257–61, 2012. With permission; [e] Modified from Yin, D. S., Zhang, E. L., and Zeng, S. Y., *Transactions of Nonferrous Metals Society of China* 18 (4):763–8, 2008. With permission; [f] Modified from Zhang, D. F., Shi, G. L., Dai, Q. W., Yuan, W., and Duan, H. L., *Transactions of Nonferrous Metals Society of China* 18:S59–63, 2008. With permission.)

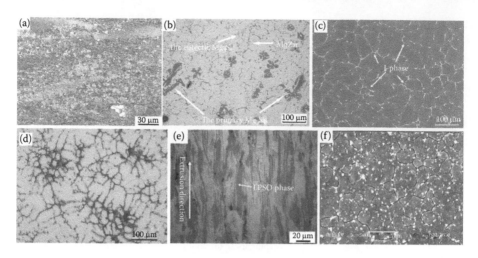

FIGURE 7.11 Typical microstructure of Mg-Zn ternary alloys: (a) as-extruded Mg-0.5Zn-2Dy, (b) as-cast Mg-6Zn-4Si, (c) Mg-7Zn-0.88Er, (d) as-cast Mg-1Zn-11Gd, (e) as-extruded Mg-4Zn-9RY, and (f) as-extruded Mg-6Zn-1Ag. ([a] Modified from Bi, G. L., Fang, D. Q., Zhao, L., Lian, J. S., Jiang, Q., and Jiang, Z. H., *Materials Science and Engineering A-Structural Materials Properties Microstructure and Processing* 528 (10–11):3609–14, 2011. With permission; [b] Modified from Cong, M. Q., Li, Z. Q., Liu, J. S., Yan, M. Y., Chen, K., Sun, Y. D., Huang, M., Wang, C., Ding, B. P., and Wang, S. L., *Journal of Alloys and Compounds* 539:168–73, 2012. With permission; [c] Modified from Li, H., Du, W. B., Li, S. B., and Wang, Z. H., *Materials and Design* 35:259–65, 2012. With permission; [d] Modified from Zhen, R., Sun, Y. S., Xue, F., Sun, J. J., and Bai, J., *Journal of Alloys and Compounds* 550:273–8, 2013. With permission; [e] Modified from Leng, Z., Zhang, J. H., Lin, H. Y., Fei, P. F., Zhang, L., Liu, S. J., Zhang, M. L., and Wu, R. Z., *Materials Science and Engineering A-Structural Materials Properties Microstructure and Processing* 576:202–6, 2013. With permission; [f] Modified from Ben-Hamu, G., Eliezer, D., Kaya, A., Na, Y. G., and Shin, K. S., *Materials Science and Engineering A-Structural Materials Properties Microstructure and Processing* 435: 579–87, 2006. With permission.)

A summary of the mechanical properties of the ternary Mg-Zn-X alloys is shown in Table 7.4.

7.3.4 DEGRADATION IN SIMULATED BODY FLUIDS

7.3.4.1 Mg-Zn-Ca Alloys

The degradation of crystalline Mg-Zn-Ca alloys are referred to in Section 6.3.4 in Chapter 6. Bulk metal glasses can possess better corrosion resistance when compared with their crystalline counterparts due to their intrinsic structure and chemical homogeneities (Q. F. Li et al. 2008). The absence of a second phase in the metal glasses even added with high Zn content exceeding its solid solubility in Mg matrix can also enhance the corrosion resistance (Gu et al. 2010c). The Mg-Zn-Ca bulk metallic glass had a more noble corrosion potential when compared with the crystalline high-purity Mg. Because the electrode potential of Zn is more positive than Mg,

TABLE 7.4

Summary of the Mechanical Properties of the Ternary Mg-Zn Alloys

Sample	Condition	Mechanical Property					Grain Size (μm)	References
		Elastic Modulus (GPA)	Yield Strength (MPa)	Ultimate Strength (MPa)	Elongation (%)	Vickers Hardness (HV)		
Mg96Zn3.4Y0.6	As-cast	N/A	210	355	23.4	N/A	N/A	Bae et al. 2002b
Mg95Zn4.3Y0.7	As-cast	N/A	220	379	19.7	N/A	N/A	Liu et al. 2013c
Mg94Zn2Y4	As-cast	N/A	96.5	130	2.9	80.8	N/A	
Mg94Zn2Y4	As-extruded	N/A	246.4	389.6	2.8	108.9	N/A	
Mg94Zn2Y4	Peak-aged	N/A	272.3	410.7	2.8	129.7	N/A	Mora et al. 2009
Mg-2.1Zn-0.26Y	As-extruded	N/A	~360	~410	~12	N/A	0.81 ± 0.03	
Mg-3.3Zn-0.43Y	As-extruded	N/A	~410	~440	~12	N/A	0.5 ± 0.02	E. Zhang et al. 2008
Mg-1.98Zn-0.36Y	As-extruded	N/A	198	260	23	N/A	N/A	
Mg-1.84Zn-0.82Y	As-extruded	N/A	213	266	25	N/A	N/A	
Mg-1.73Zn-1.54Y	As-extruded	N/A	~215	~266	27	N/A	N/A	Yin et al. 2008
Mg-1Zn-1Mn	As-extruded	N/A	246.5 ± 4.5	280.3 ± 0.9	21.8 ± 0.6	N/A	N/A	
Mg-2Zn-1Mn	As-extruded	N/A	248.8 ± 0.8	283.8 ± 1.0	20.9 ± 0.7	N/A	N/A	
Mg-3Zn-1Mn	As-extruded	N/A	275.9 ± 0.2	345.5 ± 4.3	10.5 ± 1.2	N/A	N/A	Yuan et al. 2012
Mg-5Zn-1Mn	As-extruded	N/A	229.9	299.88	11.5	N/A	30	
Mg-6Zn-1Mn	As-cast	N/A	108	335	20.3	N/A	N/A	H. J. Zhang et al. 2013b
Mg-6Zn-1Mn	Rapid solidification	N/A	154	460	20.5	N/A	N/A	
Mg-6Zn-1Mn	As-extruded	N/A	209	305	11.55	N/A	N/A	D. F. Zhang et al. 2011

(Continued)

TABLE 7.4 (CONTINUED)

Summary of the Mechanical Properties of the Ternary Mg-Zn Alloys

Sample	Condition	Mechanical Property					Grain Size (μm)	References
		Elastic Modulus (GPA)	Yield Strength (MPa)	Ultimate Strength (MPa)	Elongation (%)	Vickers Hardness (HV)		
Mg-1Zn-1Mn	As-cast	N/A	N/A	~175	~15	N/A	N/A	E. L. Zhang et al. 2009a
Mg-1Zn-1Mn	As-extruded	N/A	N/A	~275	~22	N/A	N/A	
Mg-6Zn-1Mn	As-extruded	N/A	213	312	11.1	N/A	N/A	D. F. Zhang et al. 2008
Mg-4Zn-1Pb	As-cast	N/A	84 ± 0.2	233 ± 2.4	14 ± 2.8	65	N/A	Wei et al. 2014
Mg-4Zn-1Sn	As-cast	N/A	104 ± 10	237 ± 2.4	10 ± 0.7	67	N/A	
Mg-4Zn-3Sn	As-cast	N/A	129 ± 4.2	250 ± 2.6	8 ± 1.4	70		
Mg-4Zn-0.5Zr	As-cast	64	96	176	4	N/A	N/A	Hong et al. 2013
Mg-4Zn-0.5Zr	Solution treated	68	92	134	3	N/A	N/A	
Mg-3Zn-0.6Zr	As-cast	N/A	215	300	9	N/A	N/A	Huan et al. 2010a
Mg-6Zn-0.6Zr	As-cast	N/A	235	315	8	N/A	N/A	
Mg-6Zn-0.5Zr	As-cast	N/A	227	266	15.8	N/A	N/A	Somekawa et al. 2007
Mg-6Zn-0.5Zr	As-extruded	N/A	269	315	12.3	N/A	N/A	
Mg-5.16Zn-0.57Zr	Warm rolled	N/A	187.8	270.5	20.4	N/A	N/A	Y. Kim et al. 2014
Mg-11Gd-1Zn	As-cast	N/A	159	196	3	N/A	N/A	R. Zhen et al. 2013
Mg-11Gd-1Zn	As-extruded	N/A	191	33	9	N/A	N/A	
Mg-1.92Zn-0.08Gd	As-cast	N/A	291	343	11.4	N/A	1–2	Z. H. Huang et al. 2013
Mg-2.70Zn-0.47Gd	As-cast	N/A	228	320	15.8	N/A	2–4	
Mg-1.92Zn-0.08Gd	As-extruded	N/A	410	434	14.6	N/A	<1	
Mg-2.70Zn-0.47Gd	As-extruded	N/A	417	444	13.1	N/A	<1	

Zn content in the Mg-Zn-Ca bulk metallic glasses appears to have a notable effect on the biocorrosion rate of the system. Gu et al. (2010c) reported that a $Mg_{66}Zn_{30}Ca_4$ sample exhibited a more positive OCP value than a $Mg_{70}Zn_{25}Ca_5$ sample. The higher Zn contents also led to a more steady OCP value and the diminution of the reaction of Mg with water under hydrogen evolution. Cao et al. (2012) found that more Zn content and a smaller (Ca + Mg):Zn ratio would correlate to a lower mass rate.

7.3.4.2 Mg-Zn-Zr Alloys

The addition Zr or Zn into Mg to form a binary alloy would reduce the corrosion rate in comparison with pure Mg (Gu et al. 2009b). For the ternary Mg-Zn-Zr alloys with different Zr and Zn concentrations, the corrosion rate and corrosion mode were totally different. During the immersion tests, the ZK30 alloys exhibited a similar degradation rate with WE-type alloy, and the ZK60 alloy showed an accelerated degradation rate in the first 9 weeks and completely degraded at the end of 12 weeks. Nevertheless, the ZK60 alloy showed a similar degradation mode to the WE-type alloy. During the early stages, localized pitting was the main mode of degradation. With increasing immersion time, discrete spots developed further and led to the formation of a slightly porous morphology. ZK30 alloy showed a completely different mode of degradation. A uniform shallow pitting damage can be seen on the sample surface. Pits developed very slowly, and few drop-off particles detached from the sample (Huan et al. 2010a). Gu et al. (2011c) reported that the as-extruded ZK60 alloy had higher corrosion potentials and lower corrosion current density than the as-cast alloy. The higher hydrocarbonate concentration in DMEM than that in Hank's solution also led to a faster corrosion of ZK60 alloy. These suggested that a loose external layer would form during the immersion periods (Gu et al. 2011c). Hong et al. (2013) reported that T4-treated ZK40 alloy possessed higher breakdown potential and would prevent localized corrosion by passivation phenomena.

7.3.4.3 Mg-Zn-Y Alloys

Y is regarded as a stabilizer for the surface passive films formed on the Mg alloys during the degradation procedure (Yao et al. 2003; M. Liu et al. 2010). The degradation of Mg-Zn-Y alloys, except for $Mg_{97}Zn_1Y_2$ alloy, in the corrosive medium consists of two stages. These alloys suffered from a fast degradation rate at the beginning, and then the formation of a protective layer slowed down the degradation rate. For the $Mg_{97}Zn_1Y_2$ alloy, it exhibited a steady and smaller corrosion rate (J. S. Zhang et al. 2012b).

Zhao et al. (2013a) also reported the two-stage degradation behavior of LPSO-containing Mg-Zn-Y alloys. At the stage of uniform corrosion, the dissolution of α-Mg is hindered due to the presence of the LPSO phase. Therefore, alloys with a higher volume fraction of LPSO phase exhibit better corrosion resistance. At the second stage, because the pitting corrosion occurs, the galvanic corrosion determines the degradation rate; hence a higher volume fraction of LPSO accelerates the corrosion rate. It is also suggests that both the corrosion potential and the corrosion current were increased with the increasing Zn content. The increase of Zn content may lead to the formation of a Mg_2Zn Laves phase, which is known to be a very active anodic particle and can be easily dissolved in aqueous corrosive medium (Nam et al. 2013).

7.3.4.4   Mg-Zn-Mn Alloys

The appearance of Zn in the Mg-Zn-Mn alloy can obviously shift the corrosion potential and pitting potential to the positive side due to the higher potential than Mg. It has been also reported that the Zn element can improve the stability of the passivation film and thus improve the corrosion resistance (Yin et al. 2008). However, higher Zn content, such as up to 3%, would lead to the increasing of the Mg-Zn phase and accelerate the corrosion (Yin et al. 2008). Zhang draws a similar conclusion that the corrosion resistance decreases with the increasing Zn content in the Mg-Zn-Mn alloys (E. L. Zhang et al. 2009b). At the same time, Rosalbino et al. (2013) reported that a partially protective layer was formed on the Mg-Zn-Mn alloy surface, and the extent of the current plateau significantly increased with increasing Zn and Mn content in the alloy, indicating that the corrosion resistance increases with increasing the Zn and Mn content.

Zhang reported that the RS ZM61 alloy exhibited lower corrosion current density, higher pitting potential, and a wider passive region in comparison with the as-cast ZM61 alloy. Furthermore, the corrosion penetration rate of the RS ZM61 alloy was only one 40th of the extruded ZK60 alloy (H. J. Zhang et al. 2013b).

A summary of the corrosion rates of the ternary Mg-Zn-X alloys in different corrosive media is shown in Table 7.5.

7.3.5   Cytotoxicity and Hemocompatibility

Gu et al. (2010c) evaluated the cytotoxicity of the Mg-Zn-Ca bulk glasses by both indirect and direct assays, and found that $Mg_{66}Zn_{30}Ca_4$ and $Mg_{70}Zn_{25}Ca_5$ alloy extracts improved cell viability for L929 cells (Figure 7.12b) and MG63 cell lines (Figure 7.12a) in comparison with as-rolled pure Mg extract. For the direct assay, the attached cell number was only a few on day 1, but an increased cell number with healthy cell morphologies was observed after 3 and 5 days incubation.

For Mg-Zn-Zr alloys, in vitro biocompatibility of ZK30, ZK40, and ZK60 has been studied. Zr has been reported to be suitable for orthopedic implantation when taking the biocompatibility into consideration (Huan et al. 2010a; Gu et al. 2011c). According to the calcium-Am and ethidum D-1 staining results, both the as-cast and T4-treated ZK40 alloy improved MC3T3-E1 cell viability when compared with AZ31 and pure Mg. The attached cells fixed in glutaraldehyde solution on the sample surface spread uniformly with filopodia, indicating a good health condition. The cell viability and cell morphology by direct assays are shown in Figure 7.12c, d. MTT testing revealed that the ZK40 alloy possesses good cell compatibility and showed it was noncytotoxic (Hong et al. 2013). Huan's work (2010a) suggested that the ZK30 alloy showed quite similar cytocompatibility to HA after 1 day of culture. ZK30 alloy displayed a significantly higher rBMSC cell proliferation rate with the prolonged incubation periods. Gu et al. (2011c) reported that the 100% concentration extracts of ZK60 alloy reduced the cell numbers by 40%, and increased cell numbers were observed with decreasing the concentration of the extracts. The morphology of MG63 cells and L-929 cells cultured on the sample surface is shown in Figure 7.12e, f. The ALP activity of the ZK60 alloy is significantly higher than the negative control, implying a higher bone-formation ability.

TABLE 7.5

Summary of the Corrosion Rates of the Ternary Mg-Zn Alloys in Different Corrosive Media

Sample	Condition	Corrosion Rates in Different Corrosive Media				References
		DMEM + 10% FBS	Hank's Solution	3.5% NaCl Solution	I_{corr}	
Mg-2.4Zn-0.1Ag	As-cast	N/A	N/A	1311 μm/y	60 μA/cm²	Mandal et al. 2014
Mg-2.4Zn-0.1Ag	Peak aged	N/A	N/A	874 μm/y	40 μA/cm²	
Mg97Zn1Y2	As-cast	N/A	N/A	1.15 mg/(cm²/day)	N/A	J. S. Zhang et al. 2012b
Mg-2Zn-0.2Mn	As-cast	N/A	3.7 mm/y	N/A	N/A	N. I. Z. Abidin et al. 2011a
Mg-2Zn-0.2Mn	As-cast	N/A	N/A	N/A	N/A	
Mg-6Zn-1Mn	As-cast	N/A	N/A	11 mm/y[a]	10$^{-5.44}$ μA/cm²	H. J. Zhang et al. 2013b
Mg-6Zn-1Mn	Rapid solidification	N/A	N/A	0.01 mm/y	10$^{-6.31}$ μA/cm²	
Mg-4Zn-0.5Zr	As-cast	0.84 ± 0.002 mm/y	N/A	N/A	37.19 ± 1.08 μA/cm²	Hong et al. 2013
Mg-4Zn-0.5Zr	Solution treated	0.86 ± 0.004 mm/y	N/A	N/A	38.07 ± 2.01 μA/cm²	
Mg-6Zn-0.5Zr	As-cast	30.19 μa/cm² (without FPS)	15.35 μa/cm²	N/A	N/A	Gu et al. 2011c
Mg-6Zn-0.5Zr	As-extruded	13.26 μa/cm² (without FPS)	10.17 μa/cm²	N/A	N/A	
Mg-1Zn-3Gd	As-cast	N/A	N/A	0.83 mm/y[b]	N/A	Kubasek and Vojtech 2013c
Mg-3Zn-3Gd	As-cast	N/A	N/A	5.29 mm/y[b]	N/A	

[a] Concentration was 3%.

[b] Concentration was 0.15 mol/l.

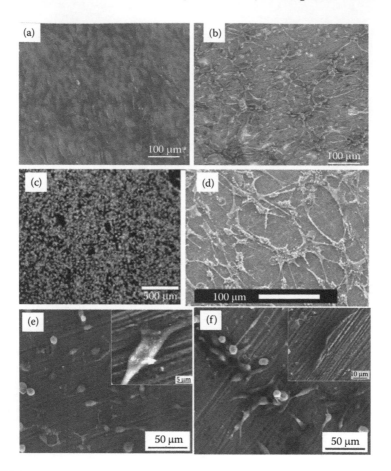

FIGURE 7.12 Cell morphology cultured with Mg-Zn ternary alloys or alloy extracts: (a) MG63 cells, (b) L929 cells cultured on Mg66Zn30Ca4 BMG for 5 days, (c) fluorescent image of live and dead MC3T3-E1 cells cultured on as-cast ZK40 alloy for 3 days, (d) MC3T3-E1 cells fixed on as-cast ZK40 alloy after 3 days of incubation, morphologies of (e) L929 and (f) MG63 cells cultured on extruded ZK60 alloy for 24 h. ([a, b] Modified from Gu, X. N., Zheng, Y. F., Zhong, S. P., Xi, T. F., Wang, J. Q., and Wang, W. H., *Biomaterials*, 31, 1093, 2010. With permission; [c, d] Modified from Hong, D., Saha, P., Chou, D. T., Lee, B., Collins, B. E., Tan, Z. Q., Dong, Z. Y., and Kumta, P. N., *Acta Biomaterialia*, 9, 8534–47, 2013. With permission; [e, f] Modified from Gu, X. N., Li, N., Zheng, Y. F., and Ruan, L Q., *Materials Science Engineering B-Advanced Functional Solid-State Materials*, 176, 1778, 2011. With permission.)

E. L. Zhang et al. (2009b) evaluated the biocompatibility of Mg-1Zn-1Mn alloy and found it was Grade 0 according to the ISO standards. Nevertheless, the hemolysis rate of bare Mg-1Zn-1Mn alloy was about 65.75%, which was much higher than the safe value of 5%. They suggested that a surface modification would be necessary before biomedical application of Mg-1Zn-1Mn alloy.

Cell morphology cultured with various Mg-Zn-X ternary alloys or alloy extracts are showed in Figure 7.12.

7.3.6 In Vivo Performance in Animal Models

Rat models were selected to investigate the in vivo biocompatibility of the Mg-1Zn-1.2Mn alloy (L. P. Xu et al. 2007). The implants were implanted in the femora of the rats with a dimension of 4.0 mm in length and 1.5 mm in diameter. The implant rods changed to an irregular shape after 9 weeks, and new bone can be observed between the degradation layer and bone tissues (Figure 7.13a). After 18 weeks of implantation, more new bone was found, and the implants were fixed tightly with no inflammation that can be seen. The newly formed bone exhibited no difference in the histological microstructure with the cortical bone. During the implantation periods, no difference was found in the blood biochemical evaluation. This implied that the Mg-Zn-Mn alloy may be a promising biomaterial for orthopedic application.

Zberg et al. (2009b) implanted $Mg_{60}Zn_{35}Ca_5$ bulk metallic glass rods into the abdominal walls and cavities of domestic pigs to investigate the tissue reaction. According to their results, no gas cavities formed by hydrogen evolution or inflammatory reaction were observed around the glass, indicating that the Mg-Zn-Ca BMG possesses good biocompatibility (Figure 7.13b, c).

Histological microstructure of the in vivo studies of ternary Mg-Zn-based alloys are shown in Figure 7.13.

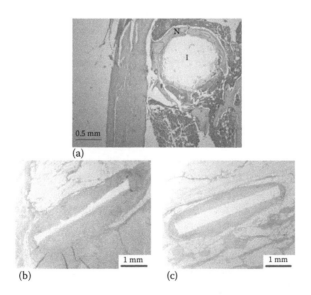

FIGURE 7.13 Histological microstructure of the in vivo studies of ternary Mg-Zn alloys: (a) new bone near the Mg-Zn-Mn implant after 9 weeks postoperation, N means newly formed bones and I means implant, Mg60Zn35Ca5 BMG in porcine (b) abdominal tissue after 27 days, (c) subcutis after 91 days. ([a] Modified from Xu, L. P., Yu, G. N., Zhang, E., Pan, F., and Yang, K., *Journal of Biomedical Materials Research Part A*, 83A, 703, 2007. With permission; [b, c] Modified from Zberg, B., Uggowitzer, P. J., and Loffler, J. F., *Nature Materials*, 8, 887, 2009. With permission.)

7.4 DEVELOPMENT OF Mg-Zn QUATERNARY ALLOYS AS DEGRADABLE METALLIC BIOMATERIALS

7.4.1 ALLOY COMPOSITION DESIGN

Rare earth elements are beneficial in terms of mechanical and corrosion properties for Mg alloys (Feyerabend et al. 2010). Feyerabend et al. (2010) suggested that highly soluble Dy and Gd and low solid solubility elements Eu, Nd, and Pr elements would be biosafe in short-term evaluation with primary cells and cell lines. Until now, large numbers of Mg-Zn-based alloys containing RE elements, such as Mg-Nd-Zn-Zr (Liao et al. 2012; Mao et al. 2012; X. G. Zhang et al. 2012b; X. B. Zhang et al. 2012d,e,f, 2013a), Mg-Zn-Dy-Zr (J. S. Zhang et al. 2013b), Mg-Zn-RE-Zr (Q. Chen et al. 2012), Mg-Zn-Y-Zr (J. Fan et al. 2013), Mg-Y-Er-Zn (L. Zhang et al. 2014), Mg-Zn-Y-RE (Perez et al. 2013), Mg-Zn-Mn-Sn (F. G. Qi et al. 2014), Mg-Zn-Mn-Y (He et al. 2010), Mg-Zn-Ca-Ce (Gao et al. 2013), Mg-Zn-Ca-Zr (Homma et al. 2010; Guan et al. 2012), Mg-Zn-Ca-Gd (M. B. Yang et al. 2013a), and Mg-Zn-Ca-Y (Jiang et al. 2013), have been widely studied.

7.4.2 FABRICATION, HOT AND COLD WORKING, AND HEAT TREATMENT

Most of the quaternary Mg-Zn-based alloys were fabricated via the conventional metallurgy method. For the fabrication of Mg-Zn-Nd-Zr alloy, pure Mg, pure Zn, commercial Mg-25Nd, and Mg-30Zr master alloys were selected. Mg and Zn were melted at 700–710°C, and then the master alloys were added at 740–760°C. Subsequently, the melts were held at 780°C for 30 min to ensure the complete dissolution of Zr. During the melting progress, argon atmosphere and a special refine flux were applied (Zong et al. 2012). Further working and heat treatment were also applied on the as-cast alloys. The temperature and solution time for the solution treatment were in the range of 525–540°C and 6–10 h, followed by quenching in water. The extrusion temperatures were selected to be 320°C or 350°C with an extrusion rate of 2 mm/s (Mao et al. 2012; X. B. Zhang et al. 2012f; Zong et al. 2012). Mg, Zn, RE elements in the form of intermediate alloys, and Mg-30Zr alloy were melted at the temperature of 760°C to fabricate the Mg-5.3Zn-1.13Nd-0.51La-0.28Pr-0.79Zr. The melt was stirred and kept at 730°C for 25 min, followed by pouring into a mold to get the as-cast ingots. Further hot-extrusion was applied at a temperature between 250°C and 450°C with a extrusion ratio of 9 to 100 (Q. Chen et al. 2012).

Fan developed the Mg-1.5Y-1.2Zn-0.44Zr alloy. Pure Mg, Zn, Mg-20Y, and Mg-30Zr master alloy were melted at 750°C, then poured into a preheated mold at 720°C. After homogenization at 450°C for 10 h, the as-cast alloys were then hot-extruded at 350°C (J. Fan et al. 2013).

Pérez developed the as-extruded Mg-2Zn-1.5Y-1.5Ce MM alloy at the temperature of 400°C and an extrusion ratio of 18:1 (Pérez et al. 2013).

A series of Mg-6Zn-1Mn-(1-10)Sn alloys were fabricated at the temperature of 730°C with pure Mg, Zn, Sn, and Mg-4.1Mn master alloy (F. G. Qi et al. 2014).

The Mg-2Zn-1Mn-(0.36,0.82,1.54)Y alloys were prepared with high-purity Mg, Zn, analysis grade $MnCl_2$, and Mg-20Y master alloy. The melting temperature varied from 698°C to 810°C, depending on the alloy composition. The as-cast ingots were subsequently extruded at 280°C with a extrusion ratio of 10:1 (He et al. 2010).

Mg-2.26Zn-0.38Mn-0.23Ce alloy was developed using pure Mg, Zn, Mg-30Ce, and Mg-11Mn master alloys by melting at about 780°C under the protection flux and poured at 720°C into the mold preheated to 200–300°C. After homogenization at a temperature of 400°C for 10 h, the as-cast alloys were rolled at 450°C with a 15% reduction per pass (Gao et al. 2013).

Homma reported that the Mg-5.7Zn-0.17Ca-0.84Zr alloy was hot-extruded at 300°C with a ram speed of 0.1 mm/s and a extrusion ratio of 20 (Homma et al. 2010).

Mg-3.8Zn2.2-Ca-(0.5-3)Gd alloy was produced by melting the pure Mg, Zn, Mg-18Ca, and Mg-25Gd master alloy at 740°C; the melts were homogenized by mechanical stirring and held at 740°C for 20 min before pouring into a mold, which was preheated to 150°C (M. B. Yang et al. 2013).

Mg-6Zn-0.5Ca-(0.5-1.5)Y alloy was produced with pure Mg, Zn, Mg-30Ca, and Mg-25.7Y master alloy at the temperature of 750°C. The as-cast alloy was further solution treated at 460°C for 24 h (Jiang et al. 2013).

Mg-1.5Zn-4.5Dy-(0.08-0.35)Zr alloys were prepared with pure Mg, Zn, Dy, and Mg-30Zr master alloy. The raw materials were melted at 760°C for about 20–30 min and poured into a mold preheated at 200°C at the temperature of 710°C. The cast alloys were solution treated at 500°C for 27–40 h, followed by quenching in hot water at 100°C (J. S. Zhang et al. 2013b).

7.4.3 MICROSTRUCTURE AND MECHANICAL PROPERTIES

The extruded Mg-Gd-Zn-Zr alloys exhibited an obviously refined structure, which was mainly composed of refinement grains, lamellar X phase in grain boundary and 14H-type LPSO structure. In addition, a small amount of β phase [$(Mg,Zn)_3Gd$] with a face-center cubic structure was also observed (Figure 7.14a) (X. B. Zhang et al. 2012d). Another study also reported that the extruded Mg-Gd-Zn-Zr alloy exhibited finer and more uniform distribution of the grain when compared with the AZ31 alloy and WE43 alloy (Figure 7.14b) (X. B. Zhang et al. 2012e). Zhang reported that a fine grain of about 1 μm, a coarse grain of about 5 μm, and a long elongated grain of more than 100 μm were observed in the extruded Mg-Gd-Zn-Zr alloy (X. Zhang et al. 2012b). The Zr concentration in the Mg-Zn-Dy-Zr alloys can tailor the phase composition as described in J. S. Zhang et al. (2013b). When the Zr level is 0.35%, the alloy mainly consisted of a Mg_8ZnDy eutectic phase and α-Mg matrix. The Mg_8ZnDy phase changed into a $Mg_{12}ZnDy$ phase when the Zr decreased to 0.17%. Furthermore, the LPSO phase decreased with the addition of Zr content, and the area fraction of the Mg_8ZnDy phase increased. It indicated that the Zr inhibited the formation of the LPSO phase but was beneficial to the formation of the Mg_8ZnDy phase. After heat treatment of the alloy, the Mg_8ZnY phase can be observed in all the alloys. The bamboo-like LPSO phases changed into acicular phases, and block LPSO phases randomly distributed can also be seen (Figure 7.14g) (J. S. Zhang et al. 2013b). The as-cast Mg-Zn-RE-Zr alloy mainly consisted of α-Mg matrix and the $MgZn_2$ phase with eutectic constituents at the grain boundaries. Wavelength dispersive X-ray analysis revealed that Zn and RE element were higher at the grain boundary than

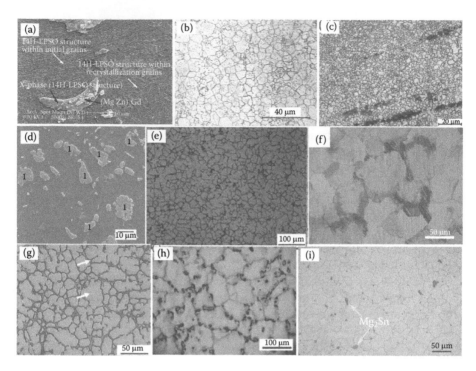

FIGURE 7.14 Typical microstructure of Mg-Zn quaternary alloys: (a) as-extruded Mg-11.3Gd-2.5Zn-0.7Zr, (b) as-extruded Mg-3.09Nd-0.22Zn-0.44Zr, (c) as-rolled Mg-5.5Zn-0.6Zr-0.8Gd, (d) as-cast Mg95Zn3Y1RE1, (e) as-cast Mg-3.8Zn-2.2Ca-2.0Gd, (f) as-cast Mg-1.5Y-1.2Zn-0.44Zr, (g) as-cast Mg-1.5Zn-4.5Y-0.35Zr, (h) as-cast Mg-Zn-Nd-La-Pr-Zr, and (i) solution treated Mg-6Zn-1Mn-4Sn. ([a] Modified from Zhang, X. B., Wu, Y. J., Xue, Y. J., Wang, Z. Z., and Yang, L., *Materials Letters*, 86, 423, 2012. With permission; [b] Modified from Zhang, X. B., Yuan, G. Y., Mao, L., Niu, J. L., and Ding, W. J., *Materials Letters*, 66, 209, 2012. With permission; [c] Modified from Yu, H. Y., Yan, H. G., Chen, J. H., Su, B., Zheng, Y., Shen, Y. J., and Ma, Z. J., *Journal of Alloys and Compounds*, 586, 757, 2014. With permission; [d] Modified from Perez, P., Onofre, E., Cabeza, S., Llorente, I., del Valle, J. A., Garcia-Alonso, M. C., Adeva, P., and Escudero, M. L., *Corrosion Science*, 69, 226, 2013. With permission; [e] Modified from Yang, M. B., Guo, T. Z., and Li, H. L., *Materials Science and Engineering A-Structural Materials Properties Microstructure and Processing*, 587, 132, 2013. With permission; [f] Modified from Fan, J., Qiu, X., Niu, X. D., Tian, Z., Sun, W., Liu, X. J., Li, Y. D., Li, W. R., and Meng, J., *Materials Science and Engineering C-Materials for Biological Applications*, 33, 2345, 2013. With permission; [g] Modified from Zhang, J. S., Xin, C., Cheng, W. L., Bian, L. P., Wang, H. X., and Xu, C. X., *Journal of Alloys and Compounds*, 558, 195, 2013. With permission; [h] Modified from Chen, Q., Shu, D. Y., Zhao, Z., Zhao, Z. X., Wang, Y. B., and Yuan, B. G., *Materials and Design*, 40, 488, 2012. With permission; [i] Modified from Qi, F. G., Zhang, D. F., Zhang, X. H., and Xu, X. X., *Journal of Alloys and Compounds*, 585, 656, 2014. With permission.)

within the grain. Zr concentration remained fairly consistent within the grain and the grain boundary. Chen found that the extrusion ratio of 9 would be sufficient for the complete dynamic recrystallization. When the extrusion ratio reached 16, fine and equiaxed recrystallized grains were uniformly distributed in the microstructure (Figure 7.14h) (Q. Chen et al. 2012). The alloying elements of Y, Zn, and Zr led to the formation of secondary phases, which existed at the triple points of the grains and distributed along the grain boundaries discontinuously and were further identified as the $Mg_{12}ZnY$ phase (Figure 7.14f). After solid solution treatment, parts of the secondary phases dissolved in the matrix. The hot extrusion process resulted in much finer grain size (J. Fan et al. 2013).

The as-cast Mg-Y-Er-Zn alloy mainly consisted of Mg dendrites and a fine lamellae intermetallic phase, which were orientated in one direction. After hot extrusion, the lamellar phase showed a long strip-like shape and was rearranged parallel to the extrusion direction. It was obvious that a refined and uniform microstructure was formed through the extrusion process (L. Zhang et al. 2014).

The addition of Sn in the Mg-Zn-Mn alloy can refine the grain size and lead to the formation of a Mg_2Sn phase at the grain boundary (Figure 7.14i). The volume fraction of the Mg_2Sn phase increased with the increasing Sn contents. After homogenization, Mg-Zn compounds at the grain boundary dissolved into the matrix, and the blocky Mg_2Sn phase was still observed. During the hot extrusion process, the undissolved block compounds were further broken into small particles and distributed as streamlined along the extrusion direction (F. G. Qi et al. 2014).

In the as-cast Mg-Zn-Ca-Gd alloy, α-Mg, Mg_2Ca, and $Ca_2Mg_6Zn_3$ were all detected, but the $Mg_3Gd_2Zn_3$ phases were only detected when the Gd content was higher than 1.49%. After the addition of Gd, the $Ca_2Mg_6Zn_3$ phases changed from the initial continuous and/or quasicontinuous net to quasicontinuous and/or disconnected shapes. Furthermore, the $Mg_3Gd_2Zn_3$ phases in the alloys with the additions of 1.49 wt.% and 2.52 wt.% Gd mainly exhibited particle and/or feather-like shapes, and increasing Gd content resulted in more and bigger $Mg_3Gd_2Zn_3$ phases along the grain boundaries (M. B. Yang et al. 2013a). Jiang reported that the increasing of Y can gradually refine the grain size of Mg-Zn-Ca-Y alloy. The spherical shape $MgZn_2$ phase precipitated in the grain interior, and $Ca_2Mg_6Zn_3$ precipitated at the grain boundary. The addition of Y resulted in the formation of the $Mg_{12}ZnY$ phase with a granular shape (Figure 7.14e) (Jiang et al. 2013).

The typical microstructure of the Mg-Zn quaternary alloys is comprehensively illustrated in Figure 7.14. A summary of the mechanical properties of the quaternary Mg-Zn alloys is listed in Table 7.6.

7.4.4 Degradation in Simulated Body Fluids

A summary of the corrosion rates of the quaternary Mg-Zn alloys in different corrosive media is given in Table 7.7.

The refined grain in the Mg-Nd-Zn-Zr alloy can be attributed to a higher fraction of grain boundaries, which accelerate the passivation kinetics and reduce the intensity of a galvanic coupling between the grain interior and grain boundary. Mao et al. (2012) reported that a Mg-Nd-Zn-Zr alloy exhibited excellent uniform corrosion behavior in artificial plasma. Another study also revealed that the Mg-Nd-Zn-Zr

TABLE 7.6

Summary of the Mechanical Properties of the Quaternary Mg-Zn Alloys

Sample	Condition	Mechanical Property			Vickers Hardness (HV)	Grain Size (μm)	References
		Yield Strength (MPa)	Ultimate Strength (MPa)	Elongation (%)			
Mg-6Zn-1Mn-4Sn	As-extruded	378	390	4.16	82	N/A	F. G. Qi et al. 2014
Mg-6Zn-0.5Ca-1Y	As-cast	166	284	14.7	N/A	N/A	Jiang et al. 2013
Mg-6Zn-0.2Ca-0.8Zr	As-extruded	310	357	18	N/A	N/A	Homma et al. 2010
Mg-3.8Zn-2.2Ca-2.0Gd	As-cast	118.9 ± 2.7	141.5 ± 1.5	4.2 ± 0.12	N/A	N/A	M. B. Yang et al. 2013a
Mg-5.5Zn-0.6Zr-0.8Gd	As-cast	242	327	22	N/A	N/A	Yu et al. 2014
Mg97Zn1Y1Nd1	As-extruded	330	358	6.6	N/A	N/A	Kim and Kawamura 2013
Mg97Zn1Y1Sm1	As-extruded	324	361	9.3	N/A	N/A	
Mg-2.26Zn-0.38Mn-0.23Ce	As-rolled	168	260	27.1	N/A	N/A	Gao et al. 2013
Mg-2Zn-8Y-1Er	As-cast	94	162	7	N/A	400	L. Zhang et al. 2014
Mg-2Zn-8Y-1Er	As-extruded	275	359	19	N/A	10	
Mg-Nd-Zn-Zr (patented)	As-extruded	308 ± 6	312±2	12.2 ± 0.6	N/A	2	X. B. Zhang et al. 2012f
Mg-1.2Zn-1.5Y-0.44Zr	As-cast	78 ± 5.29	122 ± 2.31	7.2 ± 0.62	N/A	32	J. Fan et al. 2013
Mg-1.2Zn-1.5Y-0.44Zr	As-extruded	178 ± 2.12	236 ± 1.41	28 ± 0.58	N/A	10	

TABLE 7.7

Summary of the Corrosion Rates of the Quaternary Mg-Zn Alloys in Different Corrosive Media

Sample	Condition	Corrosion Rates in Different Corrosive Media		References
		SBF	Hank's Solution	
Mg-3.09Nd-0.22Zn-0.44Zr	As-cast	N/A	0.337 mm/y	Mao et al. 2012
Mg-11.3Gd-2.5Zn-0.7Zr	As-extruded	N/A	0.17 mm/y	X. B. Zhang et al. 2012d
Mg-10.2Gd-3.3Y-0.6Zr	As-extruded	N/A	0.55 mm/y	
Mg-1.2Zn-1.5Y-0.44Zr	As-cast	9.41×10^{-3} mm/day	N/A	J. Fan et al. 2013
Mg-1.2Zn-1.5Y-0.44Zr	As-extruded	3.25×10^{-3} mm/day	N/A	

alloy exhibited more positive pitting corrosion potential and lower corrosion current density when compared with the AZ31 alloy and WE43 alloy (X. B. Zhang et al. 2012e). Zhang found that the Mg-Nd-Zn-Zr alloy with LPSO structure showed much better corrosion resistance (X. B. Zhang et al. 2012d).

The immersion test showed that during the first 24 h, the as-cast, heat-treated, and as-extruded Mg-Y-Zn-Zr alloy exhibited quit similar corrosion rates. However, with prolonged immersion time, the extruded alloy degraded more slowly. The extruded alloy also possessed a more positive corrosion potential and lower corrosion current density, and no obvious pitting corrosion point was observed (J. Fan et al. 2013). The addition of Y dropped the cathodic current density and slowed the rate of hydrogen evolution (He et al. 2010).

7.4.5 CYTOTOXICITY AND HEMOCOMPATIBILITY

Liao et al. (2012) evaluated the in vitro biocompatibility of the Mg-Nd-Zn-Zr alloys. The indirect assay showed that the chondrocytes grow well on the sample surface with both elongated and flat shapes. A higher number of cells were found on the CaHPO$_4$-coated Mg-Nd-Zn-Zr alloy, on which the cells were well distributed and clearly formed connections and confluences. Furthermore, the coated Mg-Nd-Zn-Zr alloy exhibited higher absorbance and relative express levels of Col II and aggrecan mRNA, which indicated higher chondrocyte proliferation. The morphology of chondrocyte cells cultured on the Mg-Nd-Zn-Zr alloys is shown in Figure 7.15.

X. B. Zhang et al. (2012f) evaluated the in vitro response of L-929 cells to the as-extruded Mg-Nd-Zn-Zr alloy with different extrusion ratios. The cell viability

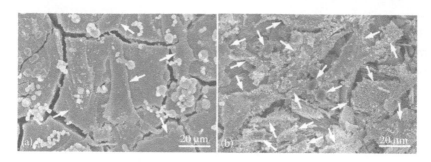

FIGURE 7.15 Morphology of chondrocyte cells cultured on the (a) JDBM (a patented Mg-Nd-Zn-Zr alloy), (b) CaHPO4-coated JDBM. (Modified from Liao, Y., Ouyang, Y. M., Niu, J. L., Zhang, J., Wang, Y. P., Zhu, Z. J., Yuan, G. Y., He, Y. H., and Jiang, Y., *Materials Letters*, 83, 206, 2012. With permission.)

of the L-929 cells after incubation in the 100% concentration extracts for 1, 3, and 5 days were more than 80%, in comparison with the negative control, and met the basic requirement of cell toxicity according to ISO 10993-5:1999. The cells exhibited a healthy morphology with a flattened spindle-like shape.

J. Fan et al. (2013) reported that the Mg-Y-Zn-Zr alloy exhibited good biocompatibility toward L-929 cells. They also suggested that the extracts of the alloy can promote the cell viability with the increasing incubation time.

7.5 CONCLUDING REMARKS

Mg-Zn system-based alloys have been proven to be biosafe for biomedical biodegradable alloys for orthopedic applications. However, the corrosion resistance and mechanical property need to be further improved to fulfill the clinical demands. The mechanisms controlling the influence of the Zn contents on the corrosion and mechanical behavior of the Mg-Zn-based alloys should be fully investigated. The further developments of Mg-Zn-based alloys may be focused on the following aspects.

7.5.1 ALLOY COMPOSITION DESIGN

Mg-Zn ternary alloys with several alloying elements were widely studied in the past, and only a small amount of research on the Mg-Zn binary and Mg-Zn quaternary or multielementary alloys for biomedical applications have been reported. Quaternary or multielementary Mg-Zn alloys provide us more choices to select the alloying elements to design Mg-Zn-based alloys specially for biomedical applications, whereas the biocompatibility of the alloys should be the first principle to consider.

7.5.2 ENHANCING MECHANICAL PROPERTIES

Alloying is an effective way to improve the mechanical behavior of the Mg-Zn alloys by adding different amounts of the various alloying elements (E. L. Zhang et al. 2009b; Cong et al. 2012; González et al. 2012; Zhao et al. 2013a; F. G. Qi et al. 2014).

In addition to the alloying methods, some other attempts were also made intended to improve the mechanical property of the Mg-Zn alloys. Zeng fabricated a carbon nanotube that contained Mg-2.0Zn alloys. The addition of 1% CNTs in the Mg-2Zn alloys significantly increased both the UTS and elongation rate (X. S. Zeng et al. 2013). Other production methods can also improve the mechanical behavior of the Mg-Zn alloys. Mora prepared the high strength Mg-Zn-Y alloys by powder metallurgy methods. The as-extruded Mg-Zn-Y alloys exhibited a UTS about 450 MPa and an elongation rate of about 14% at the temperature of 260 (Mora et al. 2009). Severe plastic deformation can be adopted to fabricate Mg alloys with ultrafine grain and excellent mechanical properties (Li and Zheng 2013).

7.5.3 ENHANCING CORROSION RESISTANCE

The Mg-Zn alloys also suffered from a fast corrosion rate in the in vitro and in vivo tests (Zhang et al. 2010b); some alloying elements, such as Ca, Zr, Y, and Mn, can tailor the microstructure and the phase distribution; thus proper amounts of the alloying elements would improve the corrosion resistance of the Mg-Zn alloys (E. L. Zhang et al. 2009b; Huan et al. 2010a; Hong et al. 2013; Yin et al. 2013). Microalloyed with Ag (Mandal et al. 2014), Pd (Gonzalez et al. 2012) can improve the corrosion behavior. Surface coating also plays an important role in increasing the corrosion resistance of the Mg-Zn alloys (Pan et al. 2013b).

7.5.4 ENHANCING BIOCOMPATIBILITY

To further improve the biocompatibility of the Mg-Zn alloys, the alloying elements and the amounts of the elements should be carefully selected. Yang reported that the micro-arc oxidation (MAO) coating contained Mg_2SiO_4 on the ZK60 alloys improved the proliferation, cell adhesion, differentiation, and hemocompatibility of murine BMSCs (X. M. Yang et al. 2013). Wang fabricated a MAO film on the Mg-1Zn-1Ca alloy. The MAO coating significantly decreased the hemolytic ratio, and the prothrombin and thrombin times were significantly increased. They concluded that the MAO coating would improve the blood compatibility of the Mg-1Zn-1Ca alloy (Wang et al. 2011).

8 Mg-Sr-Based Alloy Systems for Biomedical Applications

8.1 INTRODUCTION

8.1.1 STRONTIUM

Strontium, with a density of 2.64 g/cm³, possesses similar chemical, biological, and metallurgical properties when compared with Mg and Ca, due to the fact that they all belong to group 2 of the periodic table (Landi et al. 2008; Brar et al. 2012; Gu et al. 2012; Cipriano et al. 2013). Sr has a high alloying efficiency and is known as a grain refiner for Mg. The addition of Sr into Mg alloys has a tendency to form compounds that segregate to grain boundaries (Brar et al. 2012; Li et al. 2012). Furthermore, the Sr element improves the corrosion resistance of Mg alloys by altering its surface (Suganthi et al. 2011; Bornapour et al. 2013). It has been reported that Sr could improve both the mechanical properties and corrosion resistance of AZ91 alloy and reduce the microporosity (Zeng et al. 2006; Fan et al. 2007). From a biosafe property standpoint, Sr is an essential trace metal element in the human body, and 99% of it is stored in bones. There is about 140 mg Sr in an adult man. It is recommended that the average daily intake is 2 mg (Gu et al. 2012). Sr is also a natural bone seeking element that accumulates in the skeleton due to its close chemical and physical properties with Ca (Landi et al. 2008). It also has been known that Sr can stimulate the growth of osteoblasts and prevent bone resorption (Brar et al. 2012; Bornapour et al. 2013). Sr administration can stimulate the synthesis of bone collagen (Landi et al. 2008). And furthermore, strontium ranelate has been used in the treatment of osteoporosis to improve the bone strength and bone mineral density (Taylor 1985; Dahl et al. 2001; Marie 2005). Additionally, the degradation of Mg-Sr alloys can help to deposit Sr-substituted HA, which benefits bone mineralization (Guo et al. 2008; Tian et al. 2009).

8.1.2 PREVIOUS STUDIES ON MG-SR ALLOY SYSTEMS FOR ENGINEERING APPLICATIONS

Before its usage for biomedical applications, Sr was usually added as alloying element for creep resistance in die-casting and gravity-cast alloys to reduce microshrinkage porosity (Pekguleryuz 2000; Labelle et al. 2001; Luo and Powell 2001) (see Chapter 6, Section 6.3.2) in the past decades. Mg-Sr master alloys with different Sr contents were also fabricated to introduce elemental Sr into the casting alloys (Yang et al. 2007; Cheng et al. 2008). It had been reported that minor Sr addition between 0.01% and 1% in Mg-Al-Zn alloys could both improve the mechanical property and refine the grain size (Zeng et al. 2006; Liu et al. 2008; Sadeghi and Pekguleryuz 2011). Sr also has been reported as an effective modifier and refiner of Chinese script-shaped Mg_2Si phase in Mg-Al-Si alloys (Nam et al. 2006; Srinivasan et al. 2006). For engineering applications, it has been widely used as an alloy element in Mg-Al-Zn alloys (Hirai et al. 2005; Fan et al. 2007; Chen et al. 2011; Sadeghi et al. 2011), Mg-Al alloys (Pekguleryuz and Baril 2001; Bai et al. 2006; Nam et al. 2011), and Mg-Si alloys (Srinivasan et al. 2006; Yang et al. 2008; Samuel et al. 2014) to improve the mechanical properties, corrosion resistance, and creep properties in high-temperature situations.

8.2 DEVELOPMENT OF Mg-Sr BINARY ALLOYS AS DEGRADABLE METALLIC BIOMATERIALS

8.2.1 Alloy Composition Design

The typical binary diagram of Mg-Sr alloys is shown in Figure 8.1 (Aljarrah and Medraj 2008). As can be seen from the diagram, the element Sr has a very limited solid solubility of about 0.03% in Mg; thus the content of Sr in the Mg-Sr alloys should be carefully selected because the second phases would have detrimental effects on the mechanical and corrosion behavior. Until now, binary Mg-(1–4)Sr alloys (Gu et al. 2012), Mg-(0.5–1.5)Sr alloys (Brar et al. 2012), and Mg-(0.3–2.5)Sr alloys (Bornapour et al. 2013) were reported for biomedical applications.

8.2.2 Fabrication, Hot and Cold Working, and Heat Treatment

For the fabrication of Mg-Sr binary alloys, pure Mg, pure Sr (Brar et al. 2012; Bornapour et al. 2013), or Mg-10Sr master alloy (Gu et al. 2012) was used as raw materials with the conventional metallurgy method. Yang et al. (2012a) developed a

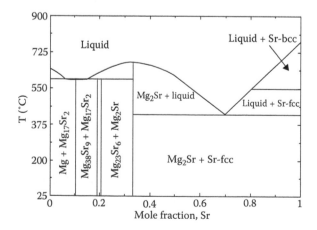

FIGURE 8.1 Typical binary diagram of Mg-Sr alloys. (Modified from Aljarrah, M. and M. Medraj, *Calphad*, 32, 240, 2008. With permission.)

metallothermic reduction method to prepare Mg-Sr master alloys by using the pure Mg and SrO. Liu et al. (2007) reported a similar electrochemical reduction method to produce a Mg-Sr alloy with Mg-base melt and a mixture of $SrCl_2$ and LiCl. The $SrCl_2$ acted as the Sr donor, and the LiCl was the flux used to adjust the density and melting point of the mixture.

Gu et al. (2012) processed as-rolled binary Mg-Sr alloy plates. The as-cast ingots were first preheated to 400°C for 3 h, then rolled with a 0.2-mm thickness reduction each single pass.

8.2.3 MICROSTRUCTURE AND MECHANICAL PROPERTIES

8.2.3.1 Microstructure

The typical metallographic microstructure of the Mg-Sr binary alloys mainly consists of α-Mg grains, and the second phase $Mg_{17}Sr_2$ precipitated along the grain boundaries as shown in Figure 8.2. The matrix was deficient in Sr, which was consistent with the low solubility of Sr in Mg. High levels of Sr can be observed in the second phase at the grain boundaries, and α-Mg can also be found in the eutectic structure (Gu et al. 2012). It is worth noting that both the as-cast and as-rolled Mg-Sr binary alloys exhibited finer grain size with increasing Sr contents, indicating that Sr

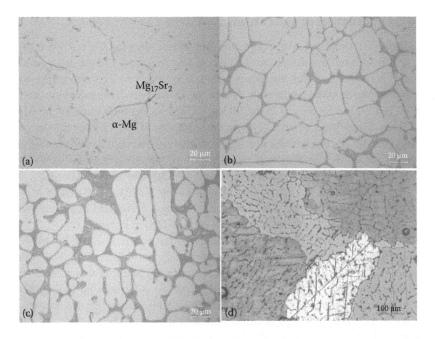

FIGURE 8.2 Optical microstructure of binary Mg-Sr alloys: (a) as-cast-Mg-0.5Sr, (b) as-cast Mg-2.5Sr, (c) as-cast Mg-6Sr, and (d) solution-treated Mg-1Sr. ([a–c] Modified from Aydin, D., Bayindir, Z., and Pekguleryuz, M., *Journal of Materials Science*, 48, 8117–32, 2013. With permission; [d] Modified from Brar, H. S., J. Wong, and M. V. Manuel, *Journal of the Mechanical Behavior of Biomedical Materials*, 7, 87–95, 2012. With permission.)

can be selected as a grain refiner for Mg alloys. And the grain refining effect of Sr can be explained by its low solubility in Mg and the concurrent effects of nonequilibrium solidification (Lee et al. 2000).

8.2.3.2 Mechanical Properties

The addition of Sr would influence the mechanical behavior of the Mg-Sr binary alloys as displayed in Table 8.1. Brar et al. (2012) reported that increasing Sr content in the alloys resulted in the increase of the hardness (HV) values of the homogenized binary Mg-Sr alloys. At the same time, the increase in Sr content had a nominal effect on the tensile strength of the binary Mg-Sr alloys. Gu et al. (2012) found that the UTS and YTS values of the as-rolled Mg-Sr alloys increased with increasing amounts of added Sr up to 2%, and further increasing the Sr contents to 4% decreased the YTS and UTS. In contrast, the elongation of the as-rolled Mg-Sr alloys decreased with increasing Sr content. Therefore, the as-rolled Mg-2Sr alloy exhibited the best combination of strength and ductility among experimental Mg-1–4Sr alloys. It has been revealed that increasing the Sr content from 3% to 4% did not significantly refine the grain size while the volume fraction of $Mg_{17}Sr_2$ phase increased. The high volume of $Mg_{17}Sr_2$ phase might be the crack source with detrimental effects on the ductility of Mg alloys (Gu et al. 2012).

8.2.4 Degradation in Simulated Body Fluids

The addition of Sr can reduce the corrosion rate when compared with the pure Mg. Gu et al. (2012) reported that when the Sr addition was less than 2%, the reduced microshrinkage porosity as well as the reduced grain size improved the corrosion resistance of the binary Mg-Sr alloys, and the alloys exhibited a uniform corrosion mode in Hank's solution. But higher Sr content in the alloys resulted in the increasing of the $Mg_{17}Sr_2$ phase. Thus, the galvanic couples between the α-Mg and $Mg_{17}Sr_2$ phases also increased. Because the $Mg_{17}Sr_2$ phase exhibited higher corrosion potential and was more stable than α-Mg, the anodic dissolution of Mg occurred when the Mg-Sr alloys were exposed in corrosive medium. The degradation of Mg-Sr binary alloys in the Hank's solution resulted in the formation of a corrosion products layer mainly composed of $Mg(OH)_2$ and HA on the sample surface. The typical morphology of as-rolled Mg-2Sr alloy after being immersed in Hank's solution for 500 h is illustrated in Figure 8.3.

Bornapour et al. (2013) reported that the corrosion potential shifted to noble values with the addition of 0.5% Sr, and the current density decreased. The cathodic hydrogen-evolution reaction was also retarded with the addition a Sr alloy element. The immersion results indicated that the Mg-Sr alloy suffered with uniform corrosion, and the changing tendency of the pH value also implied that there was formed a surface barrier layer during the immersion periods.

Brar et al. reported that the Mg-0.5Sr alloy exhibited the lowest corrosion rate according to the volume of hydrogen evolution when compared with the Mg-1Sr and Mg-1.5Sr alloys. They also found that the increasing Sr content in the alloys led to the increase of second phase precipitates, which would enhance microgalvanic corrosion, thus decreasing the corrosion rate (Brar et al. 2012).

TABLE 8.1

Summary of the Mechanical Properties of the Binary Mg-Sr Alloys

Sample	Condition	Mechanical Property				References
		Yield Strength (MPa)	Ultimate Strength (MPa)	Elongation (%)	Grain Size	
Mg-1Sr	As-rolled	~130	~165	~3.2	32.3 ± 6.7	Gu et al. 2012
Mg-2Sr	As-rolled	~148	213.3 ± 17.2	3.2 ± 0.3	25.9 ± 8.2	
Mg-3Sr	As-rolled	~115	~165	~3.0	23.0 ± 8.1	
Mg-4Sr	As-rolled	~80	~110	~2.8	20.9 ± 8.8	
Mg-0.5Sr	Homogenized	37	74	2.6	379	Brar et al. 2012
Mg-1Sr	Homogenized	33	73	3.3	390	
Mg-1.5Sr	Homogenized	40	81	2.6	145	

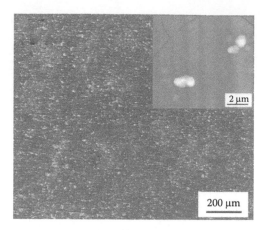

FIGURE 8.3 Morphology of as-rolled Mg-2Sr alloy after 500 h immersion in Hank's solution. The inset is the magnified structure with a smooth surface. (Modified from Gu, X. N., X. H. Xie, N. Li, Y. F. Zheng, and L. Qin, *Acta Biomaterialia*, 8, 2360–74, 2012. With permission.)

8.2.5 Cytotoxicity and Hemocompatibility

Generally speaking, Sr was regarded as a biosafe element for humans. In Gu's work (Gu et al. 2012), Mg63 cells were selected to evaluate the in vitro cytotoxicity of the Mg-(1–4)%Sr alloys. The cell viability obtained from the Mg-1Sr and Mg-2Sr alloy extracts were significantly higher than those of the Mg-3Sr and Mg-4Sr alloys. The Mg-1Sr and Mg-2Sr alloys showed Grade 1 cytotoxicity. The protein normalized ALP activity presented a similar tendency to the cell viability. Mg-3Sr and Mg-4Sr alloys exhibited a lower ALP activity.

Bornapour evaluated the viability of HUVECs in contact with the Mg-0.5Sr alloy extracts, and concluded that the Mg-0.5Sr alloy increased cell viability more than WE43 over longer exposure times to the extracts (Bornapour et al. 2013).

8.2.6 In Vivo Performance in Animal Models

Dog was chosen as the animal model to evaluate the in vivo performance of the Mg-0.5Sr stents (Bornapour et al. 2013). The stents were implanted into the right femoral artery with the WE43 stents as control. After 3 weeks implantation, all the stents maintained their tubular shape. Nevertheless, the Mg-0.5Sr stents lost about half of their initial wall thickness. Some pits caused by localized corrosion were observed on the wall surface of the retrieved Mg-0.5Sr stent. The corrosion rate of the stent was calculated to be 3 mg/day, which was well below the daily allowed limits. It should be noted that the Mg-0.5Sr stent group did not thrombose, and the WE43 stent group did. These in vivo results implied that the Mg-0.5Sr alloy had the potential to offer an optimum combination of appropriate in vivo degradation rate, good biocompatibility, and low risk of thrombosis.

Another in vivo study carried out by Gu et al. (2012) selected 3-month-old male C58BL/6 mice as the model animal. The Mg-2Sr rods were implanted into the

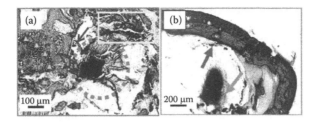

FIGURE 8.4 Histological microstructure of the as-rolled Mg-2Sr alloy implanted in mice: (a) histology of the cross-section of the mouse distal femora. Dot circle is the outline of the implant; triangle is the degraded materials, which dispersed into the bone tunnel; arrows are newly formed bones; (b) histology of cross-section of the mouse femoral shaft, thicker cortical bone was observed with dispersed implanted debris at the edge of the tunnel (black arrow) and some fibrous tissue (gray arrow) around the implant. (Modified from Gu, X. N., X. H. Xie, N. Li, Y. F. Zheng, and L. Qin, *Acta Biomaterialia*, 8, 2360–74, 2012. With permission.)

femur along the axis of the shaft from the distal femur, and in the control group, the predrilled bone tunnel was left empty for comparison. During the implantation periods, local corrosion of the implants at the rod surfaces was observed while the center maintained its integrity. Furthermore, by week 4, a significant increase in the peri-implant bone mineral density of the Mg-2Sr implant group was observed as compared with the control group. Meanwhile, the peri-implant cortical bone thickness was also significantly higher than the control group. It was worth noting that gas evolution was detected immediately after surgery and disappeared 1 week postsurgery. Histology microscopy images (Figure 8.4) revealed that more bone trabecular could be seen in the Mg-2Sr alloy group, and only the Mg-2Sr group showed endosteal new bone formation. The serum Mg and Sr concentration of the Mg-2Sr group at 2 weeks postsurgery showed no significant differences from the values of the control group, implying that the degradation of the Mg-2Sr alloys in vivo were tolerable. This implied that the Mg-2Sr alloy would be promising for orthopedic applications.

8.3 DEVELOPMENT OF Mg-Sr TERNARY ALLOYS AS DEGRADABLE METALLIC BIOMATERIALS

8.3.1 ALLOY COMPOSITION DESIGN

As was described in Gu et al. (2012), the mechanical properties and corrosion resistance of the binary Mg-Sr alloys need to be further improved for long-term implantation. From this point of view, biocompatibility elements such as Ca, Zn, Mn, Y, Zr, Sn, Li, and Si were selected as the alloying elements to fabricate the ternary Mg-Sr alloys in order to broaden its clinical applications. Mg-Ca-Sr (Berglund et al. 2012), Mg-Zn-Sr alloys (Brar et al. 2012; Cipriano et al. 2013; Guan et al. 2013), Mg-Sr-Mn alloys (Borkar et al. 2012a,b, 2013; Celikin et al. 2012; Borkar and Pekguleryuz 2013), Mg-Sr-Y alloys (Gao et al. 2011; Hu et al. 2012), Mg-Zr-Sr alloys (Li et al. 2012), Mg-Sr-Sn alloys (Liu et al. 2010; Obekcan et al. 2013), and Mg-Sr-Li alloys (Jiang et al. 2012) were reported in recent years.

8.3.2 Fabrication, Hot and Cold Working, and Heat Treatment

The fabrication and characterization of Mg-Sr-Ca alloys were described in Section 6.3.2.

The traditional metallurgical process was used to produce the Mg-Sr-Zn alloys. The pure Mg, Zn, and Sr ingots (or granules) were first melted and mechanically stirred in the crucible at temperatures between 720°C and 850°C, and the alloy melts were then held for 30–45 min. The melted mixture was finally poured into molds to get the cast ingots. During the whole melting and casting process, a protective atmosphere should applied to avoid oxidation of the melts (Brar et al. 2012; Cipriano et al. 2013; Guan et al. 2013).

For the Mg-Sr-Mn alloys, Mg-1Mn (Borkar et al. 2012a,b, 2013), Mg-5Mn (Celikin et al. 2012) master alloys were utilized to introduce the Mn element to fabricate a series of Mg-Sr-Mn alloys with Sr content ranging from 0.3% to 5%. The Mg-Mn master alloys were premelted at 700°C, and the Sr was added at 700°C. After holding for 15 min for the dissolution of the Sr, the melt was cast from a temperature of 735–740°C into the mold which was preheated at 400°C (Borkar et al. 2012a; Celikin and Pekguleryuz 2012). Further extrusion was carried out at temperatures between 350°C and 400°C with an extrusion ratio of 7 (Borkar et al. 2013; Borkar and Pekguleryuz 2013).

Pure Mg, strontium carbonate powder, and Mg-Y master alloys were used to fabricate Mg-Sr-Y alloys by melting leaching reduction methods under the protective gas of the mixture of SF_6 and N_2. The as-cast alloys were further extruded at the temperature of 400°C with a extrusion ratio of 8.3:1 (Gao et al. 2011; Hu et al. 2012).

Li prepared a series of as-cast Mg-Sr-Zr alloys. Pure Mg, Mg-30Zr, and Mg-30Sr master alloys were used as raw materials. The melt was stirred for 30 min and cast at 700°C into cylindrical steel dies, which were preheated to 250°C (Li et al. 2012).

For the Mg-Sr-Sn alloys, pure Mg, Sn, and Mg-40Sr master alloys were melted at 730–760°C with mechanical stirring. The melt was held at 700°C for 30 min, followed by casting into molds preheated at 250°C (Liu et al. 2010).

Pure Mg, Li ingots, and Mg-40Sr master alloys were selected to produce Mg-Sr-Li alloys. The raw materials were melted at 700°C and then isothermally held for 10 min, followed by solidification and cooling in the Ar atmosphere. The as-cast ingots were then extruded at 250°C with an extrusion ratio of 27 (Jiang et al. 2012).

8.3.3 Microstructure and Mechanical Properties

8.3.3.1 Microstructure of the Mg-Sr-Based Ternary Alloys

The typical microstructure of Mg-Zn-Sr alloys consists of α-Mg matrix and visible precipitate phases distributed both in the grain boundaries and inside the grains. In the as-cast Mg-4Zn-1Sr alloys, $Zn_{13}Sr$, $Mg_{17}Sr_2$, $MgZn_2$, and MgZn secondary phases were detected by XRD according to Guan's results (Guan et al. 2013), and the Zn and Sr predominantly existed in the precipitated phase.

Brar et al. (2012) reported that with increasing the Zn content from 2% to 4%, the grain size significantly decreased, but further increasing Zn content to 6% resulted in only a marginal decrease of grain size. After the aging process, MgZn phase was detected in the Mg-Zn-Sr alloys containing higher Zn content than 4%. The typical microstructure of the Mg-Zn-Sr alloys is shown in Figure 8.5.

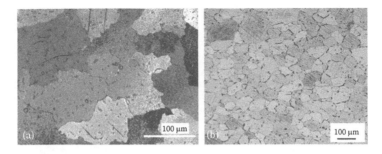

FIGURE 8.5 Typical microstructure of the Mg-Zn-Sr alloys: (a) solution-treated Mg-4Zn-0.5Sr alloy and (b) as-cast Mg-4Zn-1Sr alloy. ([a] Modified from Borkar, H., M. Hoseini, and M. Pekguleryuz, *Materials Science and Engineering A-Structural Materials Properties Microstructure and Processing*, 537, 49–57, 2012. With permission; [b] Modified from Guan, R. G., A. F. Cipriano, Z. Y. Zhao, J. Lock, D. Tie, T. Zhao, T. Cui, and H. N. Liu, *Materials Science and Engineering: C*, 33, 3661–9, 2013. With permission.)

The microstructure of the Mg-Sr-Mn alloys exhibited a typical cast structure with the interdendritic phases surrounding the α-Mg matrix. With the increasing Sr amount in the alloys, the interdendritic phases also increased. According to the XRD results, $Mg_{17}Sr_2$ phase, α-Mn, and α-Mg were detected in the Mg-Sr-Mn alloys. The Sr is concentrated in the interdendritic phase, and the α-Mn presented as polygonal bulk precipitates. Dissolved Mn was also found in the Mg matrix and the interdendritic phase (Celikin et al. 2012). The addition of 0.7% Sr led to a rapid decrease of the grain size of the Mg-Sr-Mn alloy, and the grain size changed very gradually with further increasing the Sr contents (Figure 8.6a, b) (Borkar et al. 2012a). After heat treatment, no coarsening $Mg_{17}Sr_2$ phase was observed although some coarsening α-Mn precipitates were present in the as-cast state. The dynamic precipitation of Mn phase both from the interdendritic phase and the supersaturated α-Mg matrix occurred in the intradendritic regions (Celikin et al. 2012). A typical microstructure of the Mg-Sr-Mn ternary alloys is shown in Figure 8.6.

For the as-cast Mg-Sr-Y alloys, proeutectic α-Mg, lamellae eutectic α-Mg, and $Mg_{17}Sr_2$, fine $Mg_{24}Y_5$ phase were detected. The lamellae eutectics located at the grain boundaries, and the $Mg_{24}Y_5$ particles distributed in both the dendrites and the interdendrites. The addition of Y in the alloys can be in favor of refining the grain size. After the hot extrusion, the lamellae $Mg_{17}Sr_2$ phase was redistributed in the grains and at the boundaries, and some fine recrystallized grains can also be seen at the deformed grain boundaries. The typical microstructures of the Mg-Sr-Y alloys in different states can be seen in Gao et al. (2011) and Hu et al. (2012).

For the Mg-Sr-Zr alloys, α-Mg and $Mg_{17}Sr_2$ phases were detected from the XRD characterization. The grain size of the Mg-Sr-Zr alloys decreased with increasing Zr content although there were no obvious differences in the grain size for the alloys with increasing Sr content when the Zr content was kept constant. However, the grain boundaries became rougher and broader with increasing Sr addition (Li et al. 2012). The typical microstructure of the Mg-Sr-Zr alloys is shown in Figure 8.7.

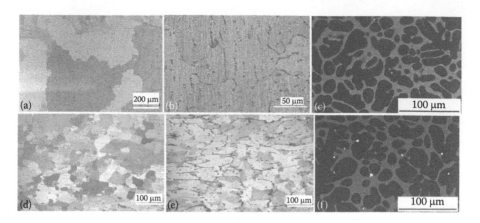

FIGURE 8.6 Typical microstructure of the Mg-Sr-Mn alloys: (a) as-cast Mg-1Mn-0.3Sr, (b) as-cast Mg-1Mn-1Sr, (c) as-cast Mg-5Sr-1.25Mn, (d) as-extruded Mg-1Mn-0.3Sr, (e) as-extruded Mg-1Mn-1Sr, and (f) Mg-5Sr-2Mn. ([a, b, d, and e] Modified from Borkar, H., M. Hoseini, and M. Pekguleryuz, *Materials Science and Engineering A-Structural Materials Properties Microstructure and Processing*, 537, 49–57, 2012. With permission; [c and f] Modified from Celikin, M. and M. Pekguleryuz, *Materials Science and Engineering: A*, 556, 911–20, 2012. With permission.)

FIGURE 8.7 Typical microstructure of the Mg-Sr-Zr alloys: (a) as-cast Mg-1Zr-2Sr and (b) as-cast Mg-2Zr-5Sr. (Modified from Li, Y. C., C. Wen, D. Mushahary, R. Sravanthi, N. Harishankar, G. Pande, and P. Hodgson, *Acta Biomaterialia*, 8, 3177–88, 2012. With permission.)

For the Mg-Sr-Sn alloys, the microstructure of the Mg-Sr-Sn alloys gradually changed from dendrite crystals to equiaxed grains with increasing Sr content. The addition of Sr can also lead to the formation of a new intermetallic phase as well as the refining of the grain size, and the Mg_2Sn phase became isolated particles. Rod-like MgSnSr phases gathered mainly in the grain boundary areas, and the volume and the size of the MgSnSr phase went up with increasing Sr additions. When the Sr content was over 2.14%, some rod-like phases grew to a bone-like phase and stretched toward the interior of the grains (Figure 8.8a,b,d,e) (Liu et al. 2010).

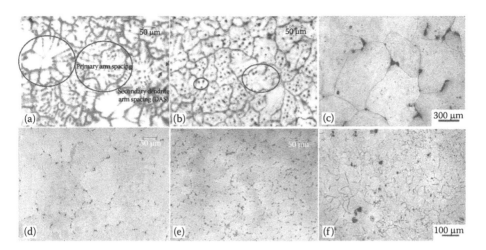

FIGURE 8.8 Typical microstructure of as-cast (a) Mg-4.84Sn-0.29Sr, (b) Mg-4.91Sn-2.14Sr, (c) Mg-14.1Li-0.19Sr, as-extruded (d) Mg-4.84Sn-0.29Sr, (e) Mg-4.91Sn-2.14Sr, and (f) Mg-14.1Li-0.19Sr. ([a, b, d, and e] Modified from Liu, H. M., Y. G. Chen, H. F. Zhao, S. H. Wei, and W. Gao, *Journal of Alloys and Compounds*, 504, 345–50, 2010. With permission; [c and f] Modified from Jiang, B., Y. Zeng, H. M. Yin, R. H. Li, and F. S. Pan, *Progress in Natural Science-Materials International*, 22, 160–8, 2012. With permission.)

The addition of Sr to the Mg-Li alloys can refine the grain size of both as-cast and ex-extruded Mg-Li-Sr alloys. The segregation power of Sr in β(Li) during solidification limited the growth of β(Li) grains in the as-cast alloys. Furthermore, the formation of the Mg_2Sr phase at the grain boundaries could impede the atomic diffusion and the movement of grain boundaries, thus restraining the growth of the β(Li) grains (Figure 8.8c). After extrusion, the fine Mg_2Sr granular structure can also act as grain refinement (Figure 8.8f) (Jiang et al. 2012). The typical microstructure of the Mg-Sr-Sn and Mg-Sr-Li ternary alloys is shown in Figure 8.8.

8.3.3.2 Mechanical Properties of Mg-Sr-Based Ternary Alloys

The addition of Zn to the Mg-Sr alloys would significantly improve the mechanical properties of the ternary alloys due to the grain refining and the increased solid solution and precipitation strengthening effect. Brar reported that the YS and UTS increased with the increasing Zn content in the Mg-0.5Sr-(2,4,6)Zn alloys, and the Mg-0.5Sr-2Zn exhibited significantly higher elongation to failure (Brar et al. 2012).

Both the YS and UTS were increased with increasing Sr content in the Mg-Sr-Mn ternary alloys during tension, and the elongation displayed irregular behavior with an initial increase as Sr increased up to 0.3%, followed by a decrease as Sr increased to 2.1%. However, the addition of Sr still exhibited higher elongation to failure when compared with Mg-Mn binary alloys. For compressive evaluation, the ultimate stresses increased with Sr addition whereas fracture strain showed the reverse trend (Borkar et al. 2012b).

TABLE 8.2

Summary of the Mechanical Properties of Mg-Sr-Based Ternary Alloys

		Mechanical Property			
Sample	Condition	Yield Strength (MPa)	Ultimate Strength (MPa)	Elongation (%)	References
Mg-3Sr-0.6Y	As-extruded	~160	~450	~28	Hu et al. 2012
Mg-3Sn-0.1Sr	As-cast	~80	~160	~6.2	Obekcan et al. 2013
Mg-4Zn-1Sr	Aged	N/A	270	12.8	Guan et al. 2013
Mg-1Mn-0.3Sr	As-extruded	~160	~230	~8.3	Borkar et al. 2012b
Mg-4.91Sn-2.14Sr	As-cast	~60	~150	~10.5	Liu et al. 2010
Mg-2Zn-0.5Sr	As-cast	62	142	8.9	Brar et al. 2012
Mg-1Zr-2Sr	As-cast	~70[a]	~250[a]	~35[a]	Li et al. 2012

[a] Compressive strength.

The precipitation strengthening from $Mg_{17}Sr_2$ and $Mg_{24}Y_5$ particles and the work hardening from the extrusion and grain strengthening from the recrystallization and Y were beneficial to the increase in mechanical properties of the Mg-Sr-Y alloys. It was reported that the Mg-3Sr-0.6Y exhibited the highest strengths and elongation to failure among the Mg-3Sr-(0.4-1)Y alloys (Hu et al. 2012).

Both the addition of Zr and Sr can affect the grain boundaries of the Mg-Sr-Zr alloys. The increasing Sr contents in the Mg-Sr-Zr alloys may lead to the increasing of $Mg_{17}Sr_2$ interphase, which was brittle and mainly located in the grain boundary zones. Thus, higher Sr content would be detrimental to the mechanical behavior of the Mg-Sr-Zr alloys. Li suggested that Mg-(1-5)Zr-2Sr exhibited better mechanical behavior than other alloys (Li et al. 2012).

As described, Sr has a very low solubility in Mg, so further increasing the Sr addition in the Mg-Sr-Sn alloys may cause a coarsening and volume fraction increase of the MgSnSr phase. Meanwhile, the volume of the Mg_2Sn phase was reduced. The cracks under applied stress would prefer initiation and propagation, and these formed large intermetallic particles, thus leading to adverse effects on both tensile strength and ductility (Liu et al. 2010). Obekcan et al. (2013) also reported that the increasing Mg_2Sn phase formed with increasing Sr contents can also decrease the YS and UTS. A summary of the mechanical properties of the Mg-Sr ternary alloys is shown in Table 8.2.

8.3.4 DEGRADATION IN SIMULATED BODY FLUIDS

Cipriano et al. (2013) evaluated the in vitro degradation of the Mg-Zn-Sr alloys and found that the corrosion resistance decreased with increasing Sr content, and the Mg-4Zn-0.15Sr alloy exhibited the lowest degradation rate in terms of total weight loss and had a significantly lower effect on pH increase.

Guan et al. (2013) reported that the degradation of Mg-4Zn-1Sr alloy in the SBF occurred in an alternating mode between pitting and localized corrosion. At the first stage, white degradation products on the sample surface served as a temporary protective layer and decelerated the corrosion progress. With increasing immersion times, the protective layer fell off. Consequently, the matrix became exposed again, resulting in new pit formation and further degradation. The preferential corrosion occurred at the grain boundaries rather than inside the grains. It is reported that lower Zn content in the Mg-Zn-Sr alloy resulted in a lower degradation rate, and the corrosion products were mainly $Mg(OH)_2$ (Brar et al. 2012).

Li et al. (2012) evaluated the in vitro corrosion behavior in both MMEM and SBF and found that the Mg-Sr-Zr alloys displayed better corrosion behavior in MMEM than in SBF. In MMEM, the Mg-Sr-Zr alloys exhibited similar corrosion current density whereas in SBF the Zr played a more important role on the corrosion resistance of the Mg-Sr-Zr alloys because the corrosion rate increased with increasing Zr content. The Mg-1Zr-2Sr alloy exhibited the lowest corrosion current density. The hydrogen evolution tests also revealed that the Mg-1Zr-2Sr alloy displayed the lowest hydrogen production rate. Moreover, higher Sr content in the alloys resulted in larger grain boundary zones with greater $Mg_{17}Sr_2$ phase content and, thus, decreased corrosion resistance.

A summary of the mechanical properties of the ternary Mg-Sr-based alloys is listed in Table 8.2.

8.3.5 Cytotoxicity and Hemocompatibility

Cipriano evaluated the cytocompatibility of four kinds of Mg-Zn-Sr alloys with H9 human embryonic stem cells (hESCs). The addition of Zn and Sr elements in the Mg improved H9 hESC cytocompatibility compared with pure Mg, and the Mg-4Zn-0.15Sr alloy showed the best cytocompatibility overall. All the Mg-based alloys displayed an initial increase in cell coverage followed by a steady decrease over time. After 30 h, all the Mg-Zn-Sr alloys exhibited coverage of visible cell colonies, and in contrast, almost no visible hESC colonies were visible on the pure Mg sample. After 48 h of culture, some viable cell colonies were observed on the Mg-4Zn-0.15Sr alloy. At the end of the 72 h incubation, the Mg-4Zn-0.15Sr alloy was the only one in which viable cell colonies were still observed in its coculture, which means that it enhanced the cell viability (Cipriano et al. 2013).

Osteoblast-like cells were selected to evaluate the cytotoxicity of the Mg-Sr-Zr alloys by indirect assays (Li et al. 2012). The additional Zr in the Mg-Sr alloys significantly improved its biocompatibility. Among the Mg-Sr-Zr ternary alloys, Mg-1Zr-2Sr and Mg-2Zr-5Sr alloys exhibited the best biocompatibility with a higher cell viability rate and lower hemolysis rate in comparison with other Mg-Sr-Zr alloys. A direct cell attachment assay revealed that the cell morphology on the Mg-1Zr-2Sr and Mg-2Zr-5Sr alloys exhibited a polygonal shape with filopodia and were more flattened than other alloy groups. Furthermore, numerous long filopodia extended from the cell bodies, and connection with distant cells was observed on the surface of the Mg-1Zr-2Sr and Mg-2Zr-5Sr alloys, which means that the surface of these two alloys provided favorable sites for initial cell attachment and growth.

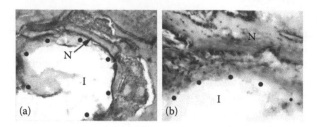

FIGURE 8.9 Histological images of bone tissue around the Mg-1Zr-2Sr alloy after 3 months of implantation stained with hematoxylin and eosin at (a) 100×, (b) 200×. The black dots were the interface between the implant and tissue, N indicates the newly formed bone, and I indicates the implant. (Modified from Li, Y. C., C. Wen, D. Mushahary, R. Sravanthi, N. Harishankar, G. Pande, and P. Hodgson, *Acta Biomaterialia*, 8, 3177–88, 2012. With permission.)

8.3.6 In Vivo Performance in Animal Models

For the in vivo evaluation of the Mg-Sr-Zr alloys, the New Zealand white rabbit was chosen as the animal model in Li's work (Li et al. 2012). Cylindrical samples with a diameter of 2 mm and length of 4 mm were implanted into the cortical region of the left femur. In the control group, the drilled hole in the femur was left bare. After 3 months implantation (see Figure 8.9), residual implants can still be seen, implying that the implant did not completely degrade. All the implants effectively induced bone formation around the implantation sites, and the BMC and BMD values were significantly higher for the alloys containing Sr. This indicates that Sr addition can improve the osteointegrative properties of the alloys. According to histochemical analysis, the Mg-1Zr-2Sr implants exhibited superior contact osteogenesis and subsequent osteointegration because it induced new trabecular bone formation. The newly formed bone was fully mineralized and remained in close contact with the existing bone. This indicated that the major phase in bone healing was complete although for the Mg-5Zr alloys the newly formed bone tissue did not integrate well with the implant surface. And for the Mg-2Zr-5Sr alloys, it displayed a slower healing effect on the wound area, and the newly developed bone around the implants was not contiguous with the implant surface.

8.4 DEVELOPMENT OF Mg-Sr QUATERNARY ALLOYS AS DEGRADABLE METALLIC BIOMATERIALS

8.4.1 Alloy Composition Design

The research on Mg-Sr quaternary or multielementary alloys in the past decades was mainly focused on the effect of additional Sr on the microstructure and mechanical properties of the Mg-Al-Zn alloys, in which the inclusion of Al made it unsuitable for biomedical applications (Suzuki et al. 2007; Kim et al. 2011; Li et al. 2011; Lou et al. 2011; Sadeghi et al. 2011; Bai et al. 2012). Only a few Mg-Sr quaternary or multielementary alloys, such as ZK60 + 0.1Sr alloy (Cheng et al. 2008),

Mg-5Gd-1.2Mn-0.4Sc-0.1Sr alloy (Yang et al. 2012b), Mg-6Zn-4Si-(0.1-2)Sr alloys (Cong et al. 2014), Mg-3Sn-2Sr-(0.1-0.3)Ti alloys (Yang et al. 2013), and Mg-1Ca-5Zr-8(Sn + Sr) alloy (Zhang et al. 2012), would be an alternative choice for the biomedical applications (Cheng et al. 2008; Yang et al. 2012b, 2013; Zhang et al. 2012; Cong et al. 2014).

8.4.2 FABRICATION, HOT AND COLD WORKING, AND HEAT TREATMENT

Yang et al. (2012b) fabricated a Sr-refined Mg-Y-Mn-Sc-Sr alloy by adding pure Mg, Mg-24.51Gd, Mg-2.7Sc, Mg-4.38Mn, and Mg-10Sr master alloys. The raw materials were held at 740°C for 20 min with mechanical stirring, followed by pouring into a mold preheated at 200°C. During the cast process, 2 wt.% RJ-2 flux was added to protect the melt from oxidation.

Cong et al. (2014) developed a series of Mg-Zn-Si-Sr alloys by adding the Mg-21.4Sr master alloy into the Mg-6Zn-4Si alloy. The melting temperature and holding time were 750°C and 15 min, respectively.

Yang produced Ti-containing Mg-Sn-Sr-Ti alloys, and Sr was added in the form of Mg-10Sr master alloy (Yang et al. 2013).

Zhang added Sr and Sn into the Mg-Zr-Ca alloy, and pure Mg, pure Tin, Mg-10Zr, Mg-40Ca, and Mg-40Sr master alloys were selected as raw materials. For the casting process, a protective atmosphere of CO_2 + SF_6 was applied (Zhang et al. 2012).

Cheng added 0.1% Sr in the form of Mg-10Sr master alloy into the ZK60 alloys. The Mg-Sr master alloys were added at the temperature of 740°C. After holding at the temperature for 80 min, the melt was poured into the mold. The as-cast alloys were further solid solution treated at 400°C for 5 h, followed by air cooling (Cheng et al. 2008).

8.4.3 MICROSTRUCTURE AND MECHANICAL PROPERTY

8.4.3.1 Microstructure

The typical microstructure of the Mg-Sr quaternary or multielementary alloys is shown in Figure 8.10.

The addition of 0.1% Sr into the Mg-Gd-Mn-Sc alloy did not influence the types of phase transformations, and the intermetallic compounds were identified as Mg_5Gd and Mn_2Sc phases. But the α-Mg phases in the Sr-containing alloy was relatively finer than those in the quaternary alloy (Figure 8.10a) (Yang et al. 2012b). The addition of Sr exceeding 0.5% led to the formation of needle-like particles, which was confirmed as the SrMgSi phase in the Mg-6Zn-4Si-0.5Sr alloy. Moreover, the Sr obviously changed the microstructure evolution. The primary Mg_2Si phase changed from dendritic to polygonal or fine block particles with increasing Sr content. And the eutectic Mg_2Si changes from Chinese script morphology to fine fiber, implying that the Mg_2Si phases were effectively modified and refined (Figure 8.10b). The average size of the primary Mg_2Si significantly decreased until the Sr addition of 0.5%, and then it gradually increased (Cong et al. 2014).

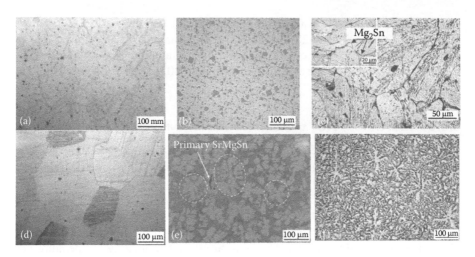

FIGURE 8.10 Typical microstructure of (a) as-cast Mg-5Gd-1.2Mn-0.4Sc-0.1Sr, (b) as-cast Mg-6Zn-4Si-0.5Sr, (c) as-cast Mg-1Ca-5Zr-8(Sn + Sr), high magnification is inset, (d) solutionized Mg-5Gd-1.2Mn-0.4Sc-0.1Sr, (e) as-cast Mg-3Sn-2Sr-0.3Ti, and (f) as-rolled ZK60 + 0.1Sr alloy. ([a and d] Modified from Yang, M. B., H. L. Li, R. J. Cheng, F. S. Pan, and H. J. Hu, *Materials Science and Engineering A-Structural Materials Properties Microstructure and Processing*, 545, 201–8, 2012. With permission; [b] Modified from Cong, M. Q., Z. Q. Li, J. S. Liu, and S. H. Li, *Materials and Design*, 53, 430–4, 2014. With permission; [c] Modified from Zhang, W. J., M. H. Li, Q. Chen, W. Y. Hu, W. M. Zhang, and W. Xin, *Materials and Design*, 39, 379–83, 2012. With permission; [e] Modified from Yang, M. B., H. L. Li, C. Y. Duan, and J. Zhang, *Journal of Alloys and Compounds*, 579, 92–9, 2013. With permission; [f] Modified from Cheng, R. J., F. S. Pan, M. B. Yang, and A. T. Tang, *Transactions of Nonferrous Metals Society of China*, 18, S50–4, 2008. With permission.)

The Ti-alloyed Mg-Sn-Sr alloys mainly consisted of α-Mg, SrMgSn, and Mg_2Sn phases, and the addition of Ti did not cause the formation of any new phases. The primary SrMgSn phases were in particle-, needle-, and short rod-like morphologies, and the eutectic SrMgSn phases were in a typical feather-like morphology. The typical $Mg_{17}Sr_2$ phase was not detected, possibly due to the relatively low electronegative difference between Mg and Sr elements (Figure 8.10e). DSC results also suggested that Ti did not influence the types of the phase transformations of the alloy (Yang et al. 2013). However, the undegradable nature of Ti in the physiological environment would make it a concern for biodegradable applications.

With the addition of Sn and Sr in the Mg-Zr-Ca alloy, a new phase Mg_2Sn was formed in the grain, A strip-shaped new phase containing Mg, Ca, Sr, and Sn was observed in the grain boundary (Figure 8.10c) (Zhang et al. 2012). After the addition of 0.1% Sr by using different Mg-Sr master alloys, ZK60 alloys exhibited a fine dendritic structure and an approximately equiaxed structure and the grain size decreased significantly (Figure 8.10f). Moreover, the as-rolled Mg-Sr master alloy had higher refinement efficiency than the as-cast, solutionized, and rapidly solidified Mg-Sr master alloys (Cheng et al. 2008).

TABLE 8.3

Summary of the Mechanical Properties of the Quaternary or Multielementary Mg-Sr Alloys

		Mechanical Property				
Sample	Condition	Yield Strength (MPa)	Ultimate Strength (MPa)	Elongation (%)	Grain Size (μm)	References
Mg-5Gd-1.2Mn-0.4Sc-0.1Sr	As-cast	184 ± 3.3	205 ± 3.1	4.1 ± 0.2	118 ± 22	Yang et al. 2012b
Mg-6Zn-4Si-0.5Sr	As-cast	~154	~160	~4.5	N/A	Cong et al. 2014
Mg-3Sn-2Sr + 0.3Ti	As-cast	104 ± 1.2	138 ± 3.2	4.1 ± 0.6	N/A	Yang et al. 2013
ZK60 + 0.1Sr	As-cast	N/A	N/A	N/A	33	Cheng et al. 2008

8.4.3.2 Mechanical Properties

A summary of the mechanical properties of these alloys is listed in Table 8.3.

The addition 0.1% Sr into the Mg-5Gd-1.2Mn-0.4Sc was beneficial to the improvement of the tensile property of the as-cast alloy. But the efficiency was lower than other alloying elements, such as Zr and Ca. Yang et al. concluded that the grain refinement effect of additional Sr may explain the improvement of mechanical behavior (Yang et al. 2012b).

Cong et al. reported that the Sr element in the Mg-6Zn-4Si alloy can change the morphology of the Mg_2Si phases and refine the eutectic and primary Mg_2Si phases, thus improving the tensile property, according to Griffith's theory. The Mg-Zn-Si alloy with 0.5% Sr addition exhibited the best UTS and elongation both in ambient and elevated temperatures (Cong et al. 2014).

Yang et al. reported that the addition of Ti into the Mg-3Sn-2Sr alloy changed the coarse needle-like shaped SrMgSn phases into refined phases, and the tensile property of the alloys was improved accordingly (Yang et al. 2013).

8.4.4 Degradation in Simulated Body Fluids

For the evaluation of the in vitro corrosion behavior of the Sr- and Sn-added Mg-Zr-Ca alloys, Zhang conducted electrochemical tests of the Mg-Zr-Ca-Sn-Sr alloys in SBF at 37°C. Compared with the Mg-Zr-Ca alloys, the addition of Sn and Sr could improve the open-circuit potential of alloys. The corrosion current density of Mg-Zr-Ca alloy was nearly tenfold that of the Mg-Zr-Ca-Sn-Sr alloy. The immersion test also released that the mass loss of the Mg-Zr-Ca-Sn-Sr alloy was stable, and the sample surface maintained integrity with only cracks appearing, and the mass loss of the Mg-Zr-Ca increased with the immersion time (Zhang et al. 2012).

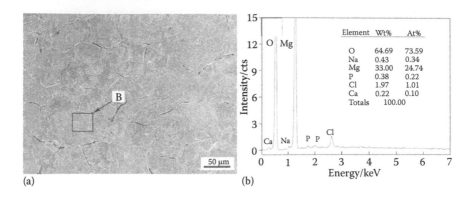

(a)   (b)

FIGURE 8.11 Illustration of morphology of the (a) Mg-Zr-Ca-Sr-Sn alloy immersed in SBF for 6 h, (b) composition of the frame B in (a). (Modified from Zhang, W. J., M. H. Li, Q. Chen, W. Y. Hu, W. M. Zhang, and W. Xin, *Materials and Design*, 39, 379–83, 2012. With permission.)

The morphology of the Mg-Zr-Ca-Sr-Sn alloy after being immersed in SBF for 6 h with an integrity surface and a few cracks is depicted in Figure 8.11.

8.5 CONCLUDING REMARKS

Various Mg-Sr-based alloys were designed in the past decade, and most of them were designed for engineering applications only, but small numbers of the reported Mg-Sr-based alloys exhibited good biocompatibility in both in vitro and in vivo experiments (Gu et al. 2012; Bornapour et al. 2013). For the consideration of biocompatibility, the alloying elements and alloying amounts should be carefully selected. The biocompatibility of the ternary and quaternary Mg-Sr alloys should be fully studied in order to screen suitable biodegradable biomaterials. Further in vivo tests also should be conducted to evaluate the in vivo compatibility of the alloys. The development of Mg-Sr-based biomaterials should be focused on the following aspects.

8.5.1 ENHANCING MECHANICAL PROPERTIES

The mechanical properties of the alloys were associated with the grain size and the distribution of the second phase. Sr is an effective grain refiner of Mg and its alloys; other alloying elements, for example, Zr, could also act as a grain refiner. Hot-working processing can change the microstructure of the alloys as well as the refinement of the grain size. The selection of proper alloying elements and suitable cold- or hot-working processing would benefit the mechanical property of the Mg-Sr alloys.

8.5.2 ENHANCING CORROSION RESISTANCE

Surface modification is an effective way to delay the corrosion of the matrix. However, up to now, little research about the surface modification of Mg-Sr alloys has been reported to the author's knowledge. Severe plastic deformation of the Mg-Sr alloys is also an option for improving the corrosion resistance.

8.5.3 ENHANCING BIOCOMPATIBILITY

Both Mg and Sr are essential metal elements for the daily metabolism and exhibit no toxicity under the daily recommended intake doses. Thus, the biocompatibility of the alloying element should be carefully considered. Bioactive coatings with biomolecular involvement with cell adherence and migration would improve the biocompatibility behavior.

9 Mg-Ag-Based Alloy Systems for Biomedical Applications

9.1 INTRODUCTION

9.1.1 Silver

Silver (Ag), with a density of 10.49 g/cm^3 at room temperature, is a precious metal chemical element. Silver is quite ductile and malleable. Silver compounds are used in photography, batteries, bactericide, catalysts, medicines, lubrication, cloud seeding, window coatings, mirrors, flower preservative, electroplating, and sanitation of swimming pools and hot tubs. Silver and silver alloys are used in jewelry, silverware, electronic components, solders, brazing alloys, bactericide, dental amalgams, bearings, coinage, heat sink, ignition-proofing, superconductors, and hydrogen storage (Wang et al. 2002; Drake 2005; Li et al. 2005; Singh et al. 2008; Zeng et al. 2010; Y. Wu et al. 2012).

The most significant trait of silver for clinical use is its antibacterial property. With such a beneficial trait, silver has a long history in biomedical treatment. Thousands of years ago, during the war in India, Alexander the Great used silver vessels to treat tropical disease in his troops, which is now thought to be attributed to the antimicrobial property of released Ag$^+$ (Silver et al. 2006). Today, Ag-containing antibacterial products are readily available in people's daily lives. The most important use of silver currently is as a biocide to prevent infections of long-term problem sites, including burns, traumatic wounds, and diabetic ulcers (Silver et al. 2006). The highly bioactive Ag$^+$ can penetrate bacterial membranes rapidly and cause cellular distortion and loss of viability due to the interaction with enzymes and other proteins in bacteria.

Considering the broad biomedical application of silver products, silver may get into the human body easily by direct ingestion, skin absorption, and other routes of entry. So far, the most common health effects associated with silver are argyria and/or argyrosis, which are long-term developments of a characteristic, irreversible pigmentation of skin and/or eyes, respectively (Drake 2005).

According to the American Conference of Governmental Industrial Hygienists, the threshold limit values for metallic silver and soluble compounds of silver are 0.1 mg/m^3 and 0.01 mg/m^3, respectively. Meanwhile, the permissible exposure limit (PEL) recommended by the Occupational Safety and Health Administration and the Mine Safety and Health Administration and the recommended exposure limit set by the National Institute for Occupational Safety and Health is 0.01 mg/m^3 for all forms of silver (Drake 2005).

Although silver is considered hazardous for human beings because of the risk of argyria and/or argyrosis, Hardes et al. (2007) and Bosetti et al. (2002) found that

269

silver exhibits no bad effects on material biocompatibility. Thus, the biocompatibility of Ag-containing materials needs further comprehensive examination.

9.1.2 PREVIOUS STUDIES ON MG-AG ALLOY SYSTEMS FOR ENGINEERING APPLICATIONS

As an alloying element, silver can refine grains in Mg alloys. The mechanical strength and ductility of the alloy will be improved with the refined grains. Silver is also a common element in facilitating the age-hardening process (Mendis et al. 2007; Gao and Nie 2008; Zhu et al. 2008, 2010; Yamada et al. 2009). Mendis et al. (2007) found that the precipitate size was reduced with the addition of Ag in a Mg-Zn-Ag alloy system. Further research revealed that the twin related ε-$Mg_{54}Ag_{17}$ precipitates, along the $[0001]_\alpha$ direction of the α-Mg matrix, played a significant role in strengthening behavior (Huang and Zhang 2012, 2014). Silver's effect on lowering the coefficient of thermal expansion (CTE) was also observed in a Mg-Sn-Ag system (Jayalakshmi et al. 2013), which is truly desired in high-temperature applications. However, the addition of Ag often increases the corrosion rate of Mg alloys (Ben-Hamu et al. 2006a,b). It is generally considered that the addition of Ag results in more microgalvanic cells between the α-Mg matrix and the MgAg second phase in the alloys.

Mg-Ag alloy systems have long been investigated for industrial/engineering applications. Typical Mg-Ag alloy systems include QE22 (Mg-Nd-Ag-Zr) (Trojanová et al. 2006; Ryspaev et al. 2008; Barucca et al. 2009; Stloukal and Čermák 2009), ZKQX (Mg-Zn-Ag-Ca-Zr) (Mendis et al. 2009, 2011), Mg-Gd-Y-Ag-Zr (H. Zhou et al. 2012), Mg-Gd-Zn-Zr-Ag (Yamada et al. 2009), and TQ50 (Mg-Sn-Ag) (Jayalakshmi et al. 2013). Lu et al. (2013) used the pseudopotential plane-waves method based on the density functional theory to investigate the mechanical property of intermetallic compounds MgAg, Mg_4Zn_8, and $Mg_4Zn_4Ag_8$ in a Mg-Zn-Ag alloy at high pressure and high temperature. The results revealed that Mg_4Zn_8 and $Mg_4Zn_4Ag_8$ are the ductility phases, and MgAg is the brittleness phase. Ryspaev et al. (2008) found that the maximum elongation of about 780% has been achieved for QE22 alloy (Mg-2Ag-2RE-Zr) at the strain rate $\varepsilon = 3 \times 10^{-4}$ s^{-1}. In addition, Wang et al. (2013c) introduced a composite material with the microstructure of the Mg-Ag-Al multilayer to improve the brittle property of the directly bonded Mg-Al interphase, which may be good for the welding application of a Mg-Al system. Moreover, the hydrogen storage material $Mg_{2-x}Ag_xNi$ (x = 0.05, 0.1, 0.2) (Wang 2002; Li et al. 2005; Zeng et al. 2010), thermoelectric material MgAgSb (Kirkham et al. 2012), and superconducting material $Mg_{1-x}Ag_x$-B_2 system (Singh et al. 2008) were also investigated.

9.2 DEVELOPMENT OF Mg-Ag BINARY ALLOYS AS DEGRADABLE METALLIC BIOMATERIALS

9.2.1 ALLOY COMPOSITION DESIGN

The binary phase diagram of the Mg-Ag system is illustrated as Figure 9.1 (Okamoto 1998). The solid line is established based on a thermodynamic assessment (Lim et al. 1997), and the other dashed line is drawn based on experimental data (Nayeb-Hashemi and Clark 1984). As shown in Figure 9.1, the solid solubility of Ag in Mg

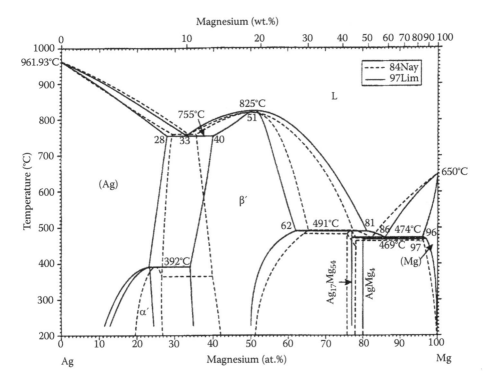

FIGURE 9.1 Mg-Ag binary phase diagram (Okamoto 1998) with solid line from thermodynamic assessment (Lim et al. 1997) and dashed line from experimental data (Nayeb-Hashemi and Clark 1984). (From Okamoto, H. *Journal of Phase Equilibria*, 19, 5, 487, 1998. With permission.)

solution is 3.83 at.% (15.14 wt.%), which is much larger compared with that of Ca, Sr, Zn, and many other common alloying elements in Mg alloys. When silver is added less than eutectic composition to pure Mg, under equilibrium conditions, Mg_4Ag will precipitate as the second phase. When more silver is added, another precipitation, $Mg_{54}Ag_{17}$, would also exist as the second phase. What needs to be noted is that the atomic radius of silver is 11% larger than that of Mg, which would result in a stronger solid solution effect. In addition, the high vacancy binding energy of the silver atom and the ability to slow down the diffusion of atoms can also eliminate the precipitations during the age-hardening process, improving the mechanical property of the Mg-Ag alloys. Thus, the strength of the Mg-Ag-based alloys would be significantly improved with Ag addition and a subsequent age-hardening process (Mendis et al. 2007; Gao and Nie 2008; Zhu et al. 2008, 2010; Yamada et al. 2009).

9.2.2 Microstructure and Mechanical Properties

For the application of Mg-based alloys as biodegradable materials, adequate mechanical property of the materials is one of the prerequisites. Thus, the microstructure should be carefully examined, and the optimization of the microstructure for better mechanical properties is essential.

Compared with the abundant data on the research of biodegradable Mg-Zn and Mg-Ca alloy systems, the scientific comprehensive statistics of a binary Mg-Ag alloy system as a degradable biomaterial are far from satisfactory.

Tie et al. (2012) investigated the binary Mg-Ag alloys as antibacterial biodegradable materials and found that the grain can be effectively refined by increasing Ag content. In addition, the secondary dendrites drastically increase inside the grains in the cast samples with higher Ag content (2 wt.%, 4 wt.%, and 6 wt.%). The EDX analysis results and X-ray diffraction pattern of the cast samples after heat treatment revealed that Mg_4Ag is the dominant secondary phase. Weak indications of $Mg_{54}Ag_{17}$ were also detected in all the samples. The results showed good correspondence to the phase diagram as mentioned.

The addition of increasing Ag can significantly improve the strength and hardness of the cast samples in comparison with pure Mg. The ultimate tensile strength of Mg-6Ag is 215.9 ± 11.3 MPa, nearly two times that of pure Mg at 108.3 ± 3.1 MPa. The increasing content of Ag also improved the tensile ductility compared with pure Mg (Tie et al. 2012). Gu et al. (2009b) also reported the increase of strength in both as-cast and as-rolled Mg-1Ag alloys compared to pure Mg.

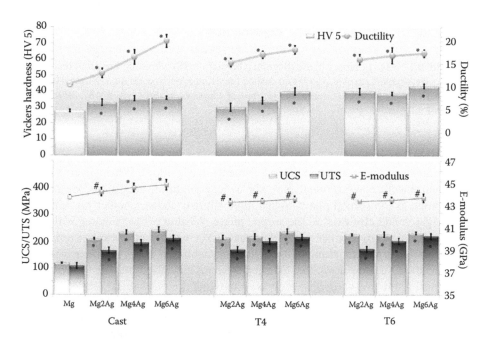

FIGURE 9.2 Vickers hardness, elongation, ultimate tensile strength (UTS), ultimate compressive strength (UCS), and E-modulus of Mg-xAg (x = 2, 4, 6 wt.%), as-cast, after T4 treatment and T6 treatment. (From Tie, D., F., Mueller W. D. et al. *European Cells and Materials*, 25, 284–98, 2012. With permission.)

The mechanical properties of the Mg-Ag alloys were dramatically improved in several parameters, especially after heat treatment, which is illustrated in Figure 9.2 (Tie et al. 2012).

9.2.3 DEGRADATION IN SIMULATED BODY FLUIDS

The addition of silver to Mg alloys often results in an increase in the corrosion rate in Mg alloys (Ben-Hamu et al. 2006b). Silver has a much higher self-corrosion potential than Mg. The addition of silver results in more microgalvanic cells between the α-Mg matrix and the MgAg second phase in the alloys, which would accelerate the corrosion rate of the alloys.

Tie et al. (2012, 2013) reported that with the addition of Ag (from 2 wt.% to 6 wt.%), the Mg-Ag alloys became more susceptible to corrosion, regardless of the heat treatment and the corrosion medium. The composition of the corrosion layer in all four T4-treated Mg-xAg alloys were also investigated. As Figure 9.3 shows, AgCl content in the corrosion layer increased with the addition of Ag. Besides, more Ca-PO_4 and Mg-PO_4 were found in Mg-Ag alloys than pure Mg, which indicated the good influence of Ag addition on osteogenesis.

Gu et al. (2009b) also found the higher corrosion rate in both as-cast and as-rolled Mg-1Ag compared with pure Mg, calculated from electrochemical measurements and immersion tests. Nevertheless, generated hydrogen and pH value of the Mg-1Ag sample are fewer and lower than pure Mg.

9.2.4 CYTOTOXICITY AND HEMOCOMPATIBILITY

For the purposes of biomedical application, the biocompatibility of implants needs to be carefully investigated. Mg is an essential bioelement of human beings. Further research reveals that Mg at a certain concentration in the human body would affect many cellular functions, including transport of potassium and calcium ions, modulating signal transduction, energy metabolism, and cell proliferation (Saris et al.

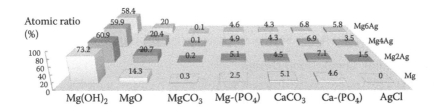

FIGURE 9.3 Compositions of corrosion products counted by Ca/Mg atoms in all four T4 treated materials derived from XPS analysis. (From Tie, D., F. Feyerabend, W. D. Mueller et al. *European Cells and Materials*, 25, 284–98, 2012. With permission.)

2000). However, considering the risk of argyria and/or argyrosis caused by silver exposure, it is really necessary to examine the cytotoxicity of Mg-Ag alloys.

To examine the cytotoxicity of binary Mg-Ag alloys, silver nitrate solutions were used at different concentrations to culture MG63 and RAW 264.7 to determine the exact LC50 of Ag$^+$ with sodium nitrate solution as a control. Further detection of the Ag$^+$ concentration from the immersion test showed it was many magnitudes lower than LC50 (Tie et al. 2012). The cell adhesion test results also revealed survival rates between 95% and nearly 100% on Mg-1Ag.

However, significantly reduced cell viability for the Mg-1Ag alloy extract has been detected in all the cultured cell lines except VSMC with the concentration of Ag at 0.9 ± 0.5 μM/l (Gu et al. 2009b). Similar cytotoxicity of Ag$^+$ (10^{-6} M/l) to the L929 and MC3T3-E1 cell lines was also observed by Yamamoto et al. (1998).

As for hemocompatibility, Gu et al. (2009b) reported a more than 30% hemolysis percentage in as-cast Mg-1Ag alloy and a more than 20% hemolysis percentage in as-rolled Mg-1Ag alloy, which they believed to be connected with a relatively higher concentration of Ag.

9.2.5 ANTIBACTERIAL PROPERTY

The antibacterial property of silver has resulted in the broad commercial application of Ag in antibacterial products and also distinguished it from other alloying elements in degradable biomedical Mg-based alloys. Ideal Mg-Ag alloys should provide both good biocompatible and antibacterial properties at the same time.

Tie et al. (2012) compared the antibacterial properties of Ti, glass, Mg, Mg-2Ag, Mg-4Ag, and Mg-6Ag. Mg-Ag alloys showed a 50–75% reduction of adherent bacteria. It is also easy to understand the tendency that increased Ag addition (from 2% to 6%) caused better antibacterial property as Figure 9.4 illustrates.

9.3 DEVELOPMENT OF Mg-Ag-BASED ALLOYS AS DEGRADABLE METALLIC BIOMATERIALS

Until now, two Ag-containing Mg alloys have been investigated as degradable biomaterials, including the Mg-Zn-Ag alloy system (Peng et al. 2013a) and the Mg-Nd-Zn-Ag-Zr alloy system (X. B. Zhang et al. 2013a).

9.3.1 MG-ZN-AG ALLOY SYSTEM AS DEGRADABLE BIOMATERIALS

9.3.1.1 Alloy Composition Design

Son et al. (2011) reported the different solidification pathway in the Mg-Zn-Ag alloy system. The solidification sequence of the Mg–6Zn–1Ag and Mg–6Zn–2Ag alloys is as follows: L → L + α-Mg → α-Mg → α-Mg + Mg$_4$Ag → α-Mg + Mg$_4$Ag + MgZn phases, and the solidification sequence of the Mg–6Zn–3Ag alloy is as follows: L → L + α-Mg → L + α-Mg + Mg$_4$Ag → α-Mg + Mg$_4$Ag → α-Mg + Mg$_4$Ag + MgZn phases.

FIGURE 9.4 CLSM image (a) and bacterial viability (b) on different materials, bacterial films were fluorescence-stained: green color for living bacteria; red color for dead bacteria. Crystal structures on alloy surface were generated by corrosion products. The differences in viability of adhering bacteria on alloys and pure magnesium were significant (*significance level $p < 0.05$). (Adapted from Tie, D., F. Feyerabend, W. D. Mueller et al. *European Cells & Materials*, 25, 284–98, 2012. With permission.)

9.3.1.2 Microstructure and Mechanical Properties

The typical microstructure of as-extruded Mg-Zn-xAg (x = 0, 0.2, 0.5, and 0.8 wt.%) is shown in Figure 9.5 (Peng et al. 2013a). A really obvious grain refinement with increasing Ag content can be observed. In addition, increasing Ag content also contributed to more coarse grain boundaries as shown in Figure 9.5a through d.

Peng et al. (2013) reported the tensile yield strength (TYS), ultimate tensile strength (UTS), compressive yield strength (CYS), ultimate compressive strength (UCS), and ductility of Mg-1Zn-xAg (x = 0.2, 0.5, and 0.8 wt.%), compared to Mg-1Zn alloys. The results revealed that the strength and ductility both increase monotonically. The better mechanical property may chiefly be the result of grain refinement mentioned here.

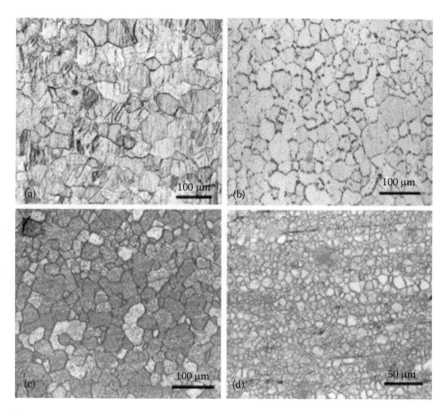

FIGURE 9.5 Microstructure of the extruded alloys perpendicular to the extruded direction: (a) Mg-1Zn alloy; (b) Mg-1Zn-0.2Ag alloy; (c) Mg-1Zn-0.5Ag alloy; (d) Mg-1Zn-0.8Ag alloy. (From Peng, Q., K. Li, Z. Han, E. Wang, Z. Xu, R. Liu, and Y. Tian, *Journal of Biomedical Materials Research.* Part A, 101 (7), 1898–906, 2013. With permission.)

9.3.1.3 Degradation in Simulated Body Fluids

The high self-corrosion potential of Ag made Ag-containing Mg alloys more susceptible to corrosion both in industrial environments and simulated body fluids. With the addition of Ag, the segregation area of Ag in the matrix is increased correspondingly, resulting in a faster degradation in SBF (Peng et al. 2013a).

Different electrochemical corrosion equivalent circuits of Mg-Zn and Mg-Zn-Ag were calculated by Peng et al. (2013a) after immersion tests in SBF. As shown in Figure 9.6, the EIS curves of the Mg-1Zn-0.8Ag alloy are mostly composed of two capacitive loops (HF and LF) and an inductive loop (LF), which is different from that of Mg-1Zn. The relationship between R_{ct} and immersion time fitting from Nyquist plots is shown in Figure 9.6e; R_{ct} is increased when retarding the immersion time in SBF.

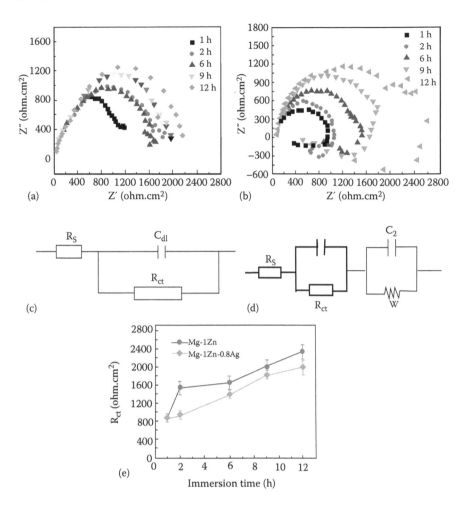

FIGURE 9.6 Electrochemical measurement in SBF at 37°C ± 2°C (Peng et al. 2013a): (a) Nyquist plots of Mg-1Zn alloy; (b) Nyquist plots of Mg-1Zn-0.8Ag alloy; (c) the equivalent circuit of Mg-1Zn alloy fitting Nyquist plots; (d) the equivalent circuit of Mg-1Zn-0.8Ag alloy fitting Nyquist plots; (e) the relationship between R_{ct} and the different immersion time. (From Peng, Q., K. Li, Z. Han, E. Wang, Z. Xu, R. Liu, and Y. Tian, *Journal of Biomedical Materials Research*. Part A, 101 (7), 1898–906, 2013. With permission.)

9.3.1.4 Cytotoxicity and Hemocompatibility

Peng et al. (2013a) studied the cytotoxicity of Mg-1Zn-0.8Ag by using HeLa cervical carcinoma cells. The results revealed that the Mg-Zn-Ag alloy specimen exhibited good cytocompatibility to HeLa cells. Typical cell morphology at different culture times is shown in Figure 9.7. For 100% extraction sample, the adhered dendrite-shaped cells started to proliferate after 72 h incubation as shown in Figure 9.7f.

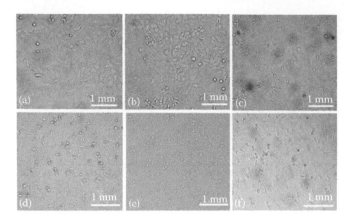

FIGURE 9.7 Relative cell growth rate of HeLa cell in solution containing Mg-1Zn-0.8Ag alloy after different incubation times, reference well (DMEM medium): (a) 24 h, (b) 48 h, and (c) 72 h; 100% extraction sample: (d) 24 h, (e) 48 h, and (f) 72 h. (From Peng, Q., K. Li, Z. Han, E. Wang, Z. Xu, R. Liu, and Y. Tian, *Journal of Biomedical Materials Research. Part A*, 101 (7), 1898–906, 2013. With permission.)

9.3.1.5 Antibacterial Property

Peng et al. (2013a) investigated the antibacterial property with interleukin-1α and nitric oxide release measurements. As for the interleukin-1α test, when the concentration of Ag is higher than 2×10^{-3} mg/l, the obvious anti-inflammatory effect was obtained. Meanwhile, the obvious anti-inflammatory effect can be detected when the concentration of Ag is higher than 3×10^{-3} mg/l with nitric oxide release measurements.

9.3.2 MG-ND-ZN-AG-ZR ALLOY SYSTEM AS DEGRADABLE BIOMATERIAL

9.3.2.1 Microstructure and Mechanical Properties

Typical microstructure and EDS results of as-cast Mg-Nd-Zn-Ag-Zr alloys are shown in Figure 9.8 (X. B. Zhang et al. 2013a). Grain refinement can be seen with the addition of Ag. A more continuous bright second phase can also be seen from Figure 9.8a through d. The EDS results revealed an increasing Ag content in the bright phase with Ag addition to the alloys.

9.3.2.2 Degradation in Simulated Body Fluids

X.B. Zhang et al. (2013a) reported that the corrosion rate of the Mg-Nd-Zn-Ag-Zr alloys accelerated with increasing Ag addition in simulated body fluid (SBF) although they believed it was a coeffect of grain refinement of the addition of Ag and discontinuous compound MgZr. The corrosion behavior of the Mg-3.0Nd-0.2Zn-xAg-0.4Zr alloy is shown in Figure 9.9 (X. B. Zhang et al. 2013a). The addition of Ag increases the corrosion rate of the alloy according to both hydrogen evolution tests and mass loss tests.

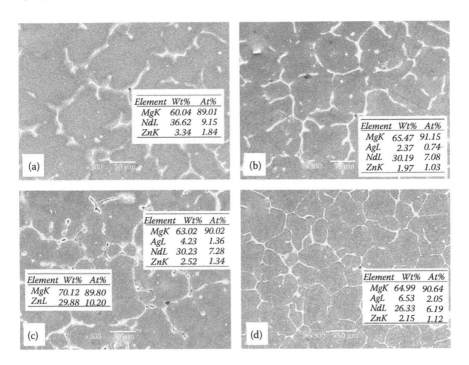

FIGURE 9.8 Microstructure and EDS results of (a) Mg-3.0Nd-0.2Zn-0.4Zr, (b) Mg-3.0Nd-0.2Zn-0.2Ag-0.4Zr, (c) Mg-3.0Nd-0.2Zn-0.4Ag-0.4Zr, and (d) Mg-3.0Nd-0.2Zn-0.8Ag-0.4Zr. (Adapted from Zhang, X., Z. Ba, Z. Wang, X. He, C. Shen, and Q. Wang, *Materials Letters*, 100, 188–91, 2013. With permission.)

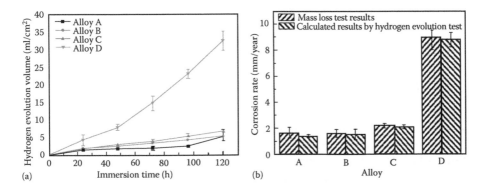

FIGURE 9.9 Corrosion behavior of Mg-3.0Nd-0.2Zn-xAg-0.4Zr alloys, (a) hydrogen evolution curves and (b) corrosion rate. (Adapted from Zhang, X., Z. Ba, Z. Wang, X. He, C. Shen, and Q. Wang, *Materials Letters*, 100, 188–91, 2013. With permission.)

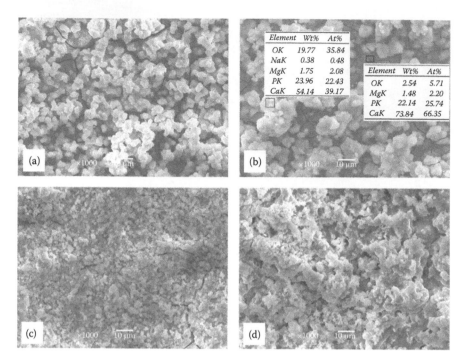

Element	Wt%	At%
OK	19.77	35.84
NaK	0.38	0.48
MgK	1.75	2.08
PK	23.96	22.43
CaK	54.14	39.17

Element	Wt%	At%
OK	2.54	5.71
MgK	1.48	2.20
PK	22.14	25.74
CaK	73.84	66.35

FIGURE 9.10 Typical corrosion morphology and EDS result of (a) Mg-3.0Nd-0.2Zn-0.4Zr, (b) Mg-3.0Nd-0.2Zn-0.2Ag-0.4Zr, (c) Mg-3.0Nd-0.2Zn-0.4Ag-0.4Zr, and (d) Mg-3.0Nd-0.2Zn-0.8Ag-0.4Zr. (Adapted from Zhang, X., Z. Ba, Z. Wang, X. He, C. Shen, and Q. Wang, *Materials Letters*, 100, 188–91, 2013. With permission.)

Figure 9.10 shows the typical corrosion morphology of the Mg-Nd-Zn-Ag-Zr alloy, and the EDS result in Figure 9.10b shows an increasing content of Ca/P after the immersion test, which would be good for the osteogenesis process (X. B. Zhang et al. 2013a).

9.4 POSSIBLE METHODS OF PROPERTY IMPROVEMENT FOR Mg-Ag-BASED ALLOY SYSTEMS

9.4.1 CURRENT TECHNIQUES FOR ENHANCING MECHANICAL PROPERTIES

Age-hardening is a common and well-understood method for improvement of Ag-containing Mg alloys for industrial/engineering applications.

Due to the accelerated occurrence of dynamic recrystallization at higher temperatures, which leads to coarsening of grains, it can be seen that at a lower extrusion temperature (275°C) the grains are smaller (Kaya et al. 2006). H. Zhou et al. (2012) compared the microstructure and mechanical properties of as-extruded Mg-8.5Gd-2.3Y-1.8Ag-0.4Zr alloy under different extrusion temperatures (400°C, 450°C, and 500°C). The results revealed that the Mg alloy under the extrusion temperature of 400°C possessed the finest grain and best mechanical properties.

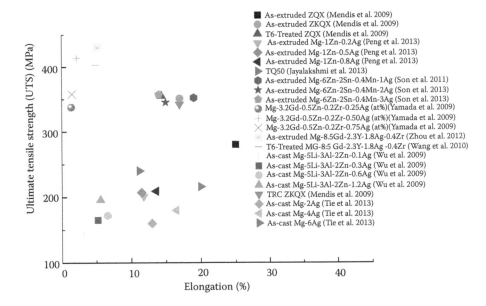

FIGURE 9.11 Mechanical property summary of current Mg-Ag alloy systems.

Figure 9.11 presents a summary of the mechanical properties of Ag-containing multielement Mg alloys, both for biomedical application and industrial use (Mendis et al. 2009, 2011; Wu and Zhang 2009; Yamada et al. 2009; Q. Wang et al. 2010a; Son et al. 2011; H. Zhou et al. 2012; Jayalakshmi et al. 2013; Peng et al. 2013a; Tie et al. 2013). The mechanical properties of industrial Ag-containing Mg alloys, the Mg-Zn-Sn-Mn-Ag alloy system, revealed good comprehensive mechanical property (Son et al. 2011).

9.4.2 Current Methods for Enhancing Corrosion Resistance

9.4.2.1 Heat Treatment

T4 treatment wiped out most of the secondary dendrites, which hinted that the solution treatment was successful. Tie et al. (2012) utilized T4 treatment on Mg-Ag alloys. Moreover, Ben-Hamu et al. (2006a) reported the corrosion rates significantly in the ZQ6X alloys in simulated body fluid (SBF) after T5D heat treatment. Because more second phases precipitate during heat treatment, the alloy would be more susceptible to corrosion. Figure 9.12 presents a summary of the corrosion rate of Ag-containing Mg alloys, both for industrial applications and biomedical use (Ben-Hamu et al. 2006a,b; Ben-Hamu 2007; Jayalakshmi et al. 2013; Peng et al. 2013a; Tie et al. 2013; X. B. Zhang et al. 2013a).

9.4.2.2 Surface Treatment

Surface modification is helpful for enhancing corrosion resistance of degradable Mg-based biomaterials. Various surface treatments, such as a MgF_2 coating (N. Li et al. 2013), microarc oxidation (MAO) (Gu et al. 2011a), alkaline heat treatment (Gu

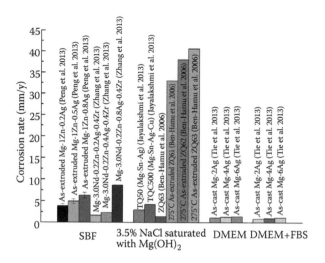

FIGURE 9.12 Corrosion rate of Ag-containing Mg alloys in different media: SBF, 3.5% NaCl saturated with $Mg(OH)_2$ with a pH = 10.5, DMEM, and DMEM + FBS.

et al. 2009a), and chitosan coating (Gu et al. 2009c), etc., have been proven to be effective for enhancing corrosion resistance, inducing calcium phosphate formation, and sustaining good biocompatibility at the same time. These surface modifications and treatments could be adoptable for Mg-Ag alloy systems and need to be studied in the future.

9.4.2.3 Amorphization

Currently investigated Mg-Ag based bulk metal glasses (BMGs) include Mg-Cu-Ag systems (Y. C. Chang et al. 2008; Y. Sun et al. 2009; Hui et al. 2010; Soubeyroux and Puech 2010; Zheng 2012) and Mg-Ni-Ag systems (Park and Kim 2011). The addition of Ag can improve the glass-forming ability, resulting in a strength of ~800 MPa, as illustrated in Figure 9.13 (Park et al. 2004, 2007, 2008, 2011; Liu et al. 2005; Qin et al. 2009; Soubeyroux et al. 2009; Y. Sun et al. 2009; Hui et al. 2010; Park and Kim 2011; Peng et al. 2011).

Because the quicker corrosion rate of Ag-containing Mg alloys is caused by the relatively big gap between the α-Mg matrix and Ag-abundant precipitations, Mg-Ag bulk metal glasses (BMGs) may be another potential way to get feasible Mg-Ag systems as degradable biomaterials with unique antibacterial properties. Subba Rao et al. (2003) reported a study on the corrosion performance of amorphous Mg65Y10Cu15Ag10 alloy in both electrolytes of pH 8.4 and 13. The presence of silver in the matrix made it more noble and stable; passive layers were also found on the amorphous alloy. Gebert et al. (2008) investigated the corrosion behavior of Mg65Cu7.5Ni7.5Ag5Zn5Gd5Y5 bulk metallic glass in borate solutions with pH 5–8.4 and in a sodium hydroxide solution of pH 13. A superior corrosion resistance was obtained in the BMG compared to pure Mg and AZ31 Mg alloy. This improved corrosion behavior is mainly attributed to the beneficial effect of the alloying components, which are homogeneously mixed in the amorphous phase. Thus,

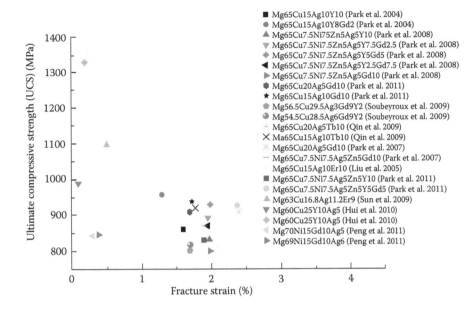

FIGURE 9.13   Summary of mechanical properties of Ag-containing Mg-based BMGs.

bulk metal glasses of Mg-Ag systems as degradable biomaterials may be a good solution for the improvement of corrosion resistance.

9.5   CONCLUDING REMARKS

To date, in vivo investigation of Ag-containing Mg alloys as degradable biomaterials have not been done. Yet Mg-Ag alloy systems as degradable biomaterials attract more and more interest for the antibacterial property of Ag, which is desired for implantation surgery.

To make a Ag-containing Mg alloy a good candidate as a degradable biomaterial, the mechanical property, corrosion resistance, and cytotoxicity need further comprehensive investigations. Heat-treatment processes are needed to obtain better mechanical properties. As for the improvement of corrosion resistance in Mg-Ag alloys, surface modification and amorphization could be possible methods. The cytotoxicity of Ag-containing Mg alloy should also be taken care of due to the risk of argyria and/or argyrosis and the uncertain cytotoxicity of Ag^+.

10 Mg-Li-Based Alloy Systems for Biomedical Applications

10.1 INTRODUCTION

10.1.1 LITHIUM

Lithium belongs to the group I alkali metal of the periodic table. As the lightest metal, Li possesses a density of 0.534 g/cm^3, which is one third that of Mg (1.74 g/cm^3), one fifth that of Al (2.70 g/cm^3), and one eighth that of Ti (4.54 g/cm^3). It has a body-centered cubic (bcc) lattice structure at room temperature with a lattice constant of 0.35 nm.

From the viewpoint of biology, Li is considered a nutrient element that can be found in plants and animals (Gu et al. 2009b). It has an effect on the central nervous system and is often used as an anti-manic-depressive drug to treat bipolar disorder. The therapeutic effects contribute to the ability of Li ions to alter ionic concentrations across cell membranes or affect enzyme systems and neurotransmitter transport (Sansone and Ziegler 1985).

10.1.2 PREVIOUS STUDIES ON MG-LI ALLOY SYSTEMS FOR INDUSTRIAL/ENGINEERING APPLICATIONS

Because Mg alloys possess a hexagonal-closed-packed (hcp) crystal structure with a high c/a ratio of 1.6236, the ductility and formability are relatively poor at ambient temperatures. This disadvantage consequently hinders the widespread use of Mg alloys in many application fields. Li is reported as the only alloying element that is able to decrease the c/a ratio of the hexagonal Mg lattice and even change the crystal structure (Witte et al. 2008a). Furthermore, adding Li into Mg is effective in decreasing the density and increasing the specific strength and specific modulus. A wide range of Li can be easily incorporated into Mg to form various Mg-Li alloys. Based on the Li content and crystal structure, Mg-Li alloys are commonly classified into three categories. Alloys containing lithium at less than 5.7 wt.% is composed of α phase with a hcp crystal structure, a solid solution of Li in Mg. When the Li content is higher than 10.3 wt.%, the alloy is composed of β phase with a bcc crystal structure, a solid solution of Mg in Li. With a content of Li between 5.7 and 10.3 wt.%, the alloy exists as a dual crystal structure ($\alpha + \beta$) (Meng et al. 2009; R. Wu et al. 2009a,b).

With a density between 1.35 and 1.65 g/cm^3, a Mg-Li alloy system is referred to as the lightest engineering alloy. It has many special properties, such as high specific strength and rigidity, excellent formability, and high energetic particle penetration resistance. Accordingly, Mg-Li alloys have attracted great interest in the fields of aerospace applications, automotive, military, electronics, and transportation industries, etc. (Trojanová et al. 2005; Zhong et al. 2005; Shao et al. 2009; Song et al. 2009a). Mg-Li alloys were introduced in the 1960s when NASA developed a series of Mg-Li-Al alloys centered on LA141A (Mg-14Li-1Al) for aerospace applications (Byrer et al. 1964). Since then, Mg-Li alloy systems kept drawing widespread attention around the world, and a series of Mg-Li alloys were developed, such as LA91 (Mg-9Li-1Al), LAZ933 (Mg-9Li-3Al-3Zn), and MA21 (Mg-7.5Li-4.3Al-1Zn) (Crawford et al. 1996; Haferkamp et al. 2000). To date, a substantial amount of research work on Mg-Li binary alloys, such as Mg-3.3Li (Liu et al. 2004), Mg-4Li (Trojanová et al. 2005), and Mg-8Li (Song et al. 2009a), have been reported for industrial purposes. To improve the mechanical properties of binary Mg-Li alloys, ternary and multielementary alloys are consequently fabricated by the addition of alloying elements, including Al, Ca, Zn, and RE, etc. (Drozd et al. 2004; Song and Kral 2005; Zhang et al. 2007a; Wu et al. 2008). Considering the chemical activity of Li, the corrosion resistance properties of Mg-Li alloys are generally weak. Various surface treatments, such as chemical conversion coating (Song et al. 2009b), anodizing (Li et al. 2006), electroless plating (L. Yang et al. 2009a), microarc oxidation (Y. Xu et al. 2009), and diamond-like carbon coating (Yamauchi et al. 2007), have been developed, and their protective effects are investigated.

10.2 DEVELOPMENT OF Mg-Li BINARY ALLOYS AS DEGRADABLE METALLIC BIOMATERIALS

10.2.1 ALLOY COMPOSITION DESIGN

According to the phase diagram of Mg-Li alloy (Figure 10.1) (Saunders 1990), the constitutional phase of Mg-Li binary alloys varies from phase α to the dual phase

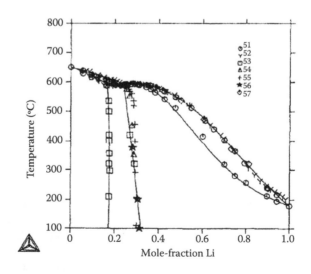

FIGURE 10.1 Equilibrium Mg-Li phase diagram. (From Saunders, N. *Calphad*, 14 (1), 61–70, 1990. With permission.)

(α + β) until it is a singular phase β with increasing Li content. The mechanical properties of the hcp α phase Mg-Li alloys are worse compared to the bcc Mg-Li alloys. The bcc Mg-Li alloys perform with lower density as well as markedly improved formability and superplasticity due to the emergence of the β phase. Experiments have shown that a rolled Mg-Li (α + β) two-phase alloy is capable of exhibiting superplastic elongations when tested at temperatures from 250°C to 350°C (Furui et al. 2005). Song et al. (2009b) also reported that the binary alloy Mg-9Li alloy has a high elongation of 460%. However, Mg-Li alloys with bcc structure are well known to possess relatively low strength, high chemical activity, and poor corrosion resistivity (Song and Kral 2005). Thus, it is of crucial importance to select the proper content of Li to meet the specific requirements of different applications.

10.2.2 Fabrication, Hot and Cold Working, and Heat Treatment

10.2.2.1 Fabrication

It is known that Mg and Li are chemically active metals; special techniques and technologies are consequently demanded for the fabrication of Mg-Li alloys. In 1982, Sahoo and Atkinson (1982) pointed out that the most successful method of melting and casting involved using the Balzers high-vacuum induction furnace under an argon atmosphere of 600 mm Hg. Single metals or master alloys are mixed and melted in a furnace under protection of a flux cover or in an inert atmosphere, and then the melts are cast into a mold. This metal casting method is now widely accepted as a convenient and fast way to produce Mg-Li-based alloys. However, many problems exist in the production process with this method, such as microsegregation, heterogeneities, complicated production process, serious metals burning, and high-energy consumption.

Recently, a molten salt electrolysis process has been developed as another feasible method to prepare Mg-Li alloys. This method can not only overcome the disadvantages of the traditional metal mixing method, but it also has many other advantages. First, the phases of alloys can be controlled by electrochemical parameters. Second, the process can be conducted with a simple apparatus. Third, mass production is easy. Figure 10.2 (Zhang et al. 2007b) illustrates a typical experimental apparatus employed in the molten salt electrolysis process. LiCl and KCl are selected as the molten chlorides with graphite being set as the anode and the magnesium rod the cathode. Argon is employed as the protection gas throughout the experiment. The temperature for the preparation of Mg-Li alloys can be adjusted in the range of 400–600°C, and the conductance is high, and as a result, energy consumption is reduced. In addition, the Mg-Li alloys obtained by electrolysis have a homogeneous composition, and the Li content is controllable. Because of these advantages, preparation of Mg-Li alloys by molten electrolysis has drawn great attention.

10.2.2.2 Hot/Cold Working and Heat Treatment

Various Mg-Li alloys have been studied, and their mechanical properties are sufficient but not impressive. So it is still hoped to improve the mechanical properties of Mg-Li alloys through rolling, extrusion, heat treatment, and solid solution treatment.

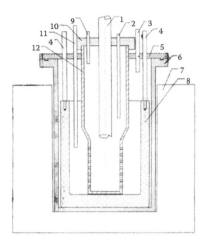

(1) Iron rod (fixing magnesium sample)
(2) Ar gas inlet
(3) Cl$_2$ gas outlet
(4) Anode extension
(5) Top cover for electrolysis cell
(6) Corundum sheath
(7) Electric furnace
(8) Graphite anode (electrolysis cell)
(9) Ar gas outlet
(10) Top cover for cathodic chamber
(11) Thermocouple
(12) Cathodic chamber (corundum crucible)

FIGURE 10.2 Schematic drawing of the experimental apparatus employed in the molten salt electrolysis process. (From Zhang M. L., Y. D. Yan, Z. Y. Hou, L. A. Fan, Z. Chen, and D. X. Tang. *Journal of Alloys and Compounds*, 440 (1–2), 362–6, 2007b. With permission.)

It is reported that a warm-extruded Mg-8.5Li sheet showed a higher elongation to failure of about 610% at 4×10^{-4} s^{-1} at 350°C, exhibiting superplasticity at conventional strain rates (Dong et al. 2007). Mg alloys containing over 10 wt.% Li have been proven to behave with good cold workability. The alloyed Li changes the crystal structure from hcp to bcc, thereby cold-working is allowed. Using cold-working, work hardening, which is impossible for ordinary hcp Mg alloys, is expected. As for single-phase alloys, including those with lower Li, intermediate annealing is required (Sahoo and Atkinson 1982).

10.2.3 MICROSTRUCTURE AND MECHANICAL PROPERTIES

It is well known that the microstructure of Mg-Li binary alloys exhibits three different kinds, depending on the Li content, thus an impact on the mechanical property will be observed. For a Mg-Li alloy containing Li lower than 5.7 wt.%, which is composed of singular α phase, its typical microstructure is highly similar to that of pure Mg as shown in Figure 10.3a (W. R. Zhou et al. 2013). Due to the decreased c/a ratio of hcp Mg, a Mg-Li alloy can slide not only in {0001}, but also in {1011} or {1010} (Ma et al. 1999). As a result, the binary alloy with single α phase shows considerably better ductility than pure Mg, and the strength is well preserved as there is no appearance of soft β phase. The Mg-Li alloy with 5.7–10.3 wt.% Li is provided with a duplex microstructure with the α phase appearing as plates in the β matrix. A simple eutectic can be seen in Figure 10.3b (Kral et al. 2007); the hcp α phase appears as the light gray interconnected lamellar phase, and the bcc β phase appears as the dark gray phase. A two-phase Mg-Li alloy is considered to have a better combination of mechanical properties at room temperature. Furthermore, different proportions of α and β in the alloy significantly change the mechanical behavior of the system. With

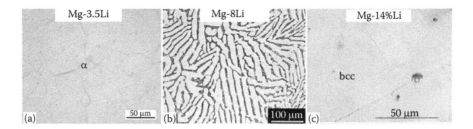

FIGURE 10.3 Microstructures of alloys (a) Mg-3.5Li, (b) Mg-8Li, and (c) Mg-14Li. ([a] From Zhou, W. R., Y. F. Zheng, M. A. Leeflang, and J. Zhou. *Acta Biomaterialia*, 9 (10), 8488–8, 2013. With permission; [b] From Kral, M. V., B. C. Muddle, and J. F. Nie. *Materials Science and Engineering: A*, 460–1, 227–32, 2007. With permission; [c] From Tsujikawa, M., S.-I. Adachi, Y. Abe, S. Oki, K. Nakata, and M. Kamita. *Plasma Processes and Polymers*, 4 (S1), S593–6, 2007. With permission.)

a high content of Li, the Mg-Li binary alloy is composed of simple β(bcc) with its optical microstructure as seen in Figure 10.3c (Tsujikawa et al. 2007). This type of Mg-Li alloy shows lower strength and excellent plasticity and are apt to creep even at room temperature.

10.2.4 DEGRADATION IN SIMULATED BODY FLUIDS

Based on its excellent ductility and formability, a Mg-Li alloy system is attractive for cardiovascular stent application because it can fulfill the mechanical specifications of radially expandable stents. Recently, Mg-3.5Li and Mg-8.5Li binary alloys have been fabricated, and their feasibilities as stent materials were investigated (W. R. Zhou et al. 2013). Electrochemical and immersion tests were conducted to evaluate the degradation behavior of these two alloys. The simulated body fluid employed in this work was Hank's solution (Kuwahara et al. 2001). Electrochemical measurements were carried out with a traditional three-electrode cell using an electrochemical analyzer. Compared to the Mg-3.5Li alloy, the Mg-8.5Li alloy was observed to have decreased corrosion potential and increased current density. Meanwhile, more hydrogen evolution was also monitored for the Mg-8.5Li alloy sample during the immersion period as illustrated in Figure 10.4 (W. R. Zhou et al. 2013). Thus, it can be concluded that the degradation rate of the Mg-Li binary alloy increases with the increasing Li content, which may be due to the presence of the more active β phase acting as a constituent for microgalvanic corrosion with the α-Mg matrix.

10.2.5 CYTOTOXICITY AND HEMOCOMPATIBILITY

The cytotoxicities of previously reported biomedical Mg-Li binary alloys Mg-3.5Li and Mg-8.5Li were evaluated by an indirect contact method using human umbilical vein endothelial cells (ECV304) and rodent vascular smooth muscle cells (VSMC) (W. R. Zhou et al. 2013). After incubating with the extracts for 1, 3, and 5 days, the cell viabilities were measured by using a microplate. The results showed slightly

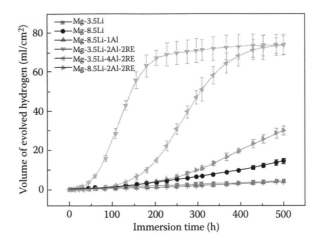

FIGURE 10.4 Hydrogen evolution volumes of Mg-Li-(Al)-(RE) alloy samples as a function of immersion time in Hank's solution. (From Zhou, W. R., Y. F. Zheng, M. A. Leeflang, and J. Zhou. *Acta Biomaterialia*, 9 (10), 8488–98, 2013. With permission.)

decreased cell viabilities to the ECV304 but significantly reduced cell viabilities to the VSMCs, which is a predominant advantage for their application as stent materials.

Hemocompatibility could reflect the anticorrosion and surface biocompatibility properties of implant biomaterials. Normally, a hemolytic ratio that is lower than 5% is required for excellent blood compatibility (ASTM. F756-00, Standard Practice for Assessment of Hemolytic Properties of Materials). For the binary Mg-3.5Li and Mg-8.5Li alloys, hemolysis and platelet adhesion tests were performed in the literature (W. R. Zhou et al. 2013). Hemolysis ratios of 3.8% and 3.4% were observed for Mg-3.5Li and Mg-8.5Li, respectively, exhibiting acceptable blood compatibility. For the platelet adhesion tests, human platelets adhering to the sample surfaces kept a nearly round shape without any pseudopodia-like structures, implying negative activation. These results indicate favorable cytotoxicity and hemocompatibility of these two Mg-Li binary alloys, meaning a greater potential to meet the requirements of stents in biocompatibility.

10.3   DEVELOPMENT OF Mg-Li-X TERNARY ALLOYS AS DEGRADABLE METALLIC BIOMATERIALS

10.3.1   ALLOY COMPOSITION DESIGN

Due to the Li addition, Mg-Li alloys are provided with excellent ductility. However, the strength of the Mg-Li alloys is generally reduced as compensation, and creep frequently occurs even at room temperature, especially for alloys with high Li contents. In order to solve these disadvantages, third alloying elements, such as Al, Zn, Ca, and RE elements, are introduced to improve the mechanical properties of Mg-Li binary alloys (Ma et al. 1999).

Among the alloying elements, Al is one of the most commonly used. The Al addition increases the tensile strength and hardness of Mg-Li alloys by means of solid solution and compound reinforcements (L. Wu et al. 2011). However, the amount of Al in Mg-Li alloys is always between 1 wt.% and 3 wt.% as the extra addition will reduce the ductility of Mg-Li alloys, and the density of alloys will be increased (M. Zhang et al. 2010).

From previous works, the addition of Zn can also improve mechanical properties because of its high solid solubility in both α and β phases (ASM 1990; Polmear 1996). Furthermore, the cold formability of Mg-Li alloys can also be improved by the Zn addition (H.-Y. Wu et al. 2009).

Ca has the potential to raise the oxidizing combustion temperature, refine the microstructure, and improve the mechanical properties at room temperature and heat resistance at an elevated temperature (Song and Kral 2005).

RE elements have been commonly recognized as favorable additives in Mg alloys as they are effective in improving mechanical properties of Mg-Li alloys. Literature has reported that the mechanical properties of Mg-Li binary alloys were improved by the addition of Y and Ce (M. L. Zhang et al. 2009; Dong et al. 2010).

10.3.2 Fabrication, Hot and Cold Working, and Heat Treatment

10.3.2.1 Fabrication

The Mg-Li-X ternary alloys are conventionally prepared by directly mixing the three metallic elements or master alloys. Zhang et al. (2009a) prepared a Mg-8.5Li-(0-3) Ce alloy system by melting commercial pure Mg, commercial pure Li, and Mg-18.26 wt.% Ce master alloy in a vacuum induction furnace. Argon gas was input as a protective gas before melting. Furthermore, the fabrication of Mg-Li-Zn (Chiu et al. 2008), Mg-Li-Y (Dong et al. 2010), Mg-Li-Al (Tsujikawa et al. 2007), and Mg-Li-Ca (Song and Kral 2005) alloy systems by this method has been reported. However, this conventional method has some demerits, such as inhomogeneous alloy composition, complicated production process, serious problems with metal oxidation, and high energy costs. As a consequence, a molten salt electro-codeposition process was proposed and has drawn great attention. Various alloy systems, including Mg-Li-Sm (Han et al. 2009), Mg-Li-Ca (Yan et al. 2009a), and Mg-Li-Al (Y. D. Yan et al. 2008), as well as Mg-Li-Zn (Yan et al. 2009b) were successfully fabricated by this electro-codeposition method in the molten system with the electrochemical formation process studied and optimal electrolysis parameters screened.

10.3.2.2 Hot/Cold Working and Heat Treatment

Enhanced tensile strengths of Mg-8Li-1Al and Mg-13Li-1Al alloys were reported through heat treatment, but the elongation presented was significantly decreased during the process (Ma et al. 1999). Chiu et al. (2008) investigated the effects of thermal and mechanical treatments on the microstructures and strengthening mechanisms on a LZ91 Mg alloy. After annealing heat treatment, the variation in hardness was believed to correspond to an unstable phase or spinodal decomposition causing an age-hardening effect.

Cold working presented a moderate hardening effect, and the hardness of the cold-rolled specimen was almost proportional to the extent of cold working. The extrusion and rolling processes have been reported to lead to an improvement in mechanical properties, and better mechanical properties were observed for the alloys after extrusion than those after rolling (M. L. Zhang et al. 2009). It is known that, during the deformation process, the defects, such as gas pores and inclusion, are decreased, and the grain size is also refined. The dislocation density of alloys also increased during deformation. Therefore, the mechanical properties of deformed alloys are improved.

10.3.3 MICROSTRUCTURE AND MECHANICAL PROPERTIES

10.3.3.1 Mg-Li-Al

Al presents a high solubility in Mg-Li alloys; thus it mainly dissolves in solid solutions. When the content of Al is larger than 3 wt.%, Al-Li compounds emerge in the alloy substrate. The addition of Al in Mg-Li alloys brings about the improvement of strength and a little increase of density, but it causes a decrease of elongation. When the addition is larger than 6 wt.%, the deterioration of elongation is serious (R. Z. Wu et al. 2010).

10.3.3.2 Mg-Li-Zn

The addition of Zn has similar effects in Mg-Li-based alloys. The deterioration of elongation is less serious than that for the addition of Al, but the increase in density is larger than that of Al. The as-cast microstructure of the Mg-9%Li-1%Zn alloy had a dual phase structure with dispersed fine particles of ZnO and MgO oxides (Chiu et al. 2008). After annealing heat treatment of the as-rolled specimen, the α phase was elongated and aligned in the rolling direction, and the β phase showed a recrystallized grain structure. A proportional relationship was observed between the hardness and percentage of cold working; the formation of texture and refined α phase during rolling might result in the work-hardening effect.

10.3.3.3 Mg-Li-Ca

The addition of Ca to the Mg-12Li alloy resulted in a transformation from a singular β phase to a microstructure of two distinct microconstituents: primary dendrites of β and a lamellar interdendritic eutectic of the β phase and $CaMg_2$ (Song and Kral 2005). In addition, the amount of the eutectic in the cast Mg-12Li-xCa alloys was observed corresponding to the Ca content as seen in Figure 10.5 (Song and Kral 2005). After cold rolling, the lamellar eutectic can be crashed. During the oxidization of the alloy, the β single phase was oxidized first, and the eutectic was hard to oxidize, indicating that the Ca in the Mg-Li base alloys has the effects of fire retarding and oxidization retarding.

10.3.3.4 Mg-Li-Ce

It has been reported that Ce obviously refines the grain size of the Mg-Li alloy and leads to the formation of an intermetallic compound ($Mg_{12}Ce$) (M. L. Zhang et al. 2009). The two aspects are both favorable for the strength of the alloys. As for the

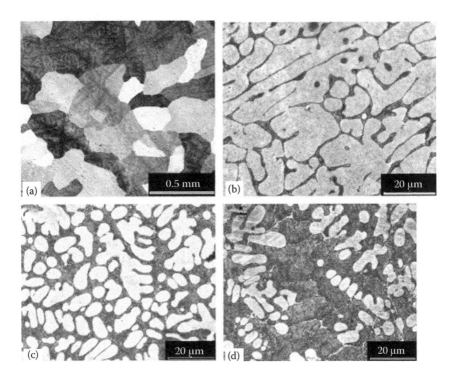

FIGURE 10.5 Optical micrographs showing microstructural evolution in the cast Mg-12Li-xCa alloys with different levels of Ca: (a) x = 0, (b) x = 5, (c) x = 10, and (d) x = 15. (From Song, G. S., and M. V. Kral. *Materials Characterization*, 54 (4–5), 279–86, 2005. With permission.)

elongation percentage of the alloys, refinement is a favorable factor, and the formation of the $Mg_{12}Ce$ compound is an unfavorable factor because $Mg_{12}Ce$ is a hard phase in Mg-Li alloys. Under the combined effects of the two factors, the Mg-8.5Li-1Ce alloy possesses the largest elongation percentage (Table 10.1).

10.3.4 DEGRADATION IN SIMULATED BODY FLUIDS

Until now, Mg-8.5Li-1Al (W. R. Zhou et al. 2013) and Mg-9Li-2Al (LA92) (Leeflang et al. 2011) have been fabricated aiming at stent application. The degradation behavior of Mg-8.5Li-1Al in simulated body fluids was investigated by electrochemical measurements and immersion tests. Hank's solution was adapted as discussed. The calculated corrosion rate of extruded Mg-8.5Li-1Al was 0.10 mm/y. During the immersion period in Hank's solution for 500 h, Mg-8.5Li-1Al maintained an extremely low hydrogen evolution tendency. Compared to the results of Mg-8.5Li, it can be concluded that the anticorrosion resistance of the Mg-Li alloy is improved by the addition of Al.

Immersion tests in the Hank's solution were also conducted to assess the degradation behavior of LA92 (Leeflang et al. 2011). This alloy displayed a steady hydrogen evolution rate over the whole period of immersion tests and even outperformed the

TABLE 10.1
Mechanical Properties of Previously Reported Mg-Li-X Ternary Alloys

Materials	Condition	Mechanical Property			References
		Yield Strength (MPa)	Ultimate Strength (MPa)	Elongation (%)	
Mg-4Li-3Al	Deformed at room temperature	~82	~236	–	Drozd et al. 2004
Mg-4Li-5Al	Deformed at room temperature	~100	~241	–	Drozd et al. 2004
Mg-8Li-1Al	As-cast	–	131	35	Ma et al. 1999
Mg-8Li-1Al	Aged	–	214	12	Ma et al. 1999
Mg-8Li-1Al	As-extruded	~105	~163	37	Liu et al. 2003
Mg-8Li-1Al	As-ECAP one pass	~169	~202	12	Liu et al. 2003
Mg-8Li-1Al	As-ECAP two passes	~185	~212	~14	Liu et al. 2003
Mg-8Li-1Al	As-ECAP three passes	~196	~219	~25	Liu et al. 2003
Mg-8Li-1Al	As-ECAP four passes	~198	~223	~27	Liu et al. 2003
Mg-8Li-7Al	As-extruded	184	239	33	Sanschagrin et al. 1996
Mg-11Li-3Al	As-cast	–	138	45	Ma et al. 1999
Mg-11Li-3Al	Aged	–	247	7	Ma et al. 1999
Mg-14Li-1Al	As-cast	–	130	12	S. J. Wang et al. 2006
Mg-7Li-1Y	As-cast	~117	~120	33	Dong et al. 2010
Mg-7Li-1Y	T4 heat-treated	~107	~112	~35	Dong et al. 2010
Mg-7Li-1Y	T6 heat-treated	~108	~115	~42	Dong et al. 2010
Mg-7Li-3Y	As-cast	144	160	20	Dong et al. 2010
Mg-7Li-5Y	As-cast	~147	~168	~7	Dong et al. 2010
Mg-7Li-7Y	As-cast	~158	~163	~5	Dong et al. 2010
Mg-7Li-7Y	T4 heat-treated	~125	~142	~5	Dong et al. 2010
Mg-7Li-7Y	T6 heat-treated	~125	~142	~7	Dong et al. 2010
Mg-5.8Li-0.46Zn	As-rolled	146.2	197.9	20.8	Lin et al. 2008
Mg-5.8Li-0.51Zn	As-rolled	146.2	197.9	20.64	H.-Y. Wu et al. 2008
Mg-6Li-1Zn	As-rolled	112	155	32.2	Takuda et al. 2002
Mg-8.5Li-1Zn	As-rolled	126	134	63.4	Takuda et al. 2000
Mg-9Li-1Zn	As-rolled	~82	~153	~24	J.-M. Song et al. 2007
Mg-9Li-1Zn	As-cast	~100	~141	~57	Chang et al. 2006
Mg-9Li-1Zn	As-ECAE	~158	~183	~31	Chang et al. 2006
Mg-9.2Li-0.47Zn	As-rolled	115.3	130.4	51.9	Lin et al. 2008

(Continued)

TABLE 10.1 (CONTINUED)
Mechanical Properties of Previously Reported Mg-Li-X Ternary Alloys

		Mechanical Property			
Materials	Condition	Yield Strength (MPa)	Ultimate Strength (MPa)	Elongation (%)	References
Mg-9.5Li-1Zn	As-rolled	121	134	71.4	Takuda et al. 2002
Mg-11Li-1Zn	As-cast	~97	~133	~62	Chang et al. 2006
Mg-11Li-1Zn	As-ECAE	~149	~173	~32	Chang et al. 2006
Mg-12Li-1Zn	As-rolled	124	125	56	Takuda et al. 2002
Mg-8.5Li-0.5Ce	As-cast	–	~102	~29.5	M. L. Zhang et al. 2009
Mg-8.5Li-0.5Ce	As-rolled	–	~108	~34	M. L. Zhang et al. 2009
Mg-8.5Li-0.5Ce	As-extruded	–	~158	~44	M. L. Zhang et al. 2009
Mg-8.5Li-1Ce	As-cast	–	~108	~45	M. L. Zhang et al. 2009
Mg-8.5Li-1Ce	As-rolled	–	~115	~48	M. L. Zhang et al. 2009
Mg-8.5Li-1Ce	As-extruded	–	~170	~64	M. L. Zhang et al. 2009
Mg-8.5Li-2Ce	As-cast	–	~114	~15	M. L. Zhang et al. 2009
Mg-8.5Li-2Ce	As-rolled	–	~120	~16	M. L. Zhang et al. 2009
Mg-8.5Li-2Ce	As-extruded	–	~173	~20	M. L. Zhang et al. 2009
Mg-8.5Li-3Ce	As-cast	–	~105	~8	M. L. Zhang et al. 2009
Mg-8.5Li-3Ce	As-rolled	–	~110	~8.7	M. L. Zhang et al. 2009
Mg-8.5Li-3Ce	As-extruded	–	~160	~8.7	M. L. Zhang et al. 2009

WE-type alloy after immersion for 94 days, which can be attributed to the alkalizing effect of lithium on the corrosion layer to stabilize the formation of magnesium hydroxides. Furthermore, Luo et al. (2013) carried out the electrochemical tests on a Mg-13Li-3Al alloy after immersion in Hank's solution with an obtained corrosion current density of 5.467×10^{-9} A/cm^2 and an associated corrosion potential of -0.50 V.

10.3.5 Cytotoxicity and Hemocompatibility

Until now, the assessment of cytotoxicity and hemocompatibility of Mg-Li-X ternary alloys could only be found for the extruded Mg-8.5Li-1Al (W. R. Zhou et al. 2013). The cytotoxicity was evaluated by indirect assay using vascular-related cell lines ECV304 and VSMC, and the hemocompatibility of Mg-8.5Li-1Al was assessed using hemolysis tests and platelet adhesion tests as mentioned in Section 10.2.5. The cytotoxicity evaluation revealed that the Mg-8.5Li-1Al alloy extracts did not induce toxicity to the ECV304 and VSMC cell lines. The hemolysis tests showed that a hemolysis ratio lower than 5% was observed for Mg-8.5Li-1Al alloy, and no platelets with any pseudopodia-like structures were detected on the surfaces, indicating a favorable hemocompatibility for stent usage.

10.4   DEVELOPMENT OF Mg-Li QUATERNARY AND MULTIELEMENTARY ALLOYS AS DEGRADABLE METALLIC BIOMATERIALS

10.4.1   ALLOY COMPOSITION DESIGN

As discussed in Section 10.3.1, various third alloying elements have been added to Mg-Li alloy systems to improve their mechanical properties, mainly including Al, Zn, and RE elements. Various Mg-Li quaternary and multielementary alloy systems, such as Mg-Li-Al-Zn (Lin et al. 2007), Mg-Li-Al-RE (C. Zhang et al. 2008), Mg-Li-Al-Cu (Saito et al. 1997), Mg-Li-Al-Zn-RE (L. Wu et al. 2011), Mg-Li-Al-Zn-Cu (Li et al. 2010c), Mg-Li-Al-Zn-Sc (H.-Y. Wu et al. 2009), and Mg-Li-Si-Ag (Matsuda et al. 1996), have been developed and investigated concerning the microstructure and mechanical and corrosion properties.

10.4.2   FABRICATION, HOT AND COLD WORKING, AND HEAT TREATMENT

The standard way to produce Mg-Li-based alloys is a vacuum melting method under an argon atmosphere. A high-frequency electric induction furnace that is mounted in a vacuum chamber is typically required. Mg-8Li-1Al-xCe was prepared by melting pure Mg ingots, pure Li ingots, pure Al ingots, and Mg-Ce master alloy (containing Ce 18.26 wt.%) in a vacuum induction-melting furnace under the protection of an argon atmosphere (M. Zhang et al. 2010). Seitz et al. (2011a) have also manufactured a LAE442 (Mg-4Li-4Al-2RE) alloy under a static atmosphere of argon at 0.2 bar with which the pre-alloy AE42 is employed. Many Mg-Li quaternary and multielementary alloys have been fabricated by this conventional method, including LAZ1010, LAZ1010Sc (H.-Y. Wu et al. 2009), Mg-14Li-1Al-xNd (B. Liu et al. 2008), LAY811, LAY831 (R. Wu et al. 2009b), Mg-15Li-5Al-0.5RE (C. Zhang et al. 2008), etc. However, there is little report about fabricating a bcc Mg-Li quaternary and multielementary alloy in air. Lin et al. successively prepared bcc Mg-11.9Li-8.5Al-0.57Zn and Mg-12Li-2.6Al-0.72Zn alloys in ambient atmosphere by electrolysis from 45 wt.% LiCl-55 wt.% KCl molten salt at 500°C (Lin et al. 2007, 2009). After fabrication, the Mg-Li alloys usually experience hot and cold working and heat treatment. Mg-5Li-3Al-2Zn-xRE (x = 0–2.5 wt.%) was reported to attain superior tensile strength and elongation after being heat treated with a protective measure at 390°C for 9 h followed by water quenching as illustrated in Figure 10.6 (L. Wu et al. 2011). The Mg-8Li-(1,3)Al-(0,1)Y alloy ingots were extruded at 280°C first; then the as-extruded specimens were rolled at 280°C (R. Wu et al. 2009b). The mechanical tests indicated that the extrusion process for these alloys has good effects on both strength and elongation, and the rolling process after extrusion can improve the strength of the alloys further, but it makes the elongation of the alloys decrease seriously. It is worthwhile to notice that cold rolling does not necessarily produce strain hardening because of concurrent offset due to dynamic recrystallization (Wang 2009). At 90% cold rolling, strain hardening plus solid-solution strengthening can produce an exceptional and unprecedented high strength in Mg-Li alloy systems.

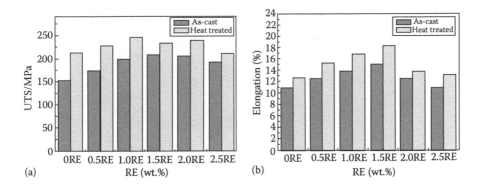

FIGURE 10.6 The tensile strength (a) and elongation (b) of as-cast and heat treated Mg-5Li-3Al-2Zn-xRE alloys. (From Wu, L., C. Cui, R. Wu, J. Li, H. Zhan, and M. Zhang. *Materials Science and Engineering: A*, 528 (4), 2174–9, 2011. With permission.)

10.4.3 MICROSTRUCTURE AND MECHANICAL PROPERTIES

Generally, the addition of Al has effective strengthening; however, it brings about the poor elongation of alloys at the same time. Figure 10.7 presents the microstructure of various Mg-8Li-(1,3)Al-(0,1)Y alloys (R. Wu et al. 2009b). The α phase (white) was observed distributing in the β phase (gray) matrix evenly in these alloys. When 1 wt.% Y is added to the alloys, some blocky compounds of AlY formed in both α and β phases. Meanwhile, the amount of α phase was reduced due to the addition of Al and Y. The mechanical tests demonstrated that the formation of AlY and the decrease of α phase presented as a contradiction both for strengthening and elongation improvement. It was reported that, with the addition of RE, Al-Ce intermetallic compounds emerged in the α-Mg matrix of Mg-3.5Li-4Al-2RE (LAE442) evenly, and markedly improved strength of LAE442 was observed (W. R. Zhou et al. 2013). L. Wu et al. (2011) also confirmed the appearance of Al-RE precipitate with the addition of RE and found that the content increased gradually whereas the Al-Li phase decreased. The room temperature tensile test revealed that the addition of RE could clearly improve the mechanical properties of Mg-5Li-3Al-2Zn (LAZ532) alloy, which are further improved after heat treatment. Furthermore, excellent tensile strength and ductility were obtained in 1.5 wt.% RE containing alloy in the as-cast state.

10.4.4 DEGRADATION IN SIMULATED BODY FLUIDS

The Mg-Li-Al-RE multielementary alloy system has been mostly evaluated as potential metallic biomaterials among the Mg-Li-based alloys. Their degradation behaviors in vitro are usually studied by immersion tests and/or electrochemical measurements in simulated body fluids. The degradation properties of Mg-Li quaternary and multielementary alloys as degradable metallic biomaterials in previous literature are reported in Table 10.2.

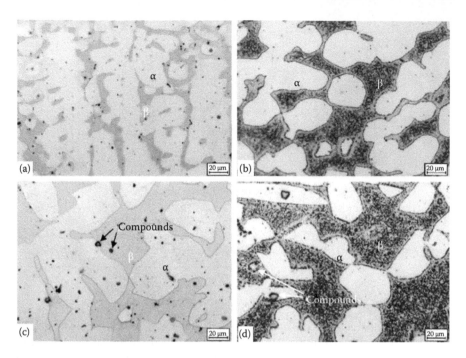

FIGURE 10.7 Microstructure of the alloys (a) LA81, (b) LA83, (c) LAY811, and (d) LAY831. (From Wu, R., Z. Qu, and M. Zhang. *Materials Science and Engineering: A*, 516 (1–2), 96–9, 2009. With permission.)

LAE442 was the first evaluated Mg-Li-based alloy system concerning the in vitro and in vivo corrosion behavior (Witte et al. 2006b). In the in vitro assessment, standardized immersion and electrochemical tests were performed in substitute ocean water and borax-phosphate buffer, respectively. The obtained results showed a stronger corrosion tendency of LAE442 than that of AZ91. However, the in vivo study presented an opposite direction from those obtained from the in vitro study. Furthermore, the electrochemical corrosion behavior of LAE442 was assessed in sodium chloride and phosphate buffer solution (PBS) electrolytes with different chloride ion and albumin concentrations (Mueller et al. 2009). The obtained results showed a complex dependence of the electrochemical parameters on chloride and albumin concentration and metal composition. Leeflang et al. (2011) conducted a long-term immersion test in Hank's balanced salt solution to observe the hydrogen evolution behavior of LAE912 and LAE922. The LAE912 alloy sample was completely degraded by day 98, and the LAE922 alloy sample degraded at a relatively low rate and stopped producing hydrogen after 330 days, as the Mg became exhausted. Compared with the results of LA92, it can be concluded that an addition of 2 wt.% rare earth elements deteriorated the corrosion resistance of the Mg-Li-Al alloy, and an increase in aluminum content from 1 to 2 wt.% would help decrease the corrosion rate. Seitz et al. (2011b) investigated the in vitro degradation of LANd442 in a stationary and flowing simulated body fluid (SBF), respectively. More homogeneous

TABLE 10.2

Degradation Properties of Mg-Li Quaternary and Multielementary Alloys as Degradable Metallic Biomaterials in Previous Literature

Materials	In Vitro			Immersion Tests	In Vivo	References
	Electrochemical Measurements					
	Current Density (A/cm²)	Corrosion Potential (V)	Corrosion Rate (mm/yr)	Corrosion Rate (mm/yr)	Corrosion Rate (mm/yr)	
Mg-13Li-3Al-1Zn-0.5Ca-0.5Sr	2.369×10^{-7}	-1.59	–	0.41	0.63	Luo et al. 2013
LAE442	–	–	6.9	5.535	1.205×10^{-4}	Witte et al. 2006
LAE442	3.54×10^{-6}	-1.33	–	–	–	Mueller et al. 2009
LAE442	1.24×10^{-5}	-1.41	–	–	–	Mueller et al. 2009
LAE442	1.77×10^{-4}	-1.43	–	–	–	Mueller et al. 2009
LAE442	1.43×10^{-5}	-1.24	–	–	–	Mueller et al. 2009
LAE442	3.95×10^{-5}	-1.43	–	–	–	Mueller et al. 2009
LAE442	5.56×10^{-5}	-1.38	–	–	–	Mueller et al. 2009
LAE442	6.10×10^{-5}	-1.42	–	–	–	Mueller et al. 2009
Mg-3.5Li-2Al-2RE	1.42×10^{-5}	-1.488	0.34	–	–	W. R. Zhou et al. 2013
Mg-3.5Li-4Al-2RE	1.04×10^{-5}	-1.507	0.24	–	–	W. R. Zhou et al. 2013
Mg-8.5Li-2Al-2RE	6.87×10^{-6}	-1.482	0.16	–	–	W. R. Zhou et al. 2013

corrosion was observed for the LANd442 samples immersed in a circulating SBF regarding the loss in mass than immersion in a noncirculating corrosive medium.

In our previous work (W. R. Zhou et al. 2013), three Mg-Li-Al-RE system alloys, Mg-3.5Li-2Al-2RE, Mg-3.5Li-4Al-2RE, and Mg-8.5Li-2Al-2RE, were fabricated, and their degradation behaviors in vitro were evaluated by immersion and electrochemical tests in Hank's solution as mentioned. The corrosion rates of these alloys presented a sequence: Mg-3.5Li-2Al-2RE > Mg-3.5Li-4Al-2RE > Mg-8.5Li-2Al-2RE.

Luo et al. (2013) investigated the degradation behavior of the Mg-13Li-3Al-1Zn-0.5Ca-0.5Sr multielementary alloy through immersion and electrochemical experiments in the Hank's solution. Compared to the control Mg-13Li-3Al alloy, this Mg-13Li-3Al-1Zn-0.5Ca-0.5Sr alloy exhibited relatively reduced corrosion resistance, which can be attributed to the second phase uniformly and discontinuously distributed at the matrix.

10.4.5 CYTOTOXICITY AND HEMOCOMPATIBILITY

The morphologies and RGR of L-929 cells cultured in Mg-13Li-3Al-1Zn-0.5Ca-0.5Sr extracts were observed using an indirect contact assay (Luo et al. 2013). The obtained results demonstrated that, after 5 days of incubation, the forged Mg-Li-based alloy exhibited cytotoxicity Grade 0-1 to the L-929 cell line, suggesting acceptable biosafety for cellular application.

W. R. Zhou et al. (2013) have assessed the cytotoxicity and hemocompatibility of three Mg-Li multielementary alloys through the cytotoxicity tests and hemolysis tests, respectively. No significantly decreased viabilities of ECV304 were observed except for with the Mg-8.5Li-2Al-2RE alloy whereas significantly reduced viabilities of VSMCs were observed, meaning a predominant advantage for their application as stent materials. Meanwhile, the hemolysis tests indicated that the Mg-3.5Li-2Al-2RE and Mg-3.5Li-4Al-2RE alloys exhibited acceptable hemolysis ratios, and no sign of thrombogenicity was observed for these Mg-Li-based alloys, indicating the feasibility of Mg-3.5Li-2Al-2RE and Mg-3.5Li-4Al-2RE multielementary alloys as potential stent materials.

10.4.6 IN VIVO PERFORMANCE IN ANIMAL MODELS

The in vivo performance of Mg-Li alloy systems was first investigated in 2005 (Witte et al. 2005), and most were conducted on LAE442. Witte et al. evaluated the corrosion property and the associated bone response of four Mg alloys (LAE442, AZ31, AZ91, and WE43). As illustrated in Figure 10.8, during implant degradation, the surface of the Mg implant materials was coated with a newly formed mineral phase, in direct contact with the surrounding bone and consisting mainly of calcium. Among these alloys, LAE442 presented the lowest degradation rate, which may be explained by the fact that the addition of lithium alkalized the corrosion layer and therefore stabilized the formed magnesium hydroxides within the corrosion layer. Later, Witte et al. performed in vivo corrosion tests by intramedullar implantation of LAE442 and AZ91D sample rods in guinea pig femura (Witte et al. 2006). Consistent with previous results, a lower degradation rate of LAE442 (1.205×10^{-4} mm/yr) was obtained

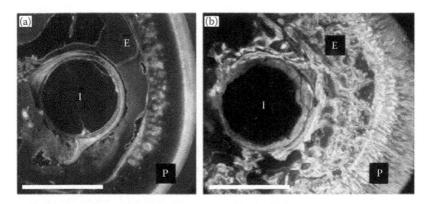

FIGURE 10.8 Flouroscopic images of cross-sections of a degradable polymer SR-PLA96 (a) and a Mg rod (b) performed 10 mm below the trochanter major in a guinea pig femur. Both specimens were harvested 18 weeks postoperatively. In vivo staining of newly formed bone by calcein green. Bar = 1.5 mm; E = endosteal bone formation; I = implant residual; P = periosteal bone formation. (From Witte, F., V. Kaese, H. Haferkamp, E. Switzer, A. Meyer-Lindenberg, C. J. Wirth, and H. Windhagen. *Biomaterials*, 26 (17), 3557–63, 2005. With permission.)

compared to that of AZ91D (3.516×10^{-4} mm/yr). The implants made of LAE442, MgCa0.8, and WE43 were inserted into the medullary cavity of rabbit tibiae to investigate the initial mechanical strength and the degradation behavior (Krause et al. 2009). The results showed that LAE442 had the best initial strength, which seemed to be sufficient for an application in weight-bearing bones, and its degradation behavior was very constant during 6 months of implantation. Bondarenko et al. (2011) examined the morphological changes of efferent lymph nodes after implantation of Mg alloys (MgCa0.8, LAE442) in comparison to the widely used implant materials titanium and PLA. LAE442 was proven to induce the lowest immunological reactions. Therefore, it was considered to be a promising candidate for implants with low immunogenic potential. Implants made of Mg alloys may be associated with skin-sensitizing reactions because of the released metal ions. Witte et al. (2008b) have studied the skin sensitizing potential of four Mg alloys (LAE442, AZ31, AZ91, and WE43) and manifested no skin sensitizing potential detected for all tested Mg alloys as illustrated in Figure 10.9. Reifenrath et al. (2010) have compared the in vivo biodegradation and biocompatibility of two very similar Mg-Li-based alloys (LAE442 and LACer442) by intramedullary implanting both alloy samples in both tibiae of New Zealand white rabbits for 3 months. The LAE442 alloy was proven to be tolerated well and showed excellent biocompatibility, and massive implant intolerance due to gas formation and fast implant degradation were observed for LACer442 (Figure 10.10). The results showed that the single rare earth element cerium cannot successfully substitute the rare earth composition metal in a Mg alloy for biomedical usage.

A newly developed Mg-13Li-3Al-1Zn-0.5Ca-0.5Sr alloy was implanted into the spinal sides of the rats (Luo et al. 2013). The degradation rate of 0.63 mm/yr was obtained by measuring the weight change after implantation, close to the data 0.41 mm/yr

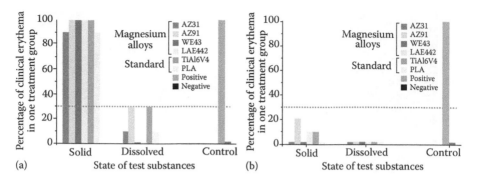

FIGURE 10.9 The percentage of the clinically observed erythema in the epicutaneous patch tests according to the Magnusson-Kligman test immediately after patch removal (a) and 24 h after patch removal (b). Initial erythema can be caused by skin irritation. If more than 30% of the animals (dashed line) in one treatment group still exhibits erythema after 24 h, the tested substance has to be considered to be an allergenic substance. (From Witte, F., I. Abeln, E. Switzer, V. Kaese, A. Meyer-Lindenberg, and H. Windhagen. *Journal of Biomedical Materials Research Part A*, 86A (4), 1041–7, 2008. With permission.)

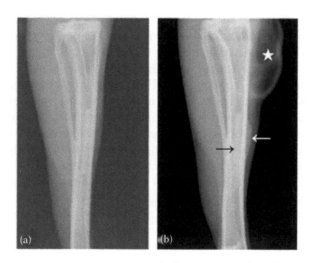

FIGURE 10.10 Radiographic differences between the Mg alloys LAE442 (a) and LACer442 (b). The LAE442 implant (a) did not show structure loss after 3 months of implantation; the LACer442 alloy (b) caused periosteal reaction (white arrow), gas formation (star), and structure loss of the implant with deformation (black arrow) after 4 weeks. (From Reifenrath, J., A. Krause, D. Bormann, B. von Rechenberg, H. Windhagen, and A. Meyer-Lindenberg. *Materialwissenschaft Und Werkstofftechnik*, 41 (12), 1054–61, 2010. With permission.)

achieved from the in vitro tests. Aiming at improving the corrosion resistance of LAE442, a MgF_2 layer has been coated at the metal surface (Witte et al. 2010). Both LAE442 and MgF_2-coated LAE442 presented an acceptable host response without the appearance of subcutaneous gas cavities and fibrous tissue encapsulations, and a reduced corrosion rate in vivo was indicated by Figure 10.11.

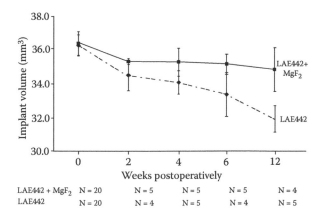

FIGURE 10.11 The implant volume of LAE442 and MgF2-coated LAE442 at different postoperative intervals. Results were obtained from SRμCT analysis. The number of analyzed samples (n) at each time point is stated below the diagram. (From Witte, F., J. Fischer, J. Nellesen, C. Vogt, J. Vogt, T. Donath, and F. Beckmann. *Acta Biomaterialia*, 6 (5), 1792–9, 2010. With permission.)

10.5 NEW ADVANCES IN PROPERTY IMPROVEMENTS FOR Mg-Li-BASED ALLOY SYSTEMS

10.5.1 NEW TECHNIQUES FOR ENHANCING MECHANICAL PROPERTIES

The wider applications of Mg-Li-based alloy systems are hindered by the relatively low strength, especially for the alloys with high lithium contents. There are two common ways to improve mechanical strength: One is an alloying element addition as discussed in Sections 10.3 and 10.4; the other approach capable of resulting in strength increase is grain refinement induced by various types of thermomechanical treatments.

Equal channel angular pressing (ECAP), one of the severe plastic deformation (SPD) methods, has been intensively employed to produce submicrometer- or even nanometer-sized materials and thus improve their strength. A Mg-8%Li-1%Al alloy has undergone a process of ECAP at 130°C for 1, 2, 3, and 4 passes, respectively (Liu et al. 2003). The tensile strength was enhanced pass by pass, and the elongation-to-failure decreased dramatically after the first ECAP pass but could be improved pass by pass during the subsequent ECAP procedures. The excellent superplastic properties of Mg-8%Li, including a maximum elongation of ~970% at 200°C, was achieved following extrusion and subsequent ECAP through two passes (Furui et al. 2005). T. Liu et al. (2007) investigated the microstructure evolution of Mg-14%Li-1%Al during the ECAP treatment. The results showed that the grains of the matrix β phase were substantially refined with the mean size decreasing from 60 μm to 200 nm. Meanwhile, the distribution of α precipitates and ternary MgLiAl$_2$ phase precipitates was homogenized by ECAP processing. It indicated that recrystallization and texture modification were caused by the ECAP processes, and the combined effects contributed to the improvement of the strength and ductility of the alloy simultaneously (Liu et al. 2004).

High-speed ratio-differential speed rolling (HRDSR) is a newly developed SPD technique with which large shear deformation is introduced uniformly along the thickness direction of a sheet during a single rolling pass. The HRDSRed Mg-9Li-1Zn alloy exhibited enhanced strength and ductility at room temperature along with excellent low-temperature superplasticity in the temperature range 150–250°C (0.49–0.6 Tm) (W. J. Kim et al. 2009). These superplastic elongations were considered comparable with those obtained by other SPD techniques or even better.

In addition, rapid solidification processing (RSP) is regarded as another possible way to activate strengthening mechanisms that could not be achieved by conventional ingot processing. The microstructures of rapidly solidified Mg-9Li, Mg-9Li-1Si, and Mg-9Li-1Ce were examined (Meschter and O'Neal 1984). The results showed that the grain refinement produced by the RSP of dispersoid-containing alloys is frequently preserved or further refined during subsequent consolidation processing, and rapidly solidified Mg-9Li-1Si and Mg-9Li-1Ce were likely to persist and give rise to significant Hall-Petch strengthening.

Recently, some interest has been reported in the enhancement of the strength and stiffness of Mg-Li alloys by using second stiffness phase reinforcement (Luo et al. 2006). S. J. Wang et al. (2006) produced a 5 wt.% Al_2Y/Mg-14Li-1Al composite by the stir-casting technique. The results showed that the tensile strength, elastic modulus, and hardness values were significantly enhanced 45.3%, 44.7%, and 58.2%, respectively, with good ductility preserved. The improved mechanical properties have also been observed for the 15 wt.% YAl_2/β-Mg-12 wt%Li composite (Luo et al. 2006), $5B_4C$/Mg-9Li (González-Doncel et al. 1990), Al_2O_3/Mg-10.3Li-6Al-6Ag-4Cd, Al_2O_3/Mg-12Li, 20 vol.% SiC/Mg-12Li (Mason et al. 1989). Trojanová et al. (2007) indicated that the mechanical properties and the deformation behavior of composites are strongly influenced by temperature. The yield strength and maximum stress were observed to decrease rapidly with increasing temperature. Although higher strength can be achieved by this composite reinforcement method, the plasticity and lightness of the alloys are always to be sacrificed. The addition of grain refiners into molten metal to inoculate heterogeneous nucleation is a simple and cost-effective method that is particularly suitable for big ingot production, compared to other techniques, such as equal channel angular pressing and the rapid solidification method (Jiang et al. 2010). After the addition of 1.25 wt.% Al-5Ti-1B master alloy to LA141 alloy, the grain size of β phase was significantly reduced from 1750 to 500 μm, resulting from the TiB_2 and Al_3Ti particles that acted as heterogeneous nucleation sites for β-Li grains (Jiang et al. 2010).

10.5.2 NEW METHODS FOR ENHANCING CORROSION RESISTANCE

It should be noted that the corrosion resistances of Mg-Li system alloys are weaker than other Mg alloys as lithium is a more chemically active metal than Mg. The poor corrosion resistance consequently limits their extensive utilization. As a result, it is necessary and of crucial importance for people to comprehend the corrosion behaviors and mechanisms of Mg-Li-based alloys to develop effective protection methods. Recently, many surface modification methods were employed to improve the corrosion resistance of Mg-Li system alloys, which were chemical conversion coating (L. Yang et al. 2008, 2009b; X. Yang et al. 2009), anodizing (Sharma et al. 1993; Li et al. 2006), electroless plating (L. Yang et al. 2009a), microarc oxidation (MAO) (Y. Xu et al. 2009), diamond-like carbon (DLC) coating (Yamauchi et al. 2007), and so on.

Among these various protection measurements, chemical conversion coating is paid more attention because it is low cost, easy to operate, and can increase the following coatings' adhesion effectively. Usually, the conversion coating as a bottom layer then painting is the most popular measurement for the industrial applications. The traditional conversion coatings are based on chromium compounds that have been shown to be highly toxic carcinogens. Consequently, novel environmentally friendly conversion coatings have been researched for Mg-Li alloys in recent years, including stannate conversion coating, phosphate conversion coating, and organic acid conversion coating, etc.

L. Yang et al. (2009b) have performed a stannate conversion coating on Mg-8Li-1Al-3Zn-0.7RE alloy by a simple immersion method. After being treated for 60 min, a uniform, dense, and corrosion-resistant surface was observed with hemispherical particles covered.

Song et al. (2009b) applied a phosphate conversion film to Mg-8.8Li alloy. This conversion film was composed of a large number of leaf-like particles to exhibit lamellar structure and contributed to the improved corrosion resistance of the Mg-8.8Li alloy during the electrochemical and immersion tests.

As rare-earth salts with low toxicity are known to inhibit the corrosion processes on several substrates, such as steel, aluminium, tin plate, and conventional Mg alloy, the rare earth conversion coatings, especially in the case of cerium and lanthanum conversion coatings, have been successfully performed on these substrates to provide good corrosion protection (L. Yang et al. 2008). A lanthanum-based conversion coating was produced on Mg-8Li alloy in Yang's work (2008). The coating was observed to be homogeneous and uniform with an aciculate-like structure. The electrochemical results revealed that the corrosion resistance of Mg-8Li alloy was improved by the lanthanum-based conversion coating. Later, a rare earth conversion coating was formed by the simple immersion on Mg-8.5Li alloy (X. Yang et al. 2009) and the results of electrochemical measurements and immersion tests revealed that this coating possessed better corrosion resistance than bare alloy and chromate conversion coating.

The phytic acid conversion coating is reported to slow down the corrosion of Mg and Mg alloys as it may generate a chemical reaction with organic coatings and increase the adherence between the alloy and organic coatings (L. Gao et al. 2009b). A uniform phytic acid conversion coating with a white flower-like deposit were prepared on the Mg-11Li-3Al-0.5RE alloy (L. Gao et al. 2009b). Lower corrosion current density, higher corrosion potential, and less hydrogen evolution during the electrochemical and immersion tests were detected for the phytic acid conversion coated alloy, indicating improved corrosion resistance using this coating.

In addition, the corrosion behavior of MAO-treated Mg-5Li-3.5Al-1.2Zn-1.2Ce alloy (Y. Xu et al. 2009), laser treated MA21 alloy (Kalimullin and Kozhevnikov 1985), galvanic anodized Mg-10.02Li-3.86Zn-2.54Al-1.76Cu alloy (Li et al. 2006), electroless nickel-plated Mg-8Li (L. Yang et al. 2009a), DLC-coated Mg-14Li alloy (Yamauchi et al. 2007), and plasma thermal-sprayed LA141 with aluminum (Tsujikawa et al. 2007) as well as PANI-coated Mg-5Li alloy (Shao et al. 2009) were investigated and proven to be effective protection to the substrate alloys. Furthermore, K. Liu et al. (2008) reported bioinspired superhydrophobic Mg-Li alloy surfaces with peony-like micro-nanoscale hierarchical structures. The obtained Mg-Li alloy surfaces presented a dramatically improved corrosion resistance and long-term stable superhydrophobic properties with a static water contact angle of about 160° and a small sliding angle of less than 5°, which may extend the practical application in industrial and high-technology fields. The degradation behavior of surface-treated and untreated Mg-Li alloys as degradable metallic biomaterials are illustrated in Table 10.3 as reported in previous literature. It can be concluded that these mentioned various surface treatments are effective in promoting the corrosion resistance of Mg-Li alloys to different extents.

As for the biomedical applications, only the MgF_2 coating has been reported for the corrosion protection of Mg-Li-based alloys (Seitz et al. 2011b; Witte et al. 2010). The in vitro immersion test performed by Seitz et al. (2011b) demonstrates that the MgF_2 coating leads to a more even degradation with less pitting corrosion in the early stages of corrosion as seen in Figure 10.12.

After being implanted into the medial femur condyle of adult rabbits for 12 weeks, localized pitting corrosion occurred in MgF_2-coated and uncoated LAE442 implants (Figure 10.13) (Witte et al. 2010). However, more uniform corrosion with singular deep pits was observed for the MgF_2-coated samples. The calculated corrosion rate of 0.13 mm/y of coated LAE442 also proved a retarded degradation behavior afforded by the MgF_2 coating, comparied to that of 0.31 mm/y of uncoated samples.

10.6 CONCLUDING REMARKS

As the lightest alloy, Mg-Li-based alloys keep drawing great attention in the fields of aerospace applications, automotive, military, electronics, and transportation industries, etc., due to the combination of high specific strength and rigidity, excellent formability, and high energetic particle penetration resistance. In recent years, their feasibilities in biomedical fields have been investigated, mainly concerning the microstructure, mechanical properties, and degradation behavior as well as biocompatibility. Most of the works are conducted on LAE442 although a few other Mg-Li systems have been studied. It has been indicated that by selecting the proper alloy system and adjusting the composition, Mg-Li-based alloys are of great potential for medical usage. Due to the chemical activity of Li, Mg-Li-based alloys are generally provided with impressive mechanical strength and corrosion resistance. Hence, new techniques, including ECAP, HRDSR, RSP, and composites have been developed to enhance the mechanical properties, and the chemical conversion coating, anodizing, electroless plating, microarc oxidation, diamond-like carbon coating are employed to improve the insufficient corrosion resistance of Mg-Li-based alloys, and their efficiencies in improving the properties have been assessed.

TABLE 10.3

Comparison of the In Vitro Degradation Behavior of Surface-Treated and Untreated Mg-Li Alloys as Degradable Metallic Biomaterials as Reported in the Previous Literature

Materials	Electrochemical Measurements			Immersion Tests	References
	Current Density (A/cm²)	Corrosion Potential (V)	Polarization Resistance	Corrosion Rate	
MA21	—	—	—	0.17 cm³/(cm²·h)	Kalimullin and Kozhevnikov 1985
Continuous laser irradiation	—	—	—	0.016 cm³/(cm²·h)	
Pulsed laser irradiation	—	—	—	0.0056 cm³/(cm²·h)	
Mg-5Li-3.5Al-1.2Zn-1.2Ce	—	~-1.57	—	—	Y. Xu et al. 2009
MAO coating (3 g/l Na₂SiO₃)	0.404×10^{-6}	~-1.52	—	—	
MAO coating (6 g/l Na₂SiO₃)	0.104×10^{-6}	~-1.50	—	—	
MAO coating (9 g/l Na₂SiO₃)	0.312×10^{-6}	~-1.41	—	—	
Mg-8Li-1Al-3Zn-0.7RE	—	-1.594	—	—	L. Yang et al. 2008
Lanthanum-based conversion coating	—	-1.315	—	—	
Mg-11Li-3Al-0.5RE	1.046×10^{-3}	-1.645	—	—	L. Gao et al. 2009b
Phytic acid conversion coating	2.633×10^{-6}	-0.905	—	—	
Mg-8.5Li-3.2Al-1.2Y-1.2Ce	1.294×10^{-4}	—	$1.942\ \Omega\ cm^{-2}$	—	X. Yang et al. 2009
Rare earth conversion coating	3.061×10^{-5}	—	$18.04\ \Omega\ cm^{-2}$	—	
Chromate conversion coating	5.587×10^{-5}	—	$14.78\ \Omega\ cm^{-2}$	—	
Mg-8Li-1Al-3Zn-0.7RE	1.046×10^{-3}	-1.594	$1.698\ \Omega\ cm^{-2}$	—	L. Yang et al. 2009b
Stannate conversion coating (30 min)	3.279×10^{-4}	-1.448	$3.628\ \Omega\ cm^{-2}$	—	
Stannate conversion coating (60 min)	1.282×10^{-5}	-1.297	$3.796\ \Omega\ cm^{-2}$	—	

(Continued)

TABLE 10.3 (CONTINUED)

Comparison of the In Vitro Degradation Behavior of Surface-Treated and Untreated Mg-Li Alloys as Degradable Metallic Biomaterials as Reported in the Previous Literature

Materials	Electrochemical Measurements			Immersion Tests		References
	Current Density (A/cm^2)	Corrosion Potential (V)	Polarization Resistance	Corrosion Rate		
Stannate conversion coating (90 min)	1.659×10^{-5}	−1.363	3.871 Ω cm^{-2}	–	–	L. Yang et al. 2009a
Mg-8Li-1Al-3Zn-0.7RE	1.046×10^{-3}	−1.594	–	–	–	
Molybdate pretreatment	1.213×10^{-4}	−1.466	–	–	–	
Electroless Ni-P coating	2.247×10^{-5}	−1.083	–	–	–	
Mg-10.02Li-3.86Zn-2.54Al-1.76Cu	1.24×10^{-3}	−1.547	21.1 Ω cm^2	–	–	Li et al. 2006
Galvanic anodization (pH 2.5, room temperature)	7.915×10^{-6}	−1.356	3296 Ω cm^2			
Galvanic anodization (pH 3.5, room temperature)	1.76×10^{-5}	−1.367	1483 Ω cm^2			
Galvanic anodization (pH 4.5, room temperature)	3.45×10^{-6}	−1.395	7551 Ω cm^2			
Galvanic anodization (pH 5.5, room temperature)	6.21×10^{-6}	−1.481	4200 Ω cm^2			
Galvanic anodization (pH 4.5, 40°C)	2.95×10^{-5}	−1.417	885.5 Ω cm^2	–		
Galvanic anodization (pH 4.5, 50°C)	1.1×10^{-4}	−1.091	236.5 Ω cm^2	–		

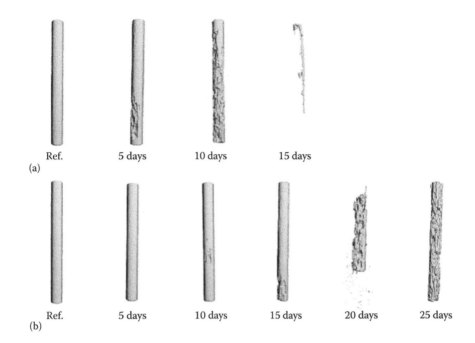

FIGURE 10.12   3-D results of the μCT analysis for the LANd442 alloys after certain times in the SBF: (a) uncoated; (b) MgF2-coated. (From Seitz, J.-M., K. Collier, E. Wulf, D. Bormann, and F.-W. Bach. *Advanced Engineering Materials*, 13 (9), B313–23, 2011. With permission.)

FIGURE 10.13   Reconstructed and visualized SRμCT data showing the morphology of in vivo corroded Mg alloy LAE442 (a) and MgF2-coated LAE442 (b) 12 weeks postoperatively. (From Witte, F., J. Fischer, J. Nellesen, C. Vogt, J. Vogt, T. Donath, and F. Beckmann. *Acta Biomaterialia*, 6 (5), 1792–9, 2010. With permission.)

11 Mg-RE-Based Alloy Systems for Biomedical Applications

11.1 GENERAL INTRODUCTION

The rare earth elements (REEs), sometimes called rare earth metals, are an abundant group of 17 elements consisting of 15 lanthanides (lanthanum, cerium, praseodymium, neodymium, promethium, samarium, europium, gadolinium, terbium, dysprosium, holmium, erbium, thulium, ytterbium, and lutetium) and two other elements: scandium and yttrium (Cordier and Hedrick 2009). Scandium and yttrium are not included in the lanthanide group, but they are considered to be REEs because they tend to occur in the same ore deposits as the lanthanide group and exhibit similar chemical properties (Hedrick 2004). In industrial applications, the lanthanide elements are traditionally divided into two groups: the light REEs (from lanthanum to europium, $Z = 57 - 63$) and the heavy REEs (from gadolinium to lutetium, $Z = 64 - 71$). Although yttrium is the lightest REE, it is placed into the heavy REE group because it is physically and chemically similar to heavy REEs.

The elemental forms of REEs are iron gray to silvery lustrous metals, which are typically soft, malleable, and ductile and usually reactive, especially at elevated temperatures or in the form of powder. Table 11.1 lists the basic physical properties of the REEs. The densities of the REEs range between 2.9 and 9.9 g/cm³. Among all the 17 elements, scandium has the lowest density of 2.985 g/cm³, and lutetium has the highest density of 9.841 g/cm³. The melting and boiling points of the REEs differ over a wide range: 790–1660°C for the melting point, 1200–3470°C for the boiling point. Elements with a hexagonal close-packed (HCP) structure are as follows: scandium, yttrium, gadolinium, terbium, dysprosium, holmium, erbium, thulium, and lutetium. Cerium and ytterbium have a face-centered cubic (FCC) structure. The crystal structures of samarium and europium are rhombohedral and body-centered cubic (BCC), respectively. The crystal structure of the remaining elements is hexagonal. The atomic structure of the REEs is similar, and, from the last column in Table 11.1, it is clear that all the REEs have a similar atomic radius. REEs are enriched in the earth's crust, and they invariably occur together naturally because all of them are in a trivalent state; Ce^{4+} and Eu^{+2} are the exceptions (Castor and Hedrick 2006).

Increasing in atomic number in the lanthanide group is not accompanied by a change in valence. The lanthanide elements occupy the same cell of the periodic table. The similar radius and the same valence in mineral allow the rare earth elements to liberally substitute each other in various crystal lattices, and that is why those elements are widely dispersed in the earth's crust, and also they co-appear in

TABLE 11.1

Basic Physical Properties of Rare Earth Elements

Atomic Number	Element	Symbol	Density (g/cm³)	Melting/Boiling Point (°C)	Crystal Structure (RT)	Radius (nm)
21	Scandium	Sc	2.985	1541/2830	HCP	0.162
39	Yttrium	Y	4.472	1526/3336	HCP	0.180
57	Lanthanum	La	6.146	920/3470	Hexagonal	0.162
58	Cerium	Ce	6.689	795/3360	FCC	0.1818
59	Praseodymium	Pr	6.640	935/3290	Hexagonal	0.1824
60	Neodymium	Nd	6.800	1024/3100	Hexagonal	0.1814
61	Promethium	Pm	7.264	1100/3000	Hexagonal	0.1834
62	Samarium	Sm	7.353	1072/1803	Rhombohedral	0.1804
63	Europium	Eu	5.244	826/1527	BCC	0.2084
64	Gadolinium	Gd	7.901	1312/3250	HCP	0.1804
65	Terbium	Tb	8.219	1356/3230	HCP	0.1773
66	Dysprosium	Dy	8.551	1407/2567	HCP	0.1781
67	Holmium	Ho	8.795	1461/2720	HCP	0.1762
68	Erbium	Er	9.066	1529/2868	HCP	0.1761
69	Thulium	Tm	9.321	1545/1950	HCP	0.1759
70	Ytterbium	Yb	6.570	824/1196	FCC	0.1933
71	Lutetium	Lu	9.841	1652/3402	HCP	0.1738

Source: Data were collected from WebElements.

one single mineral. A slight variation in the atomic/ionic radius causes physical or chemical differences among the lanthanide group and generally result in the segregation of REEs into deposits enriched in either light or heavy lanthanides.

In the 20th century, all of the REEs were finally identified, and only in the past 60 years, REEs were commercialized. The whole process of REEs in their discovery, refining, production, and use is so aggressive. Nowadays, REEs are used in a wide range of applications (Haxel et al. 2002; Castor and Hedrick 2006; Eliseeva and Bünzli 2011). Many devices that are essential to our daily life, such as computer memory, DVDs, rechargeable batteries, cell phones, magnets, fluorescent lighting, and much more are closely related to REEs. For industrial purposes, REEs are widely used as catalysts, phosphors, and polishing compounds, which are used for air pollution control, illuminated screens on electronic devices, and the polishing of optical glass. The introduction of high strength samarium–cobalt permanent magnets developed a new market for REEs. Permanent magnet technology has been revolutionized by alloys containing Nd, Sm, Gd, Dy, or Pr. This kind of tiny, lightweight, high-strength, and high-quality magnetic materials have made great contributions to the automobile industry, military equipment, and numerous electrical and electronic components. Environmental applications of REEs have dramatically increased in recent decades, and this trend won't cease. Faced with the problems of global warming and the energy crisis, researchers are working on catalysts based on

REEs that can minimize the pollution from industrial waste gas. New energy-saving fluorescent lamps (with Y, La, Ce, Eu, Gd, and Tb) are gradually becoming popular across the world. There are still many important roles REEs can play in environmental protection.

Traditionally, the addition of REEs into metallurgy is used to improve magnetic, electrical, mechanical, and some other properties of various alloys (Ferro et al. 1997; Lü et al. 2000). For Mg alloys, different REEs are added to improve various properties for engineering applications, including room temperature and elevated temperature mechanical properties and thermodynamic stability as a basic expectation and, moreover, the creep resistance, corrosion resistance, aging behavior, damping characteristics, etc.

After adding REEs into Mg as alloying elements, REE atoms can substitute Mg atoms in the lattice structure and cause strengthening effects, such as solid solution strengthening, precipitation strengthening, and grain refining strengthening. Some of the REEs (Y, Gd, Tb, Dy, Ho, Er, Tm, Yb, and Lu) have high solubilities in Mg, and they could be the proper strengthening alloying elements in Mg.

REEs are also predominantly used to improve corrosion resistance of Mg alloys. The positive effect of REEs on corrosion is attributed to two main mechanisms. One of the important effects on the corrosion resistance of Mg alloys is the so-called "scavenger effect." Impurities, such as Fe, Ni, and Co, severely deteriorate the corrosion resistance of Mg alloys. REEs can interact with those impurities and form less noble intermetallic compounds and weaken their bad influences in this way (Takenaka et al. 2007; W. Liu et al. 2009). The other reason is the change of the surface film of Mg alloys with REE addition. Rare earth atoms incorporate into the surface hydroxide layer, improving its stability and increasing its protective effect against corrosive media (Rosalbino et al. 2006; X.-W. Guo et al. 2007; Takenaka et al. 2007). Moreover, the homogenous microstructure obtained by the REE addition might also contribute to the improved corrosion resistance of Mg alloys.

For Mg alloys used for biomedical purposes, the addition of REEs or mischmetal into Mg is supposed to increase their mechanical property, enhance their corrosion resistance, and improve their biocompatibility in the physiological environment. By doing this, the biomedical Mg alloy devices can serve well in the human body. Here, the role of REEs and their mischmetals in biomedical Mg alloys will be discussed based on the recent progress in this field.

11.2   Mg-Y-BASED ALLOYS FOR BIOMEDICAL APPLICATIONS

11.2.1   Yttrium

Yttrium (Y), with a density of 4.472 g·cm^{-3}, is a silvery metal of the transition metals. It has been regarded as a rare earth element because it is chemically similar to the lanthanides. Although Y is the lightest REE, it is classified into the heavy REE group because they are physically and chemically similar. Pure Y in bulk form is relatively stable in air because of its self-passivation properties, but this is not true when Y is finely divided. The most important use of Y lies in the phosphors, such as the red phosphors used in the cathode ray tubes in televisions and phosphors in LEDs

(Ozawa and Itoh 2003; Krämer et al. 2004; Jang et al. 2007). Y and its compounds are also used in the production of electrodes, electrolytes, lasers, superconductors, and even in various medical applications (Uhrin et al. 1977; Wang et al. 1984; Dietz and Horwitz 1992; Cava et al. 1994).

The biological role of Y in the human body has no definite identification. When yttrium chloride was orally administered to male Wistar rats, a significant decrease of urine volume and creatinine excretion was observed at the dose of 58.3–116.7 mg per rat, indicating that a high Y dosage altered the glomerular function (Hayashi et al. 2006). According to the study of Drynda et al. (2009), Y did not lead to significant changes in metabolic activity of vascular smooth muscle cells over a wide concentration range; however, damage was found at high concentrations. It was also reported that the tolerant daily intake of Y is 4.2 mg·day^{-1} according to a survey on the high rare earth element background regions in China (Zhang et al. 2000).

Even though the biological role of Y in the human body is not clear, it has been added into biomedical Mg alloys to improve their performances. WE43, containing approximate 4 wt.% Y, has been intensively investigated for biomedical applications, and this alloy shows good biocompatibility. Alloys consisting of <5 wt.% Y being explored for cardiovascular stents show endothelialization, low neointima proliferation, and minimal inflammation properties (Di Mario et al. 2004; Waksman et al. 2006). Based on a previous study, Y has been demonstrated to be a suitable alloying element for biomedical Mg alloys if used in a limited amount.

11.2.2 Previous Studies on Mg-Y-Based Alloys for Industrial/Engineering Applications

Development of new advanced Mg-Y alloys that combine higher strength, higher ductility, and higher creep resistance with lower cost and less pronounced temperature dependence is of technological and economic significance.

The alloying of Mg with Y raises its strength properties at room temperature and at elevated temperatures (Bae et al. 2002; Q. Li et al. 2007; Xu et al. 2007a; Q. Wang et al. 2008a, 2012; Gao et al. 2009a; Wan et al. 2009; Dong et al. 2010; Su et al. 2010; B. L. Wu et al. 2010; Rokhlin et al. 2011; Wang et al. 2011a; Zhang et al. 2011a; Leng et al. 2013a; N. Zhou et al. 2014), and Y is considered to be one of the most effective REEs to improve mechanical properties of Mg alloys at high temperatures (Suzuki et al. 2004). Y is also beneficial to the corrosion resistance (Yao et al. 2003), creep resistance (Moreno et al. 2003; Aghion et al. 2008), and fatigue strength (Xu et al. 2007b), and addition of Y in Mg alloys can significantly improve its ignition temperature (Fan et al. 2005, 2009, 2011; X. M. Wang et al. 2008; Prasad et al. 2012a; Zhou et al. 2013b), and damping properties (Ravi Kumar et al. 2003; Somekawa et al. 2011). Here, it should be mentioned that the beneficial role of Y in Mg alloys is based on appropriate alloying composition design and proper heat treatments and manufacturing processes.

Among the engineering Mg-Y alloys, the most inspiring one is the super high strength $Mg_{97}Y_2Zn_1$ (at.%), which exhibits a tensile yield strength of about 600 MPa and an elongation of 5% at room temperature (Kawamura et al. 2001; Yamasaki and Kawamura 2009). Inspired by the super high strength $Mg_{97}Y_2Zn_1$ alloy, many efforts have been devoted to developing Mg-Y-based alloys reinforced by a long period stacking ordered structure (Hagihara et al. 2010a; Yi et al. 2011; Leng et al. 2012, 2013a; Matsumoto et al. 2012; J. S. Zhang et al. 2012b; Zhu et al. 2012a; Liu et al. 2013a,b; Mine et al. 2013; Tong et al. 2013). Much higher strengthening effects can be attained in the situation of joint alloying of Mg with two or more specific REEs. WE-type Mg alloys are the powerful proof (Rokhlin et al. 2011).

In addition to the Mg-Y-Zn and WE Mg alloys, many other novel engineering Mg-Y-based alloys have also been developed, which include Mg-Y-Ca (Zhou et al. 2013a), Mg-Y-Ni (Liu et al. 2013b), Mg-Zn-Y-Zr (Q. Yang et al. 2011), Mg-Y-Gd-Zn (Hu et al. 2012b), Mg-Y-Sm-Zr (Li et al. 2007c, 2008; Q. Wang et al. 2009; D.-Q. Li et al. 2010), Mg-Y-Zn-Cu (Shi et al. 2013), Mg-Y-Gd-Zn-Zr (Y. Gao et al. 2009a; K. Liu et al. 2010; Q. Wang et al. 2010b), Mg-Y-Nd-Gd-Zr (X. Zhao et al. 2008; Y. Li et al. 2010b; L. Wang et al. 2010; Z.-J. Liu et al. 2012) and Mg-Y-Nd-Zn-Mn (Janik et al. 2008), etc.

It is an interesting finding that the addition of Al in Mg-Y alloys can lead to in situ formation of AlY_2 particles, which can significantly reduce the average grain size. This excellent grain-refining efficiency is comparable to that of Zr for the same alloy. Moreover, the thermal stability of the grains refined by Al_2Y is much higher than those refined by Zr (Qiu and Zhang 2009; Qiu et al. 2009; H. W. Chang et al. 2013). As we know, grain refining strengthening is the optimum method among the strengthening methods in metals that will not deteriorate the elongation. So such Mg-Y Mg alloys strengthened by Al_2Y would exhibit excellent mechanical properties with large elongation.

Mg-based metallic glasses are regarded as a new family of promising materials with high specific strength. A great deal of work has been carried out on Mg-Cu-Y and Mg-Cu-Zn-Y (Men et al. 2002; Gun et al. 2006; Hojvat de Tendler et al. 2006; Lee et al. 2006; Castellero et al. 2008; Z. G. Li et al. 2008; Wang et al. 2008b) or some other Mg-Cu-Y-based systems (Linderoth et al. 2001; Baricco et al. 2007; Tian et al. 2007; Hui et al. 2010; Soubeyroux and Puech 2010; G. B. Liu et al. 2012).

11.2.3 DEVELOPMENT OF MG-Y BINARY ALLOYS AS DEGRADABLE METALLIC BIOMATERIALS

11.2.3.1 Alloy Composition Design

Elemental Y has the structure of hexagonal close packed lattice (HCP), same as Mg, but the atomic radius of Y is 12.5% larger than Mg (0.18 nm for Y and 0.16 nm for Mg). Figure 11.1 is the phase diagram of the Mg-Y binary system. Possible precipitation phases in binary Mg-Y alloys are $Mg_{24}Y_5$, Mg_2Y, and MgY, depending on the alloy composition, cooling rate, and some other factors.

Elemental Y is more effective for the strengthening of Mg alloys than the elements aluminum and manganese (Suzuki et al. 1998). The highest solubility of Y in Mg is about 11.4 wt.% at the eutectic point of 567.4°C, and the solubility decreases rapidly with the decrease of temperature. The temperature-dependent solubility and large atom radius difference between Y and Mg atoms can lead to significant solution strengthening. Precipitation hardening is another important method of enhancing

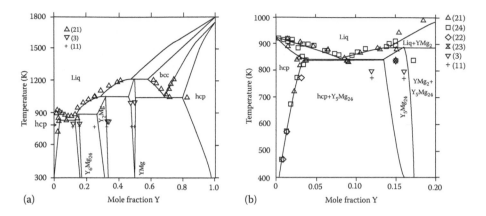

FIGURE 11.1   (a) Phase diagram of the Mg–Y system together with experimental data. (b) Enlarged part of the Mg–Y phase diagram together with all available experimental data. (From Fabrichnaya, O. B., H. L. Lukas, G. Effenberg, and F. Aldinger, *Intermetallics*, 11 (11), 1183–1188, 2003. With permission.)

the mechanical properties of Mg-Y-based alloys. Because of the ambiguous biosafety of Y, Y content in biomedical Mg alloys should be well designed. Reasonable content of Y should be added into Mg alloys to satisfy the strength requirement while maintaining the acceptable biocompatibility of the materials.

11.2.3.2   Microstructure and Mechanical Properties

The typical microstructures of as-cast and solid solution treated Mg-Y binary alloys are illustrated in Figure 11.2. The microstructure of as-cast Mg-Y alloys consists of primary α-Mg and yttrium-rich net-segregation $Mg_{24}Y_5$ phases. $Mg_{24}Y_5$ eutectic phases mainly distribute along the grain boundaries, and the amount of the eutectic $Mg_{24}Y_5$ phase increases with the increase of Y addition as shown in Figure 11.2a–c. However, the solid solution treated binary Mg-Y alloys are composed of a single α-Mg phase as can be seen in Figure 11.2d–f. No second phases can be observed in any of the studied alloys in the Y ranging from 0.8 wt.% to 6.6 wt.%. Y has little grain refining effect on the solid solution treated Mg-Y alloys.

For the solid solution treated Mg-Y alloys, which consist of a single α-Mg phase, the hardness increases with the increasing content of Y. Data from a similar study on Mg-Al and Mg-Zn alloys are included for comparison (Figure 11.3a). The yield strengths of the Mg-Y alloys show a similar trend as the hardness. From Figure 11.3b, it can be noticed that the tensile yield strength (TYS) and ultimate strength (UTS) increase gradually with increasing content of Y in the matrix while the elongation to failure declines correspondingly. The variations of tensile strength with the element Y content once again prove that Y is an effective rare earth metal in strengthening Mg, even much more effective than aluminum and manganese. Even though Y can significantly enhance the mechanical properties of Mg, generally, the strengths of solid solution treated binary Mg-Y alloys are still not satisfactory.

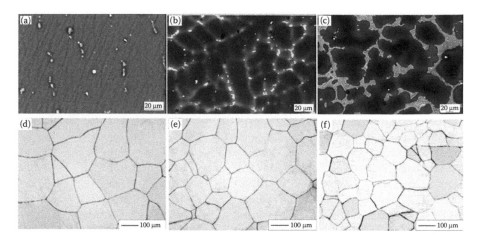

FIGURE 11.2 Typical microstructure of Mg-Y binary alloys: (a) as-cast Mg-0.25Y, (b) as-cast Mg-8Y, (c) as-cast Mg-15Y; (d) as-soluted Mg-0.8Y, (e) as-soluted Mg-3.12Y, (f) asso-luted Mg-6.6Y. (a–c) are SEM images; (d–f) are optical images. ([a–c] From Zhang, X., Y. J. Li, C. S. Wang, H. W. Li, M. L. Ma, and B. D. Zhang, *Transactions of Nonferrous Metals Society of China*, 23, 1226–1236, 2013. With permission. [d–f] From Gao, L., R. S. Chen, and E. H. Han, *Journal of Alloys and Compounds*, 472 (1), 234–240, 2009a. With permission.)

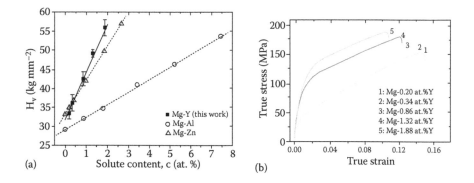

FIGURE 11.3 Mechanical properties of solution-treated Mg-xY alloy; (a) hardness of the Mg-xY alloys, (b) typical stress-strain curve for the Mg-xY alloys. (From Gao, L., R. S. Chen, and E. H. Han, *Journal of Alloys and Compounds*, 472 (1), 234–240, 2009a. With permission.)

11.2.3.3 Biodegradation Behavior

Y is a unique rare earth metal in Mg alloys because its standard electrode potential is the same as Mg, which is about −2.732 V (Sudholz et al. 2011). Table 11.2 lists the electrochemical corrosion data for various Mg-xY alloys in 3.5% NaCl solution. For the as-cast Mg-xY alloys, the corrosion potentials increase with the addition of Y up to 2.5%, and the corrosion potentials E_{corr} decrease when the Y content is above 2.5% due to the formation of the second phases. The pitting potentials E_{pit} and corrosion

TABLE 11.2

Electrochemical Corrosion Parameters of Various Mg-Y Alloys in 3.5% NaCl Solution

Alloy	Condition	E_{corr}(V)	I_{corr} ($\times 10^{-4}$ A/cm^2)	E_{pit} (V)
Mg-0.25Y (Zhang et al. 2012h)	As-cast	−1.6075	2.5098	−1.6097
Mg-2.5Y (Zhang et al. 2012h)	As-cast	−1.5499	0.7698	−1.5599
Mg-5Y (Zhang et al. 2012h)	As-cast	−1.5525	1.5639	−1.5625
Mg-8Y (Zhang et al. 2012h)	As-cast	−1.5760	1.1608	−1.5830
Mg-15Y (Zhang et al. 2012h)	As-cast	−1.5750	1.5981	−1.5810
Mg-5Y (K. Zhang et al. 2013)	As-extruded	−1.5710	1.5890	/
Mg-8Y (K. Zhang et al. 2013)	As-extruded	−1.5040	2.4600	/
Mg-14Y (K. Zhang et al. 2013)	As-extruded	−1.5310	2.4720	/
Mg-15Y (Zhang et al. 2012i)	As-cast	−1.8160	5.83	/
Mg-15Y (Zhang et al. 2012i)	As solid solution treated	−1.7450	8.37	/

current densities also show the same trend as the corrosion potentials. With the increasing of Y addition, the formation of the second phase is promoted. The Mg$_{24}$Y$_5$ phase, which acts as the anode to accelerate the corrosion, can suppress the propagation of corrosion pits when the addition of Y reaches a certain content. In the as-extruded Mg-xY alloys, it is obvious that the corrosion potentials of samples become nobler with the increasing addition of Y. The I_{corr} values increase with increasing Y, showing an accelerated corrosion with increasing Y addition. According to the results of X. Zhang et al. (2012i), the corrosion potential and corrosion rate of the as-cast Mg-15Y alloy are both lower than those of solid solution treated samples.

The electrochemical corrosion parameters (E_{corr}, I_{corr}) only reflect the corrosion rate of the initial immersion. Accompanied by the formation of the corrosion product layer, I_{corr} does not reflect the corrosion rate accurately. Therefore, direct and accurate reflection of the corrosion rate, which is measured by the weight loss or the hydrogen evolution, has been done. Gu et al. (2009) reported the corrosion of Mg-1Y in Hank's solution and simulated body fluid (SBF), and faster corrosion was observed in SBF than in Hank's solution, and this was believed to be related to the buffer in SBF. The as-rolled Mg-1Y alloy displayed less hydrogen evolution in comparison to the as-cast ones in both SBF and Hank's solution. As-cast Mg-1Y showed a higher hydrogen evolution rate in Hank's solution than that of pure Mg.

Xue et al. (2012a) tried to modify Mg-4Y with the water-based bis-[triethoxysilyl] ethane (BTSE) silane for the sake of improving corrosion resistance. The corrosion behavior of the uncoated and coated Mg-4Y alloy was evaluated in different simulated physiological environments within a self-developed corrosion probe. Figure 11.4a shows the structure of the probe whereas Figure 11.4b shows the equivalent circuit of the corrosion system. As can be found in Table 11.3, the epoxy-modified BTSE silane coating could effectively increase the corrosion resistance of the Mg-4Y

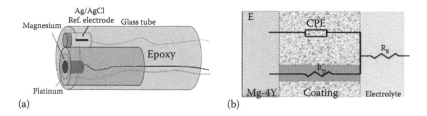

(a)   (b)

FIGURE 11.4 (a) Schematic representation of a corrosion-monitoring probe with Mg-4Y as a working electrode, platinum coiled wire as a counter electrode, and Ag/AgCl as a reference electrode; (b) the EIS equivalent circuit model used in reference. (From Xue, D., Z. Tan, M. J. Schulz, W. J. Vanooij, J. Sankar, Y. Yun, and Z. Dong, *Materials Science and Engineering: C*, 32 (5), 1230–1236, 2012. With permission.)

TABLE 11.3

Corrosion Parameters and Corresponding Estimated Parameters of the Equivalent Circuit Model of the Uncoated and BTSE Epoxy-Coated Mg–4Y Alloy Probe in Different Environments

Corrosion Environment		I_{corr} ($\mu A \cdot cm^{-2}$)	E_{corr} (V)	R_s ($\Omega \cdot cm^{-2}$)	R_p ($\Omega \cdot cm^{-2}$)	Y0 ($\mu F \cdot cm^{-2}$)
Uncoated	0.9 wt.% NaCl	32	−1.48	2072	52,290	31.752
Mg-4Y	SBF	750	−1.78	1372	7420	55.076
	Subcutaneous tissue	15	−1.67	15,428	88,060	70.966
	Body cavity	35	−1.68	7070	59,948	46.914
BTSE	0.9 wt.% NaCl	19	−1.49	1358	148,120	20.216
epoxy–coated	SBF	98	−1.76	1638	33,488	36.61
Mg-4Y	Subcutaneous tissue	9	−1.58	24,080	178,360	39.9
	Body cavity	11	−1.59	5642	165,060	25.452

Note: Y0 can be used to calculate the constant phase element impedance.

alloy, and the degradation rate under the in vivo environment in mice models were much smaller than that of in vitro environments.

For Mg-Y alloys, M. Liu et al. (2010) found Y has two contradictory effects on the corrosion of Mg. On the one hand, the Y-containing intermetallic can cause microgalvanic corrosion acceleration. From this point of view, Y can accelerate the corrosion of Mg. On the other hand, Y accumulated in the surface layer can increase the protectiveness of the surface from corrosion. How to balance the two aspects on the corrosion of Mg-Y-based alloys is still not so clear, but controlling of the content of Y in Mg is critical to the corrosion.

The corrosion behavior of Mg-Y alloys is closely related to the material itself and the corrosion medium. Johnson and Liu (2013) found the presence or absence of Y in Mg alloys, the presence or absence of surface oxides, and the presence or

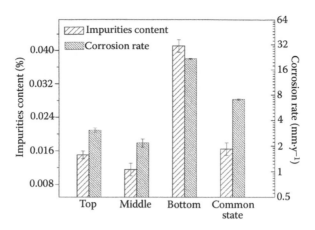

FIGURE 11.5 Impurities (Fe, Cu, and Ni) distribution and the corresponding corrosion rate in different sections of the Mg-8Y alloy. (From Peng, Q., Y. Huang, L. Zhou, N. Hort, and K. U. Kainer, *Biomaterials*, 31 (3), 398–403, 2010. With permission.)

absence of physiological ions in the immersion fluid collectively contributed to Mg degradation, and interacted with one another on influencing Mg degradation rate and mode.

During the fabrication, the cooling rates in different parts of the ingots are significantly different, and this causes different properties in different parts of the alloys, especially in the distribution of impurities. Figure 11.5 shows the impurity distribution in different parts of the zone solidification of a purified Mg-8Y alloy and the correlated corrosion rates compared to the common casting (Peng et al. 2010). The corrosion rate of the common as-cast state is about 7.11 mm·y^{-1}. However, the average corrosion rate of the purified alloy in the middle region is only about 2.17 mm·y^{-1}.

11.2.3.4 Cytotoxicity and Hemocompatibility

Gu et al. (2009b) studied the cytotoxicity of the Mg-1Y alloy on various cells, including murine fibroblast cells (L-929 and NIH3T3), murine calvarial preosteoblasts (MC3T3-E1), human umbilical vein endothelial cells (ECV304), and rodent vascular smooth muscle cells (VSMC). The pH value of the as-cast Mg-1Y extract was 8.69 ± 0.1. Except for the MC3T3-E1 cells, the Mg-1Y alloy extract showed decreased cell viability for L-929, NIH3T3, ECV304, and VSMC cells at a concentration of 2.3 ± 0.7 mM/l for Y. Generally, the cytotoxicity of the Mg-1Y alloy is not so significant, and it is better than pure Mg to some cell lines.

The hemolysis percentage of as-cast and as-rolled Mg-1Y is well below 5%, fulfilling the requirement of biomaterials. The adhered platelets show a nearly round shape and slight pseudopodia spreading, and the amount of adhered platelets is significantly reduced compared to pure Mg and Mg-1Zn.

Figure 11.6 shows the results of mesenchymal stem cell (MSC) adhesion on the surfaces of a Mg-4Y alloy (Johnson et al. 2012). Mg-4Y samples have thermal oxide

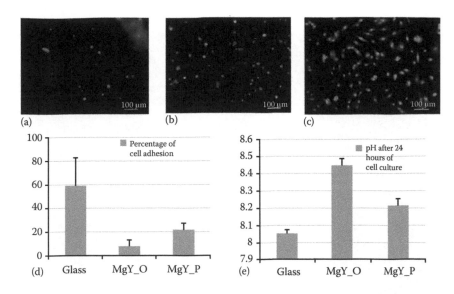

FIGURE 11.6   The results of the BMSCs adhesion on the surfaces of (a) bioactive glass, (b) MgY_O, and (c) MgY_P samples after 24 h of culture. (d) The percentage of BMSCs that adhered to the MgY and control surfaces. (e) The pH of the cell culture media after 24 h of incubation. (From Johnson, I., D. Perchy, and H. Liu, *Journal of Biomedical Materials Research Part A*, 100 (2), 477–485, 2012. With permission.)

layers on their surface after machining processes. The samples with the oxide layer were labeled MgY_O, whereas those that were simply polished were labeled as MgY_P. Bioactive glass was adopted as a comparison. Cells cultured on MgY_O had a round morphology, implying that they were unhealthy and detached from the surface. The cells cultured on MgY_P were mostly rounded, but some cells had a more elongated morphology. The percentage of cell adhesion on MgY_P was greater than on MgY_O but less than that on the bioactive glass control.

To further improve the biocompatibility of the Mg-4Y alloy, calcium phosphate (CaP) and silicon-containing calcium phosphate (Si-CaP) coatings were introduced on the surface of the Mg-4Y alloy by a sol-gel technique (Roy et al. 2011). Those coatings didn't have too much influence on the corrosion resistance of the Mg-4Y matrix because of their porous structure. The cytocompatibility study using MC3T3-E1 osteoblasts showed that the coated Mg-4Y alloy substrates were more bioactive than the uncoated Mg-4Y alloy substrates as the cells began to grow and form a matrix on the coated sample more easily than on the bare one. Figure 11.7 intuitively shows the fluorescence microscopy of the cells. Preliminary results show the potential use of sol-gel-derived calcium phosphate coatings on Mg-based degradable materials to improve their surface bioactivity.

11.2.3.5   Animal Testing

Luffy et al. (2014) evaluated the host response to a Mg-3Y alloy extraluminal stent in a canine model and assessed the in vivo biodegradation procedure. Figure 11.8

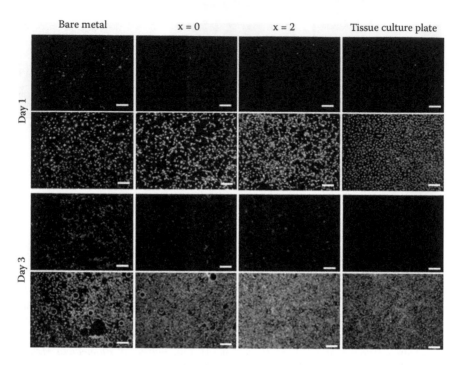

FIGURE 11.7 Fluorescence microscopy of MC3T3-E1 cells on uncoated, CaP-coated, and Si-CaP-coated substrates, tissue culture polystyrene plate as comparison (double-stained to be green for live cells, rows 2 and 4; and red for dead cells, rows 1 and 3. The scale bar for all the images is 100 µm). (From Roy, A., S. S. Singh, M. K. Datta, B. Lee, J. Ohodnicki, and P. N. Kumta, *Materials Science and Engineering: B*, 176 (20), 1679–1689, 2011. With permission.)

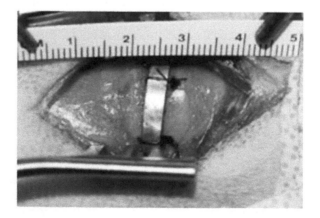

FIGURE 11.8 Surgical placement of Mg-3Y C-ring in a canine model. (The stent was secured with sutures to the abluminal surface of trachea without penetrating the lumen of the airway and persisted in vivo for 8 weeks.) (From Luffy, S. A., D. T. Chou, J. Waterman, P. D. Wearden, P. N. Kumta, and T. W. Gilbert, *Journal of Biomedical Materials Research Part A*, 102 (3), 611–620, 2014. With permission.)

shows the appearance of the Mg-4Y tracheal stent and the implantation process. Stents were machined into open C-ring structures to resemble native tracheal cartilage, and the stent was placed directly over one tracheal cartilage ring. Corrosion was assessed by microcomputed tomography. Figure 11.9 shows the radiographic images of the implantation site. Limited corrosion was observed after 8 weeks with consistent fracture patterns occurring at approximately 5 weeks. Radiographs also showed occasional formation of very small fibrous cavities in close proximity to the stent. Mechanical pressure–circumference response testing indicated that the presence of an extraluminal Mg stent didn't alter the mechanical function of the airway.

Figure 11.10 shows the morphology of the stent after removal from the mongrel dogs. Some corrosion product is visible on the surface of the stent. The

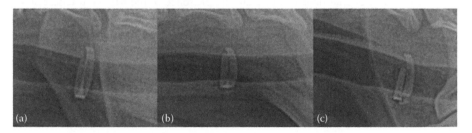

FIGURE 11.9 Radiographic analysis of W3 C-rings at (a) 1, (b) 4, and (c) 8 weeks. (From Luffy, S. A., D. T. Chou, J. Waterman, P. D. Wearden, P. N. Kumta, and T. W. Gilbert, *Journal of Biomedical Materials Research Part A*, 102 (3), 611–620, 2014. With permission.)

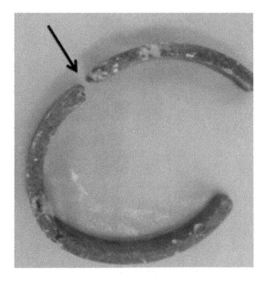

FIGURE 11.10 C-ring stent after 8-week explantation. (From Luffy, S. A., D. T. Chou, J. Waterman, P. D. Wearden, P. N. Kumta, and T. W. Gilbert, *Journal of Biomedical Materials Research Part A*, 102 (3), 611–620, 2014. With permission.)

arrow indicates the point of fracture at approximately 5 weeks. Histological analysis showed the presence of fibrous connective tissue deposited between the native trachea and the stent that was well vascularized. The connective tissue was primarily populated by mononuclear cells with only a few multinucleated giant cells (Figure 11.11a). The native tracheal cartilage was histologically normal in terms of both the structure of the tissue and the intensity of the glycosaminoglycan staining (Figure 11.11b). Herovici's staining showed that the new connective tissue was primarily composed of collagen type III (Figure 11.11c). In one specimen, calcification was observed at the site of stent–tissue interaction (Figure 11.11c), and calcification was confirmed with Alizarin Red staining (Figure 11.11d).

On the whole, despite the fracture of the stent after 5 weeks, the study of Luffy et al. (2014) showed that dogs can tolerate the presence of the Mg-3Y tracheal stent. This stent shows a mild host response in dogs and without negative effects on the native cartilage after 8 weeks. However, the stress corrosion cracking resistance and fatigue resistance should be improved to meet the requirements.

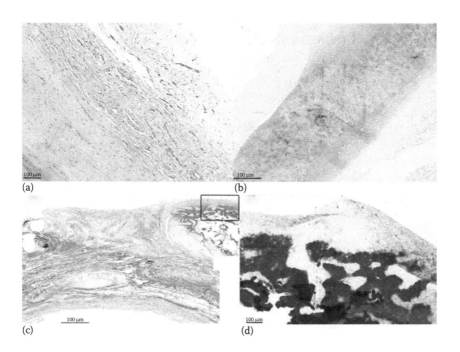

(a)

(b)

(c)

(d)

FIGURE 11.11 Histological analysis of the tissues around the tracheal stent. (a) H&E staining; (b) Alcian Blue staining; (c) Herovici's staining; (d) Alizarin Red staining. (From Luffy, S. A., D. T. Chou, J. Waterman, P. D. Wearden, P. N. Kumta, and T. W. Gilbert, *Journal of Biomedical Materials Research Part A*, 102 (3), 611–620, 2014. With permission.)

11.2.4 Development of Mg-Y-Zn-Based Alloys as Degradable Metallic Biomaterials

11.2.4.1 Alloy Composition Design

Zinc (Zn) is one of the abundant nutritionally essential elements in the human body and also an effective alloying element in strengthening Mg alloys (Tapiero and Tew 2003; S. Zhang et al. 2010). It can also reduce the harmful corrosive effect of iron and nickel in Mg alloys (Ding et al. 2014). Recently, long period stacking order (LPSO) phases in the Mg-Y-Zn system have been found to be important strengthening phases (Oñorbe et al. 2012). Based on the binary Mg-Y system, Zn is added into Mg alloys to improve the mechanical properties and to enhance the corrosion resistance.

11.2.4.2 Microstructure and Mechanical Properties

Figure 11.12a–d shows the SEM images of as-cast $Mg_{100-3x}(Zn_1Y_2)_x$ (x = 1, 2, 3) alloys (X. Zhao et al. 2013a). All of the Mg-Y-Zn alloys are composed of two phases, α-Mg and $Mg_{12}YZn$ phase. The $Mg_{12}YZn$ phase is characterized as an 18R LPSO phase with bright contrast in the metallographical images in Figure 11.12. The approximate volume fractions of the LPSO phase in the alloys increase from 25%, across 48%, to 66% with the increasing of x from 1 to 3, respectively. At x = 3, the LPSO phase even becomes the matrix. Grain sizes are around 150 µm, 110 µm, and 80 µm at x = 1, 2, 3, respectively.

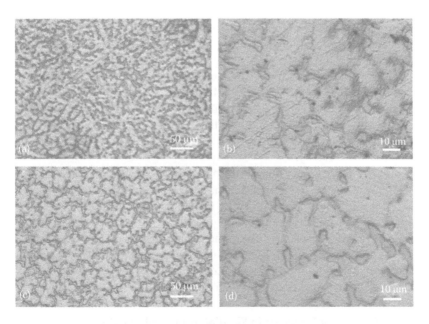

FIGURE 11.12 SEM images of $Mg_{100-3x}(Zn_1Y_2)_x$ alloys: (a) x = 1, (b) x = 2, (c) x = 3, (d) in high-magnification for x = 3. (From Zhao, X., L. L. Shi, and J. Xu, *Journal of the Mechanical Behavior of Biomedical Materials*, 18, 181–190, 2013. With permission.)

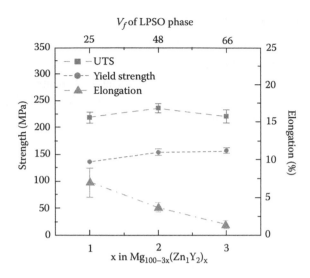

FIGURE 11.13 Mechanical properties of $Mg_{100-3x}(Zn_1Y_2)_x$ (x = 1, 2, 3) alloys. (From Zhao, X., L. L. Shi, and J. Xu, *Journal of the Mechanical Behavior of Biomedical Materials*, 18, 181–190, 2013. With permission.)

Figure 11.13 shows the strength and elongation versus volume fraction of 18R LPSO phase in as-cast $Mg_{100-3x}(Zn_1Y_2)_x$ (x = 1, 2, 3) alloys. With the increasing of Y and Zn contents, the tensile yield strength (TYS) of the alloys increases from 136 MPa at x = 1 to 157 MPa at x = 3. Ultimate tensile strength (UTS) of the alloys first increases from 218 MPa at x = 1 to 236 MPa at x = 2, then decreases to 221 MPa at x = 3. On the contrary, the elongation of the alloy decreases drastically from 7% at x = 1 to 1% at x = 3. It can be concluded that LPSO phases have a significant effect on the mechanical properties of Mg-Y-Zn alloys. LPSO phases strengthen the alloy at the cost of the plasticity.

For the extruded Mg-Y-Zn alloys with low Zn content, E. Zhang et al. (2008) found that Y improved the tensile strength and the elongation at the same time. Notably, the values of TYS and UTS can be improved to 495 MPa and 541 MPa in the Mg-Y-Zn microwires, which are about three times higher than those in the as-cast state (Peng et al. 2014a). Figure 11.14 gives the comparison of mechanical properties among various Mg-Y-Zn-based alloys. Mg-Y-Zn microwires show the highest strength but with limited elongation. Extruded Mg-Y-Zn alloys show medium strength but with excellent elongation, which can exceed 25%.

11.2.4.3 Corrosion Behavior in Simulated Physiological Environments

Table 11.4 shows the electrochemical corrosion parameters on various Mg-Y-Zn-based alloys. The corrosion data collected from the literature show that corrosion current densities of Mg-Y-Zn-based alloys in SBF are much higher than the alloys in DMEM + FBS or Hank's solution. Compared to the traditional commercial Mg alloys, such as AZ31, WE43, ZK60, and ZX60 alloys, extruded

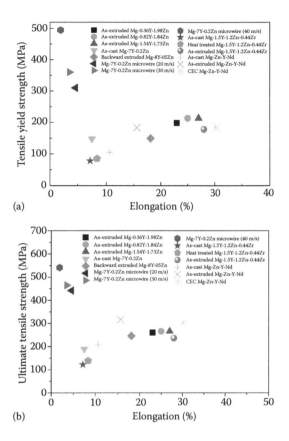

FIGURE 11.14 Tensile yield strengths, ultimate tensile strengths, and elongations of various Mg-Y-Zn-based alloys. (a) TYS-elongation, (b) UTS-elongation. (Data were collected from Zhang, E., W. W. He, H. Du, and K. Yang. *Materials Science and Engineering A-Structural Materials Properties Microstructure and Processing*, 488 (1–2), 102–111, 2008; Wu, Q., S. Zhu, L. Wang, Q. Liu, G. Yue, J. Wang, and S. Guan, *Journal of the Mechanical Behavior of Biomedical Materials*, 8, 1–7, 2012; Fan, J., X. Qiu, X. D. Niu, Z. Tian, W. Sun, X. J. Liu, Y. D. Li, W. R. Li, and J. Meng. *Materials Science and Engineering C-Materials for Biological Applications*, 33 (4), 2345–2352, 2013; Peng, Q., N. Ma, D. Fang, H. Li, R. Liu, and Y. Tian, *Journal of the Mechanical Behavior of Biomedical Materials*, 17, 176–185, 2013; Peng, Q., H. Fu, J. Pang, J. Zhang, and W. Xiao, *Journal of the Mechanical Behavior of Biomedical Materials*, 29, 375–384, 2014.)

$Mg_{100-3x}(Zn_1Y_2)_x$ ($1 \leq x \leq 3$) exhibit relatively low corrosion current densities in electrolyte of DMEM + FBS.

Izumi et al. (2009) studied the degradation behavior of the rapid solidification (RS) $Mg_{97.25}Zn_{0.75}Y_2$ (at.%) alloy and found it exhibited excellent corrosion resistance because of grain refinement and formation of a supersaturated single-phase solid solution. The improved corrosion resistance can be seen in Figure 11.15. Enhancement of microstructural and electrochemical homogeneities in the Mg-Zn-Y alloys by rapid solidification techniques results in the passivity of substrate materials. Leng et al. (2013b) observed the corrosion behavior of as-extruded

TABLE 11.4

Electrochemical Parameters Measured under Different Media of Various Mg–Y–Zn-based Alloys, Including Corrosion Potential (E_{corr}), Pitting Potential (E_{pit}), and Corrosion Current Density (I_{corr})

Material	Processing State	Electrolyte	E_{corr} (V)	I_{corr} (μA/cm^2)	E_{pit} (V)
$Mg_{97}Zn_1Y_2$ (X. Zhao et al. 2013b)	As-cast	DMEM + FBS	−1.488	7.6	−1.387
$Mg_{94}Zn_2Y_4$ (X. Zhao et al. 2013b)	As-cast	DMEM + FBS	−1.437	2.9	−1.394
$Mg_{91}Zn_3Y_6$ (X. Zhao et al. 2013b)	As-cast	DMEM + FBS	−1.474	3.3	−1.458
$Mg_{97}Zn_1Y_2Zr_{0.17}$ (X. Zhao et al. 2013b)	As-cast	DMEM + FBS	−1.638	4.5	−1.417
$Mg_{97}Zn1Y_2Zr_{0.17}$ (X. Zhao et al. 2013b)	As-extruded	DMEM + FBS	−1.702	6.7	−1.418
Mg-0.36Y-1.98Zn (E. Zhang et al. 2008)	As-extruded	Hank's	−1.498	1.883	/
Mg-0.82Y-1.84Zn (E. Zhang et al. 2008)	As-extruded	Hank's	−1.570	4.469	/
Mg-1.54Y-1.73Zn (E. Zhang et al. 2008)	As-extruded	Hank's	−1.528	2.827	/
Mg-7Y-0.2Zn (Peng et al. 2014a)	As-cast	SBF	−1.649	227	/
Mg-7Y-0.2Zn (Peng et al. 2014a)	Melt extraction microwire	SBF	−1.508	16	/
Mg-1.5Y-1.2Zn-0.44Zr (J. Fan et al. 2013)	As-cast	SBF	−1.776	122	/
Mg-1.5Y-1.2Zn-0.44Zr (J. Fan et al. 2013)	Heat-treated	SBF	−1.769	97	/
Mg-1.5Y-1.2Zn-0.44Zr (J. Fan et al. 2013)	As-extruded	SBF	−1.667	75	/

Mg-8Y-1Er-2Zn (wt.%) alloy in SBF. The corrosion rate calculated from the evolved hydrogen was 0.568 mm·y^{-1}. The corrosion products were determined to be $Mg(OH)_2$, and a $Ca(H_2PO_4)_2$ compound was also detected on the surface of the immersed samples.

Apart from the static immersion, J. Wang et al. (2010) tried to simulate the dynamic environment in vivo. Compared with the as-cast samples, the corrosion behavior of alloys in a dynamic SBF with the speed of 16 ml·min^{-1} and 800 ml·min^{-1} were investigated. The results showed that a subrapid solidification Mg-Zn-Y-Nd alloy had better corrosion resistance in dynamic SBF due to grain refinement and fine dispersion distribution of the quasicrystals and intermetallic compounds in the

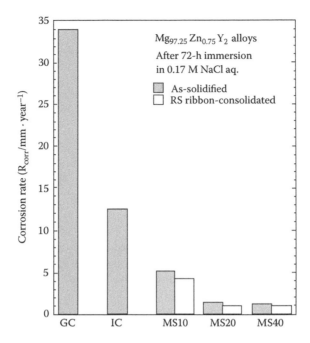

FIGURE 11.15 Corrosion rate of the as-solidified and RS ribbon-consolidated Mg97. 25Zn0.75Y2 alloys in a 0.17 M NaCl neutral aqueous solution, compared to the gravity casting and copper mold injection casting ones (GC: gravity casting; IC: copper mold injection casting; MS: RS ribbons were prepared by a single-roller melt-spinning, so MS means melt spinning. Numbers after MS mean the rotating speed). (From Izumi, S., M. Yamasaki, and Y. Kawamura, *Corrosion Science*, 51 (2), 395–402, 2009. With permission.)

α-Mg matrix. At the same time, the dynamic system made the accumulation of Cl^{-1} on the surface of the samples hard to happen. No accumulation of Cl^{-1} on the corrosion sites also contributed to the corrosion resistance in the dynamic system.

11.2.4.4 Mechanical Degeneration in Simulated Physiological Environments

When the implants are exposed to the corrosive body fluid after implantation, the erosion of the implants happened immediately. The mechanical properties of the implants decreased with the extension of postoperation time. How does the strength change? What is the rate of the decay? Can the implants sustain until the complete healing of the disease? All those problems affect whether the implantation can succeed. Now, there is no ideal way to measure the mechanical deterioration of the implants postimplantation.

Z. Leng et al. 2013b and Peng et al. (2013b) tried to find a way that could predict the mechanical deterioration of the implants postoperation. Before a tensile test, some samples were immersed in different media with different time intervals. After immersion, the samples were ultrasonically cleaned in alcohol and dried in flowing air and tested immediately. Figure 11.16 shows the Mg-8Y-1Er-2Zn alloy as an example. Corrosion preferentially occurred at the interface between the LPSO

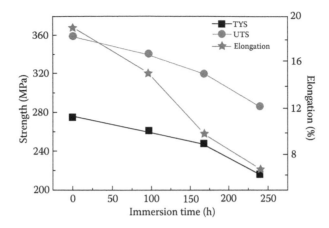

FIGURE 11.16 Tensile properties of as-extruded Mg–8Y–1Er–2Zn alloy after immersing in SBF for different time. (Data were adopted from Leng, Z., J. Zhang, T. Yin, L. Zhang, X. Guo, Q. Peng, M. Zhang, and R. Wu, *Journal of the Mechanical Behavior of Biomedical Materials*, 28, 332–339, 2013.)

phase and the Mg matrix. Before immersion, the TYS, UTS, and elongation of the alloy were 275 MPa, 359 MPa, and 19%, respectively. And these mechanical properties could be maintained at a high level even after immersion in SBF for 240 h. TYS, UTS, and elongation were 216 MPa, 286 MPa, and 6.8% after 240 h immersion, respectively. The author ascribed this to the existence of a high anticorrosion LPSO phase. However, a positive thing did not happen on the backward extruded Mg-7.25Y-0.31Zn alloy. Strengths of the Mg-7.25Y-0.31Zn alloy decreased rapidly with the extension of the immersion time. The reduction of the yield strength and elongation reached 65.8% and 78.53% after 240 h immersion, respectively.

The strength and elongation of Mg-Y-based alloys decreased with immersion time, and they are significantly influenced by the media. These results obtained from in vitro experiments offer some implications for understanding the reduced strength of Mg-based implants in the body environment. The temporary high mechanical strength in air of Mg-Y-based biomaterials is insufficient to evaluate the in vivo mechanical integrity, and further in vivo mechanical behavior evaluation is needed.

11.2.4.5 Cytotoxicity

Figure 11.17 shows the viability of L-929 cells expressed as a percentage of the negative control, cultured in 100%, 50%, and 10% Mg-1.5Y-1.2Zn-0.44Zr extracts, respectively (J. Fan et al. 2013). It is obvious that compared with the negative control and positive control, the 100%, 50%, and 10% extracts do not show significant cytotoxicity toward the cells. Moreover, it can be seen that the relative cell viability of all the extracts increases with the culture time, and cell viability is even higher than the negative control group after 3 and 5 days of culture, indicating enhanced cell proliferation. Concentrations of released Mg, Y, Zn, and Zr ions are less than 210, 0.3, 0.24, and 0.26 (μg/ml), which means, at this concentration level, the negative effects on cytocompatibility will not be significant, and moderate degradation of

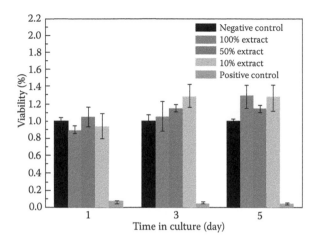

FIGURE 11.17   The viability of L-929 cells expressed as a percentage after 1, 3, and 5 days of incubation. (From Fan, J., X. Qiu, X. Niu, Z. Tian, W. Sun, X. Liu, and J. Meng, *Materials Science and Engineering: C*, 33 (4), 2345–2352, 2013. With permission.)

the Mg-1.5Y-1.2Zn-0.44Zr alloy is beneficial to the cell proliferation. In the work of Hänzi et al. (2010), the cytocompatibilities of Mg-1Y-2Zn and Mg-2Y-1Zn alloys are a little bit lower than that of WE43 alloys.

11.2.4.6   Animal Testing

The Mg-2Y-1Zn alloy showed promising mechanical performance and microstructural stability in a previous study of Hänzi et al. (2010). In order to gain insight into the in vivo performance of the Mg-2Y-1Zn alloy, a preliminary animal study on Göttingen minipigs was performed. Sample disks of 4 mm diameter and 0.4 mm thickness were implanted into four different types of tissue in the abdomen (liver, lesser omentum) and in the abdominal wall (rectus abdominis muscle, subcutaneous tissue), respectively. All animals were in good general condition until sacrifice and showed no adverse reactions. Figure 11.18 shows the histopathological preparations derived from Mg-2Y-1Zn samples in various types of tissues, and it indicates homogeneous degradation and only limited gas formation during in vivo testing. The characteristics of the tissue reactions indicate good biocompatibility, and the Mg-2Y-1Zn alloy is believed to be promising for degradable implant applications.

11.2.5   Development of WE Series Alloys as Degradable Metallic Biomaterials

11.2.5.1   Background

WE series Mg alloys, mostly WE43, initially designed for automotive applications, are basically a heat-treatable yttrium-containing cast alloy with decent mechanical properties and corrosion resistance. They are able to retain most of their mechanical properties at elevated temperatures up to 300°C (L. Wang et al. 2014). The chemical

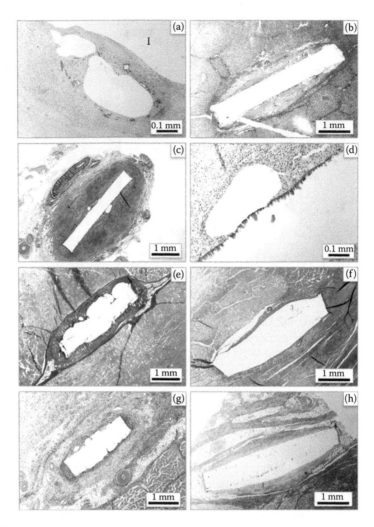

FIGURE 11.18 Histopathological preparations derived from Mg-2Y-1Zn samples. If not stated otherwise, the first column displays preparations of implants after 27 days, the second column those after 91 days. (a) Small gas bubbles adjacent to the implant (labeled as "I") in the liver; (b) limited gas formation and defined capsule around the implant in the liver; (c) very defined contours and little gas formation of the implant in the lesser omentum; (d) sector of (c) at higher magnification showing a gas bubble and characteristic degradation products at the edge of the implant after 27 days (no implant was found after 91 days); the fibrous tissue around the implant is infiltrated by eosinophilic and polymorphic granulocytes; (e) rather vivacious gas formation of the implant in the rectus abdominis muscle surrounded by a thin-walled granuloma; (f) defined contours and some gas formation (likely postmortem) of the implant in the rectus abdominis muscle; (g) thin-walled granuloma around the implant in the subcutaneous tissue, moderate gas formation; (h) vascularized tissue adjacent to the implant in the subcutaneous tissue with some gas, most likely formed post-mortem. (From Hänzi, A. C., I. Gerber, M. Schinhammer, J. F. Löffler, and P. J. Uggowitzer, *Acta Biomaterialia*, 6 (5), 1824–1833, 2010. With permission.)

compositions of the WE series alloys mainly consist of Y and some other REEs. For instance, WE43 alloys consist nominally of 4 wt.% Y and 3 wt.% RE. With excellent performance in the engineering field, naturally, WE Mg alloys have already been introduced into the biomedical field.

WE series Mg alloys are almost the first and earliest one; they have already been tested in animals, and clinical human trials have also been successfully conducted in recent decades. Moreover, WE43 (Mg-Y-RE) is regarded as the most promising candidate for biodegradable stents (Drynda et al. 2009).

11.2.5.2 Microstructure and Mechanical Properties

The typical microstructures of as-cast WE43 and as-aged WE54 alloys are illustrated in Figure 11.19. The as-cast WE43 alloy consists of solid solution α-Mg matrix with precipitates of intermetallic phases at grain boundaries as shown in Figure 11.19a. The second phases could consist of $Mg_{41}Nd_5$, MgY, $Mg_{24}Y_5$ phases and β ($Mg_{14}Nd_2Y$) compound. After solution treatment and aging, the amount of the second phases significantly decreased. As-aged WE54 consists of equiaxed grains with the size of 102 ± 8 μm as illustrated in Figure 11.19c. The residual eutectic, namely particles of the stable β phase isomorphous with Mg_5Gd containing Y and Nd instead of Gd, are also found in grain boundaries and in grain interiors.

Mechanical properties of Mg-Y-Zr and WE43 alloys in different processing states were studied by D. Liu et al. (2014). The TYS, UTS, and elongation of the original Mg-Y-Zr alloy are 116 MPa, 178 MPa, and 13%, respectively. Additions of Nd and Ce (La) elements into Mg-Y-Zr significantly increase its strength, and the elongation decreased slightly. The TYS and UTS of the as-cast WE43 are 158 MPa and 224 MPa with an elongation of 10%. The TYS, UTS, and elongation increased significantly to 190 MPa, 282 MPa, and 37% in the extruded tubes, respectively (D. Liu et al. 2014). Ultra-fine-grained WE43 alloys can have much better mechanical strength. The TYS, UTS, and elongation of ultra-fine-grained WE43 can reach 253 MPa, 298 MPa, and 25%, respectively, according to the results of Kutniy et al. (2009). According to their study, severe plastic deformation is an effective way to improve the mechanical properties of WE43 alloys. The best reported data in the

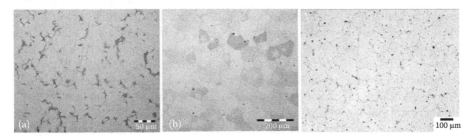

FIGURE 11.19 Typical microstructures of (a) as-cast WE43, (b) solution treated and aged WE43 and (c) as-aged WE54 alloys. ([a and b] From Ning, Z. L., J. Y. Yi, M. Qian, H. C. Sun, F. Y. Cao, H. H. Liu, and J. F. Sun, *Materials and Design* 60:218–225, 2014. With permission; [c] From Smola, B., L. Joska, V. Březina, I. Stulíková, and F. Hnilica, *Materials Science and Engineering: C*, 32 (4), 659–664, 2012. With permission.)

literature of WE43 alloys is observed after cold hydrostatic extrusion with back pressure, according to the report of Pachla et al. (2012). The TYS, UTS, and elongation of extruded WE43 with back pressure reach 350 MPa, 410 MPa, and 7%, respectively.

11.2.5.3 Expansion Test with Stent Sample

Galvin et al. (2013) carried out the expansion test on manufactured WE43 stents in order to learn the expansion performance of the stents. The stent configuration at three different phases of the expansion process correlated well with ERR data for commercially available stents (Kiousis et al. 2009). On the whole, expansion tests on WE43 stents demonstrated uniform stent expansion characteristics.

11.2.5.4 Corrosion Behavior in Simulated Physiological Environments

The pH values were recorded by D. Liu et al. (2014), when differently processed WE43 alloys were immersed in PBS solution for 7 days. The pH values of the PBS solutions increased with the extension of immersion time for all samples. The overall corrosion rates of the extruded samples were much lower than those in the as-cast state and heat treated state. The corrosion rate of the extruded microtube was in a relatively low interval value from around 4.7 mm/a to 3.8 mm/a.

The effects of different aggressive ions on the degradation behavior of WE43 Mg alloy were investigated by X. Zhou et al. (2014b). Along with the concentrations of the Cl^-, SO_4^{2-}, $H_2PO_4^-$, and HCO_3^- increasing, the open circuit potential and the corrosion potential of WE43 alloy shifted to the negative direction, and the corrosion current density increased. The order of ion corrosivity for WE43 alloy immersed in SBF was $Cl^- > SO_4^{2-} > HCO_3^- > H_2PO_4^-$. Inorganic species in body fluids also influenced the corrosion behavior of biomedical WE43 alloys. Adding calcium and phosphate to unbuffered NaCl solutions beneficially stabilized the corrosion product. But the choice of the buffer is critical for the WE43 alloy degradation mechanisms (Ott et al. 2013).

Simulated body fluid (SBF) and artificial plasma (AP) are similar in ionic content except for the higher buffering capacity of SBF and the higher carbonate content of the AP. It was reported that SBF was significantly more aggressive than AP with regard to the WE43 Mg alloys (Schmutz et al. 2007; Quach et al. 2008). Grain refinement can both improve the mechanical property and the plasticity of Mg alloys, but corrosion resistance is not always improved by grain refinement. A recent study found that the corrosion rate of ultra-fine-grained WE43 was much higher than industrial WE43 in 1% NaCl solution (Papirov et al. 2008; Kutniy et al. 2009). WE54 fabricated via powder consolidation and hot extrusion also showed decreased corrosion resistance in isotonic saline compared to commercial WE54 (Smola et al. 2012).

The corrosion layer on WE43 after being immersed in SBF with and without albumin was characterized as an amorphous layer of carbonated calcium phosphate with some calcium replaced by Mg, and Rettig and Virtanen (2009) found that calcium was only deposited in the corrosion layer if phosphates were in the solution.

11.2.5.5 Corrosion Fatigue Behavior

If Mg alloys were fabricated into implants, and this implant might bear load in the physiological environment, the implant would suffer from stress corrosion and corrosion fatigue. At this time, simple tensile testing is insufficient, and simulation and prediction of the in vivo corrosion fatigue are needed. Gu et al. (2010) reported the corrosion fatigue behavior of extruded WE43 alloy in SBF compared to die-cast AZ91D alloy. A fatigue limit of 110 MPa at 10^7 cycles in air was observed for extruded WE43 alloy compared to 40 MPa at 10^7 cycles tested in SBF at 37°C. The corrosion fatigue limit of WE43 was significantly reduced because of the corrosive SBF. Corrosion fatigue resistance of WE43 was almost two times higher than the die-cast AZ91D alloy.

11.2.5.6 Surface Coatings to Improve the Corrosion Resistance

If rapid corrosion is happening on Mg implants, rapid production of hydrogen and penetration ions may not be tolerated by host tissues. In this situation, temporary protection of the surface is necessary when the material is used in aqueous and saline media, such as the environment of the human body. Then, various surface treatment methods have been developed, and different kinds of coatings have been prepared to improve the corrosion resistance of WE series biomedical materials.

Table 11.5 lists the various coatings developed in the literature and their effects on corrosion resistance of biomedical WE Mg alloys. Except for the phosphate-coated WE43, all the other coatings improve the corrosion potential of the WE43 matrix, significantly reducing the corrosion current at one order or even more.

Additionally, Li et al. (2012) successfully fabricated an amorphous SiC film on the surface of a WE43 Mg alloy via plasma-enhanced chemical vapor deposition. The immersion test also indicated that SiC film could efficiently slow down the degradation rate of the WE43 alloy in simulated body fluid (SBF) at 37°C ± 1°C.

11.2.5.7 Cytotoxicity and Hemocompatibility

The human MG-63 osteoblast-like cells spreading and dividing in the extracts (0.28 g in 28 mL of EMEM) of WE43 alloys in different states were monitored by cinemicrography for 24 h (Smola et al. 2012). Ion concentrations in the T6-treated WE54 extracts were 543, 191, 54.6, and 43,840 ng/ml for Y, Nd, Zr, and Mg. Ion concentrations were 288, 129, 24.1, and 57,840 ng/ml for Y, Nd, Zr, and Mg in the powder consolidated and hot-extruded WE54 extracts. Concentration of Mg in extracts of all studied WE54 specimens was well below the LD50 (Hallab et al. 2002). On the whole, the MG-63 cells proliferated without cytotoxicity in all extracts.

Klocke et al. (2013) studied the cytotoxicity of electro discharge machined (EDM) WE43 alloys with different processing parameters and found that only the rough cut decreased the viability of NCTC clone 929 cells after 6 h. However, the progression of the corrosion led to an ongoing decrease of the cell viability of all the tested samples. Tissue damage indicated by an increased lactate dehydrogenase (LDH) level could not be observed in any of the tests. All the results showed that a certain toxicity of machined Mg surfaces by EDM was observed, and this could be easily avoided with optimized process parameters.

Further improvement of cytocompatibility can be obtained by phosphate coating (Ye et al. 2013) and SiC coating (M. Li et al. 2012). Figure 11.20 clearly shows the improvement of cytocompatibility after SiC coating.

Figure 11.21 shows the hemolysis percentage of the uncoated and SiC-coated WE43 alloys. Both the uncoated and SiC-coated WE43 show a hemolysis ratio lower

TABLE 11.5
Various Coatings Fabricated on WE Magnesium Alloys and Their Corresponding Electrochemical Parameters in SBF

Matrix	Coating	Preparation Method	Medium	Corrosion Parameters before Coating		Corrosion Parameters after Coating	
				E_{corr} (v)	I_{corr} (10^{-4} A/cm^2)	E_{corr} (v)	I_{corr} (10^{-4} A/cm^2)
WE43 (P. Liu et al. 2012a)	Al$_2$O$_3$–ZrO$_2$ ceramic coatings	Cathodic plasma electrolytic deposition	SBF	−1.681	5.701	−1.354	0.1044
WE43 (Jamesh et al. 2013)	Si-rich layer	Si plasma ion implantation	SBF	−1.972	6.42	−1.895	0.27
WE43 (Zhao et al. 2012)	Al$_2$O$_3$-containing layer	Al and O plasma-implantation	SBF	−1.972	6.025	−1.586	0.4468
WE43 (P. Liu et al. 2012b)	MAO coating	Micro-arc oxidation	SBF	−1.681	5.701	−1.441	0.5741
WE43 (P. Liu et al. 2012b)	MAO/LBL coating	MAO+ layer-by-layer self-assembly	SBF	−1.681	5.701	−1.295	0.02796
WE43 (Ye et al. 2013)	Phosphating coating	Immersion in phosphating solution	SBF	−1.42	1.125	−1.58	0.01583
WE43 (Zhao et al. 2013)	Ti-containing layer	Ti ion implantation	SBF	/	3.27	/	0.699
WE43 (Y. Zhao et al. 2013)	TiO$_2$-containing surface film	Dual Ti and O ion implantation	SBF	/	3.27	/	0.144

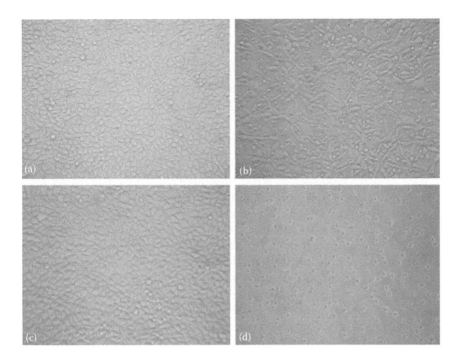

FIGURE 11.20 Optical micrographs of L929 cells that were cultured in (a) negative control, (b) uncoated, and (c) SiC-coated WE43 alloy extracts, and (d) positive control for 3 days. (From Li, M., Y. Cheng, Y. F. Zheng, X. Zhang, T. F. Xi, and S. C. Wei, *Applied Surface Science*, 258 (7), 3074–3081, 2012. With permission.)

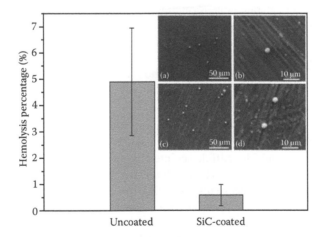

FIGURE 11.21 Hemolysis percentage of the uncoated and SiC-coated WE43 alloys. The insert SEM images show the morphology of blood platelets attached to the uncoated (a and b) and SiC-coated (c and d) WE43 alloys with different magnification. (From Li, M., Y. Cheng, Y. F. Zheng, X. Zhang, T. F. Xi, and S. C. Wei, *Applied Surface Science*, 258 (7), 3074–3081, 2012. With permission.)

than the 5% required by ISO 10993-4 standard. Obviously, the SiC-coated samples show a significant low hemolysis percentage compared to that of the uncoated one. The platelets on the uncoated WE43 alloy are at the inactivated stage with a round shape, and no spread dendritic platelets are observed, implying that WE43 has good antithrombotic properties. However, few thin pseudopods can be observed around the platelets on the SiC-coated WE43 alloys, indicating that they are slightly activated.

11.2.5.8   Animal Testing

11.2.5.8.1   Bone Implant

The bone-inducing properties of WE43 bone implants were preliminarily learned by Witte et al. (2005). Rod samples made of WE3 were implanted intramedullary into the femora of guinea pigs compared to three other kinds of Mg alloy rods and a degradable polymer. A subcutaneous gas bubble was found a week postoperation and disappeared subsequently. No adverse effects of the produced gas were observed. The corrosion layer of the WE43 alloy rods accumulated with biological calcium phosphates and was in direct contact with the surrounding bones. Rare earth elements, such as Y, were distributed homogeneously in the corrosion layer and in the remaining implant material but not found in the surrounding bone. Higher mineral apposition rates and increased bone mass were observed around the Mg rods without induced bones in the surrounding soft tissues.

A Mg-Y-RE-Zr alloy, constitutionally similar to the WE43 alloy, was machined into bone screws and implanted into the marrow cavity of the left femora of New Zealand white rabbits to investigate the acute, subacute, and chronic local effects on bone tissue as well as the systemic reactions (Waizy et al. 2014). There were no significant changes in blood values compared to normal levels. Histological examination revealed a moderate bone formation. No fibrous capsule was found, and the implant directly contacted with the surrounding tissues. Histopathological evaluation of lung, liver, intestine, kidneys, pancreas, and spleen tissue samples showed no abnormalities. Preliminary results showed that the Mg-Y-RE-Zr screws had good biocompatibility and osteoconductivity without acute, subacute, or chronic toxicity.

11.2.5.8.2   Vascular Stent

The first generation of bioabsorbable Mg stents was made from a WE43 Mg alloy named as AMS-1 or Lekton Magic (Loos et al. 2007). The AMS-1 was a tubular, slotted, balloon-expandable scaffold sculpted by laser from a tube of a bioabsorbable Mg alloy without drug elution. The mechanical characteristics of the Mg scaffolds were similar to stainless steel stents, including low elastic recoil (less than 8%), high collapse pressure (0.8 bar), and a minimum amount of shortening after inflation (less than 5%) (Erbel et al. 2007).

The safety and efficacy of bioabsorbable Mg alloy stents (AMS-1) in coronary arteries of domestic or minipigs were studied by Waksman et al. (2006). The histologic evaluation of AMS-1 showed that none of the arteries analyzed had incomplete stent apposition, excess of intimal thickening at the stent edges, or intraluminal thrombus. The neointimal tissue proliferation was significantly less in the stented segments of the Mg alloy scaffold as compared to a control group of stainless steel stents. This could be seen in the intravascular ultrasound images (IVUS) of the coronary arteries. Here, the intravascular ultrasound images of coronary arteries are presented in Figure 11.22. The reduction of neointima formation was not translated to larger vessel lumen, and the overall stented segment was significantly smaller when compared to the stainless steel stent. Representative histological photomicrographs are presented in Figure 11.23.

Although statistically it is not significant, the extent of fibrin deposition and inflammation for stented segments of stainless steel stents was slightly higher than those with Mg bioabsorbable stents. The AMS-1 was largely biodegraded into inorganic ions within 60 days of implantation. In total, AMS-1 was safe and was associated with less neointima formation; however, reduced neointima did not result in larger lumen.

In order to solve the modest degree of late recoil and intimal hyperplasia of the AMS-1 stents, adjunct vascular brachytherapy (VBT) was adopted as a combined treatment (Waksman et al. 2007). The results showed that VBT as an adjunct to AMS further reduced the intimal hyperplasia and improved the lumen area when compared to AMS alone but did not have any impact on late recoil. Consistently, all implantation studies suggested a desirable prolonged degradation time of AMS-1.

AMS-1 was redesigned to prolong the vessel scaffolding. To reduce neointimal growth, the newly designed stent was coated with a 1-μm bioabsorbable poly(lactide-co-glycolide) polymer matrix (PLGA) containing the antiproliferative drug paclitaxel ($0.07 \mu g/mm^2$). The newly designed stent was known as DREAMS (Wittchow et al. 2013). In vivo evaluation of DREAMS was performed on a porcine model and tried to find the best lactide-to-glycolide ratio for the PLGA polymer formulation (Wittchow et al. 2013). This formulation regulates the resorption rate of the drug-carrying PLGA polymer and, in this way, controls the release of paclitaxel. The best-performing Mg scaffold with high molecular weight and a ratio of 85/15 is equivalent

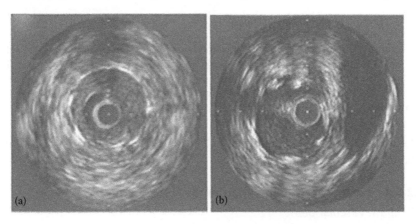

FIGURE 11.22 Intravascular ultrasound images of coronary arteries 28 days after stainless steel stent (a) or magnesium alloy stent (b) implantation. Note the intima hyperplasia in the vessels implanted with stainless steel stents. (From Waksman, R., R. Pakala, P. K. Kuchulakanti, R. Baffour, D. Hellinga, R. Seabron, and A. Haverich, *Catheterization and Cardiovascular Interventions*, 68 (4), 607–617, 2006. With permission.)

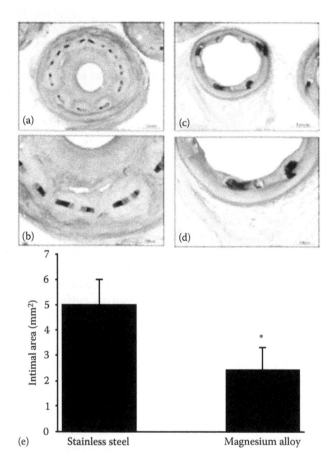

FIGURE 11.23 Representative photomicrographs of hematoxylin and eosin-stained sections of porcine coronary arteries 28 days after stainless steel stent (a) 40× and (b) 100×, and magnesium alloy stent (c) 40× and (d) 100×, implantation. (e) Bar graph showing the intimal area. Note the large intimal area in the vessels implanted with stainless steel stents. (From Waksman, R., R. Pakala, P. K. Kuchulakanti, R. Baffour, D. Hellinga, R. Seabron, and A. Haverich, *Catheterization and Cardiovascular Interventions*, 68 (4), 607–617, 2006. With permission.)

to TAXUS Liberté and superior to euca TAX regarding late luminal loss, intimal area, fibrin score, and endothelialization.

11.2.5.9 Preliminary Clinical Trials in the Human Body

11.2.5.9.1 Vascular Stent

Until now, most of the preliminary clinical trails on the human body were about the absorbable Mg stents (AMS). In 2005, Peeters et al. (2005) reported a preliminary 3-month follow-up of the AMS-1 stent for treatment of infrapopliteal lesions in 20 patients with critical limb ischemia (CLI), and the results were encouraging, yielding a limb salvage rate of 100%. The primary clinical patency and limb salvage rates suggested a potentially promising performance of these AMS-1 devices in the treatment of below-knee lesions in CLI patients.

Compared to single case reports, results of clinical trials based on large patient populations are much more reliable. In 2007, it was reported that 71 AMS-1 stents were evaluated in a prospective, nonrandomized, multicenter, clinical trial in 63 patients (Erbel et al. 2007). However, angiography results at 4 months showed an increased diameter stenosis of 48.4 (17.0%). A follow-up report in 2009 revealed that the main mechanism of restenosis was a faster than expected stent degradation with an early loss of radial force and consequent vessel recoil (Waksman et al. 2009).

The first-in-man trial of the developed AMS named DREAMS, which was the first generation of drug-eluting Mg stent developed to overcome the drawbacks of AMS-1, was carried out in 46 patients with 47 lesions at five European centers (Haude et al. 2013). However, results showed the late lumen loss with DREAMS still did not match the excellent results of currently available drug-eluting stents.

Considering the specificity in neonates and infants, the use of conventional stents is limited because of further vessel growth, the need of redilation, and later surgical removal. The AMS will not need further stent expansion with the child's growth, and the potential risks of conventional metal stents can be avoided for further surgical approaches. Cases of the use of AMS in infants and children were also reported.

Implantation of AMS to open an occlusion in the left pulmonary artery of a preterm baby was reported in 2005 (Zartner et al. 2005). All those pathological and histological findings show minimal alteration of the vessel wall and an increase of the arterial diameter after stent degradation and give us a new and deeper cognition of the AMS in small infants (Zartner et al. 2007). Later in 2006, the first use of AMS for acute treatment of a 3-week-old male newborn with severely impaired heart function due to a long segment recoarctation after a complex surgical repair was also reported (Schranz et al. 2006). Despite the small size of the baby, the degradation of implanted AMS was clinically well tolerated, and AMS had already shown clinical feasibility in pediatrics.

The third generation of AMS based on the first generation of drug-eluting bioabsorbable Mg stents (DREAMS) has already been developed. The DREAMS 2 G stent is made of a WE43 alloy with a 6-crown 2-link design and a strut thickness of 150 µm with radiopaque markers at both ends, resulting in slower dismantling and resorption rate. To further reduce the neointima formation, the DREAMS 2G is coated with a bioresorbable polylactic acid polymer (7 µm) featuring sirolimus at a dose of 1.4 µg/mm^2. DREAMS 2G has completed preclinical assessment and is currently under evaluation in BIOSOLVE-II trial (Campos et al. 2013).

11.2.5.9.2 Bone Screw

Mg-based MgYREZr screws, similar to WE43, have recently received the CE marking of medical devices for medical applications within Europe. Positive results were obtained from their clinical pilot trial (Windhagen et al. 2013). The clinical results of this prospective controlled study demonstrate that degradable Mg-based screws are equivalent to titanium screws for the treatment of mild hallux valgus deformities.

More details about the clinical trials in humans on WE Mg alloys can be found in Chapter 14.

11.2.6 Conclusions on Mg-Y-Based Biomedical Materials

Y contributes to the strength of Mg via two main ways: solid solution strengthening and precipitation strengthening from aging. Long period stacking ordered structure (LPSO) in the Mg-Y system has a great impact on the mechanical properties of the alloys. Mechanical properties of Mg-Y and Mg-Y-based alloys can be adjusted by various heat treatments and deformation processes in a wide range. Other alloying additions also give us another way to adjust the mechanical properties to some extent. All those properties ensure that the mechanical compatibility of Mg-Y alloys can preliminarily satisfy the requirement of some applications, such as vessel stents.

The mechanical properties of the Mg-Y alloys can preliminarily satisfy the requirement under some circumstances. Degradation (corrosion) becomes the dominating factor that determines the viability of the material. The concomitant gas problem can also be solved by controlling the degradation process. From the discussion, we find that the degradation of Mg-Y alloys is faster than we expected, especially for absorbable stent application. However, the Mg-3Y extraluminal tracheal stent is an exception. Degradation of the Mg-3Y tracheal stent seems quite reasonable for the healing of the lesioned trachea, but early fracture of the stent is closely related to the corrosion cracking resistance and fatigue resistance of the material. Similar fracturing may happen in orthopedic implants or intravascular stents because of the cyclical loading from the physiological environment. So mechanical behavior or corrosion behavior in vitro simulation should concern those factors. If possible, add those factors, such as load and blood flow, into in vitro simulation experiments to well predict the in vivo behavior.

Proper coating on Mg alloys can improve the corrosion resistance in a limited implantation time. It can contribute to the mechanical integrity of the implants and avoid the implantation failure in the early implantation period. Some bioactive layers even help the diseased tissue to reconstruct. However, from the viewpoint of the author, the shielding layer should be absorbable. The reason we are developing Mg implants is mainly because of the degradation properties of Mg. If we add a layer that can't be degraded, why choose the Mg alloys as the substrate? Other inert metals, such as titanium, can work better in this situation.

Because the toxicity of rare earth elements are not well known, Mg-Y-Zn alloys consisting of one single rare earth element seem to be suitable because more rare earth elements make the case more complicated and interactions between different rare earth elements are not easy to clarify. If the addition of one of the rare earth elements into Mg—here it is Y—can reach the requirement, no more is needed. Mg-Y-Zn alloys with mediate strength exhibit relatively low corrosion current density in vitro. Those alloys seem to be good biomaterials in orthopedic implants bearing no load or light load. Preliminary results of the cytotoxicity and animal testing have already showed good biocompatibility and uniform degradation of this material.

Plenty of study has been carried out on the Mg-Y-RE (WE) alloy systems for biomedical applications, especially on WE43. The mechanical properties and corrosion resistance of WE43 Mg alloys are much better than many other Mg alloys, and this has been proven in the engineering field. Of the biodegradable metals that have been intensively investigated in animal tests and also, in some practical cases, in humans,

WE43 is among those with the most potential. Compression screws made of a Mg-Y-RE-Zr alloy similar to WE43 have recently received the CE marking of medical devices for medical applications within Europe. We are waiting for the approval of absorbable Mg stents.

A series of clinical tests with absorbable Mg stents made of WE43 have already demonstrated WE43 to be a suitable Mg alloy for cardiovascular stents, which shows endothelialization, low neointima proliferation, minimal inflammation, and estimated complete stent degradation. Excessive neointima proliferation, possible thrombosis and restenosis can be further suppressed by drug-loading on the stent. DREAMS 2G has completed preclinical assessment and is currently under evaluation in a BIOSOLVE-II trial. The viability of the second generation of drug-eluting Mg stents (DREAMS 2G) will be seen in the near future.

11.3   Mg-Nd-BASED ALLOYS FOR BIOMEDICAL APPLICATIONS

11.3.1   Neodymium

Neodymium (Nd), with a density of 6.8 g/cm^3, is a soft silvery metal that tarnishes in air. The most successful application of Nd is in the NdFeB permanent magnets, which can be used in computers (disks, voice coil motors, printer hammer, etc.), automobiles (sensors, electric steering, electric fuel pumps, etc.), and consumer electronics (cameras, microphones, loudspeakers, cell phones, etc.) (Brown et al. 2002; Ma et al. 2002; Matsuura 2006). In addition, Nd and its compounds are also used in the Nd-doped lasers, colorant in glasses, additives in rubber, and as an alloying element added into aluminum and Mg alloys (Mears et al. 1985; Zhou et al. 1985; Campbell and Suratwala 2000).

Nd is one of the light rare earth elements and doesn't exhibit obvious cell toxicity (Feyerabend et al. 2010), and the anticarcinogenic effect of Nd has also been reported (Ji et al. 2000; Witte et al. 2008). The median lethal dose for Nd (LD50) is about 0.6 kg^{-1} according to the result of Haley et al. (1964). If used at low concentrations, Nd appears to be a suitable alloying element for biomedical Mg alloys causing no negative effects on cell viability, inflammatory response, and apoptosis in the in vitro test system.

11.3.2   Previous Studies on Mg-Nd-Based Alloys
for Industrial/Engineering Applications

Nd is a common rare earth alloying element used in engineering Mg due to the optimum combination of its wide solubility range in solid Mg and its relatively low cost (Bronfin and Moscovitch 2006). The maximum solubility of Nd in Mg is 3.6 wt.% at eutectic temperature, and Mg-Nd binary alloys have already shown significant strengthening effects (X. B. Zhang et al. 2012g, 2013b). Study on the binary Mg-Nd alloys reveals that Nd in Mg can improve the creep resistance (J. Yan et al. 2008, 2009), corrosion resistance (Takenaka et al. 2007; Birbilis et al. 2009), and antioxidation properties at high temperature (Aydin et al. 2013, 2014). Nd can also refine the grains and weaken the textures (Hadorn et al. 2012). Those effects of Nd in binary Mg-Nd alloys are also observed in ternary and multielementary Mg-Nd alloys.

Based on the benefits of Nd in Mg alloys, various Mg-Nd-based alloy systems have been developed. The most popular one is the Mg-Nd-Zn-Zr alloy system. Effects of different alloy composition designs, mechanical processes, heat treatments on the microstructure, mechanical properties, creep resistance, fatigue behaviors, age-hardening properties, and corrosion resistance of the Mg-Nd-Zn-Zr quaternary alloys have been widely studied (Chang et al. 2007a,b; Q. Li et al. 2007; Ding et al. 2008; Penghuai et al. 2008; Wen et al. 2008; Ji et al. 2009; Ning et al. 2009, 2010; X. Zheng et al. 2010; Liang et al. 2012; L. Ma et al. 2012; Z. M. Li et al. 2013; Zheng et al. 2013). From the previous reports, it can be concluded that Nd plays an important role, mostly a positive role, in the mechanical properties, age hardening behavior, and corrosion resistance in the engineering Mg-Nd-Zn-Zr alloys. Z. M. Li et al. (2013) did some exploratory study to find the optimized composition of this system. In their work, the effect of Zn and Nd addition on the microstructure and mechanical properties in different heat-treated conditions were discussed, and when Zn addition was 0.2 wt.% and Nd addition was 3.0 wt.%, the resulting alloy revealed the best combination of strength and elongation. In order to further improve the corrosion resistance of Mg-Nd-Zn-Zr alloys, some surface treatment methods are adopted, such as anodic coating in alkaline electrolyte (J.-W. Chang et al. 2008), plasma immersion ion implantation and deposition (G. Wu et al. 2014), microarc combined with sol-gel method (S. Wang et al. 2012), etc. What is more, it is an interesting finding that the machined chips of Mg-Nd-Zn-Zr can be recycled and directly processed into compact alloys via a solid recycling process without smelting, and this recycled alloy shows decent mechanical properties (Ji et al. 2009).

Apart from the Mg-Nd-Zn-Zr system, some other Mg-Nd-based alloys have also been developed, including Mg-Nd-Al (Lu et al. 2004; Bettles et al. 2009), Mg-Nd-Zn (Wilson et al. 2003; J. Yang et al. 2008; J. Zhang et al. 2009a; Ma et al. 2011; Qi et al. 2011; D. Wu et al. 2014), Mg-Nd-Zr (Chang et al. 2007c; Y.-D. Kim et al. 2008; Betsofen et al. 2011), Mg-Nd-Al-Mn (J. Zhang et al. 2008a; Wang et al. 2008c; N. Liu et al. 2009), Mg-Nd-Al-Zn (Hyuk Kim et al. 1997; Song et al. 2007a), and some other alloys with more than two rare earth elements, such as Mg-Nd-Y (Rokhlin et al. 2004; Smola and Stulíková 2004; Meng et al. 2007b; Aghion et al. 2008), Mg-Nd-Y-Zn (Q. Wu et al. 2012), Mg-Nd-Gd-Zn-Zr (J.-H. Li et al. 2008; Sha et al. 2010; Yongdong et al. 2011), Mg-Nd-Gd-Dy (Li et al. 2007b), Mg-Nd-Y-Gd-Zr (Peng et al. 2006; Z.-J. Liu et al. 2012), etc., and so on.

In the fabrication and mechanical processing of engineering Mg-Nd-based alloys, some advanced technologies are employed. Bettles et al. (2009) fabricated a Mg-5Al-2Nd alloy via a powder metallurgical route using a blend of two dissimilar alloy powders, which has a homogeneous distribution of nanoscaled precipitates, and such microstructure is believed to be beneficial to the creep resistance. Severe plastic deformation is another effective way to modify the microstructure and improve the mechanical properties of Mg alloys. Sha et al. (2010) found that high temperature equal-channel angular pressing was an effective way to modify precipitate microstructure to obtain precipitation strengthening in the Mg-Nd-Gd-Zn-Zr alloy.

11.3.3   DEVELOPMENT OF MG-ND BINARY ALLOYS
AS DEGRADABLE METALLIC BIOMATERIALS

11.3.3.1   Alloy Composition Design

Figure 11.24 is the phase diagram of a Mg-Nd binary system. Possible phases in Mg-Nd alloys could be $Mg_{12}Nd$, $Mg_{41}Nd_5$, Mg_3Nd, MgNd, or their coexistence. According to the results of Meng et al. (2007a), Mg_3Nd has the greatest relative stability in the Mg-Nd system compared with other possible compounds. However, Can's results (2011) indicate that among all the intermetallic compounds formed in this system, MgNd is the most stable, and $Mg_{12}Nd$ is the most unstable. Chang et al. (2007a) reported that $Mg_{12}Nd$ phase was corrosion resistant, and its corrosion potential was only a little more positive than that of pure Mg.

The maximum solubility of Nd in Mg is about 3.6 wt.% at the eutectic temperature of 545°C, and it decreases drastically with decreasing temperature. Too much Nd in Mg alloys may lead to the formation of second phases. So the content of Nd in Mg-Nd-based alloys should be well designed and under strict control.

11.3.3.2   Microstructure and Mechanical Properties

The typical microstructure of an as-cast Mg-Nd binary alloy is illustrated in Figure 11.25. In this range of Nd content, the alloys consist of dendritic α-Mg matrix surrounded by $Mg_{12}Nd$ interdendritic phase. This two-phase microstructure is formed during the nonequilibrium solidification. The amount of the intermetallic phase increases with increasing Nd levels.

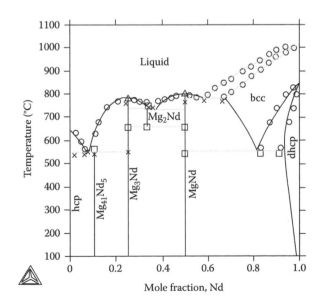

FIGURE 11.24   Phase diagram of Mg-Nd. (From Gorsse, S., C. R. Hutchinson, B. Chevalier, and J. F. Nie, *Journal of Alloys and Compounds*, 392 (1–2), 253–262, 2005. With permission.)

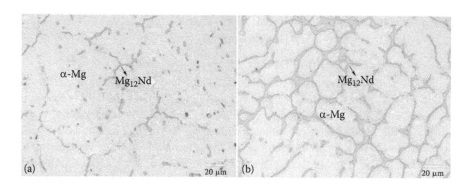

FIGURE 11.25 Typical microstructure of as-cast Mg-Nd alloys: (a) Mg–2.5%Nd and (b) Mg–6%Nd. (From Aydin, D. S., Z. Bayindir, and M. O. Pekguleryuz, *Journal of Alloys and Compounds*, 584, 558–565, 2014. With permission.)

Too much Nd in Mg will lead to the formation of complicated second phases. In the work of Gorsse et al. (2005), the second phase in the slowly cooled Mg-7Nd (at.%) alloy is $Mg_{41}Nd_5$, and it is $Mg_{12}Nd$ in the samples that quenched from the melt, and the $Mg_{12}Nd$ phase changes into $Mg_{41}Nd_5$ during the following annealing. In Mg-9Nd (at.%) alloy, the phase constitution is totally different. Mg_3Nd and $Mg_{41}Nd_5$ form in the slowly cooled state, and those phases turn into $Mg_{12}Nd$ and Mg_3Nd in as-quenched samples. Phases in the annealed state coincide with that in annealed Mg-7Nd (at.%). It can conclude that the second phases in as-cast Mg-Nd binary alloys is closely related to Nd content, cooling rate during solidification, and subsequent heat treatment.

In the study of J. Yan et al. (2008), Mg-xNd (x = 1, 2, 4) alloys were fabricated via casting. The room temperature tensile properties revealed that Nd addition caused the increase of both yield (TYS) and ultimate strength (UTS) but a decrease of ductility. The as-cast Mg-4Nd exhibited TYS and UTS of 141.2 MPa and 153.6 MPa, respectively, but with a poor elongation of 1.08%.

Seitz et al. (2012) developed a Mg-2Nd (in their work, this alloy was named Nd2) alloy aimed at the potential applications in bioabsorbable implantable devices. Tensile results revealed a yield strength of 77 MPa and a tensile strength of 193 MPa for Mg-2Nd in the hot-extruded condition with the elongation reaching 30%. For the compression test, a yield stress of 102 MPa and a compressive strength of 327 MPa were reached. Subsequent T5 and T6 heat treatments could adjust the tensile and compressive properties of Mg-2Nd in a limited range.

11.3.3.3 Corrosion Behavior in Simulated Physiological Environments

Seitz et al. (2012) studied the corrosion behavior of biomedical Mg-2Nd alloys in both 0.9% NaCl solution and simulated body fluid (SBF). The open circuit potentials for Mg-2Nd in NaCl solution and in SBF at 36.5°C were −1.75 V and −1.94 V, respectively. The corrosion current densities determined by Tafel curve were 3.5×10^{-5} A·cm^{-2} and 1.4×10^{-5} A·cm^{-2} for Mg-2Nd exposed to NaCl solution and SBF, respectively. Average mass losses calculated from the corrosion current density by

TABLE 11.6

Corrosion Parameters of Binary Mg-Nd Alloys

Alloy	Condition	OCP (V)	I_{corr} ($\mu A \cdot cm^{-2}$)
Pure Mg (>99.9%)	As-cast	−1.88	23
Mg-1Nd	As-cast	−1.63	11
Mg-4Nd	As-cast	−1.62	9
Mg-4Nd	T4	−1.84	18
Mg-9Nd	As-cast	−1.56	20

Note: I_{corr}, corrosion current density; OCP, open circuit potential.

using Faraday's law were 8.4×10^{-5} g·day^{-1} in NaCl solution and 2.1×10^{-4} g·day^{-1} in SBF. The corrosion rate of Mg-2Nd decreased with the increase of immersion time, owing to the protection of the corrosion products layer.

In the work of Kubasek and Vojtech (2013), they strictly controlled the contents of Fe, Ni, Co, and Cu in the Mg-Nd alloys, aiming at eliminating the impact of those impurities on corrosion. Table 11.6 lists the results of corrosion parameters determined from potentiodynamic curves in their work. The corrosion potential of the as-cast alloys increased with the increasing amount of Nd. Additions of 1% and 4% of Nd caused significant decrease of the corrosion rate compared to pure Mg in the as-cast state. However, 9% Nd addition had a slight influence on the corrosion behavior. This trend was in line with the results of immersion tests.

11.3.3.4 Cytotoxicity

Figure 11.26 depicts the cell viability data after 24 h incubation in the extract of uncoated, MgF$_2$-coated, and bioglass-coated Mg-2Nd specimens (Seitz et al. 2012). All the tested Mg-2Nd samples show no significant effect on cell viability of murine fibroblasts or keratinocytes. The cell proliferation assay demonstrates a significant inhibition of cell proliferation for the uncoated Mg-2Nd specimen with small extract ratio (volume of the extract medium/surface area of the specimen). Murine keratinocytes show the same trend but without statistical significance. Neither bioglass- nor MgF$_2$-coated Mg-2Nd specimens affect cell proliferation of murine fibroblasts or keratinocytes in vitro. Larger quantities of the extract medium led to lower concentrations of the leaching ions; thus, proliferation is not affected to such a degree.

11.3.3.5 Animal Testing

Some Mg-2Nd alloy medical devices have been designed (see Figure 11.27). Stents made of this alloy (Figure 11.27a and b) had been implanted in cases of (postoperative) trauma and stenosis of tissue to maintain the circulation of physiological fluid. Figure 11.27c and d show a fixing element aimed at suture applications. Small tubes enable the Mg sutures to be connected. The tube in which the sutures to be connected are inserted is finally cold deformed by tongs and enables the resulting pinch as shown in Figure 11.27c. The slow, steady degradation rates of the Mg-2Nd alloy enable stability to be maintained in the diseased tissue until it has completely healed, and the degradation

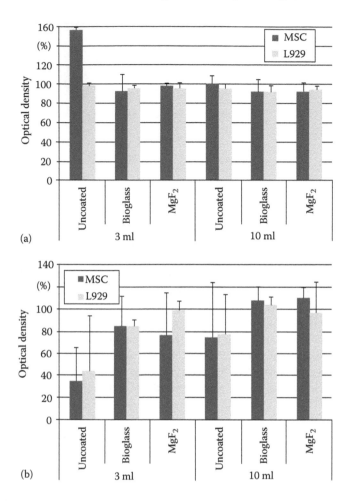

FIGURE 11.26 (a) MTS cell viability results on L929 and MSC cells for Mg-2Nd alloy in different coated states; (b) BrdU cell proliferation results on L929 and MSC cells for Mg-2Nd ally in different coated states; 3 or 10 ml means different extract ratio. (From Seitz, J. M., R. Eifler, J. Stahl, M. Kietzmann, and F. W. Bach, *Acta Biomaterialia*, 8 (10), 3852–3864, 2012. With permission.)

can be adjusted by the coating (MgF$_2$). The manufacture, testing, and implantation of stents or fixing/clamping elements have already shown promising results.

11.3.4 Development of Mg-Nd-Zn-Zr Quaternary Alloys as Degradable Metallic Biomaterials

11.3.4.1 Alloy Composition Design

Based on the binary Mg-Nd system, some other alloying elements are added into Mg alloys to optimize the mechanical properties and improve the corrosion resistance.

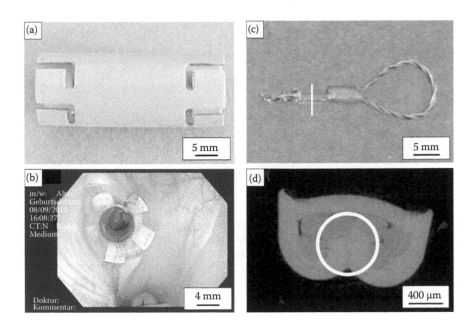

FIGURE 11.27 Prototypes of two medical devices: (a) Mg-2Nd stent before implantation; (b) Mg-2Nd stent postimplantation; (c) Mg-2Nd fixing/clamping element. The initial tube was deformed at its lower part, which is marked by the line. (b) μCT analysis of the tube's cold deformation area. The μCT analyses were carried out on the vertical section marked by the line in (a). It is shown that the tube experiences a single central crack (circle); as a result, the wires were easily clamped. (From Seitz, J. M., R. Eifler, J. Stahl, M. Kietzmann, and F. W. Bach, *Acta Biomaterialia*, 8 (10), 3852–3864, 2012. With permission.)

Zinc (Zn) is one of the abundant nutritionally essential elements in the human body and also an effective alloying element in strengthening Mg alloys. It can also reduce the harmful corrosive effect of iron and nickel in Mg alloys. Sometimes, it is added into Mg alloys in a combination with Nd. Zirconium (Zr) is usually added into Mg alloys as an effective grain-refining agent; in addition, it can also remove the Fe impurity element from the melt for Mg alloys (Prasad et al. 2012b). The biocompatibility of a small amount of zirconium in a Mg alloy has been verified by Ye et al. (2010).

11.3.4.2 Microstructure and Mechanical Properties

The typical microstructure of biomedical Mg-Nd-Zn-Zr alloys with low Zn and low Zr (0.2–0.4 wt.%) is illustrated in Figure 11.28. The microstructure of the as-cast alloy consists of the α-Mg matrix and eutectic compounds distributed along the grain boundary. The eutectic compounds are $Mg_{12}Nd$ identified by XRD. Figure 11.28b displays the microstructure of Mg-Nd-Zn-Zr alloy after solution and aging thermal treatments. Judged by the naked eye, it seems the alloy has a uniform microstructure

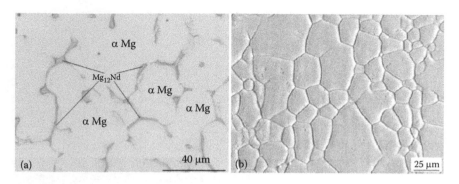

FIGURE 11.28 Typical microstructure of biomedical Mg-Nd-Zn-Zr alloys: (a) optical microstructure in as-cast state. (b) SEM image in aged state. ([a] From Zhang, X., G. Yuan, J. Niu, P. Fu, and W. Ding, *Journal of the Mechanical Behavior of Biomedical Materials*, 9, 153–162, 2012. With permission; [b] From Zhang, X. B., Y. J. Xue, and Z. Z. Wang, *Transactions of Nonferrous Metals Society of China*, 22 (10), 2343–2350, 2012. With permission.)

without second phases. But it actually is composed of fine nanoscaled Nd-rich platelets in the Mg-matrix, which can be verified by TEM.

Zhang et al. (2013d) reported that as-cast Mg-3.08Nd-0.27Zn-0.46Zr alloy had moderate strength with TYS and UTS of 102 MPa and 194 MPa, respectively. In addition, the elongation of this alloy exceeded 11%, higher than many other as-cast Mg alloys. Different extrusion ratios, extrusion temperatures, and subsequent heat treatments and their combinations were applied, and their enhancing effects on the mechanical properties are listed in Table 11.7. Under identical extrusion temperature and extrusion speed, both the TYS and the ultimate tensile strength (UTS) increase first and then decrease with increasing the extrusion ratio. Specimens with an extrusion ratio of 18 show the highest TYS and UTS, significantly higher than those in the as-cast state. The strength of the alloy extruded at different temperatures, with an extrusion ratio of 25, shows that the ultimate tensile strength has no obvious difference, but the yield strength decreases apparently with increasing extrusion temperature. When those alloys are subsequently aged, both the TYS and UTS are further improved, but the elongation decreases slightly. The extruded alloys own high yield ratios, and the yield ratio can be adjusted by solution treatment. For biodegradable stent application, a high yield ratio (TYS is closer to UTS) of the manufacturing material means the stent can be easily recoiled and even fractured during the expanding process.

Zhang et al. (2012j) reported the mechanical properties of an Mg-Nd-Zn-Zr alloy extruded at 250°C with the extrusion rate of 8. They found the extruded Mg-2.7Nd-0.2Zn-0.4Zr had a TYS and an UTS of 363 MPa and 376 MPa, respectively. The elongation of this alloy reached 8.4%. Mechanical properties could even be further improved by aging after extrusion. The TYS and UTS were even a little higher than those in Table 11.7 (extruded at 280°C, with the extrusion ratio of 18), and the elongation of this alloy was comparable.

TABLE 11.7

Different Treatment Parameters on the Mechanical Properties of Mg-Nd-Zn-Zr Alloys

	Treatment Parameters		TYS (MPa)	UTS (MPa)	Elongation (%)	Approximate Yield Ratio (%)
Mg-3.08Nd-0.27Zn-0.46Zr (X. Zhang et al. 2013d)	Casting	/	102 ± 1.2	194 ± 0.8	11.1 ± 0.01	52.58
Mg-3.08Nd-0.27Zn-0.46Zr (X. Zhang et al. 2013d)	Extrusion ratio	R8[a]	306 ± 2.8	307 ± 3.5	15.3 ± 0.14	99.67
		R18[a]	360 ± 8.6	365 ± 10.1	14.6 ± 2.92	98.63
		R25[a]	246 ± 10.8	269 ± 2.4	26.5 ± 0.67	91.45
Mg-3Nd-0.2Zn-0.4Zr (Zhang et al. 2012g)	Extrusion temperature (°C)	E250[b]	162 ± 1	234 ± 1	26 ± 0.7	69.23
		E350[b]	145 ± 2	229 ± 3	25 ± 1.5	63.32
		E450[b]	124 ± 2	226 ± 2	22 ± 0.3	54.87
Mg-2.7Nd-0.22Zn-0.4Zr (X. Zhang et al. 2013c)	Solution temperature (°C)	S425[c]	267 ± 14	322 ± 2	11.4 ± 4.3	82.92
		S450[c]	255 ± 10	313 ± 16	14.9 ± 2.1	81.99
		S475[c]	178 ± 3	268 ± 4	18.7 ± 0.8	66.42
		S500[c]	163 ± 9	287 ± 2	25.1 ± 4.9	56.79
Mg-2.7Nd-0.22Zn-0.4Zr (X. B. Zhang et al. 2012f)	Extrusion ratio + aging temperature (°C)	R8 + Aging[d]	333 ± 4	334 ± 4	7.9 ± 0.2	99.70
		R25 + Aging[d]	177 ± 2	238 ± 3	20.4 ± 0.3	74.37
Mg-3Nd-0.2Zn-0.4Zr (Zhang et al. 2012g)	Extrusion temperature (°C) + aging	E250 + Aging[e]	189 ± 2	243 ± 3	21 ± 0.9	77.78
		E350 + Aging[e]	159 ± 2	240 ± 2	19 ± 1.1	66.25
		E450 + Aging[e]	137 ± 3	236 ± 5	17 ± 1.0	58.05

Note: Capital letters in front of the numbers in the second column mean the corresponding treatment parameters.

[a] Extruded a 280°C.

[b] Extrusion ratio of 25.

[c] Before solution treatment, the specimens were extruded at 320°C with extrusion ratio of 8.

[d] Extruded at 320°C, aging means aged at 200°C for 8 h in an oil bath.

[e] Extruded with a ratio of 25, and then aged at 200°C for 10 h in an oil bath. Approximate yield ratio was calculated from the average value of TYS dividing UTS.

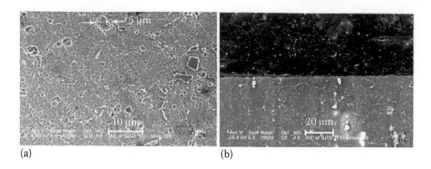

(a)   (b)

FIGURE 11.29 (a) Surface and (b) cross-section corrosion morphologies of Mg-Nd-Zn-Zr alloy. (From Mao, L., G. Yuan, S. Wang, J. Niu, G. Wu, and W. Ding, *Materials Letters*, 88, 1–4, 2012. With permission.)

11.3.4.3 Corrosion Behavior in Simulated Body Fluids

For Mg-3.09Nd-0.22Zn-0.44Zr alloy extruded at 350°C with an extrusion ratio of 25 and annealed at 300°C for 30 min, the in vitro corrosion rate in artificial plasma was 0.337 mm·y^{-1} (0.165 mg·cm^{-2}·day^{-1}) (Mao et al. 2012). Figure 11.29 shows the surface and cross-section morphologies of this kind of Mg-Nd-Zn-Zr alloy within removal of the corrosion layer after a 10-day immersion. Apparently, the surface and cross-section are very smooth, indicating that this alloy suffers a uniform corrosion. X. Zhang et al. (2013d) studied the effects of different extrusion ratios and thermal treatment parameters or their combinations on the corrosion behavior of Mg-Nd-Zn-Zr alloy. Corrosion results in Hank's solution revealed that the corrosion morphologies of both as-cast and extruded Mg-Nd-Zn-Zr alloy are uniform corrosion, and corrosion resistance of extruded ones are better than that in the as-cast condition. Higher extrusion ratios result in better corrosion resistance.

In another work of Yuan et al. (2013), the corrosion resistance of the Mg-2.7Nd-0.2Zn-0.4Zr alloy in artificial plasma after solution treatment is improved slightly when the solution temperature is below 475°C. Moreover, aging treatment on the extruded Mg-3Nd-0.2Zn-0.4Zr alloy shows much better biocorrosion resistance than that at solution state in simulated body fluid (SBF) (X. B. Zhang et al. 2012g).

11.3.4.4 Cytotoxicity

Y. Wang et al. (2012) studied the cytotoxicity of Mg-Nd-Zn-Zr alloy on different kind of cells. The cytotoxicity results with MC$_3$T$_3$-E$_1$ cells indicated that Mg-Nd-Zn-Zr alloy extracts had no significant reduction in the cell viability. In another work of X. Zhang et al. (2012f), an indirect assay was performed to evaluate the L-929 cell response to extruded Mg-Nd-Zn-Zr alloy with different extrusion ratios (R 8 and R 25). Cell viabilities of all the extracts were more than 80% during 1, 3, and 5 days of culture. The cytotoxicity of the Mg-Nd-Zn-Zr alloy met the requirement of cell toxicity, according to ISO 10993-5:1999. Typical cell morphologies are illustrated in Figure 11.30. Liao et al. (2012) also reported the effects of Mg-Nd-Zn-Zr alloy on the viability of chondrocytes. A direct adhesion experiment showed good cell spreading and adhesion on Mg-Nd-Zn-Zr alloy. Indirect cytotoxicity and RT-PCR

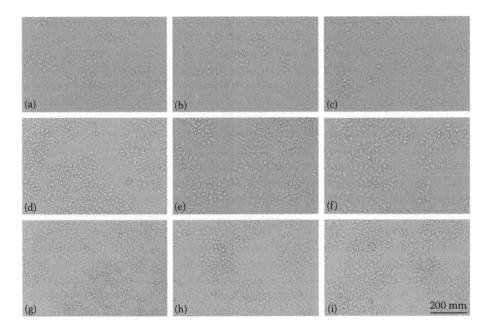

FIGURE 11.30 Optical morphologies of L-929 cells that were cultured in the control (a, d, and g), 100% R8 (b, e, and h), and 100% R25 (c, f, and i) alloy extracts for 1 (a, b, and c), 3 (d, e, and f), and 5 (g, h, and i) days. (From Zhang, X., G. Yuan, J. Niu, P. Fu, and W. Ding, *Journal of the Mechanical Behavior of Biomedical Materials*, 9, 153–162, 2012. With permission.)

test results found the expression of collagen II mRNA and aggrecan mRNA. All the results showed that Mg-Nd-Zn-Zr alloy exhibited an in vitro cytocompatibility with chondrocytes similar to that of pure Mg.

11.3.4.5 Hemocompatibility

In the study of Wang et al. (2013d), the in vitro blood biocompatibility of Mg-Nd-Zn-Zr alloy was investigated with rabbit blood. The results showed that the Mg-Nd-Zn-Zr alloy did not induce significant blood cell aggregation, platelet adhesion, and protein absorption, compared to the clinically used Ti-6Al-4V alloy. Red blood corpuscle (RBC) aggregation and platelet adhesion on the surface of Mg-Nd-Zn-Zr alloys were even lighter than the Ti-6Al-4V alloy, indicating that the Mg-Nd-Zn-Zr alloy had excellent in vitro blood compatibility.

11.3.4.6 Coatings on Biomedical Mg-Nd-Zn-Zr Alloys

Anodic coating (J.-W. Chang et al. 2008), plasma immersion ion implantation combined deposition (G. Wu et al. 2014), and microarc oxidation and sol-gel method (S. Wang et al. 2012) are employed to obtain surface coatings that can enhance the corrosion resistance of engineering Mg-Nd-Zn-Zr alloys. Various kinds of surface coatings are also adopted to improve the corrosion resistance and biocompatibility of biomedical Mg-Nd-Zn-Zr alloys.

To further improve the corrosion resistance and biocompatibility of the Mg-Nd-Zn-Zr alloy, a biodegradable calcium phosphate coating (Ca-P coating) was developed via a chemical deposition method, according to the study of Niu et al. (2013). The bonding strength of the layer was over 10 MPa. In vitro corrosion tests indicated that the Ca-P coating improved the corrosion resistance of the Mg-Nd-Zn-Zr alloy in Hank's solution. Ca-P coating significantly reduced the hemolysis rate of the Mg-Nd-Zn-Zr alloy and induced no toxicity to MC3T3-E1 cells.

Fluoride treatment is another method of surface treatment of Mg-Nd-Zn-Zr alloys. A compact MgF_2 film was obtained on the Mg-Nd-Zn-Zr alloy substrate by the hydrofluoric acid chemical conversion method by Mao et al. (2013). Due to the protective MgF_2 film, the corrosion rate in artificial plasma decreased from 0.337 mm·y^{-1} to 0.253 mm·y^{-1}. Meanwhile, the hemolysis ratio of Mg-Nd-Zn-Zr alloy decreased from 52.0% to 10.1%, and the Mg-Nd-Zn-Zr alloy showed a good antiplatelet adhesion property. A direct cell adhesion test showed many more live HUVECs retained than with the bare Mg-Nd-Zn-Zr alloy. Judging from cytoskeleton staining images, HUVECs stretched out and appeared in normal spindle shape similar to that of the negative control, indicating no adverse effect of untreated and HF-treated Mg-Nd-Zn-Zr alloys on HUVEC spreading. Both HF-treated and untreated Mg-Nd-Zn-Zr alloys showed no adverse effect on HUVEC viability and spreading morphology. In brief, cytotoxicity results met the requirement of cellular application after HF treatment.

11.3.4.7 Animal Testing

Bone plates and screws made of bare Mg-Nd-Zn-Zr alloy and Ca-P-coated Mg-Nd-Zn-Zr alloy were fabricated and implanted into the tibias of New Zealand rabbits with the same shape of Ti plates and screws as a control (Niu et al. 2013). Appearance of the implants and schematic figure of the implantation process are presented in Figure 11.31.

Figure 11.32 shows the radiographs of bone implants in the tibias of New Zealand rabbits 4 weeks after implantation. No obvious degradation of either coated or uncoated Mg-Nd-Zn-Zr alloys was observed. Only a small amount of hydrogen gas was observed around the uncoated Mg-Nd-Zn-Zr alloy implants, indicating that Ca-P coating could protect the Mg-Nd-Zn-Zr alloy matrix effectively. Plenty of bony calluses were observed around both Ca-P coated and uncoated Mg-Nd-Zn-Zr alloy implants; meanwhile, few bony calluses could be observed around the Ti implant. It was clear that both coated and uncoated Mg-Nd-Zn-Zr alloy implants could stimulate bone formation.

Figure 11.33 shows the morphologies of implants after removal from the rabbits. The uncoated Mg-Nd-Zn-Zr alloy screw suffered severe corrosion after 8 weeks in vivo; however, the coated Mg-Nd-Zn-Zr alloy only degraded slightly at the same postimplantation time. Up to 18 weeks postimplantation, only the end of the coated Mg-Nd-Zn-Zr alloy screw degraded out.

Both coated and uncoated Mg-Nd-Zn-Zr alloy implants show good biocompatibility and bone-inducing ability in vivo. The degradation of naked Mg-Nd-Zn-Zr alloy implants is still too fast, and better results were obtained on the Ca-P-coated Mg-Nd-Zn-Zr alloy implants. It can be inferred that the present Ca-P coating technique can prolong the in vivo degradation time of Mg-Nd-Zn-Zr alloy for at least 10 weeks, which is very important for the future clinical applications.

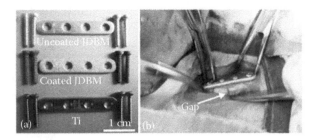

FIGURE 11.31 (a) Appearance of uncoated and Ca-P-coated Mg-Nd-Zn-Zr alloy bone plates and screws; (b) operation on the tibia of a rabbit. (From Niu, J., G. Yuan, Y. Liao, L. Mao, J. Zhang, Y. Wang, and W. Ding, *Materials Science and Engineering: C*, 33 (8), 4833–4841, 2013. With permission.)

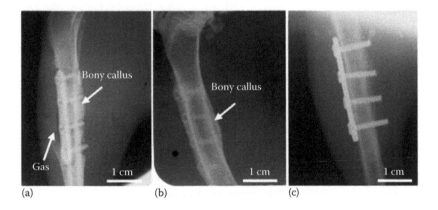

FIGURE 11.32 Radiographs of Mg-Nd-Zn-Zr alloy implants in tibias of New Zealand rabbits 4 weeks postoperation. (From Niu, J., G. Yuan, Y. Liao, L. Mao, J. Zhang, Y. Wang, and W. Ding, *Materials Science and Engineering: C*, 33 (8), 4833–4841, 2013. With permission.)

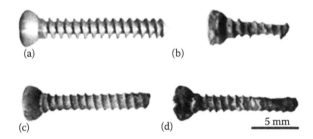

FIGURE 11.33 Optical morphology of the implants after removing from the rabbits and CrO_3 solution cleaning: (a) JDBM screw preimplantation; (b) 8 weeks postimplantation for uncoated Mg-Nd-Zn-Zr alloy; (c) 8 weeks postimplantation for coated Mg-Nd-Zn-Zr alloy; (d) 18 weeks postimplantation for coated Mg-Nd-Zn-Zr alloy. (From Niu, J., G. Yuan, Y. Liao, L. Mao, J. Zhang, Y. Wang, and W. Ding, *Materials Science and Engineering: C*, 33 (8), 4833–4841, 2013. With permission.)

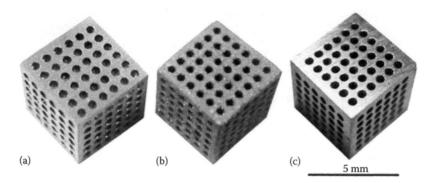

FIGURE 11.34 Porous magnesium alloy scaffolds made of different magnesium alloys: (a) Mg-Nd-Zn-Zr alloy, (b) CaHPO4-coated Mg-Nd-Zn-Zr alloy, and (c) WE43. (From Liao, Y., D. Chen, J. Niu, J. Zhang, Y. Wang, Z. Zhu, and Y. Jiang, *Materials Letters*, 100, 306–308, 2013. With permission.)

11.3.4.8 Mg-Nd-Zn-Zr Alloy as a Potential Tissue Engineering Scaffold

Mg-Nd-Zn-Zr alloy was also developed as tissue engineering scaffold by Liao et al. (2013). Figure 11.34 presents the appearance of polyporous Mg-Nd-Zn-Zr alloy scaffolds fabricated by using a computer numerical control machine. Some scaffolds were CaHPO$_4$-coated subsequently. The porosities were 61.1% ± 2.81% for the noncoated Mg-Nd-Zn-Zr alloy scaffolds and 49.74% ± 2.11% for the coated Mg-Nd-Zn-Zr alloy scaffolds. The polyporous Mg-Nd-Zn-Zr alloy scaffolds possessed similar elastic modulus and compressive strength to those of human cancellous bones. CaHPO$_4$-coated porous Mg-Nd-Zn-Zr alloy scaffolds maintained mechanical integrity, and noncoated Mg-Nd-Zn-Zr alloy scaffolds disaggregated after immersion for 8 weeks. CaHPO$_4$-coated Mg-Nd-Zn-Zr alloy scaffolds showed greater potential in the in vivo applications compared with uncoated Mg-Nd-Zn-Zr alloy scaffolds and WE43 scaffolds.

11.3.5 DEVELOPMENT OF OTHER MG-ND-BASED ALLOYS AS DEGRADABLE METALLIC BIOMATERIALS

11.3.5.1 Mg-Nd-Zr-Sr Alloy System

Strontium (Sr) is a trace metal in the human body (99% of which is located in the bones), and biocompatibility of Sr had already been identified (Meunier et al. 2004, 2009). Sr in biomedical Mg alloys was reported to stimulate new bone formation (X. N. Gu et al. 2012). Addition of Sr into Mg alloys can refine the microstructure and improve the mechanical properties and corrosion resistance (Fan et al. 2007).

Mg-2.2Nd-0.3Zr-xSr alloys (x = 0, 0.4, 0.7, 2.0) were prepared by traditional gravity casting, and the effects of Sr addition on the microstructure and corrosion of Mg-Nd-Zr alloys in SBF were studied by X. Zhang et al. (2013b). Mg$_{17}$Sr$_2$ phase was detected in as-cast Mg-2.2Nd-0.3Zr-xSr alloys, and the grain size decreased with additional Sr content. Corrosion rates calculated from mass loss in SBF were 0.89, 0.83, 0.77, and 56.49 mm·y^{-1} for Mg-2.2Nd-0.3Zr, Mg-2.2Nd-0.3Zr-0.4Sr,

Mg-2.2Nd-0.3Zr-0.7Sr, and Mg-2.2Nd-0.3Zr-2.0Sr, respectively. Elements in the corrosion layer were O, Ca, P, and Mg, probably with some Na.

11.3.5.2 Mg-Zn-Nd-Y Alloy System

Cyclic extrusion was applied on the Mg-Zn-Nd-Y alloy. The grain size was greatly refined to 1 μm, and the second phase distributed along the grain boundaries, and nanosized particles uniformly distributed in grains. The elongation of cyclic extruded alloy was 30.2%, and the TYS and UTS reached 185 MPa and 303 MPa, respectively. The cyclic extruded alloy showed a uniform corrosion, and the corrosion current density decreased from 2.8×10^{-4} A·cm^2 to 6.6×10^{-5} A·cm^2. The improved mechanical properties, uniform corrosion, and reduced corrosion rate were critical to the alloy for stent applications (Q. Wu et al. 2012).

11.3.5.3 Mg-Zn-Mn-Ca-Nd Alloy System

According to previous studies, Ca, Mn, and Zn could be suitable alloying elements in biomedical Mg alloys (Song 2007; Z. Li et al. 2008; E. Zhang et al. 2009b). Wang et al. (2013b) reported the microstructure and in vitro corrosion of the Mg-6Zn-1Mn-0.5Ca-0.4Nd alloy. It was found that the Mg-6Zn-1Mn-0.5Ca-0.4Nd alloy was composed of α-Mg, Ca$_2$Mg$_6$Zn$_3$, and Mg$_{41}$Nd$_5$ phases, and the degradation rate in SBF was about 10.975 mm/a. After immersion in SBF, Ca$_4$O(PO$_4$)$_2$, CaPO$_3$(OH)·2H$_2$O, and Ca$_3$(PO$_4$)$_2$ were detected in the corrosion products.

11.3.5.4 Mg-Nd-Zn-Zr-Ag Alloy System

Silver (Ag) exhibits antibacterial properties and biocompatibility in the human body and can enhance mechanical properties in Mg alloys. X. Zhang et al. (2013a) added Ag into biomedical Mg-Nd-Zn-Zr and studied the effects of Ag addition on the microstructure and corrosion resistance. The amount of Ag in the second phase increased and the amount of the MgZr compound increased with increasing Ag addition. It seemed that Ag addition did not have any benefits on the corrosion resistance. Unfortunately, whether this alloy had the antibacterial property or not was not mentioned.

11.3.5.5 Mg-Nd-Gd-Sr-Zn-Zr Alloy System

For as-cast Mg-(4-x)Nd-xGd-0.3Sr-0.2Zn-0.4Zr (x = 0, 1, 2, and 3) alloys, α-Mg and Mg$_{41}$Nd$_5$ phases were found in the as-cast alloys with 4% and 3% Nd addition, and the new Mg$_3$Gd phase was found in the alloy with 3% Gd addition. Microstructure was refined by Gd addition. Corrosion results obtained by different methods showed that the corrosion rate of the alloys decreased with increasing Gd addition. The alloys exhibited a uniform corrosion (X. B. Zhang et al. 2014).

11.3.5.6 Mg-1.2Nd-0.5Y-0.5Zr-0.4Ca

The addition of 0.4 wt.% Ca into the Mg-1.2Nd-0.5Y-0.5Zr resulted in the formation of Mg$_2$Ca phase, and another precipitate in this alloy was Mg$_{41}$Nd$_5$. The corrosion resistance was improved by the Ca addition. But contrary to the positive effect on the corrosion, Ca addition had a damaging effect on the stress corrosion behavior in terms of reduced UTS and ductility (Aghion and Levy 2010).

Animal testing in Wister male rats was applied to evaluate the in vivo behavior of Mg-1.2Nd-0.5Y-0.5Zr and Mg-1.2Nd-0.5Y-0.5Zr-0.4Ca alloys (Aghion et al. 2012). Cylindrical disks were implanted within the subcutaneous layer at the back midline of Wister male rats at different locations. The postoperative performance, behavior, and surgical wounds of all the experimental rats appeared normal with no signs of ill health. Weekly body weight gains of rats with both Mg-1.2Nd-0.5Y-0.5Zr and Mg-1.2Nd-0.5Y-0.5Zr-0.4Ca were normal and similar to that of the Ti-6Al-4V control alloy. The blood biochemical parameters obtained before and after the implantation of all the materials were within the normal values for rats, indicating that both of the two Mg alloys did not have any detrimental effects on Mg metabolism or renal function. Judging from the morphology of the explanted materials, degradation of the implants at the ventral aspect was greater than at the dorsal side. Implants at the interscapular tended to degrade faster than implants at the lumbar, which led to much hydrogen gas at the interscapular site. In addition, the addition of 4% Ca improved the corrosion resistance of the Mg-1.2Nd-0.5Y-0.5Zr alloy in vivo, and this was well in accordance with the results in vitro. Histological analysis showed that there were no remarkable differences in the histological lesions of the skin and subcutis around the implants among the two Mg alloys and the inert titanium control alloy. In all cases, there was none to very mild inflammatory changes that did not cause marked tissue reaction. All the results obtained from the in vivo experiment clearly indicated, apart from the natural degradation characteristics of biomedical Mg alloys, the in vivo behaviors of Mg-1.2Nd-0.5Y-0.5Zr-(0.4Ca) alloys were adequate and comparable to that of the Ti-6Al-4V alloy, which was under clinical use.

11.3.6 NEW METHODS TO IMPROVE THE PROPERTIES OF BIOMEDICAL MG-ND-BASED ALLOYS

Up to now, insufficient mechanical properties and fast corrosion of biomedical Mg alloys are still two crucial obstacles to their use. Some new manufacturing and processing technologies in various fields may provide new ways to design and process biomedical Mg-Nd alloys that can improve the performance of those alloys.

Rapid solidification processing has the potential for producing substantial improvement in the strength and corrosion resistance of Mg alloys, and this metallurgical method could be applied in the fabrication of biomedical Mg-Nd-based alloys. Similarly, fabrication of biomedical Mg alloys via powder metallurgy is also interesting, and the availability of this method needs experiments to verify it.

Extrusion is an effective way to refine the grains of Mg, and grain refinement is a suitable way to improve the mechanical properties that will not deteriorate the plasticity. However, the fine and homogeneous microstructure is difficult to obtain by one-time extrusion. Double extrusion was applied in the preparation of biomedical Mg-Nd-Zn-Zr alloy by X. Zhang et al. (2012c). Microstructures of double-extruded Mg-2.25Nd-0.11Zn-0.43Zr and Mg-2.70Nd-0.20Zn-0.41Zr alloys became much finer and more homogeneous than those under single extrusion. The TYS, UTS, and elongation of the alloys under double extrusion were more than 270 MPa, 300 MPa, and 32%, respectively. Results of immersion tests showed that double extrusion had little influence on the corrosion resistance in SBF compared to single extrusion. Generally, double extrusion is an effective way to refine grains and to improve mechanical properties.

Cyclic extrusion and compression processing (CEC) is another way that can be adopted to refine grains, to improve the mechanical properties and corrosion resistance. TYS, UTS, and elongation of Mg-Nd-Zn-Zr alloys under CEC have ~71%, ~28%, and ~154% improvement, respectively, according to the results of X. Zhang et al. (2012b). Corrosion resistance of CEC-processed alloys is also improved apparently. CEC-processed Mg-Zn-Y-Nd alloys also exhibit improved mechanical properties, uniform corrosion, and reduced corrosion rates (Q. Wu et al. 2012).

11.3.7 CONCLUSIONS ON MG-ND-BASED BIOMEDICAL ALLOYS

Nd seems to be a suitable alloying element in biomedical Mg alloys causing no negative effects on cell viability, inflammatory response and apoptosis in the in vivo and in vitro tests. In addition, it can refine the microstructure, improve the mechanical properties, and benefit the corrosion resistance if used in a proper content and suitable way.

Mg-Nd binary and Mg-Nd-Zn-Zr quaternary alloys and some other Mg-Nd-based alloys, such as Mg-Nd-Zr-Sr, Mg-1.2Nd-0.5Y-0.5Zr-(0.4Ca), and even more complicated Mg-Nd-Gd-Sr-Zn-Zr, have already been developed. Mg-Nd binary alloys are the base of other Mg-Nd-based alloys, but their mechanical properties are not so satisfactory. Figure 11.35 lists the mechanical properties of previously learned Mg-Nd binary alloys. Extrusion can significantly improve the elongation of Mg-2Nd, and this elongation (30%) can meet the criteria for a biodegradable stent proposed by Erinc et al. (2009). However, the strength of the binary Mg-Nd alloys is still not so sufficient and needs to be improved.

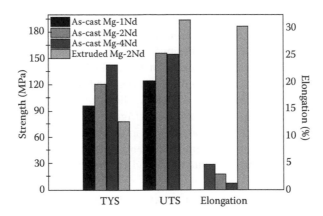

FIGURE 11.35 Mechanical properties of Mg-Nd binary alloys. (Data were collected from Seitz, J. M., R. Eifler, J. Stahl, M. Kietzmann, and F. W. Bach, *Acta Biomaterialia*, 8(10), 3852–3864, 2012; Yan, J., Y. Sun, F. Xue, S. Xue, and W. Tao, *Materials Science and Engineering: A*, 476 (1–2), 366–371, 2008.)

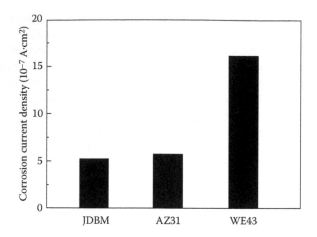

FIGURE 11.36 Corrosion current density of Mg-Nd-Zn-Zr alloy and commercial AZ31 and WE43 alloys. (Data were from Zhang, X. B., G. Y. Yuan, L. Mao, J. L. Niu, and W. J. Ding, *Materials Letters*, 66 (1), 209–211, 2012.)

Compared to Mg-Nd binary alloys, Mg-Nd-Zn-Zr alloys seem more suitable for biomedical applications because of their excellent mechanical properties. Strength of the Mg-Nd-Zn-Zr system can even exceed 300 MPa. These alloys can be used in stents or some orthopedic implants with a light load. Biocompatibility of biomedical Mg-Nd-Zn-Zr alloys has already been proven by both in vitro and in vivo experiments. Apart from the mechanical properties and biocompatibility, degradation rate is another factor that closely relates to the practical use of biodegradable Mg alloys. Figure 11.36 shows the corrosion current density of Mg-Nd-Zn-Zr alloy in SBF compared to AZ31 and WE43. The corrosion resistance of the Mg-Nd-Zn-Zr alloy is much better than the other two commercial Mg alloys. The corrosion rate of the Mg-Nd-Zn-Zr alloy in Hank's solution is about 0.28 mm·y^{-1}, much slower than that of AZ31 (1.02 mm·y^{-1}), and this corrosion rate is well below the criteria proposed by Erinc.

The environment around the implants in animals is closer to the practical physiological environment in the human body. Positive results in vivo put forward the materials much closer to clinical use. All the results of the animal testing show Mg-Nd-Zn-Zr alloys have good biocompatibility, no inflammation, and bone stimulation properties. This alloy system could be an excellent one among the ocean of biodegradable Mg alloys.

It is worth mentioning that biomedical Mg-Nd-Zn-Zr alloys have excellent mechanical properties, especially for elongation and blood compatibility, and those properties are quite desirable for degradable stents. But until now, there have been no experiments putting this material into the blood vessels in animals.

11.4 Mg-Dy-BASED ALLOYS FOR BIOMEDICAL APPLICATIONS

11.4.1 Dysprosium

Dysprosium (Dy), with an atomic number of 66, belongs to the heavy REE group. It has a HCP crystal lattice with a radius of 0.1781 nm, a bit larger than that of Mg

(0.16 nm). Dy and dysprosium compounds are used in some control rods at nuclear power plants (Risovany et al. 2000, 2006) and also used in certain kinds of laser, high-intensity lighting (Langenscheidt et al. 2008) and magnetostrictive alloys, such as Terfenol-D, to raise the coercivity (Hirosawa et al. 1990; Stepankin 1995; Fang et al. 1998). In medical science, Dy is also used as a magnetic resonance imaging (MRI) contrast agent and in the treatment of synovectomy (McLaren et al. 1990; Kattel et al. 2012). Dy is a nonabsorbable element, presenting at no higher than trace amounts in the diet (Sheng et al. 2005). In the literature, Dy shows good cytocompatibility according to the in vitro study of cytotoxicity and inflammatory response (Feyerabend et al. 2010). A half-lethal dose of dysprosium chlorides is 585 mg/kg (Haley et al. 1966). L. Yang et al. (2013a) found that 4000 µM $DyCl_3$ had no adverse influence on the cell viability of SaoS-2 cells. Furthermore, Dy was also used in the MRI contrast agents and also showed good cytotoxicity up to 100 µM (McLaren et al. 1990; Kattel et al. 2012).

11.4.2 PREVIOUS STUDIES ON MG-DY-BASED ALLOYS FOR INDUSTRIAL/ENGINEERING APPLICATIONS

Previous studies on Mg-Dy-based alloys for industrial purposes focused mainly on the development of high-strength, high-stability, and heat-resistant Mg alloys that are relatively cheap. Even though dysprosium atoms can substitute some Mg atoms at the lattice site and cause solid solution strengthening, the strengthening efficiency is limited. People don't pay much attention to the practical use of Mg-Dy binary alloys in engineering, and several papers referred to the aging and recovery precipitation characteristics and deformation behavior of Mg-Dy binary alloys (Rokhlin and Nikitina 2001; Miura et al. 2008; Saito et al. 2011). Additional elements were chosen to improve the mechanical properties of Mg-Dy alloys. Ternary Mg-Dy-Zn (Bi et al. 2011a), Mg-Dy-Gd (Peng et al. 2009), Mg-Dy-Nd (D. Li et al. 2010), and quaternary Mg-Dy-Gd-Nd (Li et al. 2007a,b), Mg-Dy-Nd-Zr (X. Q. Zeng et al. 2008), Mg-Dy-Gd-Zn (Zhang et al. 2011b), Mg-Dy-Gd-Zr (Peng et al. 2008), Mg-Dy-Zn-Zr (Z. H. Huang et al. 2013; J. Zhang et al. 2013b) were developed.

Bi et al. (2011b) found that the Mg-2Dy-0.5Zn alloy with high volume fraction LPSO phase showed high thermal stability and excellent elevated temperature properties. Tensile tests revealed that the yield and ultimate tensile strengths of this alloy reached 245 MPa and 260 MPa at 300°C, respectively. In another work of Bi et al. (2011b), the double-peak aging behavior of Mg-2Dy-0.5Zn alloy was found (see Figure 11.37). Two significant aging peaks were observed at 36 h and 80 h, respectively. A large amount of fine $(Mg, Zn)_xDy$ particles precipitated from the matrix contributed to the first aging peak. The second aging peak was mainly attributed to the strengthening of a high volume fraction and well-dispersed LPSO phase.

Kawamura and Yamasaki (2007) also investigated the LPSO structure in the Mg-Zn-Dy alloy and discussed the criteria for Dy that forms an LPSO phase in Mg-Zn-Dy alloys. Zhang et al. (2011b) improved the ultimate tensile strength and elongation of Mg-6.5Gd-2.5Dy-1.8Zn from 276 MPa 10.8% to 392 MPa and 6.1% by aging treatment.

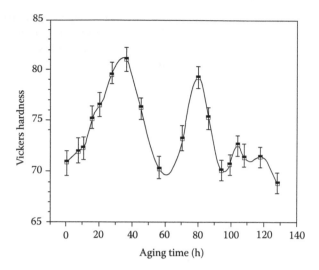

FIGURE 11.37 Age hardening curve of the Mg-2Dy-0.5Zn alloy at 180°C. (From Bi, G., D. Fang, L. Zhao, Q. Zhang, J. Lian, Q. Jiang, and Z. Jiang, *Journal of Alloys and Compounds*, 509 (32), 8268–8275, 2011. With permission.)

Recently, some new methods for the fabrication of Mg-Dy-based alloys have been tried. Gibbs energy showed that $MgCl_2$ can chloridize Dy_2O_3 and release Dy ions in the $LiCl$-KCl-$MgCl_2$-Dy_2O_3 melts. Based on this principle, M. Zhang et al. (2013) fabricated a ternary Mg-Li-Dy alloy by using electrodeposition. With the attractive properties of bulk metallic glasses (BMG), Dy was also applied to prepare Mg-Cu-Dy BMG (X. F. Wu et al. 2011).

11.4.3 ALLOY COMPOSITION DESIGN OF MG-DY-BASED ALLOYS

Because Mg and Dy share the same crystal lattice and similar radius, Dy can considerably solute in Mg. The maximum dissolution of Dy in Mg is 25.3 wt.% at the eutectic temperature of 561°C (L. Yang et al. 2011a); moreover, the solubility decreases with the temperature, and this characteristic can contribute to the strength and aging behavior of Mg alloys (Rokhlin and Nikitina 1999). Addition of appropriate Dy into Mg is an effective way to improve the corrosion resistance. In a proper composition range, Dy can change the corrosion morphology and improve the corrosion resistance of Mg alloys.

Figure 11.38 is the binary phase diagram of a Mg-Dy alloy system. Possible second phases in Mg-Dy binary alloys could be $Mg_{24}Dy_5$, Mg_2Dy, and $MgDy$ or their coexistence. With the content of Dy less than 25.34 wt.%, the main phases are α-Mg solid solution with the secondary $Mg_{24}Dy_5$ phase. When the Dy content is between 25.34 and 58 wt.%, the main phases are α-Mg solid solution and eutectic phase composed of mixture of α-Mg matrix and $Mg_{24}Dy_5$ phase. However, the exact phases in the Mg-Dy alloys could be adjusted by the metallurgical methods, alloying elements, etc.

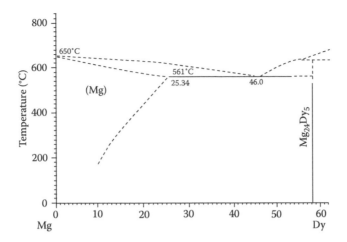

FIGURE 11.38 Phase diagram of Mg-Dy. (From Yang, L., Y. Huang, F. Feyerabend, R. Willumeit, K. U. Kainer, and N. Hort, *Journal of the Mechanical Behavior of Biomedical Materials*, 13, 36–44, 2012. With permission.)

11.4.4 Microstructure and Phase Constitution of Mg-Dy-Based Alloys

11.4.4.1 Microstructure and Phase Constitution of Binary Mg-Dy Alloys

Figure 11.39 is the typical microstructure of Mg-Dy binary alloys. The scanning electron microscopic images show that a very inhomogeneous microstructure is formed due to the rapid cooling during the casting process. The grain size

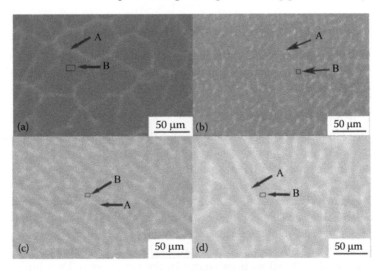

FIGURE 11.39 Typical microstructure of as-cast Mg-Dy alloys under SEM: (a) Mg-5Dy; (b) Mg-10Dy; (c) Mg-15Dy; (d) Mg-20Dy (arrow A: segregation area of Dy; arrow B: Mg matrix; and arrow C: phase). (From Yang, L., Y. Huang, Q. Peng, F. Feyerabend, K. U. Kainer, R. Willumeit, and N. Hort, *Materials Science and Engineering: B*, 176 (20), 1827–1834, 2011. With permission.)

significantly decreases with the increasing of Dy content, but when Dy addition exceeds 10 wt.%, further addition of Dy has less influence on the grain size. Dy segregation areas are found in all of these alloys, and this increases with the increment of Dy content. The amount of the second phases has a similar trend. It is interesting that Dy content in the matrix is about half of the Dy content in these alloys. According to the EDX results, the composition of the second phase is about 85% Mg, 14% Dy, and balanced O in atomic ratio. Based on the phase diagram of Mg-Dy system (Figure 11.38), it can be inferred that the second phase in Mg-Dy binary alloys is $Mg_{24}Dy_5$.

11.4.4.2 Microstructure and Phase Constitution of Mg-Dy-Based Alloys

The microstructure of Mg-Dy-based alloys is similar to that of Mg-Dy binary alloys, which mainly consists of α-Mg matrix with dissolved Dy and other Mg-RE phases. Taking the Mg-Dy-Zn ternary alloy as an example, in Bi's work (2011a), a lot of secondary phases could be observed at the grain boundaries and in the grain interior of the α-Mg matrix. XRD results revealed that those secondary phases were mainly the $Mg_{12}ZnDy$ phase with a 18R LPSO structure. Such 18R LPSO structure was usually found in as-cast Mg-Dy-Zn alloys (Kawamura and Yamasaki 2007). In order to observe the phase transformation in this Mg-2Dy-0.5Zn (at.%) alloy, heat treatment at 525°C for 10 h was applied to the Mg-2Dy-0.5Zn ingots. The microstructure morphology showed that fine lamellar structure formed in all the α-Mg grains, and some precipitates also formed both at the grain boundaries and in the grains. Latter TEM analysis revealed that this lamellar structure is the 14H LPSO phase, and the precipitates were (Mg, Zn)xDy phases.

11.4.5 MECHANICAL PROPERTIES OF MG-DY-BASED ALLOYS

It was proven that Dy can improve the mechanical properties both at room temperature and elevated temperatures via simulation (Wu and Hu 2008) and experimental methods (Lorimer et al. 2003).

11.4.5.1 Mg-Dy Binary Alloys

Rokhlin and Nikitina (2001) studied the aging behavior and recovery characteristics of Mg-23.6% Dy alloy. The hardness at solution state was 80.8 HB and this value reached its maximum of 120 HB after aging at 200°C for 100 h. The recovery was observed at the peak-aged specimens followed by elevated temperature heat treatment. The hardness decreased to 81.7 HB after peak-aged specimens were treated at 300°C for 0.5 h. Elevated temperature heat treatment caused the dissolution of Dy in the solid Mg and resulted in the decrease of hardness.

According to Yang's work (L. Yang et al. 2011b, 2012), the tensile yield strength (TYS) of as-cast Mg-Dy alloys is enhanced by the increment of Dy addition (varies from 5 to 20 wt.%) with the maximum value of 112 MPa for as-cast Mg-20Dy. Both the compressive yield strength (CYS) and TYS are enhanced with the increase of Dy content, and those values more or less decrease after solution treatment. In the as-cast condition, Mg-5Dy shows the lowest ultimate tensile strength (UTS) with a value of about 77 MPa. The highest UTS only reaches about 140 MPa for the Mg-20Dy. The TYS after solution shows the same trend compared to the as-cast condition but is lower than its counterparts. Little change is observed for the UTS of Mg-xDy (x = 5, 15, 20) except for the remarkable decrease in Mg-10Dy. The highest TYS is about 112 MPa, and the highest UTS is about 142 MPa in all the as-cast or solution-treated alloys; in addition, no elongation higher than 6% is obtained. Various aging treatments were taken to improve the mechanical properties of binary Mg-Dy alloys. The highest TYS and UTS are 167 MPa and 219 MPa after aging at 200°C for 168 h, respectively.

11.4.5.2 Mg-Dy-Based Alloys

As mentioned, Dy in Mg can enhance the Mg matrix via solid solution strengthening and precipitation strengthening, but the strengthening effect cannot satisfy the requirements of practical use. A lot of other alloying elements were selected to add into the Mg-Dy alloys to solve this problem, and then various Mg-Dy ternary and multielementary alloys have been developed. Table 11.8 lists some representative mechanical properties of the Mg-Dy ternary and multielementary alloys for structural applications. For Mg-2Dy-0.5Zn alloys, the mechanical properties are much higher than the Mg-Dy binary alloys. The TYS and YTS can reach 287 MPa and 321 MPa with the elongation of >11% at peak-aged condition. TYS and UTS decrease only slightly, and the elongation increases slightly with the increasing of tested temperatures below 200°C both for the as-extruded and peak-aged conditions. At the temperature range of 200–300°C, TYS and UTS of extruded Mg-2Dy-0.5Zn drop rapidly, and the corresponding elongation rises rapidly with increasing temperatures. The peak-aged Mg-2Dy-0.5Zn alloy still has a relatively high strength even at 300°C with the TYS and UTS of 245 MPa and 260 MPa, respectively. Elongation of the peak-aged Mg-2Dy-0.5Zn alloy has a dramatic increase at 300°C, and it reaches 36%. This excellent elevated temperature strength of the peak-aged alloy is attributed to the LPSO phase and the grain refinement strengthening. Wrought $Mg_{97}Zn_1Dy_2$ (at.%) fabricated by Kawamura et al. also exhibited excellent mechanical properties (Kawamura and Yamasaki 2007). The yield strength, ultimate tensile strength, and elongation of the wrought $Mg_{97}Zn_1Dy_2$ alloy are more than 342 MPa, 372 MPa, and 3%, respectively. The wrought $Mg_{97}Zn_1Dy_2$ alloy exhibits a yield strength above 292 MPa, ultimate tensile strength above 322 MPa, and elongation of >4% at 200°C. It is clear that the mechanical properties of wrought Mg-Zn-Dy alloys display high strength and good ductility both at ambient and elevated temperatures.

For the Mg-Dy quaternary alloys, there seems to be a trend that the strength of the peak-aged condition is always higher than the as-cast or solution-treated condition, and the stability of the alloys is also better than the other two conditions. This can be seen from the change of TYS and UTS toward the elevated temperature in Table 11.8 and acts as a guide for which heat treatment condition the alloy would serve better.

Although some Mg-Dy-based alloys were developed and exhibited excellent mechanical properties, the majority of previously studied Mg-Dy-based alloys were intended for structural uses. Quite a few were aimed at biomedical applications,

TABLE 11.8
Mechanical Properties of Mg-Dy Ternary and Multielementary Alloys

Alloy	Condition	Phase Constitution	Grain Size (μm)	Strain Rate	TYS (MPa)	UTS (MPa)	Elongation (%)
Mg-2Dy-0.52Zn (at.%) (Bi et al. 2012)	As-extruded	α + 14H LPSO	~2	1×10^{-3} s^{-1}	262 (RT)	320	11.8
					249 (100°C)	310	12
	Peak-aged	α + 14H LPSO	~3		230 (200°C)	282	14
					182 (300°C)	227	42
					287 (RT)	321	11.6
					250 (100°C)	316	13
					248 (200°C)	283	17
					245 (300°C)	260	36
Mg-6.5Gd-2.5Dy-1.8Zn (Zhang et al. 2011)	As-cast	α + Mg$_3$ (Gd, Dy, Zn) + 14H LPSO	60–80	8.33×10^{-4} s^{-1}	145 (RT)	276	10.8
					130 (200°C)	240	33.8
	Peak-aged	α + Mg$_3$ (Gd, Dy, Zn)	>100		295 (RT)	392	6.1
					152 (200°C)	247	12.0
Mg-8.31Gd-1.12Dy-0.38Zr (Peng et al. 2008)	As-cast	α + Mg$_5$RE	60	1.7×10^{-3} s^{-1}	131 (RT)	210	5.7
					116 (250°C)	187	7.9
	T4	α + Mg$_5$RE + Mg$_{15}$RE$_3$	100		135 (RT)	226	6.9
					118 (250°C)	191	8.2
	T6	α + Mg$_5$RE + Mg$_{15}$RE$_3$	100		261 (RT)	355	3.8
					174 (250°C)	230	7.4
ZK60-4.32Dy (Z. H. Huang et al. 2013)	As-extruded	α + Mg-Zn-Dy phase	1	2 mm/min	355 (RT)	395	>12
Mg-3.5Dy-4.0Gd-3.1Nd-0.4Zr (De-Hui et al. 2006)	Peak-aged	/	/	1 mm/min	/	298.25 (RT)	3.31
		/	/		/	285.23 (200°C)	11.6

including binary Mg-Dy alloys and quaternary Mg-Dy-Gd-Zr alloys, which are discussed in more detail in the next section.

11.4.6 Corrosion Behavior of Biomedical Mg-Dy-Based Alloys

11.4.6.1 Corrosion of Binary Mg-Dy Alloys

For as-cast binary Mg-Dy alloys, the total amount of the second phase increased, filiform corrosion was reduced, and pitting corrosion was increased with the increment of Dy from 5 wt.% to 20 wt.% (see Figure 11.40). L. Yang et al. (2011a) observed the corrosion process of as-cast Mg-Dy alloys in 0.9 wt.% NaCl solution. After removing the corrosion products, Mg-5Dy showed serious filiform corrosion, but Mg-20Dy showed a severe pitting corrosion after being immersed for 72 h. Corrosion morphology of the initial stage revealed that the preliminary corrosion mechanism of Mg-5Dy was pitting, and this corrosion mechanism gradually changed into filiform corrosion with prolonged immersion time. Here, it can be inferred that when the as-cast Mg-xDy alloys were initially immersed in saline solution, the corrosion mechanism was pitting, and then the corrosion mechanism differed with increased immersion time of different Dy contents. The corrosion rate calculated from the

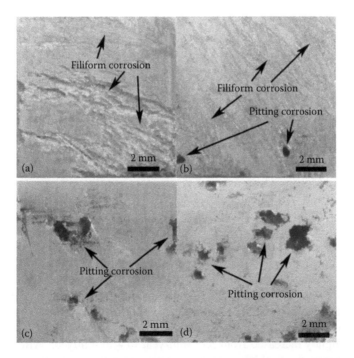

FIGURE 11.40 Typical corrosion morphology of as-cast Mg-Dy alloys after immersion in 0.9 wt.% NaCl solution at room temperature for 72 h (corrosion products have been removed). (a) Mg-5Dy alloy; (b) Mg-10Dy alloy; (c) Mg-15Dy alloy; and (d) Mg-20Dy alloy. (From Yang, L., Y. Huang, Q. Peng, F. Feyerabend, K. U. Kainer, R. Willumeit, and N. Hort, *Materials Science and Engineering: B*, 176 (20), 1827–1834, 2011. With permission.)

evolved hydrogen showed that the Mg-5Dy alloy exhibited the highest corrosion rate, and the Mg-10Dy alloy exhibited the lowest corrosion rate (3 mm/year at 72 h). The trend of corrosion rate for all the alloys tended to increase with increased immersion time.

For the solution-treated Mg-10Dy alloy, L. Yang et al. (2013a) carried out the corrosion experiment under sterile conditions in a cell culture medium (CCM) consisting of Dulbecco's modified Eagle's medium (DMEM) and 10% fetal bovine serum (FBS). The surface of the tested samples were covered by a corrosion layer, and a large amount of cracks formed on the samples due to the dehydration during drying. The corrosion morphology of the solution-treated Mg-10Dy alloy remained uniform even after 14 days of immersion, and no localized corrosion was found after removing the corrosion products. Corrosion morphology showed that the Mg-10Dy alloy suffered from a uniform corrosion. After immersion for 3 days, the corrosion rate of Mg-10Dy was about 1 mm/year, and this value dropped to 0.56 mm/year after 14 days. The corrosion rate of solution-treated Mg-10Dy alloy decreased with immersion time, inversely to the variation of the as-cast Mg-10Dy alloy.

In the solution-treated condition, the corrosion medium was cell culture medium, not identical to the 0.9% NaCl solution in the as-cast state. It is hard to judge whether the heat-treated condition or the corrosion medium resulted in the different variation of corrosion rate toward immersion time or they both contributed.

Yang et al. (2013a) also studied the element distribution and phase constitution in the corrosion layer of Mg-10Dy. Dy was enriched on the surface of the corrosion layer; it was about seven times higher than that contained in the substrate. EDX analysis showed that N and C mainly distributed on the surface of the corrosion layer. The amount of P and Ca in the surface layer was higher than that near the bottom of the corrosion layer. The O content was higher at the interface between the corrosion layer and the substrate. Combined EDX, X-ray photoelectron spectroscopy (XPS), and XRD analysis revealed that the corrosion layers were likely composed of Dy_2O_3, $Dy(OH)_3$, and $MgCO_3 \cdot 3H_2O$ phases.

In Yang et al.'s work (2012), half of the Mg-xDy (x = 10, 15, 20) specimens were solution-treated and aged at 250°C for 16 h (T6-1). Another half were aged at 200°C for 168 h (T6-2). Specimens aged at 200°C apparently showed age hardening behavior; however, specimens aged at 250°C showed less age hardening behavior. After immersion for 3 days in 0.9 wt.% NaCl solution at ambient temperatures and compared with the solid-solution condition (T4), the corrosion rate slightly decreased for Mg-10Dy and Mg-15Dy alloys after T6-1 treatment. It decreased by around 28% and 25% from 1.34 mm/year and 1.37 mm/year to 0.97 mm/year and 1.03 mm/year for Mg-10Dy and Mg-15Dy, respectively. However, the corrosion rate of T6-1-treated Mg-20Dy slightly increased from 3.13 mm/year to 4.16 mm/year. After T6-2 treatment, the corrosion rate of Mg-10Dy alloy remained at the same level in comparison with that in the T4 condition; however, it increased significantly for the Mg-15Dy and Mg-20Dy alloys. Especially for Mg-20Dy, the corrosion rate increased from 3.13 mm/year to 21.53 mm/year, which was about seven times that in the T4 condition.

In the corrosion medium of CCM under cell culture condition for 14 days, there was little change in the corrosion rate for all the alloys after T6-1 treatment compared to the T4 condition. After T6-2 treatment, the corrosion rate remained the same for the Mg-20Dy alloy, and it was even reduced for Mg-10Dy and Mg-15Dy alloys. The corrosion of Mg-10Dy was reduced from 0.75 mm/year to 0.48 mm/year compared with that of the T4 condition.

Macrocorrosion morphology of T6-2-treated Mg-20Dy alloy in 0.9 wt.% NaCl solution after removing the corrosion products showed that it suffered serious pitting corrosion. In MCC medium, the corrosion morphology was similar to that of the T4 condition, which was severe localized corrosion. O, Mg, P, Ca, and Dy elements were detected in the corrosion layer by EDX analysis, and Dy was enriched in the corrosion products.

Both T6-1 and T6-2 were aging treatments, but specimens after T6-1 and T6-2 treatments showed different corrosion rates even in the same corrosion medium. The age hardening response of T6-2-treated Mg-20Dy was much more obvious than the T6-1-treated one. Clearly, the precipitates in T6-2-treated Mg-20Dy were more than that of the T6-1-treated one. With more precipitates after aging, the galvanic couples in the T6-2-treated Mg-20Dy were far more than that in the T6-1-treated one, and this resulted in the differences in corrosion. In short, the amount, size, and morphology of the precipitated phases in the matrix after aging dramatically affected the corrosion behavior of the aged Mg-Dy binary alloy.

11.4.6.2 Corrosion of Mg-Dy-Zn Alloys with Different LPSO Phases

LPSO phases in Mg-Dy alloys are mostly divided into 18R-type and 14H-type, and they can be obtained by different heat treatments. Currently, the effect of LPSO phases on corrosion behavior of Mg alloys is still unclear. From the standpoint of electrochemistry, the LPSO phases act as a heterogeneous spot, and then the galvanic corrosion might be accelerated. However, recent study showed that corrosion resistance of an as-extruded Mg-Gd-Zn-Zr alloy in Hank's solution was enhanced owing to the presence of 14H-LPSO phase (X. B. Zhang et al. 2012d). For the effect of the 18R-LPSO phase toward corrosion, X. Zhao et al. (2013b) also verified that the corrosion resistance of the Mg-Zn-Y alloy-containing 18R-LPSO phase in simulated body fluid was better than conventional engineering Mg alloys.

Peng et al. (2014b) found that LPSO phases played an essential role in the corrosion of biomedical Mg-Dy-based alloys, and those phases had a direct effect on the mechanical and corrosion properties of the alloys. In 0.9 wt.% NaCl solution, it was demonstrated that a fine metastable 14H-LPSO phase in the grain interior was more effective at improving resistance due to the presence of a homogeneous oxidation film and the rapid film remediation ability. It is clear that the corrosion mechanisms are different between alloys with 14H-LPSO phase and 18H-LPSO phase as illustrated in Figure 11.41. Two main factors should be responsible for the decreased corrosion rate. One is whether the exposed part can form a homogeneous and compact oxidation film at the initial corrosion stage. The other is the

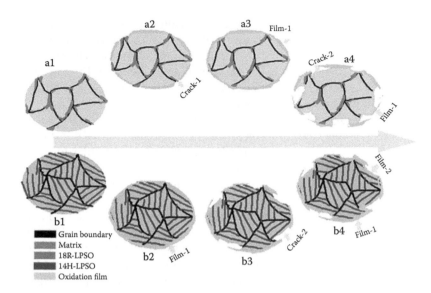

FIGURE 11.41 Schematic diagrams of corrosion process, a1–a4 graphs correspond to alloys with 18H-LPSO structure; b1–b4 graphs correspond to alloys with 14H-LPSO structure. (From Peng, Q., J. Guo, H. Fu, X. Cai, Y. Wang, B. Liu, and Z. Xu, *Scientific Reports*, 4, 2014. With permission.)

remediation ability, that is, when the protective corrosion is broken, whether it is easily self-repaired.

11.4.6.3 Corrosion of Biomedical Quaternary Mg-Dy-Gd-Zr Alloys

L. Yang et al. (2013b) developed a new biomedical quaternary Mg-xDy-(10-x) Gd-Zr alloy based on their previous study of binary Mg-Dy alloys. The corrosion of Mg-Dy-Gd-Zr alloys were performed in CCM under cell culture conditions with a ratio of 1.5 ml/cm². Weight loss was used to calculate the corrosion rate. In the as-cast condition, D10K and DG82K had similar corrosion rates of ~0.55 mm/year. The corrosion rate increased to 0.7 and 0.85 mm/year when the Gd content increased to 5 and 8 wt.% in the DG55K and DG28K alloys. After T4 treatment, the corrosion rate of D10K and DG82K changed little, and it reduced to ~0.5 mm/year for the DG55K and DG28K alloys. All the alloys have similar corrosion rates in the T4 condition. Aging treatment had no adverse effect on the corrosion rate or even reduced it slightly. All the samples in the T4 and T6 conditions showed no localized corrosion, and the corrosion is quite uniform.

11.4.7 Cytotoxicity of Biomedical Mg-Dy Alloys

L. Yang et al. (2013a) calculated the released ion concentration in the extract of solution-treated Mg-10Dy based on the corrosion data after being incubated under cell culture conditions for 72 h and found the concentrations of released Mg^{2+} and Dy^{3+} in the extracts were about 37.5 and 0.62 mM/L. Then the relative cell viability in the indirect cell assay was applied to evaluate the in vitro cytotoxicity of the Mg-10Dy alloy. Osteoblast cells were cultured in Mg-10Dy extracts with different concentrations for 2 days and 6 days. Similar cell viabilities were observed for the 25% and 50% extracts and 25%, 50%, and 75% extracts showed no cytotoxicity to the osteoblasts during the whole culture period. The cell viability of 100% extract reduced to ~80% at 6 days. There was a decrease in the cell viability accompanying with the increasing of the extract concentration, but only 100% extract showed slight cytotoxicity.

Cell morphology on the surface of the specimens revealed that a large number of cells were observed, and the cells were well spread after 3 days of culture. After being cultured for 7 days, no negative effect on cells was observed. The live/dead staining method was also used to further examine the influence of the corrosion layer on cell viability. Before cells were seeded on the surface of the specimens, the specimens were immersed in CCM for 3 days to form a stable corrosion layer. A large number of live cells and only a few dead cells were observed on Mg-10Dy after culture for 3 days. Even after 7 days, the trend remained the same as indicated by Figure 11.42. The quantification of cell viability was about 91.6% and 93% at 3 days and 7 days, respectively. These results confirmed that the corrosion layer had no cytotoxicity to osteoblast cells.

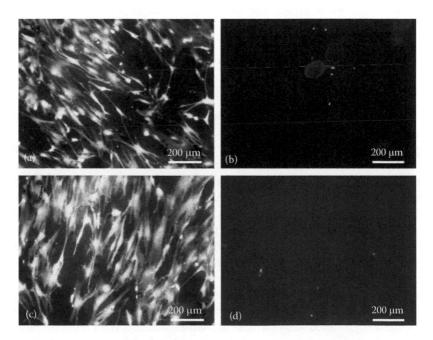

FIGURE 11.42 Live (left) and dead (right) cells on the surface of specimens after culture for 7 days: (a and b) pure Mg; (c and d) Mg–10Dy alloy. The images of live and dead cells were taken from the same location. (From Yang, L., N. Hort, D. Laipple, D. Höche, Y. Huang, K. U. Kainer, and F. Feyerabend, *Acta Biomaterialia*, 9 (10), 8475–8487, 2013. With permission.)

11.4.8 CONCLUSIONS ON MG-DY-BASED ALLOYS

Dy can contribute a lot to the strength and aging behavior of Mg alloys. Addition of REEs into Mg is an effective way to improve the mechanical properties and, in proper content, it can also improve the corrosion resistance in various simulated body fluids. Elevated temperature performance of industrial Mg alloys can also be enhanced by Dy addition. In addition, Dy shows good cytocompatibility according to the in vitro study of cytotoxicity and inflammatory response. Considered the mechanical compatibility, corrosion compatibility, and biocompatibility, Dy seems to be a suitable alloying element in biomedical Mg alloys.

Fabrication methods of biomedical Mg-Dy and Mg-Dy-based alloys are similar to those of industrial Mg-Dy alloys, but a high quality of raw materials is required. Heat treatment of Mg-Dy alloys should also be well designed because heat treatment affects the microstructure, and it indirectly affects the mechanical properties and corrosion. An excellent combination of mechanical properties and corrosion resistance can ensure safe and reliable service of biodegradable implants.

Currently, several works on biomedical Mg-Dy and Mg-Dy-based alloys only focus on the mechanical properties, in vitro corrosion behavior, and cytotoxicity evaluations. A preliminary study on biomedical Mg-Dy and Mg-Dy-based alloys have just come to the forefront, and further study, such as gene expression and animal experiments, are needed.

11.5 OTHER RARE EARTH METALS IN MG ALLOYS

11.5.1 MG-GD-BASED ALLOY SYSTEM

The Mg-Gd system is one of the most promising ultra-high-strength Mg systems in engineering applications due to its remarkable age-hardening response and very good thermal stability. A recent study on Mg-Gd systems has shown that this alloy system has excellent strength and toughness, and some alloys even show ultra high strength compared to other Mg alloys (Yamasaki et al. 2005; Yamada et al. 2006; K. Zhang et al. 2008; C. Xu et al. 2012a, 2013a,b; K. Liu et al. 2013). A Mg-8.2Gd-3.8Y-1.0Zn-0.4Zr alloy sheet exhibits excellent tensile properties at ambient temperature with an ultimate tensile strength of 517 MPa, 0.2% proof stress of 426 MPa, and elongation to failure of 4.5% (Xu et al. 2012b). Because of the excellent mechanical properties of Mg-Gd-based alloys, some alloys in the Mg-Gd system are brought into the biomedical field.

A few reports about Mg-Gd-based biomedical alloys have come into our sights, including Mg-Gd binary alloys, Mg-Gd-Zn ternary alloys, and some Mg-Gd-Zn-Zr alloys. Currently, limited study on those Mg-Gd-based alloys is all about the mechanical properties and in vitro corrosion behavior, and no further biocompatibility study can be found. However, this system has shown a promising future. The mechanical properties of Mg-Gd alloys can be adjusted over a wide range, and this makes them promising for the future design of various degradable metallic implants (Hort et al. 2010). High-strength Mg-11.3Gd-2.5Zn-0.7Zr alloys exhibit uniform corrosion in

Hank's solution with a corrosion rate of 0.17 mm/y, and it shows grade 1 cell toxicity (X. B. Zhang et al. 2012d).

Compatibility of elemental Gd is still in doubt. Some authors state that Gd is highly toxic; the acute toxicity is only moderate. The intraperitoneal LD_{50} dose of $GdCl_3$ was 550 mg/kg in mice, and $GdNO_3$ induced acute toxicity at the concentration of 300 mg/kg in mice and 230 mg/kg in rats, respectively (Haley et al. 1961; Bruce et al. 1963). At the same time, Gd-based contrast agents are widely used as the contrast media in magnetic resonance imaging (Hifumi et al. 2006, 2009). Tests regarding the cytotoxicity of gadolinium chlorides in osteoblast-like cells showed that it could be a suitable element in biomedical Mg alloys (Feyerabend et al. 2010).

The excellent mechanical properties of Mg-Gd systems are the driving force to develop high-strength-toughness Mg-Gd-based degradable alloys, and if possible, this alloy system can be used in orthopedic applications bearing load.

11.5.2 MG-LA AND MG-CE BINARY ALLOYS

REEs are usually added to Mg alloys as mischmetal of various compositions, and these mischmetals are commonly rich in Ce, La, and Nd. In our recent work (Willbold et al. 2015), we systematically investigated the microstructures, in vitro corrosion behavior, cytotoxicity, and in vivo biocompatibility of binary Mg-La and Mg-Ce alloys with the commonly used RE (La, Ce) concentrations.

The two alloys and their corrosion products caused no systemic or local cytotoxicologic effects, clinically and histologically. In vivo degradation quantified by microtomography showed low corrosion rates. Low concentrations of rare earth elements had no influence on bone growth inside a 750-µm-broad area around the implant. However, increased bone growth was observed at further distances from the degrading alloys. Typical histological images are presented in Figure 11.43.

11.5.3 OTHER MG-REE ALLOY SYSTEMS

Except for Y, Nd, Dy, and Gd, other rare earth metals are rarely reported in biomedical Mg alloys. However, some of them are widely used in the industrial/ engineering field. Proper amounts of Sc, La, Ce, Pr, Sm, more or less contribute to the mechanical properties (Fan et al. 2006; Wang and Chou 2008; Xiao et al. 2008; K. Li et al. 2009; Son et al. 2009; J. Zhang et al. 2009c,d; M. L. Zhang et al. 2009; Cui et al. 2010; Pan et al. 2011a,b; Sha et al. 2011; Tamura et al. 2011; Wang et al. 2011b; Ma et al. 2013; Su et al. 2013), corrosion resistance (Song et al. 2007b; Takenaka et al. 2007; J. Zhang et al. 2008b; W. Liu et al. 2009; Q. Li et al. 2010; Zhang et al. 2011c; N. Ma et al. 2012), creep resistance (Mordike 2002; Mordike et al. 2005; Wei et al. 2009; M.-B. Yang et al. 2009; J. Zheng et al. 2010a,b; Bai et al. 2012a), thermal stability (Li et al. 2009; J. Zhang et al. 2010a) in various Mg alloys. Normally, they are used in the form of mischmetals for the sake of cost. La is also widely used in the fabrication of various bulk metallic glasses (González et al. 2009a,b) and hydrogen storage materials (Ouyang et al.

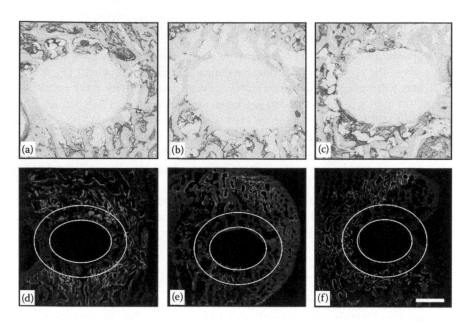

FIGURE 11.43   Reaction of the bone tissue in a 750-μm broad area around the drill hole as shown by Toluidin Blue staining and Xylenol Orange fluorescence around MgLa (a and d), MgNd (b and e), and MgCe (c and f) implants. Bar = 750 μm (a, b, and c); 1500 μm (d, e, and f). (From Willbold, E., X. Gu, D. Albert, K. Kalla, K. Bobe, M. Brauneis, C. Janning, J. Nellesen, W. Czayka, W. Tillmann, Y. Zheng, and F. Witte, *Acta Biomaterialia*, 11, 554–562, 2015. With permission.)

2006, 2009; Tarasov et al. 2007; Gu et al. 2008; Song et al. 2008; Tanaka 2008; D. Wu et al. 2010; Capurso et al. 2013; Y. Fan et al. 2014).

Promethium (Pm) is a special rare earth metal. It is an artificial radioactive element, and no reports have talked about this special element in Mg alloys. At the same time, the author cannot find any other papers working on europium (Eu) and thulium (Tm) in Mg except for their phase diagrams. Only limited studies report on the microstructure, mechanical properties, and corrosion behaviors of Mg-Yb (Dobromyslov et al. 2007; Yamasaki et al. 2007; Yu et al. 2008), and Mg-Tb (Drits et al. 1978; Silonov et al. 1996; Neubert et al. 2007; Čížek et al. 2012; Bryła et al. 2013). Phase diagrams of Mg-Ho, Mg-Er, and Mg-Lu are similar to each other; in addition, Er and Ho additions have been reported to significantly improve the corrosion resistance of Mg-Al and Mg-Al-Zn alloys, respectively (Rosalbino et al. 2005; Zhou et al. 2006).

The roles of REEs in Mg are quite unique. A small addition of REEs can even significantly influence the properties of Mg alloys. Feyerabend et al. (2010) inferred that suitable elements in biomedical with low solid solubility could be Eu and Pr. If not avoidable, La and Ce should be used cautiously. However, the underlying damage that REEs can cause to the human body remains controversial. Whether the remaining REEs are suitable alloying elements in biomedical Mg alloys requires further research.

11.6 CONCLUSIONS ON Mg-RE-BASED BIOMEDICAL Mg ALLOYS

In this chapter, we have reviewed the current research progress on the Mg-RE-based biomedical Mg alloys, and we have discussed their advantages and disadvantages. Major work has been done on the Mg-Y-based biomedical Mg alloys, especially in the WE series (Mg-Y-RE) Mg alloys. The Mg-Nd system includes another popular biomedical Mg-RE Mg alloy that has attracted wide attention. Some researchers also did some exploratory research on the Mg-Dy-based alloys. Recently, Mg-Gd-based biomedical Mg alloys have just emerged into view.

The mechanical properties of Mg-RE-based biomedical Mg alloys can preliminarily satisfy the basic requirements of intravascular stents. No cytotoxicity to mild cytotoxicity of Mg-RE alloys have been proven by a large amount of in vitro and in vivo experiments, indicating that the cytocompatibility of various Mg-RE-based alloys is acceptable. The serious problem remaining is the controlling of degradation of Mg-RE-based alloys. Only in the case of treating an occluded left pulmonary artery in a preterm baby, have the mechanical and degradation characteristics of AMS-1 been proven to be adequate to secure reperfusion of the occluded left pulmonary artery. The residual clinical data all reflect that degradation of Mg-RE-based stents is still too fast to allow the complete healing of the tissues before disintegration. Future work should vitally focus on how to improve the corrosion resistance and degradation behavior of Mg-RE-based stents.

Putting the AMS aside, even though the mechanical properties of Mg-RE-based alloys are much better than many other biodegradable Mg alloys, if we want to apply this kind of alloy in the orthopedic field, improved mechanical properties, such as tensile strength and fatigue resistance are also needed. The stress corrosion cracking fracture is another problem that needs addressing. Compared to other commercial orthopedic implants, maybe the first step toward Mg-RE-based alloys in the orthopedic field should concentrate upon the implants bearing no or light load.

Basically, the toxicity of rare earth elements remains in debate. The clear metabolic pathways, safe dose, organ toxicity, etc., are still unknown. At the same time as developing novel Mg-RE-based alloys, those basic questions should be well studied and made clear. We still want to stress, though, that, even though REEs in biomedical Mg alloys do not show observable adverse effects on animals or on human beings, the long-term safety of REEs still needs to be clarified.

Among so many biomedical Mg alloys, Mg-Y-based Mg alloys are the first type that have been fabricated into practical devices and tested in the human body and even in the treatment of some typical diseases. Those kinds of Mg alloys have attracted attention around the world. Some biomedical manufacturers are devoted to the exploitation of new devices of Mg-Y-based alloys.

The absorbable Mg stents (AMS) have already been developed to the third generation-DREAMS 2G. The first generation of AMS is a bare metallic stent. Late lumen loss because of fast degradation is the major problem of AMS-1. Modified AMS with paclitaxel (DREAMS 1G) significantly decreased the late lumen loss compared to the bare stent AMS-1, but DREAMS 1G still cannot match the excellent results of currently available drug-eluting stents. That is the key reason why DREAMS 1G has not been approved in Europe. DREAMS 2G, the second generation of drug eluting bioabsorbable Mg stents, has been optimized in structural design and drug loading, hoping to further reduce neointima formation and degradation rate. Results from a porcine coronary model have confirmed the improvement in degradation period. Now, DREAMS 2G has completed a preclinical assessment, and existing experimental data suggest that it may be able to compete with the drug-eluting metallic stents in terms of safety and efficacy. DREAMS 2G is currently under BIOSOLVE-II trial. We hope the DREAMS 2G can bring us good news in the near future.

Despite hundreds of studies devoted to Mg-based biodegradable alloys, almost no commercial implants are approved and available at present. Recently, the most exciting progress in the biodegradable metals field is that Mg-based MgYREZr screws, chemically similar to WE43, have recently received the CE marking of medical devices for medical applications within Europe because of their positive clinical results. This is the first one to be approved. Researchers around the world will be inspired and a bright future for biodegradable implants based on Mg-RE alloys can be expected.

12 Fabrication and Processing of Biodegradable Mg and Its Alloys

From Raw Materials to Final Medical Devices

12.1 INTRODUCTION

All steps are inseparable from fabricating the materials, processing them into semi-products, designing devices, and manufacturing to the performance of the device in its final form as an implant. In this chapter, we will discuss the fabrication process of biodegradable Mg and its alloys from raw materials to the final medical devices, which is schematically illustrated in Figure 12.1. To produce the raw material forms, such as ingots, magnesium and its alloys are vacuum melted, alloyed, refined, and then cast. Then the castings are further processed into semiproducts such as plates, rods, wires, and tubes through hot or cold working. Subsequent manufacturing processes, such as machining, cutting, finishing operation, joining, sterilization, and packaging may be applied as needed to produce the final medical devices. Due to the high activity of Mg alloys and their limited ductility and poor workability at room temperature, resulting from their hexagonal crystal lattice, the processing of biodegradable Mg alloy devices is quite different from other bioinert metallic materials, such as stainless steel and Ti alloys.

12.2 FABRICATION OF BIODEGRADABLE Mg ALLOY RAW MATERIALS

Mg alloy melts must be protected against oxidation and burning during melting, alloying, and casting. Fluxes or fluxless melting techniques can be used, depending on alloy compositions and processing conditions.

The flux is comprised of a mixture of chlorides and fluorides, such as $MgCl_2$, KCl, and CaF_2. The mixture melts below the melting point of the alloys so that they spread out and protect the solid metal from oxidation prior to melting. The other function of the flux is to remove oxides and nonmetallic inclusions from the melt, that is, refining. The flux with $MgCl_2$ as the main component can be used for the

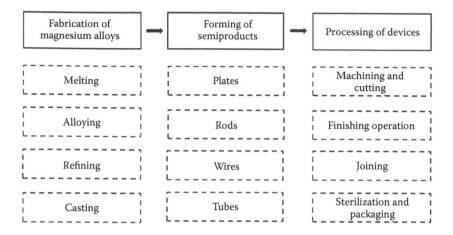

FIGURE 12.1 Flowchart of fabrication process of biodegradable Mg alloys from raw materials to devices.

melting of Mg-Al-Zn, Mg-Mn, and Mg-Zn-Zr alloys. For Mg alloys that contain Ca, Sr, La, Ce, and Nd, flux without $MgCl_2$ should be used. $MgCl_2$ can react with these elements and form inclusions.

Flux inclusions in the castings are not uncommon, which creates a major hindrance to the development of Mg. The invention of a fluxless process is a significant breakthrough. Mixtures of CO_2, CO_2 and air, or air alone combined with 0.3% to 0.5% SF_6 have been successfully used for melt protection in casting operations, thus essentially eliminating problems due to flux inclusions. Due to the global warming potential (GWP) of SF_6, which is 24,000 times that of CO_2, alternative protective gases are currently under development. Several potential substitutes, including sulfuryl fluoride (SO_2F_2), boron trifluoride (BF_3), HFC-134a, hydrofluoroethers (e.g., HFE-7100 and HFE-7200), and a fluorinated ketone (Novec™ 612) have been investigated (Bartos et al. 2007).

Alloys can be produced directly from virgin materials but many users prefer to use a prealloyed ingot as a starting material as control of composition is simpler. The alloying parameters are summarized in Table 12.1.

Mg alloy castings can be produced by nearly all of the conventional casting methods, namely, gravity and low-pressure casting (which include sand casting, permanent mold casting, semipermanent mold casting, and investment mold casting), and die-casting. The choice of a casting method for a particular part depends upon factors such as the configuration of the proposed design, the application, the properties required, the total number of castings required, and the properties of the alloy. Currently investigated biodegradable Mg alloy castings are usually produced by gravity and low-pressure casting. Die-casting Mg alloys are mainly Mg-Al-based alloys, which have good mold-filling capacity but are not suitable for biomedical applications.

In addition, novel fabrication technologies are applied to Mg alloys. For example, rapid solidification (RS) processing can result in the refinement of both matrix

TABLE 12.1

Alloying Parameters of Some Mg Alloys

Element	Addition Form	Protection Method	Addition Temperature (°C)	Recovery (%)	Remarks
Al	Pure metal	Flux or fluxless	680–750	70–90	
Zn	Pure metal	Flux or fluxless	680–750	70–90	
Si	Pure material	Flux or fluxless	700–750 (flux); 800–850 (fluxless)	70–90	Requires longer time and agitation; should be added prior to Mn
Mn	Anhydrous MnCl₂ flake or electrolytic Mn flake or briquettes	Flux	700–730 (MnCl₂) 750–800 (Mn)	35–90	HCl emissions during the reagent addition
RE	Lumps of mischmetal or pure metal	Fluxless	600–900	90–98	
Ca	Master alloy	Fluxless	680–720	80–90	Dissolution or mixing time of 15–30 min after the addition is recommended
Sr	Master alloy	Fluxless	675–700	85–95	
Li	Pure metal	Flux and/or fluxless	680–700		Add Li last if there are other alloying elements
Zr	Hardener	Flux			
Ag	Pure metal			100	
Y	Hardener	Inert gas	~800	92–95	

Source: Friedrich, H. E., and B. L. Mordike. Melting, alloying and refining. In *Magnesium Technology*, eds. Friedrich, H. E., and B. L. Mordike, Berlin Heidelberg: Springer, 2006a.

grains and intermetallic particles, extension of solid solubility, and improved chemical homogeneity with a consequent suppression of internal galvanic corrosion and thus enhancement of strength and corrosion resistance. Mg-3Ca alloy ribbons prepared by the melt-spinning technique have a very fine grain size, 200–500 nm, compared to the coarse grain size, 50–100 μm, of the as-cast Mg-3Ca alloy sample (Gu et al. 2010a) as shown in Figure 12.2. Some nanoscale precipitates were observed along the grain boundary of the RS45 Mg-3Ca alloy ribbon. The corrosion resistance for the as-spun Mg-3Ca alloy ribbons in SBF is improved compared with the as-cast Mg-3Ca alloy ingot, and the corrosion rates decreased with increasing wheel-rotating speeds. Also, the cytotoxicity evaluation revealed that the three experimental as-spun Mg-3Ca alloy ribbon extracts do not induce toxicity to the

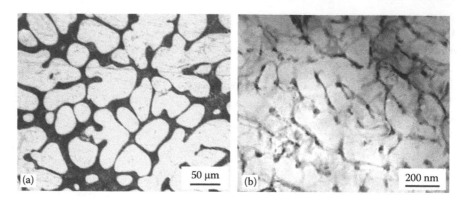

FIGURE 12.2 (a) Optical micrograph of the as-cast Mg-3Ca alloy ingot; (b) TEM micrograph of the as-spun RS45 Mg-3Ca alloy ribbon. (From Gu, X. N., Li, X. L., Zhou, W. R., Cheng, Y., and Zheng, Y. F., *Biomedical Materials*, 5, 2010. With permission.)

L-929 cells whereas the as-cast Mg-3Ca alloy ingot extract does (Gu et al. 2010a). J. Wang et al. (2010) found the subrapid solidification Mg-Zn-Y-Nd alloy has better corrosion resistance than the as-cast sample in dynamic SBF due to grain refinement and fine dispersion distribution of the quasicrystals and intermetallic compounds in the α-Mg matrix.

12.3 FORMING OF BIODEGRADABLE Mg ALLOY SEMIPRODUCTS

12.3.1 SHEET PROCESSING OF BIODEGRADABLE MG ALLOYS

Due to the low workability of Mg alloy, the fabrication technique of sheet or plate was once described as "the best-kept secret in the nonferrous industry." Mg plates are commonly manufactured by rolling and more recently by twin-roll strip casting. The rolled products are usually time-consuming and costly because the slabs have to be reheated and annealed multiple times, especially before each pass of 5–20% reduction in the final finish rolling mill. Twin-roll strip casting combines the casting and the forming processes permitting the production of semifinished products with a thickness of approximately 3.5 mm in only one process step (Bian et al. 2009). The fast solidification of the melt between the rolls leads to fine and homogeneously distributed grain sizes and precipitates, which improve the final product properties. However, in the case of products with reduced thicknesses, a further step using the conventional rolling mill after the twin-roll casting process is necessary.

A typical basal texture with the *c*-axis of most grains parallel to the normal sheet is usually formed in the conventionally rolled or extruded Mg alloy sheet. This strong texture induces a poor deformation capability of sheet thinning and a greater anisotropy and consequently results in low ductility at room temperature. Rolling at high temperature, which is about 100°C higher than that of a conventional rolling process (below 400°C) can control the crystal orientation. Wide-width commercial Mg alloy (AZ31 alloy, etc.) sheets with high room-temperature formability are produced by this method (AIST 2010).

There have been extensive efforts to process Mg alloys with weakened and tilted basal texture by advanced technologies, for example, cross-rolling (CR) (X. Li et al. 2011), three-directional rolling (TDR) (H. Zhang et al. 2013a), equal channel angular rolling (ECAR) (Cheng et al. 2007; Hassani et al. 2011), different speed rolling (DSR) (Kim et al. 2005, 2010; Huang et al. 2009), asymmetric extrusion (L. L. Chang et al. 2008; Q. Yang et al. 2013), cyclic extrusion compression (CEC) (Y. J. Chen et al. 2008), friction stir process (FSP) (Darras et al. 2007; H.-W. Lee et al. 2009), repeated unidirectional bending (RUB) (Song et al. 2010), and accumulative roll bonding (ARB) (Zhan et al. 2011). These technologies can effectively improve the mechanical properties and formability of Mg alloy sheets by changing the basal texture.

It also has been found that the addition of rare earth (RE) elements, such as yttrium, cerium, or neodymium, significantly weakens the rolling or extrusion texture of Mg alloys (Bohlen et al. 2007; Hantzsche et al. 2010; Sanjari et al. 2013). The texture-weakening mechanism has been associated with different mechanisms, such as particle-stimulated mechanism (PSN) or deformation or shear bands containing twins or retardation of dynamic recrystallization (DRX). The amount of the RE addition required for sufficient texture weakening is connected with the solid solubility of the respective element. Gd demonstrated the highest potential to modify the sheet texture of rolled and annealed Mg although the light RE elements, Ce, La, Nd, and mischmetal, depicted a common rare earth sheet texture characterized by a weak basal component and a broad scatter of basal poles toward the sheet transverse direction. Elemental Gd is of particular interest because it gave rise to a desired Mg sheet texture despite its coarse grain size, resulting in promising mechanical properties (Al-Samman and Li 2011).

12.3.2 Wire Processing of Biodegradable Mg Alloys

Mg wires with small diameter are needed for resorbable suture applications as well as for knitting of stents. Mg electrode wires or filler wires are also used in the welding of Mg.

Pure Mg (99.95% purity) and AZ31 alloy wires, with the smallest diameter of 0.4 mm, have been fabricated through the extrusion and drawing process (Xu 2006; He 2010; H. F. Sun et al. 2011). With the decrease of the original grain size of the billet, the recrystallization temperature decreases, and the cumulative deformation increases (Xu 2006). The smaller the grain size of the wire, the better the corrosion resistance exhibited (He 2010; H. F. Sun et al. 2011). Chu et al. fabricated AZ31 wires with the smallest diameter of 0.2 mm (as shown in Figure 12.3) and employed MAO and sealing treatment to improve the corrosion resistance (Chu et al. 2012, 2013).

Seitz et al. (2010) produced Mg alloy wires, including ZEK100 (Zn 1 wt.%, RE 0.5 wt.%, and Zr 0.5 wt.%), AX30 (Al 3 wt.% and Ca 0.8 wt.%), AL36 (Li 6 wt.% and Al 3 wt.%), and Mg-0.8Ca (Ca 0.8 wt.% with diameters of 0.5, 0.4, or 0.3 mm) by a single wire-extrusion pass from a pre-extruded bar (30 mm in diameter) at 300–450°C. The extruded wires exhibited dramatically refined grain sizes and improved tensile strengths of 300–400 MPa. Meanwhile, dramatically reduced ductilities were observed for ZEK100 and Mg-0.8Ca wires, but the extruded AX30 and AL36 wires exhibited elongation values close to those wires form extruded profiles (~10%), which were able to form tight knots (Figure 12.4). The diameter of the

FIGURE 12.3   AZ31 wires with a diameter of 0.2 mm. (Courtesy of Prof. Jing Bai, School of Materials Science and Engineering, Southeast University, Nanjing, China.)

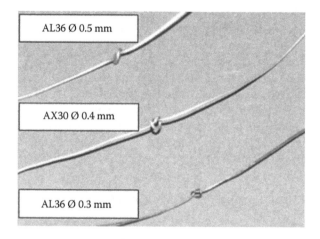

FIGURE 12.4   Knotted wires (AL36 and AX30) with diameters between 0.5 and 0.3 mm. (From Seitz, J. M., E. Wulf, P. Freytag, D. Bormann, and F. W. Bach, *Advanced Engineering Materials*, 12, 1099–105, 2010. With permission.)

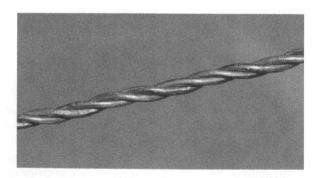

FIGURE 12.5   Stranded wires of alloy ZEK100, each with a diameter of 0.396 mm. (From Seitz, J. M., D. Utermohlen, E. Wulf, C. Klose, and F. W. Bach, *Advanced Engineering Materials*, 13, 1087–95, 2011b. With permission.)

extruded ZEK100, Mg-0.8Ca, and AL36 wires can be further reduced to 0.114 mm by a 14-pass drawing process integrated with a four-pass annealing (200–270°C; 16 h) (Seitz et al. 2011c). Among the monofilaments, ZEK100 exhibits the highest tensile strengths (up to 540 MPa), and Mg-0.8Ca shows the best combination of strength and elongation. These filaments are also twisted into poly-filament suture material using stranding (Figure 12.5), possessing significantly lower tensile strengths but higher fracture strains.

Additionally, amorphous Mg67Zn28Ca5 BMG wires with great surface quality were produced via a melt-extraction setup that was designed in-house (Zberg et al. 2009a). These wires exhibited a tensile strength of 675–894 MPa with a characteristic strength of 817 MPa and a Weibull modulus of 20.6 GPa.

12.3.3 MINITUBE PROCESSING OF BIODEGRADABLE MG ALLOYS

The requirement in stent applications challenges the manufacturing of minitubes made of Mg alloys, which typically have an outside diameter of 2.5–4.0 mm and a wall thickness of only 0.1–0.2 mm. The general fabrication processes of Mg alloy minitubes consist of (1) hot extrusion of cast feedstock to obtain a solid rod, (2) deep drilling of a hole in the rod to obtain a tube, (3) tube indirect extrusion, and (4) multiple drawing and intermediate annealing of the tube to reduce the diameter and wall thickness. Although Mg alloys exhibit better formability at elevated temperatures, hot drawing leads to uncontrollable increases in tube wall thickness and possibilities of unsatisfactory microstructure and surface quality of drawn tubes (Fang et al. 2013). Cold drawing is a precision forming technology for the fabrication of minitubes.

Fang et al. (2013) developed a multipass cold drawing process of Mg minitubes with a moving mandrel to define the inside diameter of the tube. The ZM21 tubes with an outside diameter of 2.9 mm and a wall thickness of 0.217 mm were obtained through five passes of cold drawing with an interpass annealing procedure after the fourth pass, which is shown in Figure 12.6. The interpass annealing at 300°C for 1 h changes a twinned microstructure into a recrystallized grain structure, so that the Mg alloy can be further drawn through the fifth pass with a reduced drawing force to achieve a cumulative area reduction of 37.68%.

Hanada et al. (2013) propose a fixed mandrel drawing method to enable high-accuracy cold drawing of the Mg alloys. Postprocessing is not needed to remove the mandrel unlike the moving mandrel method. Mg-0.8%Ca and AZ61 alloy tubes with 1.5- to 1.8-mm outer diameter and 150-μm thickness were fabricated. They also found the machine oil showed better lubrication performance than mineral oil, boron nitride (BN), molybdenum disulfide (MoS$_2$), and water-based lubricant, making it possible to draw tubes with highly smooth surfaces with an average roughness of 0.24 μm.

Ge et al. (2013) fabricated ultra-fine-grained (UFG) ZM21 tubes by ECAP and low-temperature extrusion. ECAP is performed according to a two-step strategy aimed at achieving a first refining of the structure at 200°C, and then reaching the submicrometer grain size range by lowering the processing temperature down to 150°C. The billets are then extruded into small tubes with 4-mm outer and 2-mm inner diameters. The grain size after extrusion remains in the submicrometer range

FIGURE 12.6 ZM21 minitubes. (From Fang, G., W.-J. Ai, S. Leeflang, J. Duszczyk, and J. Zhou, *Materials Science and Engineering: C*, 33, 3481–8, 2013. With permission.)

owing to low-temperature processing and to the contribution of dynamic recrystallization, which contributes to the improvement of hardness.

12.4 MACHINING AND CUTTING OF BIODEGRADABLE Mg ALLOYS

12.4.1 Machining of Biodegradable Mg Alloys

The machining of functional elements, such as holes and threads for orthopedic implants, is usually required using wire electro discharge machining (EDM), turning, or milling techniques. Klocke et al. (2011) reported the wire electro discharge machining technique for WE43 alloy with ASP23 steel as a comparison. The applied trim cut with sequential decreasing discharge energies could effectively improve the surface integrity and reduce the surface roughness. Due to the lower melting temperature and higher thermal conductivity of the Mg alloy, the gap between the tool electrode and the WE43 alloy work piece is larger than that of ASP23 steel, which should be taken into consideration when setting process parameters.

Guo and Salahshoor (2010) reported safely performed high-speed dry milling for a Mg-0.8Ca alloy using polycrystalline diamond (PCD) tools. Superior surface integrity of low surface roughness, highly compressive residual stress, increased microhardness, and microstructure free of phase transformation can be produced in the presence of a certain amount of flank build-up. In turning the AZ91 alloy, no adhesion on the flank face occurred using carbide tools coated with PCD compared to uncoated and TiN-coated carbide tools (Salahshoor and Guo 2011e). This is related to lower friction between the PCD coating and Mg plus the higher thermal conductivity of PCD, which cause lower heat generation and faster heat conduction from the cutting-edge zone, respectively. Denkena and Lucas (2007) reported that the high residual compressive stresses generated in the subsurface via a deep rolling process after turning reduced the corrosion rate of Mg-3.0Ca alloy by a factor of approximately 100.

The generated heat is one of the main problems for Mg alloy machining, which may lead to ignition of the chips. Several kinds of coolants are used to reduce the surface temperature of the Mg alloy. In contact with water-based coolants, Mg tends to have chemical reactions and form hydrogen, which is extremely flammable. Oil-based lubricants introduce the danger of oil mist explosions. However, Salahshoor and Guo (2013) used dual-purpose oil, serving as both lubricant and coolant during the ball-burnishing process on Mg-0.8Ca alloy, which can effectively reduce the temperature, indicating only a 5–6°C increase in the applied pressure range. Pu et al. (2012a) compared the influence of dry and cryogenic machining (spraying the liquid nitrogen on the machined surface) using different cutting-edge radius tools on the surface integrity of AZ31B alloy, and the cryogenic machining with a large cutting-edge radius could lead to the desired surface finish.

12.4.2 LASER CUTTING OF BIODEGRADABLE MG ALLOYS

Although stent mesh manufacturing can involve various methods, such as photo-chemical etching, water jet cutting, and electric discharge machining, the vast major-ity of the stents are produced via laser beam machining of minitubes (Stoeckel et al. 2002; Demir et al. 2012, 2013). Stent machining requires good accuracy and control of the stent mesh, which should also involve less thermal affection for the integrity of the mechanical properties and high surface quality. The short pulse Nd-YAG laser with higher pulse repetition rate has shown promising results with respect to quality and economic viability (Kathuria 2005). Short pulse width is advantageous in reduc-ing the heat-affected zone.

Demir et al. (2012, 2013) reported the use of a pulsed fiber laser system to cut a novel stent mesh on AZ31 alloy tubes to manufacture biodegradable stents as shown in Figure 12.7. Due to the high reactivity of Mg, the appropriate process gas for Mg alloy tubes is Ar whereas the better machining condition for AISI 316L was with O_2.

12.5 FINISHING OF BIODEGRADABLE Mg ALLOYS

The surface characteristics of biomedical implants determine the nature of immedi-ate and long-term tissue responses. So the finishing process of the surface is impor-tant to increase the quality of products of biomedical implants. Chemical etching with HNO_3 ethanol can clean the deposited spatter on the stent surface and clean the kerf (Demir et al. 2012, 2013). Electropolishing is usually performed to remove burrs and mechanical defects resulting from the heat ablation of the laser cutting, etching, and forming steps and to achieve a smooth product surface. Magnetic abrasive pol-ishing (MAP) has also been proposed with flexible tools, including iron powder and abrasive particles. Not only are the cutting force and generated temperature lower than with other machining processes, but it is also possible to polish the free surface of products. MAP has been verified on AZ31 with an improving strategy of the mag-netic force with a permanent magnet (Kim and Kwak 2008).

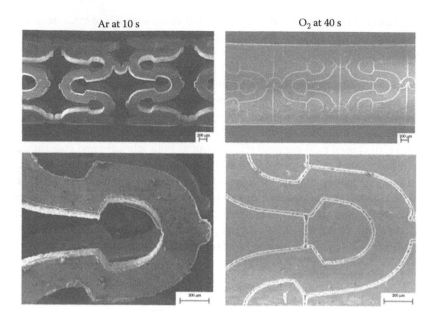

FIGURE 12.7 SEM images of the laser microcut stents using Ar and O_2 after chemical etching. (From Demir, A. G., B. Previtali, and C. A. Biffi, *Advances in Materials Science and Engineering*, 1–11, 2013. With permission.)

12.6 JOINING OF BIODEGRADABLE Mg ALLOYS

Currently investigated Mg-based devices (such as cardiovascular stents, orthopedic implants, and sutures) are quite simple and usually made from a whole piece of Mg alloy. However, some complex medical devices consist of components and materials that must be joined together. A Mg alloy may be joined with itself or other biomaterial using methods similar to those used for other metallic materials, that is, mechanical joining, welding, and adhesive bonding.

12.6.1 MECHANICAL METHODS

Mechanical joining is a process for joining parts through mechanical methods, which clamp or fasten the parts of the assembly together (e.g., nails, screws, bolts, rivets). The advantages of mechanical joining include versatility and ease of use and possible dismantling of the product, making it ideal for cases in which periodic maintenance requires disassembly. The possibility of the joining of dissimilar materials is also an advantage. It allows movement of components relative to one another, for example, by the use of hinges and bearings. Mechanical joints also have disadvantages: The joint is achieved through discrete points, thus not creating a continuous connection between the parts, and the holes employed for mechanical joining are vulnerable to fractures.

In order to minimize the notch effect and improve the resistance to fatigue, the screw threads should be coarse with flat or rounded tops. Studs must be tight fitting, and therefore, the holes to take them should be tapped undersize to obtain a slight interference fit (Magnesium Elektron n.d.). A difference in physicochemical properties of dissimilar joint components creates challenges for mechanically bolted assemblies as well. Due to their very low electronegative potential, Mg alloys are susceptible to galvanic corrosion, thus affecting performance of mechanical joints in conductive environments.

12.6.2 Welding of Biodegradable Mg Alloys

To date, Mg alloys have not usually been welded because of the occurrence of defects, such as oxide films, cracks, and cavities (Cao et al. 2006). However, with wider application of Mg alloys, a predictable, repeatable, consistent, and reliable welding process will be needed.

12.6.2.1 Arc Welding

There are two basic methods of arc welding, that is, inert gas tungsten arc welding (TIG) and inert gas metal arc welding (MIG). TIG-welding is the most commonly used welding for Mg alloys. In the TIG process, an arc is generated between a non-consumable tungsten electrode and the welded metal. The electrode and welded metal are shielded with an inert gas, typically argon. In general, a weld can be made with or without filler. In MIG-welding, the arc is formed between the consumable electrode and the part to be welded. The electrode is continuously provided from the spool. Both the welded area and the arc zone are protected by a gas shield. TIG is recommended for thinner materials, and MIG is recommended for thicker materials; however, there is considerable overlap.

TIG welding has a good weld appearance and high quantity but relatively shallow penetration. There are many methods to improve TIG welding penetration and, consequently, production. One of the most notable examples is the use of activating flux. MgO, CaO, TiO_2, MnO_2, and Cr_2O_3 fluxes all can increase the penetration of TIG welding of the Mg alloy. However, the as-weld fusion zone with oxide flux exhibited a rather larger crystal grain in the weld interior than that without flux.

Weldability varies with Mg alloy type from excellent to limited. Wrought alloys are usually weldable more easily than certain cast alloys. Zinc is a common alloying element in Mg alloys. However, zinc content of more than 1% increases hot shortness, which may cause weld cracking. So high-zinc alloys are not recommended for arc welding. Mg-Zr-RE alloys have good weldability. Mg-Mn alloys also have good weldability and are not subject to stress corrosion.

The choice of electrode wire or filler metal is governed by the composition of the base metal. For Mg alloys, filler rods may be of the same chemistry as the welded part or of a lower melting range. The latter allows the weld to remain liquid until other parts of the weld are solid, thus reducing the probability of cracking.

12.6.2.2 Laser-Beam Welding

Two main types of lasers, CO_2 and Nd:YAG with wavelengths of 10.6 and 1.06 μm, respectively, have been used to weld Mg alloys (Cao et al. 2006). The weldability of Mg alloys is significantly better with the Nd:YAG laser due to its shorter wavelength. Nd:YAG laser beams also have a higher welding efficiency compared with CO_2 lasers.

Laser welding is controlled by many parameters, including power, beam characteristics, welding speed, focal position, gas flow, and material characteristics. Generally, the penetration depth and weld width both increase with increasing laser power and decreasing welding speed. Figures 12.8 and 12.9 show the influence of the laser power and welding speed on the penetration depth, respectively. The welding speed should be adjusted to provide the penetration depth required at a given power level. For the WE43 alloy at speeds less than 2 m/min, no further improvement in the penetration depth but an increase in the width of the bead and the heat-affected

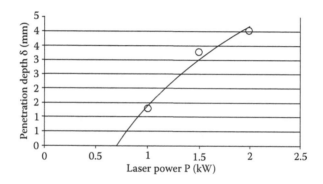

FIGURE 12.8 Influence of the laser power on the penetration depth: WE43, 4 mm thickness; welding speed: $v = 2$ m/min; focal position: $M_f = \pm 1$; gas flow, helium: $\mu = 50$ l/min. (From Dhahri, M., J. E. Masse, J. F. Mathieu, G. Barreau, and M. Autric, *Advanced Engineering Materials*, 3, 504–7, 2001. With permission.)

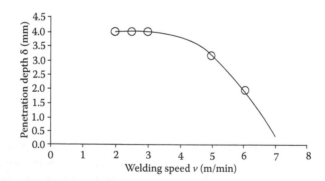

FIGURE 12.9 Influence of the welding speed on the penetration depth: WE43, 4 mm thickness; laser power: $P = 2$ kW; focal position: $M_f = \pm 1$; gas flow, helium: $\mu = 50$ l/min. (From Dhahri, M., J. E. Masse, J. F. Mathieu, G. Barreau, and M. Autric, *Advanced Engineering Materials*, 3, 504–7, 2001. With permission.)

zone can be observed (Dhahri et al. 2001). For welding speeds faster than 3.5 m/min, the width becomes narrow, leading to a lack of penetration depth (Dhahri et al. 2001). For welding of thicker plates, a higher power and a lower speed are necessary (Weisheit et al. 1998). The penetration depth and the weld profile are affected by the variation of the focused spot size and the focal position, respectively. For thin plates (2.5 and 3 mm), an adequate weld can be obtained for a focal position on the surface or less than 1 mm under (Weisheit et al. 1998; Dhahri et al. 2001), whereas for thick plates (5 and 8 mm) a position of 2 mm below the surface of the work piece has proven to be the best (Weisheit et al. 1998).

At present, there are no commercially available filler wires developed specifically for the laser welding of Mg alloys. The small fusion zone during laser welding means that smaller wire diameters should be used at high welding speeds. Due to the hexagonal lattice structure of Mg, commercially available Mg weld wires have rather high production costs and large sizes. Wires of less than 1.6 mm in diameter are rare, and below 1.2 mm, nothing is available at all (Kainer 2003). Therefore, different commercial filler wires with smaller diameters or even with various sectional shapes should be developed for laser welding of various Mg alloys in the future.

12.6.2.3 Electron-Beam Welding

Electron beam welding (EBW) is a method of fusion welding that employs a dense stream of high-velocity electrons to bombard, heat, and melt the materials being joined. The electron beam is generated by an electron gun composed of a cathode made of tungsten and an anode placed in a high vacuum. When the electron beam moves forward, the melted and evaporated alloy flows from the front to the back of the keyhole. It is suitable for difficult welding in which high repeatability is required. Nevertheless, the volatilization of the Mg will contaminate the vacuum chamber. Electron-beam welding under nonvacuum conditions is more suitable for Mg alloys (Bach et al. 2003).

With increasing Al content, the Mg-based materials were found to be more easily fusion welded, and the weld zone aspect ratio became higher (Su et al. 2002). But the $Mg_{17}(Al,Zn)_{12}$ phase containing oxygen can be dissolved and form micropores (Chi et al. 2007). The AZ91 alloy plate with a thickness of 30 mm could be electron beam welded using a power of 2200 W and a weld speed of 16 mm/s, resulting in a weld depth of 29 mm with a fusion-zone aspect ratio of 8.2 (Su et al. 2002).

12.6.2.4 Friction Stir Welding

Friction stir welding (FSW) is a relatively new joining technique that uses a rotating, nonconsumable, cylindrical shouldered tool to deform the surrounding material without melting. Thus, the joint is essentially formed in a solid state, which does not result in solute loss by evaporation or segregation during solidification, resulting in homogeneous distribution of solutes in the weld. Also, many Mg alloys in the cast condition contain porosity, which can be healed during FSW. Due to frictional contact of the tool and welded parts, the heat generated plasticizes metal. As the tool moves forward, its special profile forces the plasticized material to the back, and due to substantial forging force, consolidates the material so the joint is formed. The half-plate where the direction of rotation is the same as that of the welding is called the advancing side with the other side designated as the retreating side. The process is accompanied by severe plastic deformation, involving dynamic recrystallization of the base metal.

Properties of the friction stir welded alloys depend on the composition and microstructure of the base metal. For Mg–Zn–Y–Zr (Xie et al. 2007), ZK60 (Mironov et al. 2007), and AZ91 (Lee et al. 2003) alloys, FSW leads to a considerable refinement of the grain structure as well as dispersedness and dissolution of the intermetallic compounds, both of which contribute to the increase of microhardness. The strength of these alloys can be retained or even improved. FSW is feasible for 6-mm-thick Mg–Zn–Y–Zr plates (Xie et al. 2007), and defects associated with an inability to supply sufficient heat during welding appear in 6-mm-thick ZK60 alloys (Mironov et al. 2007). The microstructure of the Mg–Zn–Y–Zr alloy after FSW is shown in Figure 12.10. Grain refinement is also observed in the AZ31 alloy, but there is no significant precipitation hardening (Esparza et al. 2002). For AZ31B-H24 alloy with fine grains less than 10 μm, the grain size in the stir zone and thermomechanically affected zone (TMAZ) increased after recrystallization (Figure 12.11), and lower hardness was observed (Afrin et al. 2008).

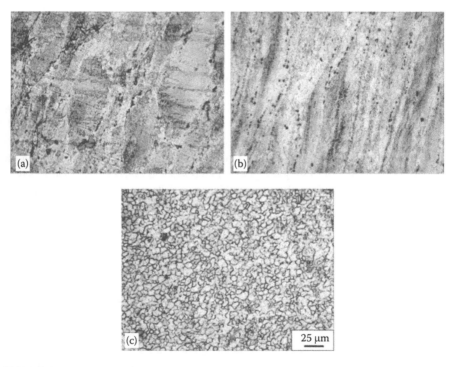

FIGURE 12.10 Microstructure of Mg–Zn–Y–Zr alloy: (a) base metal, (b) thermomechanically affected zone, and (c) nugget zone. (From Xie, G. M., Z. Y. Ma, L. Geng, and R. S. Chen, *Materials Science and Engineering: A*, 471, 63–8, 2007. With permission.)

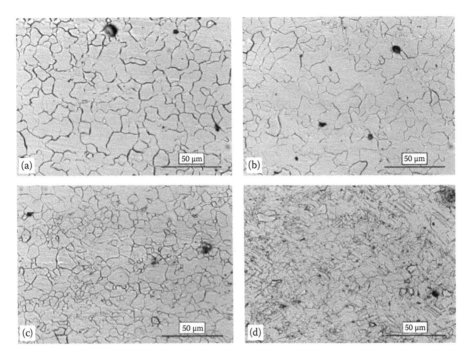

FIGURE 12.11 Optical microscope images of AZ31B-H24 alloy: (a) weld nugget, (b) thermomechanically affected zone (TMAZ), (c) heat-affected zone (HAZ), and (d) base metal. (From Afrin, N., D. L. Chen, X. Cao, and M. Jahazi, *Materials Science and Engineering: A*, 472, 179–86, 2008. With permission.)

12.6.2.5 Other Welding Processes

The resistance welding processes can be used for welding Mg alloys, including spot welding, seam welding, and flash welding. Mg alloys can also be joined by brazing. Most of the different brazing techniques can be used. In all cases, brazing flux is required, and the flux residue must be completely removed from the finished part. Soldering is not too popular because the strength of the joint is relatively low.

Mg alloys can be stud welded, gas welded, and plasma welded. Finely divided pieces of Mg alloys, such as shavings, fillings, etc., should not be in the welding area because they will burn. Mg alloy castings or wrought materials do not create a safety hazard because the possibility of fire caused by welding on these sections is very remote.

12.6.3 ADHESIVE BONDING OF BIODEGRADABLE MG ALLOYS

Adhesive bonding is defined as joining using a nonmetallic substance (adhesive) that undergoes hardening due to a physical or chemical reaction. As a result, parts are joined together through surface adherence (adhesion) and internal strength (cohesion). During adhesive bonding of Mg alloys and polypropylene, satisfactory joint strength was achieved only after chemical pretreatment of both components (Liu and Xie 2007). An addition of 1% of SiO_2 particles into the adhesive increased the joint strength. The mechanism of bonding between polypropylene and the adhesive is based on diffusion and mechanical interaction. The bonding between a Mg alloy and adhesive is controlled by the coordinative bond forces. Gluing is one of the most flexible processes for joining components of similar and dissimilar materials. The poor aging and heat resistance and the low strength are rather bad compared to welding. The main cause of failure, however, is not the poor strength but an insufficient preparation of the surfaces before gluing.

12.7 STERILIZATION OF BIODEGRADABLE Mg ALLOYS

Sterilization is the final step in manufacturing any implant device. The process of sterilization could alter surface properties of materials more or less, for which reason the sterilization method should be chosen cautiously.

Steam autoclave sterilization is the most widely used process in hospitals and laboratories, and its advantages are that it is nontoxic, highly effective, and inexpensive. However, it is deleterious for heat-sensitive and moisture-sensitive instruments.

Dry-heat sterilization can be used to sterilize heat-stable, nonaqueous materials that cannot be sterilized by steam because of its corrosive effects or failure to penetrate but is not recommended for use in hospitals because it is inefficient compared with autoclaves, and many devices cannot withstand the high temperatures involved (Lewis and McIndoe 2004).

Glutaraldehyde is widely used as a cold sterilant to disinfect a variety of heat-sensitive instruments, such as endoscopes, bronchoscopes, and dialysis equipment. But exposure to glutaraldehyde could bring about respiratory effects and anaphylactic reaction (U.S. Department of Labor 2006).

Compared with steam autoclave sterilization and dry-heat sterilization, ethylene oxide (ETO) sterilization is compatible with heat- or moisture-sensitive materials. The main disadvantages associated with ETO are the lengthy cycle time, the cost, and its potential hazards to patients and staff.

One common method of irradiation sterilization is using γ rays emitted from cobalt-60 at more than 25,000 Gy to produce sterility. Co60 γ-ray radiation sterilization is ideal for prepacked heat-labile single-use items and leaves no chemical residue (Lewis and McIndoe 2004). This method is costly and requires highly specialized equipment.

Seitz et al. (2011a) and X. L. Liu et al. (2013) have carried out two comparative studies on the effect of various sterilization methods on mechanical properties and surface characteristics of Mg and Mg alloys, respectively. Because recrystallization of Mg alloys is initiated at temperatures around 200°C, and they corrode in a hydrous environment, steam autoclave, dry-heat, and glutaraldehyde sterilization are not compatible with Mg and its alloys. ETO sterilization causes the least change in the mechanical strength of the Mg alloys (Seitz et al. 2011a). However, because of the toxicity of ETO, ETO sterilization inhibits cell adhesion and increases the hemolysis rate of Mg alloys (X. L. Liu et al. 2013). Co60 γ-ray radiation sterilization comprehensively minimizes the effects of the sterilization process on the surface chemistry and consequent biocompatibility, and is believed to be the optimal sterilization method for Mg-based biodegradable materials.

12.8 CONCLUDING REMARKS

Techniques have been developed that allow most metal-working processes to be applied to Mg and its alloys, and a wide range of properties can be obtained. Some challenges still exist; thus the optimization of current techniques and the development of new techniques are expected. For example, it is reported that three-dimensional printing has been applied to produce biodegradable iron–manganese-based scaffolds (Magnesium Elektron 2014). We believe the emergence of three-dimensional printed Mg devices are in prospect. As technology advances, biodegradable Mg alloys will be used for sophisticated new biomedical devices in the future.

13 Design of Biodegradable Mg Alloy Implants with Finite Element Analysis

13.1 INTRODUCTION

In mathematics, the finite element method (FEM) is a numerical technique for finding approximate solutions to boundary value problems for differential equations. It uses variational methods (the calculus of variations) to minimize an error function and produce a stable solution. Analogous to the idea that connecting many tiny straight lines can approximate a larger circle, FEM encompasses all the methods for connecting many simple element equations over many small subdomains, named finite elements, to approximate a more complex equation over a larger domain. Finite element analysis (FEA) has been widely used for the testing and optimization of biomedical device designs for its low costs and high efficiency compared to the conventional prototype testing. Abaqus, Ansys, Adina, Hypermesh, Femap/NX Nastran, and some other FEA software are often used in this area.

13.2 FEA IN THE DESIGN OF BONE IMPLANTS

Much FEA work has been done on the design of bone implants, such as bone screws, bone plates, and spinal products. For the screw design and its influence on the pullout force, many numerical (Gefen 2002; Zhang et al. 2004, 2006; Hsu et al. 2005; Chao et al. 2008; Kourkoulis and Chatzistergos 2009) studies have been carried out. Avery et al. (2013) and others (Lovald et al. 2010; Fouad 2011; Izaham et al. 2012; Wieding et al. 2013) used FEA methods for the optimization of bone plates. As most of the FEA work on the design of Mg alloy bone implants are based on bone screws, this section will be focused on the design of bone screws.

13.2.1 DESIGN OF BIOINERT ALLOY BONE SCREWS WITH THE FINITE ELEMENT METHOD

In the dental and orthopedic fields, a screw is a threaded device to tighten interconnective components and fix bone segments. In the literature, screw performance is often evaluated in terms of structural stiffness, fatigue strength, holding power, and insertion resistance (Yerby et al. 2001; Pfeiffer et al. 2006; Pedroza et al. 2007; International Organization for Standardization, Standard 14801; ASTM F543-07; ASTM F1717-04; ASTM F1264-03; Lee et al. 2012).

The holding power of the bone–screw interface is one of the key factors in the clinical performance of screw design (Lee et al. 2012). In order to increase efficiency and avoid laborious tests, researchers tend to use the FEM to evaluate the holding power of various screws (dental, traumatic, spinal) (Verheyen et al. 1991; Dhert et al. 1992; Dar et al. 2002; Chao and Hsiao 2006; Zhang et al. 2006).

For locking screws, the design factors include length, outer diameter, inner diameter, pitch, thread width, half angle (distal/proximal), and root radius (distal/proximal) as shown in Figure 13.1. Besides these eight factors, the materials of the screws also affect their mechanical properties.

Y. Wang et al. (2009) tested the pullout strength of three kinds of stainless-steel screws. Their proximal half angle was set at 0°, 30°, and 60°. The screws were inserted into porcine cancellous bone. Experimental results are shown in Figure 13.2.

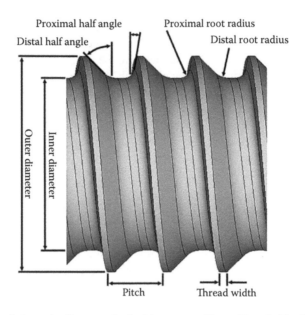

FIGURE 13.1 Schematic diagram of a locking screw. (From Hou, S.-M., C.-C. Hsu, J.-L. Wang, C.-K. Chao, and J. Lin. *Clinical Biomechanics*, 19, 738–745, 2004. With permission.)

	Proximal half angle		
	0°	30°	60°
Metal screw	301.9 (35.9)	348.8 (44.1)	126.5 (39.0)
Bone screw	278.2 (30.6)	326.6 (39.4)	174.8 (29.7)

Mean (SD). For each screw type, $n = 40$. **$P < 0.01$. *$P < 0.05$. ns, not significant.

FIGURE 13.2 Pull-out strength (N). (From Wang, Y., R. Mori, N. Ozoe, T. Nakai, and Y. Uchio. *Clinical Biomechanics*, 24, 781–785, 2009. With permission.)

The pullout strength of the screws was maximal at 30°. A proximal half angle of around 30° is appropriate because the pullout force is applied to the recipient bone evenly. Commercial cancellous screws can be improved by changing the thread shape to minimize the damage to the recipient bone.

Hou et al. (2004) used the Taguchi method to do factorial analysis of bone holding power, and the testing results were closely related to those of FEA. The results showed that increasing the outer diameter or decreasing the inner diameter, pitch, and half angle could significantly increase the bone holding power. The outer diameter had the highest contribution (46.40%) to the holding power of the screw, and then the descending order of contribution was pitch (29.01%), half angle (16.39%), and inner diameter (4.52%). The effects of root radius and the thread width were minimal (Table 13.1).

Chatzistergos et al. (2010) compared experimental and numerical results and proved that the major parameter influencing the pullout force is the outer radius, which corresponds with Hou's work.

Lin et al. (2010) also used the Taguchi method combined with FEA and found that bone strain decreased obviously when the screw material had the high elastic modulus of stainless/titanium alloys, a small exposure length, and a large diameter. The results indicated that using a strong stainless/titanium alloy as a screw material is advantageous, and an increase in mechanical stability can be achieved by reducing the screw exposure length (Table 13.2).

It is indicated that the fixation strength of a bone screw can be improved by the use of screws with a conical core (Kwok et al. 1996; Abshire et al. 2001; Hsu et al. 2005; Chao et al. 2008; Krenn et al. 2008; Kim et al. 2012). Chatzistergos et al. (2014) used FEA methods to study the influence of under-tapping and conical profile on pullout force. The numerical results indicated that pretension could indeed increase a screw's pullout force but only up to a certain degree. Under-tapping increased cylindrical screws' pullout force up to 12%, 15%, and 17% for synthetic bones of

TABLE 13.1
ANOVA Table for Total Strain Energy

Factor	Sum of Squares	Degrees of Freedom	Mean Square	F Value	Contribution (%)
Outer diameter	2.22	2	1.11	48.63	46.40
Inner diameter	0.22	2	0.11	4.74	4.52
Root radius	0.04	2	0.02	0.80	0.77
Pitch	1.39	2	0.69	30.41	29.01
Half angle	0.78	2	0.39	17.18	16.39
Thread width	0.03	2	0.01	0.56	0.53
Error	0.11	5	0.02	–	2.38
Total	4.79	17	–	–	100

Source: Hou, S.-M., C.-C. Hsu, J.-L. Wang, C.-K. Chao, and J. Lin. *Clinical Biomechanics*, 19, 738–745, 2004. With permission.

TABLE 13.2
Summary of the ANOVA Showed the Statistical Results of Maximum Strain with Respect to Cortical Bone

Design Factor	Degrees of Freedom	Sum of Squares	Mean Sum of Squares	Total Sum of Squares (%)
Osseointegration	1	0	0.0	0
Screw length	2	16	8	1
Screw diameter	2	101	50.5	7
Thread shape	2	30	15	2
Thread depth	2	20	10	1
Screw material	2	917	458.5	63
Head diameter	2	33	16.5	2
Head exposure length	2	353	176.5	24
Total	15	1470	735	100

Source: Lin, C.-L., J.-H. Yu, H.-L. Liu, C.-H. Lin, and Y.-S. Lin. *Journal of Biomechanics*, 43, 2174–2181, 2010. With permission.

density equal to 0.08 g/cm^3, 0.16 g/cm^3, and 0.28 g/cm^3, respectively. Inserting a conical screw into a cylindrical hole increased pullout force up to 11%. In any case, an optimum level of screw-induced pretension exists.

13.2.2 Design of Biodegradable Polymer Bone Screws with the Finite Element Method

In the clinic, bone plates and screws have been fabricated from stainless steel, vitallium, chromium-cobalt, and most importantly titanium and its alloys, which have become the preferred permanent metal of choice due to their ability to osteointegrate (Meslemani and Kellman 2012). However, permanent implants may cause infection, exposure, pain, and discomfort.

Resorbable polymer plates and screws are becoming more popular because they allow for fixation and stabilization but are not permanent. Wozniak et al. (2005) compared the insertion-torque and load-at-failure characteristics of two types of screws (Figure 13.3) during retrograde fixation of a soft-tissue graft. These two types of screws are used for surgical reconstruction of the anterior cruciate ligament (ACL). The top one is a small buttress thread large-taper screw design, and the bottom one has a large buttress thread small-taper screw design. The results (Table 13.3) showed that mean maximum load at failure and stiffness was similar between constructs fixed with a large buttress thread small-taper screw and a small buttress thread large-taper screw, respectively. The small buttress thread screw with a large taper displayed greater mean insertion torque at one third insertion whereas the large buttress thread screw with a small taper displayed greater mean insertion

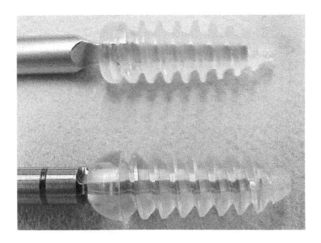

FIGURE 13.3   Two types of screws. (From Wozniak, T. D., Y. Kocabey, S. Klein, J. Nyland, and D. N. M. Caborn. *Arthroscopy*: *The Journal of Arthroscopic and Related Surgery*, 21 (7), 815–819, 2005. With permission.)

TABLE 13.3
Results by Sample Pair

Sample Pair	Thread Type	Insertion Torque at One Third (in lbs.)	Insertion Torque at Full Insertion (in lbs.)	Insertion Torque at Two Thirds (in lbs.)	Load at Failure (N)
1	LB	2.4	5.7	11.8	254.0
1	SB	4.3	8.2	10.9	309.5
2	LB	3.5	5.7	11.5	440.4
2	SB	2.9	6.4	10.4	394.8
3	LB	3.4	5.4	10.3	417.6
3	SB	4.7	5.7	7.9	426.6
4	LB	2.9	6.0	10.8	421.9
4	SB	4.1	5.5	9.2	353.9
5	LB	3.7	5.9	11.7	339.2
5	SB	4.6	6.8	11.0	371.9
6	LB	3.8	5.4	10.0	348.7
6	SB	4.2	6.6	9.5	314.3
7	LB	2.7	5.7	10.4	384.4
7	SB	4.3	6.0	7.7	322.9
8	LB	3.3	6.2	11.8	278.1
8	SB	3.7	5.7	8.9	240.2

Source: Wozniak, T. D., Y. Kocabey, S. Klein, J. Nyland, and D. N. M. Caborn. *Arthroscopy: The Journal of Arthroscopic and Related Surgery*, 21 (7), 815–819, 2005. With permission.
Note: LB, large buttress; SB, small buttress.

torque at full insertion. So large buttress thread small-taper screws displayed biomechanical fixation characteristics comparable to small buttress thread large-taper
screws.

Hunt et al. (2005) developed a perforated biodegradable poly-(L-co-D,L-lactide)
interference screw to allow for enhanced osseous implant integration without impair-
ing screw stability during insertion. The four kinds of biodegradable interference
screws are shown in Figure 13.4. They are unperforated control screws, clockwise
perforation (group 1), counterclockwise perforation (group 2), and parallel perforation (group 3). By applying these screws to the proximal tibia of 12 sheep for
24 weeks, Table 13.4 indicates that perforated, "cage-like" interference screws may
be promising for the acceleration of osseous implant integration into the bone with a
very low risk of screw breakage during insertion.

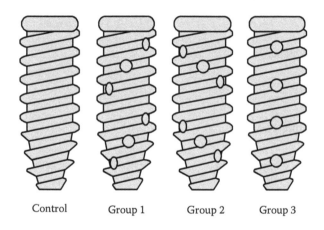

Control   Group 1   Group 2   Group 3

FIGURE 13.4 Unperforated screw and perforated screws. (From Hunt, P., F. N. Unterhauser,
M. J. Strobel, and A. Weiler. *Arthroscopy: The Journal of Arthroscopic and Related Surgery*,
21 (3), 258–265, 2005. With permission.)

TABLE 13.4
Results of Mechanical Tests of the Three Differently Perforated Screws

		Perforation				
	Insertion in Block	Unperforated	Clockwise	Counterclockwise	Parallel Manual	Parallel Molded
Test 1	23 mm	7.74 ± 0.48	6.48 ± 0.49	5.29 ± 1.09	6.03 ± 0.85	7.03 ± 0.43
Test 2	11.5 mm	3.95 ± 0.66	3.4 ± 0.19	2.74 ± 0.25	3.67 ± 0.26	4.43 ± 0.26

Source: Hunt, P., F. N. Unterhauser, M. J. Strobel, and A. Weiler. *Arthroscopy: The Journal of Arthroscopic
and Related Surgery*, 21 (3), 258–265, 2005. With permission.

Note: Values are given as mean ± SD.

However, biodegradable polymers, such as poly-L-lactide, are biomechanically inferior to their metal counterparts (Suuronen et al. 1992). Another two short-comings of the polymer implants include the need for a heating device to provide implant malleability and the need to tap the bone prior to screw placement (Bell and Kindsfater 2006).

13.2.3 Design of Biodegradable Mg Alloy Bone Screws with the Finite Element Method

The elastic modulus of Mg alloys, which is approximate to human bone, is much lower than stainless steel and titanium alloy. As a result, Mg alloy bone implants would cause much less stress shielding in fixed bone segments. Moreover, during degradation, the Mg ions being released have induction and guiding reactions in bone growth. Thus Mg alloys seem to be promising biodegradable materials for bone implants. Some designs of Mg alloy screws are shown in Figure 13.5.

Erdmann et al. (2011) compared the biomechanical properties of degradable magnesium calcium alloy (MgCa0.8) screws and commonly used stainless steel (S316L) screws and assessed the in vivo degradation behavior of MgCa0.8 by implanting MgCa0.8 screws and S316L screws into both tibiae of 40 adult rabbits for a follow-up of 2, 4, 6, and 8 weeks. SEM images of MgCa0.8 screws and a magnification of the corresponding first thread are shown in Figure 13.6. No significant differences could be noted between the pullout forces of MgCa0.8 and S316L 2 weeks after surgery. Six weeks after surgery, the pullout force of MgCa0.8 decreased slightly. Because MgCa0.8 showed good biocompatibility and biomechanical properties, comparable with those of S316L in the first 2–3 weeks of implantation, its application as a biodegradable implant is conceivable.

Henderson et al. (2014) created a finite-element model of the Mg craniofacial screw using the pullout test data and compared the simulation results with experimental results. Synthetic bone was modeled as a cylinder with a diameter of 5 mm with the screw inserted along the longitudinal axis (Figure 13.7ai). The screw and the cylindrical bone were discretized with 3731 and 14,254 four-noded tetrahedral finite elements, respectively. Details of the finite-element discretization are shown in Figure 13.7aii. The simulated pullout force profile along with representative experimental data for pure Mg experimental curves were utilized to extract the model parameters as shown in Figure 13.7b. Finite element modeling can also be used to help design future screws for specific applications. The results of this study show promise for the future use of degradable Mg alloys for craniofacial fixation applications.

Tan et al. (2014) studied the loss of mechanical properties and the interface strength of coated AZ31B Mg alloy (a Mg–Al alloy) screws with the surrounding host tissues being investigated and compared with noncoated AZ31B, degradable polymer, and biostable titanium alloy screws in a rabbit animal model after 1, 4, 12, and 21 weeks of implantation. The coating plays an important part during the degradation: The degradation of the mechanical load-bearing ability of Mg screws with coatings was slower, and the bone–implant interface strength of coated AZ31B increased with implantation time, higher than that of the noncoated screws after 21

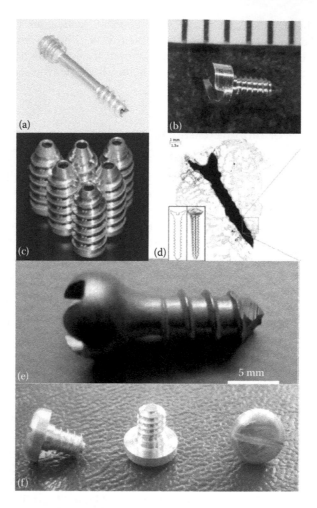

FIGURE 13.5 (a) MAGNEZIX® Compression Screw (Syntellix AG Schiffgraben 11, 30159 Hannover, Germany), (b) Mg alloy craniofacial bone screws, (c) Mg-based interference screw, (d) Extruded Mg-5 wt.%Ca-1 wt.%Zn alloy bone screw, (e) The AZ31B screw with Si-containing coating, and (f) MgCa0.8 alloy bone screws. ([a] From Windhagen, H., K. Radtke, A. Weizbauer, J. Diekmann, Y. Noll, U. Kreimeyer, R. Schavan, C. Stukenborg-Colsman, and H. Waizy. *BioMedical Engineering OnLine*, 12, 62, 2013. With permission; [b] From Henderson, S. E., K. Verdelis, S. Maiti, S. Pal, W. L. Chung, D.-T. Chou, P. N. Kumta, and A. J. Almarza. *Acta Biomaterialia*, 10 (5), 2323–2332, 2014. With permission; [c] From Farraro, K. F., K. E. Kim, S. L.-Y. Woo, J. R. Flowers, and M. B. McCullough. *Journal of Biomechanics*, 47 (9), 1979–1986, 2014. With permission; [d] From Cha, P.-R., H.-S. Han, G.-F. Yang, Y.-C. Kim, K.-H. Hong, S.-C. Lee, J.-Y. Jung, J.-P. Ahn, Y.-Y. Kim, S.-Y. Cho, J. Y. Byun, K.-S. Lee, S.-J. Yang, and H.-K. Seok. *Scientific Reports*, 3, 2367, 2013. With permission; [e] From Tan, L., Q. Wang, X. Lin, P. Wan, G. Zhang, Q. Zhang, and K. Yang. *Acta Biomaterialia*, 10 (5), 2333–2340, 2014. With permission; [f] From Erdmann, N., N. Angrisani, J. Reifenrath, A. Lucas, F. Thorey, D. Bormann, and A. Meyer-Lindenberg. *Acta Biomaterialia*, 7 (3), 1421–1428, 2011. With permission.)

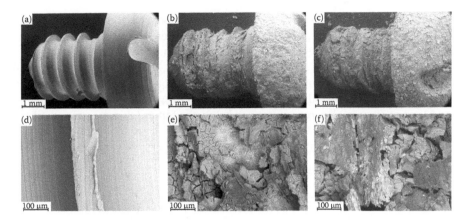

FIGURE 13.6   (a, d) MgCa0.8 before implantation; (b, e) after 2 weeks; (c, f) after 8 weeks. (From Erdmann, N., N. Angrisani, J. Reifenrath, A. Lucas, F. Thorey, D. Bormann, and A. Meyer-Lindenberg. *Acta Biomaterialia*, 7 (3), 1421–1428, 2011. With permission.)

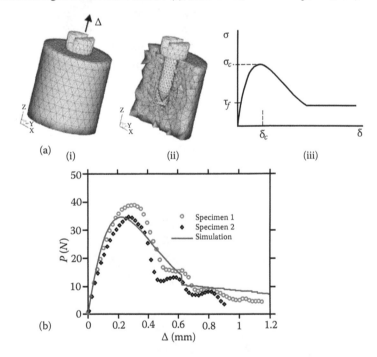

FIGURE 13.7   Pull-out testing. (a) (i) Pull-out experiment simulation domain. (ii) Details of the finite element discretization. (iii) Interfacial constitutive law used to simulate the pull-out tests with three model parameters indicated. (b) Simulated pull-out force profile along with two representative experimental data for pure Mg. Experimental curves were utilized to extract the model parameters as shown in (iii). (From Henderson, S. E., K. Verdelis, S. Maiti, S. Pal, W. L. Chung, D.-T. Chou, P. N. Kumta, and A. J. Almarza. *Acta Biomaterialia*, 10 (5), 2323–2332, 2014. With permission.)

weeks of implantation (Figure 13.8). The corrosion biodegradation mechanism and the coating to control the degradation rate of AZ31B were thought to affect both the loss of mechanical properties and the interface strength.

In order to optimize Mg alloy screws through FEA, Pal et al. (2013) considered four screw designs, each with a tapered head and unique thread profile. A pullout force of 50N was used for the simulation. It was observed that resulting maximum von-Mises stresses in all screws were well below the failure/ultimate strength of the

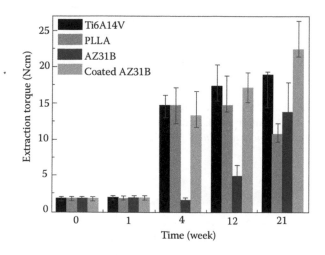

FIGURE 13.8 Results of extraction torque measurements on noncoated AZ31B, coated AZ31B, PLLA, and Ti6Al4V implanted in vivo for 1, 4, 12, and 21 weeks, respectively. (From Tan, L., Q. Wang, X. Lin, P. Wan, G. Zhang, Q. Zhang, and K. Yang. *Acta Biomaterialia*, 10 (5), 2333–2340, 2014. With permission.)

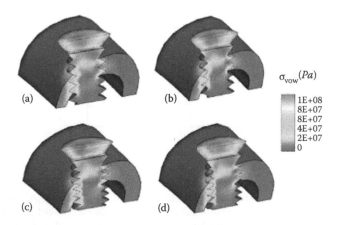

FIGURE 13.9 Stress profiles for screw design 1 (a), design 2 (b), design 3 (c), and design 4 (d). (From Pal, S., A. Chaya, S. Yoshizawa, D.-T. Chou, D. Hong, S. Maiti, P. N. Kumta, and C. Sfeir. *Proceedings of the ASME 2013 Conference on Frontiers in Medical Devices: Applications of Computer Modeling and Simulation*. 2013. With permission.)

alloy. Designs 2 and 4 provided more thread surface area compared to designs 1 and 3. Therefore, it can be envisaged that these designs would perform better. Designs 3 and 4 allow more threads to engage in the bone when compared to designs 1 and 2. Therefore, designs 3 and 4 are considered to be superior (Figure 13.9).

Unlike stents, the applications of Mg alloys in bone implants have just started. Last year, a privately held start-up company, Syntellix, announced the first Mg alloy (WE43-based) screw for bone correction osteotomy in foot surgery (Windhagen et al. 2013).

13.3 FEA IN THE DESIGN OF STENTS

13.3.1 DESIGN OF BIOINERT ALLOY STENTS WITH THE FINITE ELEMENT METHOD

For bioinert alloy stents, the most widely used materials are stainless steel, nitinol, and Co-Cr alloy. According to their constitutive relationships, stainless steel, and Co-Cr alloy are elastoplastic materials, and nitinol is a SMA material.

The choice of a particular type of stent design depends on the possible performance of that device in the lesion that is going to be treated. For example, laser-cut tube stents made of stainless steel are mainly used for the treatment of coronary heart disease; on the other hand, stents made of nickel-titanium alloy are more appropriate for the treatment of peripheral vascular disease located in the femoral, renal, carotid, or abdominal artery because of their super-elasticity and shape memory properties (Azaouzi et al. 2012).

Vascular stents are tubular cellular structures available in numerous cell shapes and sizes and in balloon-expandable and self-expanding varieties. Stent design parameters include thickness, strut shape, and bridging links.

13.3.1.1 Stainless Steel Stents

13.3.1.1.1 Thickness

The performance of the stent may be improved by an increase in thickness, such as an increase in radio visibility, radial strength, and arterial wall support although increase in thickness means more inflation pressure for expansion, which may provoke a higher risk of restenosis than thinner stents. Zahedmanesh and Lally (2009) investigated the mechanical atmosphere in arteries stented with thinner and thicker strut stents by using numerical modeling techniques. Idealized stenosed vessel geometries were developed, and a finite element model of stent was deployed in order to get and compare the mechanical behavior of different stents. Figure 13.10 shows the finite element stress analysis obtained for the numerical simulation of the stenosed vessel with the thick and thin stent models. Case 1 shows that both stents are expanded to 3.22 mm to get the final stresses on the vessel loading and unloading, and it is observed that the thicker stent induces 15% more stress than the thinner stent. As shown in Figure 13.10, that is due to the high recoil of the thin stent having fewer vessel stresses as compared to the thick stent. The case 2 stents are expanded on the same final diameter, and in this case, the thin stent expanded more to get a final diameter similar to the thick stent because of high stress; the results show that the thicker stent induced 23% higher stresses than the thin stent.

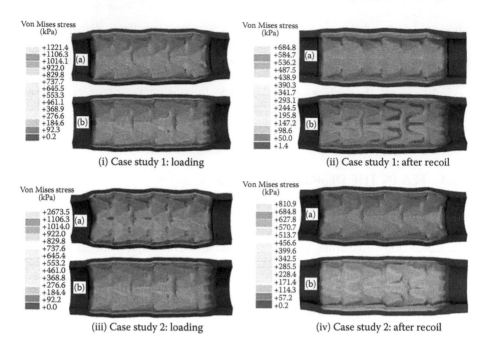

(i) Case study 1: loading

(ii) Case study 1: after recoil

(iii) Case study 2: loading

(iv) Case study 2: after recoil

FIGURE 13.10 Von Mises stresses in the stenosed vessel with stent by (a) Multilink$_{thin}$ (50 µm strut thickness) and (b) Multilink$_{thick}$ (140 µm strut thickness). (From Zahedmanesh, H., and C. Lally. *Medical and Biological Engineering and Computing*, 47 (4), 385–393, 2009. With permission.)

13.3.1.1.2 Strut Shape

Pant et al. (2011) measured the performance of different CYPHER-type stent designs by six figures of merit (objectives/metrics) representing (i) acute recoil, (ii) tissue stresses, (iii) hemodynamic disturbance, (iv) drug delivery, (v) uniformity of drug distribution, and (vi) flexibility. The results demonstrated that a change in one parameter that leads to an improvement in one of the objectives often leads to a compromise in one or more of the other objectives. It is found that strut width and the length of the circumferential rings most affect the volume average stress, and the length of the links in the cross-flow direction significantly affect volume average stress, flexibility, and the flow index. Despite this complex interplay between stent design and stent performance, they presented a methodology to perform design optimization studies on stents and the process of choosing different stent designs appropriate to different needs.

Douglas et al. (2014) found that stent designs with positive, negative, or zero foreshortening over the expansion phase can be designed by tailoring unit cell geometries, as shown in Figure 13.11, and hence obtain desired length–diameter and pressure–diameter characteristics. This analysis was for balloon-expandable stents but can be modified by changing the loading and material model for self-expanding stents.

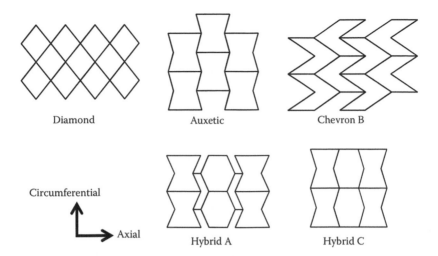

Diamond   Auxetic   Chevron B

Circumferential

Axial   Hybrid A   Hybrid C

FIGURE 13.11   Idealized geometries used in Douglas's study. (From Douglas, G. R., A. S. Phani, and J. Gagnon. *Journal of Biomechanics*, 47 (6), 1438–1446, 2014. With permission.)

W.-Q. Wang et al. (2006) studied the transient expansion process of a stent/balloon system with different stent structures and balloon lengths under internal pressure. The FEA results (see Figure 13.12) showed that it is not the changes in dimension but the configuration of the connector that mainly determines the foreshortening of stents when they are expanded. The dog-boning of a stent/balloon system is closely correlated with not only the distal geometry of the stent but also the overall length of the balloon. The dog-boning of a stent can be weakened by both decreasing the overall length of the balloon and increasing the distal stent strut width, and the cooperation of these two methods would produce a more apparent result.

Fatigue life is another important factor of stent design. Fatigue life prediction of these devices is critical for the designer. Azaouzi et al. (2013c) conducted numerical investigations using an effective approach based on the well-known stress-based Goodman diagram (GD) for the FLP of a given stent design. The results demonstrated that the dimensions of strut and the percentage of an artery's expansion both have a noticeable impact on the fatigue behavior of cardiovascular stents.

13.3.1.1.3   Bridging Links

M. Azaouzi et al. (2013a) studied the structural behavior (bridge) of the stent (Figure 13.13) in order to show the effect of the stent design in terms of flexibility, torsion, and expansion. The numerical results indicated that the geometry of the bridges that connect two adjacent rings together play an important role in the improvement of the stent behavior, especially in terms of flexibility. The unsymmetrical N-shaped Flex-connectors (Figure 13.14) have proven better mechanical properties regarding elastic recoil, flexibility, torsion, and radial strength.

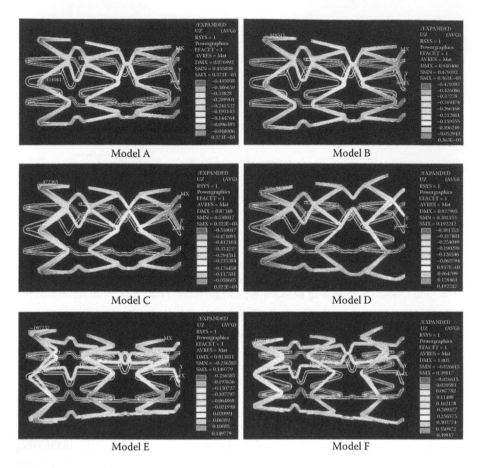

FIGURE 13.12 Simulation results of the axial displacements of six models with no display of balloons. (From Wang, W.-Q., D.-K. Liang, D.-Z. Yang, and M. Qi. *Journal of Biomechanics*, 39 (1), 21–32, 2006. With permission.)

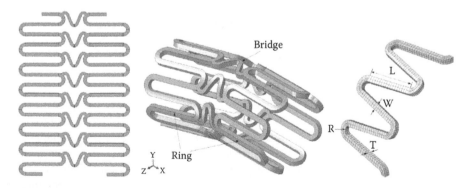

FIGURE 13.13 Typical BE stent design (ring and connecting elements "bridges"). (From Azaouzi, M., A. Makradi, and S. Belouettar. *Computational Materials Science*, 72, 54–61, 2013. With permission.)

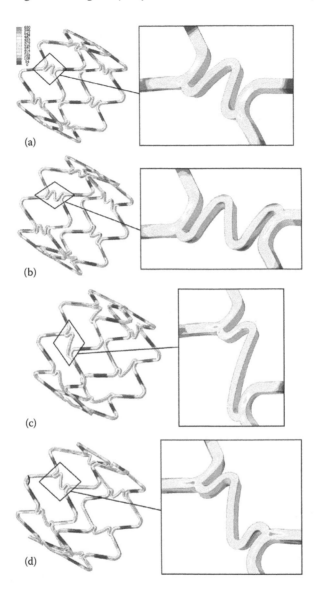

FIGURE 13.14 Von-Mises stresses over the stent obtained after expansion. (From Azaouzi, M., A. Makradi, and S. Belouettar. *Computational Materials Science*, 72, 54–61, 2013. With permission.)

Wu et al. (2007b) proposed a FEM to study the expansion of a stent in a curved vessel (the CV model) and its interactions. A model of the same stent in a straight vessel (the SV model) was also studied. The results (Figure 13.15) indicated that the nonconformity of the stent to the vessel is mainly due to insufficient deformation of links, especially at stent extremes. So stent design should consider softer links at stent extremes.

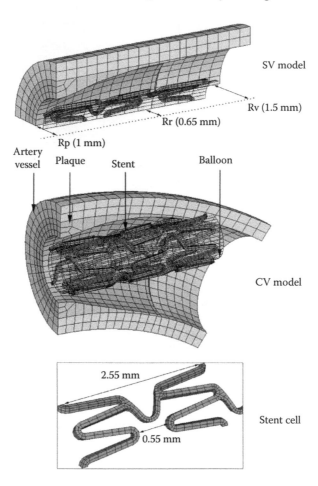

FIGURE 13.15 Finite element models of the SV and CV model. (From Wu, W., W.-Q. Wang, D.-Z. Yang, and M. Qi. *Journal of Biomechanics*, 40 (11), 2580–2585, 2007. With permission.)

13.3.1.2 Nitinol Stent

Self-expanding Nitinol (nickel–titanium alloy) stents are tubular, often mesh-like in structure. Nitinol stents are widely used in peripheral vessel interventions for their superelasticity, good corrosion resistance, and biocompatibility (Stoeckel et al. 2004). As Nitinol stents are subjected to a long-term cyclic pulsating load due to the heart beating (typically 4×10^7 cycles/year), fatigue fracture may occur. Nitinol stents should be designed to withstand at least 380 million cycles, which is equivalent to 10 years of cyclic fatigue. The fatigue endurance is a major design requirement that could be achieved by optimizing the stent design and manufacturing process.

Wu et al. (2007a) built a FEM model that is composed of stenotic carotid tissue, a segmented design nitinol stent, and a sheath. Results (Figure 13.16) show that a stent with shorter struts may have better clinical results, and the different stent designs can cause different carotid vessel geometry changes.

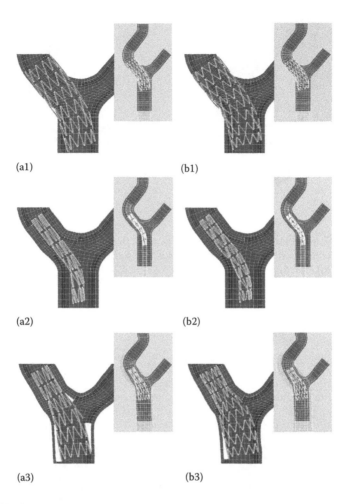

FIGURE 13.16 The stent delivery and release procedures of the Sori and Smod (both full view and local view are shown for every procedure, and the sheath is not shown in local view). (a1 and b1) The Sori and Smod were pushed forward in the sheath and delivered to the location of the plaque when the displacement of the two proximal links reached −19 mm in the global Z direction. (a2 and b2) The Sori and Smod were compressed into the stenotic ICA lumen after the diameter of the sheath was reduced to 3 mm. (a3 and b3) When the sheath was expanded and separated completely with the stent, the Sori and Smod were released from the sheath and fully contacted with the tissue. The stent malapposition areas of the Sori and Smod are highlighted in (a3 and b3), respectively. (From Wu, W., M. Qi, X.-P. Liu, D.-Z. Yang, and W.-Q. Wang. *Journal of Biomechanics*, 40 (13), 3034–3040, 2007. With permission.)

Azaouzi et al. (2013b) established a reliable procedure of FEA to provide quantitative measures of a stent's strain amplitude ($\Delta\varepsilon_{eq}$) and mean strain (ε_{mean}), which are generated by the cyclic pulsating load. An optimization-based simulation methodology was developed in order to improve the fatigue endurance of the stent. Numerical results (Figure 13.17) of the simulation indicate that the strut dimensions have an impact on $\Delta\varepsilon_{eq}$ and ε_{mean} and, consequently, the fatigue lifetime of Nitinol stents. An

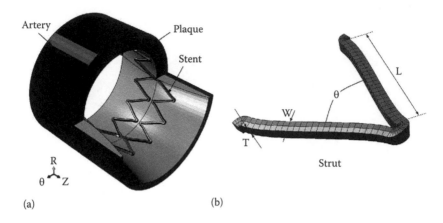

(a)   (b)

FIGURE 13.17 (a) Finite element model of the stent deployment and (b) parameters of a strut. (From Azaouzi, M., N. Lebaal, A. Makradi, and S. Belouettar. *Materials and Design*, 50, 917–928, 2013. With permission.)

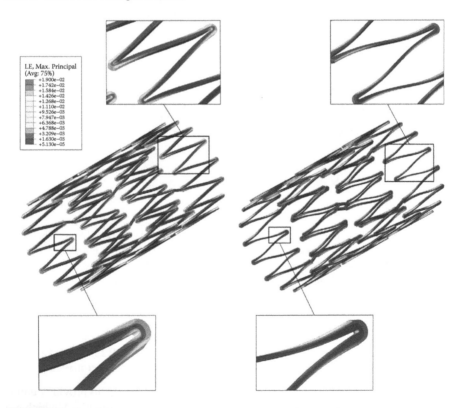

FIGURE 13.18 Contour plot comparison of strain distribution between standard stent (left) and tapered-strut stent at 50% strut width reduction via curved-line approach (right). (From Hsiao, H.-M., L.-W. Wu, M.-T. Yin, C.-H. Lin, and H. Chen. *Computational Materials Science*, 86, 57–63, 2014. With permission.)

increase in L or decrease in W results in a decrease of $\Delta\varepsilon_{eq}$ and ε_{mean}. On the other hand, a decrease in T results in a moderate increase of $\Delta\varepsilon_{eq}$ and ε_{mean}. Moreover, contrary to L and W, the decrease in T leads to an increase of $\Delta\varepsilon_{eq}$. The relationship between stent oversizing, strut dimensions, and fatigue life of Nitinol stents was demonstrated.

Hsiao et al. (2014) investigated a simple and interesting concept of stent design aimed at enhancing pulsatile fatigue life. This can be achieved by tapering the stent's strut width so the highly concentrated stresses/strains will be shifted away from the crown and redistributed along the stress-free bar arm. This kind of design (see Figure 13.18) allows the stress-free bar arm to carry more of the load by narrowing its strut width at the midpoint. Simulation results show that when the strut width was reduced to 50% of its original dimension at the midpoint of the bar arm, the fatigue safety factor could be increased up to 5.4 times that of its standard counterpart via the curved-line approach.

13.3.2 DESIGN OF BIODEGRADABLE MG ALLOY STENTS WITH THE FINITE ELEMENT METHOD

Different from bioinert alloy stents, biodegradable Mg alloy stents (MAS; as illustrated in Figure 13.19) provide a temporary scaffolding to target vessels, degrade gradually, and are ultimately absorbed by the body (Waksman 2007). During degradation, the stents lose their scaffolding abilities. So the most important difference between the designs of MAS and traditional bioinert alloy stents is corrosion time (mainly controlled by both uniform and stress corrosion processes) (Winzer et al. 2007). This can be achieved by improving alloy composition, shape optimization (which requires, at the same time, an increase in mass and a reduction in maximum stress during scaffolding), and the application of coatings. In addition to degradation, there are two controversial requirements in mechanical properties compared to a

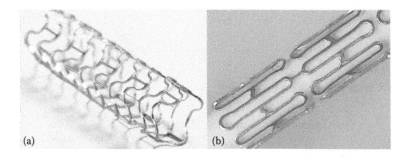

(a)   (b)

FIGURE 13.19   (a) Magic stent (Biotronik, Berlin, Germany) and (b) the absorbable metallic stent. ([a] From Erbel, R., C. Di Mario, J. Bartunek, J. Bonnier, B. de Bruyne, F. R. Eberli, P. Erne, M. Haude, B. Heublein, M. Horrigan, C. Ilsley, D. Böse, J. Koolen, T. F. Lüscher, N. Weissman, and R. Waksman. *The Lancet*, 369 (9576), 1869–1875, 2007. With permission; [b] From Garg, S. and P. W. Serruys. *Journal of the American College of Cardiology*, 56, S43–S78, 2010. With permission.)

conventional stainless steel stents: First, SS stents undergo a large local strain (about 0.4–0.5) during stent expansion (McGarry et al. 2004), and most Mg alloys have a much lower ultimate elongation (usually below 0.2) (Dalla Torre et al. 2008). Second, the elastic modulus of Mg alloys is about 25% of SS; thus a MAS needs more material to provide adequate scaffolding. The future direction for the development of Mg alloy biodegradable devices in both orthopedic and intravascular applications is the control of the corrosion rate (Song 2007).

Finite element stent models are currently a vital component of the design process. Regulatory agencies, such as the U.S. Food and Drug Administration (FDA), require a detailed stress analysis before the actual practice of a new stent design. The influence of various design variables on properties is illustrated in Table 13.5.

The mechanical behavior of Mg stent struts was investigated by Grogan et al. (2014) using computational micromechanics, based on FEA and crystal plasticity theory. The plastic deformation in tension and bending of textured and nontextured Mg stent struts with different numbers of grains through the strut dimension was investigated. It is predicted that, unlike 316LSS and L605, the failure risk and load-bearing capacity of Mg stent struts during expansion is not strongly affected by the number of grains across the strut dimensions; however, texturing, which may be introduced and controlled in the manufacturing process, is predicted to have a significant influence on these measures of strut performance.

W. Wu et al. (2010) utilized a morphing procedure to facilitate the optimization of MAS. The optimized designs were compared to an existing MAS by means of three-dimensional FEA. The strain distributions of the original design and optimized design are shown in Figure 13.20. The change of mass distribution along the strut can be a useful way to reduce strain and stress concentration during stent expansion and scaffolding.

To better optimize the design of a MAS, the corrosion process should be taken into consideration. There are five kinds of mechanisms affecting the corrosion behavior of MAS: microgalvanic corrosion, pitting corrosion, filiform corrosion, stress corrosion cracking (SCC), and fatigue corrosion (Song and Atrens 1999, 2007; Zeng et al. 2008b).

TABLE 13.5
Design Variable Effect on Properties

Design Variable	Radial Recoil (%)	Foreshortening (%)	Deployment Pressure	Visibility	Stent Stress (Mpa)	Vessel Scaffolding
Strut thickness	↑	↓	↓	↑	↑	↑	↑
Number crest	↑	↑	↓	↓	–	↑	↑
Strut length	↑	↑	↑	↓	–	↓	↓
Bridging links	↑	–	↑	↑	–	↑	↑

Source: Sabir, M. I. Doctoral thesis, 2011. With permission.

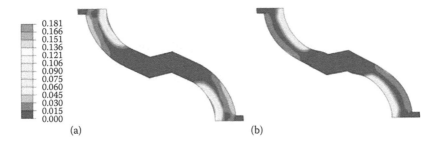

FIGURE 13.20 (a) Original design and (b) optimized design. (From Wu, W., L. Petrini, D. Gastaldi, T. Villa, M. Vedani, E. Lesma, B. Previtali, and F. Migliavacca. *Annals of Biomedical Engineering*, 38 (9), 2829–2840, 2010. With permission.)

Gastaldi et al. (2011) developed a model of Mg degradation (the continuum damage model, CDM) that is able to predict the corrosion rate, thus providing a valuable tool for the design of bioresorbable stents. In this model, the damage is assumed to be the superposition of stress corrosion and uniform microgalvanic corrosion processes. The effect of the different damage contributions is clarified in Figure 13.21. D was a scalar field that modeled the loss of integrity due to the degradation process. D is monotonically increasing, and, in particular, D = 0 when the material is undamaged, and D = 1 when it is completely damaged and loses its ability to sustain loads.

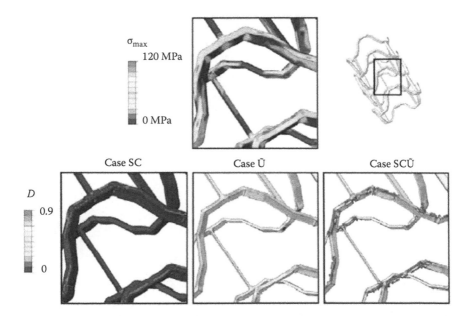

FIGURE 13.21 Maximum principal stress is mapped on a portion of the stent (top). At the bottom, the damage after 50 time units is plotted with SC only (case SC) with reduced uniform corrosion alone (case Ũ) and with both mechanisms activated (case SCŨ). (From Gastaldi, D., V. Sassi, L. Petrini, M. Vedani, S. Trasatti, and F. Migliavacca. *Journal of the Mechanical Behavior of Biomedical Materials*, 4 (3), 352–365, 2011. With permission.)

Uniform corrosion alone produced an even loss of material from the surface (case Ũ), and SC concentrated where the maximum principal stress (top) exceeded σ_{th} (case SC), forming pitting that was further amplified when both mechanisms were present (case SCŨ).

Grogan et al. (2011) also developed a numerical model based on the continuum damage theory (Lemaitre 1996) to predict the effects of corrosion on the mechanical integrity of bioabsorbable metallic stents. The major difference between Grogan's and Gastaldi's work is that Grogan enhanced the damage evolution in this work by the introduction of an element-specific dimensionless pitting parameter, λ_e. The model was calibrated by assessing the effects of corrosion on the integrity of bio-degradable metallic foils (struts) experimentally. This corrosion model captured both the experimentally observed corrosion rate and the resulting corrosion-induced reduction in foil (strut) integrity.

Grogan et al. (2013) used the continuum damage model to study the effects of corrosion on the mechanical performance of a range of stent designs. In the simulations, the Mg alloy stents underwent four processes as shown in Figure 13.22: deploy, recoil, corrode, and crush. It is predicted that the radial stiffness in stents undergoing uniform corrosion decreases as an initially linear function of the quantity corrosion

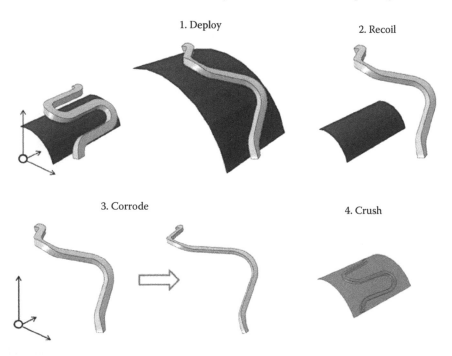

FIGURE 13.22 The hinge is deployed by expanding a semirigid inner cylinder. It then undergoes recoil and corrosion and is subsequently crushed using an outer cylinder. Symmetry boundary conditions are used so only quarter of the stent unit is modeled. (From Grogan, J. A., S. B. Leen, and P. E. McHugh. *Biomaterials*, 34 (33), 8049–8060, 2013. With permission.)

rate multiplied by immersion time. For AMSs undergoing uniform corrosion, it was predicted that initial design studies could avoid modeling corrosion in improving long-term scaffolding performance by focusing on maximizing radial stiffness and forces. The corrosion results are shown in Figure 13.23.

W. Wu et al. (2011) applied CDM to three different MAS designs, including an already implanted stent (stent A), an optimized design (stent B), and a patented stent design (stent C). The locations where the stent structure was severely damaged are indicated by circles in Figure 13.24. With the empirical material parameters, the results numerically verified the expectation that MAS design with more mass and optimized mechanical properties can increase scaffolding time: Stent B increased the half normalized recoil time of the vessel by nearly 120% compared to stent A. Although stent C has more materials than stent B, it only increased the half normalized recoil time of the vessel by nearly 50% compared to stent A because of much higher stress concentration than stent B.

Later on, W. Wu et al. (2012) carried out an experimental validation for the developed FEA model. Twelve stent samples of AZ31B were manufactured according to two MAS designs (an optimized one and a conventional one) with six samples of each design. All the samples were balloon-expanded and subsequently immersed in D-Hank's solution for a degradation test lasting 14 days. With a good match between the simulation and the experimental results (Figure 13.25), the work shows that the FEA numerical modeling constitutes an effective tool for design and thus the improvement of novel biodegradable MAS.

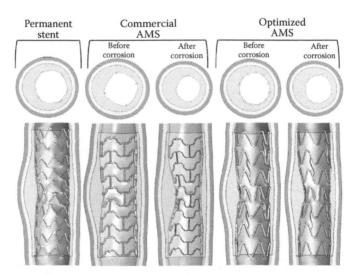

FIGURE 13.23 Cross-sections of predicted stent configurations before and after uniform corrosion. (From Grogan, J. A., S. B. Leen, and P. E. McHugh. *Biomaterials*, 34 (33), 8049–8060, 2013. With permission.)

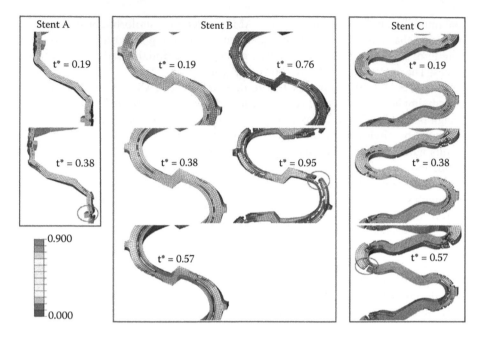

FIGURE 13.24 Damage evolution of the three stent designs. (From Wu, W., D. Gastaldi, K. Yang, L. Tan, L. Petrini, and F. Migliavacca. *Materials Science and Engineering B*, 176 (20), 1733–1740, 2011. With permission.)

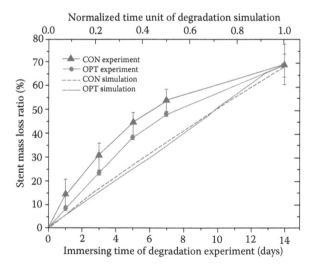

FIGURE 13.25 Mass loss ratio of the CON and OPT samples for the experiment and the corresponding models for the simulations. (From Wu, W., S. Chen, D. Gastaldi, L. Petrini, D. Mantovani, K. Yang, L. Tan, and F. Migliavacca. *Acta Biomaterialia*, 9 (10), 8730–8739, 2012. With permission.)

13.4   CONCLUDING REMARKS: FUTURE

In FEA, many assumptions have been made, such as regarding material behavior, loading conditions, solver algorithm, and post-processing index. So the correlation between FEA results and mechanical tests varied while these assumptions changed. The correctness of the numerical result highly depends on these assumptions. Only if the assumptions are accurate is the FEA method an efficient tool in the analysis and design of biomedical devices.

In literature with animal experiments, Mg and its alloys have shown great advantages and benefits as promising biodegradable materials for biomedical devices. But corrosion is a critical factor for Mg alloy devices. The future direction for the development of biodegradable Mg alloy devices in both orthopedic and intravascular applications is the control of the corrosion rate (Song 2007).

In order to control the corrosion rate, the design of Mg devices, including both stents and bone implants, should reduce strain and stress concentration in order to avoid severe local stress corrosion. For Mg stents, much research and simulations can be found in the literature, but as far as the authors know, little work on the design of Mg bone implants (screws and plates) has been done. Especially for Mg screws, during degradation, threads almost disappeared after a few weeks (Erdmann et al. 2011). As a result, thread width should be designed a little thicker than SS and other bioinert screws.

For the design of new biodegradable implants, Nguyen (2011) proposed a prototype porous Mg alloy screw as shown in Figure 13.26. Porous Mg alloys have the advantages of being strong enough to provide suitable mechanical properties for the implanted site while remaining lightweight. More importantly, the porous architecture helps facilitate tissue growth and anchoring of the implant to the recipient tissue (e.g., bone) as well as supporting the flow of tissue fluid carrying nutrients, which are two of the most important requirements for successful application of porous

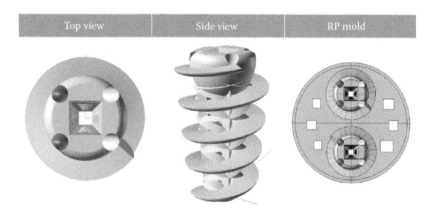

FIGURE 13.26   CAD models of the screw in top and side view as well as the RP model consisting of two-screw mold for TOPM manufacture. (From Nguyen, T. L. Doctoral thesis, 2011. With permission.)

biomaterial scaffolds in the repair or replacement of damaged or diseased tissues. However, the control of the degradation mode and rate is still a big concern.

Parameters such as materials, strut shape, strut thickness, and links determine the mechanical properties and corrosion life of Mg alloy stents. The change of one parameter may lead to the opposite effects on different properties. In order to optimize the design of Mg alloy stents, all the parameters should be taken into consideration and a balance found.

For finite element simulations, the future direction should be the establishment of the corrosion model for Mg and its alloys. Improving the purity of pure Mg and alloying of Mg can both decelerate the corrosion rate, but the pure Mg degraded with a different profile to that of the Mg alloys (Henderson et al. 2014). With the establishment of the corrosion models, FEA methods will be better used to help design future biomedical Mg devices for specific applications.

14 In Vivo Testing of Biodegradable Mg Alloy Implants

14.1 IN VIVO TESTING OF BIODEGRADABLE Mg ALLOY IMPLANTS WITHIN BONE

14.1.1 BONE HEALING PROCESS

There are three major phases of fracture healing: inflammation, repair, and remodeling (Rüedi and Murphy 2001; Johnson et al. 2006) as shown in Figure 14.1.

The inflammation stage begins the moment the bone is broken and lasts for around 1–7 days. As the inflammation response subsides, the repair phase begins and gradually becomes the predominant pattern. As the first step in the reparative phase, the hematoma is organized. The hematoma serves primarily as a fibrin scaffold over which repair cells perform their function. During the first 2–3 weeks postfracture, the soft callus is formed, which corresponds roughly to the time when the fragments are no longer moving freely. This early soft callus can resist compression, but shows similar tensile properties to fibrous tissue (Johnson et al. 2006). Hence the mineralization of the soft callus proceeds from the fragment ends toward the center of the fracture site and forms a hard callus, which has regained enough strength and rigidity to allow low-impact exercise at the end of the repair phases (Rüedi and Murphy 2001; Johnson et al. 2006). The time to achieve the hard bone union varies greatly according to the fracture configuration and location and status of the adjacent soft tissues as well as patient characteristics (age, health status, concurrent injuries/diseases). According to Perkin's classification of fracture healing, a spiral fracture in the upper limb unites in 3 weeks and consolidates in 6 weeks. The fracture healing time doubles for a transverse fracture and doubles again for the lower limb. Table 14.1 lists a rough estimate of bone healing time for different fractures. Hence, the mechanical support should be sustained for 12–24 weeks, depending on the clinical conditions.

14.1.2 ANIMAL TESTING OF BIODEGRADABLE Mg ALLOY BONE IMPLANTS

More than 20 kinds of Mg-based implants have been evaluated within the bones of small (rats, guinea pigs, and rabbits) and large animal (minipigs, sheep) models to characterize the in vivo corrosion as well as the bone response. The results are summarized in Table 14.2.

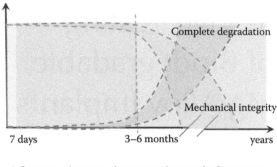

7 days   3–6 months   years

Inflammation; hematoma formation with a typical inflammatory
response lasting 1–7 days

Repair; hematoma→granulation tissue→connective tissue→
cartilage→mineralization→woven bone; continues for 3–6 months
depending on the fracture position and type

Remodeling; woven bone is replaced by cortical bone and the
medullary cavity is restored, which persists for several years

FIGURE 14.1 The schematic diagram of degradation behavior and the change of mechanical integrity of BM implant during the bone healing process. (From Zheng, Y. F., X. N. Gu, and F. Witte, *Materials Science Engineering Reports*, 77, 1, 2014. With permission.)

14.1.2.1 Animal Models

Although nonhuman primates are often considered to be the most appropriate model for human bone (Wang et al. 1998; Turner 2001), there are clear ethical concerns in using this species for medical research as well as cost, zoonotic disease risks, and handling difficulties. Rat is one of the most commonly used species in medical research,

TABLE 14.1
Bone Healing Time for Different Fractures

Fracture		Healing Time (Weeks)
Upper limb	Clavicle	5–7
	Fingers	4–8
	Scaphoid	>10
	Humerus	5–8
	Radius and ulna	8–12
	Distal radius	3–4
Lower limb	Pelvis	6–10
	Femur	8–14
	Neck of femur	12–24
	Tibia	8–12
	Calcaneus	6
	Toes	6–8

Source: Zheng, Y. F., X. N. Gu, and F. Witte, *Materials Science and Engineering Reports*, 77, 1, 2014. With permission.

TABLE 14.2
Summary of Animal Tests of Mg Alloy Implants within Bone

Implants	Implantation Site	Period/ Weeks	New Bone	Bone Contact	Subcutaneous Gas Cavity	Degradation Rate/mm/yr or Weight/Volume Loss %	Strength Loss	References
Pure Mg (99.99%)/cast, rod	Femoral diaphysis, rabbit	12	+	+	–	N/A	N/A	Gao et al. 2010a
Pure Mg (99.99%) + heat-organic-film/cast, rod			+	+	–	N/A	N/A	
Pure Mg (99.9%)/rod	Femora, rat	6	+	N/A	N/A	29.41 wt.%	N/A	W. Yang et al. 2011
Pure Mg (99.99%)/screw	Tibia, rabbit	12	+	+	N/A	~34 vol.%	~50% (tensile)	S.-M. Kim et al. 2014
Pure Mg (99.99%) + HA/screw			+	+	N/A	~24 vol.%	~10% (tensile)	

(Continued)

TABLE 14.2 (CONTINUED)
Summary of Animal Tests of Mg Alloy Implants within Bone

Implants	Implantation Site	Period/ Weeks	New Bone	Bone Contact	Subcutaneous Gas Cavity	Degradation Rate/mm/yr or Weight/Volume Loss %		Strength Loss	References
Pure Mg (99.9%)/screw	Mandible, rabbit	12	+	+	N/A	Total screw	~65 vol.%	N/A	Henderson et al. 2014
						Screw head	68.7 vol.%	N/A	
						Screw shaft in cortical bone	28.5 vol.%		
						Screw shaft in bone marrow	90.4 vol.%		
AZ31/screw			+	+	N/A	Total screw	~50 vol.%	N/A	
						Screw head	38.5 vol.%		
						Screw shaft in cortical bone	55.8 vol.%		
						Screw shaft in bone marrow	80 vol.%		

(Continued)

TABLE 14.2 (CONTINUED)
Summary of Animal Tests of Mg Alloy Implants within Bone

Implants	Implantation Site	Period/ Weeks	New Bone	Bone Contact	Subcutaneous Gas Cavity	Degradation Rate/mm/yr or Weight/Volume Loss %	Strength Loss	References
AZ31/gravity cast, rod	Marrow cavity, guinea pig	18	+	+	+	N/A	N/A	Witte et al. 2005, 2006b
AZ91/gravity cast, rod						100 vol.%	N/A	
WE43/gravity cast, rod						N/A	N/A	
LAE442/gravity cast, rod						70 vol.%	N/A	
AZ31B/rolled, stabilization splint	Mandible, rabbit	26	+	+	–	N/A	N/A	Hong et al. 2008
AZ31B/rolled, dictyo- plate			+	–	+	N/A	N/A	
AZ31B/extruded, screw	Femoral diaphysis, rabbit	21	N/A	N/A	N/A	N/A	42.4% (bending)	Tan et al. 2014
AZ31B + Si coating/ extruded, screw			N/A	N/A	N/A	N/A	29.9% (bending)	
AZ31/cast, rod	Femur, rat	12	+	N/A	N/A	33% cross-section area	N/A	Chai et al. 2012
AZ31 + β-TCP/cast, rod			+	N/A	N/A	17% cross-section area	N/A	
AZ60/cast, rod	Femoral shaft, rabbit	12	N/A	N/A	–	~1.9 mm/yr	N/A	Xiao et al. 2013
AZ60 + Ca-P/cast, rod			N/A	N/A	–	~0.6 mm/yr	N/A	

(Continued)

TABLE 14.2 (CONTINUED)
Summary of Animal Tests of Mg Alloy Implants within Bone

Implants	Implantation Site	Period/Weeks	New Bone	Bone Contact	Subcutaneous Gas Cavity	Degradation Rate/mm/yr or Weight/Volume Loss %	Strength Loss	References
AZ91/cast, rod	Femoral diaphysis, rabbit	8	+	+	–	33 vol.%	N/A	Wong et al. 2010
AZ91 + PCL/cast, rod						0–5 vol.%	N/A	
AZ91/cast, rod	Femoral epicondyle, rat	8	+	+	N/A	~7 vol.%	N/A	Wong et al. 2013
AZ91 + Al_2O_3/cast, rod			+	+	N/A	~2 vol.%	N/A	
AZ31/extruded, screw	Hip bone, sheep	12	+	+	N/A	main body	N/A	Duygulu et al. 2007
Mg–Sr/rolled, rod	Marrow cavity, rat	4	+	–	+	1.01 mm/yr	N/A	X. N. Gu et al. 2012
LAE442/extruded, rod	Femoral condyle, rabbit	12	N/A	+	–	0.31 mm/yr 11 vol.%	N/A	Witte et al. 2010
LAE442 + MgF_2/extruded, rod			N/A	+	–	0.13 mm/yr	N/A	
LACer442/extruded, rod	Marrow cavity, rabbit	4	+	–	++	N/A	N/A	Reifenrath et al. 2010
LANd442/extruded, rod	Marrow cavity, rabbit	26	+	+	+	0.072 mm/yr 5.5 vol.%	38.3%	Ullmann et al. 2011; Hampp et al. 2012
LAE442/extruded, rod	Marrow cavity, rabbit	52	+	–	–	21 vol.%	~46%	Thomann et al. 2009
Mg–0.8Ca/extruded, rod, smooth			+	–	–	40 vol.%	~72%	

(Continued)

TABLE 14.2 (CONTINUED)
Summary of Animal Tests of Mg Alloy Implants within Bone

Implants	Implantation Site	Period/Weeks	New Bone	Bone Contact	Subcutaneous Gas Cavity	Degradation Rate/mm/yr or Weight/Volume Loss %	Strength Loss	References
Mg-0.8Ca + MgF₂/ extruded, rod	Marrow cavity, rabbit	26	+	+	–	25.33 vol.%	~55%	Thomann et al. 2010
Mg-0.8Ca/extruded, rod Smooth Sand blast Threaded	Femoral epicondyle, rabbit	26	+ + +	+ + +	+ + +	V_{corr}, sand blast > threaded > smooth	N/A N/A N/A	Von Der Höh et al. 2009a,b
Mg-0.8Ca/extruded, screw	Tibia, rabbit	8	+	N/A	+	Screw head: 8.77 vol.% Thread within the cortex: 1.37 vol.% Thread within the marrow cavity: 8.82 vol.%	Decreased pull out force to 123.37N	Erdmann et al. 2011
Mg-Ca/cast, screw	Femoral diaphysis, rabbit	12	+	–	+	1.27 mm/yr 80 mass.%	N/A	Z. J. Li et al. 2008
Mg-Ca-Zn/extruded, screw	Femoral condyle, rabbit	52	N/A	N/A	–	0.32 mm/yr 100 vol%	N/A	Cho et al. 2013
Mg-6Zn/extruded, rod	Femoral diaphysis, rabbit	14	+	–	+	2.32 mm/yr 87 wt.%	N/A	Zhang et al. 2010b

(Continued)

TABLE 14.2 (CONTINUED)
Summary of Animal Tests of Mg Alloy Implants within Bone

Implants	Implantation Site	Period/Weeks	New Bone	Bone Contact	Subcutaneous Gas Cavity	Degradation Rate/mm/yr or Weight/Volume Loss %	Strength Loss	References
Mg-Mn-Zn/extruded, rod	Femoral diaphysis, rat	18	+	–	N/A	54% cross-section area	N/A	L. P. Xu et al. 2007; Zhang et al. 2009a
Mg-Mn-Zn + Ca-P coating/extruded, rod	Femoral diaphysis, rabbit	4	+	+	–	Main body	N/A	Xu et al. 2009a
Mg-Mn-Zn/disk	Mandible, rabbit	12	+	N/A	N/A	N/A	N/A	Miao and Jinag 2009
Mg-Mn-Zn + MAO (Ca-P)/disk			+	N/A	N/A	N/A	N/A	
Mg-Mn-Zn + MAO (Ca-P-Ag)/disk			+	N/A	N/A	N/A	N/A	
Mg-2Zn-0.2Ca/extruded, rod	Femoral diaphysis, rabbit	50	+	–	–	2.15 mm/yr	N/A	S. Chen et al. 2012
Mg-2Zn-0.2Ca + MAO + DCPD/extruded, rod			+	+	–	1.24 mm/yr	N/A	
Mg-5Zr/cast, rod	Tibia, rabbit	12	+	–	N/A	V_{corr}, Mg-1Zr-2Sr > Mg-5Zr and Mg-2Sr-5Sr	N/A	Y. C. Li et al. 2012
Mg-1Zr-2Sr/cast, rod			+	+	N/A			
Mg-2Zr-5Sr/cast, rod			+	–	N/A			

(Continued)

TABLE 14.2 (CONTINUED)
Summary of Animal Tests of Mg Alloy Implants within Bone

Implants	Implantation Site	Period/Weeks	New Bone	Bone Contact	Subcutaneous Gas Cavity	Degradation Rate/mm/yr or Weight/Volume Loss %	Strength Loss	References
Mg-Y-Nd-HRE*/—, rod	Femoral diaphysis, rat	24	+	+	–	Main body	Increasing bonding strength with time	Castellani et al. 2011; Lindtner et al. 2013
MgYREZr/PM, screw	Femoral condyle, rabbit	52	+	+	–	N/A	N/A	Waizy et al. 2014
WE43/extruded, plate	Nasal, minipig	24	N/A	N/A	N/A	~13 wt.%	13% (bending)	Imwinkelried et al. 2013
WE43 + plasma electrolytic coating/ extruded, plate			N/A	N/A	N/A	~14 wt.%	8%	
ZEK100/extruded, rod	Marrow cavity, rabbit	26	+	–	–	0.154 mm/yr	58.7%	Huehnerschulte et al. 2011; Reifenrath et al. 2011; Huehnerschulte et al. 2012
AX30/extruded, rod			+	–	–	0.11 mm/yr	70.3%	
ZEK100/extruded, rod	Marrow cavity, rabbit	52	+	+	+	1.285 mm/yr	N/A	Dziuba et al. 2013

(Continued)

TABLE 14.2 (CONTINUED)
Summary of Animal Tests of Mg Alloy Implants within Bone

Implants	Implantation Site	Period/Weeks	New Bone	Bone Contact	Subcutaneous Gas Cavity	Degradation Rate/mm/yr or Weight/Volume Loss %	Strength Loss	References
ZEK100/extruded, rod	Middle ear, rabbit	12	+	+	N/A	N/A	N/A	Lensing et al. 2014
ZK60/extruded, rod	Femoral diaphysis, rabbit	12	+	+	+	100 vol%	N/A	Lin et al. 2013
ZK60 + MAO/extruded, rod			+	+	+	100 vol%	N/A	
ZK60/extruded, rod	Femoral condyle, rat	26	+	–	–	0.67 mm/yr	N/A	Z. R. Qi et al. 2014
ZK60 + MAO/extruded, rod			+	–	–	0.58 mm/yr	N/A	
WZ21/extruded, pin	Femoral diaphysis, rat	24	+	+	–	~70 vol%	Increasing bonding strength with time	Celarek et al. 2012; Kraus et al. 2012; Fischerauer et al. 2013
ZX50/extruded, rod			+	–	+	4 mm/yr at 12 weeks, 100 vol% at 12 weeks	100% loss at 12 weeks	
ZX50 + MAO/extruded, rod			+	+	+	6 mm/yr at 8 weeks, 100 vol% at 8 weeks	N/A	

(Continued)

TABLE 14.2 (CONTINUED)
Summary of Animal Tests of Mg Alloy Implants within Bone

Implants	Implantation Site	Period/Weeks	New Bone	Bone Contact	Subcutaneous Gas Cavity	Degradation Rate/mm/yr or Weight/Volume Loss %	Strength Loss	References
MgZnCa BMG/cast, rod			+	+	N/A	N/A	Fracture at 3 months	Remennik et al. 2011
Mg-5Bi-1Ca/RS, rod	Femoral condyle, rabbit	4	+	–	–	1.85 mm/yr	N/A	Willbold et al. 2013
Mg-6Zn-1Y-0.6Ce-0.6Zr/RS, rod	Rabbit, Femoral condyle	8	+	–	–	5.45 mm/yr 71.61 vol%	N/A	
	Lumbar musculature				+	3.04 mm/yr 80.84 vol%	N/A	
	Subcutaneously				+	3.46 mm/yr 90.4 vol%	N/A	
AX30 + MgF_2/scaffold, porosity 63% ± 6%	Femoral head, rabbit	24	+	+	+	76 vol%	N/A	Lalk et al. 2013
AX30 + Ca-P/scaffold, porosity 6% ± 4%			+	–	+	93 vol%	N/A	
AZ91D/scaffold, porosity 72–76%	Condyle, rabbit	12	+	–	+	100 vol%	N/A	Witte et al. 2007a,b
MgY4/scaffold	Femoral condyle, rabbit	12	+	+	–	93 vol%	N/A	Bobe et al. 2013

and approximately a quarter of the reported in vivo research of biodegradable Mg orthopedic implants is carried out in rats. However, there are significant dissimilarities between rat and human bone (Pearce et al. 2007). The rabbit is the most commonly used model for implant biomaterial research in bone, which is in part due to ease of handling. Nevertheless, this species still shows limited similarities to human bone. Rats and rabbits are more suitable for screening implant materials prior to testing in a larger animal model. The dog is described as perhaps having the most similar bone composition (ash weight, hydroxyproline, extractable proteins, and IGF-1 content) to human bone; however, there are ethical implications of using companion animals for medical research. Although species such as the sheep and pig are not as ethically emotive and are of a more similar body weight to humans, they may pose housing, handling, and availability issues.

Although no species fulfills the requirements of an ideal animal model, an understanding of the differences in bone macroscopic, microscopic, and remodeling attributes is likely to improve the choice of animal species and interpretation of results from these in vivo studies.

14.1.2.2 Methods for the Analysis of In Vivo Biocompatibility and Corrosion

Many studies use radiography throughout the experimental period and before euthanasia as a tool to identify gas pockets within the soft tissue associated with the implants and the formation of new bone (Z. J. Li et al. 2008; Erdmann et al. 2011; Dziuba et al. 2013). Figure 14.2 shows the radiographs of rabbit femora with Mg-1Ca alloy pins at 1 month and 3 months postoperation.

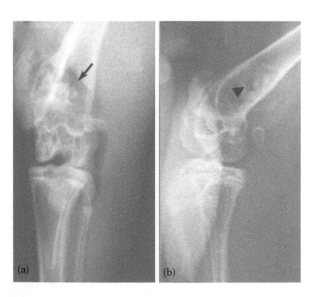

FIGURE 14.2 Femora radiographs of rabbit with Mg-1Ca alloy pins (a) at 1 month postoperation (the arrow marks the gas shadows) and (b) at 3 months postoperation (the black triangle marks the circumferential osteogenesis). (From Li, Z. J., X. N. Gu, S. Q. Lou, and Y. F. Zheng, *Biomaterials*, 29, 1329, 2008. With permission.)

The weight loss method is a basic and commonly used method to reliably quantify the degradation of an implant after explantation (Z. J. Li et al. 2008; Thomann et al. 2009; Zhang et al. 2010b; Erdmann et al. 2011; W. Yang et al. 2011; Aghion et al. 2012; Imwinkelried et al. 2013). The results can be compared directly with weight loss data gathered from in vitro studies. The removed implants can be analyzed by SEM and EDS (Z. J. Li et al. 2008; Erdmann et al. 2011; Tan et al. 2014). However, the necessity to remove the implant from the surrounding tissue destroys the bone–implant complex and eliminates the opportunity to investigate the implant–tissue interface.

Another method is to analyze the remaining cross-sectional area of explanted Mg samples after embedding in methylmethacrylate (Witte et al. 2005; L. P. Xu et al. 2007; E. L. Zhang et al. 2009a; Chai et al. 2012; Willbold et al. 2013). However, the limitation of this method is that the implant is usually not uniformly corroded so that cross-sections only provide local information on the sample.

The most common method is the analysis of the remaining volume of the implanted material, new bone growth, bone–implant contact, and osseointegration using microcomputed tomography (µ-CT) or synchrotron radiation-based microtomography (SRµCT). Micro-CT scanning can be performed both in vivo (Erdmann et al. 2011; Huehnerschulte et al. 2011; X. N. Gu et al. 2012; Kraus et al. 2012; Dziuba et al. 2013; Henderson et al. 2014) and ex vivo (Dziuba et al. 2013; Henderson et al. 2014). In vivo µ-CT applies to small animals, such as rats and rabbits. It allows continuous monitoring of one animal over time without interfering with the ongoing experiment. The 3-D micro-CT reconstruction can represent both the new bone formation and 3-D degradation of the implants, and 2-D cross-sectional images provide information about new bone formation, the bone–implant contact, and cavity formation as shown in Figures 14.3 and 14.4. Typical in vivo µ-CT scanners have resolutions ranging from 100 to 30 µm, and ex vivo scanners have resolutions from 30 to 1 µm (Zagorchev et al. 2010). Ex vivo µ-CT is generally performed

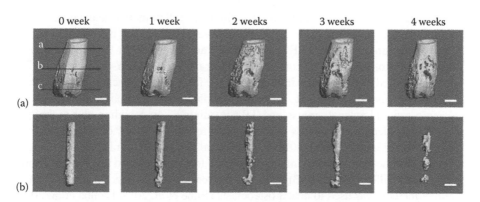

FIGURE 14.3 3-D micro-CT topographies of a representative mouse femur with an intramedullary as-rolled Mg–2Sr alloy implant scanned in vivo at different postimplantation time points. The three lines in the first image are the locations used for 2-D cross-sectional imaging of the mouse femur shown in Figure 14.4. (a) Complete outline of the distal femur of a mouse; (b) complete outline of the Mg–2Sr alloy implant. Bar 1.0 mm. (From Gu, X. N., X. H. Xie, N. Li, Y. F. Zheng, and L. Qin, *Acta Biomaterialia*, 8, 2360, 2012. With permission.)

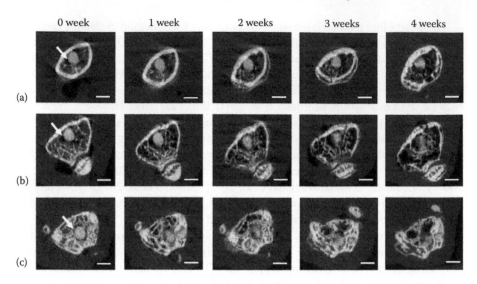

FIGURE 14.4 2-D cross-sectional images from micro-CT at three different regions of a representative mouse femur as specified in the first image of Figure 14.3 with an as-rolled Mg–2Sr alloy implant (arrows) at different postimplantation time points. (a) Proximal part of the distal femur, (b) middle part of the distal femur, (c) distal part of the distal femur. Bar 1.0 mm. (From Gu, X. N., X. H. Xie, N. Li, Y. F. Zheng, and L. Qin, *Acta Biomaterialia*, 8, 2360, 2012. With permission.)

on explants of tissue containing an implant that has been fixed and embedded in resin. After µ-CT imaging, the samples can be used for further histological analysis (Dziuba et al. 2013). Additionally, element-specific SRµCT can be utilized as a specific and sensitive method for detecting rare earth elements among other elements in bone and in Mg implants (Witte et al. 2006a).

Histological techniques can be used to indicate inflammation and neovascularization in addition to orthopedic-specific responses, such as new bone formation. In addition to commonly used H&E staining, there are various other techniques. Figure 14.5 shows histological analysis of the RS66 (Mg-6.0% Zn-1.0% Y-0.6% Ce-0.6% Zr, rapidly solidified) implant–bone interface and the foreign body reaction with toluidine blue staining, von Kossa staining, polysequential fluorescence labeling technique with calcein green and xylenol orange, and TRAP staining. The toluidine blue technique was used to identify osteoblasts and characterize the degree of osteoblast activity (Bobe et al. 2013; Dziuba et al. 2013; Willbold et al. 2013). The von Kossa technique is used to mark the extent of new mineralized bone formation next to the implant (Bobe et al. 2013; Willbold et al. 2013). The tartrate-resistant acidic phosphatase (TRAP) staining method was used to detect osteoclasts (Bobe et al. 2013; Dziuba et al. 2013; Willbold et al. 2013). Some research only provides qualitative descriptions of the resultant tissue response, and others adopt a scoring system for the semiquantitative assessment of local effects after implantation according to ISO 10993-6 (Dziuba et al. 2013).

Additional methods are also used to assess the general toxicity of the implanted materials. Most commonly, this involves the analysis of serum for the presence of Mg ions or alloying components (Gu et al. 2012; Dziuba et al. 2013) but can also be

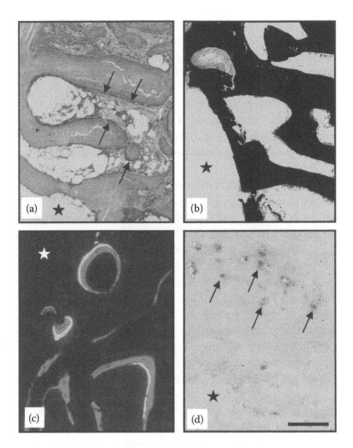

FIGURE 14.5   Histological analysis of the implant–bone interface and the foreign body reaction with different staining methods. The original drill hole is marked in each sub-image by a black or white star. Toluidine blue staining (a) nicely shows uncalcified osteoid and osteoblasts (marked by black arrows), indicating bone growth. The von Kossa stain (b) marks the calcified bone areas and was used for the determination of the bone volume. The polysequential fluorescence labeling technique (c) with calcein green (lightest part) and xylenol orange (medium grey) provides valuable information about the viability and the spatiotemporal pattern of bone growth. TRAP staining (d) selectively marks osteoclasts (black arrows) and provides information about the bone-degradation processes. Bar = 100 µm (a–c), 200 µm (d). (From Willbold, E., K. Kalla, I. Bartsch et al., *Acta Biomaterialia*, 9, 8509, 2013. With permission.)

used to measure general markers for the health of the animal (such as serum creatinine, blood urea nitrogen, and cell counts). Histological samples of visceral organs have also been analyzed to identify any systemic pathology that could be due to the implantation of Mg-based biomaterials.

14.1.2.3   In Vivo Corrosion of Mg Alloy Bone Implants

The bare Mg alloys exhibited a wide range of degradation period from 4 to 52 weeks (1 year), depending on their chemical composition and processing history (see

Table 14.2). Furthermore, the anatomical implantation site has a remarkable effect on the rate of corrosion, which is of great importance for the design of implants. The Mg alloy implant exhibited a faster degradation within the bone tissue or areas with high blood supply, that is, the marrow cavity or the cancellous bone, than within the cortical bone tissue (Erdmann et al. 2011; Huehnerschulte et al. 2011; Z. R. Qi et al. 2014). Zhang et al. (2009a) found that up to 95% of the extruded Mg-Mn-Zn alloy implants degraded in the bone marrow in comparison with 5% in the cortical bone 6 weeks postsurgery. Remennik et al. (2011) revealed that depending on the area of implantation, the degradation rate could be ranked in the following order: subcutaneous > muscle > bone.

The design of the implants also influences the corrosion of the implants. The most common orthopedic implant designs used in animal models are either the screw type (threaded) or cylindrical (rod-shaped). The screw type implants provide good initial stability, and analysis of rod or cylindrically shaped implants may be less complicated, and the exact fit into the bone gives accurate results regarding their effect on bone integration (Carlsson et al. 1988). Several designs of Mg implants, including screws, interference screws, plates, and intramedullary nails, are shown in Figure 14.6. Von Der Höh et al. (2009b) found obvious pitting corrosion occurred in threaded cylinders of MgCa0.8 alloy but not for smooth and sandblasted implants. In addition, smooth implants showed the best integration into the bone compared to sandblasted and threaded cylinders. Willbold et al. (2011) found an accelerated corrosion of the screw heads when compared to the screw threads. Han et al. (2014) reported the cusps are the most vulnerable place compared to other parts in the pin model.

The subcutaneous gas cavities were clinically observed in approximately 30% of the reported cases for bare Mg alloy tests (see Table 14.2), which usually initiated in the early stage (7~30 days) of implantation and gradually disappeared without special treatment with moderate inflammatory response. In some cases, the gas cavity is not visible clinically but can be found in the 2-D images of the μ-CT scans (Huehnerschulte et al. 2011; Bobe et al. 2013). There are two studies published in which clinically massive gas formation was observed. Reifenrath et al. (2010) compared in vivo degradation and biocompatibility of LAE442 and LACer442 in a rabbit model. The LAE442-alloys were tolerated well, and the in vivo corrosion process of the LACer442 implants was much too fast and caused massive clinical problems with gas formation and bone reactions to the point of bone fracture and, in one animal, a spontaneous death. In another study, massive gas formation was caused by LAE442 plate-screw systems with large soft tissue contact, leading to severe lameness and bone reaction/lysis (Wolters et al. 2013). Witte et al. (2008) considered the formation of gas cavities to be related to the anatomical implantation sites because of different hydrogen solubility and diffusion coefficient in different tissues. Therefore, the local gas cavities should be avoided for the Mg alloys exhibiting slower degradation rates with less hydrogen generation per time interval. There was an exceptional case in which no gas formation was clinically observed in a fast-degrading RS/PM Mg-5Bi-1Ca alloy that almost completely dissolved in 4 weeks (Remennik et al. 2011). Remennik et al. (2011) attributed the lack of local gas cavities to the fine microstructure (grain size ~2 μm), which restricted the coalescence of hydrogen gas

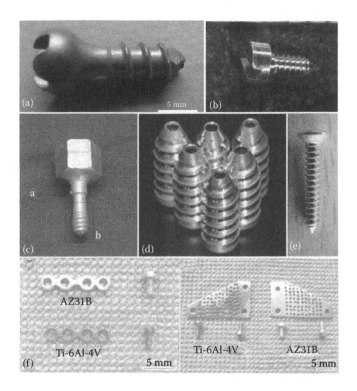

FIGURE 14.6 Mg-based orthopedic implants. (a) AZ31B screw with Si-containing coating, (b) AZ31 craniofacial bone screw, (c) threaded cylinder with hexagonal bolt head, which was pinched off with a pair of pliers at (a, b) remained in the femoral epicondyle, (d) AZ31 interference screw, (e) Mg-Ca-Zn screw, (f) left: stabilization splint systems and right: dictyo-plate systems, made of AZ31B and Ti-6Al-4V. (From Tan, L., Q. Wang, X. Lin et al., *Acta Biomaterialia*, 10, 2333, 2014; Henderson, S. E., K. Verdelis, S. Maiti et al., *Acta Biomaterialia*, 10, 2323, 2014; Von Der Höh, N., D. Bormann, A. Lucas et al., *Advanced Engineering Materials*, 11, B47, 2009; Farraro, K. F., K. E. Kim, S. L. Y. Woo, J. R. Flowers, and M. B. McCullough, *Journal of Biomechanics*, 2013; Cho, S. Y., S. W. Chae, K. W. Choi et al., *Journal of Biomechanics Research Part B*, 101, 201, 2014; Hong, Y. S., K. Yang, G. D. Zhang et al., *Acta Metallurgica Sinica*, 44, 1035, 2008. With permission.)

bubbles on the metal surface. Thus, the large gas–liquid surface enhanced hydrogen dissolution and local removal from the corrosion front.

The fast degeneration of mechanical integrity of bare Mg alloys needs to be improved independent of the implantation site. From Table 14.2, surface coating could provide the protection of Mg alloy substrate in the early stage of implantation, exhibiting slower degradation and higher corrosion uniformity and thus slowing down the strength decay. The Mg-0.8Ca alloy indicated a 40% loss in volume and resulted in a 70% loss in bending strength, and a 25% volume and 55% decay in strength were observed when coated with the MgF_2 layer over a 6-month period of implantation (Thomann et al. 2010). The gas cavities were not observed in most cases of the surface-modified Mg alloys. However, MAO coatings only delay the initial corrosion process and accelerate degradation in the later stage due to the high level of porosity (Fischerauer et al. 2013; Lin et al. 2013).

14.1.2.4 Biocompatibility of Mg Implants in Bone

All of the research work until now reported enhanced new bone formation around the implants of Mg alloys and enhanced local periosteal and endosteal bone formation in the vicinity. The degradation product layer on experimental Mg alloy implants revealed high deposition of calcium phosphate-based mineral. Half of the cases in which bare Mg alloys were implanted showed direct contact between the layer of degradation product and new bone, and the rest indicated the presence of a fibrotic layer or a gap between the degradation product layer from bare Mg-based implants and the new bone as listed in Table 14.2. Zhang et al. (2009a) indicated that the layer comprised of two distinct membrane structures with many fibroblasts is one that is closer to the bone. This layer was still present 6 months postsurgery. All of the cases with modified surfaces demonstrated direct contact between Mg-based implants and the new bone without the fibrotic layer.

Most studies published on Mg implants report beneficial osteoinductive effects of Mg alloys. However, there are controversies. Some researchers (Hampp et al. 2012; Huehnerschulte et al. 2012; Dziuba et al. 2013) reported that Mg alloys induce an osteoclast-driven resorption of bone, and they believe resorption cavities cause locally heightened strain levels in the remaining adjacent bone and thereby stimulate the periosteal formation of new bone. Dziuba et al. (2013) claimed that favorable in vivo degradation behavior is not necessarily associated with good biocompatibility. Furthermore, the absence of general pathological disorders does not definitely indicate acceptable local biocompatibility of Mg implants. Interestingly, Hampp et al. (2013) found the tibiae of rabbits, which were only subjected to the surgery but received no implant, also reacted by forming cavities. Hence, it is assumed that mere manipulation of the bone under surgical conditions leads to cell activation and remodeling processes and can thus not be assessed as an exclusive effect of the implant material.

14.1.3 CLINICAL TRIAL OF BIODEGRADABLE MG ALLOY BONE IMPLANTS

The MAGNEZIX® Compression Screw is a compression screw that can be used for fixing small bones and bone fragments. It is the first Mg screw product to get a CE mark. There was a prospective, randomized, clinical pilot trial (Windhagen et al. 2013) to determine if they are equivalent to standard titanium screws for fixation during chevron osteotomy in 26 patients with a mild hallux valgus. The implants are made of a powder metallurgically processed Mg alloy MgYREZr, which is similar to WE43. The design of this compression screw is shown in Figure 14.7. The cannulated screws had a total length of 20 mm, a shaft diameter of 2.0 mm, and a bore diameter of 1.3 mm. The design of the screw includes two threads with different pitches (3.0 and 4.0 mm) to gain interfragment compression. The postoperative radiographs (Figure 14.8) show a bony healing in both Mg and titanium groups. Six-month follow-up, including clinical, laboratory, and radiographic assessments, revealed that the degradable magnesium-based screw is equivalent to the

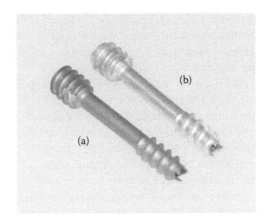

FIGURE 14.7 The two cannulated screws with the same design. (a) The titanium screw (Fracture compressing screw, Königsee Implantate GmbH, Am Sand 4, 07426 Allendorf, Germany), (b) MAGNEZIX® Compression Screw (Syntellix AG Schiffgraben 11, 30159 Hannover, Germany). (From Windhagen, H., K. Radtke, A. Weizbauer et al., *Biomedical Engineering Online*, 12, 1–10, 2013.)

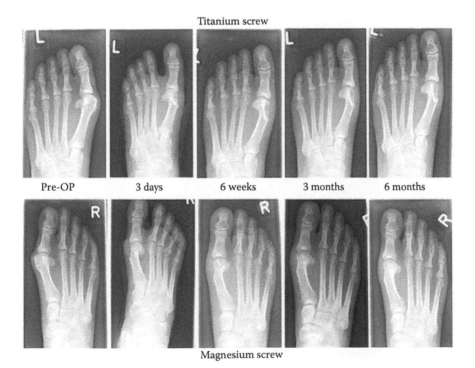

FIGURE 14.8 Preoperative radiographs (posterior–anterior) of a mild hallux valgus deformity. The correction is achieved by a chevron osteotomy. (From Windhagen, H., K. Radtke, A. Weizbauer et al., *Biomedical Engineering Online*, 12, 62, 2013. With permission.)

conventional titanium screw, and both groups showed good to excellent clinical and radiographic results with a high satisfaction rate. No foreign body reaction, osteolysis, or systemic inflammatory reaction, palpable gas cavity, and significant elevations in blood Mg levels are observed.

Zhao and Huang (2013) used a pure Mg screw (99.99% purity) in conservative treatment in 18 patients with avascular necrosis of the femoral head. A vascular pedicled iliac bone flap graft has been reported to be a successful conservative method for the treatment of young patients. However, a bone graft alone may not provide enough support, causing further collapse. Zhao and Huang use a biodegradable Mg screw to fix and strengthen the bone flap. The postoperative radiographs show bone flap ingrowth, and Harris scores increased from preoperative values of 41.7 ± 11.2 to 89.1 ± 8.9 at 8 weeks follow-up. No adverse reactions were detected. The pure Mg screws degraded completely after 8 weeks. No increase of necrotic area or collapse occurred.

14.2 IN VIVO TESTING OF BIODEGRADABLE Mg ALLOY IMPLANTS WITHIN BLOOD VESSELS

14.2.1 HEALING PROCEDURE OF BLOOD VESSELS

For cardiovascular applications, vascular injury is common and often extensive following balloon angioplasty. The injured vessels exhibit a wound-healing response that can be described in three overlapping phases: inflammation, granulation, and remodeling (Forrester et al. 1991) as shown in Figure 14.9. A complete degradation is expected to occur after the vessel remodel phase, which is usually completed at

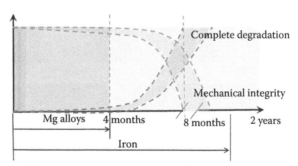

Inflammation; platelet deposition and infiltration of inflammatory cells which lasts for several days
Granulation; endothelial cells migrate to cover the injured surface, and smooth muscle cell modulate and proliferate lasting for 1–2 weeks
Remodeling; extracellular matrix deposition and remodeling continues for months. This phase is variable in duration but largely complete at 90–120 days

FIGURE 14.9 Schematic diagram of degradation behavior and the change of mechanical integrity of biodegradable magnesium stent during the vascular healing process. (From Zheng, Y. F., X. N. Gu, and F. Witte, *Materials Science and Engineering Reports*, 77, 1–34, 2014. With permission.)

90–120 days. Serruys et al. (1988) observed that almost all lesions deteriorated to some extent by 120 days post-coronary angioplasty, and the minimal luminal diameter tended to stabilize after the 3-months follow-up in 342 patients. Hence, a very slow degeneration of mechanical support is expected in the first 4 months, providing sufficient support to the injured vessels. Even though experts still debate whether full mechanical support in stented lesions is mandatory during the first 4 months after implantation, it would certainly be wise to use this clinical opinion as a safety design parameter and a benchmark for market approval evaluations based on the fact that there is insufficient human in vivo data available, especially for mechanical vessel wall properties during the healing/remodeling phases.

14.2.2 METHODS FOR THE ASSESSMENT OF BIODEGRADABLE MG CARDIOVASCULAR STENTS

Commonly used medical imaging techniques include angiography, intravascular ultrasonographic (IVUS), and optical coherence tomography (OCT). Angiography is done by injecting a radio-opaque contrast agent into the blood vessel and imaging using X-ray-based techniques, such as fluoroscopy. Traditional angiography provides information only about the contour of the vascular lumen. The components of the vascular wall are not visualized. What is more, Mg stents are radiolucent and cannot be imaged by X-rays. Intravascular ultrasonography produces images based on the reflected amplitude of ultrasound pulses, showing the lumen geometry and the structure of the vascular wall. For a 20 to 40 MHz IVUS transducer, the typical resolution is 80 microns axially and 200 to 250 microns laterally (Mintz et al. 2001). Grayscale IVUS allows robust quantitative measurements, including lumen, vessel, stent, and plaque area; qualitative assessment of lesions preintervention; and quantitative assessment and complications of lesions postintervention. OCT is an optical analogue of IVUS with a resolution of approximately 10 to 20 μm, which is about 10 times higher than IVUS (Kubo et al. 2007). Cross-sectional images of plaque microarchitecture, stent placement, apposition, and strut coverage can be provided by OCT. With its high resolution capability, OCT allowed measurements of stent strut width and was an effective tool for quantitative assessment of stent degradation. The major disadvantage of OCT is limited tissue penetration.

After the animals are sacrificed, histomorphometry and histopathology can be performed with the treated vessel segments removed and embedded in resin. Computer-assisted morphometry can be performed to determine intimal area and medial area. Semiquantitative and descriptive histopathology included scores for inflammation, intimal fibrin content, and endothelialization.

14.2.3 ANIMAL TESTING OF BIODEGRADABLE MG ALLOY STENTS

Although started earlier, far fewer animal studies of Mg-based stents have been carried out than the Mg-based orthopedic implants, as shown in Table 14.3, which is mainly due to the difficulty of fabricating the stent. The stents are usually tested in a porcine model because the vessels of pigs are similar to those of humans.

TABLE 14.3

Summary of Animal Tests of Mg Alloy Stent within Blood Vessels

Alloy	Biocompatibility	Degradation	References
AE21 stent (pig, coronary artery)	40% loss of perfused lumen diameter between days 10 and 35 due to neointima formation; a 25% re-enlargement between days 35 and 56 caused by vascular remodeling	~89 days (extrapolated)	Heublein et al. 2003
WE43 stent (minipig, coronary artery)	The struts are covered by neointima after 6 days; higher minimal lumen diameter on weeks 4 and 12 than the 316L stent group	~98 days (extrapolated)	Di Mario et al. 2004
AMS (pig, coronary artery)	Show signs of degradation 28 days postsurgery; less neointima but no significant increase in the lumen area compare to 316L stent	–	Waksman et al. 2006
AMS (minipig, coronary artery)	A significantly higher minimal lumen diameter than 316L at 28 and 56 days; significantly reduced neointimal areas than 316L; all stent struts were completely covered by a dense cellular layer after 7 days	–	Loos et al. 2007
AMS (pig, coronary artery)	Lumen narrowing at 28 days and positive remodeling of the lumen and vessel area at 3 months; Mean stent strut width (0.24 ± 0.032 mm) decreased to 0.12 ± 0.007 mm at 28 days and to 0.151 ± 0.032 mm at 3 months	–	Slottow et al. 2008
AMS (pig, coronary artery)	Compared to two nonabsorbable stents, neointima formation was smallest in the AMS group, but the AMS group also has smallest lumen area at 3-month follow-up because of negative vascular remodeling	–	Maeng et al. 2009
AMS + VBT (pig, coronary artery)	VBT reduces the intimal hyperplasia and improves the lumen area, but does not have impact on late recoil	–	Waksman et al. 2007
AMS-3.0, drug-eluting (pig, coronary artery)	Equivalent to TAXUS Liberté and superior to eucaTAX regarding late luminal loss, intimal area, fibrin score, and endothelialization	180 days	Wittchow et al. 2013
AZ31B (rabbit, infrarenal abdominal aorta)	Sirolimus-eluting AZ31B stents reduces intimal hyperplasia and improves the lumen area when compared to uncoated AZ31B stents, but delays vascular healing and endothelialization	~120 days	H. Li et al. 2011
MZX (pig, coronary artery)	Neointimal thickness and percent stenosis were significantly higher when compared to platinum chromium bare metal stents at 30 days	~120 days	Deng et al. 2011

Note: MZX is a series of noncommercial alloys, which contains magnesium, zinc, and other elements.

The first animal study of biodegradable Mg cardiovascular stents was carried out by Heublein et al. (2003). They selected AE21 alloy, which was expected to have only 50% mass loss during the first half year of the implantation. Twenty AE21 stents were implanted into coronary arteries of 11 domestic pigs. Neointima formation caused 40% loss of perfused lumen diameter between days 10 and 35. The loss of stent integrity between days 35 and 56 allowed the local vessel tissue to have a positive remodeling response with a 25% re-enlargement of perfused lumen diameter. No platelet deposition or thrombus formation was found at the endothelial sites after any of the assessment intervals. However, the AE21 stent degraded faster than expected. Therefore, further improvements are necessary with respect to prolongation of the degradation and mechanical stability over a defined time.

Di Mario et al. (2004) reported the results of experimental implantation of Lekton Magic coronary stent (Biotronik, Bulach, Switzerland) made from WE43 Mg alloy in the coronary artery of 33 minipigs. The stent has a novel design characterized by circumferential noose-shaped elements connected by unbowed cross-links along its longitudinal axis as shown in Figure 14.10. The minimal lumen diameter of the Mg stent group is higher than the 316L stent group on weeks 4 and 12, and significant positive remodeling of the Mg stent group was found from week 4 to week 12. The Mg stents not only inhibited the growth of smooth muscle cells, but also induced homogenous and rapid endothelialization. The struts were covered by neointima after 6 days.

This Lekton Magic coronary stent was further developed and denominated as absorbable metallic stents (AMS) later. The first generation, AMS-1, started to show signs of degradation by 28 days, as shown by radiographs (Figure 14.11) that demonstrated reduced opacity of the stents with multiple areas of discontinuity and

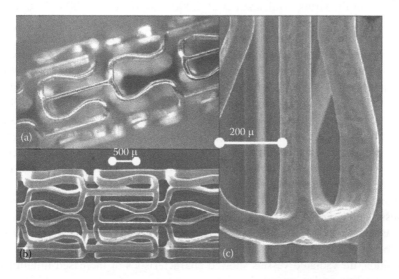

FIGURE 14.10   (a) Photograph of the tubular slot balloon expandable magnesium alloy stent after electropolishing. (b, c) Electron microscopy at low (b) and high (c) magnification. (From Di Mario, C., H. U. W. Griffiths, and O. Goktekin, *Journal Interventional Cardiology*, 17, 391–5, 2004. With permission.)

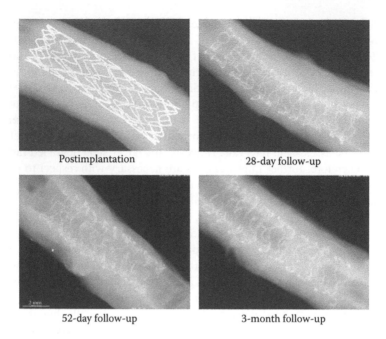

Postimplantation   28-day follow-up

52-day follow-up   3-month follow-up

FIGURE 14.11 Low-kilovoltage radiographs (60 kV/25 s with Kodak X-Omat V film) of porcine coronary arteries implanted with AMS taken at noted time points. (From Slottow, T. L. P., R. Pakala, T. Okabe et al., *Cardiovascular Revascularization Medicine*, 9, 248–54, 2008. With permission.)

dispersion of the struts (Waksman et al. 2006; Slottow et al. 2008). The decrease in radiopacity, the areas of stent discontinuity, and the strut dispersion progressed with time. There was no evidence of stent particle embolization, thrombosis, excess inflammation, or fibrin deposition (Waksman et al. 2006). As shown in Figure 14.12, the neointimal area was significantly less in the AMS group as compared with non-absorbable stents (Waksman et al. 2006; Loos et al. 2007). However, although Loos et al. (2007) found a significantly higher minimal lumen diameter in the AMS group than 316L at 28 and 56 days, Waksman et al. (2006) discovered that reduction of neo-intima formation did not result in a larger vessel lumen. Maeng et al. (2009) reported absorbable Mg stents, compared to two nonabsorbable stents, was associated with the smallest lumen area at 3-month follow-up owing to negative vascular remodeling rather than neointima formation. Waksman et al. (2007) evaluated the impact of adjunct vascular brachytherapy (VBT) on restenosis and positive remodeling of the vessels following AMS implantation in porcine coronaries. The results indicated VBT as adjunct to AMS could reduce the formation of neointima from 1.30 ± 0.62 mm^2 to 0.49 ± 0.34 mm^2 at 28 days follow-up but does not have impact on late recoil. In addition, the radiated vessels showed characteristics of increased inflammation, delayed healing, and re-endothelialization.

To prolong vessel scaffolding, AMS-2.1 was developed with several alloy and design iterations that resulted in the increase of radial force and a slower degradation rate as demonstrated on the bench and in animal studies. AMS-2.1 serves as the

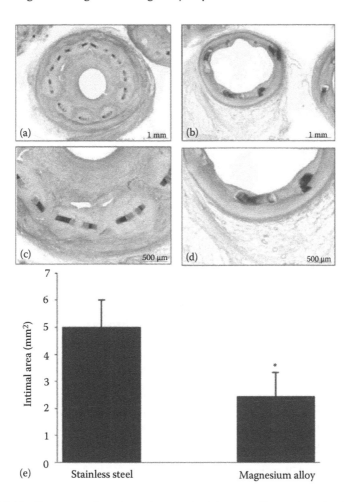

FIGURE 14.12 Representative photomicrographs of hematoxylin–eosin stained sections of porcine coronary arteries 28 days after stainless steel stent (a: 40× and b: 100×) and Mg alloy stent (c: 40× and d: 100×) implantation. (e) Bar graph showing the intimal area. Note the large intimal area in the vessels implanted with stainless steel stents. (From Waksman, R., R. Pakala, P. K. Kuchulakanti et al., *Catheterization and Cardiovascular Interventions*, 68, 607–17, 2006. With permission.)

stent platform for AMS-3, the drug-eluting AMS (DREAMS). The collapse pressure of DREAMS is higher than AMS-1 (1.5 vs. 0.8 bar). The cross-sectional profile of scaffold struts in DREAMS was redesigned to be square-shaped as opposed to the rectangular shape in AMS-1 (Figure 14.13a and b). As a preferential attack occurs at the lateral sides of the strut (Figure 14.14), the square-shaped struts can preserve radial strength during anisotropic scaffold degradation over a longer time period. The reduced strut thickness from 165 to 120 μm should also facilitate endothelialization and reduce restenosis.

Drug-elution kinetics of AMS-3 can be modulated by varying the composition of their bioresorbable poly(lactide-co-glycolide) coating loaded with paclitaxel.

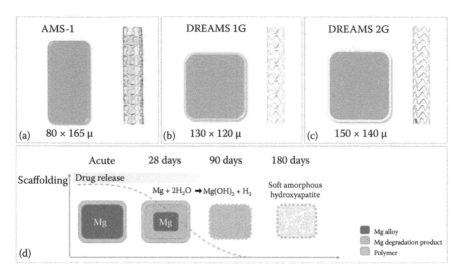

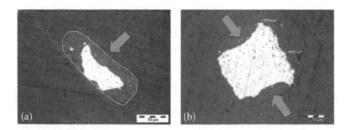

FIGURE 14.13 (a) Schematic cross-sectional profile of Mg scaffolds struts of (a) uncoated, noneluting, AMS-1 with 80×165 μ; (b) DREAMS 1st Generation (DREAMS 1G) with 130×120 μ struts; and (c) DREAMS 2nd generation (2G) with 150×140 μ struts. The poly(lactide-co-glycolide)-coating with paclitaxel elution of the DREAMS 1G scaffold is indicated by the thin light orange layer. The PLA-coating with sirolimus elution of the DREAMS 2G scaffold is indicated by the thin dark orange layer; and (d) schematic representation of the resorption process in the drug-eluting absorbable magnesium scaffold. The release of the antiproliferative drug occurs within the first 3 months after device implantation. Hydrolysis of the scaffold affects the radial strength of the scaffold, resulting in a gradual resorption of the device into a soft amorphous hydroxyapatite at 9 months follow-up. AMS-1, first-generation bare absorbable metal scaffold; DREAMS, drug-eluting absorbable metal scaffold. (From Campos, C. M., T. Muramatsu, J. Iqbal et al., *International Journal of Molecular Science*, 14, 24492–500, 2013. With permission.)

FIGURE 14.14 Appearance of struts of (a) AMS-1 and (b) DREAMS under the light microscope (unstained ground sections) 28 days after implantation in a porcine coronary model. (From Wittchow, E., N. Adden, J. Riedmuller et al., *EuroIntervention*, 8, 1441–50, 2013. With permission.)

Wittchow et al. (2013) compared three versions of AMS-3 with established, pacli-taxel-eluting, permanent stents TAXUS Liberté and eucaTAX. They found the best-performing AMS-3, 85/15H, with an 85:15 ratio lactide to glycolide and high-molecular-weight polymer, was equivalent to TAXUS Liberté and superior to eucaTAX regarding late luminal loss, intimal area, fibrin score, and endothelializa-tion. Representative histological sections for 85/15H, TAXUS Liberté, and eucaTAX, taken at 28–180 days, are presented in Figure 14.15. Intimal inflammation score was higher in 85/15H than in the control stents at 28 days, but this effect disappeared at later time points. The Mg-alloy is not decayed completely. Figure 14.13d describes the dynamics and byproducts of Mg scaffold resorption at 28, 90, and 180 days. At first, a Mg-rich compound containing a large amount of oxygen is formed, possibly representing a mixture of Mg hydroxide. Then the compound converted to amor-phous calcium phosphate, taking exactly the morphology of the dissolved scaffold struts at 180 days. Measured at 28 days, the average in vivo degradation rates for the three DREAMS versions ranged from 0.036–0.072 mg/cm^2 day.

DREAMS was further modified to create the next generation: the DREAMS 2nd generation (DREAMS 2G, Figure 14.13c), which is made of a WE43 alloy with a six-crown, two-link design and a strut thickness of 150 μm with radiopaque markers at both ends (made from tantalum), resulting in slower dismantling and resorption rates (Figure 14.16). The distal markers were added to make scaffold implantation

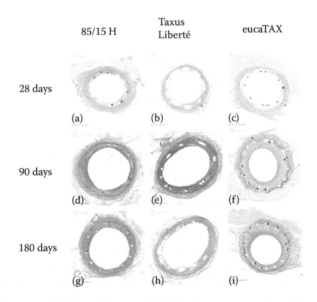

FIGURE 14.15 Representative histological sections removed at 28 days (a–c), 90 days (d–f), and 180 days (g–i) and stained with Verhoeff–van Gieson. Note small or even absent media in (f, g, and i). (From Wittchow, E., N. Adden, J. Riedmuller et al., *EuroIntervention*, 8, 1441–50, 2013. With permission.)

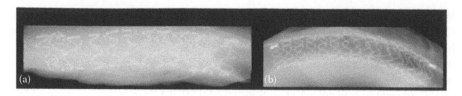

FIGURE 14.16 High-resolution faxitron evaluation from a porcine coronary model after 90 days of implantation. At this time point, faster dismantling rate and resorption of the scaffold DREAMS 1G (a) than its latest development, the DREAMS 2G (b) could be detected. (From Campos, C. M., T. Muramatsu, J. Iqbal et al., *International Journal of Molecular Sciences*, 14, 24492–500, 2013. With permission.)

and possible postdilation more precise. Previous studies have shown that the use of sirolimus-eluting stents resulted in fewer major adverse cardiac events (Windecker et al. 2005; Daemen et al. 2007; Stone et al. 2007). The DREAMS 2G was coated with a bioresorbable polylactic acid polymer (7 μm) featuring sirolimus at a dose of 1.4 μg/mm². DREAMS 2G has completed preclinical assessment and is currently being evaluated in a BIOSOLVE-II trial.

In addition to the AMS series, a few other biodegradable Mg stents were investigated. The MZX stent, which was made of a noncommercial alloy, contains Mg, zinc, and other elements and caused significantly higher neointimal thickness and percentage of stenosis when compared to platinum chromium bare metal stents at 30 days (Deng et al. 2011). H. Li et al. (2011) developed an AZ31-based sirolimus-eluting stent, which degraded completely after 120 days in a rabbit model. Sirolimus-eluting AZ31B stents reduced intimal hyperplasia and improved the lumen area when compared to uncoated AZ31B stents, but delayed vascular healing and endothelialization.

14.2.4 CLINICAL TRIAL OF BIODEGRADABLE MG ALLOY STENTS

The AMS-1 has been evaluated in three clinical trials for the treatment of lower limb lesions and coronary lesions. Another application of AMS-1 is the treatment of pulmonary atresia and aortic coarctation of newborns. The safety and performance of the first-generation DREAMS have also been assessed. The results are summarized in Table 14.4.

Traditionally, the implantation of stents in young children with small blood vessels has been avoided as the presence of a metallic or fixed stent structure limits the ability to further dilate the stent with vessel growth. The biodegradable stent can address this problem. The first successful implantation of a biodegradable metal stent in a human was performed by Zartner et al. in the left pulmonary artery of a preterm baby with a congenital heart disease (Zartner et al. 2005). Reperfusion of the left lung was established and persisted throughout the 4-month follow-up period, and the degradation process of the stent was clinically well tolerated despite the small size of the baby. Unfortunately, the patient died from multiple organ failure caused by severe pneumonia about 5 months after implantation. At the autopsy, no evidence was found of an adverse effect of the stent on the cardiac situation.

TABLE 14.4
Summary of Clinical Trial of Mg Alloy Stents within Blood Vessels

	Alloy	Biocompatibility	Degradation	References
Pediatric use	AMS (preterm baby, pulmonary artery)	Normal serum Mg level at 72 h; persistent left lung perfusion throughout the 4-month follow-up; the stent struts were substituted by a jelly-like $CaPO_4$ compound and fibrotic structure	~5 months	Zartner et al. 2005
	AMS (newborn baby, aortic arch)	Restenosis after 3 weeks implantation; implantation of a 2nd AMS; only residual Mg metal struts without stability forces were visible after 6 weeks after the implantation of the second stent; normal Mg levels in serum	—	Schranz et al. 2006; Zartner et al. 2007
	AMS (2-month-old girl, aorto-pulmonary collateral)	Restenosis after 4 months of implantation of AMS	—	McMahon et al. 2007
Clinical trial	AMS (20 patients, 23 stents, lower limb vascular)	A low immediate elastic recoil; 89.5% primary clinical patency after 3 months; no blood or vessel toxicity	6 weeks	Peeters et al. 2005; Bosiers et al. 2006
	AMS INSIGHT (60 patients, 74 stents, lower limb vascular)	Lower angiographic patency rate 31.8% for AMS treatment and comparable complication rate 5% with PTA treatment (patency 58%, complication rate 5.3%)	~4 months	Bosiers et al. 2009
	PROGRESS-AMS (63 patients, 71 stents, eight centers, coronary artery)	Restenosis caused by stent recoil and intra- and extra-stent neointima at 4 months; the neointima decreased after over 12 months; no myocardial infarction, subacute or late thrombosis, or death occurred	~4 months	Erbel et al. 2007; Waksman et al. 2009
	BIOSOLVE-I DREAMS (46 patients, two cohorts, cohort 1 for 6 months and 2 for 12 months)	7% rate of target lesion failure, 4.7% revascularization rate, no significant change of vasoreactivity between 6 and 12 months, reduced lumen loss from 6 month to 12 month, no death and no thrombosis	—	Haude et al. 2013

Degradation of the Mg stent and disintegration of the metal struts were completed in this specimen, leaving small bulks of jelly-like calcium-phosphate compounds as shown in Figure 14.17. Some infiltration of neointimal cells into the stent strut relicts indicated progression of cellular substitution of the stent material by cells. Only a mild intimal proliferation with no stent-related inflammatory reaction was observed. The diameter of the stented vessel segment was 3.7 mm, which is larger than the diameter of the stent, demonstrating some vessel growth after degradation of the resorbable metal stent without further intervention.

Schranz et al. (2006) reported the use of an AMS for the treatment of aortic coarctation in a newborn, which was not very successful. At first, an AMS with a length of 15 mm premounted on a 3.5-mm balloon was implanted in the descending aortic arch of this 3-week-old male newborn. Restenosis occurred after 3 weeks implantation, resulting from the degradation of the stent. Then, a second $4 \times 15\text{-mm}^2$ AMS was implanted. Only residual Mg metal struts without stability forces were visible intraoperatively when ventricular septal defect closure was performed at the age of 3 months (6 weeks after the implantation of the second stent). Despite the use of two metal stents, pathological Mg levels in serum of the patient were not detected.

In another case, McMahon et al. (2007) reported a 2-month-old girl with pulmonary atresia, VSD, and multiple aorto-pulmonary collaterals with severely hypoplastic pulmonary arteries who underwent placement of a biodegradeable Mg stent within a stenotic aorto-pulmonary collateral. Although there was an initial

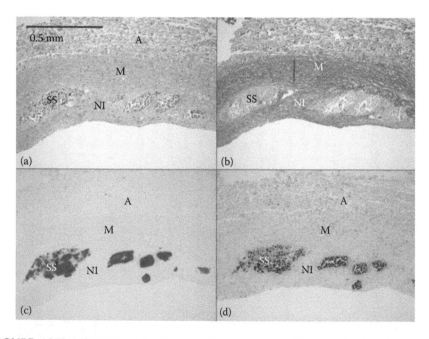

FIGURE 14.17 Hematoxylin/eosin stain (a), elastica van Gieson stain (b), alizarin red stain (c), and von Kossa's stain (d) of A, adventitia; M, media; SS, stent strut; NI, neointima. (From Zartner, P., M. Buettner, H. Singer, and M. Sigler, *Catheterization and Cardiovascular Interventions*, 69, 443–6, 2007.)

significant increase in vessel diameter, significant restenosis occurred 4 months after stent placement.

The first clinical trial to test the feasibility and safety of the Mg bioabsorbable stent was performed in 20 patients with critical limb ischemia (CLI). After 3 months, the primary clinical patency and limb salvage rates were 89.5% and 100%, respectively, suggesting a potentially promising performance of these AMS devices in the treatment of below-knee lesions in CLI patients. Morphological analysis indicated that the stent was almost completely degraded 6 weeks after implantation.

The AMS INSIGHT is a prospective, multicenter, randomized study, designed to evaluate the safety and performance of the first-generation AMS for the treatment of infrapopliteal lesions in patients with critical limb ischemia (CLI). The 6-month angiographic patency rate for lesions treated with AMS (31.8%) was significantly lower than the rate for those treated with percutaneous transluminal angioplasty (PTA, 58.0%). Although the study indicates that the AMS technology can be safely applied, it did not demonstrate efficacy in long-term patency over standard PTA in the infrapopliteal vessels.

PROGRESS AMS is the first-in-man study of AMS in coronary arteries. It included 63 patients at eight international clinical sites. The study's primary endpoint was major adverse cardiac events (MACEs) at 4 months defined as cardiac death, nonfatal myocardial infarction (MI), and ischemia driven TLR. Briefly, 71 stents, 10–15 mm in length and 3.0–3.5 mm in diameter, were successfully implanted after predilatation in 63 patients. Diameter stenosis (DS) was reduced from 61.5% ± 13.1% to 12.6% ± 5.6% with an acute gain of 1.41 ± 0.46 mm. No MI, subacute or late thrombosis, or death occurred. The ischemia-driven TLR rate was 23.8%. Four-month angiography showed an increased DS to 48.4% ± 17.0%. On the basis of the serial IVUS examinations, it was found that only small remnants of the original struts were visible and well embedded into the intima. The main cause of restenosis in this study was attributed to recoil, which suggested that longer degradation rates are needed to provide support to the vessel after intervention.

The first-in-man BIOSOLVE-I trial assessed the safety and performance of the first-generation drug-eluting absorbable metal scaffold (DREAMS) in 46 patients with 47 lesions at five European centers. Overall device and procedural success was 100%. Two of the 46 (4%) patients had target lesion failure at 6 months (both clinically driven target lesion revascularizations), which rose to three of 43 (7%) at 12 months (one periprocedural target vessel myocardial infarction occurred during angiography at the 12-month follow-up visit). An illustrative case of DREAMS with OCT images is shown in Figure 14.18. No cardiac death or scaffold thrombosis happened. The in-scaffold late lumen loss was reduced at 6 months (0.65 ± 0.5 mm) and at 12 months (0.52 ± 0.39 mm) compared to 1.08 ± 0.49 mm of the prior generation bare AMS-1 Mg scaffold. In addition, the natural vessel angulation was observed in the 6-month follow-up and maintained at 12 months, suggesting the real restoration of the vessel's natural architecture. The rate of target lesion failure with DREAMS at 6-month and 12-month follow-up is similar to that with contemporary drug-eluting stents and the bioabsorbable everolimus-eluting coronary scaffold system. However, the late lumen loss with DREAMS still did not match the excellent results of currently available drug-eluting stents. The study suffers from "the absence of a direct

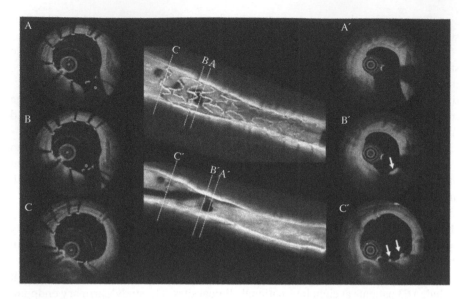

FIGURE 14.18 Postimplantation and 12-month follow-up optical coherence tomography (OCT; LightLab Imaging, Westford, MA, USA) of a percutaneous coronary intervention of the left anterior descending coronary artery, whereby a 3.25 × 16 mm paclitaxel-eluting absorbable metal scaffold (DREAMS 1G; Biotronik, Bülach, Switzerland) was implanted. Postprocedurally, a side branch was jailed by the struts of DREAMS 1G (A and B). At 12 months follow-up, OCT showed a smooth luminal surface with moderate neointimal hyperplasia in the scaffolded segment. Just a few remnants of struts were still visible with shadows (C', marked by arrows). The struts overhanging a side branch ostium were partially replaced by a neointimal membranous bridge (B', arrow), and three-dimensional OCT revealed unobstructed and widely opened ostium of side branch. (From Campos, C. M., T. Muramatsu, J. Iqbal et al., *International Journal of Molecular Sciences*, 14, 24492–500, 2013. With permission.)

comparison with other permanent stents or scaffolds." Furthermore, it included only a few patients with simple lesions, thus the results cannot be generalized to other types of lesions, and findings need to be confirmed in larger studies (Lim 2013).

14.3 IN VIVO TESTING OF BIODEGRADABLE Mg ALLOY IMPLANTS WITHIN THE NONVASCULAR LUMEN

There are limitations of currently used stents for the treatment of biliary strictures. Permanent metal stents (often made of titanium) may induce sludge accumulation and epithelial hyperplasia, and polymeric stents (made of PE, PVC, PU, or PTFE) lack radial force. Biodegradable biliary stents made of Mg and its alloys may address the problems. Chen et al. (2014b) implanted Mg-6Zn stents in the common bile duct (CBD) of rabbits. The results revealed no significant differences ($P > 0.05$) in serum Mg, CREA, BUN, LPS, TB, or GPT before and after the operation. The Mg-6Zn stents did not harm the function or morphology of the CBD, kidney, pancreas, or liver. However, only 9% of the original weight remained after 3 weeks, which cannot meet the clinical requirement.

Mg-6Zn was also evaluated for use in the pins of circular staplers for gastrointestinal reconstruction. Mg-6Zn alloy pins, 5 × 1 × 1 mm, were embedded in the cecum incision of rats. The edge Mg-6Zn of the implants became fuzzy in the radiographs (Figure 14.19), indicating the degradation of Mg-6Zn. The degradation did not have an impact on serum Mg. Superior to the Ti-3Al-2.5V alloy, the Mg-6Zn alloy enhanced the expression of transforming growth factor-β1 in healing tissue and promoted the expression of both the vascular endothelial growth factor and the basic fibroblast growth factor, which helped angiogenesis and healing. The Mg-6Zn alloy reduced the expression of the tumor necrosis factor (TNF-α) at different stages and decreased inflammatory response, which may have been related to the zinc inhibiting TNF-α. In general, the Mg-6Zn alloy performed better than Ti-3Al-2.5V at promoting healing and reducing inflammation.

Mg alloys are also considered for ureteral stent applications. The bacterial colony formation coupled with the encrustation on the stent surface from extended use often leads to clinical complications and contributes to the failure of indwelling medical devices. Lock et al. (2014) demonstrated that Mg4Y alloy decreased Escherichia coli viability and reduced the colony-forming units over a 3-day incubation period in an artificial urine (AU) solution when compared with a currently used commercial polyurethane stent. Further studies are needed for clinical translation of biodegradable metallic ureteral stents.

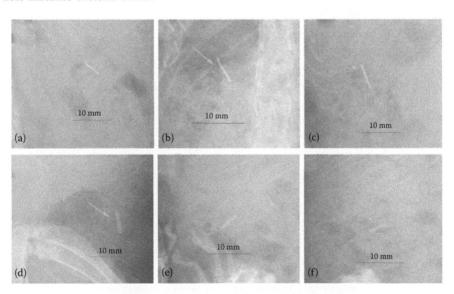

FIGURE 14.19 Radiography assessment of rats: no change was radiographically detected in the Ti group; the Mg–6Zn implant started to degrade in the third week as depicted. No gas bubbles appeared around the implants. Titanium stick: (a) first week; (b) second week; (c) third week. Mg–6Zn alloy: (d) first week; (e) second week; (f) third week. (From Yan, J., Y. Chen, Q. Yuan et al., *Biomedical Materials*, 9, 025011, 2014. With permission.)

14.4 CONCLUDING REMARKS

First-generation biomaterials are bioinert materials that achieve a suitable combination of physical properties to match those of the replaced tissue with a minimal toxic response in the host. Whereas second-generation biomaterials were designed to be either resorbable or bioactive, the next generation of biomaterials is combining these two properties with the aim of developing materials that, once implanted, will help the body heal itself (Hench and Polak 2002). Undoubtedly, biodegradable Mg alloys are the rising stars as the next generation of metallic biomaterials.

The successes of the animal tests and clinical trials of Mg-based implants within bone, blood vessels, and the nonvascular lumen give us great confidence in the future of Mg-based implants and devices. In addition to the implants and devices mentioned above, Mg-based interbody fusion cages, rib fixation plates, and staplers for gastrointestinal anastomosis have been developed as shown in Figures 14.20 and 14.21. The results of in vivo tests can be expected in the near future. Mg alloys are also considered to be materials for biodegradable septal occluders, which should provide benefits, such as a decreased risk of long-term complications when compared

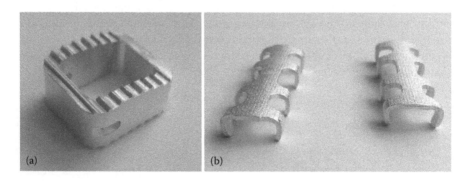

FIGURE 14.20 Mg-based (a) interbody fusion cage and (b) rib fixation plate. (Courtesy Suzhou Origin Medical Technology Co., Ltd.)

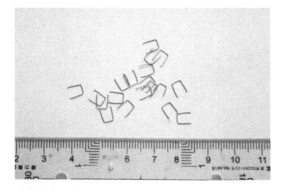

FIGURE 14.21 Mg-based stapler for gastrointestinal anastomosis. (Courtesy Suzhou Origin Medical Technology Co., Ltd.)

with their metallic counterparts (i.e., late erosion), potential improvements in MR safety (heating and artifact), and no risk of nickel allergy (Hijazi et al. 2010). An attempt failed because of the insufficient mechanical properties of Mg (Weishaupt et al. 2013). Clearly, improvement of the mechanical properties and more predictable degradation behaviors of Mg-based devices will be helpful for broadening their use.

The choices among degradable metallic devices and nondegradable metallic degradable polymer devices must be carefully weighed considering many factors, such as the patient age (child or adult) and personal physical condition, the type of lesion, the risk of infection, etc. But it can be conjectured that, at the beginning, the competition between permanent metallic materials and biodegradable Mg alloys is relatively low; however, the competition between biodegradable Mg alloys and biodegradable polymers might be relatively high (Zheng et al. 2014).

The combination/integration with other biomaterials, especially the biodegradable ceramics and polymers with advanced techniques for materials fabrication, will be a promising direction for future development. Figure 14.22 shows a biodegradable poly-lactic acid based-composite reinforced unidirectionally with high-strength magnesium alloy wires (MAWs) which could provide a great impact-bearing ability for bone fracture fixation (Li et al. 2015). Implantable biodegradable electronic devices are another area wide open for exploration and innovation. A pioneer work (Hwang et al. 2012) uses Mg for the conductors, magnesium oxide (MgO) (silicon dioxide, SiO_2, is also possible) for the dielectrics, monocrystalline silicon (Si) nanomembranes (NMs) for the semiconductors, and silk (which is water soluble and enzymatically degradable) for the substrate and packaging material. Inductive coils of Mg combined with resistive microheaters of doped Si NMs, integrated on silk

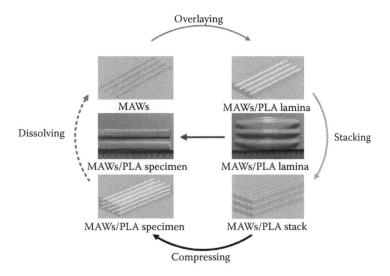

FIGURE 14.22 Illustration of preparation process of the MAWs/PLA composites and the pictures in the center are those of the as-prepared specimens. (From Li, X., C. L. Chu, L. Liu et al., *Biomaterials*, 49, 135, 2015. With permission.)

substrates and housed in silk packages, provided transient thermal therapy to control surgical site infections as a programmable nonantibiotic bacteriocide. In the future, various transient electronic components, circuits, and sensors, including simple integrated circuits and sensor arrays, might be designed with biodegradable Mg alloys, and they would function for medically useful time frames but then completely disappear via resorption by the body.

There is still a long way to go until biodegradable Mg alloys are used widely in clinical applications. There are still a lot of unknowns about biodegradable alloys, such as the exact metabolic pathways of the alloying elements. We expect improvement in the mechanical properties, better control of the corrosion rate, deeper understanding of the biological functions, and more ingenious design of the devices.

References

Abdal-hay, A., N. A. M. Barakat, and J. K. Lim. 2013. Hydroxyapatite-doped poly(lactic acid) porous film coating for enhanced bioactivity and corrosion behavior of AZ31 Mg alloy for orthopedic applications. *Ceramics International* 39 (1):183–95.

Abdal-hay, A., M. Dewidar, J. Lim, and J. K. Lim. 2014. Enhanced biocorrosion resistance of surface modified magnesium alloys using inorganic/organic composite layer for biomedical applications. *Ceramics International* 40 (1):2237–47.

Abidin, N. I. Z., D. Martin, and A. Atrens. 2011a. Corrosion of high purity Mg, AZ91, ZE41 and Mg2Zn0.2Mn in Hank's solution at room temperature. *Corrosion Science* 53 (3):862–72.

Abidin, N. I. Z., A. D. Atrens, D. Martin, and A. Atrens. 2011b. Corrosion of high purity Mg, Mg2Zn0.2Mn, ZE41 and AZ91 in Hank's solution at 37 degrees C. *Corrosion Science* 53 (11):3542–56.

Abidin, N. I. Z., B. Rolfe, H. Owen, J. Malisano, D. Martin, J. Hofstetter, P. J. Uggowitzer, and A. Atrens. 2013. The in vivo and in vitro corrosion of high-purity magnesium and magnesium alloys WZ21 and AZ91. *Corrosion Science* 75:354–66.

Abshire, B. B., R. F. McLain, A. Valdevit, and H. E. Kambic. 2001. Characteristics of pullout failure in conical and cylindrical pedicle screws after full insertion and back out. *The Spine Journal* 1:408–414.

Abu Leil, T., N. Hort, W. Dietzel, C. Blawert, Y. Huang, K. U. Kainer, and K. P. Rao. 2009. Microstructure and corrosion behavior of Mg-Sn-Ca alloys after extrusion. *Transactions of Nonferrous Metals Society of China* 19 (1):40–4.

Afrin, N., D. L. Chen, X. Cao, and M. Jahazi. 2008. Microstructure and tensile properties of friction stir welded AZ31B magnesium alloy. *Materials Science and Engineering: A* 472 (1–2):179–86.

Aghion, E., and G. Levy. 2010. The effect of Ca on the in vitro corrosion performance of biodegradable Mg–Nd–Y–Zr alloy. *Journal of Materials Science* 45 (11):3096–101.

Aghion, E., Y. Gueta, N. Moscovitch, and B. Bronfin. 2008. Effect of yttrium additions on the properties of grain-refined Mg–3% Nd alloy. *Journal of Materials Science* 43 (14):4870–5.

Aghion, E., T. Yered, Y. Perez, and Y. Gueta. 2010. The prospects of carrying and releasing drugs via biodegradable magnesium foam. *Advanced Engineering Materials* 12 (8):B374–9.

Aghion, E., G. Levy, and S. Ovadia. 2012. In vivo behavior of biodegradable Mg-Nd-Y-Zr-Ca alloy. *Journal of Materials Science. Materials in Medicine* 23 (3):805–12.

Ai, Y. L., C. P. Luo, and J. W. Liu. 2007. Twinning of CaMgSi phase in a cast Mg-1.0Ca-0.5Si-0.3Zr alloy. *Acta Materialia* 55 (2):531–8.

AIST (National Institute of Advanced Industrial Science and Technology). 2010. Development of a new rolling process for commercial magnesium alloy sheets with high room-temperature formability. Press release, January 26. http://www.aist.go.jp/aist_e/latest _research/2010/20100217/20100217.html (accessed May 30, 2015).

Alabbasi, A., A. Mehjabeen, M. B. Kannan, Q. Ye, and C. Blawert. 2014a. Biodegradable polymer for sealing porous PEO layer on pure magnesium: An in vitro degradation study. *Applied Surface Science* 301:463–7.

Alabbasi, A., M. B. Kannan, and C. Blawert. 2014b. Dual layer inorganic coating on magnesium for delaying the biodegradation for bone fixation implants. *Materials Letters* 124:188–91.

Al-Abdullat, Y., S. Tsutsumi, N. Nakajima, M. Ohta, H. Kuwahara, and K. Ikeuchi. 2001. Surface modification of magnesium by $NaHCO_3$ and corrosion behavior in Hank's solution for new biomaterial applications. *Materials Transactions-JIM* 42 (8):1777–80.

Aljarrah, M., and M. Medraj. 2008. Thermodynamic modelling of the Mg-Ca, Mg-Sr, Ca-Sr and Mg-Ca-Sr systems using the modified quasichemical model. *Calphad-Computer Coupling of Phase Diagrams and Thermochemistry* 32 (2):240–51.

Al-Samman, T., and X. Li. 2011. Sheet texture modification in magnesium-based alloys by selective rare earth alloying. *Materials Science and Engineering: A* 528 (10–11):3809–22.

Alvarez-Lopez, M., M. D. Pereda, J. A. del Valle, M. Fernandez-Lorenzo, M. C. Garcia-Alonso, O. A. Ruano, and M. L. Escudero. 2010. Corrosion behaviour of AZ31 magnesium alloy with different grain sizes in simulated biological fluids. *Acta Biomaterialia* 6 (5):1763–71.

Amaravathy, P., C. Rose, S. Sathiyanarayanan, and N. Rajendran. 2012. Evaluation of in vitro bioactivity and MG63 oesteoblast cell response for TiO_2 coated magnesium alloys. *Journal of Sol-Gel Science and Technology* 64 (3):694–703.

American Society for Testing Materials, Standard F1264-03: Standard Specification and Test Methods for Intramedullary Fixation Devices. Philadelphia, PA.

American Society for Testing Materials, Standard F1717-04: Standard Test Methods for Spinal Implant Constructs in a Vertebrectomy Model. Philadelphia, PA.

American Society for Testing Materials, Standard F543-07: Standard Specification and Test Methods for Metallic Medical Bone Screw. Philadelphia, PA.

Andrews, E. W. 1917. Absorbable metal clips as substitutes for ligatures and deep sutures in wound closure. *Journal of the American Medical Association* 69 (4):278–81.

Anilchandra, A. R., and M. K. Surappa. 2012. Microstructure and damping behaviour of consolidated magnesium chips. *Materials Science and Engineering: A* 542:94–103.

Antunes, R. A., and M. C. de Oliveira. 2012. Corrosion fatigue of biomedical metallic alloys: Mechanisms and mitigation. *Acta Biomaterialia* 8 (3):937–62.

Asgharzadeh, H., E. Y. Yoon, H. J. Chae, T. S. Kim, J. W. Lee, and H. S. Kim. 2014. Microstructure and mechanical properties of a Mg-Zn-Y alloy produced by a powder metallurgy route. *Journal of Alloys and Compounds* 586:S95–100.

ASM. 1990. *Metals Handbook*, 10th Edition. ASM International, Russell Township, Geauga County, OH. 2:455–79.

Assadian, M., M. H. Idris, S. Mehraban, S. Jafari, and M. R. A. Kadir. 2012. Effect of HF concentration on corrosion resistance of biomedical implants. *Advanced Materials Research* 463–4:837–40.

ASTM. 2004. G31-72, Standard practice for laboratory immersion corrosion testing of metals. *Book of ASTM Standards* 3.

ASTM. F 756-00, Standard Practice for Assessment of Hemolytic Properties of Materials.

Atchison, G. J., and W. H. Beamer. 1952. Determination of trace impurities in magnesium by activation analysis. *Analytical Chemistry* 24 (11):1812–15.

Atrens, A., M. Liu, and N. I. Zainal Abidin. 2011a. Corrosion mechanism applicable to biodegradable magnesium implants. *Materials Science and Engineering: B* 176 (20):1609–36.

Atrens, A., M. Liu, N. I. Zainal Abidin, and G.-L. Song. 2011b. Corrosion of magnesium (Mg) alloys and metallurgical influence. In G.-L. Song (ed.), *Corrosion of Magnesium Alloys*, pp. 117–65. Woodhead Publishing, Cambridge, UK.

Aurbach, D., G. S. Suresh, E. Levi, A. Mitelman, O. Mizrahi, O. Chusid, and M. Brunelli. 2007. Progress in rechargeable magnesium battery technology. *Advanced Materials* 19 (23):4260–7.

Avedesian, M. M., and H. Baker. 1999. *ASM Specialty Handbook: Magnesium and Magnesium Alloys Materials Park*, The Materials Information Society, ASM International, Russell Township, Geauga County, OH.

Avery, C. M. E., P. Bujtár, J. Simonovics, T. Dézsi, K. Váradi, G. K. B. Sándor, and J. Pan. 2013. A finite element analysis of bone plates available for prophylactic internal fixation of the radial osteocutaneous donor site using the sheep tibia model. *Medical Engineering and Physics* 35:1421–30.

Aydin, D. S., Z. Bayindir, M. Hoseini, and M. O. Pekguleryuz. 2013. The high temperature oxidation and ignition behavior of Mg–Nd alloys. Part I: The oxidation of dilute alloys. *Journal of Alloys and Compounds* 569:35–44.

Aydin, D. S., Z. Bayindir, and M. O. Pekguleryuz. 2014. The high temperature oxidation behavior of Mg–Nd alloys. Part II: The effect of the two-phase microstructure on the on-set of oxidation and on oxide morphology. *Journal of Alloys and Compounds* 584:558–65.

Azaouzi, M., A. Makradi, and S. Belouettar. 2012. Deployment of a self-expanding stent inside an artery: A finite element analysis. *Materials and Design* 41:410–20.

Azaouzi, M., A. Makradi, and S. Belouettar. 2013a. Numerical investigations of the structural behavior of a balloon expandable stent design using finite element method. *Computational Materials Science* 72:54–61.

Azaouzi, M., N. Lebaal, A. Makradi, and S. Belouettar. 2013b. Optimization based simulation of self-expanding Nitinol stent. *Materials and Design* 50:917–28.

Azaouzi, M., A. Makradi, J. Petit, S. Belouettar, and O. Polit. 2013c. On the numerical investigation of cardiovascular balloon-expandable stent using finite element method. *Computational Materials Science* 79:326–35.

Azevedo, C. R. F. 2003. Failure analysis of a commercially pure titanium plate for osteosynthesis. *Engineering Failure Analysis* 10 (2):153–64.

Bach, F. W., A. Szelagowski, R. Versemann, and M. Zelt. 2003. Welding of magnesium alloys by means of non-vacuum electron-beam welding. *Welding in the World* 47 (3–4):4–10.

Bae, D. H., S. H. Kim, D. H. Kim, and W. T. Kim. 2002a. Deformation behavior of Mg-Zn-Y alloys reinforced by icosahedral quasicrystalline particles. *Acta Materialia* 50 (9):2343–56.

Bae, D. H., M. H. Lee, K. T. Kim, W. T. Kim, and D. H. Kim. 2002b. Application of quasicrystalline particles as a strengthening phase in Mg-Zn-Y alloys. *Journal of Alloys and Compounds* 342 (1–2):445–50.

Bai, J., Y. S. Sun, S. Xun, F. Xue, and T. B. Zhu. 2006. Microstructure and tensile creep behavior of Mg-4Al based magnesium alloys with alkaline-earth elements Sr and Ca additions. *Materials Science and Engineering A-Structural Materials Properties Microstructure and Processing* 419 (1–2):181–8.

Bai, J., Y. Sun, F. Xue, and J. Qiang. 2012a. Microstructures and creep properties of Mg–4Al–(1–4) La alloys produced by different casting techniques. *Materials Science and Engineering: A* 552:472–80.

Bai, J., Y. S. Sun, F. Xue, and J. Zhou. 2012b. Microstructures and creep behavior of as-cast and annealed heat-resistant Mg-4Al-2Sr-1Ca alloy. *Materials Science and Engineering A-Structural Materials Properties Microstructure and Processing* 531:130–40.

Bai, K., Y. Zhang, Z. Fu, C. Zhang, X. Cui, E. Meng, S. Guan, and J. Hu. 2012. Fabrication of chitosan/magnesium phosphate composite coating and the in vitro degradation properties of coated magnesium alloy. *Materials Letters* 73:59–61.

Bakhsheshi-Rad, H. R., M. R. Abdul-Kadir, M. H. Idris, and S. Farahany. 2012a. Relationship between the corrosion behavior and the thermal characteristics and microstructure of Mg-0.5Ca-xZn alloys. *Corrosion Science* 64:184–97.

Bakhsheshi-Rad, H. R., M. H. Idris, M. R. A. Kadir, and S. Farahany. 2012b. Microstructure analysis and corrosion behavior of biodegradable Mg-Ca implant alloys. *Materials and Design* 33:88–97.

Bakhsheshi-Rad, H. R., M. H. Idris, and M. R. Abdul-Kadir. 2013a. Synthesis and in vitro degradation evaluation of the nano-HA/MgF$_2$ and DCPD/MgF$_2$ composite coating on biodegradable Mg–Ca–Zn alloy. *Surface and Coatings Technology* 222:79–89.

Bakhsheshi-Rad, H. R., M. H. Idris, M. R. Abdul-Kadir, and M. Daroonparvar. 2013b. Effect of fluoride treatment on corrosion behavior of Mg–Ca binary alloy for implant application. *Transactions of Nonferrous Metals Society of China* 23 (3):699–710.

Bakhsheshi-Rad, H. R., E. Hamzah, M. Daroonparvar, R. Ebrahimi-Kahrizsangi, and M. Medraj. 2014a. In-vitro corrosion inhibition mechanism of fluorine-doped hydroxyapatite and brushite coated Mg–Ca alloys for biomedical applications. *Ceramics International* 40 (6):7971–82.

Bakhsheshi-Rad, H. R., M. H. Idris, M. R. Abdul-Kadir, A. Ourdjini, M. Medraj, M. Daroonparvar, and E. Hamzah. 2014b. Mechanical and bio-corrosion properties of quaternary Mg-Ca-Mn-Zn alloys compared with binary Mg-Ca alloys. *Materials and Design* 53:283–92.

Baricco, M., A. Castellero, M. Di Chio, Z. S. Kovacs, P. Rizzi, M. Satta, and A. Ziggiotti. 2007. Thermal stability and hardness of Mg–Cu–Au–Y amorphous alloys. *Journal of Alloys and Compounds* 434:183–6.

Barlis, P., J. Tanigawa, and C. Di Mario. 2007. Coronary bioabsorbable magnesium stent: 15-month intravascular ultrasound and optical coherence tomography findings. *European Heart Journal* 28 (19):2319.

Bartos, S., C. Laush, J. Scharfenberg, and R. Kantamaneni. 2007. Reducing greenhouse gas emissions from magnesium die casting. *Journal of Cleaner Production* 15 (10): 979–87.

Barucca, G., R. Ferragut, D. Lussana, P. Mengucci, F. Moia, and G. Riontino. 2009. Phase transformations in QE22 Mg alloy. *Acta Materialia* 57 (15):4416–25.

Beausir, B., S. Biswas, D. I. Kim, L. S. Tóth, and S. Suwas. 2009. Analysis of microstructure and texture evolution in pure magnesium during symmetric and asymmetric rolling. *Acta Materialia* 57 (17):5061–77.

Bell, R. B., and C. S. Kindsfater. 2006. The use of biodegradable plates and screws to stabilize facial fractures. *Journal of Oral and Maxillofacial Surgery* 64:31–9.

Ben-Hamu, G. E. A. 2007. The influence of Ag and Si additions on the electrochemical behavior in extruded Mg-Zn alloys. *Israel Journal of Chemistry* 47 (3–4):309–17.

Ben-Hamu, G., D. Eliezer, A. Kaya, Y. G. Na, and K. S. Shin. 2006a. Microstructure and corrosion behavior of Mg–Zn–Ag alloys. *Materials Science and Engineering A-Structural Materials Properties Microstructure and Processing* 435–6:579–87.

Ben-Hamu, G., D. Eliezer, and K. S. Shin. 2006b. Influence of Si, Ca and Ag addition on corrosion behaviour of new wrought Mg–Zn alloys. *Materials Science and Technology* 22 (10):1213–18.

Berglund, I. S., H. S. Brar, N. Dolgova, A. P. Acharya, B. G. Keselowsky, M. Sarntinoranont, and M. V. Manuel. 2012. Synthesis and characterization of Mg-Ca-Sr alloys for biodegradable orthopedic implant applications. *Journal of Biomedical Materials Research Part B—Applied Biomaterials* 100B (6):1524–34.

Betsofen, S. Y., E. F., Volkova, and A. A. Shaforostov. 2011. Effect of alloying elements on the formation of rolling texture in Mg-Nd-Zr and Mg-Li alloys. *Russian Metallurgy (Metally)* 2011 (1):66–71.

Bettles, C. J., M. H. Moss, and R. Lapovok. 2009. A Mg–Al–Nd alloy produced via a powder metallurgical route. *Materials Science and Engineering: A* 515 (1–2):26–31.

Beyerlein, I. J., R. J. McCabe, and C. N. Tomé. 2011. Effect of microstructure on the nucleation of deformation twins in polycrystalline high-purity magnesium: A multi-scale modeling study. *Journal of the Mechanics and Physics of Solids* 59 (5):988–1003.

Bhan, S., and A. Lal. 1993. The Mg-Zn-Zr system (magnesium-zinc-zirconium). *Journal of Phase Equilibria* 14 (5):634–7.

Bhattacharjee, T., C. L. Mendis, T. T. Sasaki, T. Ohkubo, and K. Hono. 2012. Effect of Zr addition on the precipitation in Mg-Zn-based alloy. *Scripta Materialia* 67 (12):967–70.

Bi, G., D. Fang, L. Zhao, J. Lian, Q. Jiang, and Z. Jiang. 2011a. An elevated temperature Mg–Dy–Zn alloy with long period stacking ordered phase by extrusion. *Materials Science and Engineering: A* 528 (10–11):3609–14.

Bi, G., D. Fang, L. Zhao, Q. Zhang, J. Lian, Q. Jiang, and Z. Jiang. 2011b. Double-peak ageing behavior of Mg–2Dy–0.5 Zn alloy. *Journal of Alloys and Compounds* 509 (32):8268–75.

Bi, G., D. Fang, W. Zhang, J. Sudagar, Q. Zhang, J. Lian, and Z. Jiang. 2012. Microstructure and mechanical properties of an extruded Mg-2Dy-0.5 Zn alloy. *Journal of Materials Science and Technology* 28 (6):543–51.

Bian, Z., I. Bayandorian, H. W. Zhang, G. Scamans, and Z. Fan. 2009. Extremely fine and uniform microstructure of magnesium AZ91D alloy sheets produced by melt conditioned twin roll casting. *Materials Science and Technology* 25 (5):599–606.

Birbilis, N., M. A. Easton, A. D. Sudholz, S. M. Zhu, and M. A. Gibson. 2009. On the corrosion of binary magnesium-rare earth alloys. *Corrosion Science* 51 (3):683–9.

Birbilis, N., G. Williams, K. Gusieva, A. Samaniego, M. A. Gibson, and H. N. McMurray. 2013. Poisoning the corrosion of magnesium. *Electrochemistry Communications* 34:295–8.

Biswas, S., S. S. Dhinwal, and S. Suwas. 2010. Room-temperature equal channel angular extrusion of pure magnesium. *Acta Materialia* 58 (9):3247–61.

Bobe, K., E. Willbold, I. Morgenthal et al. 2013. In vitro and in vivo evaluation of biodegradable, open-porous scaffolds made of sintered magnesium W4 short fibres. *Acta Biomaterialia* 9 (10):8611–23.

Boehlert, C. J., and K. Knittel. 2006. The microstructure, tensile properties, and creep behavior of Mg-Zn alloys containing 0-4.4 wt.% Zn. *Materials Science and Engineering A-Structural Materials Properties Microstructure and Processing* 417 (1–2):315–21.

Bohlen, J., M. R. Nürnberg, J. W. Senn, D. Letzig, and S. R. Agnew. 2007. The texture and anisotropy of magnesium–zinc–rare earth alloy sheets. *Acta Materialia* 55 (6): 2101–12.

Boivin, G., P. Deloffre, B. Perrat et al. 1996. Strontium distribution and interactions with bone mineral in monkey iliac bone after strontium salt (S 12911) administration. *Journal of Bone and Mineral Research* 11 (9):1302–11.

Bonarski, B. J., E. Schafler, B. Mingler, W. Skrotzki, B. Mikulowski, and M. J. Zehetbauer. 2008. Texture evolution of Mg during high-pressure torsion. *Journal of Materials Science* 43 (23–24):7513–18.

Bonarski, B. J., E. Schafler, B. Mikulowski, and M. J. Zehetbauer. 2010. Effects of recrystallization on texture, microstructure and mechanical properties in HPT-deformed pure Mg. Paper read at Journal of Physics: Conference Series.

Bondarenko, A., M. Hewicker-Trautwein, N. Erdmann, N. Angrisani, J. Reifenrath, and A. Meyer-Lindenberg. 2011. Comparison of morphological changes in efferent lymph nodes after implantation of resorbable and non-resorbable implants in rabbits. *BioMedical Engineering OnLine* 10 (1):32.

Borkar, H., and M. Pekguleryuz. 2013. Microstructure and texture evolution in Mg-1%Mn-Sr alloys during extrusion. *Journal of Materials Science* 48 (4):1436–47.

Borkar, H., M. Hoseini, and M. Pekguleryuz. 2012a. Effect of strontium on flow behavior and texture evolution during the hot deformation of Mg-1 wt%Mn alloy. *Materials Science and Engineering A-Structural Materials Properties Microstructure and Processing* 537:49–57.

Borkar, H., M. Hoseini, and M. Pekguleryuz. 2012b. Effect of strontium on the texture and mechanical properties of extruded Mg-1%Mn alloys. *Materials Science and Engineering A-Structural Materials Properties Microstructure and Processing* 549:168–75.

Borkar, H., R. Gauvin, and M. Pekguleryuz. 2013. Effect of extrusion temperature on texture evolution and recrystallization in extruded Mg-1% Mn and Mg-1% Mn-1.6% Sr alloys. *Journal of Alloys and Compounds* 555:219–24.

Bornapour, M., N. Muja, P. Shum-Tim, M. Cerruti, and M. Pekguleryuz. 2013. Biocompatibility and biodegradability of Mg-Sr alloys: The formation of Sr-substituted hydroxyapatite. *Acta Biomaterialia* 9 (2):5319–30.

Bose, S., and S. Tarafder. 2012. Calcium phosphate ceramic systems in growth factor and drug delivery for bone tissue engineering: A review. *Acta Biomaterialia* 8 (4):1401–21.

Bosetti, M., A. Massé, E. Tobin, and M. Cannas. 2002. Silver coated materials for external fixation devices in vitro biocompatibility and genotoxicity. *Biomaterials* 23:887–92.

Bosiers, M., K. Deloose, J. Verbist, and P. Peeters. 2006. Percutaneous transluminal angioplasty for treatment of "below-the-knee" critical limb ischemia: Early outcomes following the use of sirolimus-eluting stents. *The Journal of Cardiovascular Surgery* 47 (2):171–6.

Bosiers, M., P. Peeters, O. D'Archambeau et al. 2009. AMS INSIGHT—absorbable metal stent implantation for treatment of below-the-knee critical limb ischemia: 6-month analysis. *Cardiovascular and Interventional Radiology* 32 (3):424–35.

Bowen, P. K., J. Drelich, and J. Goldman. 2013a. A new in vitro–in vivo correlation for bioabsorbable magnesium stents from mechanical behavior. *Materials Science and Engineering: C* 33 (8):5064–70.

Bowen, P. K., J. Drelich, and J. Goldman. 2013b. Zinc exhibits ideal physiological corrosion behavior for bioabsorbable stents. *Advanced Materials* 25 (18):2577–82.

Bowen, P. K., A. Drelich, J. Drelich, and J. Goldman. 2014a. Rates of in vivo (arterial) and in vitro biocorrosion for pure magnesium. *Journal of Biomedical Materials Research Part A* 103 (1):341–49.

Bowen, P. K., J. Drelich, and J. Goldman. 2014b. Magnesium in the murine artery: Probing the products of corrosion. *Acta Biomaterialia* 10 (3):1475–83.

Brar, H. S., M. O. Platt, M. Sarntinoranont, P. I. Martin, and M. V. Manuel. 2009. Magnesium as a biodegradable and bioabsorbable material for medical implants. *JOM* 61 (9):31–4.

Brar, H. S., J. Wong, and M. V. Manuel. 2012. Investigation of the mechanical and degradation properties of Mg-Sr and Mg-Zn-Sr alloys for use as potential biodegradable implant materials. *Journal of the Mechanical Behavior of Biomedical Materials* 7:87–95.

Bronfin, B., and N. Moscovitch. 2006. New magnesium alloys for transmission parts. *Metal Science and Heat Treatment* 48 (11–12):479–86.

Brown, D., B.-M. Ma, and Z. Chen. 2002. Developments in the processing and properties of NdFeB-type permanent magnets. *Journal of Magnetism and Magnetic Materials* 248 (3):432–40.

Brubaker, C. O., and Z. K. Liu. 2004. A computational thermodynamic model of the Ca-Mg-Zn system. *Journal of Alloys and Compounds* 370 (1–2):114–22.

Bruce, D. W., B. E. Hietbrink, and K. P. DuBois. 1963. The acute mammalian toxicity of rare earth nitrates and oxides. *Toxicology and Applied Pharmacology* 5 (6):750–9.

Bruckner, J., R. Gunzel, E. Richter, and W. Moller. 1998. Metal plasma immersion ion implantation and deposition (MPIIID): Chromium on magnesium. *Surface and Coatings Technology* 104:227–30.

Brunelli, K., M. Dabala, I. Calliari, and M. Magrini. 2005. Effect of HCl pre-treatment on corrosion resistance of cerium-based conversion coatings on magnesium and magnesium alloys. *Corrosion Science* 47 (4):989–1000.

Bryła, K., J. Dutkiewicz, L. L. Rokhlin, L. Litynska-Dobrzynska, K. Mroczka, and P. Kurtyka. 2013. Microstructure and mechanical properties of Mg-2.5% Tb-0.78% Sm alloy after ECAP and ageing. *Archives of Metallurgy and Materials* 58 (2):481–7.

Buchholz, B. M., D. J. Kaczorowski, R. Sugimoto et al. 2008. Hydrogen inhalation ameliorates oxidative stress in transplantation induced intestinal graft injury. *American Journal of Transplantation* 8 (10):2015–24.

Buchholz, B. M., K. Masutani, T. Kawamura et al. 2011. Hydrogen-enriched preservation protects the isogeneic intestinal graft and amends recipient gastric function during transplantation. *Transplantation* 92 (9):985–92.

Busk, R. S. 1987. *Magnesium Products Design*: Taylor & Francis, New York.

Byrer, T. G., E. L. White, and P. D. Frost. 1964. The development of magnesium-lithium alloys for structural applications. DTIC Document.

Cabrini, M., P. Colombi, S. Lorenzi, and T. Pastore. 2014. Evaluation of corrosion resistance of biocompatible coatings on magnesium. *Metallurgia Italiana* 7–8:23–8.

Cai, J., Z. Kang, W. W. Liu et al. 2008. Hydrogen therapy reduces apoptosis in neonatal hypoxia–ischemia rat model. *Neuroscience Letters* 441 (2):167–72.

Cai, J., Z. Kang, K. Liu et al. 2009. Neuroprotective effects of hydrogen saline in neonatal hypoxia–ischemia rat model. *Brain Research* 1256:129–37.

Cai, K., X. Sui, Y. Hu, L. Zhao, M. Lai, Z. Luo, P. Liu, and W. Yang. 2011. Fabrication of anti-corrosive multilayer onto magnesium alloy substrates via spin-assisted layer-by-layer technique. *Materials Science and Engineering: C* 31 (8):1800–8.

Cai, S. H., T. Lei, N. F. Li, and F. F. Feng. 2012. Effects of Zn on microstructure, mechanical properties and corrosion behavior of Mg-Zn alloys. *Materials Science and Engineering C-Materials for Biological Applications* 32 (8):2570–7.

Campbell, J. H., and T. I. Suratwala. 2000. Nd-doped phosphate glasses for high-energy/high-peak-power lasers. *Journal of Non-Crystalline Solids* 263:318–41.

Campos, C. M., T. Muramatsu, J. Iqbal, Y.-J. Zhang, Y. Onuma, H. M. Garcia-Garcia, M. Haude, P. A. Lemos, B. Warnack, and P. W. Serruys. 2013. Bioresorbable drug-eluting magnesium-alloy scaffold for treatment of coronary artery disease. *International Journal of Molecular Sciences* 14 (12):24492–500.

Can, W., H. Peide, Z. Lu, Z. Caili, and X. Bingshe. 2011. First-principles study on the stabilities of the intermetallic compounds in Mg-Nd alloys. *Rare Metal Materials and Engineering* 40 (4):590–4.

Cao, X., M. Jahazi, J. P. Immarigeon, and W. Wallace. 2006. A review of laser welding techniques for magnesium alloys. *Journal of Materials Processing Technology* 171 (2):188–204.

Cao, J. D., N. T. Kirkland, K. J. Laws, N. Birbilis, and M. Ferry. 2012. Ca-Mg-Zn bulk metallic glasses as bioresorbable metals. *Acta Biomaterialia* 8 (6):2375–83.

Capurso, G., S. Lo Russo, A. Maddalena, A. Saccone, F. Gastaldo, and S. De Negri. 2013. Study on La–Mg based ternary system for hydrogen storage. *Journal of Alloys and Compounds* 580:S159–62.

Carboneras, M., L. S. Hernández, J. A. del Valle, M. C. García-Alonso, and M. L. Escudero. 2010. Corrosion protection of different environmentally friendly coatings on powder metallurgy magnesium. *Journal of Alloys and Compounds* 496 (1):442–8.

Carboneras, M., M. C. García-Alonso, and M. L. Escudero. 2011a. Biodegradation kinetics of modified magnesium-based materials in cell culture medium. *Corrosion Science* 53 (4):1433–9.

Carboneras, M., B. T. Peréz-Maceda, J. A. del Valle, M. C. García-Alonso, R. M. Lozano, and M. L. Escudero. 2011b. In vitro performance of magnesium processed by different routes for bone regeneration applications. *Materials Letters* 65 (19–20):3020–3.

Cardinal, J. S., J. Zhan, Y. Wang et al. 2009. Oral hydrogen water prevents chronic allograft nephropathy in rats. *Kidney International* 77 (2):101–9.

Carlsson, L., T. Röstlund, B. Albrektsson, and T. Albrektsson. 1988. Implant fixation improved by close fit cylindrical implant—Bone interface studied in rabbits. *Acta Orthopaedica* 59 (3):272–5.

Castellani, C., R. A. Lindtner, P. Hausbrandt et al. 2011. Bone-implant interface strength and osseointegration: Biodegradable magnesium alloy versus standard titanium control. *Acta Biomaterialia* 7 (1):432–40.

Castellero, A., B. Moser, D. I. Uhlenhaut, F. H. Dalla Torre, and J. F. Löffler. 2008. Room-temperature creep and structural relaxation of Mg–Cu–Y metallic glasses. *Acta Materialia* 56 (15):3777–85.

Castor, S. B., and J. B. Hedrick. 2006. Rare earth elements. *Industrial Minerals Volume*, 7th edition. Littleton, CO: Society for Mining, Metallurgy, and Exploration, 769–92.

Cava, R. J., H. Takagi, B. Batlogg, H. W. Zandbergen, J. J. Krajewski, W. F. Peck, R. B. Van
    Dover, R. J. Felder, T. Siegrist, and K. Mizuhashi. 1994. Superconductivity at 23 K in
    yttrium palladium boride carbide. *Nature* 367 (6459):146–8.
Celarek, A., T. Kraus, E. K. Tschegg et al. 2012. PHB, crystalline and amorphous magne-
    sium alloys: Promising candidates for bioresorbable osteosynthesis implants? *Materials
    Science and Engineering: C* 32 (6):1503–10.
Celikin, M., and M. Pekguleryuz. 2012. The role of α-Mn precipitation on the creep mecha-
    nisms of Mg–Sr–Mn. *Materials Science and Engineering: A* 556:911–20.
Celikin, M., A. A. Kaya, and M. Pekguleryuz. 2012. Microstructural investigation and the
    creep behavior of Mg-Sr-Mn alloys. *Materials Science and Engineering A-Structural
    Materials Properties Microstructure and Processing* 550:39–50.
Cha, P.-R., H.-S. Han, G.-F. Yang, Y.-C. Kim, K.-H. Hong, S.-C. Lee, J.-Y. Jung, J.-P.
    Ahn, Y.-Y. Kim, S.-Y. Cho, J. Y. Byun, K.-S. Lee, S.-J. Yang, and H.-K. Seok. 2013.
    Biodegradability engineering of biodegradable Mg alloys: Tailoring the electrochemical
    properties and microstructure of constituent phases. *Scientific Reports* 3:2367.
Chai, H. W., L. Guo, X. T. Wang et al. 2012. In vitro and in vivo evaluations on osteogenesis
    and biodegradability of a ss-tricalcium phosphate coated magnesium alloy. *Journal of
    Biomedical Materials Research Part A* 100A (2):293–304.
Chakrapani, D. G., and E. N. Pugh. 1975. The transgranular SCC of a Mg-Al alloy: Crystal-
    lographic, fractographic and acoustic-emission studies. *Metallurgical Transactions A*
    6 (6):1155–63.
Champagne, C. M. 2008. Magnesium in hypertension, cardiovascular disease, metabolic syn-
    drome, and other conditions: A review. *Nutrition in Clinical Practice* 23 (2):142–51.
Chang, S. Y., H. Tezuka, and A. Kamio. 1997. Mechanical properties and structure of
    ignition-proof Mg-Ca-Zr alloys produced by squeeze casting. *Materials Transactions*
    38 (6):526–35.
Chang, T.-C., J.-Y. Wang, C.-L. Chu, and S. Lee. 2006. Mechanical properties and microstruc-
    tures of various Mg–Li alloys. *Materials Letters* 60 (27):3272–6.
Chang, J.-W., P.-H. Fu, X.-W. Guo, L.-M. Peng, and W.-J. Ding. 2007a. The effects of heat
    treatment and zirconium on the corrosion behaviour of Mg–3Nd–0.2 Zn–0.4 Zr (wt.%)
    alloy. *Corrosion Science* 49 (6):2612–27.
Chang, J.-W., X.-W. Guo, P.-H. Fu, L.-M. Peng, and W.-J. Ding. 2007b. Effect of heat treat-
    ment on corrosion and electrochemical behaviour of Mg–3Nd–0.2 Zn–0.4 Zr (wt.%)
    alloy. *Electrochimica Acta* 52 (9):3160–7.
Chang, J.-W., X.-W. Guo, P.-H. Fu, L.-M. Peng, and W.-J. Ding. 2007c. Relationship between
    heat treatment and corrosion behaviour of Mg-3.0% Nd-0.4% Zr magnesium alloy.
    Transactions of Nonferrous Metals Society of China 17 (6):1152–7.
Chang, J.-W., X.-W. Guo, L.-M. Peng, W.-J. Ding, and Y.-H. Peng. 2008. Characterization
    of anodic coating formed on Mg-3Nd-0.2 Zn-0.4 Zr Mg alloy in alkaline electrolyte.
    Transactions of Nonferrous Metals Society of China 18:s318–22.
Chang, L. L., Y. N. Wang, X. Zhao, and J. C. Huang. 2008. Microstructure and mechanical
    properties in an AZ31 magnesium alloy sheet fabricated by asymmetric hot extrusion.
    Materials Science and Engineering: A 496 (1–2):512–16.
Chang, Y. C., J. C. Huang, C. W. Tang, C. I. Chang, and J. S. C. Jang. 2008. Viscous flow behav-
    ior and workability of Mg-Cu-(Ag)-Gd bulk metallic glasses. *Materials Transactions* 49
    (11):2605–10.
Chang, H., M. Y. Zheng, K. Wu, W. M. Gan, L. B. Tong, and H.-G. Brokmeier. 2010.
    Microstructure and mechanical properties of the accumulative roll bonded (ARBed)
    pure magnesium sheet. *Materials Science and Engineering: A* 527 (27):7176–83.
Chang, H. W., D. Qiu, J. A. Taylor, M. A. Easton, and M. X. Zhang. 2013. The role of Al2Y
    in grain refinement in MgeAleY alloy system. *Journal of Magnesium and Alloys*
    1 (2013):115–21.

Chang, L., L. Tian, W. Liu, and X. Duan. 2013. Formation of dicalcium phosphate dihydrate on magnesium alloy by micro-arc oxidation coupled with hydrothermal treatment. *Corrosion Science* 72:118–24.

Chao, C. K., and C. C. Hsiao. 2006. Parametric study on bone screw designs for holing power. *Journal of Mechanics* 22 (1):13–18.

Chao, C. K., C. C. Hsu, J. L. Wang, and J. Lin. 2008. Increasing bending strength and pull-out strength in conical pedicle screws: Biomechanical tests and finite element analyses. *Journal of Spinal Disorders and Techniques* 21:130–18.

Chatzistergos, P. E., E. A. Magnissalis, and S. K. Kourkoulis. 2010. A parametric study of cylindrical pedicle screw design implications on the pullout performance using an experimentally validated finite-element model. *Medical Engineering and Physics* 32:145–54.

Chatzistergos, P. E., E. A. Magnissalis, and S. K. Kourkoulis. 2014. Numerical simulation of bone screw induced pretension: The cases of under-tapping and conical profile. *Medical Engineering and Physics* 36 (3):378–86.

Chen, A. H., W. J. Huang, and Z. F. Li. 2006. Fretting wear behavior of AZ91D magnesium alloy. *Transactions of Nonferrous Metals Society of China* 16:S1879–83.

Chen, J., R.-C. Zeng, W.-J. Huang, Z.-Q. Zheng, Z.-L. Wang, and J. Wang. 2008. Characterization and wear resistance of macro-arc oxidation coating on magnesium alloy AZ91 in simulated body fluids. *Transactions of Nonferrous Metals Society of China* 18:s361–4.

Chen, Y. J., Q. D. Wang, H. J. Roven et al. 2008. Microstructure evolution in magnesium alloy AZ31 during cyclic extrusion compression. *Journal of Alloys and Compounds* 462 (1–2):192–200.

Chen, C., A. Manaenko, Y. Zhan et al. 2010a. Hydrogen gas reduced acute hyperglycemia-enhanced hemorrhagic transformation in a focal ischemia rat model. *Neuroscience* 169 (1):402–14.

Chen, C., Q. Chen, Y. Mao et al. 2010b. Hydrogen-rich saline protects against spinal cord injury in rats. *Neurochemical Research* 35 (7):1111–18.

Chen, H., Y. P. Sun, Y. Li et al. 2010. Hydrogen-rich saline ameliorates the severity of l-arginine-induced acute pancreatitis in rats. *Biochemical and Biophysical Research Communications* 393 (2):308–13.

Chen, W., W. Gao, and Y. He. 2010. A novel electroless plating of Ni–P–TiO$_2$ nano-composite coatings. *Surface and Coatings Technology* 204 (15):2493–8.

Chen, Y., S. Zhang, J. Li et al. 2010. Dynamic degradation behavior of MgZn alloy in circulating m-SBF. *Materials Letters* 64 (18):1996–9.

Chen, D. Y., Y. H. He, H. R. Tao, Y. Zhang, Y. Jiang, X. N. Zhang, and S. X. Zhang. 2011. Biocompatibility of magnesium-zinc alloy in biodegradable orthopedic implants. *International Journal of Molecular Medicine* 28 (3):343–8.

Chen, G., X. D. Peng, P. G. Fan, W. D. Xie, Q. Y. Wei, H. Ma, and Y. Yang. 2011. Effects of Sr and Y on microstructure and corrosion resistance of AZ31 magnesium alloy. *Transactions of Nonferrous Metals Society of China* 21 (4):725–31.

Chen, H., Y. P. Sun, P. F. Hu et al. 2011. The effects of hydrogen-rich saline on the contractile and structural changes of intestine induced by ischemia-reperfusion in rats. *Journal of Surgical Research* 167 (2):316–22.

Chen, S., S. Guan, B. Chen, W. Li, J. Wang, L. Wang, S. Zhu, and J. Hu. 2011. Corrosion behavior of TiO2 films on Mg–Zn alloy in simulated body fluid. *Applied Surface Science* 257 (9):4464–7.

Chen, X. B., N. Birbilis, and T. B. Abbott. 2011. A simple route towards a hydroxyapatite–Mg(OH)$_2$ conversion coating for magnesium. *Corrosion Science* 53 (6):2263–8.

Chen, Y., S. Zhang, J. Li, Y. Song, C. Zhao, H. Wang, and X. Zhang. 2011. Influence of Mg2+ concentration, pH value and specimen parameter on the hemolytic property of biodegradable magnesium. *Materials Science and Engineering: B* 176 (20):1823–6.

Chen, S., S. K. Guan, W. Li, H. X. Wang, J. Chen, Y. S. Wang, and H. T. Wang. 2012. In vivo degradation and bone response of a composite coating on Mg-Zn-Ca alloy prepared by microarc oxidation and electrochemical deposition. *Journal of Biomedical Materials Research Part B-Applied Biomaterials* 100 (2):533–43.

Chen, Q., D. Y. Shu, Z. Zhao, Z. X. Zhao, Y. B. Wang, and B. G. Yuan. 2012. Microstructure development and tensile mechanical properties of Mg-Zn-RE-Zr magnesium alloy. *Materials and Design* 40:488–96.

Chen, X. B., N. Birbilis, and T. B. Abbott. 2012a. Effect of [Ca^{2+}] and [PO$_4{}^{3-}$] levels on the formation of calcium phosphate conversion coatings on die-cast magnesium alloy AZ91D. *Corrosion Science* 55:226–32.

Chen, X. B., N. T. Kirkland, H. Krebs, M. A. Thiriat, S. Virtanen, D. Nisbet, and N. Birbilis. 2012b. In vitro corrosion survey of Mg-xCa and Mg-3Zn-yCa alloys with and without calcium phosphate conversion coatings. *Corrosion Engineering Science and Technology* 47 (5):365–73.

Chen, M.-A., N. Cheng, Y.-C. Ou, and J.-M. Li. 2013. Corrosion performance of electroless Ni–P on polymer coating of MAO coated AZ31 magnesium alloy. *Surface and Coatings Technology* 232:726–33.

Chen, Y., G. Wan, J. Wang, S. Zhao, Y. Zhao, and N. Huang. 2013. Covalent immobilization of phytic acid on Mg by alkaline pre-treatment: Corrosion and degradation behavior in phosphate buffered saline. *Corrosion Science* 75:280–6.

Chen, Y., Z. Xu, C. Smith, and J. Sankar. 2014a. Recent advances on the development of magnesium alloys for biodegradable implants. *Acta Biomaterialia* 10 (11):4561–73.

Chen, Y., J. Yan, C. Zhao et al. 2014b. In vitro and in vivo assessment of the biocompatibility of an Mg-6Z(n) alloy in the bile. *Journal of Materials Science. Materials in Medicine* 25 (2):471–80.

Cheng, Y. Q., Z. H. Chen, W. J. Xia, and T. Zhou. 2007. Effect of channel clearance on crystal orientation development in AZ31 magnesium alloy sheet produced by equal channel angular rolling. *Journal of Materials Processing Technology* 184 (1–3):97–101.

Cheng, R. J., F. S. Pan, M. B. Yang, and A. T. Tang. 2008. Effects of various Mg-Sr master alloys on microstructural refinement of ZK60 magnesium alloy. *Transactions of Nonferrous Metals Society of China* 18:S50–4.

Cheng, J., B. Liu, Y. H. Wu, and Y. F. Zheng. 2013. Comparative in vitro study on pure metals (Fe, Mn, Mg, Zn and W) as biodegradable metals. *Journal of Materials Science and Technology* 29 (7):619–27.

Chi, C.-T., C.-G. Chao, T.-F. Liu, and C.-H. Lee. 2007. Aluminum element effect for electron beam welding of similar and dissimilar magnesium–aluminum–zinc alloys. *Scripta Materialia* 56 (9):733–6.

Chino, Y., M. Kobata, H. Iwasaki, and M. Mabuchi. 2002. Tensile properties from room temperature to 673 K of Mg-0.9 mass% Ca alloy containing lamella Mg2Ca. *Materials Transactions* 43 (10):2643–6.

Chino, Y., A. Yamamoto, H. Iwasaki, M. Mabuchi, and H. Tsubakino. 2003. Solid recycling of an AZ31 Mg alloy with a vapor deposition coating layer of high purity Mg. *Materials Transactions* 44 (4):578–82.

Chino, Y., T. Hoshika, and M. Mabuchi. 2006. Enhanced corrosion properties of pure Mg and AZ31Mg alloy recycled by solid-state process. *Materials Science and Engineering: A* 435:275–81.

Chiu, K. Y., M. H. Wong, F. T. Cheng, and H. C. Man. 2007. Characterization and corrosion studies of fluoride conversion coating on degradable Mg implants. *Surface and Coatings Technology* 202 (3):590–8.

Chiu, C.-H., H.-Y. Wu, J.-Y. Wang, and S. Lee. 2008. Microstructure and mechanical behavior of LZ91 Mg alloy processed by rolling and heat treatments. *Journal of Alloys and Compounds* 460 (1–2):246–52.

Chlumsky, V. 1900. Über die Wiederherstellung der Beweglichkeit des Gelenkes bei Ankylose. *Zentralbl Chir* 37:37.

Cho, S. Y., S. W. Chae, K. W. Choi et al. 2013. Biocompatibility and strength retention of biodegradable Mg-Ca-Zn alloy bone implants. *Journal of Biomedical Materials Research Part B-Applied Biomaterials* 101 (2):201–12.

Chou, D. T., D. Hong, P. Saha, J. Ferrero, B. Lee, Z. Q. Tan, Z. Y. Dong, and P. N. Kumta. 2013. In vitro and in vivo corrosion, cytocompatibility and mechanical properties of biodegradable Mg-Y-Ca-Zr alloys as implant materials. *Acta Biomaterialia* 9 (10):8518–33.

Christoffersen, J., M. R. Christoffersen, N. Kolthoff, and O. Bärenholdt. 1997. Effects of strontium ions on growth and dissolution of hydroxyapatite and on bone mineral detection. *Bone* 20 (1):47–54.

Chu, C. L., X. Han, J. Bai, F. Xue, and P. K. Chu. 2012. Fabrication and degradation behavior of micro-arc oxidized biomedical magnesium alloy wires. *Surface and Coatings Technology* 213:307–12.

Chu, C. L., X. Han, F. Xue, J. Bai, and P. K. Chu. 2013. Effects of sealing treatment on corrosion resistance and degradation behavior of micro-arc oxidized magnesium alloy wires. *Applied Surface Science* 271:271–5.

Chuai, Y., L. Zhao, J. Ni et al. 2011. A possible prevention strategy of radiation pneumonitis: Combine radiotherapy with aerosol inhalation of hydrogen-rich solution. *Medical Science Monitor* 17 (4):HY1–4.

Chung, S. J., A. Roy, D. H. Hong, J. P. Leonard, and P. N. Kumta. 2011. Microstructure of Mg-Zn-Ca thin film derived by pulsed laser deposition. *Materials Science and Engineering B-Advanced Functional Solid-State Materials* 176 (20):1690–4.

Cipriano, A. F., T. Zhao, I. Johnson, R. G. Guan, S. Garcia, and H. N. Liu. 2013. In vitro degradation of four magnesium-zinc-strontium alloys and their cytocompatibility with human embryonic stem cells. *Journal of Materials Science-Materials in Medicine* 24 (4):989–1003.

Čížek, J., B. Smola, I. Stulíková, P. Hruška, M. Vlach, M. Vlček, O. Melikhova, and I. Prochazka. 2012. Natural aging of Mg–Gd and Mg–Tb alloys. *Physica Status Solidi (a)* 209 (11):2135–41.

Cong, M. Q., Z. Q. Li, J. S. Liu, M. Y. Yan, K. Chen, Y. D. Sun, M. Huang, C. Wang, B. P. Ding, and S. L. Wang. 2012. Effect of Ca on the microstructure and tensile properties of Mg-Zn-Si alloys at ambient and elevated temperature. *Journal of Alloys and Compounds* 539:168–73.

Cong, M. Q., Z. Q. Li, J. S. Liu, and S. H. Li. 2014. Effect of Sr on microstructure, tensile properties and wear behavior of as-cast Mg-6Zn-4Si alloy. *Materials and Design* 53:430–4.

Cooke, M. N., J. P. Fisher, D. Dean, C. Rimnac, and A. G. Mikos. 2003. Use of stereolithography to manufacture critical-sized 3D biodegradable scaffolds for bone ingrowth. *Journal of Biomedical Materials Research Part B: Applied Biomaterials* 64 (2):65–9.

Cordier, D. J., and J. B. Hedrick. 2009. *Minerals Yearbook Rare Earths* [Advance Release], Tech. Rep. US Geological Survey, Reston, VA.

Correa, P. S., C. F. Malfatti, and D. S. Azambuja. 2011. Corrosion behavior study of AZ91 magnesium alloy coated with methyltriethoxysilane doped with cerium ions. *Progress in Organic Coatings* 72 (4):739–47.

Cramer, S. D., and B. S. Covino. 2003. *ASM Handbook Vol. 13 A Corrosion: Fundamentals, Testing, and Protection.* Materials Park, OH: ASM International.

Crawford, P., R. Barrosa, J. Mendez, J. Foyos, and O. S. Es-Said. 1996. On the transformation characteristics of LA141A (Mg-Li-Al) alloy. *Journal of Materials Processing Technology* 56 (1):108–18.

Cui, X.-P., H.-F. Liu, J. Meng, and D.-P. Zhang. 2010. Microstructure and mechanical properties of die-cast AZ91D magnesium alloy by Pr additions. *Transactions of Nonferrous Metals Society of China* 20:s435–8.

Cui, X., Y. Yang, E. Liu, G. Jin, J. Zhong, and Q. Li. 2011. Corrosion behaviors in physiological solution of cerium conversion coatings on AZ31 magnesium alloy. *Applied Surface Science* 257 (23):9703–9.

Cui, W., E. Beniash, E. Gawalt, Z. Xu, and C. Sfeir. 2013. Biomimetic coating of magnesium alloy for enhanced corrosion resistance and calcium phosphate deposition. *Acta Biomaterialia* 9 (10):8650–9.

da Conceicao, T. F., N. Scharnagl, C. Blawert, W. Dietzel, and K. U. Kainer. 2010. Surface modification of magnesium alloy AZ31 by hydrofluoric acid treatment and its effect on the corrosion behaviour. *Thin Solid Films* 518 (18):5209–18.

da Conceicao, T. F., N. Scharnagl, W. Dietzel, and K. U. Kainer. 2011. Corrosion protection of magnesium AZ31 alloy using poly(ether imide) [PEI] coatings prepared by the dip coating method: Influence of solvent and substrate pre-treatment. *Corrosion Science* 53 (1):338–46.

Da Forno, A., M. Bestetti, and N. Lecis. 2013. Effect of anodising electrolyte on performance of AZ31 and AM60 magnesium alloys microarc anodic oxides. *Transactions of the IMF* 91 (6):336–41.

Daemen, J., P. Wenaweser, K. Tsuchida et al. 2007. Early and late coronary stent thrombosis of sirolimus-eluting and paclitaxel-eluting stents in routine clinical practice: Data from a large two-institutional cohort study. *The Lancet* 369 (9562):667–78.

Dahifar, H., A. Faraji, S. Yassobi, and A. Ghorbani. 2007. Asymptomatic rickets in adolescent girls. *The Indian Journal of Pediatrics* 74 (6):571–5.

Dahl, S. G., P. Allain, P. J. Marie, Y. Mauras, G. Boivin, P. Ammann, Y. Tsouderos, P. D. Delmas, and C. Christiansen. 2001. Incorporation and distribution of strontium in bone. *Bone* 28 (4):446–53.

Dalla Torre, F. H., A. C. Hanzi, and P. J. Uggowitzer. 2008. Microstructure and mechanical properties of microalloyed and equal channel angular extruded Mg alloys. *Scripta Materialia* 59:207–10.

Dar, F. H., Meakin, J. R., and Aspden, R. M. 2002. Statistical methods in finite element analysis. *Journal of Biomechanics* 35 (9):1155–61.

Darras, B. M., M. K. Khraisheh, F. K. Abu-Farha, and M. A. Omar. 2007. Friction stir processing of commercial AZ31 magnesium alloy. *Journal of Materials Processing Technology* 191 (1–3):77–81.

Datta, M. K., D. T. Chou, D. H. Hong, P. Saha, S. J. Chung, B. Lee, A. Sirinterlikci, M. Ramanathan, A. Roy, and P. N. Kumta. 2011. Structure and thermal stability of biodegradable Mg-Zn-Ca based amorphous alloys synthesized by mechanical alloying. *Materials Science and Engineering B-Advanced Functional Solid-State Materials* 176 (20):1637–43.

De-Hui, L., D. Jie, Z. Xiao-Qin, L. Chen, and D. Wen-Jiang. 2006. Age hardening characteristics and mechanical properties of Mg-3.5 Dy-4.0 Gd-3.1 Nd-0.4 Zr alloy. *Transactions of Nonferrous Metals Society of China* 16:s1694–7.

Demir, A. G., B. Previtali, D. Colombo et al. 2012. Fiber laser micromachining of magnesium alloy tubes for biocompatible and biodegradable cardiovascular stents. *SPIE LASE.* International Society for Optics and Photonics 823730-823730-9.

Demir, A. G., B. Previtali, and C. A. Biffi. 2013. Fibre laser cutting and chemical etching of AZ31 for manufacturing biodegradable stents. *Advances in Materials Science and Engineering* 2013:1–11.

Deng, C. Z., R. Radhakrishnan, S. R. Larsen et al. 2011. Magnesium alloys for bioabsorbable stents: A feasibility assessment. In W. H. Sillekens, S. R. Agnew, N. R. Neelameggham and S. N. Mathaudhu (eds.), *Magnesium Technology* 413–8.

Denkena, B., and A. Lucas. 2007. Biocompatible magnesium alloys as absorbable implant materials—Adjusted surface and subsurface properties by machining processes. *CIRP Annals—Manufacturing Technology* 56 (1):113–16.

Dhahri, M., J. E. Masse, J. F. Mathieu, G. Barreau, and M. Autric. 2001. Laser welding of AZ91 and WE43 magnesium alloys for automotive and aerospace industries. *Advanced Engineering Materials* 3 (7):504–7.

Dhert, W. J. A., C. C. P. M. Verheyen, L. H. Braak, C. P. A. T. Klein, K. de Groot, and P. M. Rozing. 1992. A finite element analysis of the push-out test: Influence of test condition. *Journal of Biomedical Materials Research* 26 (1):119–30.

Di Mario, C., H. U. W. Griffiths, O. Goktekin et al. 2004. Drug-eluting bioabsorbable magnesium stent. *Journal of Interventional Cardiology* 17 (6):391–5.

Dietz, M. L., and E. P. Horwitz. 1992. Improved chemistry for the production of yttrium-90 for medical applications. *International Journal of Radiation Applications and Instrumentation. Part A. Applied Radiation and Isotopes* 43 (9):1093–101.

Diez, M., H.-E. Kim, V. Serebryany, S. Dobatkin, and Y. Estrin. 2014. Improving the mechanical properties of pure magnesium by three-roll planetary milling. *Materials Science and Engineering: A* 612:287–92.

Ding, W., D. Li, Q. Wang, and Q. Li. 2008. Microstructure and mechanical properties of hot-rolled Mg–Zn–Nd–Zr alloys. *Materials Science and Engineering: A* 483–4:228–30.

Ding, H. L., K. Hirai, T. Homma, and S. Kamado. 2010. Numerical simulation for microstructure evolution in AM50 Mg alloy during hot rolling. *Computational Materials Science* 47 (4):919–25.

Ding, Y., C. Wen, P. Hodgson, and Y. Li. 2014. Effects of alloying elements on the corrosion behavior and biocompatibility of biodegradable magnesium alloys: A review. *Journal of Materials Chemistry B* 2 (14):1912–33.

Dobromyslov, A. V., L. I. Kaigorodova, V. D. Sukhanov, and T. V. Dobatkina. 2007. Decomposition of a supersaturated solid solution in the Mg-3.3 wt.% Yb alloy. *The Physics of Metals and Metallography* 103 (1):64–71.

Doepke, A., J. Kuhlmann, X. Guo, R. T. Voorhees, and W. R. Heineman. 2013. A system for characterizing Mg corrosion in aqueous solutions using electrochemical sensors and impedance spectroscopy. *Acta Biomaterialia* 9 (11):9211–19.

Doi, Y., S. Kitamura, and H. Abe. 1995. Microbial synthesis and characterization of poly (3-hydroxybutyrate-co-3-hydroxyhexanoate). *Macromolecules* 28 (14):4822–8.

Domoki, F., O. Oláh, A. Zimmermann et al. 2010. Hydrogen is neuroprotective and preserves cerebrovascular reactivity in asphyxiated newborn pigs. *Pediatric Research* 68:387–92.

Dong, J., H. Kojima, T. Uemura, M. Kikuchi, T. Tateishi, and J. Tanaka. 2001. In vivo evaluation of a novel porous hydroxyapatite to sustain osteogenesis of transplanted bone marrow-derived osteoblastic cells. *Journal of Biomedical Materials Research* 57 (2):208–16.

Dong, S. L., T. Imai, S.-W. Lim, N. Kanetake, and N. Saito. 2007. Superplasticity evaluation in an extruded Mg–8.5Li alloy. *Journal of Materials Science* 42 (13):5296–8.

Dong, H., L. Wang, Y. Wu, and L. Wang. 2010. Effect of Y on microstructure and mechanical properties of duplex Mg–7Li alloys. *Journal of Alloys and Compounds* 506 (1):468–74.

Dou, Y., S. Cai, X. Ye, G. Xu, K. Huang, X. Wang, and M. Ren. 2013. 45S5 bioactive glass–ceramic coated AZ31 magnesium alloy with improved corrosion resistance. *Surface and Coatings Technology* 228:154–61.

Douglas, G. R., A. S. Phani, and J. Gagnon. 2014. Analyses and design of expansion mechanisms of balloon expandable vascular stents. *Journal of Biomechanics* 47 (6):1438–46.

Drake, P. L. 2005. Exposure-related health effects of silver and silver compounds: A review. *Annals of Occupational Hygiene* 49 (7):575–85.

Drits, M. E., L. L. Rokhlin, E. M. Padezhnova, and L. S. Guzei. 1978. Phase diagram and mechanical properties of Mg–Tb alloys. *Metal Science and Heat Treatment* 20 (9):771–4.

Drozd, Z., Z. Trojanová, and S. Kúdela. 2004. Deformation behaviour of Mg–Li–Al alloys. *Journal of Alloys and Compounds* 378 (1–2):192–5.

Drynda, A., N. Deinet, N. Braun, and M. Peuster. 2009. Rare earth metals used in biodegradable magnesium-based stents do not interfere with proliferation of smooth muscle cells but do induce the upregulation of inflammatory genes. *Journal of Biomedical Materials Research Part A* 91 (2):360–9.

Drynda, A., T. Hassel, R. Hoehn, A. Perz, F. W. Bach, and M. Peuster. 2010. Development and biocompatibility of a novel corrodible fluoride-coated magnesium-calcium alloy with improved degradation kinetics and adequate mechanical properties for cardiovascular applications. *Journal of Biomedical Materials Research. Part A* 93 (2):763–75.

Du, H., Z. J. Wei, X. W. Liu, and E. L. Zhang. 2011a. Effects of Zn on the microstructure, mechanical property and bio-corrosion property of Mg-3Ca alloys for biomedical application. *Materials Chemistry and Physics* 125 (3):568–75.

Du, H., Z. Wei, H. Wang, E. Zhang, L. Zuo, and L. Du. 2011b. Surface microstructure and cell compatibility of calcium silicate and calcium phosphate composite coatings on Mg-Zn-Mn-Ca alloys for biomedical application. *Colloids and Surface B Biointerfaces* 83 (1):96–102.

Du, Y. Z., M. Y. Zheng, X. G. Qiao, K. Wu, X. D. Liu, G. J. Wang, and X. Y. Lv. 2013. Microstructure and mechanical properties of Mg-Zn-Ca-Ce alloy processed by semi-continuous casting. *Materials Science and Engineering A-Structural Materials Properties Microstructure and Processing* 582:134–9.

Du, J., Z. Wang, Y. Niu, W. Duan, and Z. Wu. 2014. Double liquid electrolyte for primary Mg batteries. *Journal of Power Sources* 247:840–4.

Dumas, L., C. Chatillon, and E. Quesnel. 2001. Thermodynamic calculations of congruent vaporization and interactions with residual water during magnesium fluoride vacuum deposition. *Journal of Crystal Growth* 222 (1–2):215–34.

Durdu, S., A. Aytaç, and M. Usta. 2011. Characterization and corrosion behavior of ceramic coating on magnesium by micro-arc oxidation. *Journal of Alloys and Compounds* 509 (34):8601–6.

Duygulu, O., R. A. Kaya, G. Oktay, and A. A. Kaya. 2007. Investigation on the potential of magnesium alloy AZ31 as a bone implant. *Materials Science Forum* 546–9:421–4.

Dziuba, D., A. Meyer-Lindenberg, J. M. Seitz et al. 2013. Long-term in vivo degradation behaviour and biocompatibility of the magnesium alloy ZEK100 for use as a biodegradable bone implant. *Acta Biomaterialia* 9 (10):8548–60.

Eckermann, J. M., W. Chen, V. Jadhav et al. 2011. Hydrogen is neuroprotective against surgically induced brain injury. *Medical Gas Research* 1 (1):1–7.

Edalati, K., A. Yamamoto, Z. Horita, and T. Ishihara. 2011. High-pressure torsion of pure magnesium: Evolution of mechanical properties, microstructures and hydrogen storage capacity with equivalent strain. *Scripta Materialia* 64 (9):880–3.

Ehrenberger, S. I., S. A. Schmid, S. Song, and H. E. Friedrich. 2008. Status and potentials of magnesium production in China: Life cycle analysis focusing on CO_2eq emissions. Paper read at 65th Annual World Magnesium Conference, Warsaw, Poland.

Eliezer, D., E. Aghion, and F. H. S. Froes. 1998. Magnesium science, technology and applications. *Advanced Performance Materials* 5 (3):201–12.

Eliseeva, S. V., and J.-C. G. Bünzli. 2011. Rare earths: Jewels for functional materials of the future. *New Journal of Chemistry* 35 (6):1165–76.

El-Rahman, S. S. A. 2003. Neuropathology of aluminum toxicity in rats (glutamate and GABA impairment). *Pharmacological Research* 47 (3):189–94.

Emsley, J. 2011. *Nature's Building Blocks: An AZ Guide to the Elements*. Oxford University Press, Oxford.

Erbel, R., C. Di Mario, J. Bartunek, J. Bonnier, B. de Bruyne, F. R. Eberli, P. Erne, M. Haude, B. Heublein, M. Horrigan, C. Ilsley, D. Böse, J. Koolen, T. F. Lüscher, N. Weissman, and R. Waksman. 2007. Temporary scaffolding of coronary arteries with bioabsorbable magnesium stents: A prospective, non-randomised multicentre trial. *The Lancet* 369 (9576):1869–75.

Erdmann, N., N. Angrisani, J. Reifenrath et al. 2011. Biomechanical testing and degradation analysis of MgCa0.8 alloy screws: A comparative in vivo study in rabbits. *Acta Biomaterialia* 7 (3):1421–8.

Erinc, M., W. H. Sillekens, R. G. T. M. Mannens, and R. J. Werkhoven. 2009. Applicability of existing magnesium alloys as biomedical implant materials. *Magnesium Technology*: 209–14.

Esparza, J. A., W. C. Davis, E. A. Trillo, and L. E. Murr. 2002. Friction-stir welding of magnesium alloy AZ31B. *Journal of Materials Science Letters* 21 (12):917–20.

Estrin, Y., E. P. Ivanova, A. Michalska, V. K. Truong, R. Lapovok, and R. Boyd. 2011. Accelerated stem cell attachment to ultrafine grained titanium. *Acta Biomaterialia* 7 (2):900–6.

Fabrichnaya, O. B., H. L. Lukas, G. Effenberg, and F. Aldinger. 2003. Thermodynamic optimization in the Mg–Y system. *Intermetallics* 11 (11):1183–8.

Faghihi, S., A. P. Zhilyaev, J. A. Szpunar, F. Azari, H. Vali, and M. Tabrizian. 2007. Nanostructuring of a titanium material by high-pressure torsion improves pre-osteoblast attachment. *Advanced Materials* 19 (8):1069–73.

Fan, J. F., S. L. Cheng, H. Xie, W. X. Hao, M. Wang, G. C. Yang, and Y. H. Zhou. 2005. Surface oxidation behavior of Mg-Y-Ce alloys at high temperature. *Metallurgical and Materials Transactions A* 36 (1):235–9.

Fan, Y., G. Wu, and C. Zhai. 2006. Influence of cerium on the microstructure, mechanical properties and corrosion resistance of magnesium alloy. *Materials Science and Engineering: A* 433 (1):208–15.

Fan, Y., G. Wu, and C. Zhai. 2007. Effect of strontium on mechanical properties and corrosion resistance of AZ91D. *Materials Science Forum* 546–9:567–70.

Fan, J. F., G. C. Yang, Y. H. Zhou, Y. H. Wei, and B. S. Xu. 2009. Selective oxidation and the third-element effect on the oxidation of Mg-Y alloys at high temperatures. *Metallurgical and Materials Transactions A* 40 (9):2184–9.

Fan, J. F., C. L. Yang, G. Han, S. Fang, W. D. Yang, and B. S. Xu. 2011. Oxidation behavior of ignition-proof magnesium alloys with rare earth addition. *Journal of Alloys and Compounds* 509 (5):2137–42.

Fan, G. D., M. Y. Zheng, X. S. Hu, C. Xu, K. Wu, and I. S. Golovin. 2012. Improved mechanical property and internal friction of pure Mg processed by ECAP. *Materials Science and Engineering: A* 556:588–94.

Fan, G. D., M. Y. Zheng, X. S. Hu, K. Wu, W. M. Gan, and H. G. Brokmeier. 2013. Internal friction and microplastic deformation behavior of pure magnesium processed by equal channel angular pressing. *Materials Science and Engineering: A* 561:100–8.

Fan, J., X. Qiu, X. D. Niu, Z. Tian, W. Sun, X. J. Liu, Y. D. Li, W. R. Li, and J. Meng. 2013. Microstructure, mechanical properties, in vitro degradation and cytotoxicity evaluations of Mg-1.5Y-1.2Zn-0.44Zr alloys for biodegradable metallic implants. *Materials Science and Engineering C-Materials for Biological Applications* 33 (4):2345–52.

Fan, X., Y. Wang, B. Zou, L. Gu, W. Huang, and X. Cao. 2013. Preparation and corrosion resistance of MAO/Ni–P composite coat on Mg alloy. *Applied Surface Science* 277:272–80.

Fan, T., L. Luo, L. Ma, B. Tang, L. Peng, and W. Ding. 2014. First-principles study of full a-dislocations in pure magnesium. *Journal of Applied Mechanics and Technical Physics* 55 (4):672–81.

Fan, Y., X. Peng, T. Su, H. Bala, and B. Liu. 2014. Modifying microstructures and hydrogen storage properties of 85mass% Mg–10mass% Ni–5mass% La alloy by ultra-high pressure. *Journal of Alloys and Compounds* 596:113–17.

Fang, X., Y. Shi, and D. C. Jiles. 1998. Modeling of magnetic properties of heat treated Dy-doped NdFeB particles bonded in isotropic and anisotropic arrangements. *Magnetics, IEEE Transactions on* 34 (4):1291–3.

Fang, Y., X.-J. Fu, C. Gu et al. 2011. Hydrogen-rich saline protects against acute lung injury induced by extensive burn in rat model. *Journal of Burn Care and Research* 32 (3):e82–91.

Fang, G., W.-J. Ai, S. Leeflang, J. Duszczyk, and J. Zhou. 2013. Multipass cold drawing of magnesium alloy minitubes for biodegradable vascular stents. *Materials Science and Engineering: C* 33 (6):3481–8.

Farahany, S., H. R. Bakhsheshi-Rad, M. H. Idris, M. R. A. Kadir, A. F. Lotfabadi, and A. Ourdjini. 2012. In-situ thermal analysis and macroscopical characterization of Mg-xCa and Mg-0.5Ca-xZn alloy systems. *Thermochimica Acta* 527:180–9.

Farraro, K. F., K. E. Kim, S. L.-Y. Woo, J. R. Flowers, and M. B. McCullough. 2014. Revolutionizing orthopaedic biomaterials: The potential of biodegradable and bioresorbable magnesium-based materials for functional tissue engineering. *Journal of Biomechanics* 47 (9):1979–86.

Fathi, M., M. Meratian, and M. Razavi. 2011. Novel magnesium-nanofluorapatite metal matrix nanocomposite with improved biodegradation behavior. *Journal of Biomedical Nanotechnology* 7 (3):441–5.

Feng, A., and Y. Han. 2010. The microstructure, mechanical and corrosion properties of calcium polyphosphate reinforced ZK60A magnesium alloy composites. *Journal of Alloys and Compounds* 504 (2):585–93.

Feng, A., and Y. Han. 2011. Mechanical and in vitro degradation behavior of ultrafine calcium polyphosphate reinforced magnesium-alloy composites. *Materials and Design* 32 (5):2813–20.

Ferrando, W. A. 1989. Review of corrosion and corrosion control of magnesium alloys and composites. *Journal of Materials Engineering* 11 (4):299–313.

Ferro, R., A. Saccone, and G. Borzone. 1997. Rare earth metals in light alloys. *Journal of Rare Earths* 15 (1):45–61.

Feyerabend, F., J. Fischer, J. Holtz et al. 2010. Evaluation of short-term effects of rare earth and other elements used in magnesium alloys on primary cells and cell lines. *Acta Biomaterialia* 6 (5):1834–42.

Feyerabend, F., H. Drücker, D. Laipple, C. Vogt, M. Stekker, N. Hort, and R. Willumeit. 2012. Ion release from magnesium materials in physiological solutions under different oxygen tensions. *Journal of Materials Science: Materials in Medicine* 23 (1):9–24.

Fischer, J., M. H. Prosenc, M. Wolff, N. Hort, R. Willumeit, and F. Feyerabend. 2010. Interference of magnesium corrosion with tetrazolium-based cytotoxicity assays. *Acta Biomaterialia* 6 (5):1813–23.

Fischer, J., D. Pröfrock, N. Hort, R. Willumeit, and F. Feyerabend. 2011. Reprint of: Improved cytotoxicity testing of magnesium materials. *Materials Science and Engineering: B* 176 (20):1773–7.

Fischerauer, S. F., T. Kraus, X. Wu et al. 2013. In vivo degradation performance of microarc-oxidized magnesium implants: A micro-CT study in rats. *Acta Biomaterialia* 9 (2):5411–20.

Flaten, T. P. 2001. Aluminium as a risk factor in Alzheimer's disease, with emphasis on drinking water. *Brain Research Bulletin* 55 (2):187–96.

Fontana, M. G., and W. Stactile. 1970. *Corrosion Science and Technology.* London: Plenum Press.

Fontenier, G., R. Freschard, and M. Mourot. 1975. Study of the corrosion in vitro and in vivo of magnesium anodes involved in an implantable bioelectric battery. *Medical and Biological Engineering* 13 (5):683–9.

Forrester, J., M. Fishbein, R. Helfant, and J. Fagin. 1991. A paradigm for restenosis based on cell biology: Clues for the development of new preventive therapies. *Journal of the American College of Cardiology* 17 (3):758–69.

Fouad, H. 2011. Assessment of function-graded materials as fracture fixation bone-plates under combined loading conditions using finite element modelling. *Medical Engineering and Physics* 33:456–63.

Friedrich, H. E., and B. L. Mordike. 2006a. Melting, alloying and refining. In H. E. Friedrich, and B. L. Mordike, (eds.). *Magnesium Technology*. Berlin: Springer Berlin Heidelberg.

Friedrich, H. E., and B. L. Mordike. 2006b. In H. E. Friedrich, and B. L. Mordike, (eds.). *Metallurgy, Design Data, Applications*. Berlin: Springer Berlin Heidelberg.

Froats, A., T. K. Aune, D. Hawke, W. Unsworth, and J. Hillis. 1987. Corrosion of magnesium and magnesium alloys. *ASM Handbook* 13:740–54.

Froes, F. H., D. Eliezer, and E. Aghion. 1998. The science, technology, and applications of magnesium. *JOM* 50 (9):30–4.

Fu, J. W., and Y. S. Yang. 2011. Formation of the solidified microstructure in Mg-Sn binary alloy. *Journal of Crystal Growth* 322 (1):84–90.

Fu, Y., M. Ito, Y. Fujita et al. 2009. Molecular hydrogen is protective against 6-hydroxy-dopamine-induced nigrostriatal degeneration in a rat model of Parkinson's disease. *Neuroscience Letters* 453 (2):81–5.

Fujita, K., T. Seike, N. Yutsudo et al. 2009. Hydrogen in drinking water reduces dopaminergic neuronal loss in the 1-methyl-4-phenyl-1, 2, 3, 6-tetrahydropyridine mouse model of Parkinson's disease. *PLoS One* 4 (9):e7247.

Fukuda, K.-I., S. Asoh, M. Ishikawa et al. 2007. Inhalation of hydrogen gas suppresses hepatic injury caused by ischemia/reperfusion through reducing oxidative stress. *Biochemical and Biophysical Research Communications* 361 (3):670–4.

Furui, M., C. Xu, T. Aida, M. Inoue, H. Anada, and T. G. Langdon. 2005. Improving the superplastic properties of a two-phase Mg–8% Li alloy through processing by ECAP. *Materials Science and Engineering: A* 410–1:439–42.

Galvin, E., M. M. Morshed, C. Cummins, S. Daniels, C. Lally, and B. MacDonald. 2013. Surface modification of absorbable magnesium stents by reactive ion etching. *Plasma Chemistry and Plasma Processing* 33 (6):1137–52.

Gan, W. M., M. Y. Zheng, H. Chang, X. J. Wang, X. G. Qiao, K. Wu, B. Schwebke, and H.-G. Brokmeier. 2009. Microstructure and tensile property of the ECAPed pure magnesium. *Journal of Alloys and Compounds* 470 (1):256–62.

Gan, J., L. Tan, K. Yang, Z. Hu, Q. Zhang, X. Fan, Y. Li, and W. Li. 2013. Bioactive Ca–P coating with self-sealing structure on pure magnesium. *Journal of Materials Science: Materials in Medicine* 24 (4):889–901.

Gao, X., and J. F. Nie. 2007. Characterization of strengthening precipitate phases in a Mg-Zn alloy. *Scripta Materialia* 56 (8):645–8.

Gao, X., and J. F. Nie. 2008. Enhanced precipitation-hardening in Mg–Gd alloys containing Ag and Zn. *Scripta Materialia* 58 (8):619–22.

Gao, J.-C., L.-C. Li, and Y. Wang. 2004. Surface modification on magnesium by alkali-heat-treatment and its corrosion behaviors in SBF. *Chinese Journal of Nonferrous Metals* 14 (9):1508–13.

Gao, X., S. M. Zhu, B. C. Muddle, and J. F. Nie. 2005. Precipitation-hardened Mg-Ca-Zn alloys with superior creep resistance. *Scripta Materialia* 53 (12):1321–6.

Gao, J.-C., L.-Y. Qiao, L.-C. Li, and Y. Wang. 2006. Hemolysis effect and calcium-phosphate precipitation of heat-organic-film treated magnesium. *Transactions of Nonferrous Metals Society of China* 16 (3):539–44.

Gao, L., R. S. Chen, and E. H. Han. 2009a. Effects of rare-earth elements Gd and Y on the solid solution strengthening of Mg alloys. *Journal of Alloys and Compounds* 481 (1):379–84.

Gao, L., C. Zhang, M. Zhang, X. Huang, and X. Jiang. 2009b. Phytic acid conversion coating on Mg–Li alloy. *Journal of Alloys and Compounds* 485 (1–2):789–93.

Gao, Y., Q. Wang, J. Gu, Y. Zhao, Y. Tong, and D. Yin. 2009a. Comparison of microstructure in Mg–10Y–5Gd–0.5 Zr and Mg–10Y–5Gd–2Zn–0.5 Zr alloys by conventional casting. *Journal of Alloys and Compounds* 477 (1):374–8.

Gao, Y., D. Xiong, C. Wang, and X. Pan. 2009b. Laser cladding hydroxypatite coating on magnesium alloy for biomaterials application. *Special Casting and Nonferrous Alloys* 4:010.

Gao, J. C., L. Y. Qiao, Y. Wang, and R. L. Xin. 2010a. Research on bone inducement of magnesium in vivo. *Rare Metal Materials and Engineering* 39 (2):296–9.

Gao, J. C., L.-Y. Qiao, and R.-L. Xin. 2010b. Corrosion and bone response of magnesium implants after surface modification by heat-self-assembled monolayer. *Frontiers of Materials Science in China* 4 (2):120–5.

Gao, J. H., S. K. Guan, J. Chen, L. G. Wang, S. J. Zhu, J. H. Hu, and Z. W. Ren. 2011a. Fabrication and characterization of rod-like nano-hydroxyapatite on MAO coating supported on Mg–Zn–Ca alloy. *Applied Surface Science* 257 (6):2231–7.

Gao, J. H., S. K. Guan, Z. W. Ren, Y. F. Sun, S. J. Zhu, and B. Wang. 2011b. Homogeneous corrosion of high pressure torsion treated Mg-Zn-Ca alloy in simulated body fluid. *Materials Letters* 65 (4):691–3.

Gao, J. H., X. Y. Shi, B. Yang, S. S. Hou, E. C. Meng, F. X. Guan, and S. K. Guan. 2011c. Fabrication and characterization of bioactive composite coatings on Mg-Zn-Ca alloy by MAO/sol-gel. *Journal of Materials Science. Materials in Medicine* 22 (7):1681–7.

Gao, J., H. C. Yu, F. F. Zhao, and M. Hu. 2011. A study on morphologies of second phases in the Mg-Sr and Mg-Sr-Y alloys. *Advanced Materials Research* 146:336–9.

Gao, L., H. Yan, J. Luo, A. A. Luo, and R. Chen. 2013. Microstructure and mechanical properties of a high ductility Mg–Zn–Mn–Ce magnesium alloy. *Journal of Magnesium and Alloys* 1 (4):283–91.

Garces, G., A. Muller, E. Onorbe, P. Perez, and P. Adeva. 2008. Effect of hot forging on the microstructure and mechanical properties of Mg-Zn-Y alloy. *Journal of Materials Processing Technology* 206 (1–3):99–105.

Garg, S., and P. W. Serruys. 2010. Coronary stents: Looking forward. *Journal of the American College of Cardiology* 56:S43–78.

Gastaldi, D., V. Sassi, L. Petrini, M. Vedani, S. Trasatti, and F. Migliavacca. 2011. Continuum damage model for bioresorbable magnesium alloy devices—Application to coronary stents. *Journal of the Mechanical Behavior of Biomedical Materials* 4 (3):352–65.

Ge, Q., D. Dellasega, A. G. Demir, and M. Vedani. 2013. The processing of ultrafine-grained Mg tubes for biodegradable stents. *Acta Biomaterialia* 9 (10):8604–10.

Gebert, A., V. Haehnel, E. S. Park, D. H. Kim, and L. Schultz. 2008. Corrosion behaviour of Mg65Cu7.5Ni7.5Ag5Zn5Gd5Y5 bulk metallic glass in aqueous environments. *Electrochimica Acta* 53 (8):3403–11.

Gefen, A. 2002. Optimizing the biomechanical compatibility of orthopaedic screws for bone fracture fixation. *Medical Engineering and Physics* 24:337–47.

Geng, F., L. L. Tan, X. X. Jin, J. Y. Yang, and K. Yang. 2009a. The preparation, cytocompatibility, and in vitro biodegradation study of pure beta-TCP on magnesium. *Journal of Materials Science. Materials in Medicine* 20 (5):1149–57.

Geng, F., L. Tan, B. Zhang, C. Wu, Y. He, J. Yang, and K. Yang. 2009b. Study on beta-TCP coated porous Mg as a bone tissue engineering scaffold material. *Journal of Material Science and Technology* 25 (1):123–9.

Geng, L., B. P. Zhang, A. B. Li, and C. C. Dong. 2009. Microstructure and mechanical properties of Mg-4.0Zn-0.5Ca alloy. *Materials Letters* 63 (5):557–9.

George, J. F., and A. Agarwal. 2010. Hydrogen: Another gas with therapeutic potential. *Kidney International* 77 (2):85–7.

Geringer, J., B. Forest, and P. Combrade. 2006. Wear analysis of materials used as orthopaedic implants. *Wear* 261 (9):971–9.

Ghali, E., W. Dietzel, and K. U. Kainer. 2013. General and localized corrosion of magnesium alloys: A critical review. *Journal of Materials Engineering and Performance* 22 (10):2875–91.

Gharib, B., S. Hanna, O. Abdallahi et al. 2001. Anti-inflammatory properties of molecular hydrogen: Investigation on parasite-induced liver inflammation. *Comptes Rendus de l'Académie des Sciences-Series III-Sciences de la Vie* 324 (8):719–24.

Gil Posada, J. O., and P. J. Hall. 2014. SANS characterization of porous magnesium for hydrogen storage. *International Journal of Hydrogen Energy* 39 (16):8321–30.

Giles, J. J., and J. G. Bannigan. 2006. Teratogenic and developmental effects of lithium. *Current Pharmaceutical Design* 12 (12):1531–41.

Glass, E. 1926. Klinische und experimentelle Untersuchungen über die Payrsche Magnesiumpfeilbehandlung von Angiomen. *Langenbeck's Archives of Surgery* 194 (5):352–66.

Golubev, S. V., O. S. Pokrovsky, and V. S. Savenko. 1999. Unseeded precipitation of calcium and magnesium phosphates from modified seawater solutions. *Journal of Crystal Growth* 205 (3):354–60.

Gong, W., A. Abdelouas, and W. Lutze. 2001. Porous bioactive glass and glass–ceramics made by reaction sintering under pressure. *Journal of Biomedical Materials Research* 54 (3):320–7.

González, S., I. A. Figueroa, and I. Todd. 2009a. Influence of minor alloying additions on the glass-forming ability of Mg–Ni–La bulk metallic glasses. *Journal of Alloys and Compounds* 484 (1):612–18.

González, S., I. A. Figueroa, H. Zhao, H. A. Davies, I. Todd, and P. Adeva. 2009b. Effect of mischmetal substitution on the glass-forming ability of Mg–Ni–La bulk metallic glasses. *Intermetallics* 17 (11):968–71.

González, S., E. Pellicer, J. Fornell, A. Blanquer, L. Barrios, E. Ibanez, P. Solsona, S. Surinach, M. D. Baro, C. Nogues, and J. Sort. 2012. Improved mechanical performance and delayed corrosion phenomena in biodegradable Mg-Zn-Ca alloys through Pd-alloying. *Journal of the Mechanical Behavior of Biomedical Materials* 6:53–62.

González-Doncel, G., J. Wolfenstine, P. Metenier, O. A. Ruano, and O. D. Sherby. 1990. The use of foil metallurgy processing to achieve ultrafine grained Mg-9 Li laminates and Mg-9Li-5B4C particulate composites. *Journal of Materials Science* 25 (10):4535–40.

Gorsse, S., C. R. Hutchinson, B. Chevalier, and J. F. Nie. 2005. A thermodynamic assessment of the Mg–Nd binary system using random solution and associate models for the liquid phase. *Journal of Alloys and Compounds* 392 (1–2):253–62.

Gouzman, I., M. Dubey, M. D. Carolus, J. Schwartz, and S. L. Bernasek. 2006. Monolayer vs. multilayer self-assembled alkylphosphonate films: X-ray photoelectron spectroscopy studies. *Surface Science* 600 (4):773–81.

Grandfield, K., E. McNally, A. Palmquist, G. Botton, P. Thomsen, and H. Engqvist. 2010. Visualizing biointerfaces in three dimensions: Electron tomography of the bone–hydroxyapatite interface. *Journal of the Royal Society Interface* 7 (51):1497–501.

Gray-Munro, J. E., and M. Strong. 2009. The mechanism of deposition of calcium phosphate coatings from solution onto magnesium alloy AZ31. *Journal of Biomedical Materials Research. Part A* 90 (2):339–50.

Gray-Munro, J. E., C. Seguin, and M. Strong. 2009. Influence of surface modification on the in vitro corrosion rate of magnesium alloy AZ31. *Journal of Biomedical Materials Research. Part A* 91 (1):221–30.

Grogan, J. A., B. J. O'Brien, S. B. Leen, and P. E. McHugh. 2011. A corrosion model for bioabsorbable metallic stents. *Acta Biomaterialia* 7:3523–33.

Grogan, J. A., S. B. Leen, and P. E. McHugh. 2013. Optimizing the design of a bioabsorbable metal stent using computer simulation methods. *Biomaterials* 34 (33):8049–60.

Grogan, J. A., S. B. Leen, and P. E. McHugh. 2014. Computational micromechanics of bioabsorbable magnesium stents. *Journal of the Mechanical Behavior of Biomedical Materials* 4:93–105.

Groves, E. W. H. 1913. An experimental study of the operative treatment of fractures. *British Journal of Surgery* 1 (3):438–501.

Grubač, Z., M. Metikoš-Huković, and R. Babić. 2013a. Electrocrystallization, growth and characterization of calcium phosphate ceramics on magnesium alloys. *Electrochimica Acta* 109:694–700.

Grubač, Z., M. Metikos-Hukovic, R. Babic, I. S. Roncevic, M. Petravic, and R. Peter. 2013b. Functionalization of biodegradable magnesium alloy implants with alkylphosphonate self-assembled films. *Materials Science and Engineering. C, Materials for Biological Applications* 33 (4):2152–8.

Gu, X. N. 2011. *Biodegradation and biocompatibility evaluation of magnesium based materials.* (doctoral thesis), Peking University.

Gu, H., Y. Zhu, and L. Li. 2008. Effect of La/Ni ratio on hydrogen storage properties of Mg–Ni–La system prepared by hydriding combustion synthesis followed by mechanical milling. *International Journal of Hydrogen Energy* 33 (12):2970–4.

Gu, X., G. J. Shiflet, F. Q. Guo, and S. J. Poon. 2005. Mg-Ca-Zn bulk metallic glasses with high strength and significant ductility. *Journal of Materials Research* 20 (8):1935–8.

Gu, X. N., W. Zheng, Y. Cheng, and Y. F. Zheng. 2009a. A study on alkaline heat treated Mg-Ca alloy for the control of the biocorrosion rate. *Acta Biomaterialia* 5 (7):2790–9.

Gu, X. N., Y. F. Zheng, Y. Cheng, S. P. Zhong, and T. F. Xi. 2009b. In vitro corrosion and biocompatibility of binary magnesium alloys. *Biomaterials* 30 (4):484–98.

Gu, X. N., Y. F. Zheng, Q. X. Lan, Y. Cheng, Z. X. Zhang, T. F. Xi, and D. Y. Zhang. 2009c. Surface modification of an Mg-1Ca alloy to slow down its biocorrosion by chitosan. *Biomedical Materials* 4 (4):044109.

Gu, X. N., Y. F. Zheng, and L. J. Chen. 2009d. Influence of artificial biological fluid composition on the biocorrosion of potential orthopedic Mg-Ca, AZ31, AZ91 alloys. *Biomedical Materials* 4 (6).

Gu, X. N., X. L. Li, W. R. Zhou, Y. Cheng, and Y. F. Zheng. 2010a. Microstructure, biocorrosion and cytotoxicity evaluations of rapid solidified Mg-3Ca alloy ribbons as a biodegradable material. *Biomedical Materials* 5 (3):035013.

Gu, X. N., W. R. Zhou, Y. F. Zheng et al. 2010b. Corrosion fatigue behaviors of two biomedical Mg alloys—AZ91D and WE43—in simulated body fluid. *Acta Biomaterialia* 6 (12):4605–13.

Gu, X. N., Y. F. Zheng, S. P. Zhong, T. F. Xi, J. Q. Wang, and W. H. Wang. 2010c. Corrosion of, and cellular responses to Mg-Zn-Ca bulk metallic glasses. *Biomaterials* 31 (6):1093–103.

Gu, X. N., W. Zhou, Y. Zheng, L. Dong, Y. Xi, and D. Chai. 2010d. Microstructure, mechanical property, bio-corrosion and cytotoxicity evaluations of Mg/HA composites. *Materials Science and Engineering: C* 30 (6):827–32.

Gu, X. N., W. R. Zhou, Y. F. Zheng, Y. Liu, and Y. X. Li. 2010e. Degradation and cytotoxicity of lotus-type porous pure magnesium as potential tissue engineering scaffold material. *Materials Letters* 64 (17):1871–4.

Gu, Y., C.-S. Huang, T. Inoue et al. 2010. Drinking hydrogen water ameliorated cognitive impairment in senescence-accelerated mice. *Journal of Clinical Biochemistry and Nutrition* 46 (3):269–76.

Gu, X. N., N. Li, W. R. Zhou, Y. F. Zheng, X. Zhao, Q. Z. Cai, and L. Q. Ruan. 2011a. Corrosion resistance and surface biocompatibility of a microarc oxidation coating on a Mg-Ca alloy. *Acta Biomaterialia* 7 (4):1880–9.

Gu, X. N., N. Li, Y. Zheng, F. Kang, J. Wang, and L. Ruan. 2011b. In vitro study on equal channel angular pressing AZ31 magnesium alloy with and without back pressure. *Materials Science and Engineering: B* 176 (20):1802–6.

Gu, X. N., N. Li, Y. F. Zheng, and L. Q. Ruan. 2011c. In vitro degradation performance and biological response of a Mg-Zn-Zr alloy. *Materials Science and Engineering B-Advanced Functional Solid-State Materials* 176 (20):1778–84.

Gu, X. N., X. Wang, N. Li, L. Li, Y. F. Zheng, and X. G. Miao. 2011d. Microstructure and characteristics of the metal-ceramic composite (MgCa-HA/TCP) fabricated by liquid metal infiltration. *Journal of Biomedical Materials Research Part B-Applied Biomaterials* 99 (1):127–34.

Gu, X. N., X. H. Xie, N. Li, Y. F. Zheng, and L. Qin. 2012. In vitro and in vivo studies on a Mg-Sr binary alloy system developed as a new kind of biodegradable metal. *Acta Biomaterialia* 8 (6):2360–74.

Gu, Y., S. Bandopadhyay, C.-F. Chen, Y. Guo, and C. Ning. 2012a. Effect of oxidation time on the corrosion behavior of micro-arc oxidation produced AZ31 magnesium alloys in simulated body fluid. *Journal of Alloys and Compounds* 543:109–17.

Gu, Y., C.-F. Chen, S. Bandopadhyay, C. Ning, Y. Zhang, and Y. Guo. 2012b. Corrosion mechanism and model of pulsed DC microarc oxidation treated AZ31 alloy in simulated body fluid. *Applied Surface Science* 258 (16):6116–26.

Gu, X.-N., S.-S. Li, X.-M. Li, and Y.-B. Fan. 2014. Magnesium based degradable biomaterials: A review. *Frontiers of Materials Science* 8 (3):200–18.

Guan, Y. C., W. Zhou, and H. Y. Zheng. 2009. Effect of laser surface melting on corrosion behaviour of AZ91D Mg alloy in simulated-modified body fluid. *Journal of Applied Electrochemistry* 39 (9):1457–64.

Guan, R. G., I. Johnson, T. Cui, T. Zhao, Z. Y. Zhao, X. Li, and H. N. Liu. 2012. Electrodeposition of hydroxyapatite coating on Mg-4.0Zn-1.0Ca-0.6Zr alloy and in vitro evaluation of degradation, hemolysis, and cytotoxicity. *Journal of Biomedical Materials Research. Part A* 100 (4):999–1015.

Guan, R.-G., A. F. Cipriano, Z.-Y. Zhao, J. Lock, D. Tie, T. Zhao, T. Cui, and H. Liu. 2013. Development and evaluation of a magnesium–zinc–strontium alloy for biomedical applications—Alloy processing, microstructure, mechanical properties, and biodegradation. *Materials Science and Engineering: C* 33 (7):3661–9.

Gun, B., K. J. Laws, and M. Ferry. 2006. Static and dynamic crystallization in Mg–Cu–Y bulk metallic glass. *Journal of Non-Crystalline Solids* 352 (36):3887–95.

Gunatillake, P. A., and R. Adhikari. 2003. Biodegradable synthetic polymers for tissue engineering. *European Cells and Materials* 5 (1):1–16.

Guo, X., and D. Shechtman. 2007. Reciprocating extrusion of rapidly solidified Mg–6Zn–1Y–0.6 Ce–0.6 Zr alloy. *Journal of Materials Processing Technology* 187:640–4.

Guo, Y. B., and M. Salahshoor. 2010. Process mechanics and surface integrity by high-speed dry milling of biodegradable magnesium–calcium implant alloys. *CIRP Annals—Manufacturing Technology* 59 (1):151–4.

Guo, C. P., Z. M. Du, and C. R. Li. 2007. A thermodynamic description of the Gd-Mg-Y system. *Calphad-Computer Coupling of Phase Diagrams and Thermochemistry* 31 (1):75–88.

Guo, X.-W., J.-W. C., S.-M. He, W.-J. Ding, and X. Wang. 2007. Investigation of corrosion behaviors of Mg-6Gd–3Y–0.4 Zr alloy in NaCl aqueous solutions. *Electrochimica Acta* 52 (7):2570–9.

Guo, D. G., K. W. Xu, and Y. Han. 2008. The influence of Sr doses on the in vitro biocompatibility and in vivo degradability of single-phase Sr-incorporated HAP cement. *Journal of Biomedical Materials Research Part A* 86A (4):947–58.

Guo, M., L. Cao, P. Lu, Y. Liu, and X. Xu. 2011. Anticorrosion and cytocompatibility behavior of MAO/PLLA modified magnesium alloy WE42. *Journal of Materials Science. Materials in Medicine* 22 (7):1735–40.

Guo, Y., M. P. Sealy, and C. Guo. 2012. Significant improvement of corrosion resistance of biodegradable metallic implants processed by laser shock peening. *CIRP Annals—Manufacturing Technology* 61 (1):583–6.

Guo, X., K. Du, Q. Guo, Y. Wang, R. Wang, and F. Wang. 2013a. Effect of phytic acid on corrosion inhibition of composite film coated on Mg–Gd–Y alloy. *Corrosion Sci. 70:129–41.

Guo, X., K. Du, Q. Guo, Y. Wang, R. Wang, and F. Wang. 2013b. Study of barrier property of composite film coated on Mg-Gd-Y alloy by water diffusion. *ECS Electrochemistry Letters* 2 (8):C27–30.

Gupta, U. C., and S. C. Gupta. 2014. Sources and deficiency diseases of mineral nutrients in human health and nutrition: A review. *Pedosphere* 24 (1):13–38.

Gupta, R. K., K. Mensah-Darkwa, J. Sankar, and D. Kumar. 2013. Enhanced corrosion resistance of phytic acid coated magnesium by stearic acid treatment. *Transactions of Nonferrous Metals Society of China* 23 (5):1237–44.

Habibnejad-Korayem, M., R. Mahmudi, H. M. Ghasemi, and W. J. Poole. 2010. Tribological behavior of pure Mg and AZ31 magnesium alloy strengthened by Al2O3 nano-particles. *Wear* 268 (3):405–12.

Hadorn, J. P., K. Hantzsche, S. Yi, J. Bohlen, D. Letzig, and S. R. Agnew. 2012. Effects of solute and second-phase particles on the texture of Nd-containing Mg alloys. *Metallurgical and Materials Transactions A* 43 (4):1363–75.

Haferkamp, H., M. Niemeyer, R. Boehm, U. Holzkamp, C. Jaschik, and V. Kaese. 2000. Development, processing and applications range of magnesium lithium alloys. *Magnesium Alloys* 350 (3):31–41.

Hagihara, K., A. Kinoshita, Y. Sugino, M. Yamasaki, Y. Kawamura, H. Y. Yasuda, and Y. Umakoshi. 2010a. Effect of long-period stacking ordered phase on mechanical properties of Mg97Zn1Y2 extruded alloy. *Acta Materialia* 58 (19):6282–93.

Hagihara, K., N. Yokotani, and Y. Umakoshi. 2010b. Plastic deformation behavior of Mg12YZn with 18R long-period stacking ordered structure. *Intermetallics* 18 (2):267–76.

Haley, T. J., K. Raymond, N. Komesu, and H. C. Upham. 1961. Toxicological and pharmacological effects of gadolinium and samarium chlorides. *British Journal of Pharmacology and Chemotherapy* 17 (3):526–32.

Haley, T. J., N. Komesu, M. Efros, L. Koste, and H. C. Upham. 1964. Pharmacology and toxicology of praseodymium and neodymium chlorides. *Toxicology and Applied Pharmacology* 6 (5):614–20.

Haley, T. J., L. Koste, N. Komesu, M. Efros, and H. C. Upham. 1966. Pharmacology and toxicology of dysprosium, holmium, and erbium chlorides. *Toxicology and Applied Pharmacology* 8 (1):37–43.

Hallab, N. J., C. Vermes, C. Messina, K. A. Roebuck, T. T. Glant, and J. J. Jacobs. 2002. Concentration- and composition-dependent effects of metal ions on human MG-63 osteoblasts. *Journal of Biomedical Materials Research* 60 (3):420–33.

Hampl, M., J. Gröbner, and R. Schmid-Fetzer. 2007. Experimental study of phase equilibria and solidification microstructures of Mg-Ca-Ce alloys combined with thermodynamic modeling. *Journal of Materials Science* 42 (24):10023–31.

Hampp, C., B. Ullmann, J. Reifenrath et al. 2012. Research on the biocompatibility of the new magnesium alloy LANd442—An in vivo study in the rabbit tibia over 26 weeks. *Advanced Engineering Materials* 14 (3):B28–37.

Hampp, C., N. Angrisani, J. Reifenrath et al. 2013. Evaluation of the biocompatibility of two magnesium alloys as degradable implant materials in comparison to titanium as non-resorbable material in the rabbit. *Materials Science and Engineering: C* 33 (1): 317–26.

Han, W., Y. Tian, M. Zhang, Y. Yan, and X. Jing. 2009. Preparation of Mg-Li-Sm alloys by electrocodeposition in molten salt. *Journal of Rare Earths* 27 (6):1046–50.

Han, P., M. Tan, S. Zhang et al. 2014. Shape and site dependent in vivo degradation of Mg-Zn pins in rabbit femoral condyle. *International Journal of Molecular Sciences* 15 (2):2959–70.

Hada, K., K. Matsuzaki, X. Huang, and Y. Chino. 2013. Fabrication of Mg alloy tubes for biodegradable stent application. *Materials Science and Engineering. C, Materials for Biological Applications* 33 (8):4146–50.

Hanaoka, T., N. Kamimura, T. Yokota, S. Takai, and S. Ohta. 2011. Molecular hydrogen protects chondrocytes from oxidative stress and indirectly alters gene expressions through reducing peroxynitrite derived from nitric oxide. *Medical Gas Research* 1 (1):18.

Hanawalt, J. D., C. E. Nelson, and J. A. Peloubet. 1942. Corrosion studies of magnesium and its alloys. *Transactions of AIME* 147:273–99.

Hantzsche, K., J. Bohlen, J. Wendt et al. 2010. Effect of rare earth additions on microstructure and texture development of magnesium alloy sheets. *Scripta Materialia* 63 (7):725–30.

Hänzi, A. C., I. Gerber, M. Schinhammer, J. F. Löffler, and P. J. Uggowitzer. 2010. On the in vitro and in vivo degradation performance and biological response of new biodegradable Mg–Y–Zn alloys. *Acta Biomaterialia* 6 (5):1824–33.

Harandi, S. E., M. H. Idris, and H. Jafari. 2011. Effect of forging process on microstructure, mechanical and corrosion properties of biodegradable Mg-1Ca alloy. *Materials and Design* 32 (5):2596–603.

Harandi, S. E., M. Mirshahi, S. Koleini, M. H. Idris, H. Jafari, and M. R. A. Kadir. 2013. Effect of calcium content on the microstructure, hardness and in-vitro corrosion behavior of biodegradable Mg-Ca binary alloy. *Materials Research-Ibero-American Journal of Materials* 16 (1):11–18.

Hardes, J., A. Streitburger, H. Ahrens, T. Nusselt, C. Gebert, W. Winkelmann, A. Battmann, and G. Gosheger. 2007. The influence of elementary silver versus titanium on osteoblasts behaviour in vitro using human osteosarcoma cell lines. *Sarcoma* 2007:26539.

Hartwig, A. 2001. Role of magnesium in genomic stability. *Mutation Research/Fundamental and Molecular Mechanisms of Mutagenesis* 475 (1–2):113–21.

Hashimoto, M., M. Katakura, T. Nabika et al. 2011. Effects of hydrogen-rich water on abnormalities in a SHR. Cg-Leprcp/NDmcr rat-a metabolic syndrome rat model. *Medical Gas Research* 1 (1):26.

Hassan, S., and M. Gupta. 2002. Development of ductile magnesium composite materials using titanium as reinforcement. *Journal of Alloys and Compounds* 345 (1):246–51.

Hassani, F. Z., M. Ketabchi, and M. T. Hassani. 2011. Effect of twins and non-basal planes activated by equal channel angular rolling process on properties of AZ31 magnesium alloy. *Journal of Materials Science* 46 (24):7689–95.

Hassel, T., F. W. Bach, C. Krause, and P. Wilk. 2005. Corrosion protection and repassivation after the deformation of magnesium alloys coated with a protective magnesium fluoride layer. In *Magnesium Technology 2005*, edited by N. Neelameggham, H. I. Kaplan, and B. R. Powell, Berlin: Springer Berlin Heidelberg.

Haude, M., R. Erbel, P. Erne et al. 2013. Safety and performance of the drug-eluting absorbable metal scaffold (DREAMS) in patients with de-novo coronary lesions: 12 month results of the prospective, multicentre, first-in-man BIOSOLVE-I trial. *The Lancet* 381 (9869):836–44.

Hawke, D. 1975. Corrosion and wear resistance of magnesium die castings. Paper read at SYCE 8th International Die Casting Exposition and Congress.

Haxel, G. B., J. B. Hedrick, and G. J. Orris. 2002. Rare earth elements: Critical resources for high technology. Fact sheet 087-02. Reston, VA: US Geological Survey. pubs.er.usgs .gov, No. 087-02. 2002.

Hayashi, S., K. Usuda, G. Mitsui, T. Shibutani, E. Dote, K. Adachi, M. Fujihara, Y. Shimbo, W. Sun, and R. Kono. 2006. Urinary yttrium excretion and effects of yttrium chloride on renal function in rats. *Biological Trace Element Research* 114 (1–3):225–35.

Hayashi, T., T. Yoshioka, K. Hasegawa et al. 2011. Inhalation of hydrogen gas attenuates left ventricular remodeling induced by intermittent hypoxia in mice. *American Journal of Physiology-Heart and Circulatory Physiology* 301 (3):H1062–9.

Hayashida, K., M. Sano, I. Ohsawa et al. 2008. Inhalation of hydrogen gas reduces infarct size in the rat model of myocardial ischemia–reperfusion injury. *Biochemical and Biophysical Research Communications* 373 (1):30–5.

He, L. X. 2010. *Research on optimization of extrusion and drawing process of pure magnesium wire and influences on biological corrosion.* Masters Thesis, Harbin Institute of Technology, Harbin, China.

He, Y. H., H. R. Tao, Y. Zhang, Y. Jiang, S. X. Zhang, C. L. Zhao, J. N. Li, B. L. Zhang, Y. Song, and X. N. Zhang. 2009. Biocompatibility of bio-Mg-Zn alloy within bone with heart, liver, kidney and spleen. *Chinese Science Bulletin* 54 (3):484–91.

He, W. W., E. L. Zhang, and K. Yang. 2010. Effect of Y on the bio-corrosion behavior of extruded Mg-Zn-Mn alloy in Hank's solution. *Materials Science and Engineering C-Materials for Biological Applications* 30 (1):167–74.

Hedrick, J. B. 2004. Rare earths. *Minerals Yearbook* 1:60.1–15.

Hench, L. L. 1998. Bioceramics. *Journal of the American Ceramic Society* 81 (7):1705–28.

Hench, L. L., and J. M. Polak. 2002. Third-generation biomedical materials. *Science* 295 (5557):1014–17.

Henderson, S. E., K. Verdelis, S. Maiti et al. 2014. Magnesium alloys as a biomaterial for degradable craniofacial screws. *Acta Biomaterialia* 10 (5):2323–32.

Hermawan, H., D. Dubé, and D. Mantovani. 2010. Developments in metallic biodegradable stents. *Acta Biomaterialia* 6 (5):1693–7.

Heublein, B., R. Rohde, V. Kaese, M. Niemeyer, W. Hartung, and A. Haverich. 2003. Biocorrosion of magnesium alloys: A new principle in cardiovascular implant technology? *Heart* 89 (6):651–6.

Hifumi, H., S. Yamaoka, A. Tanimoto, D. Citterio, and K. Suzuki. 2006. Gadolinium-based hybrid nanoparticles as a positive MR contrast agent. *Journal of the American Chemical Society* 128 (47):15090–1.

Hifumi, H., S. Yamaoka, A. Tanimoto, T. Akatsu, Y. Shindo, A. Honda, D. Citterio, K. Oka, S. Kuribayashi, and K. Suzuki. 2009. Dextran coated gadolinium phosphate nanoparticles for magnetic resonance tumor imaging. *Journal of Materials Chemistry* 19 (35):6393–9.

Hijazi, Z. M., T. Feldman, M. H. A. Al-Qbandi, and H. Sievert. 2010. *Transcatheter Closure of ASDs and PFOs: A Comprehensive Assessment.* Minneapolis, MN: Cardiotext Publishing.

Hillis, J. E. 1983. The effects of heavy metal contamination on magnesium corrosion performance. SAE Technical Paper.

Hirai, K., H. Somekawa, Y. Takigawa, and K. Higashi. 2005. Effects of Ca and Sr addition on mechanical properties of a cast AZ91 magnesium alloy at room and elevated temperature. *Materials Science and Engineering A-Structural Materials Properties Microstructure and Processing* 403 (1–2):276–80.

Hirano, S., and K. T. Suzuki. 1996. Exposure, metabolism, and toxicity of rare earths and related compounds. *Environmental Health Perspectives* 104 (Suppl 1):85–95.

Hiromoto, S., and S. Mischler. 2006. The influence of proteins on the fretting–corrosion behaviour of a Ti6Al4V alloy. *Wear* 261 (9):1002–11.

Hiromoto, S., and M. Tomozawa. 2010a. Corrosion behavior of magnesium with hydroxyapatite coatings formed by hydrothermal treatment. *Materials Transactions* 51 (11):2080–7.

Hiromoto, S., and A. Yamamoto. 2010b. Control of degradation rate of bioabsorbable magnesium by anodization and steam treatment. *Materials Science and Engineering: C* 30 (8):1085–93.

Hiromoto, S., T. Shishido, A. Yamamoto, N. Maruyama, H. Somekawa, and T. Mukai. 2008a. Precipitation control of calcium phosphate on pure magnesium by anodization. *Corrosion Science* 50 (10):2906–13.

Hiromoto, S., A. Yamamoto, N. Maruyama, H. Somekawa, and T. Mukai. 2008b. Influence of pH and flow on the polarisation behaviour of pure magnesium in borate buffer solutions. *Corrosion Science* 50 (12):3561–8.

Hirosawa, S., H. Tomizawa, S. Mino, and A. Hamamura. 1990. High-coercivity Nd-Fe-B-type permanent magnets with less dysprosium. *Magnetics, IEEE Transactions on* 26 (5):1960–2.

Hoeppner, D. W., and V. Chandrasekaran. 1994. Fretting in orthopedic implants—A review. *Wear* 173 (1–2):189–97.

Hoey, G. R., and M. Cohen. 1958. Corrosion of anodically and cathodically polarized magnesium in aqueous media. *Journal of the Electrochemical Society* 105 (5):245–50.

Hofstetter, J., E. Martinelli, A. M. Weinberg, M. Becker, B. Mingler, P. J. Uggowitzer, and J. F. Löffler. 2014. Assessing the degradation performance of ultrahigh-purity magnesium in vitro and in vivo. *Corrosion Science* 91:29–36.

Hojvat de Tendler, R., M. R. Soriano, M. E. Pepe, J. A. Kovacs, E. E. Vicente, and J. A. Alonso. 2006. Calculation of metastable free-energy diagrams and glass formation in the Mg–Cu–Y alloy and its boundary binaries using the Miedema model. *Intermetallics* 14 (3):297–307.

Hollister, S. J. 2005. Porous scaffold design for tissue engineering. *Nature Materials* 4 (7):518–24.

Homma, T., C. L. Mendis, K. Hono, and S. Kamado. 2010. Effect of Zr addition on the mechanical properties of as-extruded Mg-Zn-Ca-Zr alloys. *Materials Science and Engineering A-Structural Materials Properties Microstructure and Processing* 527 (9):2356–62.

Hong, Y. S., K. Yang, G. D. Zhang et al. 2008. The role of bone induction of a biodegradable magnesium alloy. *Acta Metallurgica Sinica* 44 (9):1035–41.

Hong, D., P. Saha, D. T. Chou, B. Lee, B. E. Collins, Z. Q. Tan, Z. Y. Dong, and P. N. Kumta. 2013. In vitro degradation and cytotoxicity response of Mg-4% Zn-0.5% Zr (ZK40) alloy as a potential biodegradable material. *Acta Biomaterialia* 9 (10):8534–47.

Höpfner, E. 1903. Ueber Gefässnaht, Gefässtransplantationen und Replantation von amputirten Extremitäten: Inaugural-Dissertation welche zur Erlangung der Doktorwürde in der Medicin und Chirurgie: Mit Zustimmung der medicinischen Fakultät der Königl. Friedrich-Wilhelms-Universität zu Berlin: Am 15. Mai 1903 nebst den angeführten Thesen: öffentlich verteidigen wird, Druck von L. Schumacher.

Hort, N., Y. D. Huang, T. Abu Leil, P. Maier, and K. U. Kainer. 2006. Microstructural investigations of the Mg-Sn-xCa system. *Advanced Engineering Materials* 8 (5):359–64.

Hort, N., Y. Huang, D. Fechner, M. Stormer, C. Blawert, F. Witte, C. Vogt, H. Drucker, R. Willumeit, K. U. Kainer, and F. Feyerabend. 2010. Magnesium alloys as implant materials—Principles of property design for Mg-RE alloys. *Acta Biomaterialia* 6 (5):1714–25.

Hou, S.-M., C.-C. Hsu, J.-L. Wang, C.-K. Chao, and J. Lin. 2004. Mechanical tests and finite element models for bone holding power of tibial locking screws. *Clinical Biomechanics* 19:738–45.

Hou, S. S., R. R. Zhang, S. K. Guan, C. X. Ren, J. H. Gao, Q. B. Lu, and X. Z. Cui. 2012. In vitro corrosion behavior of Ti-O film deposited on fluoride-treated Mg–Zn–Y–Nd alloy. *Applied Surface Science* 258 (8):3571–7.

Hradilova, M., D. Vojtech, J. Kubasek, J. Capek, and M. Vlach. 2013. Structural and mechanical characteristics of Mg-4Zn and Mg-4Zn-0.4Ca alloys after different thermal and mechanical processing routes. *Materials Science and Engineering A-Structural Materials Properties Microstructure and Processing* 586:284–91.

Hsiao, H.-M., L.-W. Wu, M.-T. Yin, C.-H. Lin, and H. Chen. 2014. Quintupling fatigue resistance of intravascular stents via a simple design concept. *Computational Materials Science* 86:57–63.

Hsu, C. C., C. K. Chao, J. L. Wang, S. M. Hou, Y. T. Tsai, and J. Lin. 2005. Increase of pullout strength of spinal pedicle screws with conical core: Biomechanical tests and finite element analyses. *Journal of Orthopaedic Research* 23:788–94.

Hu, X. S., Y. K. Zhang, M. Y. Zheng, and K. Wu. 2005. A study of damping capacities in pure Mg and Mg–Ni alloys. *Scripta Materialia* 52 (11):1141–5.

Hu, J., C. Zhang, B. Cui, K. Bai, S. Guan, L. Wang, and S. Zhu. 2011. In vitro degradation of AZ31 magnesium alloy coated with nano TiO_2 film by sol-gel method. *Applied Surface Science* 257 (21):8772–7.

Hu, M., H. Fei, J. Gao, and F. F. Zhao. 2012a. A study on microstructures and creep behaviors in the Mg-3Sr-xY alloys. *Advanced Materials Research* 418:602–5.

Hu, M.-L., Q.-D. Wang, C. Li, and W.-J. Ding. 2012b. Dry sliding wear behavior of cast Mg–11Y–5Gd–2Zn magnesium alloy. *Transactions of Nonferrous Metals Society of China* 22 (8):1918–23.

Hu, X.-S., X.-J. Wang, X.-D. He, K. Wu, and M.-Y. Zheng. 2012. Low frequency damping capacities of commercial pure magnesium. *Transactions of Nonferrous Metals Society of China* 22 (8):1907–11.

Huaiying, Z., G. Tianlong, Y. Daoguo, J. Zhengyi, and Z. Jianmin. 2011. Compressive properties of hot-rolled Mg-Zr-Ca alloys for biomedical applications. *Advanced Materials Research* 197–8:56–9.

Huan, Z. G., M. A. Leeflang, J. Zhou, L. E. Fratila-Apachitei, and J. Duszczyk. 2010a. In vitro degradation behavior and cytocompatibility of Mg-Zn-Zr alloys. *Journal of Materials Science-Materials in Medicine* 21 (9):2623–35.

Huan, Z., J. Zhou, and J. Duszczyk. 2010b. Magnesium-based composites with improved in vitro surface biocompatibility. *Journal of Materials Science: Materials in Medicine* 21 (12):3163–9.

Huan, Z., M. Leeflang, J. Zhou, and J. Duszczyk. 2011. ZK30-bioactive glass composites for orthopedic applications: A comparative study on fabrication method and characteristics. *Materials Science and Engineering: B* 176 (20):1644–52.

Huan, Z., S. Leeflang, J. Zhou, W. Zhai, J. Chang, and J. Duszczyk. 2012. In vitro degradation behavior and bioactivity of magnesium-Bioglass® composites for orthopedic applications. *Journal of Biomedical Materials Research Part B: Applied Biomaterials* 100 (2):437–46.

Huang, X. F., and W. Z. Zhang. 2012. Improved age-hardening behavior of Mg–Sn–Mn alloy by addition of Ag and Zn. *Materials Science and Engineering: A* 552:211–21.

Huang, X., and W. Zhang. 2014. Characterization and interpretation on the new crystallographic features of a twin-related ε′-Mg54Ag17 precipitates in an Mg–Sn–Mn–Ag–Zn alloy. *Journal of Alloys and Compounds* 582:764–8.

Huang, W. J., B. Hou, Y. X. Pang, and Z. R. Zhou. 2006. Fretting wear behavior of AZ91D and AM60B magnesium alloys. *Wear* 260 (11–2):1173–8.

Huang, X. S., K. Suzuki, and N. Saito. 2009. Microstructure and mechanical properties of AZ80 magnesium alloy sheet processed by differential speed rolling. *Materials Science and Engineering: A* 508 (1–2):226–33.

Huang, C.-S., T. Kawamura, S. Lee et al. 2010. Hydrogen inhalation ameliorates ventilator-induced lung injury. *Crit Care* 14 (6):R234.

Huang, C.-S., T. Kawamura, X. Peng et al. 2011. Hydrogen inhalation reduced epithelial apoptosis in ventilator-induced lung injury via a mechanism involving nuclear factor-kappa B activation. *Biochemical and Biophysical Research Communications* 408 (2):253–8.

Huang, Y., K. Xie, J. Li et al. 2011. Beneficial effects of hydrogen gas against spinal cord ischemia–reperfusion injury in rabbits. *Brain Research* 1378:125–36.

Huang, H., G. Y. Yuan, Z. H. Chu, and W. J. Ding. 2013. Microstructure and mechanical properties of double continuously extruded Mg-Zn-Gd-based magnesium alloys. *Materials Science and Engineering A-Structural Materials Properties Microstructure and Processing* 560:241–8.

Huang, K., S. Cai, G. Xu, X. Ye, Y. Dou, M. Ren, and X. Wang. 2013a. Preparation and characterization of mesoporous 45S5 bioactive glass–ceramic coatings on magnesium alloy for corrosion protection. *Journal of Alloys and Compounds* 580:290–7.

Huang, K., X. Lin, C. Xie, and T. Yue. 2013b. Laser cladding of Zr-based coating on AZ91D magnesium alloy for improvement of wear and corrosion resistance. *Bulletin of Materials Science* 36 (1):99–105.

Huang, Z. H., W. J. Qi, K. H. Zheng, X. M. Zhang, M. Liu, Z. M. Yu, and J. Xu. 2013. Microstructures and mechanical properties of Mg–Zn–Zr–Dy wrought magnesium alloys. *Bulletin of Materials Science* 36 (3):437–45.

Huang, K., S. Cai, G. Xu, M. Ren, X. Wang, R. Zhang, S. Niu, and H. Zhao. 2014. Sol-gel derived mesoporous 58S bioactive glass coatings on AZ31 magnesium alloy and in vitro degradation behavior. *Surface and Coatings Technology* 240:137–44.

Huehnerschulte, T. A., N. Angrisani, D. Rittershaus et al. 2011. In vivo corrosion of two novel magnesium alloys ZEK100 and AX30 and their mechanical suitability as biodegradable implants. *Materials* 4 (12):1144–67.

Huehnerschulte, T. A., J. Reifenrath, B. von Rechenberg et al. 2012. In vivo assessment of the host reactions to the biodegradation of the two novel magnesium alloys ZEK100 and AX30 in an animal model. *Biomedical Engineering Online* 11:14.

Hugyecz, M., É. Mracskó, P. Hertelendy et al. 2011. Hydrogen supplemented air inhalation reduces changes of prooxidant enzyme and gap junction protein levels after transient global cerebral ischemia in the rat hippocampus. *Brain Research* 1404:31–8.

Hui, X., G. Y. Sun, C. M. Zhang, S. N. Liu, E. R. Wang, M. L. Wang, and G. L. Chen. 2010. Mg–Cu–Y–Ag bulk metallic glasses with enhanced compressive strength and plasticity. *Journal of Alloys and Compounds* 504:S6–9.

Hunt, P., F. N. Unterhauser, M. J. Strobel, and A. Weiler. 2005. Development of a perforated biodegradable interference screw. *Arthroscopy: The Journal of Arthroscopic and Related Surgery* 21 (3):258–65.

Huse, E. 1878. A new ligature? *Chicago Medical Journal and Examiner* 37:171–2.

Hussl, H., C. Papp, I. Höpfel-Kreiner, E. Rumpl, and P. Wilflingseder. 1981. Resorption time and tissue reactions with magnesium rods in rats and rabbits. *Chirurgia Plastica* 6 (2):117–26.

Hwang, I. J., B. U. Lee, Y. G. Ko, and D. H. Shin. 2011. Corrosion response of annealed oxide film of pure Mg via plasma electrolytic oxidation in an electrolyte containing KMnO4. *Journal of Alloys and Compounds* 509:S473–7.

Hwang, S. W., H. Tao, D. H. Kim et al. 2012. A physically transient form of silicon electronics. *Science* 337 (6102):1640–4.

Hynes, R. O. 1992. Integrins—Versatility, modulation, and signaling in cell-adhesion. *Cell* 69 (1):11–25.

Hyuk Kim, S., D. H. Kim, and N. J. Kim. 1997. Structure and properties of rapidly solidified Mg-Al-Zn-Nd alloys. *Materials Science and Engineering: A* 226:1030–4.

Imwinkelried, T., S. Beck, T. Iizuka, and B. Schaller. 2013. Effect of a plasma electrolytic coating on the strength retention of in vivo and in vitro degraded magnesium implants. *Acta Biomaterialia* 9 (10):8643–9.

Inoue, M., M. Iwai, K. Matuzawa, S. Kamado, and Y. Kojima. 1998. Effect of impurities on corrosion behavior of pure magnesium in salt water environment. *Journal of Japan Institute of Light Metals* 48 (6):257–62.

Inoue, H., K. Sugahara, A. Yamamoto, and H. Tsubakino. 2002. Corrosion rate of magnesium and its alloys in buffered chloride solutions. *Corrosion Science* 44 (3):603–10.

International Organization for Standardization, Standard 14801: Dentistry – Implants – Dynamic fatigue test for endosseous dental implants.

Ishizaki, T., K. Teshima, Y. Masuda, and M. Sakamoto. 2011. Liquid phase formation of alkyl- and perfluoro-phosphonic acid derived monolayers on magnesium alloy AZ31 and their chemical properties. *Journal of Colloid and Interface Science* 360 (1):280–8.

Ismail, A. A., Y. Ismail, and A. A. Ismail. 2013. Clinical assessment of magnesium status in the adult: An overview. In R. R. Watson, V. R. Preedy, and S. Zibadi (eds.). Humana Press, Totowa, NJ. *Magnesium in Human Health and Disease.*

ISO, E. 2000. 10993-15. Biological evaluation of medical devices–Part 15: Identification and quantification of degradation products from metals and alloys.

ISO, E. 2002. 10993-4: 2007–Biological evaluation of medical devices–Part 4: Selection of tests for interactions with blood. ISO.

ISO, E. 2008. 10993-12: 2008–Biological evaluation of medical devices–Part 12: Sample preparation and reference materials (ISO 10993-12: 2007). *German version: DIN EN ISO*:10993-12.

Ito, M., T. Ibi, K. Sahashi et al. 2011. Open-label trial and randomized, double-blind, placebo-controlled, crossover trial of hydrogen-enriched water for mitochondrial and inflammatory myopathies. *Medical Gas Research* 1 (1):24.

Itoh, T., Y. Fujita, M. Ito et al. 2009. Molecular hydrogen suppresses FcεRI-mediated signal transduction and prevents degranulation of mast cells. *Biochemical and Biophysical Research Communications* 389 (4):651–6.

Itoh, T., N. Hamada, R. Terazawa et al. 2011. Molecular hydrogen inhibits lipopolysaccharide/interferon γ-induced nitric oxide production through modulation of signal transduction in macrophages. *Biochemical and Biophysical Research Communications* 411 (1):143–9.

Itoi, T., T. Seimiya, Y. Kawamura, and M. Hirohashi. 2004. Long period stacking structures observed in $Mg_{97}Zn_1Y_2$ alloy. *Scripta Materialia* 51 (2):107–11.

Izaham, R., R. M. Aizat, M. R. A. Kadir, A. H. A. Rashid, M. G. Hossain, and T. Kamarul. 2012. Finite element analysis of Puddu and Tomofix plate fixation for open wedge high tibial osteotomy. *Injury* 43 (6): 898–902.

Izumi, S., M. Yamasaki, and Y. Kawamura. 2009. Relation between corrosion behavior and microstructure of Mg–Zn–Y alloys prepared by rapid solidification at various cooling rates. *Corrosion Science* 51 (2):395–402.

Jain, I. P., Chhagan L., and A. Jain. 2010. Hydrogen storage in Mg: A most promising material. *International Journal of Hydrogen Energy* 35 (10):5133–44.

James, D. W. 1969. High damping metals for engineering applications. *Materials Science and Engineering* 4 (1):1–8.

James, B. A., and R. A. Sire. 2010. Fatigue-life assessment and validation techniques for metallic vascular implants. *Biomaterials* 31 (2):181–6.

Jamesh, M., S. Kumar, and T. S. N. Sankara Narayanan. 2011a. Corrosion behavior of commercially pure Mg and ZM21 Mg alloy in Ringer's solution–long term evaluation by EIS. *Corrosion Science* 53 (2):645–54.

Jamesh, M., S. Kumar, and T. S. N. Sankara Narayanan. 2011b. Electrodeposition of hydroxyapatite coating on magnesium for biomedical applications. *Journal of Coatings Technology and Research* 9 (4):495–502.

Jamesh, M., G. Wu, Y. Zhao, and P. K. Chu. 2013. Effects of silicon plasma ion implantation on electrochemical corrosion behavior of biodegradable Mg–Y–RE Alloy. *Corrosion Science* 69:158–63.

Jang, H. S., W. B. Im, D. C. Lee, D. Y. Jeon, and S. S. Kim. 2007. Enhancement of red spectral emission intensity of Y3Al5O12: Ce3+ phosphor via Pr co-doping and Tb substitution for the application to white LEDs. *Journal of Luminescence* 126 (2):371–7.

Jang, J. S., J. Ciou, T. Li, J. Huang, and T. Nieh. 2010. Dispersion toughening of Mg-based bulk metallic glass reinforced with porous Mo particles. *Intermetallics* 18 (4):451–8.

Jang, J., T. Li, S. Jian, J. Huang, and T. Nieh. 2011. Effects of characteristics of Mo dispersions on the plasticity of Mg-based bulk metallic glass composites. *Intermetallics* 19 (5):738–43.

Jang, J., J. Li, S. Lee, Y. Chang, S. Jian, J. Huang, and T. Nieh. 2012. Prominent plasticity of Mg-based bulk metallic glass composites by ex-situ spherical Ti particles. *Intermetallics* 30:25–9.

Janik, V., F. Hnilica, P. Zuna, V. Očenášek, and I. Stulíková. 2008. Cavitation and grain boundary sliding during creep of Mg-Y-Nd-Zn-Mn alloy. *Transactions of Nonferrous Metals Society of China* 18:s64–8.

Janning, C., E. Willbold, C. Vogt et al. 2010. Magnesium hydroxide temporarily enhancing osteoblast activity and decreasing the osteoclast number in peri-implant bone remodelling. *Acta Biomaterialia* 6 (5):1861–8.

Jayalakshmi, S., S. Sankaranarayanan, S. P. X. Koh, and M. Gupta. 2013. Effect of Ag and Cu trace additions on the microstructural evolution and mechanical properties of Mg–5Sn alloy. *Journal of Alloys and Compounds* 565:56–65.

Jayaraj, J., C. L. Mendis, T. Ohkubo, K. Oh-ishi, and K. Hono. 2010. Enhanced precipitation hardening of Mg-Ca alloy by Al addition. *Scripta Materialia* 63 (8):831–4.

Jeon, H., H. Hidai, D. J. Hwang, K. E. Healy, and C. P. Grigoropoulos. 2010. The effect of micronscale anisotropic cross patterns on fibroblast migration. *Biomaterials* 31 (15):4286–95.

Ji, Y. J., B. Xiao, Z. H. Wang, M. Z. Cui, and Y. Y. Lu. 2000. The suppression effect of light rare earth elements on proliferation of two cancer cell lines. *Biomedical and Environmental Sciences: BES* 13 (4):287–92.

Ji, Z. S., L. H. Wen, and X. L. Li. 2009. Mechanical properties and fracture behavior of Mg–2.4Nd–0.6Zn–0.6Zr alloys fabricated by solid recycling process. *Journal of Materials Processing Technology* 209 (4):2128–34.

Ji, X., W. Liu, K. Xie et al. 2010. Beneficial effects of hydrogen gas in a rat model of traumatic brain injury via reducing oxidative stress. *Brain Research* 1354:196–205.

Ji, Q., K. Hui, L. Zhang et al. 2011. The effect of hydrogen-rich saline on the brain of rats with transient ischemia. *Journal of Surgical Research* 168 (1):e95–101.

Jia, Z. J., M. Li, Q. Liu, X. C. Xu, Y. Cheng, Y. F. Zheng, T. F. Xi, and S. C. Wei. 2014. Microarc oxidization of a novel Mg-1Ca alloy in three alkaline KF electrolytes: Corrosion resistance and cytotoxicity. *Applied Surface Science* 292:1030–9.

Jiang, Y.-F., H.-T. Zhou, and S.-M. Zeng. 2009. Microstructure and properties of oxalate conversion coating on AZ91D magnesium alloy. *Transactions of Nonferrous Metals Society of China* 19 (6):1416–22.

Jiang, B., D. Qiu, M. X. Zhang, P. D. Ding, and L. Gao. 2010. A new approach to grain refinement of an Mg-Li–Al cast alloy. *Journal of Alloys and Compounds* 492 (1–2):95–8.

Jiang, B., Y. Zeng, H. M. Yin, R. H. Li, and F. S. Pan. 2012. Effect of Sr on microstructure and aging behavior of Mg-14Li alloys. *Progress in Natural Science-Materials International* 22 (2):160–8.

Jiang, D. M., Z. Y. Cao, X. Sun, L. Guo, and J. G. Liu. 2013. Effect of yttrium addition on microstructure and mechanical properties of Mg-Zn-Ca alloy. *Materials Research Innovations* 17:S33–8.

Jiang, H. B., Y. K. Kim, J. H. Ji, I. S. Park, T. S. Bae, and M. H. Lee. 2014. Surface modification of anodized Mg in ammonium hydrogen fluoride by various voltages. *Surface and Coatings Technology* 259:310–7.

Jin, Q. M., H. Takita, T. Kohgo, K. Atsumi, H. Itoh, and Y. Kuboki. 2000. Effects of geometry of hydroxyapatite as a cell substratum in BMP-induced ectopic bone formation. *Journal of Biomedical Materials Research* 52 (4):841–51.

Jo, J. H., B. G. Kang, K. S. Shin, H. E. Kim, B. D. Hahn, D. S. Park, and Y. H. Koh. 2011. Hydroxyapatite coating on magnesium with MgF(2) interlayer for enhanced corrosion resistance and biocompatibility. *Journal of Materials Science. Materials in Medicine* 22 (11):2437–47.

Jo, J.-H., J.-Y. Hong, K.-S. Shin, H.-E. Kim, and Y.-H. Koh. 2012. Enhancing biocompatibility and corrosion resistance of Mg implants via surface treatments. *Journal of Biomaterials Applications* 27 (4):469–76.

Jo, J. H., Y. Li, S. M. Kim, H. E. Kim, and Y. H. Koh. 2013. Hydroxyapatite/poly(epsilon-caprolactone) double coating on magnesium for enhanced corrosion resistance and coating flexibility. *Journal of Biomaterials Applications* 28 (4):617–25.

Johnson, I., and H. Liu. 2013. A study on factors affecting the degradation of magnesium and a magnesium-yttrium alloy for biomedical applications. *PLoS One* 8 (6):e65603.

Johnson, A. L., J. E. F. Houlton, and R. Vannini. 2006. *AO Principles of Fracture Management in the Dog and Cat.* Davos, Switzerland: AO Publishing.

Johnson, I., D. Perchy, and H. Liu. 2012. In vitro evaluation of the surface effects on magnesium-yttrium alloy degradation and mesenchymal stem cell adhesion. *Journal of Biomedical Materials Research Part A* 100 (2):477–85.

Jones, D. A. 1992. *Principles and Prevention of Corrosion.* Basingstoke, New York: Macmillan.

Jones, J. R., L. M. Ehrenfried, and L. L. Hench. 2006. Optimising bioactive glass scaffolds for bone tissue engineering. *Biomaterials* 27 (7):964–73.

Jung, J. Y., S. J. Kwon, H. S. Han, J. Y. Lee, J. P. Ahn, S. J. Yang, S. Y. Cho, P. R. Cha, Y. C. Kim, and H. K. Seok. 2012. In vivo corrosion mechanism by elemental interdiffusion of biodegradable Mg-Ca alloy. *Journal of Biomedical Materials Research Part B-Applied Biomaterials* 100B (8):2251–60.

Kainer, K. U. 2003. *Magnesium Alloys and Technology.* Weinheim, Germany: Wiley-VCH.

Kainer, K. U., and B. L. Mordike. 2000. *Magnesium Alloys and Their Applications,* Weinheim, Germany: Wiley-VCH.

Kajiya, M., K. Sato, M. J. Silva et al. 2009a. Hydrogen from intestinal bacteria is protective for Concanavalin A-induced hepatitis. *Biochemical and Biophysical Research Communications* 386 (2):316–21.

Kajiya, M., M. J. Silva, K. Sato, K. Ouhara, and T. Kawai. 2009b. Hydrogen mediates suppression of colon inflammation induced by dextran sodium sulfate. *Biochemical and Biophysical Research Communications* 386 (1):11–15.

Kajiyama, S., G. Hasegawa, M. Asano et al. 2008. Supplementation of hydrogen-rich water improves lipid and glucose metabolism in patients with type 2 diabetes or impaired glucose tolerance. *Nutrition Research* 28 (3):137–43.

Kalimullin, R. K., and Y. Y. Kozhevnikov. 1985. Structure and corrosion resistance of an Mg-Li base alloy after laser treatment. *Metal Science and Heat Treatment* 27 (4):272–4.

Kamimura, N., K. Nishimaki, I. Ohsawa, and S. Ohta. 2011. Molecular hydrogen improves obesity and diabetes by inducing hepatic FGF21 and stimulating energy metabolism in db/db mice. *Obesity* 19 (7):1396–403.

Kamitakahara, M., C. Ohtsuki, and T. Miyazaki. 2008. Review paper: Behavior of ceramic biomaterials derived from tricalcium phosphate in physiological condition. *Journal of Biomaterials Applications* 23 (3):197–212.

Kang, F., J. Q. Liu, J. T. Wang, and X. Zhao. 2010. Equal channel angular pressing of a Mg–3Al–1Zn alloy with back pressure. *Advanced Engineering Materials* 12 (8):730–4.

Kang, K.-M., Y.-N. Kang, I.-B. Choi et al. 2011. Effects of drinking hydrogen-rich water on the quality of life of patients treated with radiotherapy for liver tumors. *Medical Gas Research* 1 (1):11.

Kang, Z., X. Lai, J. Sang, and Y. Li. 2011. Fabrication of hydrophobic/super-hydrophobic nanofilms on magnesium alloys by polymer plating. *Thin Solid Films* 520 (2):800–6.

Kannan, M. B. 2010. Influence of microstructure on the in vitro degradation behaviour of magnesium alloys. *Materials Letters* 64 (6):739–42.

Kannan, M. B. 2012. Enhancing the performance of calcium phosphate coating on a magnesium alloy for bioimplant applications. *Materials Letters* 76:109–12.

Kannan, M. B., and R. K. S. Raman. 2008. In vitro degradation and mechanical integrity of calcium-containing magnesium alloys in modified-simulated body fluid. *Biomaterials* 29 (15):2306–14.

Kannan, M. B., and L. Orr. 2011. In vitro mechanical integrity of hydroxyapatite coated magnesium alloy. *Biomedical Materials* 6 (4):045003.

Kannan, M. B., and S. Liyanaarachchi. 2013. Hybrid coating on a magnesium alloy for minimizing the localized degradation for load-bearing biodegradable mini-implant applications. *Materials Chemistry and Physics* 142 (1):350–4.

Kannan, M. B., and O. Wallipa. 2013. Potentiostatic pulse-deposition of calcium phosphate on magnesium alloy for temporary implant applications—An in vitro corrosion study. *Materials Science and Engineering: C* 33 (2):675–9.

Kannan, M. B., W. Dietzel, and R. Zettler. 2011. In vitro degradation behaviour of a friction stir processed magnesium alloy. *Journal of Materials Science. Materials in Medicine* 22 (11):2397–401.

Karageorgiou, V., and D. Kaplan. 2005. Porosity of 3D biomaterial scaffolds and osteogenesis. *Biomaterials* 26 (27):5474–91.

Karavai, O. V., A. C. Bastos, M. L. Zheludkevich, M. G. Taryba, S. V. Lamaka, and M. G. S. Ferreira. 2010. Localized electrochemical study of corrosion inhibition in microdefects on coated AZ31 magnesium alloy. *Electrochimica Acta* 55 (19):5401–6.

Kartsonakis, I. A., E. P. Koumoulos, C. A. Charitidis, and G. Kordas. 2013. Hybrid organic–inorganic coatings including nanocontainers for corrosion protection of magnesium alloy ZK30. *Journal of Nanoparticle Research* 15 (8):1871.

Kathuria, Y. P. 2005. Laser microprocessing of metallic stent for medical therapy. *Journal of Materials Processing Technology* 170 (3):545–50.

Kato, K., A. Yamamoto, S. Ochiai, Y. Daigo, T. Isobe, S. Matano, and K. Omori. 2012. Cell proliferation, corrosion resistance and mechanical properties of novel titanium foam with sheet shape. *Materials Transactions* 53 (4):724–32.

Kattel, K., J. Y. Park, W. Xu, H. G. Kim, E. J. Lee, B. A. Bony, W. C. Heo, S. Jin, J. S. Baeck, and Y. Chang. 2012. Paramagnetic dysprosium oxide nanoparticles and dysprosium hydroxide nanorods as T_2 MRI contrast agents. *Biomaterials* 33 (11):3254–61.

Kaur, G., O. P. Pandey, K. Singh, D. Homa, B. Scott, and G. Pickrell. 2014. A review of bioactive glasses: Their structure, properties, fabrication, and apatite formation. *Journal of Biomedical Materials Research Part A* 102 (1):254–74.

Kawamura, Y., and M. Yamasaki. 2007. Formation and mechanical properties of Mg~ 9~ 7Zn~ 1RE~ 2 alloys with long-period stacking ordered structure. *Materials Transactions* 48 (11):2986.

Kawamura, Y., K. Hayashi, A. Inoue, and T. Masumoto. 2001. Rapidly solidified powder metallurgy Mg97Zn1Y2 alloys with excellent tensile yield strength above 600 MPa. *Materials Transactions* 42:1171–4.

Kawamura, T., C.-S. Huang, N. Tochigi et al. 2010. Inhaled hydrogen gas therapy for prevention of lung transplant-induced ischemia/reperfusion injury in rats. *Transplantation* 90 (12):1344–51.

Kawasaki, H., J. Guan, and K. Tamama. 2010. Hydrogen gas treatment prolongs replicative lifespan of bone marrow multipotential stromal cells in vitro while preserving differentiation and paracrine potentials. *Biochemical and Biophysical Research Communications* 397 (3):608–13.

Kaya, A. E. D., G. Ben-Hamu et al. 2006. Microstructure and corrosion resistance of alloys of the Mg-Zn-Ag system. *Metal Science and Heat Treatment* 48 (11–12):524–30.

Keim, S., J. G. Brunner, B. Fabry, and S. Virtanen. 2011. Control of magnesium corrosion and biocompatibility with biomimetic coatings. *Journal of Biomedical Materials Research Part B: Applied Biomaterials* 96 (1):84–90.

Khakbaz, H., R. Walter, T. Gordon, and M. B. Kannan. 2014. Self-dissolution assisted coating on magnesium metal for biodegradable bone fixation devices. *Materials Research Express* 1 (4):045406.

Khanra, A. K., H. C. Jung, K. S. Hong, and K. S. Shin. 2010. Comparative property study on extruded Mg–HAP and ZM61–HAP composites. *Materials Science and Engineering: A* 527 (23):6283–8.

Kikkawa, Y. S., T. Nakagawa, R. T. Horie, and J. Ito. 2009. Hydrogen protects auditory hair cells from free radicals. *Neuroreport* 20 (7):689–94.

Kim, Y., and D. Bae. 2004. Superplasticity in a fine-grained Mg-Zn-Y-Zr alloy containing quasicrystalline particles. *Materials Transactions* 45 (12):3298–303.

Kim, S. O., and J. S. Kwak. 2008. Magnetic force improvement and parameter optimization for magnetic abrasive polishing of AZ31 magnesium alloy. *Transactions of Nonferrous Metals Society of China* 18 (Suppl 1):s369–73.

Kim, J., and Y. Kawamura. 2013. Influence of rare earth elements on microstructure and mechanical properties of Mg97Zn1Y1RE1 alloys. *Materials Science and Engineering A-Structural Materials Properties Microstructure and Processing* 573:62–6.

Kim, S., J. Lee, Y. Kim et al. 2003. Synthesis of Si, Mg substituted hydroxyapatites and their sintering behaviors. *Biomaterials* 24 (8):1389–98.

Kim, S.-H., B.-S. You, C. Dong Yim, and Y.-M. Seo. 2005. Texture and microstructure changes in asymmetrically hot rolled AZ31 magnesium alloy sheets. *Materials Letters* 59 (29–30):3876–80.

Kim, W. C., J. G. Kim, J. Y. Lee, and H. K. Seok. 2008. Influence of Ca on the corrosion properties of magnesium for biomaterials. *Materials Letters* 62 (25):4146–8.

Kim, Y.-D., N.-H. Kang, I.-G. Jo, K.-H. Kim, and I.-B. Kim. 2008. Aging Behavior of Mg-Y-Zr and Mg-Nd-Zr cast alloys. *Journal of Materials Science and Technology* 24 (1):80–4.

Kim, D. H., C. H. Seo, K. Han, K. W. Kwon, A. Levchenko, and K. Y. Suh. 2009. Guided cell migration on microtextured substrates with variable local density and anisotropy. *Advanced Functional Materials* 19 (10):1579–86.

Kim, J. H., N. E. Kang, C. D. Yim, and B. K. Kim. 2009. Effect of calcium content on the microstructural evolution and mechanical properties of wrought Mg-3Al-1Zn alloy. *Materials Science and Engineering A-Structural Materials Properties Microstructure and Processing* 525 (1–2):18–29.

Kim, W. J., M. J. Kim, and J. Y. Wang. 2009. Ultrafine-grained Mg–9Li–1Zn alloy sheets exhibiting low temperature superplasticity. *Materials Science and Engineering: A* 516 (1–2):17–22.

Kim, W. J., M. J. Lee, B. H. Lee, and Y. B. Park. 2010. A strategy for creating ultrafine-grained microstructure in magnesium alloy sheets. *Materials Letters* 64 (6):647–9.

Kim, B. H., K. C. Park, Y. H. Park, and I. M. Park. 2011. Effect of Ca and Sr additions on high temperature and corrosion properties of Mg-4Al-2Sn based alloys. *Materials Science and Engineering A-Structural Materials Properties Microstructure and Processing* 528 (3):808–14.

Kim, Y. Y., W. S. Choi, and K. W. Rhyu. 2012. Assessment of pedicle screw pullout strength basedon various screw designs and bone densities—An ex vivo biomechanical study. *The Spine Journal* 12:164–8.

Kim, S. B., J. H. Jo, S. M. Lee, H. E. Kim, K. H. Shin, and Y. H. Koh. 2013. Use of a poly(ether imide) coating to improve corrosion resistance and biocompatibility of magnesium (Mg) implant for orthopedic applications. *Journal of Biomedical Materials Research. Part A* 101 (6):1708–15.

Kim, S.-M., J.-H. Jo, S.-M. Lee et al. 2014. Hydroxyapatite-coated magnesium implants with improved in vitro and in vivo biocorrosion, biocompatibility, and bone response. *Journal of Biomedical Materials Research Part A* 102 (2):429–41.

Kim, Y., H. Sohn, and W. Kim. 2014. Ductility enhancement through texture control and strength restoration through subsequent age-hardening in Mg-Zn-Zr alloys. *Materials Science and Engineering: A* 597:157–63.

King, A. D., N. Birbilis, and J. R. Scully. 2014. Accurate electrochemical measurement of magnesium corrosion rates: A combined impedance, mass-loss and hydrogen collection study. *Electrochimica Acta* 121:394–406.

Kiousis, D. E., A. R. Wulff, and G. A. Holzapfel. 2009. Experimental studies and numerical analysis of the inflation and interaction of vascular balloon catheter-stent systems. *Annals of Biomedical Engineering* 37 (2):315–30.

Kirkham, M. J., A. M. dos Santos, C. J. Rawn, E. Lara-Curzio, J. W. Sharp, and A. J. Thompson. 2012. Ab initio determination of crystal structures of the thermoelectric material MgAgSb. *Physical Review B* 85 (14):144120.

Kirkland, N. T., and N. Birbilis. 2013. *Magnesium Biomaterials: Design, Testing, and Best Practice*. Heidelberg, Berlin: Springer.

Kirkland, N. T., N. Birbilis, J. Walker, T. Woodfield, G. J. Dias, and M. P. Staiger. 2010. In vitro dissolution of magnesium–calcium binary alloys: Clarifying the unique role of calcium additions in bioresorbable magnesium implant alloys. *Journal of Biomedical Materials Research Part B: Applied Biomaterials* 95 (1):91–100.

Kirkland, N. T., N. Birbilis, and M. P. Staiger. 2012. Assessing the corrosion of biodegradable magnesium implants: A critical review of current methodologies and their limitations. *Acta Biomaterialia* 8 (3):925–36.

Kitahara, H., F. Maruno, M. Tsushida, and S. Ando. 2014. Deformation behavior of Mg single crystals during a single ECAP pass at room temperature. *Materials Science and Engineering: A* 590:274–80.

Kitamura, A., S. Kobayashi, T. Matsushita, H. Fujinawa, and K. Murase. 2014. Experimental verification of protective effect of hydrogen-rich water against cisplatin-induced nephrotoxicity in rats using dynamic contrast-enhanced CT. *The British Journal of Radiology* 83 (990):509–14.

Klocke, F., M. Schwade, A. Klink, and A. Kopp. 2011. EDM machining capabilities of magnesium (Mg) alloy WE43 for medical applications. *Procedia Engineering* 19:190–5.

Klocke, F., M. Schwade, A. Klink, D. Veselovac, and A. Kopp. 2013. Influence of electro discharge Machining of biodegradable magnesium on the biocompatibility. *Procedia CIRP* 5:88–93.

Kokubo, T., and H. Takadama. 2006. How useful is SBF in predicting in vivo bone bioactivity? *Biomaterials* 27 (15):2907–15.

Koleini, S., M. H. Idris, and H. Jafari. 2012. Influence of hot rolling parameters on microstructure and biodegradability of Mg-1Ca alloy in simulated body fluid. *Materials and Design* 33:20–5.

Kourkoulis, S. K., and E. P. Chatzistergos. 2009. The influence of the "penetration-" and the "filling-ratios" on the pull-out strength of transpedicular screws. *Journal of Mechanics in Medicine and Biology* 9:283–300.

Kozlov, A., M. Ohno, T. Abu Leil, N. Hort, K. U. Kainer, and R. Schmid-Fetzer. 2008. Phase equilibria, thermodynamics and solidification microstructures of Mg-Sn-Ca alloys, Part 2: Prediction of phase formation in Mg-rich Mg-Sn-Ca cast alloys. *Intermetallics* 16 (2):316–21.

Kozlov, A., J. Grobner, and R. Schmid-Fetzer. 2011. Phase formation in Mg-Sn-Si and Mg-Sn-Si-Ca alloys. *Journal of Alloys and Compounds* 509 (7):3326–37.

Kozov, A., M. Ohno, R. Arroyave, Z. K. Liu, and R. Schmid-Fetzer. 2008. Phase equilibria, thermodynamics and solidification microstructures of Mg-Sn-Ca alloys, Part 1: Experimental investigation and thermodynamic modeling of the ternary Mg-Sn-Ca system. *Intermetallics* 16 (2):299–315.

Kral, M. V., B. C. Muddle, and J. F. Nie. 2007. Crystallography of the bcc/hcp transformation in a Mg–8Li alloy. *Materials Science and Engineering: A* 460–1:227–32.

Krämer, K. W., D. Biner, G. Frei, H. U. Güdel, M. P. Hehlen, and S. R. Lüthi. 2004. Hexagonal sodium yttrium fluoride based green and blue emitting upconversion phosphors. *Chemistry of Materials* 16 (7):1244–51.

Kraus, T., S. F. Fischerauer, A. C. Hanzi et al. 2012. Magnesium alloys for temporary implants in osteosynthesis: In vivo studies of their degradation and interaction with bone. *Acta Biomaterialia* 8 (3):1230–8.

Krause, A., N. Höh, D. Bormann, C. Krause, F.-W. Bach, H. Windhagen, and A. Meyer-Lindenberg. 2009. Degradation behaviour and mechanical properties of magnesium implants in rabbit tibiae. *Journal of Materials Science* 45 (3):624–32.

Krenn, M. H., W. P. Piotrowski, R. Penzkofer, and P. Augat. 2008. Influence of thread design on pedicle screw fixation. *Journal of Neurosurgery: Spine* 9 (1):90–5.

Kroll, W. J., W. F. Hergert, and L. A. Yerkes. 1950. Improvements in methods for the reduction of zirconium chloride with magnesium. *Journal of the Electrochemical Society* 97 (10):305–10.

Kubasek, J., D. Vojtech, and I. Pospisilova. 2012. Structural and corrosion characterization of biodegradable Mg-Zn alloy castings. *Kovove Materialy-Metallic Materials* 50 (6):415–24.

Kubasek, J., and D. Vojtech. 2013a. Structural and corrosion characterization of biodegradable Mg–RE (RE = Gd, Y, Nd) alloys. *Transactions of Nonferrous Metals Society of China* 23 (5):1215–25.

Kubasek, J., and D. Vojtech. 2013b. Structural characteristics and corrosion behavior of biodegradable Mg-Zn, Mg-Zn-Gd alloys. *Journal of Materials Science. Materials in Medicine* 24 (7):1615–26.

Kubo, T., T. Imanishi, S. Takarada et al. 2007. Assessment of culprit lesion morphology in acute myocardial infarction ability of optical coherence tomography compared with intravascular ultrasound and coronary angioscopy. *Journal of the American College of Cardiology* 50 (10):933–9.

Kubota, M., S. Shimmura, S. Kubota et al. 2011. Hydrogen and N-acetyl-L-cysteine rescue oxidative stress-induced angiogenesis in a mouse corneal alkali-burn model. *Investigative Ophthalmology and Visual Science* 52 (1):427–33.

Kulekci, M. K. 2008. Magnesium and its alloys applications in automotive industry. *International Journal of Advanced Manufacturing Technology* 39 (9–10):851–65.

Kumar, S., T. S. N. Sankara Narayanan, S. Ganesh Sundara Raman, and S. K. Seshadri. 2010. Evaluation of fretting corrosion behaviour of CP-Ti for orthopaedic implant applications. *Tribology International* 43 (7):1245–52.

Kumar, A., S. Dhara, K. Biswas, and B. Basu. 2013. In vitro bioactivity and cytocompatibility properties of spark plasma sintered HA-Ti composites. *Journal of Biomedical Materials Research Part B: Applied Biomaterials* 101 (2):223–36.

Kutniy, K. V., I. I. Papirov, M. A. Tikhonovsky, A. I. Pikalov, S. V. Sivtzov, L. A. Pirozhenko, V. S. Shokurov, and V. A. Shkuropatenko. 2009. Influence of grain size on mechanical and corrosion properties of magnesium alloy for medical implants. *Materialwissenschaft und Werkstofftechnik* 40 (4):242–6.

Kuwahara, H., Y. Al-Abdullat, M. Ohta, S. Tsutsumi, K. Ikeuchi, N. Mazaki, and T. Aizawa. 2000. Surface reaction of magnesium in Hank's solutions. *Materials Science Forum* 350–1:349–58.

Kuwahara, H., Y. Al-Abdullat, N. Mazaki, S. Tsutsumi, and T. Aizawa. 2001. Precipitation of magnesium apatite on pure magnesium surface during immersing in Hank's solution. *Materials Transactions (Japan)* 42 (7):1317–21.

Kwok, A. W., J. A. Finkelstein, T. Woodside, T. C. Hearn, and R. W. Hu. 1996. Insertional torque and pull-out strengths of conical and cylindrical pedicle screws in cadaveric bone. *Spine* 21:2429–34.

Labelle, P., M. Pekguleryuz, D. Argo, M. Dierks, T. Sparks, and T. Waltematte. 2001. Heat resistant magnesium alloys for power-train applications. *Training* 2012:3–12.

Lal, C., and I. P. Jain. 2012. Effect of ball milling on structural and hydrogen storage properties of Mg – x wt% FeTi (x = 2 & 5) solid solutions. *International Journal of Hydrogen Energy* 37 (4):3761–6.

Lalk, M., J. Reifenrath, N. Angrisani, A. Bondarenko, J. M. Seitz, P. P. Mueller, and A. Meyer-Lindenberg. 2013. Fluoride and calcium-phosphate coated sponges of the magnesium alloy AX30 as bone grafts: A comparative study in rabbits. *Journal of Materials Science. Materials in Medicine* 24 (2):417–36.

Lam, C. X. F., X. Mo, S.-H. Teoh, and D. Hutmacher. 2002. Scaffold development using 3D printing with a starch-based polymer. *Materials Science and Engineering: C* 20 (1):49–56.

Lamaka, S. V., M. F. Montemor, A. F. Galio, M. L. Zheludkevich, C. Trindade, L. F. Dick, and M. G. S. Ferreira. 2008. Novel hybrid sol-gel coatings for corrosion protection of AZ31B magnesium alloy. *Electrochimica Acta* 53 (14):4773–83.

Lamaka, S. V., G. Knörnschild, D. V. Snihirova, M. G. Taryba, M. L. Zheludkevich, and M. G. S. Ferreira. 2009. Complex anticorrosion coating for ZK30 magnesium alloy. *Electrochimica Acta* 55 (1):131–41.

Lambotte, A. 1932. L'utilisation du magnésium comme matérial perdu dans l'ostéosynthése. *Bull Mem Soc Nat Chir* (28):1325–34.

Landi, E., S. Sprio, M. Sandri, G. Celotti, and A. Tampieri. 2008. Development of Sr and CO3 co-substituted hydroxyapatites for biomedical applications. *Acta Biomaterialia* 4 (3):656–63.

Lang, J., and J. Huot. 2011. A new approach to the processing of metal hydrides. *Journal of Alloys and Compounds* 509 (3):L18–22.

Langelier, B., X. Wang, and S. Esmaeili. 2012. Evolution of precipitation during non-isothermal ageing of an Mg-Ca-Zn alloy with high Ca content. *Materials Science and Engineering A-Structural Materials Properties Microstructure and Processing* 538:246–51.

Langenscheidt, O., M. Westermeier, J. Reinelt, J. Mentel, and P. Awakowicz. 2008. Investigation of the gas-phase emitter effect of dysprosium in ceramic metal halide lamps. *Journal of Physics D: Applied Physics* 41 (14):144005.

Larionova, T. V., W. W. Park, and B. S. You. 2001. A ternary phase observed in rapidly solidified Mg-Ca-Zn alloys. *Scripta Materialia* 45 (1):7–12.

Laws, K. J., B. Gun, and M. Ferry. 2008a. Large-scale production of Ca65Mg15Zn20 bulk metallic glass samples by low-pressure die-casting. *Materials Science and Engineering A-Structural Materials Properties Microstructure and Processing* 475 (1–2):348–54.

Laws, K. J., B. Gun, and M. Ferry. 2008b. Mechanical stability of Ca65Mg15Zn20 bulk metallic glass during deformation in the supercooled liquid region. *Materials Science and Engineering A-Structural Materials Properties Microstructure and Processing* 480 (1–2):198–204.

Lee, D. W., and B. K. Kim. 2003. Synthesis of nano-structured titanium carbide by Mg-thermal reduction. *Scripta Materialia* 48 (11):1513–18.

Lee, Y. C., A. K. Dahle, and D. H. St. John. 2000. The role of solute in grain refinement of magnesium. *Metallurgical and Materials Transactions A-Physical Metallurgy and Materials Science* 31 (11):2895–906.

Lee, Y. M., Y. J. Seol, Y. T. Lim, S. Kim, S. B. Han, I. C. Rhyu, S. H. Baek, S. J. Heo, J. Y. Choi, and P. R. Klokkevold. 2001. Tissue-engineered growth of bone by marrow cell transplantation using porous calcium metaphosphate matrices. *Journal of Biomedical Materials Research* 54 (2):216–23.

Lee, W. B., J. W. Kim, Y. M. Yeon, and S. B. Jung. 2003. The joint characteristics of friction stir welded AZ91D magnesium alloy. *Materials Transactions* 44 (5):917–23.

Lee, J. Y., H. K. Do, H. K. Lim, and D. H. Kim. 2005. Effects of Zn/Y ratio on microstructure and mechanical properties of Mg-Zn-Y alloys. *Materials Letters* 59 (29–30):3801–5.

Lee, P. Y., M. C. Kao, C. K. Lin, and J. C. Huang. 2006. Mg–Y–Cu bulk metallic glass prepared by mechanical alloying and vacuum hot pressing. *Intermetallics* 14 (8):994–9.

Lee, H.-W., T.-S. Lui, and L.-H. Chen. 2009. Studies on the improvement of tensile ductility of hot-extrusion AZ31 alloy by subsequent friction stir process. *Journal of Alloys and Compounds* 475 (1–2):139–44.

Lee, J. Y., G. Han, Y. C. Kim et al. 2009. Effects of impurities on the biodegradation behavior of pure magnesium. *Metals and Materials International* 15 (6):955–61.

Lee, C.-C., S.-C. Lin, S.-W. Wu, Y.-C. Li, and P.-Y. Fu. 2012. Correlation of the experimental and numerical results for the holding power of dental, traumatic, and spinal screws. *Medical Engineering and Physics* 34:1123–31.

Lee, T., Y.-U. Heo, and C. S. Lee. 2013. Microstructure tailoring to enhance strength and ductility in Ti–13Nb–13Zr for biomedical applications. *Scripta Materialia* 69 (11):785–8.

Leeflang, M. A., J. S. Dzwonczyk, J. Zhou, and J. Duszczyk. 2011. Long-term biodegradation and associated hydrogen evolution of duplex-structured Mg–Li–Al–(RE) alloys and their mechanical properties. *Materials Science and Engineering: B* 176 (20):1741–5.

Lei, T., C. Ouyang, W. Tang, L.-F. Li, and L.-S. Zhou. 2010a. Enhanced corrosion protection of MgO coatings on magnesium alloy deposited by an anodic electrodeposition process. *Corrosion Science* 52 (10):3504–8.

Lei, T., C. Ouyang, W. Tang, L.-F. Li, and L.-S. Zhou. 2010b. Preparation of MgO coatings on magnesium alloys for corrosion protection. *Surface and Coatings Technology* 204 (23):3798–803.

Lei, T., W. Tang, S.-H. Cai, F.-F. Feng, and N.-F. Li. 2012. On the corrosion behaviour of newly developed biodegradable Mg-based metal matrix composites produced by in situ reaction. *Corrosion Science* 54:270–7.

Lemaitre, J. 1996. *A Course on Damage Mechanics*. Berlin: Springer.

Leng, Z., J. Zhang, M. Zhang, X. Liu, H. Zhan, and R. Wu. 2012. Microstructure and high mechanical properties of Mg–9RY–4Zn (RY: Y-rich misch metal) alloy with long period stacking ordered phase. *Materials Science and Engineering: A* 540:38–45.

Leng, Z., J. Zhang, C. Cui, J. Sun, S. Liu, R. Wu, and M. Zhang. 2013a. Compression properties at different loading directions of as-extruded Mg–9RY–4Zn (RY: Y-rich misch metal) alloy with long period stacking ordered phase. *Materials and Design* 51:561–6.

Leng, Z., J. Zhang, T. Yin, L. Zhang, X. Guo, Q. Peng, M. Zhang, and R. Wu. 2013b. Influence of biocorrosion on microstructure and mechanical properties of deformed Mg–Y–Er–Zn biomaterial containing 18R-LPSO phase. *Journal of the Mechanical Behavior of Biomedical Materials* 28:332–9.

Leng, Z., J. H. Zhang, H. Y. Lin, P. F. Fei, L. Zhang, S. J. Liu, M. L. Zhang, and R. Z. Wu. 2013c. Superplastic behavior of extruded Mg–9RY–4Zn alloy containing long period stacking ordered phase. *Materials Science and Engineering a–Structural Materials Properties Microstructure and Processing* 576:202–6.

Lensing, R., P. Behrens, P. P. Muller, T. Lenarz, and M. Stieve. 2014. In vivo testing of a bioabsorbable magnesium alloy serving as total ossicular replacement prostheses. *Journal of Biomaterials Applications* 28 (5):688–96.

Lespinasse, V., G. C. Fisher, and J. Eisenstaedt. 1910. A practical mechanical method of end-to-end anastomosis of blood-vessels: Using absorbable magnesium rings. *Journal of the American Medical Association* 55 (21):1785–90.

Leukers, B., H. Gülkan, S. H. Irsen, S. Milz, C. Tille, M. Schieker, and H. Seitz. 2005. Hydroxyapatite scaffolds for bone tissue engineering made by 3D printing. *Journal of Materials Science: Materials in Medicine* 16 (12):1121–4.

Levesque, J., H. Hermawan, D. Dube, and D. Mantovani. 2008. Design of a pseudo-physiological test bench specific to the development of biodegradable metallic biomaterials. *Acta Biomaterialia* 4 (2):284–95.

Lewis, S., and A. K. McIndoe. 2004. Cleaning, disinfection and sterilization of equipment. *Anaesthesia and Intensive Care Medicine* 5 (11):360–3.

Li, N., and Y. F. Zheng. 2013. Novel magnesium alloys developed for biomedical application: A review. *Journal of Materials Science and Technology* 29 (6):489–502.

Li, L., J. Gao, and Y. Wang. 2004. Evaluation of cyto-toxicity and corrosion behavior of alkali-heat-treated magnesium in simulated body fluid. *Surface and Coatings Technology* 185 (1):92–8.

Li, Q., L.-J. Jiang, K.-C. Chou, Q. Lin, F. Zhan, K.-D. Xu, X.-G. Lu, and J.-Y. Zhang. 2005. Effect of hydrogen pressure on hydriding kinetics in the Mg2–xAgxNi–H (x = 0.05, 0.1) system. *Journal of Alloys and Compounds* 399 (1–2):101–5.

Li, J. F., Z. Q. Zheng, S. C. Li, W. D. Ren, and Z. Zhang. 2006. Preparation and galvanic anodizing of a Mg–Li alloy. *Materials Science and Engineering: A* 433 (1–2):233–40.

Li, D., J. Dong, X. Zeng, C. Lu, and W. Ding. 2007a. Characterization of precipitate phases in a Mg–Dy–Gd–Nd alloy. *Journal of Alloys and Compounds* 439 (1–2):254–7.

Li, D. H., J. Dong, X. Q. Zeng, C. Lu, and W. J. Ding. 2007b. Characterization of β″ precipitate phase in a Mg–Dy–Gd–Nd alloy. *Materials Characterization* 58 (10):1025–8.

Li, D., Q. Wang, and W. Ding. 2007c. Effects of heat treatments on microstructure and mechanical properties of Mg–4Y–4Sm–0.5 Zr alloy. *Materials Science and Engineering: A* 448 (1):165–70.

Li, Q., Q. Wang, Y. Wang, X. Zeng, and W. Ding. 2007. Effect of Nd and Y addition on microstructure and mechanical properties of as-cast Mg–Zn–Zr alloy. *Journal of Alloys and Compounds* 427 (1–2):115–23.

Li, D., Q. Wang, and W. Ding. 2008. Precipitate phases in the Mg-4Y-4Sm-0.5 Zr alloy. *Journal of Alloys and Compounds* 465 (1):119–26.

Li, J.-H., W.-Q. Jie, and G.-Y. Yang. 2008. Effect of gadolinium on aged hardening behavior, microstructure and mechanical properties of Mg-Nd-Zn-Zr alloy. *Transactions of Nonferrous Metals Society of China* 18:s27–32.

Li, Q. F., H. R. Weng, Z. Y. Suo, Y. L. Ren, X. G. Yuan, and K. Q. Qiu. 2008. Microstructure and mechanical properties of bulk Mg-Zn-Ca amorphous alloys and amorphous matrix composites. *Materials Science and Engineering A-Structural Materials Properties Microstructure and Processing* 487 (1–2):301–8.

Li, Z., X. Gu, S. Lou, and Y. Zheng. 2008. The development of binary Mg–Ca alloys for use as biodegradable materials within bone. *Biomaterials* 29 (10):1329–44.

Li, K., Q. Li, X. Jing, J. Chen, X. Zhang, and Q. Zhang. 2009. Effects of Sm addition on microstructure and mechanical properties of Mg–6Al–0.6 Zn alloy. *Scripta Materialia* 60 (12):1101–4.

Li, D., Q. Wang, and W. Ding. 2009. Effects of samarium on microstructure and mechanical properties of Mg–Y–Sm–Zr alloys during thermo-mechanical treatments. *Journal of Materials Science* 44 (12):3049–56.

Li, Z. G., X. Hui, C. M. Zhang, and G. L. Chen. 2008. Formation of Mg–Cu–Zn–Y bulk metallic glasses with compressive strength over gigapascal. *Journal of Alloys and Compounds* 454 (1):168–73.

Li, Z. J., X. N. Gu, S. Q. Lou, and Y. F. Zheng. 2008. The development of binary Mg-Ca alloys for use as biodegradable materials within bone. *Biomaterials* 29 (10):1329–44.

Li, D., J. Dong, X. Zeng, and C. Lu. 2010. Transmission electron microscopic investigation of the β1→β phase transformation in a Mg–Dy–Nd alloy. *Materials Characterization* 61 (8):818–23.

Li, D.-Q., Q.-D. Wang, W.-J. Ding, J. J. Blandin, and M. Suery. 2010. Influence of extrusion temperature on microstructure and mechanical properties of Mg-4Y-4Sm-0.5 Zr alloy. *Transactions of Nonferrous Metals Society of China* 20 (7):1311–15.

Li, J., Y. Song, S. Zhang, C. Zhao, F. Zhang, X. Zhang, L. Cao, Q. Fan, and T. Tang. 2010a. In vitro responses of human bone marrow stromal cells to a fluoridated hydroxyapatite coated biodegradable Mg-Zn alloy. *Biomaterials* 31 (22):5782–8.

Li, J., P. Cao, X. N. Zhang, S. X. Zhang, and Y. H. He. 2010b. In vitro degradation and cell attachment of a PLGA coated biodegradable Mg–6Zn based alloy. *Journal of Materials Science* 45 (22):6038–45.

Li, J., Z. Qu, R. Wu, and M. Zhang. 2010c. Effects of Cu addition on the microstructure and hardness of Mg–5Li–3Al–2Zn alloy. *Materials Science and Engineering: A* 527 (10–11):2780–3.

Li, J., C. Wang, J. H. Zhang et al. 2010d. Hydrogen-rich saline improves memory function in a rat model of amyloid-beta-induced Alzheimer's disease by reduction of oxidative stress. *Brain Research* 1328:152–61.

Li, Q., X. Li, Q. Zhang, and J. Chen. 2010. Effect of rare-earth element Sm on the corrosion behavior of Mg-6Al-1.2 Y-0.9 Nd alloy. *Rare Metals* 29 (6):557–60.

Li, Y., M. H. Li, W. Y. Hu, P. D. Hodgson, and C. E. Wen. 2010a. Biodegradable Mg-Ca and Mg-Ca-Y alloys for regenerative medicine. Paper read at Materials Science Forum.

Li, Y., K. Zhang, Y. Zhang, X. Li, and M. Ma. 2010b. Microstructural evolution and mechanical properties of Mg-5Y-5Gd-xNd-0.5 Zr magnesium alloys at different states. *Rare Metals* 29 (3):317–22.

Li, H., H. Zhong, K. Xu et al. 2011. Enhanced efficacy of sirolimus-eluting bioabsorbable magnesium alloy stents in the prevention of restenosis. *Journal of Endovascular Therapy* 18 (3):407–15.

Li, J., P. Han, W. Ji, Y. Song, S. Zhang, Y. Chen, C. Zhao, F. Zhang, X. Zhang, and Y. Jiang. 2011a. The in vitro indirect cytotoxicity test and in vivo interface bioactivity evaluation of biodegradable FHA coated Mg–Zn alloys. *Materials Science and Engineering: B* 176 (20):1785–8.

Li, J., W. Xu, X. Wu, H. Ding, and K. Xia. 2011b. Effects of grain size on compressive behaviour in ultrafine grained pure Mg processed by equal channel angular pressing at room temperature. *Materials Science and Engineering: A* 528 (18):5993–8.

Li, M. H., Q. A. Chen, W. J. Zhang, W. Y. Hu, and Y. Su. 2011. Corrosion behavior in SBF for titania coatings on Mg-Ca alloy. *Journal of Materials Science* 46 (7):2365–9.

Li, R. H., F. S. Pan, B. Jiang, H. M. Yin, and T. T. Liu. 2011. Effects of yttrium and strontium additions on as-cast microstructure of Mg-14Li-1Al alloys. *Transactions of Nonferrous Metals Society of China* 21 (4):778–83.

Li, W., S. K. Guan, J. Chen, J. H. Hu, S. Chen, L. G. Wang, and S. J. Zhu. 2011. Preparation and in vitro degradation of the composite coating with high adhesion strength on biodegradable Mg-Zn-Ca alloy. *Materials Characterization* 62 (12):1158–65.

Li, X., T. Al-Samman, and G. Gottstein. 2011. Mechanical properties and anisotropy of ME20 magnesium sheet produced by unidirectional and cross rolling. *Materials and Design* 32 (8–9):4385–93.

Li, Y., T. Hamasaki, N. Nakamichi et al. 2011a. Suppressive effects of electrolyzed reduced water on alloxan-induced apoptosis and type 1 diabetes mellitus. *Cytotechnology* 63 (2):119–31.

Li, Y. C., P. D. Hodgson, and C. E. Wen. 2011b. The effects of calcium and yttrium additions on the microstructure, mechanical properties and biocompatibility of biodegradable magnesium alloys. *Journal of Materials Science* 46 (2):365–71.

Li, H., W. B. Du, S. B. Li, and Z. H. Wang. 2012. Effect of Zn/Er weight ratio on phase formation and mechanical properties of as-cast Mg-Zn-Er alloys. *Materials and Design* 35:259–65.

Li, M., Y. Cheng, Y. F. Zheng, X. Zhang, T. F. Xi, and S. C. Wei. 2012. Surface characteristics and corrosion behaviour of WE43 magnesium alloy coated by SiC film. *Applied Surface Science* 258 (7):3074–81.

Li, Y. C., C. Wen, D. Mushahary, R. Sravanthi, N. Harishankar, G. Pande, and P. Hodgson. 2012. Mg-Zr-Sr alloys as biodegradable implant materials. *Acta Biomaterialia* 8 (8):3177–88.

Li, N., Y. D. Li, Y. B. Wang, M. Li, Y. Cheng, Y. H. Wu, and Y. F. Zheng. 2013. Corrosion resistance and cytotoxicity of a MgF_2 coating on biomedical Mg-1Ca alloy via vacuum evaporation deposition method. *Surface and Interface Analysis* 45 (8):1217–22.

Li, Z. M., P. H. Fu, L. M. Peng, Y. X. Wang, H. Y. Jiang, and G. H. Wu. 2013. Comparison of high cycle fatigue behaviors of Mg–3Nd–0.2Zn–Zr alloy prepared by different casting processes. *Materials Science and Engineering: A* 579:170–9.

Li, H., Y. Zheng, and L. Qin. 2014. Progress of biodegradable metals. *Progress in Natural Science: Materials International* 24 (5):414–22.

Li, L.-H., T. S. N. Sankara Narayanan, Y. K. Kim, J. Y. Kang, I. S. Park, T. S. Bae, and M. H. Lee. 2014a. Characterization and corrosion resistance of pure Mg modified by micro-arc oxidation using phosphate electrolyte with/without NaOH. *Surface and Interface Analysis* 46 (1):7–15.

Li, L.-H., T. S. N. Sankara Narayanan, Y. K. Kim, Y.-M. Kong, G.-S. Shin, S.-K. Lyu, I. S. Park, and M. H. Lee. 2014b. Coloring and corrosion resistance of pure Mg modified by micro-arc oxidation method. *International Journal of Precision Engineering and Manufacturing* 15 (8):1625–30.

Li, R. W., N. T. Kirkland, J. Truong, J. Wang, P. N. Smith, N. Birbilis, and D. R. Nisbet. 2014. The influence of biodegradable magnesium alloys on the osteogenic differentiation of human mesenchymal stem cells. *Journal of Biomedical Materials Research Part A* 102 (12):4346–57.

Li, Y., G. Liu, Z. Zhai et al. 2014. Antibacterial properties of magnesium in vitro and in an in vivo model of implant-associated methicillin-resistant staphylococcus aureus infection. *Antimicrobial Agents and Chemotherapy* 58 (12):7586–91.

Li, Z., G.-L. Song, and S. Song. 2014. Effect of bicarbonate on biodegradation behaviour of pure magnesium in a simulated body fluid. *Electrochimica Acta* 115:56–65.

Li, X., C.L. Chu, L. Liu et al. 2015. Biodegradable poly-lactic acid based-composite reinforced unidirectionally with high-strength magnesium alloy wires. *Biomaterials* 49:135–44.

Liang, M.-J., H.-H. Liao, W.-J. Ding, L.-M. Peng, and P.-H. Fu. 2012. Microstructure characterization on Mg-2Nd-4Zn-1Zr alloy during heat treatment. *Transactions of Nonferrous Metals Society of China* 22 (10):2327–33.

Liao, Y., Y. M. Ouyang, J. L. Niu, J. Zhang, Y. P. Wang, Z. J. Zhu, G. Y. Yuan, Y. H. He, and Y. Jiang. 2012. In vitro response of chondrocytes to a biodegradable Mg-Nd-Zn-Zr alloy. *Materials Letters* 83:206–8.

Liao, Y., D. Chen, J. Niu, J. Zhang, Y. Wang, Z. Zhu, G. Yuan, Y. He, and Y. Jiang. 2013. In vitro degradation and mechanical properties of polyporous CaHPO$_4$-coated Mg–Nd–Zn–Zr alloy as potential tissue engineering scaffold. *Materials Letters* 100:306–8.

Lietaert, K., L. Weber, J. Van Humbeeck, A. Mortensen, J. Luyten, and J. Schrooten. 2013. Open cellular magnesium alloys for biodegradable orthopaedic implants. *Journal of Magnesium and Alloys* 1 (4):303–11.

Lim, G. B. 2013. Interventional cardiology: DREAMS of a bioabsorbable stent coming true. *Nature Reviews Cardiology* 10 (3):120.

Lim, M., J. E. Tibballs, and P. L. Rossiter. 1997. Thermodynamic assessment of the Ag-Mg binary system. *Zeitschrift Fur Metallkunde* 88 (2):162–9.

Lin, M. C., C. Y. Tsai, and J. Y. Uan. 2007. Converting hcp Mg–Al–Zn alloy into bcc Mg–Li–Al–Zn alloy by electrolytic deposition and diffusion of reduced lithium atoms in a molten salt electrolyte LiCl–KCl. *Scripta Materialia* 56 (7):597–600.

Lin, Y.-N., H.-Y. Wu, G.-Z. Zhou, C.-H. Chiu, and S. Lee. 2008. Mechanical and anisotropic behaviors of Mg–Li–Zn alloy thin sheets. *Materials and Design* 29 (10):2061–5.

Lin, M. C., C. Y. Tsai, and J. Y. Uan. 2009. Electrochemical behaviour and corrosion performance of Mg–Li–Al–Zn anodes with high Al composition. *Corrosion Science* 51 (10):2463–72.

Lin, C.-L., J.-H. Yu, H.-L. Liu, C.-H. Lin, and Y.-S. Lin. 2010. Evaluation of contributions of orthodontic mini-screw design factors based on FE analysis and the Taguchi method. *Journal of Biomechanics* 43:2174–81.

Lin, Y., A. Kashio, T. Sakamoto et al. 2011. Hydrogen in drinking water attenuates noise-induced hearing loss in guinea pigs. *Neuroscience Letters* 487 (1):12–16.

Lin, X., L. Tan, P. Wan, X. Yu, K. Yang, Z. Hu, Y. Li, and W. Li. 2013a. Characterization of micro-arc oxidation coating post-treated by hydrofluoric acid on biodegradable ZK60 magnesium alloy. *Surface and Coatings Technology* 232:899–905.

Lin, X., L. Tan, Q. Wang, G. Zhang, B. Zhang, and K. Yang. 2013b. In vivo degradation and tissue compatibility of ZK60 magnesium alloy with micro-arc oxidation coating in a transcortical model. *Materials Science and Engineering. C, Materials for Biological Applications* 33 (7):3881–8.

Lin, X., L. L. Tan, Q. Zhang, K. Yang, Z. Q. Hu, J. H. Qiu, and Y. Cai. 2013c. The in vitro degradation process and biocompatibility of a ZK60 magnesium alloy with a forsterite-containing micro-arc oxidation coating. *Acta Biomaterialia* 9 (10):8631–42.

Lin, X., X. Yang, L. Tan, M. Li, X. Wang, Y. Zhang, K. Yang, Z. Hu, and J. Qiu. 2014. In vitro degradation and biocompatibility of a strontium-containing micro-arc oxidation coating on the biodegradable ZK60 magnesium alloy. *Applied Surface Science* 288:718–26.

Linderoth, S., N. H. Pryds, M. Ohnuma, A. Schrøder Pedersen, M. Eldrup, N. Nishiyama, and A. Inoue. 2001. On the stability and crystallisation of bulk amorphous Mg–Cu–Y–Al alloys. *Materials Science and Engineering: A* 304:656–9.

Lindtner, R. A., C. Castellani, S. Tangl et al. 2013. Comparative biomechanical and radiological characterization of osseointegration of a biodegradable magnesium alloy pin and a copolymeric control for osteosynthesis. *Journal of the Mechanical Behavior of Biomedical Materials* 28:232–43.

Liu, X., and P. X. Ma. 2004. Polymeric scaffolds for bone tissue engineering. *Annals of Biomedical Engineering* 32 (3):477–86.

Liu, L. M., and L. Xie. 2007. Adhesive bonding between Mg alloys and polypropylene. *Materials Technology* 22 (2):76–80.

Liu, B., and Y. Zheng. 2011. Effects of alloying elements (Mn, Co, Al, W, Sn, B, C and S) on biodegradability and in vitro biocompatibility of pure iron. *Acta Biomaterialia* 7 (3):1407–20.

Liu, T., W. Zhang, S. D. Wu, C. B. Jiang, S. X. Li, and Y. B. Xu. 2003. Mechanical properties of a two-phase alloy Mg–8%Li–1%Al processed by equal channel angular pressing. *Materials Science and Engineering: A* 360 (1–2):345–9.

Liu, T., Y. D. Wang, S. D. Wu, R. Lin Peng, C. X. Huang, C. B. Jiang, and S. X. Li. 2004. Textures and mechanical behavior of Mg–3.3%Li alloy after ECAP. *Scripta Materialia* 51 (11):1057–61.

Liu, W. Y., H. F. Zhang, Z. Q. Hu, and H. Wang. 2005. Formation and mechanical properties of Mg65Cu25Er10 and Mg65Cu15Ag10Er10 bulk amorphous alloys. *Journal of Alloys and Compounds* 397 (1–2):202–6.

Liu, C., Y. Xin, X. Tian, and P. K. Chu. 2007a. Corrosion behavior of AZ91 magnesium alloy treated by plasma immersion ion implantation and deposition in artificial physiological fluids. *Thin Solid Films* 516 (2–4):422–7.

Liu, C., Y. Xin, X. Tian, J. Zhao, and P. K. Chu. 2007b. Corrosion resistance of titanium ion implanted AZ91 magnesium alloy. *Journal of Vacuum Science and Technology A: Vacuum, Surfaces, and Films* 25 (2):334–9.

Liu, C. L., Y. C. Xin, X. B. Tian, and P. K. Chu. 2007c. Degradation susceptibility of surgical magnesium alloy in artificial biological fluid containing albumin. *Journal of Materials Research* 22 (7):1806–14.

Liu, H. M., Y. G. Chen, Y. B. Tang, S. H. Wei, and G. Niu. 2007. The microstructure, tensile properties, and creep behavior of as-cast Mg-(1-10)% Sn alloys. *Journal of Alloys and Compounds* 440 (1–2):122–6.

Liu, N. C., W. D. Xie, X. D. Peng, Q. Y. Wei, and H. D. Li. 2007. Preparation of Mg-Sr alloy using electrochemical reduction. *Advanced Materials Research* 26:119–23.

Liu, T., S. D. Wu, S. X. Li, and P. J. Li. 2007. Microstructure evolution of Mg–14% Li–1% Al alloy during the process of equal channel angular pressing. *Materials Science and Engineering: A* 460–1:499–503.

Liu, B., M. Zhang, and R. Wu. 2008. Effects of Nd on microstructure and mechanical properties of as-cast LA141 alloys. *Materials Science and Engineering: A* 487 (1–2):347–51.

Liu, K., M. Zhang, J. Zhai, J. Wang, and L. Jiang. 2008. Bioinspired construction of Mg–Li alloys surfaces with stable superhydrophobicity and improved corrosion resistance. *Applied Physics Letters* 92 (18):183103.

Liu, L., Z. Liu, K. Chan, H. Luo, Q. Cai, and S. Zhang. 2008. Surface modification and biocompatibility of Ni-free Zr-based bulk metallic glass. *Scripta Materialia* 58 (3):231–4.

Liu, M., P. J. Uggowitzer, P. Schmutz, and A. Atrens. 2008. Calculated phase diagrams, iron tolerance limits, and corrosion of Mg-Al alloys. *JOM* 60 (12):39–44.

Liu, S. F., L. Liu, and L. G. Kang. 2008. Refinement role of electromagnetic stirring and strontium in AZ91 magnesium alloy. *Journal of Alloys and Compounds* 450 (1–2):546–50.

Liu, M., P. J. Uggowitzer, A. V. Nagasekhar, P. Schmutz, M. Easton, G.-L. Song, and A. Atrens. 2009. Calculated phase diagrams and the corrosion of die-cast Mg–Al alloys. *Corrosion Science* 51 (3):602–19.

Liu, N., J. Wang, L. Wang, Y. Wu, and L. Wang. 2009. Electrochemical corrosion behavior of Mg–5Al–0.4Mn–xNd in NaCl solution. *Corrosion Science* 51 (6):1328–33.

Liu, W., F. Cao, L. Chang, Z. Zhang, and J. Zhang. 2009. Effect of rare earth element Ce and La on corrosion behavior of AM60 magnesium alloy. *Corrosion Science* 51 (6):1334–43.

Liu, C., Y. Wang, R. Zeng, X. Zhang, W. Huang, and P. Chu. 2010. In vitro corrosion degradation behaviour of Mg–Ca alloy in the presence of albumin. *Corrosion Science* 52 (10):3341–7.

Liu, D.-B., M.-F. Chen, and X.-Y. Ye. 2010. Fabrication and corrosion behavior of HA/Mg-Zn biocomposites. *Frontiers of Materials Science in China* 4 (2):139–44.

Liu, H. M., Y. G. Chen, H. F. Zhao, S. H. Wei, and W. Gao. 2010. Effects of strontium on microstructure and mechanical properties of as-cast Mg-5 wt.%Sn alloy. *Journal of Alloys and Compounds* 504 (2):345–50.

Liu, K., L. L. Rokhlin, F. M. Elkin, D. Tang, and J. Meng. 2010. Effect of ageing treatment on the microstructures and mechanical properties of the extruded Mg–7Y–4Gd–1.5 Zn–0.4 Zr alloy. *Materials Science and Engineering: A* 527 (3):828–34.

Liu, M., P. Schmutz, P. J. Uggowitzer, G. L. Song, and A. Atrens. 2010. The influence of yttrium (Y) on the corrosion of Mg–Y binary alloys. *Corrosion Science* 52 (11):3687–701.

Liu, Q., W. F. Shen, H. Y. Sun et al. 2010. Hydrogen-rich saline protects against liver injury in rats with obstructive jaundice. *Liver International* 30 (7):958–68.

Liu, X. B., D. Y. Shan, Y. W. Song, and E. H. Han. 2010. Effects of heat treatment on corrosion behaviors of Mg-3Zn magnesium alloy. *Transactions of Nonferrous Metals Society of China* 20 (7):1345–50.

Liu, G. Y., J. Hu, Z. K. Ding, and C. Wang. 2011a. Bioactive calcium phosphate coating formed on micro-arc oxidized magnesium by chemical deposition. *Applied Surface Science* 257 (6):2051–7.

Liu, G. Y., J. Hu, Z. K. Ding, and C. Wang. 2011b. Formation mechanism of calcium phosphate coating on micro-arc oxidized magnesium. *Materials Chemistry and Physics* 130 (3):1118–24.

Liu, S., K. Liu, Q. Sun et al. 2011. Consumption of hydrogen water reduces paraquat-induced acute lung injury in rats. *BioMed Research International* 2011:305086.

Liu, Y., W. Liu, X. Sun et al. 2011. Hydrogen saline offers neuroprotection by reducing oxidative stress in a focal cerebral ischemia-reperfusion rat model. *Medical Gas Research* 1 (1):15.

Liu, D., Y. Zuo, W. Meng, M. Chen, and Z. Fan. 2012. Fabrication of biodegradable nano-sized β-TCP/Mg composite by a novel melt shearing technology. *Materials Science and Engineering: C* 32 (5):1253–8.

Liu, G. B., P. Gao, Z. Xue, S. Q. Yang, and M. L. Zhang. 2012. Study on the formation of new Mg–Cu–Ti–Y quaternary bulk metallic glasses with high mechanical strength. *Journal of Non-Crystalline Solids* 358 (23):3084–8.

Liu, J.-R., Y.-N. Guo, and W.-D. Huang. 2012. Formation process and properties of phytic acid conversion coatings on magnesium. *Anti-Corrosion Methods and Materials* 59 (5):225–30.

Liu, P., X. Pan, W. Yang, K. Cai, and Y. Chen. 2012a. Al$_2$O$_3$–ZrO$_2$ ceramic coatings fabricated on WE43 magnesium alloy by cathodic plasma electrolytic deposition. *Materials Letters* 70:16–18.

Liu, P., X. Pan, W. Yang, K. Cai, and Y. Chen. 2012b. Improved anticorrosion of magnesium alloy via layer-by-layer self-assembly technique combined with micro-arc oxidation. *Materials Letters* 75:118–21.

Liu, X. H., R. Z. Wu, Z. Y. Niu, J. H. Zhang, and M. L. Zhang. 2012. Superplasticity at elevated temperature of an Mg-8%Li-2%Zn alloy. *Journal of Alloys and Compounds* 541:372–5.

Liu, Z.-J., G.-H. Wu, W.-C. Liu, S. Pang, and W.-J. Ding. 2012. Effect of heat treatment on microstructures and mechanical properties of sand-cast Mg–4Y–2Nd–1Gd–0.4Zr magnesium alloy. *Transactions of Nonferrous Metals Society of China* 22 (7):1540–8.

Liu, H., F. Xue, J. Bai, and Y. Sun. 2013a. Effect of heat treatments on the microstructure and mechanical properties of an extruded Mg95.5Y3Zn1.5 alloy. *Materials Science and Engineering: A* 585:261–7.

Liu, H., F. Xue, J. Bai, and J. Zhou. 2013b. Microstructure and mechanical properties of a Mg94Y4Ni2 alloy with long period stacking ordered structure. *Journal of Materials Engineering and Performance* 22 (11):3500–6.

Liu, H., F. Xue, J. Bai, J. Zhou, and Y. S. Sun. 2013c. Microstructure and mechanical properties of Mg94Zn2Y4 extruded alloy with long-period stacking ordered structure. *Transactions of Nonferrous Metals Society of China* 23 (12):3598–603.

Liu, K., X. Wang, and W. Du. 2013. Development of extraordinary high-strength-toughness Mg alloy via combined processes of repeated plastic working and hot extrusion. *Materials Science and Engineering: A* 573:127–31.

Liu, X. L., W. R. Zhou, Y. H. Wu, Y. Cheng, and Y. F. Zheng. 2013. Effect of sterilization process on surface characteristics and biocompatibility of pure Mg and MgCa alloys. *Materials Science and Engineering. C, Materials for Biological Applications* 33 (7):4144–54.

Liu, C.-L., Y. Zhang, C.-Y. Zhang, W. Wang, W.-J. Huang, and P. K. Chu. 2014. Synergistic effect of chloride ion and albumin on the corrosion of pure magnesium. *Frontiers of Materials Science* 8 (3):244–55.

Liu, D., Y. Ding, T. Guo, X. Qin, C. Guo, S. Yu, and S. Lin. 2014. Influence of fine-grain and solid-solution strengthening on mechanical properties and in vitro degradation of WE43 alloy. *Biomedical Materials* 9 (1):015014.

Lock, J. Y., E. Wyatt, S. Upadhyayula et al. 2014. Degradation and antibacterial properties of magnesium alloys in artificial urine for potential resorbable ureteral stent applications. *Journal of Biomedical Materials Research. Part A* 102 (3):781–92.

Loghin, F., A. Olinic, D.-S. Popa, C. Socaciu, and S. E. Leucuta. 1999. Effects of long-term administration of lithium and hydrochlorothiazide in rats. *Metal-Based Drugs* 6 (2):87–93.

Loos, A., R. Rohde, A. Haverich, and S. Barlach. 2007. In vitro and in vivo biocompatibility testing of absorbable metal stents. *Macromolecular Symposia* 253 (1):103–8.

López, A. J., E. Otero, and J. Rams. 2010. Sol–gel silica coatings on ZE41 magnesium alloy for corrosion protection. *Surface and Coatings Technology* 205 (7):2375–85.

Lorenz, C., J. G. Brunner, P. Kollmannsberger, L. Jaafar, B. Fabry, and S. Virtanen. 2009. Effect of surface pre-treatments on biocompatibility of magnesium. *Acta Biomaterialia* 5 (7):2783–9.

Lorimer, G. W., P. J. Apps, H. Karimzadeh, and J. F. King. 2003. Improving the performance of Mg-rare earth alloys by the use of Gd or Dy additions. *Materials Science Forum* 419:279–84.

Lou, Y., X. Bai, and L. X. Li. 2011. Effect of Sr addition on microstructure of as-cast Mg-Al-Ca alloy. *Transactions of Nonferrous Metals Society of China* 21 (6):1247–52.

Lovald, S., B. Baack, C. Gaball, G. Olson, and A. Hoard. 2010. Biomechanical optimization of bone plates used in rigid fixation of mandibular symphysis fractures. *Journal of Oral and Maxillofacial Surgery* 68 (8):1833–41.

Lu, L. C., and Mikos, A. G. 1999. Poly(lactic acid). In: Mark J. E., editor. Polymer data handbook. Oxford: Oxford Press. pp. 527–633.

Lu, L., K. Raviprasad, and M. O. Lai. 2004. Nanostructured Mg–5%Al–x%Nd alloys. *Materials Science and Engineering: A* 368 (1–2):117–25.

Lü, Y., Q. Wang, X. Zeng, W. Ding, C. Zhai, and Y. Zhu. 2000. Effects of rare earths on the microstructure, properties and fracture behavior of Mg–Al alloys. *Materials Science and Engineering: A* 278 (1):66–76.

Lu, P., H. Fan, Y. Liu, L. Cao, X. Wu, and X. Xu. 2011a. Controllable biodegradability, drug release behavior and hemocompatibility of PTX-eluting magnesium stents. *Colloids and Surfaces B Biointerfaces* 83 (1):23–8.

Lu, P., L. Cao, Y. Liu, X. Xu, and X. Wu. 2011b. Evaluation of magnesium ions release, biocorrosion, and hemocompatibility of MAO/PLLA-modified magnesium alloy WE42. *Journal of Biomedical Materials Research Part B: Applied Biomaterials* 96 (1):101–9.

Lu, L., T. Liu, M.-J. Tan, J. Chen, and Z. Wang. 2012. Effect of annealing on microstructure evolution and mechanical property of cold forged magnesium pipes. *Materials and Design* 39:131–9.

Lu, T.-F., K.-Y. Yin, B.-Y. Sun, Q. Dong, and B. Chen. 2012. Effects of micro-arc oxidation coating on corrosion behavior of Mg-Y-Zn in simulated body fluid. *Journal of Shanghai Jiaotong University (Science)* 17 (6):668–72.

Lu, W., Z. Chen, P. Huang, and B. Yan. 2012. Microstructure, corrosion resistance and biocompatibility of biomimetic HA-based Ca-P coatings on ZK60 magnesium alloy. *International Journal of Electrochemical Science* 7 (2012):12668–79.

Lu, W., C. Ou, Z. Zhan, P. Huang, B. Yan, and M. Chen. 2013. Microstructure and in vitro corrosion properties of ZK60 magnesium alloy coated with calcium phosphate by electrodeposition at different temperatures. *International Journal of Electrochemical Science* 8 (2013):10746–57.

Lu, Z.-W., D.-W. Zhou, J.-P. Bai, C. Lu, Z.-G. Zhong, and G.-Q. Li. 2013. Theoretical investigation on structural and thermodynamic properties of the intermetallic compound in Mg–Zn–Ag alloy under high pressure and high temperature. *Journal of Alloys and Compounds* 550:406–11.

Lu, Y., P. Wan, L. Tan, B. Zhang, K. Yang, and J. Lin. 2014a. Preliminary study on a bioactive Sr containing Ca–P coating on pure magnesium by a two-step procedure. *Surface and Coatings Technology* 252:79–86.

Lu, Y., L. Tan, B. Zhang, J. Lin, and K. Yang. 2014b. Synthesis and characterization of Ca–Sr–P coating on pure magnesium for biomedical application. *Ceramics International* 40 (3):4559–65.

Luffy, S. A., D.-T. Chou, J. Waterman, P. D. Wearden, P. N. Kumta, and T. W. Gilbert. 2014. Evaluation of magnesium-yttrium alloy as an extraluminal tracheal stent. *Journal of Biomedical Materials Research Part A* 102 (3):611–20.

Lunder, O., J. Lein, S. Hesjevik, T. K. Aune, and K. Nişancıoğlu. 1994. Corrosion morphologies on magnesium alloy AZ 91. *Materials and Corrosion* 45 (6):331–40.

Luo, A., and M. O. Pekguleryuz. 1994. Cast magnesium alloys for elevated-temperature applications. *Journal of Materials Science* 29 (20):5259–71.

Luo, Z. P., and S. Q. Zhang. 2000. High-resolution electron microscopy on the X-Mg12ZnY phase in a high strength Mg-Zn-Zr-Y magnesium alloy. *Journal of Materials Science Letters* 19 (9):813–15.

Luo, A. A., and B. R. Powell. 2001. Tensile and compressive creep of magnesium–aluminum–calcium based alloys. *Magnesium Technology* 2001:137–44.

Luo, X., and X. T. Cui. 2011. Electrochemical deposition of conducting polymer coatings on magnesium surfaces in ionic liquid. *Acta Biomaterialia* 7 (1):441–6.

Luo, G. X., G. Q. Wu, S. J. Wang, R. H. Li, and Z. Huang. 2006. Effects of YAl2 particulates on microstructure and mechanical properties of β-Mg–Li alloy. *Journal of Materials Science* 41 (17):5556–8.

Luo, S., Q. Q. Zhang, Y. C. Zhang, C. Li, X. Q. Xu, and T. T. Zhou. 2013. In vitro and in vivo studies on a MgLi-X alloy system developed as a new kind of biological metal. *Materials Science Forum* 747–8:257–63.

Lynch, S., and P. Trevena. 1988. Stress corrosion cracking and liquid metal embrittlement in pure magnesium. *Corrosion* 44 (2):113–24.

Ma, C., D. Zhang, J. Qin, W. Hu, and Z. Shi. 1999. Aging behavior of Mg-Li-Al alloys. *Transactions-Nonferrous Metals Society of China-English Edition* 9 (4):772–7.

Ma, B. M., J. W. Herchenroeder, B. Smith, M. Suda, D. N. Brown, and Z. Chen. 2002. Recent development in bonded NdFeB magnets. *Journal of Magnetism and Magnetic Materials* 239 (1):418–23.

Ma, L., R. K. Mishra, L. Peng, A. A. Luo, W. Ding, and A. K. Sachdev. 2011. Texture and mechanical behavior evolution of age-hardenable Mg–Nd–Zn extrusions during aging treatment. *Materials Science and Engineering: A* 529:151–5.

Ma, L., R. K. Mishra, M. P. Balogh, L. Peng, A. A. Luo, A. K. Sachdev, and W. Ding. 2012. Effect of Zn on the microstructure evolution of extruded Mg–3Nd (–Zn)–Zr (wt.%) alloys. *Materials Science and Engineering: A* 543:12–21.

Ma, N., Q. Peng, X. Li, H. Li, J. Zhang, and Y. Tian. 2012. Influence of scandium on corrosion properties and electrochemical behaviour of Mg alloys in different media. *International Journal of Electrochemical Science* 7 (9):8020.

Ma, Y., N. Li, D. Li, M. Zhang, and X. Huang. 2012. Characteristics and corrosion studies of vanadate conversion coating formed on Mg–14wt%Li–1wt%Al–0.1wt%Ce alloy. *Applied Surface Science* 261:59–67.

Ma, N., H. Li, J. Pan, D. Fang, Y. Tian, and Q. Peng. 2013. Effects of cerium on microstructures, recovery behavior and mechanical properties of backward extruded Mg–0.5 Mn alloys. *Materials Science and Engineering: A* 564:310–16.

Ma, X., S. Zhu, L. Wang, C. Ji, C. Ren, and S. Guan. 2014. Synthesis and properties of a biocomposite coating formed on magnesium alloy by one-step method of micro-arc oxidation. *Journal of Alloys and Compounds* 590:247–53.

Maeng, M., L. O. Jensen, E. Falk, H. R. Andersen, and L. Thuesen. 2009. Negative vascular remodelling after implantation of bioabsorbable magnesium alloy stents in porcine coronary arteries: A randomised comparison with bare-metal and sirolimus-eluting stents. *Heart* 95 (3):241–6.

Magnesium Elektron n.d. *Joining Magnesium Alloys.* Available from http://www.magnesium -elektron.com/data/downloads/ds250jo.pdf.

Magnesium Elektron. 2014. Researchers continue to investigate 3D printed bone & tissue. March 12. http://3dprinterplans.info/tag/magnesium-elektron/ (accessed May 30, 2015).

Magnissalis, E., S. Zinelis, T. Karachalios, and G. Hartofilakidis. 2003. Failure analysis of two Ti-alloy total hip arthroplasty femoral stems fractured in vivo. *Journal of Biomedical Materials Research Part B: Applied Biomaterials* 66 (1):299–305.

Maguire, M. E., and J. A. Cowan. 2002. Magnesium chemistry and biochemistry. *Biometals* 15 (3):203–10.

Maier, O. 1940. Über die Verwendbarkeit von Leichtmetallen in der Chirurgie (metallisches Magnesium als Reizmittel zur Knochenneubildung). *Langenbeck's Archives of Surgery* 253 (8):552–6.

Maier, J. A., D. Bernardini, Y. Rayssiguier, and A. Mazur. 2004. High concentrations of magnesium modulate vascular endothelial cell behaviour in vitro. *Biochimica et Biophysica Acta* 1689 (1):6–12.

Makar, G. L., and J. Kruger. 1990. Corrosion studies of rapidly solidified magnesium alloys. *Journal of the Electrochemical Society* 137 (2):414–21.

Makar, G. L., and J. Kruger. 1993. Corrosion of magnesium. *International Materials Reviews* 38 (3):138–53.

Makau, F., K. Morsi, N. Gude, R. Alvarez, M. Sussman, and K. May-Newman. 2013. Viability of titanium-titanium boride composite as a biomaterial. *ISRN Biomaterials* 2013:970535.

Mandal, M., A. P. Moon, G. Deo, C. L. Mendis, and K. Mondal. 2014. Corrosion behavior of Mg-2.4Zn alloy micro-alloyed with Ag and Ca. *Corrosion Science* 78:172–82.

Mao, Y.-F., X.-F. Zheng, J.-M. Cai et al. 2009. Hydrogen-rich saline reduces lung injury induced by intestinal ischemia/reperfusion in rats. *Biochemical and Biophysical Research Communications* 381 (4):602–5.

Mao, L. H., Y. L. Wang, Y. Z. Wan, F. He, and Y. A. Huang. 2010. Corrosion resistance of Ag-ion implanted Mg-Ca-Zn alloys in SBF. *Rare Metal Materials and Engineering* 39 (12):2075–8.

Mao, L., G. Y. Yuan, S. H. Wang, J. L. Niu, G. H. Wu, and W. J. Ding. 2012. A novel bio-degradable Mg-Nd-Zn-Zr alloy with uniform corrosion behavior in artificial plasma. *Materials Letters* 88:1–4.

Mao, L., G. Yuan, J. Niu, Y. Zong, and W. Ding. 2013. In vitro degradation behavior and bio-compatibility of Mg–Nd–Zn–Zr alloy by hydrofluoric acid treatment. *Materials Science and Engineering: C* 33 (1):242–50.

Marie, P. J. 2005. Strontium ranelate: A novel mode of action optimizing bone formation and resorption. *Osteoporosis International* 16:S7–10.

Mason, J. F., C. M. Warwick, P. J. Smith, J. A. Charles, and T. W. Clyne. 1989. Magnesium-lithium alloys in metal matrix composites—A preliminary report. *Journal of Materials Science* 24 (11):3934–46.

Matsubara, H., Y. Ichige, K. Fujita, H. Nishiyama, and K. Hodouchi. 2013. Effect of impurity Fe on corrosion behavior of AM50 and AM60 magnesium alloys. *Corrosion Science* 66:203–10.

Matsuda, A., C. C. Wan, J.-M. Yang, and W. H. Kao. 1996. Rapid solidification processing of a Mg-Li-Si-Ag alloy. *Metallurgical and Materials Transactions A* 27 (5):1363–70.

Matsuda, M., S. Ii, Y. Kawamura, Y. Ikuhara, and M. Nishida. 2005. Variation of long-period stacking order structures in rapidly solidified Mg97Zn1Y2 alloy. *Materials Science and Engineering A-Structural Materials Properties Microstructure and Processing* 393 (1–2):269–74.

Matsumoto, R., M. Otsu, M. Yamasaki, T. Mayama, H. Utsunomiya, and Y. Kawamura. 2012. Application of mixture rule to finite element analysis for forging of cast Mg–Zn–Y alloys with long period stacking ordered structure. *Materials Science and Engineering: A* 548:75–82.

Matsushita, T., Y. Kusakabe, A. Kitamura, S. Okada, and K. Murase. 2011. Investigation of protective effect of hydrogen-rich water against cisplatininduced nephrotoxicity in rats using blood oxygenation level-dependent magnetic resonance imaging. *Japanese Journal of Radiology* 29 (7):503–12.

Matsuura, Y. 2006. Recent development of Nd–Fe–B sintered magnets and their applications. *Journal of Magnetism and Magnetic Materials* 303 (2):344–7.

McBride, E. D. 1938a. Absorbable metal in bone surgery: A further report on the use of magnesium alloys. *Journal of the American Medical Association* 111 (27):2464–7.

McBride, E. D. 1938b. Magnesium screw and nail transfixion in fractures. *Southern Medical Journal* 31 (5):508–14.

McGarry, J. P., B. P. O'Donnell, P. E. McHugh, and J. G. McGarry. 2004. Analysis of the mechanical performance of a cardiovascular stent design based on micromechanical modelling. *Computational Materials Science* 31:421–38.

McKnight, R. F., M. Adida, K. Budge, S. Stockton, G. M. Goodwin, and J. R. Geddes. 2012. Lithium toxicity profile: A systematic review and meta-analysis. *The Lancet* 379 (9817):721–8.

McLaren, A., E. Hetherington, D. Maddalena, and G. Snowdon. 1990. Dysprosium (165Dy) hydroxide macroaggregates for radiation synovectomy—animal studies. *European Journal of Nuclear Medicine* 16 (8–10):627–32.

McMahon, C. J., P. Oslizlok, and K. P. Walsh. 2007. Early restenosis following biodegradable stent implantation in an aortopulmonary collateral of a patient with pulmonary atresia and hypoplastic pulmonary arteries. *Catheterization and Cardiovascular Interventions* 69 (5):735–8.

McNulty, R. E., and J. D. Hanawalt. 1942. Some corrosion characteristics of high purity magnesium alloys. *Transactions of The Electrochemical Society* 81 (1):423–33.

Mears, R. J., L. Reekie, S. B. Poole, and D. N. Payne. 1985. Neodymium-doped silica single-mode fibre lasers. *Electronics Letters* 21 (17):738–40.

Meletis, E., and R. Hochman. 1984. Crystallography of stress corrosion cracking in pure magnesium. *Corrosion* 40 (1):39–45.

Men, H., Z. Q. Hu, and J. Xu. 2002. Bulk metallic glass formation in the Mg–Cu–Zn–Y system. *Scripta Materialia* 46 (10):699–703.

Mendis, C. L., K. Oh-ishi, and K. Hono. 2007. Enhanced age hardening in a Mg–2.4at.% Zn alloy by trace additions of Ag and Ca. *Scripta Materialia* 57 (6):485–8.

Mendis, C., K. Ohishi, Y. Kawamura, T. Honma, S. Kamado, and K. Hono. 2009. Precipitation-hardenable Mg–2.4Zn–0.1Ag–0.1Ca–0.16Zr (at.%) wrought magnesium alloy. *Acta Materialia* 57 (3):749–60.

Mendis, C. L., J. H. Bae, N. J. Kim, and K. Hono. 2011. Microstructures and tensile properties of a twin roll cast and heat-treated Mg–2.4Zn–0.1Ag–0.1Ca–0.1Zr alloy. *Scripta Materialia* 64 (4):335–8.

Meng, F.-G., H.-S. Liu, L.-B. Liu, and Z.-P. Jin. 2007a. Thermodynamic optimization of Mg–Nd system. *Transactions of Nonferrous Metals Society of China* 17 (1):77–81.

Meng, F. G., J. Wang, H. S. Liu, L. B. Liu, and Z. P. Jin. 2007b. Experimental investigation and thermodynamic calculation of phase relations in the Mg–Nd–Y ternary system. *Materials Science and Engineering: A* 454–5:266–73.

Meng, X., R. Wu, M. Zhang, L. Wu, and C. Cui. 2009. Microstructures and properties of superlight Mg–Li–Al–Zn wrought alloys. *Journal of Alloys and Compounds* 486 (1–2):722–5.

Meng, E. C., S. K. Guan, H. X. Wang, L. G. Wang, S. J. Zhu, J. H. Hu, C. X. Ren, J. H. Gao, and Y. S. Feng. 2011. Effect of electrodeposition modes on surface characteristics and corrosion properties of fluorine-doped hydroxyapatite coatings on Mg–Zn–Ca alloy. *Applied Surface Science* 257 (11):4811–16.

Merino, M. C., A. Pardo, R. Arrabal et al. 2010. Influence of chloride ion concentration and temperature on the corrosion of Mg–Al alloys in salt fog. *Corrosion Science* 52 (5):1696–704.

Meschter, P. J., and J. E. O'Neal. 1984. Rapid solidification processing of magnesium-lithium alloys. *Metallurgical and Materials Transactions A* 15 (1):237–40.

Meslemani, D., and R. M. Kellman. 2012. Recent advances in fixation of the craniomaxillofacial skeleton. *Current Opinion in Otolaryngology and Head and Neck Surgery* 20:304–9.

Meunier, P. J., C. Roux, E. Seeman, S. Ortolani, J. E. Badurski, T. D. Spector, J. Cannata, A. Balogh, E.-M. Lemmel, and S. Pors-Nielsen. 2004. The effects of strontium ranelate on the risk of vertebral fracture in women with postmenopausal osteoporosis. *New England Journal of Medicine* 350 (5):459–68.

Meunier, P. J., C. Roux, S. Ortolani, M. Diaz-Curiel, J. Compston, P. Marquis, C. Cormier, G. Isaia, J. Badurski, and J. D. Wark. 2009. Effects of long-term strontium ranelate treatment on vertebral fracture risk in postmenopausal women with osteoporosis. *Osteoporosis International* 20 (10):1663–73.

Mhaede, M., F. Pastorek, and B. Hadzima. 2014. Influence of shot peening on corrosion properties of biocompatible magnesium alloy AZ31 coated by dicalcium phosphate dihidrate (DCPD). *Materials Science and Engineering: C* 39 (1):330–5.

Miao, B., and D. Z. Jinag. 2009. Bone inductivity and antibacterial property of MAO-treated magnesium alloys. *Heilongjiang Medicine and Pharmacy* 32 (5):7–8.

Middleton, J. C., and A. J. Tipton. 2000. Synthetic biodegradable polymers as orthopedic devices. *Biomaterials* 21 (23):2335–46.

Mine, Y., H. Yoshimura, M. Matsuda, K. Takashima, and Y. Kawamura. 2013. Microfracture behaviour of extruded Mg–Zn–Y alloys containing long-period stacking ordered structure at room and elevated temperatures. *Materials Science and Engineering: A* 570:63–9.

Mintz, G. S., S. E. Nissen, W. D. Anderson et al. 2001. American College of Cardiology clinical expert consensus document on standards for acquisition, measurement and reporting of intravascular ultrasound studies (ivus)33: A report of the American College of Cardiology task force on clinical expert consensus documents developed in collaboration with the European Society of Cardiology endorsed by the society of cardiac angiography and interventions. *Journal of the American College of Cardiology* 37 (5):1478–92.

Mironov, S., Y. Motohashi, T. Ito et al. 2007. Feasibility of friction stir welding for joining and microstructure refinement in a ZK60 magnesium alloy. *Materials Transactions* 48 (12):3140–8.

Misra, R., W. Thein-Han, T. Pesacreta, K. Hasenstein, M. Somani, and L. Karjalainen. 2009a. Cellular response of preosteoblasts to nanograined/ultrafine-grained structures. *Acta Biomaterialia* 5 (5):1455–67.

Misra, R., W. W. Thein-Han, T. Pesacreta, K. Hasenstein, M. Somani, and L. Karjalainen. 2009b. Favorable modulation of pre-osteoblast response to nanograined/ultrafine-grained structures in austenitic stainless steel. *Advanced Materials* 21 (12):1280–5.

Miura, S., S. Imagawa, T. Toyoda, K. Ohkubo, and T. Mohri. 2008. Effect of rare-earth elements Y and Dy on the deformation behavior of Mg alloy single crystals. *Materials Transactions* 49 (5):952–6.

Mizutani, Y., T. Tamura, and K. Miwa. 2005. Microstructural refinement process of pure magnesium by electromagnetic vibrations. *Materials Science and Engineering: A* 413:205–10.

Mora, E., G. Garces, E. Onorbe, P. Perez, and P. Adeva. 2009. High-strength Mg-Zn-Y alloys produced by powder metallurgy. *Scripta Materialia* 60 (9):776–9.

Mordike, B. L. 2002. Creep-resistant magnesium alloys. *Materials Science and Engineering: A* 324 (1):103–12.

Mordike, B. L., I. Stulikova, and B. Smola. 2005. Mechanisms of creep deformation in Mg-Sc-based alloys. *Metallurgical and Materials Transactions A* 36 (7):1729–36.

Moreno, I. P., T. K. Nandy, J. W. Jones, J. E. Allison, and T. M. Pollock. 2003. Microstructural stability and creep of rare-earth containing magnesium alloys. *Scripta Materialia* 48 (8):1029–34.

Moulton, S. E., M. D. Imisides, R. L. Shepherd, and G. G. Wallace. 2008. Galvanic coupling conducting polymers to biodegradable Mg initiates autonomously powered drug release. *Journal of Materials Chemistry* 18 (30):3608–13.

Mueller, W. D., M. F. Lorenzo De Mele, M. L. Nascimento, and M. Zeddies. 2009. Degradation of magnesium and its alloys: Dependence on the composition of the synthetic biological media. *Journal of Biomedical Materials Research—Part A* 90 (2):487–95.

Muller, A., G. Garces, P. Perez, and P. Adeva. 2007. Grain refinement of Mg-Zn-Y alloy reinforced by an icosahedral quasicrystalline phase by severe hot rolling. *Journal of Alloys and Compounds* 443 (1–2):L1–5.

Nagase, T., K. Kinoshita, and Y. Umakoshi. 2008. Preparation of Zr-based metallic glass wires for biomaterials by arc-melting type melt-extraction method. *Materials Transactions* 49 (6):1385–94.

Nagata, M., and B. Lönnerdal. 2011. Role of zinc in cellular zinc trafficking and mineralization in a murine osteoblast-like cell line. *The Journal of Nutritional Biochemistry* 22 (2):172–8.

Nagata, K., N. Nakashima-Kamimura, T. Mikami, I. Ohsawa, and S. Ohta. 2008. Consumption of molecular hydrogen prevents the stress-induced impairments in hippocampus-dependent learning tasks during chronic physical restraint in mice. *Neuropsychopharmacology* 34 (2):501–8.

Nakao, A., D. J. Kaczorowski, Y. Wang et al. 2010a. Amelioration of rat cardiac cold ischemia/reperfusion injury with inhaled hydrogen or carbon monoxide, or both. *The Journal of Heart and Lung Transplantation* 29 (5):544–53.

Nakao, A., Y. Toyoda, P. Sharma, M. Evans, and N. Guthrie. 2010b. Effectiveness of hydrogen rich water on antioxidant status of subjects with potential metabolic syndrome—An open label pilot study. *Journal of Clinical Biochemistry and Nutrition* 46 (2):140–9.

Nakashima-Kamimura, N., T. Mori, I. Ohsawa, S. Asoh, and S. Ohta. 2009. Molecular hydrogen alleviates nephrotoxicity induced by an anti-cancer drug cisplatin without compromising anti-tumor activity in mice. *Cancer Chemotherapy and Pharmacology* 64 (4):753–61.

Nakayama, M., S. Kabayama, H. Nakano et al. 2009. Biological effects of electrolyzed water in hemodialysis. *Nephron Clinical Practice* 112 (1):c9–15.

Nakayama, M., H. Nakano, H. Hamada et al. 2010. A novel bioactive haemodialysis system using dissolved dihydrogen (H2) produced by water electrolysis: A clinical trial. *Nephrology Dialysis Transplantation* 25 (9):3026–33.

Nam, K. Y., D. H. Song, C. W. Lee, S. W. Lee, Y. H. Park, K. M. Cho, and I. M. Park. 2006. Modification of Mg2Si morphology in as-cast Mg-Al-Si alloys with strontium and antimony. *Eco-Materials Processing and Design VII* 510–1:238–41.

Nam, N. D., W. C. Kim, J. G. Kim, K. S. Shin, and H. C. Jung. 2011. Corrosion resistance of Mg-5Al-xSr alloys. *Journal of Alloys and Compounds* 509 (14):4839–47.

Nam, S. W., W. T. Kim, D. H. Kim, and T. S. Kim. 2013. Microstructure and corrosion behavior of rapidly solidified Mg-Zn-Y alloys. *Metals and Materials International* 19 (2):205–9.

Nayeb-Hashemi, A. 1988. *Phase Diagrams of Binary Magnesium Alloys*. Metals Park, OH: ASM International.

Nayeb-Hashemi, A. A. and J. B. Clark. 1984. The Ag-Mg (silver-magnesium) system. *Bulletin of Alloy Phase Diagrams* 5 (4):348–58.

Nazarov, R., H.-J. Jin, and D. L. Kaplan. 2004. Porous 3-D scaffolds from regenerated silk fibroin. *Biomacromolecules* 5 (3):718–26.

Neubert, V., I. Stulíková, B. Smola, B. L. Mordike, M. Vlach, A. Bakkar, and J. Pelcová. 2007. Thermal stability and corrosion behaviour of Mg–Y–Nd and Mg–Tb–Nd alloys. *Materials Science and Engineering: A* 462 (1–2):329–33.

Ng, W. F., M. H. Wong, and F. T. Cheng. 2010a. Cerium-based coating for enhancing the corrosion resistance of bio-degradable Mg implants. *Materials Chemistry and Physics* 119 (3):384–8.

Ng, W. F., M. H. Wong, and F. T. Cheng. 2010b. Stearic acid coating on magnesium for enhancing corrosion resistance in Hank's solution. *Surface and Coatings Technology* 204 (11):1823–30.

Ng, C. C., M. M. Savalani, M. L. Lau, and H. C. Man. 2011. Microstructure and mechanical properties of selective laser melted magnesium. *Applied Surface Science* 257 (17):7447–54.

Nguyen, T. L. 2011. *Synthesis of topologically ordered porous magnesium*. Doctoral thesis, New Zealand: University of Canterbury, Canterbury, Kent, UK.

Nguyen, T. L., A. Blanquet, M. P. Staiger, G. J. Dias, and T. B. Woodfield. 2012. On the role of surface roughness in the corrosion of pure magnesium in vitro. *Journal of Biomedical Materials Research Part B: Applied Biomaterials* 100 (5):1310–18.

Nguyen, T. Y., C. G. Liew, and H. Liu. 2013. An in vitro mechanism study on the proliferation and pluripotency of human embryonic stems cells in response to magnesium degradation. *PLoS One* 8 (10):e76547.

Ni, X.-X., Z.-Y. Cai, D.-F. Fan et al. 2011. Protective effect of hydrogen-rich saline on decompression sickness in rats. *Aviation, Space, and Environmental Medicine* 82 (6):604–9.

Nie, J. F., and B. C. Muddle. 1997. Precipitation hardening of Mg-Ca(-Zn) alloys. *Scripta Materialia* 37 (10):1475–81.

Ning, C., and Y. Zhou. 2002. In vitro bioactivity of a biocomposite fabricated from HA and Ti powders by powder metallurgy method. *Biomaterials* 23 (14):2909–15.

Ning, C., and Y. Zhou. 2004. On the microstructure of biocomposites sintered from Ti, HA and bioactive glass. *Biomaterials* 25 (17):3379–87.

Ning, Z. L., G. J. Wang, F. Y. Cao, J. F. Sun, and J. F. Du. 2009. Tensile deformation of a Mg–2.54Nd–0.26Zn–0.32Zr alloy at elevated temperature. *Journal of Materials Science* 44 (16):4264–9.

Ning, Z. L., H. Wang, H. H. Liu, F. Y. Cao, S. T. Wang, and J. F. Sun. 2010. Effects of Nd on microstructures and properties at the elevated temperature of a Mg–0.3Zn–0.32Zr alloy. *Materials and Design* 31 (9):4438–44.

Ning, Z. L., J. Y. Yi, M. Qian, H. C. Sun, F. Y. Cao, H. H. Liu, and J. F. Sun. 2014. Microstructure and elevated temperature mechanical and creep properties of Mg–4Y–3Nd–0.5 Zr alloy in the product form of a large structural casting. *Materials and Design* 60:218–25.

Nishida, M., Y. Kawamura, and T. Yamamuro. 2004. Formation process of unique microstructure in rapidly solidified Mg97Zn1Y2 alloy. *Materials Science and Engineering A-Structural Materials Properties Microstructure and Processing* 375:1217–23.

Niu, J., G. Yuan, Y. Liao, L. Mao, J. Zhang, Y. Wang, F. Huang, Y. Jiang, Y. He, and W. Ding. 2013. Enhanced biocorrosion resistance and biocompatibility of degradable Mg–Nd–Zn–Zr alloy by brushite coating. *Materials Science and Engineering: C* 33 (8):4833–41.

Nogara, G. 1939. Sulla tolleranza dell'osso verso i metalli riassorbibili magnesio ed electron. *Arch Ital Chir* 56 (5):459–78.

Nunez-Lopez, C., H. Habazaki, P. Skeldon, G. Thompson, H. Karimzadeh, P. Lyon, and T. Wilks. 1996. An investigation of microgalvanic corrosion using a model magnesium-silicon carbide metal matrix composite. *Corrosion Science* 38 (10):1721–9.

Oak, J.-J., D. V. Louzguine-Luzgin, and A. Inoue. 2007. Fabrication of Ni-free Ti-based bulk-metallic glassy alloy having potential for application as biomaterial, and investigation of its mechanical properties, corrosion, and crystallization behavior. *Journal of Materials Research* 22 (5):1346–53.

Obekcan, M., A. Ayday, H. Sevik, and S. C. Kurnaz. 2013. Addition of strontium to an Mg-3Sn alloy and investigation of its properties. *Materiali in Tehnologije* 47 (3):299–301.

Oguchi, H., K. Ishikawa, K. Mizoue, K. Seto, and G. Eguchi. 1995. Long-term histological evaluation of hydroxyapatite ceramics in humans. *Biomaterials* 16 (1):33–8.

Oharazawa, H., T. Igarashi, T. Yokota et al. 2010. Protection of the retina by rapid diffusion of hydrogen: Administration of hydrogen-loaded eye drops in retinal ischemia–reperfusion injury. *Investigative Ophthalmology and Visual Science* 51 (1):487–92.

Oh-ishi, K., R. Watanabe, C. L. Mendis, and K. Hono. 2009. Age-hardening response of Mg-0.3 at.%Ca alloys with different Zn contents. *Materials Science and Engineering A-Structural Materials Properties Microstructure and Processing* 526 (1–2):177–84.

Ohno, K., M. Ito, M. Ichihara, and M. Ito. 2012. Molecular hydrogen as an emerging therapeutic medical gas for neurodegenerative and other diseases. *Oxidative Medicine and Cellular Longevity* 2012:353152.

Ohsawa, I., M. Ishikawa, K. Takahashi et al. 2007. Hydrogen acts as a therapeutic antioxidant by selectively reducing cytotoxic oxygen radicals. *Nature Medicine* 13 (6):688–94.

Ohsawa, I., K. Nishimaki, K. Yamagata, M. Ishikawa, and S. Ohta. 2008. Consumption of hydrogen water prevents atherosclerosis in apolipoprotein E knockout mice. *Biochemical and Biophysical Research Communications* 377 (4):1195–8.

Ohta, S. 2011. Recent progress toward hydrogen medicine: Potential of molecular hydrogen for preventive and therapeutic applications. *Current Pharmaceutical Design* 17 (22):2241–52.

Oka, H., W. Guo, T. Wada, and H. Kato. 2013. Mg-based metallic glass matrix composite with in situ porous titanium dispersoids by dealloying in metallic melt. *Materials Science and Engineering: A* 582:76–83.

Okamoto, H. 1998. Ag–Mg (silver–magnesium). *Journal of Phase Equilibria* 19 (5):487.

Okamoto, H. 2012. Mg-Mn (magnesium-manganese). *Journal of Phase Equilibria and Diffusion* 33 (6):496.

Olguín-González, M., D. Hernández-Silva, M. García-Bernal, and V. Sauce-Rangel. 2014. Hot deformation behavior of hot-rolled AZ31 and AZ61 magnesium alloys. *Materials Science and Engineering: A* 597:82–8.

Omanovic, S., and S. G. Roscoe. 1999. Electrochemical studies of the adsorption behavior of bovine serum albumin on stainless steel. *Langmuir* 15 (23):8315–21.

Ono, H., Y. Nishijima, N. Adachi et al. 2011. Improved brain MRI indices in the acute brain stem infarct sites treated with hydroxyl radical scavengers, Edaravone and hydrogen, as compared to Edaravone alone. A non-controlled study. *Medical Gas Research* 1 (1):12.

Onoki, T., and S. Yamamoto. 2010. Hydroxyapatite ceramics coating on magnesium alloy via a double layered capsule hydrothermal hot-pressing. *Journal of the Ceramic Society of Japan* 118 (1380):749–52.

Oñorbe, E., G. Garcés, P. Pérez, and P. Adeva. 2012. Effect of the LPSO volume fraction on the microstructure and mechanical properties of Mg–Y2X–Zn X alloys. *Journal of Materials Science* 47 (2):1085–93.

Ortega, Y., and J. del Rio. 2005. Study of Mg-Ca alloys by positron annihilation technique. *Scripta Materialia* 52 (3):181–6.

Ostrowski, N., B. Lee, N. Enick, B. Carlson, S. Kunjukunju, A. Roy, and P. N. Kumta. 2013. Corrosion protection and improved cytocompatibility of biodegradable polymeric layer-by-layer coatings on AZ31 magnesium alloys. *Acta Biomaterialia* 9 (10):8704–13.

Ott, N., P. Schmutz, C. Ludwig, and A. Ulrich. 2013. Local, element-specific and time-resolved dissolution processes on a Mg–Y–RE alloy–influence of inorganic species and buffering systems. *Corrosion Science* 75:201–11.

Ou, C., W. Lu, Z. Zhan, P. Huang, P. Yan, B. Yan, and M. Chen. 2013. Effect of Ca and P ion concentrations on the structural and corrosion properties of biomimetic Ca-P coatings on ZK60 magnesium alloy. *International Journal of Electrochemical Science* 8 (2013):9518–30.

Ouyang, L. Z., F. X. Qin, and M. Zhu. 2006. The hydrogen storage behavior of Mg_3La and $Mg_3LaNi_{0.1}$. *Scripta Materialia* 55 (12):1075–8.

Ouyang, L. Z., Y. J. Xu, H. W. Dong, L. X. Sun, and M. Zhu. 2009. Production of hydrogen via hydrolysis of hydrides in Mg–La system. *International Journal of Hydrogen Energy* 34 (24):9671–6.

Ozawa, L., and M. Itoh. 2003. Cathode ray tube phosphors. *Chemical Reviews* 103 (10):3835–56.

Pachla, W., A. Mazur, J. Skiba, M. Kulczyk, and S. Przybysz. 2012. Wrought magnesium alloys ZM21, ZW3 and WE43 processed by hydrostatic extrusion with back pressure. *Archives of Metallurgy and Materials* 57 (2):485–93.

Pal, S., A. Chaya, S. Yoshizawa, D.-T. Chou, D. Hong, S. Maiti, P. N. Kumta, and C. Sfeir. 2013. Finite element analysis of magnesium alloy based bone fixation devices. *Proceedings of the ASME 2013 Conference on Frontiers in Medical Devices: Applications of Computer Modeling and Simulation.*

Pan, F.-S., T.-T. Liu, X.-Y. Zhang, A.-T. Tang, and W.-Q. Wang. 2011a. Effects of scandium addition on microstructure and mechanical properties of ZK60 alloy. *Progress in Natural Science: Materials International* 21 (1):59–65.

Pan, F., M. Yang, J. Shen, and L. Wu. 2011b. Effects of minor Zr and Sr on as-cast microstructure and mechanical properties of Mg–3Ce–1.2 Mn–0.9 Sc (wt.%) magnesium alloy. *Materials Science and Engineering: A* 528 (13):4292–9.

Pan, Y. K., C. Z. Chen, D. G. Wang, and X. Yu. 2012. Microstructure and biological properties of micro-arc oxidation coatings on ZK60 magnesium alloy. *Journal of Biomedical Materials Research Part B: Applied Biomaterials* 100 (6):1574–86.

Pan, Y. K., C. Z. Chen, D. G. Wang, and Z. Q. Lin. 2013a. Preparation and bioactivity of micro-arc oxidized calcium phosphate coatings. *Materials Chemistry and Physics* 141 (2–3):842–9.

Pan, Y. K., C. Z. Chen, D. G. Wang, and T. G. Zhao. 2013b. Effects of phosphates on microstructure and bioactivity of micro-arc oxidized calcium phosphate coatings on Mg-Zn-Zr magnesium alloy. *Colloids and Surfaces B Biointerfaces* 109:1–9.

Pan, Y., S. He, D. Wang, D. Huang, T. Zheng, S. Wang, P. Dong, and C. Chen. 2015. In vitro degradation and electrochemical corrosion evaluations of microarc oxidized pure Mg,Mg-Ca and Mg-Ca-Zn alloys for biomedical applications. *Materials Science and Engineering: C* 47:85–96.

Pang, S., T. Zhang, K. Asami, and A. Inoue. 2002. Synthesis of Fe–Cr–Mo–C–B–P bulk metallic glasses with high corrosion resistance. *Acta Materialia* 50 (3):489–97.

Pant, S., G. Limbert, N. P. Curzen, and N. W. Bressloff. 2011. Multiobjective design optimisation of coronary stents. *Biomaterials* 32:7755–73.

Papirov, I. I., M. A. Tikhonovsky, K. V. Kutniy, A. I. Pikalov, S. V. Sivtsov, L. A. Pirozhenko, and V. S. Shokurov. 2008. Biodegradable magnesium alloys for medical application. *Functional Materials* 15 (1):139–43.

Pardo, A., M. C. Merino, A. E. Coy, R. Arrabal, F. Viejo, and E. Matykina. 2008. Corrosion behaviour of magnesium/aluminium alloys in 3.5 wt.% NaCl. *Corrosion Science* 50 (3):823–34.

Park, E., and D. Kim. 2005. Design of bulk metallic glasses with high glass forming ability and enhancement of plasticity in metallic glass matrix composites: A review. *Metals and Materials International* 11 (1):19–27.

Park, E. S., and D. H. Kim. 2011. Formation of Mg–Cu–Ni–Ag–Zn–Y–Gd bulk glassy alloy by casting into cone-shaped copper mold in air atmosphere. *Journal of Materials Research* 20 (6):1465–9.

Park, W. W., B. S. You, and H. R. Lee. 2002. Precipitation hardening and microstructures of rapidly solidified Mg-Zn-Ca-X alloys. *Metals and Materials International* 8 (2):135–8.

Park, E. S., W. T. Kim, and D. H. Kim. 2004. Bulk glass formation in Mg-Cu-Ag-Y-Gd alloy. *Materials Transactions* 45 (7):2474–7.

Park, E. S., J. S. Kyeong, and D. H. Kim. 2007. Enhanced glass forming ability and plasticity in Mg-based bulk metallic glasses. *Materials Science and Engineering: A* 449–51:225–9.

Park, E. S., J. Y. Lee, D. H. Kim, A. Gebert, and L. Schultz. 2008. Correlation between plasticity and fragility in Mg-based bulk metallic glasses with modulated heterogeneity. *Journal of Applied Physics* 104 (2):023520.

Park, J., M. Kim, U. Yoon, and W. Kim. 2009. Microstructures and mechanical properties of Mg-Al-Zn-Ca alloys fabricated by high frequency electromagnetic casting method. *Journal of Materials Science* 44 (1):47–54.

Park, E. S., J. Y. Lee, and D. H. Kim. 2011. Effect of Ag addition on the improvement of glass-forming ability and plasticity of Mg–Cu–Gd bulk metallic glass. *Journal of Materials Research* 20 (9):2379–85.

Park, J.-W., H.-J. Ko, J.-H. Jang, H. Kang, and J.-Y. Suh. 2012. Increased new bone formation with a surface magnesium-incorporated deproteinized porcine bone substitute in rabbit calvarial defects. *Journal of Biomedical Materials Research Part A* 100A (4):834–40.

Paul, W., and C. P. Sharma. 2006. Nanoceramic matrices: Biomedical applications. *American Journal of Biochemistry and Biotechnology* 2 (2):41–8.

Payr, E. 1901a. Zur Verwendung des Magnesiums für resorbirbare Darmknöpfe und andere chirurgisch-technische Zwecke. *Centralblatt für Chirurgie* 28 (20):513–15.

Payr, E. 1901b. Blutgefäß und Nervennaht (nebst Mittheilung über die Verwendung eines resorbirbaren Metalles in der Chirurgie). *Centralblatt für Chirurgie* 28.

Payr, E. 1902. Ueber Verwendung von Magnesium zur Behandlung von Blutgefässerkrankungen. *Langenbeck's Archives of Surgery* 63 (5):503–11.

Payr, E. 1903. Zur Technik der Behandlung kavernöser Tumoren. *Zentralbl Chir* 47:233–7.

Payr, E. 1905. Weitere Erfahrungen über die Behandlung von Blutgefäßgeschwülsten mit Magnesiumpfeilen. *Zentralbl Chir* 49:1335–40.

Payr, E., and A. Martina. 1905. Experimentelle und klinische Beiträge zur Lebernaht und leberresection (Magnesiumplattennaht). *Arch Klin Chir* 77 (4):962–98.

Pearce, A. I., R. G. Richards, S. Milz, E. Schneider, and S. G. Pearce. 2007. Animal models for implant biomaterial research in bone: A review. *European Cells and Materials* 13:1–10.

Pedroza, J. E., Y. Torrealba, A. Elias, and W. Psoter. 2007. Comparison of the compressive strength of 3 different implant design systems. *Journal of Oral Implantology* 33 (1):1–7.

Peeters, P., M. Bosiers, J. Verbist, K. Deloose, and B. Heublein. 2005. Preliminary results after application of absorbable metal stents in patients with critical limb ischemia. *Journal of Endovascular Therapy* 12 (1):1–5.

Pekguleryuz, M. O. 2000. Development of creep resistant magnesium diecasting alloys. Paper read at Materials Science Forum.

Pekguleryuz, M. O., and E. Baril. 2001. Development of creep resistant Mg-Al-Sr alloys. *Magnesium Technology* 2001:119–25.

Peng, Q., J. Wang, Y. Wu, and L. Wang. 2006. Microstructures and tensile properties of Mg–8Gd–0.6 Zr–xNd–yY (x+ y = 3, mass%) alloys. *Materials Science and Engineering: A* 433 (1):133–8.

Peng, Q., H. Dong, L. Wang, Y. Wu, and L. Wang. 2008. Microstructure and mechanical property of Mg–8.31 Gd–1.12 Dy–0.38 Zr alloy. *Materials Science and Engineering: A* 477 (1):193–7.

Peng, Q., L. Wang, Y. Wu, and L. Wang. 2009. Structure stability and strengthening mechanism of die-cast Mg–Gd–Dy based alloy. *Journal of Alloys and Compounds* 469 (1):587–92.

Peng, Q., Y. Huang, L. Zhou, N. Hort, and K. U. Kainer. 2010. Preparation and properties of high purity Mg–Y biomaterials. *Biomaterials* 31 (3):398–403.

Peng, H., S. S. Li, Y. P. Qi, and T. Y. Huang. 2011. Mg–Ni–Gd–Ag bulk metallic glass with improved glass-forming ability and mechanical properties. *Intermetallics* 19 (7):829–32.

Peng, Q. M., X. J. Li, N. Ma, R. P. Liu, and H. J. Zhang. 2012. Effects of backward extrusion on mechanical and degradation properties of Mg-Zn biomaterial. *Journal of the Mechanical Behavior of Biomedical Materials* 10:128–37.

Peng, Q., K. Li, Z. Han, E. Wang, Z. Xu, R. Liu, and Y. Tian. 2013a. Degradable magnesium-based implant materials with anti-inflammatory activity. *Journal of Biomedical Materials Research. Part A* 101 (7):1898–906.

Peng, Q., N. Ma, D. Fang, H. Li, R. Liu, and Y. Tian. 2013b. Microstructures, aging behaviour and mechanical properties in hydrogen and chloride media of backward extruded Mg–Y based biomaterials. *Journal of the Mechanical Behavior of Biomedical Materials* 17:176–85.

Peng, Q., H. Fu, J. Pang, J. Zhang, and W. Xiao. 2014a. Preparation, mechanical and degradation properties of Mg–Y-based microwire. *Journal of the Mechanical Behavior of Biomedical Materials* 29:375–84.

Peng, Q., J. Guo, H. Fu, X. Cai, Y. Wang, B. Liu, and Z. Xu. 2014b. Degradation behavior of Mg-based biomaterials containing different long-period stacking ordered phases. *Scientific Reports* 4:3620.

Penghuai, F., P. Liming, J. Haiyan, C. Jianwei, and Z. Chunquan. 2008. Effects of heat treatments on the microstructures and mechanical properties of Mg–3Nd–0.2 Zn–0.4 Zr (wt.%) alloy. *Materials Science and Engineering: A* 486 (1):183–92.

Pereda, M. D., C. Alonso, L. Burgos-Asperilla, J. A. del Valle, O. A. Ruano, P. Perez, and M. A. Fernández Lorenzo de Mele. 2010. Corrosion inhibition of powder metallurgy Mg by fluoride treatments. *Acta Biomaterialia* 6 (5):1772–82.

Pereda, M. D., C. Alonso, M. Gamero, J. A. del Valle, and M. Fernández Lorenzo de Mele. 2011. Comparative study of fluoride conversion coatings formed on biodegradable powder metallurgy Mg: The effect of chlorides at physiological level. *Materials Science and Engineering: C* 31 (5):858–65.

Pérez, P., E. Onofre, S. Cabeza, I. Llorente, J. A. del Valle, M. C. Garcia-Alonso, P. Adeva, and M. L. Escudero. 2013. Corrosion behaviour of Mg-Zn-Y-Mischmetal alloys in phosphate buffer saline solution. *Corrosion Science* 69:226–35.

Persaud, D., and A. McGoran. 2011. Biodegradable magnesium alloys: A review of material development and applications. *Journal of Biomimetics, Biomaterials and Tissue Engineering* 12:25–39.

Pfann, W. G. 1957. Zone melting. *Metallurgical Reviews* 2 (1):29–76.

Pfeiffer, F. M., D. L. Abernathie, and D. E. Smith. 2006. A comparison of pullout strength for pedicle screws of different designs: A study using tapped and untapped pilot holes. *Spine* 31 (23):E867–70.

Pfister, A., R. Landers, A. Laib, U. Hübner, R. Schmelzeisen, and R. Mülhaupt. 2004. Biofunctional rapid prototyping for tissue-engineering applications: 3D bioplotting versus 3D printing. *Journal of Polymer Science Part A: Polymer Chemistry* 42 (3): 624–38.

Poggiali, F. S. J., R. B. Figueiredo, M. T. P. Aguilar, and P. R. Cetlin. 2012. Grain refinement of commercial purity magnesium processed by ECAP (equal channel angular pressing). *Materials Research* 15 (2):312–16.

Poggiali, F. S. J., R. B. Figueiredo, M. T. P. Aguilar, and P. R. Cetlin. 2013. Effect of grain size on compression behavior of magnesium processed by equal channel angular pressing. *Journal of Materials Research and Technology* 2 (1):30–5.

Poinern, G. E. J., S. Brundavanam, and D. Fawcett. 2012. Biomedical magnesium alloys: A review of material properties, surface modifications and potential as a biodegradable orthopaedic implant. *American Journal of Biomedical Engineering* 2 (6):218–40.

Polmear, I. J. 1994. Magnesium alloys and applications. *Materials Science and Technology* 10 (1):1–16.

Polmear, I. J. 1996. Recent developments in light alloys. *JIM, Materials Transactions* 37 (1):12–31.

Post, I. J., J. K. Eibl, and G. M. Ross. 2008. Zinc induces motor neuron death via a selective inhibition of brain-derived neurotrophic factor activity. *Amyotrophic Lateral Sclerosis* 9 (3):149–55.

Prasad, Y. V. R. K., K. P. Rao, N. Hort, and K. U. Kainer. 2009. Optimum parameters and rate-controlling mechanisms for hot working of extruded Mg-3Sn-1Ca alloy. *Materials Science and Engineering A-Structural Materials Properties Microstructure and Processing* 502 (1–2):25–31.

Prasad, A., Z. Shi, and A. Atrens. 2012a. Influence of Al and Y on the ignition and flammability of Mg alloys. *Corrosion Science* 55:153–63.

Prasad, A., P. J. Uggowitzer, Z. Shi, and A. Atrens. 2012b. Production of high purity magnesium alloys by melt purification with Zr. *Advanced Engineering Materials* 14 (7):477–90.

Pravahan Salunke, M. J., F. Witte, M. Schulz, M. Pink, B. Collins, and V. Shanov. 2013. Growing and characterization of magnesium single crystal for biodegradable implant applications. *TMS Annual Meeting and Exhibition*.

Pu, Z., O. Dillon, I. Jawahir, and D. Puleo. 2010. Microstructural changes of AZ31 magnesium alloys induced by cryogenic machining and its influence on corrosion resistance in simulated body fluid for biomedical applications. Paper read at ASME 2010 International Manufacturing Science and Engineering Conference.

Pu, Z., J. C. Outeiro, A. C. Batista, O. W. Dillon, D. A. Puleo, and I. S. Jawahir. 2011a. Surface integrity in dry and cryogenic machining of AZ31B Mg alloy with varying cutting edge radius tools. *Procedia Engineering* 19:282–7.

Pu, Z., D. A. Puleo, O. W. Dillon, and I. S. Jawahir. 2011b. Controlling the biodegradation rate of magnesium-based implants through surface nanocrystallization induced by cryogenic machining. In W. H. Sillekens, S. R. Agnew, N. R. Neelameggham and S. N. Mathaudhu (eds.). *Magnesium Technology 2011*. Springer: Berlin, Heidelberg.

Pu, Z., S. Yang, G. L. Song, O. W. Dillon, D. A. Puleo, and I. S. Jawahir. 2011c. Ultrafine-grained surface layer on Mg–Al–Zn alloy produced by cryogenic burnishing for enhanced corrosion resistance. *Scripta Materialia* 65 (6):520–3.

Pu, Z., J. C. Outeiro, A. C. Batista, O. W. Dillon, D. A. Puleo, and I. S. Jawahir. 2012a. Enhanced surface integrity of AZ31B Mg alloy by cryogenic machining towards improved functional performance of machined components. *International Journal of Machine Tools and Manufacture* 56:17–27.

Pu, Z., G. L. Song, S. Yang, J. C. Outeiro, O. W. Dillon, D. A. Puleo, and I. S. Jawahir. 2012b. Grain refined and basal textured surface produced by burnishing for improved corrosion performance of AZ31B Mg alloy. *Corrosion Science* 57:192–201.

Qi, H. Y., G. X. Huang, H. Bo, G. L. Xu, L. B. Liu, and Z. P. Jin. 2011. Thermodynamic description of the Mg–Nd–Zn ternary system. *Journal of Alloys and Compounds* 509 (7):3274–81.

Qi, F. G., D. F. Zhang, X. H. Zhang, and X. X. Xu. 2014. Effect of Sn addition on the microstructure and mechanical properties of Mg-6Zn-1Mn (wt.%) alloy. *Journal of Alloys and Compounds* 585:656–66.

Qi, Z. R., Q. Zhang, L. L. Tan et al. 2014. Comparison of degradation behavior and the associated bone response of ZK60 and PLLA in vivo. *Journal of Biomedical Materials Research. Part A* 102 (5):1255–63.

Qian, M., A. Ramirez, and A. Das. 2009. Ultrasonic refinement of magnesium by cavitation: Clarifying the role of wall crystals. *Journal of Crystal Growth* 311 (14):3708–15.

Qian, L., F. Cao, J. Cui et al. 2010a. The potential cardioprotective effects of hydrogen in irradiated mice. *Journal of Radiation Research* 51 (6):741–7.

Qian, L., F. Cao, J. Cui et al. 2010b. Radioprotective effect of hydrogen in cultured cells and mice. *Free Radical Research* 44 (3):275–82.

Qian, L., B. Li, F. Cao et al. 2010c. Hydrogen-rich PBS protects cultured human cells from ionizing radiation-induced cellular damage. *Nuclear Technology and Radiation Protection* 25 (1):23–9.

Qiao, Z., Z. Shi, N. Hort, N. I. Zainal Abidin, and A. Atrens. 2012. Corrosion behaviour of a nominally high purity Mg ingot produced by permanent mould direct chill casting. *Corrosion Science* 61:185–207.

Qiao, Y., X. Wang, Z. Liu, and E. Wang. 2013a. Effect of temperature on microstructures, texture and mechanical properties of hot rolled pure Mg sheets. *Materials Science and Engineering: A* 568:202–5.

Qiao, Y., X. Wang, Z. Liu, and E. Wang. 2013b. Effects of grain size, texture and twinning on mechanical properties and work-hardening behaviors of pure Mg. *Materials Science and Engineering: A* 578:240–6.

Qiao, X. G., Y. W. Zhao, W. M. Gan, Y. Chen, M. Y. Zheng, K. Wu, N. Gao, and M. J. Starink. 2014. Hardening mechanism of commercially pure Mg processed by high pressure torsion at room temperature. *Materials Science and Engineering: A* 619:95–106.

Qin, W., J. Li, H. Kou, X. Gu, X. Xue, and L. Zhou. 2009. Effects of alloy addition on the improvement of glass forming ability and plasticity of Mg–Cu–Tb bulk metallic glass. *Intermetallics* 17 (4):253–5.

Qin, F. X., G. Q. Xie, Z. Dan, S. L. Zhu, and I. Seki. 2013. Corrosion behavior and mechanical properties of Mg-Zn-Ca amorphous alloys. *Intermetallics* 42:9–13.

Qiu, D., and M.-X. Zhang. 2009. Effect of active heterogeneous nucleation particles on the grain refining efficiency in an Mg–10wt.% Y cast alloy. *Journal of Alloys and Compounds* 488 (1):260–4.

Qiu, D., M.-X. Zhang, J. A. Taylor, and P. M. Kelly. 2009. A new approach to designing a grain refiner for Mg casting alloys and its use in Mg–Y-based alloys. *Acta Materialia* 57 (10):3052–9.

Quach, N.-C., P. J. Uggowitzer, and P. Schmutz. 2008. Corrosion behaviour of an Mg-Y-RE alloy used in biomedical applications studied by electrochemical techniques. *Comptes Rendus Chimie* 11 (9):1043–54.

Ramakrishna, S. 2004. *An Introduction to Biocomposites.* London: Imperial College Press.

Rashad, M., F. Pan, A. Tang, M. Asif, and M. Aamir. 2014. Synergetic effect of graphene nanoplatelets (GNPs) and multi-walled carbon nanotube (MW-CNTs) on mechanical properties of pure magnesium. *Journal of Alloys and Compounds* 603:111–18.

Ravi Kumar, N. V., J. J. Blandin, M. Suery, and E. Grosjean. 2003. Effect of alloying elements on the ignition resistance of magnesium alloys. *Scripta Materialia* 49 (3):225–30.

Razavi, M., M. Fathi, and M. Meratian. 2010. Microstructure, mechanical properties and bio-corrosion evaluation of biodegradable AZ91-FA nanocomposites for biomedical applications. *Materials Science and Engineering: A* 527 (26):6938–44.

Razavi, M., M. Fathi, O. Savabi, S. Mohammad Razavi, B. Hashemi Beni, D. Vashaee, and L. Tayebi. 2013. Surface modification of magnesium alloy implants by nanostructured bredigite coating. *Materials Letters* 113:174–8.

Razavi, M., M. Fathi, O. Savabi, B. H. Beni, S. M. Razavi, D. Vashaee, and L. Tayebi. 2014. Coating of biodegradable magnesium alloy bone implants using nanostructured diopside (CaMgSi2O6). *Applied Surface Science* 288:130–7.

Reifenrath, J., A. Krause, D. Bormann et al. 2010. Profound differences in the in-vivo-degradation and biocompatibility of two very similar rare-earth containing Mg-alloys in a rabbit model. Massive Unterschiede im in-vivo-Degradationsverhalten und in der Biokompatibilität zweier sehr ähnlicher Seltene-Erden enthaltender Magnesiumlegierungen im Kaninchenmodell. *Materialwissenschaft und Werkstofftechnik* 41 (12):1054–61.

Reifenrath, J., D. Bormann, and A. Meyer-Lindenberg. 2011. Magnesium alloys as promising degradable implant materials in orthopaedic research. In *Magnesium Alloys—Corrosion and Surface Treatments*, edited by F. Czerwinski: InTech. Springer: Berlin, Heidelberg.

Remennik, S., I. Bartsch, E. Willbold, F. Witte, and D. Shechtman. 2011. New, fast corroding high ductility Mg-Bi-Ca and Mg-Bi-Si alloys, with no clinically observable gas formation in bone implants. *Materials Science and Engineering B-Advanced Functional Solid-State Materials* 176 (20):1653–9.

Ren, L., X. Lin, L. Tan, and K. Yang. 2011. Effect of surface coating on antibacterial behavior of magnesium based metals. *Materials Letters* 65 (23–4):3509–11.

Ren, M., S. Cai, G. Xu, X. Ye, Y. Dou, K. Huang, and X. Wang. 2013. Influence of heat treatment on crystallization and corrosion behavior of calcium phosphate glass coated AZ31 magnesium alloy by sol-gel method. *Journal of Non-Crystalline Solids* 369: 69–75.

Ren, M., S. Cai, T. Liu, K. Huang, X. Wang, H. Zhao, S. Niu, R. Zhang, and X. Wu. 2014. Calcium phosphate glass/MgF$_2$ double layered composite coating for improving the corrosion resistance of magnesium alloy. *Journal of Alloys and Compounds* 591: 34–40.

Rengier, F., A. Mehndiratta, H. von Tengg-Kobligk, C. M. Zechmann, R. Unterhinninghofen, H.-U. Kauczor, and F. L. Giesel. 2010. 3D printing based on imaging data: Review of medical applications. *International Journal of Computer Assisted Radiology and Surgery* 5 (4):335–41.

Rettig, R., and S. Virtanen. 2008. Time-dependent electrochemical characterization of the corrosion of a magnesium rare-earth alloy in simulated body fluids. *Journal of Biomedical Materials Research Part A* 85 (1):167–75.

Rettig, R., and S. Virtanen. 2009. Composition of corrosion layers on a magnesium rare-earth alloy in simulated body fluids. *Journal of Biomedical Materials Research Part A* 88 (2):359–69.

Revel, G., J.-L. Pastol, J.-C. Rouchaud, and R. Fromageau. 1978. Purification of magnesium by vacuum distillation. *Metallurgical Transactions B* 9 (4):665–72.

Rezwan, K., Q. Z. Chen, J. J. Blaker, and A. R. Boccaccini. 2006. Biodegradable and bioactive porous polymer/inorganic composite scaffolds for bone tissue engineering. *Biomaterials* 27 (18):3413–31.

Risovany, V. D., E. E. Varlashova, and D. N. Suslov. 2000. Dysprosium titanate as an absorber material for control rods. *Journal of Nuclear Materials* 281 (1):84–9.

Risovany, V. D., A. V. Zakharov, E. M. Muraleva, V. M. Kosenkov, and R. N. Latypov. 2006. Dysprosium hafnate as absorbing material for control rods. *Journal of Nuclear Materials* 355 (1):163–70.

Robinson, D. A., R. W. Griffith, D. Shechtman, R. B. Evans, and M. G. Conzemius. 2010. In vitro antibacterial properties of magnesium metal against *Escherichia coli, Pseudomonas aeruginosa* and *Staphylococcus aureus. Acta Biomaterialia* 6 (5):1869–77.

Robson, J. D., D. T. Henry, and B. Davis. 2011. Particle effects on recrystallization in magnesium-manganese alloys: Particle pinning. *Materials Science and Engineering A-Structural Materials Properties Microstructure and Processing* 528 (12):4239–47.

Rocca, E., C. Juers, and J. Steinmetz. 2010. Corrosion behaviour of chemical conversion treatments on as-cast Mg–Al alloys: Electrochemical and non-electrochemical methods. *Corrosion Science* 52 (6):2172–8.

Rocha, L. B., G. Goissis, and M. A. Rossi. 2002. Biocompatibility of anionic collagen matrix as scaffold for bone healing. *Biomaterials* 23 (2):449–56.

Rojaee, R., M. Fathi, and K. Raeissi. 2013a. Controlling the degradation rate of AZ91 magnesium alloy via sol-gel derived nanostructured hydroxyapatite coating. *Materials Science and Engineering. C, Materials for Biological Applications* 33 (7):3817–25.

Rojaee, R., M. Fathi, and K. Raeissi. 2013b. Electrophoretic deposition of nanostructured hydroxyapatite coating on AZ91 magnesium alloy implants with different surface treatments. *Applied Surface Science* 285:664–73.

Rokhlin, L. L., and N. I. Nikitina. 1999. Mechanical properties of magnesium alloys with dysprosium. *Metal Science and Heat Treatment* 41 (6):271–3.

Rokhlin, L. L., and N. I. Nikitina. 2001. A study of recovery in aged magnesium alloys with dysprosium. *Metal Science and Heat Treatment* 43 (5):209–11.

Rokhlin, L. L., T. V. Dobatkina, I. E. Tarytina, V. N. Timofeev, and E. E. Balakhchi. 2004. Peculiarities of the phase relations in Mg-rich alloys of the Mg–Nd–Y system. *Journal of Alloys and Compounds* 367 (1–2):17–19.

Rokhlin, L. L., T. V. Dobatkina, N. I. Nikitina, and I. E. Tarytina. 2011. A study of properties of high-strength magnesium alloy of the Mg–Y–Gd–Zr system. *Metal Science and Heat Treatment* 52 (11–12):588–91.

Rosalbino, F., E. Angelini, S. De Negri, A. Saccone, and S. Delfino. 2005. Effect of erbium addition on the corrosion behaviour of Mg–Al alloys. *Intermetallics* 13 (1):55–60.

Rosalbino, F., E. Angelini, S. De Negri, A. Saccone, and S. Delfino. 2006. Electrochemical behaviour assessment of novel Mg-rich Mg–Al–RE alloys (RE = Ce, Er). *Intermetallics* 14 (12):1487–92.

Rosalbino, F., S. De Negri, A. Saccone, E. Angelini, and S. Delfino. 2010. Bio-corrosion characterization of Mg-Zn-X (X = Ca, Mn, Si) alloys for biomedical applications. *Journal of Materials Science-Materials in Medicine* 21 (4):1091–8.

Rosalbino, F., S. De Negri, G. Scavino, and A. Saccone. 2013. Microstructure and in vitro degradation performance of Mg-Zn-Mn alloys for biomedical application. *Journal of Biomedical Materials Research Part A* 101A (3):704–11.

Rosalie, J. M., H. Somekawa, A. Singh, and T. Mukai. 2013. Effect of precipitation on strength and ductility in a Mg-Zn-Y alloy. *Journal of Alloys and Compounds* 550:114–23.

Rosemann, P., J. Schmidt, and A. Heyn. 2013. Short and long term degradation behaviour of Mg-1Ca magnesium alloys and protective coatings based on plasma-chemical oxidation and biodegradable polymer coating in synthetic body fluid. *Materials and Corrosion* 64 (8):714–22.

Roy, A., S. S. Singh, M. K. Datta, B. Lee, J. Ohodnicki, and P. N. Kumta. 2011. Novel sol-gel derived calcium phosphate coatings on Mg4Y alloy. *Materials Science and Engineering: B* 176 (20):1679–89.

Rudd, A. L., C. B. Breslin, and F. Mansfeld. 2000. The corrosion protection afforded by rare earth conversion coatings applied to magnesium. *Corrosion Science* 42 (2):275–88.

Rüedi, T. P., and W. M. Murphy. 2001. *AO Principles of Fracture Management*. Davos, Switzerland: AO Publishing.

Ruggeri, R. T., and T. R. Beck. 1983. An analysis of mass transfer in filiform corrosion. *Corrosion* 39 (11):452–65.

Russo, J. 2012. The effects of laser shock peening on the residual stress and corrosion characteristics of magnesium alloy AZ91D for use as biodegradable implants, Doctoral dissertation. University of Cincinnati.

Ryspaev, T., Z. Trojanová, O. Padalka, and V. Wesling. 2008. Microstructure of superplastic QE22 and EZ33 magnesium alloys. *Materials Letters* 62 (24):4041–3.

Sabir, M. I. 2011. Biodegradable magnesium coronary stent structure design and optimization by finite element method. Doctoral thesis. *Material Science and Engineering*, Harbin Engineering University.

Sadeghi, A., and M. Pekguleryuz. 2011. Microstructural investigation and thermodynamic calculations on the precipitation of Mg-Al-Zn-Sr alloys. *Journal of Materials Research* 26 (7):896–903.

Sadeghi, A., M. Hoseini, and M. Pekguleryuz. 2011. Effect of Sr addition on texture evolution of Mg-3Al-1Zn (AZ31) alloy during extrusion. *Materials Science and Engineering A-Structural Materials Properties Microstructure and Processing* 528 (7–8):3096–104.

Saengsai, A., Y. Miyashita, and Y. Mutoh. 2009. Effects of humidity and contact material on fretting fatigue behavior of an extruded AZ61 magnesium alloy. *Tribology International* 42 (9):1346–51.

Sahin, O., O. Sulak, Y. Yavuz, E. Uz, I. Eren, H. Ramazan Yilmaz, M. A. Malas, I. Altuntas, and A. Songur. 2006. Lithium-induced lung toxicity in rats: The effect of caffeic acid phenethyl ester (CAPE). *Pathology* 38 (1):58–62.

Sahoo, M., and J. T. N. Atkinson. 1982. Magnesium-lithium-alloys—constitution and fabrication for use in batteries. *Journal of Materials Science* 17 (12):3564–74.

Saito, N., M. Mabuchi, M. Nakanishi, K. Kubota, and K. Higashi. 1997. The aging behavior and the mechanical properties of the Mg-Li-Al-Cu alloy. *Scripta Materialia* 36 (5):551–5.

Saito, K., A. Yasuhara, M. Nishijima, and K. Hiraga. 2011. Structural changes of precipitates by aging of an Mg-4 at% Dy solid solution studied by atomic-scaled transmission electron microscopy. *Keikinzoku/Journal of Japan Institute of Light Metals* 61 (5):199–205.

Saitoh, Y., H. Okayasu, L. Xiao, Y. Harata, and N. Miwa. 2008. Neutral pH hydrogen-enriched electrolyzed water achieves tumor-preferential clonal growth inhibition over normal cells and tumor invasion inhibition concurrently with intracellular oxidant repression. *Oncology Research Featuring Preclinical and Clinical Cancer Therapeutics* 17 (6):247–55.

Salahshoor, M., and Y. Guo. 2011a. Process mechanics in deep rolling of magnesium-calcium (MgCa) biomaterial. Paper read at ASME 2011 International Manufacturing Science and Engineering Conference.

Salahshoor, M., and Y. B. Guo. 2011b. Surface integrity of biodegradable orthopedic magnesium–calcium alloy by high-speed dry face milling. *Production Engineering* 5 (6):641–50.

Salahshoor, M., and Y. B. Guo. 2011c. Surface integrity of magnesium-calcium implants processed by synergistic dry cutting-finish burnishing. *Procedia Engineering* 19:288–93.

Salahshoor, M., and Y. B. Guo. 2011d. Surface integrity of biodegradable magnesium-calcium orthopedic implant by burnishing. *Journal of the Mechanical Behavior of Biomedical Materials* 4 (8):1888–904.

Salahshoor, M., and Y. B. Guo. 2011e. Cutting mechanics in high speed dry machining of biomedical magnesium–calcium alloy using internal state variable plasticity model. *International Journal of Machine Tools and Manufacture* 51 (7–8):579–90.

Salahshoor, M., and Y. B. Guo. 2013. Process mechanics in ball burnishing biomedical magnesium–calcium alloy. *The International Journal of Advanced Manufacturing Technology* 64 (1–4):133–44.

Saldaña, L., A. Méndez-Vilas, L. Jiang, M. Multigner, J. L. González-Carrasco, M. T. Pérez-Prado, M. L. González-Martín, L. Munuera, and N. Vilaboa. 2007. In vitro biocompatibility of an ultrafine grained zirconium. *Biomaterials* 28 (30):4343–54.

Salman, S. A., K. Kuroda, and M. Okido. 2013. Preparation and characterization of hydroxyapatite coating on AZ31 Mg alloy for implant applications. *Bioinorganic Chemistry and Applications* 2013:175756.

Salunke, P., V. Shanov, and F. Witte. 2011. High purity biodegradable magnesium coating for implant application. *Materials Science and Engineering: B* 176 (20):1711–17.

Samuel, A. M., H. W. Doty, S. Valtierra, and F. H. Samuel. 2014. Effect of grain refining and Sr-modification interactions on the impact toughness of Al-Si-Mg cast alloys. *Materials and Design* 56:264–73.

Sanchez, A. H. M., B. J. Luthringer, F. Feyerabend, and R. Willumeit. 2015. Mg and Mg alloys: How comparable are in vitro and in vivo corrosion rates? A review. *Acta Biomaterialia* 13:16–31.

Sandlöbes, S., S. Zaefferer, I. Schestakow, S. Yi, and R. Gonzalez-Martinez. 2011. On the role of non-basal deformation mechanisms for the ductility of Mg and Mg–Y alloys. *Acta Materialia* 59 (2):429–39.

Sandlöbes, S., M. Friák, S. Zaefferer, A. Dick, S. Yi, D. Letzig, Z. Pei, L.-F. Zhu, J. Neugebauer, and D. Raabe. 2012. The relation between ductility and stacking fault energies in Mg and Mg–Y alloys. *Acta Materialia* 60 (6):3011–21.

Sanjari, M., S. F. Farzadfar, T. Sakai et al. 2013. Microstructure and texture evolution of Mg3Zn3Ce magnesium alloys sheets and associated restoration mechanisms during annealing. *Materials Science and Engineering: A* 561:191–202.

Sankara Narayanan, S., S. Jayalakshmi, and M. Gupta. 2011. Effect of addition of mutually soluble and insoluble metallic elements on the microstructure, tensile and compressive properties of pure magnesium. *Materials Science and Engineering: A* 530:149–60.

Sankara Narayanan, T. S. N., I. S. Park, and M. H. Lee. 2014a. Strategies to improve the corrosion resistance of microarc oxidation (MAO) coated magnesium alloys for degradable implants: Prospects and challenges. *Progress in Materials Science* 60:1–71.

Sankara Narayanan, T. S. N., I. S. Park, and M. H. Lee. 2014b. Tailoring the composition of fluoride conversion coatings to achieve better corrosion protection of magnesium for biomedical applications. *Journal of Materials Chemistry B* 2014 (2):3365–82.

Sankara Narayanan, S., R. K. Sabat, S. Jayalakshmi, S. Suwas, and M. Gupta. 2014c. Effect of nanoscale boron carbide particle addition on the microstructural evolution and mechanical response of pure magnesium. *Materials and Design* 56:428–36.

Sanschagrin, A., R. Tremblay, R. Angers, and D. Dube. 1996. Mechanical properties and microstructure of new magnesium—Lithium base alloys. *Materials Science and Engineering: A* 220 (1):69–77.

Sansone, M. E. G., and D. K. Ziegler. 1985. Lithium toxicity: A review of neurologic complications. *Clinical neuropharmacology* 8 (3):242–8.

Santhanakrishnan, S., Y. H. Ho, and N. B. Dahotre. 2012. Laser coating of hydroxyapatite on Mg for enhanced physiological corrosion resistance and biodegradability. *Materials Technology: Advanced Performance Materials* 27 (4):273–7.

Saris, N. E. L., E. Mervaala, H. Karppanen, J. A. Khawaja, and A. Lewenstam. 2000. Magnesium: An update on physiological, clinical and analytical aspects. *Clinica Chimica Acta* 294 (1–2):1–26.

Sasaki, T. T., J. D. Ju, K. Hono, and K. S. Shin. 2009. Heat-treatable Mg-Sn-Zn wrought alloy. *Scripta Materialia* 61 (1):80–3.

Sasaki, T. T., K. Oh-ishi, T. Ohkubo, and K. Hono. 2011. Effect of double aging and microalloying on the age hardening behavior of a Mg-Sn-Zn alloy. *Materials Science and Engineering A-Structural Materials Properties Microstructure and Processing* 530:1–8.

Sasha, O., and Roscoe, S. G. 1999. Electrochemical studies of the adsorption behavior of bovine serum albumin on stainless steel. *Langmuir* 15 (23):8315–21.

Sato, T. 1999. Precipitation behavior of heat-resistant magnesium alloys containing rare earth elements. *Materia Japan (Japan)* 38 (4):294–7.

Sato, Y., S. Kajiyama, A. Amano et al. 2008. Hydrogen-rich pure water prevents superoxide formation in brain slices of vitamin C-depleted SMP30/GNL knockout mice. *Biochemical and Biophysical Research Communications* 375 (3):346–50.

Saunders, N. 1990. A review and thermodynamic assessment of the Al-Mg and Mg-Li Systems. *Calphad* 14 (1):61–70.

Scheideler, L., C. Fuger, C. Schille, F. Rupp, H. P. Wendel, N. Hort, H. P. Reichel, and J. Geis-Gerstorfer. 2013. Comparison of different in vitro tests for biocompatibility screening of Mg alloys. *Acta Biomaterialia* 9 (10):8740–5.

Schmutz, P., V. Guillaumin, R. Lillard, J. Lillard, and G. Frankel. 2003. Influence of dichromate ions on corrosion processes on pure magnesium. *Journal of the Electrochemical Society* 150 (4):B99–110.

Schmutz, P., N. Chang Quach-Vu, O. Guseva, P. Gunde, and P. Uggowitzer. 2007. Corrosion mechanisms and protection of WE43 Mg alloys in "biologically relevant" environments. *Meeting Abstracts*: 900.

Schranz, D., P. Zartner, I. Michel-Behnke, and H. Akintürk. 2006. Bioabsorbable metal stents for percutaneous treatment of critical recoarctation of the aorta in a newborn. *Catheterization and Cardiovascular Interventions* 67 (5):671–3.

Schroers, J., G. Kumar, T. M. Hodges, S. Chan, and T. R. Kyriakides. 2009. Bulk metallic glasses for biomedical applications. *JOM* 61 (9):21–9.

Schumacher, S., I. Roth, J. Stahl, W. Bäumer, and M. Kietzmann. 2014. Biodegradation of metallic magnesium elicits an inflammatory response in primary nasal epithelial cells. *Acta Biomaterialia* 10 (2):996–1004.

Seal, B., T. Otero, and A. Panitch. 2001. Polymeric biomaterials for tissue and organ regeneration. *Materials Science and Engineering: R: Reports* 34 (4):147–230.

Sealy, M. P., and Y. B. Guo. 2010. Surface integrity and process mechanics of laser shock peening of novel biodegradable magnesium-calcium (Mg-Ca) alloy. *Journal of the Mechanical Behavior of Biomedical Materials* 3 (7):488–96.

Sealy, M. P., and Y. B. Guo. 2011. Fabrication and characterization of surface texture for bone ingrowth by sequential laser peening biodegradable orthopedic magnesium-calcium implants. *Journal of Medical Devices-Transactions of the ASME* 5 (1):011003.

Seelig, M. 1924. A study of magnesium wire as an absorbable suture and ligature material. *Archives of Surgery* 8 (2):669–80.

Segura, T., B. C. Anderson, P. H. Chung, R. E. Webber, K. R. Shull, and L. D. Shea. 2005. Crosslinked hyaluronic acid hydrogels: A strategy to functionalize and pattern. *Biomaterials* 26 (4):359–71.

Seiler, H. G., and H. Sigel. 1988. *Handbook of Toxicity of Inorganic Compounds*. New York: Marcel Dekker.

Seitz, J. M., E. Wulf, P. Freytag, D. Bormann, and F. W. Bach. 2010. The manufacture of resorbable suture material from magnesium. *Advanced Engineering Materials* 12 (11):1099–105.

Seitz, J. M., K. Collier, E. Wulf et al. 2011a. The effect of different sterilization methods on the mechanical strength of magnesium based implant materials. *Advanced Engineering Materials* 13 (12):1146–51.

Seitz, J.-M., K. Collier, E. Wulf, D. Bormann, and F.-W. Bach. 2011b. Comparison of the corrosion behavior of coated and uncoated magnesium alloys in an in vitro corrosion environment. *Advanced Engineering Materials* 13 (9):B313–23.

Seitz, J. M., D. Utermöhlen, E. Wulf, C. Klose, and F.-W. Bach. 2011c. The manufacture of resorbable suture material from magnesium—Drawing and stranding of thin wires. *Advanced Engineering Materials* 13 (12):1087–95.

Seitz, J. M., R. Eifler, J. Stahl, M. Kietzmann, and F. W. Bach. 2012. Characterization of MgNd2 alloy for potential applications in bioresorbable implantable devices. *Acta Biomaterialia* 8 (10):3852–64.

Sepulveda, P., J. R. Jones, and L. L. Hench. 2002. Bioactive sol-gel foams for tissue repair. *Journal of Biomedical Materials Research* 59 (2):340–8.

Serre, C. M., M. Papillard, P. Chavassieux, J. C. Voegel, and G. Boivin. 1998. Influence of magnesium substitution on a collagen-apatite biomaterial on the production of a calcifying matrix by human osteoblasts. *Journal of Biomedical Materials Research* 42 (4):626–33.

Serruys, P., H. Luijten, K. Beatt et al. 1988. Incidence of restenosis after successful coronary angioplasty: A time-related phenomenon. A quantitative angiographic study in 342 consecutive patients at 1, 2, 3, and 4 months. *Circulation* 77 (2):361–71.

Serruys, P. W., J. A. Ormiston, Y. Onuma, E. Regar, N. Gonzalo, H. M. Garcia-Garcia, K. Nieman, N. Bruining, C. Dorange, and K. Miquel-Hébert. 2009. A bioabsorbable everolimus-eluting coronary stent system (ABSORB): 2-year outcomes and results from multiple imaging methods. *The Lancet* 373 (9667):897–910.

Seyfoori, A., S. Mirdamadi, A. Khavandi, and Z. S. Raufi. 2012. Biodegradation behavior of micro-arc oxidized AZ31 magnesium alloys formed in two different electrolytes. *Applied Surface Science* 261:92–100.

Seyfoori, A., S. Mirdamadi, M. Mehrjoo, and A. Khavandi. 2013a. In-vitro assessments of micro arc oxidized ceramic films on AZ31 magnesium implant: Degradation and cell-surface response. *Progress in Natural Science: Materials International* 23 (4):425–33.

Seyfoori, A., S. Mirdamadi, Z. S. Seyedraoufi, A. Khavandi, and M. Aliofkhazraei. 2013b. Synthesis of biphasic calcium phosphate containing nanostructured films by micro arc oxidation on magnesium alloy. *Materials Chemistry and Physics* 142 (1):87–94.

Sha, G., J. H. Li, W. Xu, K. Xia, W. Q. Jie, and S. P. Ringer. 2010. Hardening and microstructural reactions in high-temperature equal-channel angular pressed Mg–Nd–Gd–Zn–Zr alloy. *Materials Science and Engineering: A* 527 (20):5092–9.

Sha, G., X. Sun, T. Liu, Y. Zhu, and T. Yu. 2011. Effects of Sc addition and annealing treatment on the microstructure and mechanical properties of the as-rolled Mg-3Li alloy. *Journal of Materials Science and Technology* 27 (8):753–8.

Shadanbaz, S., J. Walker, M. P. Staiger, G. J. Dias, and A. Pietak. 2013. Growth of calcium phosphates on magnesium substrates for corrosion control in biomedical applications via immersion techniques. *Journal of Biomedical Materials Research Part B: Applied Biomaterials* 101 (1):162–72.

Shadanbaz, S., J. Walker, T. B. Woodfield, M. P. Staiger, and G. J. Dias. 2014. Monetite and brushite coated magnesium: In vivo and in vitro models for degradation analysis. *Journal of Materials Science. Materials in Medicine* 25 (1):173–83.

Shahzad, M., and L. Wagner. 2009a. Microstructure development during extrusion in a wrought Mg-Zn-Zr alloy. *Scripta Materialia* 60 (7):536–8.

Shahzad, M., and L. Wagner. 2009b. The role of Zr-rich cores in strength differential effect in an extruded Mg-Zn-Zr alloy. *Journal of Alloys and Compounds* 486 (1–2):103–8.

Shao, Y., H. Huang, T. Zhang, G. Meng, and F. Wang. 2009. Corrosion protection of Mg–5Li alloy with epoxy coatings containing polyaniline. *Corrosion Science* 51 (12):2906–15.

Sharma, A. K., R. Uma Rani, H. Bhojaraj, and H. Narayanamurthy. 1993. Galvanic black anodizing on Mg-Li alloys. *Journal of Applied Electrochemistry* 23 (5):500–7.

Shen, L., J. Wang, K. Liu et al. 2011. Hydrogen-rich saline is cerebroprotective in a rat model of deep hypothermic circulatory arrest. *Neurochemical Research* 36 (8):1501–11.

Sheng, X.-Y., K. M. Hambidge, N. F. Krebs, S. Lei, J. E. Westcott, and L. V. Miller. 2005. Dysprosium as a nonabsorbable fecal marker in studies of zinc homeostasis. *The American Journal of Clinical Nutrition* 82 (5):1017–23.

Shepelev, D., M. Bamberger, and A. Katsman. 2009. Precipitation hardening of Zr-modified Mg-Ca-Zn alloy. *Journal of Materials Science* 44 (20):5627–35.

Shi, Z., G. Song, and A. Atrens. 2006. The corrosion performance of anodised magnesium alloys. *Corrosion Science* 48 (11):3531–46.

Shi, P., W. F. Ng, M. H. Wong, and F. T. Cheng. 2009. Improvement of corrosion resistance of pure magnesium in Hanks' solution by microarc oxidation with sol–gel TiO_2 sealing. *Journal of Alloys and Compounds* 469 (1–2):286–92.

Shi, Z., M. Liu, and A. Atrens. 2010. Measurement of the corrosion rate of magnesium alloys using Tafel extrapolation. *Corrosion Science* 52 (2):579–88.

Shi, Y., M. Qi, Y. Chen, and P. Shi. 2011. MAO-DCPD composite coating on Mg alloy for degradable implant applications. *Materials Letters* 65 (14):2201–4.

Shi, B. Q., R. S. Chen, and W. Ke. 2013. Effects of Solid Solution Treatments On Microstructures And Tensile Properties Of a Mg-Y-Zn-Cu Alloy. *Materials Science Forum* 747:449–56.

Shimaya, M., T. Muneta, S. Ichinose, K. Tsuji, and I. Sekiya. 2010. Magnesium enhances adherence and cartilage formation of synovial mesenchymal stem cells through integrins. *Osteoarthritis Cartilage* 18 (10):1300–9.

Sikavitsas, V. I., J. V. D. Dolder, G. N. Bancroft, J. A. Jansen, and A. G. Mikos. 2003. Influence of the in vitro culture period on the in vivo performance of cell/titanium bone tissue-engineered constructs using a rat cranial critical size defect model. *Journal of Biomedical Materials Research Part A* 67 (3):944–51.

Silonov, V. M., E. V. Evlyukhina, and L. L. Rokhlin. 1996. Investigation of short-range order in magnesium-terbium alloy. *Russian Physics Journal* 39 (7):622–5.

Silva, D. N., M. Gerhardt de Oliveira, E. Meurer, M. I. Meurer, J. V. Lopes da Silva, and A. Santa-Bárbara. 2008. Dimensional error in selective laser sintering and 3D-printing of models for craniomaxillary anatomy reconstruction. *Journal of Cranio-Maxillofacial Surgery* 36 (8):443–9.

Silver, S., L. T. Phung, and G. Silver. 2006. Silver as biocides in burn and wound dressings and bacterial resistance to silver compounds. *Journal of Industrial Microbiology and Biotechnology* 33 (7):627–34.

Singh, A., M. Watanabe, A. Kato, and A. P. Tsai. 2004. Formation of icosahedral-hexagonal H phase nano-composites in Mg-Zn-Y alloys. *Scripta Materialia* 51 (10):955–60.

Singh, K., R. Mohan, V. Shelke et al. 2008. Superconductivity in Mg1-xMxB2 (M = Cu and Ag) system. *Indian Journal of Pure and Applied Physics* 46 (6):420–2.

Singh, S. S., A. Roy, B. Lee, and P. N. Kumta. 2011. Aqueous deposition of calcium phosphates and silicate substituted calcium phosphates on magnesium alloys. *Materials Science and Engineering: B* 176 (20):1695–702.

Slottow, T. L. P., R. Pakala, T. Okabe et al. 2008. Optical coherence tomography and intra-vascular ultrasound imaging of bioabsorbable magnesium stent degradation in porcine coronary arteries. *Cardiovascular Revascularization Medicine* 9 (4):248–54.

Slutsky, I., N. Abumaria, L.-J. Wu et al. 2010. Enhancement of learning and memory by elevating brain magnesium. *Neuron* 65 (2):165–77.

Smola, B., and I. Stulíková. 2004. Equilibrium and transient phases in Mg–Y–Nd ternary alloys. *Journal of Alloys and Compounds* 381 (1–2):L1–L2.

Smola, B., L. Joska, V. Březina, I. Stulíková, and F. Hnilica. 2012. Microstructure, corrosion resistance and cytocompatibility of Mg–5Y–4Rare Earth–0.5 Zr (WE54) alloy. *Materials Science and Engineering: C* 32 (4):659–64.

Somekawa, H., and T. Mukai. 2007. High strength and fracture toughness balance on the extruded Mg-Ca-Zn alloy. *Materials Science and Engineering A-Structural Materials Properties Microstructure and Processing* 459 (1–2):366–70.

Somekawa, H., A. Singh, and T. Mukai. 2007. Effect of precipitate shapes on fracture toughness in extruded Mg-Zn-Zr magnesium alloys. *Journal of Materials Research* 22 (4):965–73.

Somekawa, H., H. Watanabe, and T. Mukai. 2011. Damping properties in Mg–Zn–Y alloy with dispersion of quasicrystal phase particle. *Materials Letters* 65 (21):3251–3.

Son, H.-T., J.-S. Lee, D.-G. Kim, K. Yoshimi, and K. Maruyama. 2009. Effects of samarium (Sm) additions on the microstructure and mechanical properties of as-cast and hot-extruded Mg-5wt% Al-3wt% Ca-based alloys. *Journal of Alloys and Compounds* 473 (1):446–52.

Son, H.-T., D.-G. Kim, and J. S. Park. 2011. Effects of Ag addition on microstructures and mechanical properties of Mg–6Zn–2Sn–0.4Mn-based alloy system. *Materials Letters* 65 (19–20):3150–3.

Song, G. L. 2005. Recent progress in corrosion and protection of magnesium alloys. *Advanced Engineering Materials* 7 (7):563–86.

Song, G. 2007. Control of biodegradation of biocompatable magnesium alloys. *Corrosion Science* 49 (4):1696–701.

Song, G. 2011. Corrosion electrochemistry of magnesium (Mg) and its alloys. In G. Song (ed.) *Corrosion of Magnesium Alloys*. Cambridge: Elsevier.

Song, G. L., and A. Atrens. 1999. Corrosion mechanisms of magnesium alloys. *Advanced Engineering Materials* 1 (1):11–33.

Song, G. L., and A. Atrens. 2003. Understanding magnesium corrosion: A framework for improved alloy performance. *Advanced Engineering Materials* 5 (12):837–58.

Song, G. S., and M. V. Kral. 2005. Characterization of cast Mg-Li-Ca alloys. *Materials Characterization* 54 (4–5):279–86.

Song, G., and A. Atrens. 2007. Recent insights into the mechanism of magnesium corrosion and research suggestions. *Advanced Engineering Materials* 9 (3):177–83.

Song, J.-M., T.-X. Wen, and J.-Y. Wang. 2007. Vibration fracture properties of a lightweight Mg–Li–Zn alloy. *Scripta Materialia* 56 (6):529–32.

Song, Y. L., Y. H. Liu, S. R. Yu, X. Y. Zhu, and S. H. Wang. 2007a. Effect of neodymium on microstructure and corrosion resistance of AZ91 magnesium alloy. *Journal of Materials Science* 42 (12):4435–40.

Song, Y. L., Y. H. Liu, S. H. Wang, S. R. Yu, and X. Y. Zhu. 2007b. Effect of cerium addition on microstructure and corrosion resistance of die cast AZ91 magnesium alloy. *Materials and Corrosion* 58 (3):189–92.

Song, M. Y., C.-D. Yim, J.-S. Bae, D. R. Mumm, and S.-H. Hong. 2008. Preparation by gravity casting and hydrogen-storage properties of Mg–23.5 wt.% Ni–(5, 10 and 15wt.%) La. *Journal of Alloys and Compounds* 463 (1):143–7.

Song, Y., D. Shan, R. Chen, and E.-H. Han. 2009a. Corrosion characterization of Mg–8Li alloy in NaCl solution. *Corrosion Science* 51 (5):1087–94.

Song, Y., D. Shan, R. Chen, F. Zhang, and E.-H. Han. 2009b. A novel phosphate conversion film on Mg–8.8Li alloy. *Surface and Coatings Technology* 203 (9):1107–13.

Song, Y., D. Shan, R. Chen, F. Zhang, and E. H. Han. 2009c. Biodegradable behaviors of AZ31 magnesium alloy in simulated body fluid. *Materials Science and Engineering: C* 29 (3):1039–45.

Song, B., G. Huang, H. Li et al. 2010. Texture evolution and mechanical properties of AZ31B magnesium alloy sheets processed by repeated unidirectional bending. *Journal of Alloys and Compounds* 489 (2):475–81.

Song, D., A. Ma, J. Jiang, P. Lin, D. Yang, and J. Fan. 2010. Corrosion behavior of equal-channel-angular-pressed pure magnesium in NaCl aqueous solution. *Corrosion Science* 52 (2):481–90.

Song, Y., S. Zhang, J. Li, C. Zhao, and X. Zhang. 2010. Electrodeposition of Ca-P coatings on biodegradable Mg alloy: In vitro biomineralization behavior. *Acta Biomaterialia* 6 (5):1736–42.

Song, G., H. Tian, J. Liu et al. 2011. H2 inhibits TNF-α-induced lectin-like oxidized LDL receptor-1 expression by inhibiting nuclear factor κB activation in endothelial cells. *Biotechnology Letters* 33 (9):1715–22.

Song, Y., D. Shan, and E.-H. Han. 2013. A novel biodegradable nicotinic acid/calcium phosphate composite coating on Mg–3Zn alloy. *Materials Science and Engineering: C* 33 (1):78–84.

Song, M. Y., Y. J. Kwak, S. H. Lee, and H. R. Park. 2014. Comparison of hydrogen storage properties of pure Mg and milled pure Mg. *Bulletin of Materials Science* 37 (4):831–5.

Soubeyroux, J.-L., and S. Puech. 2010. Phases formation during heating of Mg–Cu–Ag–Y bulk metallic glasses. *Journal of Alloys and Compounds* 495 (2):330–3.

Soubeyroux, J. L., S. Puech, and J. J. Blandin. 2009. Effect of silver on the glass forming ability of MgCuGdY bulk metallic glasses. *Journal of Alloys and Compounds* 483 (1–2):107–11.

Spoerke, E. D., N. G. Murray, H. Li, L. C. Brinson, D. C. Dunand, and S. I. Stupp. 2005. A bioactive titanium foam scaffold for bone repair. *Acta Biomaterialia* 1 (5):523–33.

Sreekanth, D., and N. Rameshbabu. 2012. Development and characterization of MgO/hydroxyapatite composite coating on AZ31 magnesium alloy by plasma electrolytic oxidation coupled with electrophoretic deposition. *Materials Letters* 68:439–42.

Srinivasan, A., U. T. S. Pillai, J. Swaminathan, S. K. Das, and B. C. Pai. 2006. Observations of microstructural refinement in Mg-Al-Si alloys containing strontium. *Journal of Materials Science* 41 (18):6087–9.

Srinivasan, P. B., J. Liang, C. Blawert, M. Störmer, and W. Dietzel. 2010. Characterization of calcium containing plasma electrolytic oxidation coatings on AM50 magnesium alloy. *Applied Surface Science* 256 (12):4017–22.

Sripanyakorn, S., R. Jugdaohsingh, H. Elliott et al. 2007. The silicon content of beer and its bioavailability in healthy volunteers. *British Journal of Nutrition* 91 (3):403–9.

Staiger, M. P., A. M. Pietak, J. Huadmai, and G. Dias. 2006. Magnesium and its alloys as orthopedic biomaterials: A review. *Biomaterials* 27 (9):1728–34.

Staiger, M. P., I. Kolbeinsson, N. T. Kirkland, T. Nguyen, G. Dias, and T. B. Woodfield. 2010. Synthesis of topologically ordered open-cell porous magnesium. *Materials Letters* 64 (23):2572–4.

Staiger, M. P., F. Feyerabend, R. Willumeit, C. S. Sfeir, Y. F. Zheng, S. Virtanen, W. D. Müeller, A. Atrens, M. Peuster, P. N. Kumta, D. Mantovani, and F. Witte. 2011. Summary of the panel discussions at the 2nd Symposium on Biodegradable Metals, Maratea, Italy, 2010. *Materials Science and Engineering: B* 176 (20):1596–9.

Stalmann, A., W. Sebastian, H. Friedrich, S. Schumann, and K. Dröder. 2001. Properties and processing of magnesium wrought products for automotive applications. *Advanced Engineering Materials* 3 (12):969–74.

Stampella, R., R. Procter, and V. Ashworth. 1984. Environmentally induced cracking of magnesium. *Corrosion Science* 24 (4):325–41.

Stanford, N. 2010. The effect of calcium on the texture, microstructure and mechanical properties of extruded Mg-Mn-Ca alloys. *Materials Science and Engineering A-Structural Materials Properties Microstructure and Processing* 528 (1):314–22.

Stepankin, V. 1995. Magnetically aligned polycrystalline dysprosium as ultimate saturation ferromagnet for high magnetic field polepieces. *Physica B: Condensed Matter* 211 (1):345–7.

Stipanuk, M. H., and M. A. Caudill. 2013. *Biochemical, Physiological, and Molecular Aspects of Human Nutrition.* Elsevier Health Sciences: Oxford.

Stippich, F., E. Vera, G. K. Wolf, G. Berg, and C. Friedrich. 1998. Enhanced corrosion protection of magnesium oxide coatings on magnesium deposited by ion beam-assisted evaporation. *Surface and Coatings Technology* 104:29–35.

Stloukal, I., and J. Čermák. 2009. Tracer diffusion of 65Zn, 110mAg and 88Y in QE22 metal matrix composite. *Journal of Alloys and Compounds* 471 (1–2):83–9.

Stoeckel, D., C. Bonsignore, and S. Duda. 2002. A survey of stent designs. *Minimally Invasive Therapy and Allied Technologies* 11 (4):137–47.

Stoeckel, D., A. Pelton, and T. Duerig. 2004. Self-expanding nitinol stents: Material and design considerations. *European Radiology* 14:292–301.

Stone, P., and J. Lord Jr. 1951. An experimental study of the thrombogenic properties of magnesium and magnesium-aluminum wire in the dog's aorta. *Surgery* 30 (6):987–93.

Stone, G. W., J. W. Moses, S. G. Ellis et al. 2007. Safety and efficacy of sirolimus- and paclitaxel-eluting coronary stents. *New England Journal of Medicine* 356 (10):998–1008.

Su, S. F., H. K. Lin, J. C. Huang, and N. J. Ho. 2002. Electron-beam welding behavior in Mg-Al-based alloys. *Metallurgical and Materials Transactions A* 33 (5):1461–73.

Su, G.-H., L. Zhang, L.-R. Cheng, Y.-B. Liu, and Z.-Y. Cao. 2010. Microstructure and mechanical properties of Mg-6Al-0.3Mn-xY alloys prepared by casting and hot rolling. *Transactions of Nonferrous Metals Society of China* 20 (3):383–9.

Su, P., X. Wu, Z. Jiang, and Y. Guo. 2011. Effects of working frequency on the structure and corrosion resistance of plasma electrolytic oxidation coatings formed on a ZK60 Mg alloy. *International Journal of Applied Ceramic Technology* 8 (1):112–19.

Su, X., D. J. Li, Y. C. Xie, X. Q. Zeng, and W. J. Ding. 2013. Effect of Sm on the microstructure and mechanical property of Mg-xSm-0.4 Zn-0.3 Zr alloys. *Materials Science Forum* 747:238–44.

Subba Rao, R. V., U. Wolff, S. Baunack et al. 2003. Corrosion behaviour of the amorphous Mg65Y10Cu15Ag10 alloy. *Corrosion Science* 45 (4):817–32.

Sudholz, A. D., K. Gusieva, X. B. Chen, B. C. Muddle, M. A. Gibson, and N. Birbilis. 2011. Electrochemical behaviour and corrosion of Mg–Y alloys. *Corrosion Science* 53 (6):2277–82.

Suganthi, R. V., K. Elayaraja, M. I. A. Joshy, V. S. Chandra, E. K. Girija, and S. N. Kalkura. 2011. Fibrous growth of strontium substituted hydroxyapatite and its drug release. *Materials Science and Engineering C-Materials for Biological Applications* 31 (3):593–9.

Sugiyama, N., H. Xu, T. Onoki, Y. Hoshikawa, T. Watanabe, N. Matsushita, X. Wang, F. Qin, M. Fukuhara, and M. Tsukamoto. 2009. Bioactive titanate nanomesh layer on the Ti-based bulk metallic glass by hydrothermal–electrochemical technique. *Acta Biomaterialia* 5 (4):1367–73.

Sun, Q., Z. Kang, J. Cai et al. 2009. Hydrogen-rich saline protects myocardium against ischemia/reperfusion injury in rats. *Experimental Biology and Medicine* 234 (10):1212–19.

Sun, Y., H. F. Zhang, H. M. Fu, A. M. Wang, and Z. Q. Hu. 2009. Mg–Cu–Ag–Er bulk metallic glasses with high glass forming ability and compressive strength. *Materials Science and Engineering: A* 502 (1–2):148–52.

Sun, H., L. Chen, W. Zhou et al. 2011a. The protective role of hydrogen-rich saline in experimental liver injury in mice. *Journal of Hepatology* 54 (3):471–80.

Sun, H. F., C. J. Li, and W. B. Fang. 2011. Corrosion behavior of extrusion-drawn pure Mg wire immersed in simulated body fluid. *Transactions of Nonferrous Metals Society of China* 21:s258–61.

Sun, Q., J. Cai, J. Zhou et al. 2011a. Hydrogen-rich saline reduces delayed neurologic sequelae in experimental carbon monoxide toxicity. *Critical Care Medicine* 39 (4):765–9.

Sun, Q., J. Cai, S. Liu et al. 2011b. Hydrogen-rich saline provides protection against hyperoxic lung injury. *Journal of Surgical Research* 165 (1):e43–9.

Sun, H.-F., C.-J. Li, Y. Xie, and W.-B. Fang. 2012. Microstructures and mechanical properties of pure magnesium bars by high ratio extrusion and its subsequent annealing treatment. *Transactions of Nonferrous Metals Society of China* 22:s445–9.

Sun, Y., B. P. Zhang, Y. Wang, L. Geng, and X. H. Jiao. 2012. Preparation and characterization of a new biomedical Mg-Zn-Ca alloy. *Materials and Design* 34:58–64.

Sun, F., C. Shi, K. Y. Rhee, and N. Zhao. 2013. In situ synthesis of CNTs in Mg powder at low temperature for fabricating reinforced Mg composites. *Journal of Alloys and Compounds* 551:496–501.

Suuronen, R., T. Pohjonen, L. Tech, J. Vasenius, and S. Vainionpaa. 1992. Comparison of absorbable self-reinforced multilayer poly-l-lactide and metallic plates for the fixation of mandibular body osteotomies: An experimental study in sheep. *Journal of Oral and Maxillofacial Surgery* 50:255–62.

Suzuki, M., H. Sato, K. Maruyama, and H. Oikawa. 1998. Creep behavior and deformation microstructures of Mg–Y alloys at 550 K. *Materials Science and Engineering: A* 252 (2):248–55.

Suzuki, M., T. Kimura, J. Koike, and K. Maruyama. 2004. Effects of zinc on creep strength and deformation substructures in Mg–Y alloy. *Materials Science and Engineering: A* 387:706–9.

Suzuki, A., N. D. Saddock, L. Riester, E. Lara-Curzio, J. W. Jones, and T. M. Pollock. 2007. Effect of Sr additions on the microstructure and strength of a Mg-Al-Ca ternary alloy. *Metallurgical and Materials Transactions A-Physical Metallurgy and Materials Science* 38A (2):420–7.

Taboas, J., R. Maddox, P. Krebsbach, and S. Hollister. 2003. Indirect solid free form fabrication of local and global porous, biomimetic and composite 3D polymer-ceramic scaffolds. *Biomaterials* 24 (1):181–94.

Takenaka, T., T. Ono, Y. Narazaki, Y. Naka, and M. Kawakami. 2007. Improvement of corrosion resistance of magnesium metal by rare earth elements. *Electrochimica Acta* 53 (1):117–21.

Takuda, H., T. Enami, K. Kubota, and N. Hatta. 2000. The formability of a thin sheet of Mg–8.5 Li–1Zn alloy. *Journal of Materials Processing Technology* 101 (1):281–6.

Takuda, H., H. Matsusaka, S. Kikuchi, and K. Kubota. 2002. Tensile properties of a few Mg-Li-Zn alloy thin sheets. *Journal of Materials Science* 37 (1):51–7.

Tamura, Y., S. Kawamoto, H. Soda, and A. McLean. 2011. Effects of lanthanum and zirconium on cast structure and room temperature mechanical properties of Mg-La-Zr alloys. *Materials Transactions* 52 (9):1777–86.

Tan, L., M. Gong, F. Zheng, B. Zhang, and K. Yang. 2009. Study on compression behavior of porous magnesium used as bone tissue engineering scaffolds. *Biomedical Materials* 4 (1):015016.

Tan, L., Q. Wang, X. Lin, P. Wan, G. Zhang, Q. Zhang, and K. Yang. 2014. Loss of mechanical properties in vivo and bone-implant interface strength of AZ31B magnesium alloy screws with Si-containing coating. *Acta Biomaterialia* 10 (5):2333–40.

Tanaka, K. 2008. Hydride stability and hydrogen desorption characteristics in melt-spun and nanocrystallized Mg–Ni–La alloy. *Journal of Alloys and Compounds* 450 (1):432–9.

Tang, H., and F. Wang. 2013. Synthesis and properties of $CaTiO_3$-containing coating on AZ31 magnesium alloy by micro-arc oxidation. *Materials Letters* 93:427–30.

Tang, H., T. Z. Xin, Y. Luo, and F. P. Wang. 2013a. In vitro degradation of AZ31 magnesium alloy coated with hydroxyapatite by sol–gel method. *Materials Science and Technology* 29 (5):547–52.

Tang, H., D. Yu, Y. Luo, and F. Wang. 2013b. Preparation and characterization of HA microflowers coating on AZ31 magnesium alloy by micro-arc oxidation and a solution treatment. *Applied Surface Science* 264:816–22.

Tapiero, H., and K. D. Tew. 2003. Trace elements in human physiology and pathology: Zinc and metallothioneins. *Biomedicine and Pharmacotherapy* 57 (9):399–411.

Tarasov, B. P., P. V. Fursikov, D. N. Borisov, M. V. Lototsky, V. A. Yartys, and A. Schrøder Pedersen. 2007. Metallography and hydrogenation behaviour of the alloy Mg-72mass%– Ni-20mass%–La-8mass%. *Journal of Alloys and Compounds* 446:183–7.

Tathgar, H. S. 2001. Solubility of nickel in Mg-Al, Mg-Al-Fe, and Mg-Al-Mn systems, Fakultet for naturvitenskap og teknologi.

Tathgar, H. S., P. Bakke, and T. A. Engh. 2006. Impurities in magnesium and magnesium based alloys and their removal. *Magnesium Alloys and Their Applications*: 767–79.

Taura, A., Y. S. Kikkawa, T. Nakagawa, and J. Ito. 2010. Hydrogen protects vestibular hair cells from free radicals. *Acta Oto-Laryngologica* 130 (S563):95–100.

Taylor, A. 1985. Therapeutic uses of trace elements. *Clinics in Endocrinology and Metabolism* 14 (3):703–24.

Teoh, S. 2000. Fatigue of biomaterials: A review. *International Journal of Fatigue* 22 (10):825–37.

Terasaki, Y., I. Ohsawa, M. Terasaki et al. 2011. Hydrogen therapy attenuates irradiation-induced lung damage by reducing oxidative stress. *American Journal of Physiology-Lung Cellular and Molecular Physiology* 301 (4):L415–26.

Thomann, M., C. Krause, D. Bormann et al. 2009. Comparison of the resorbable magnesium. alloys LAE442 und MgCa0.8 concerning their mechanical properties, their progress of degradation and the bone-implant-contact after 12 months implantation duration in a rabbit model. *Materialwissenschaft und Werkstofftechnik* 40 (1–2):82–7.

Thomann, M., C. Krause, N. Angrisani et al. 2010. Influence of a magnesium-fluoride coating of magnesium-based implants (MgCa0.8) on degradation in a rabbit model. *Journal of Biomedical Materials Research Part A* 93A (4):1609–19.

Tian, W., S. Pang, H. Men, C. Ma, and T. Zhang. 2007. Mg-Cu-Zn-Y-Zr bulk metallic glassy composite with high strength and plasticity. *Journal of University of Science and Technology Beijing, Mineral, Metallurgy, Material* 14:43–5.

Tian, M., F. Chen, W. Song, Y. C. Song, Y. W. Chen, C. X. Wan, X. X. Yu, and X. H. Zhang. 2009. In vivo study of porous strontium-doped calcium polyphosphate scaffolds for bone substitute applications. *Journal of Materials Science-Materials in Medicine* 20 (7):1505–12.

Tie, D., F. Feyerabend, W. D. Mueller et al. 2012. Antibacterial biodegradable Mg-Ag alloys. *European Cells and Materials* 25:284–98.

Tie, D., F. Feyerabend, N. Hort, D. Hoeche, K. U. Kainer, R. Willumeit, and W. D. Mueller. 2013. In vitro mechanical and corrosion properties of biodegradable Mg-Ag alloys. *Materials and Corrosion* 65 (6):569–76.

Tilocca, A. 2010. Models of structure, dynamics and reactivity of bioglasses: A review. *Journal of Materials Chemistry* 20 (33):6848–58.

Tomozawa, M., and S. Hiromoto. 2011. Microstructure of hydroxyapatite- and octacalcium phosphate-coatings formed on magnesium by a hydrothermal treatment at various pH values. *Acta Materialia* 59 (1):355–63.

Tong, L. B., M. Y. Zheng, S. W. Xu, X. S. Hu, K. Wu, S. Kamado, G. J. Wang, and X. Y. Lv. 2010. Room-temperature compressive deformation behavior of Mg-Zn-Ca alloy processed by equal channel angular pressing. *Materials Science and Engineering A-Structural Materials Properties Microstructure and Processing* 528 (2):672–9.

Tong, L. B., X. H. Li, and H. J. Zhang. 2013. Effect of long period stacking ordered phase on the microstructure, texture and mechanical properties of extruded Mg–Y–Zn alloy. *Materials Science and Engineering: A* 563:177–83.

Trantolo, D. J., S. T. Sonis, B. Thompson, D. L. Wise, K.-U. Lewandrowski, and D. D. Hile. 2003. Evaluation of a porous, biodegradable biopolymer scaffold for mandibular reconstruction. *The International Journal of Oral and Maxillofacial Implants* 18 (2):182–8.

Tritschler, B., B. Forest, and J. Rieu. 1999. Fretting corrosion of materials for orthopaedic implants: A study of a metal/polymer contact in an artificial physiological medium. *Tribology International* 32 (10):587–96.

Trojanová, Z., Z. Drozd, P. Lukáč, and F. Chmelík. 2005. Deformation behaviour of Mg–Li alloys at elevated temperatures. *Materials Science and Engineering: A* 410–1:148–51.

Trojanová, Z., B. Weidenfeller, and W. Riehemann. 2006. Thermal stresses in Mg–Ag–Nd alloy reinforced by short Saffil fibers studied by internal friction. *Materials Science and Engineering: A* 442 (1–2):480–3.

Trojanová, Z., Z. Drozd, S. Kudela, Z. Szaraz, and P. Lukac. 2007. Strengthening in Mg–Li matrix composites. *Composites Science and Technology* 67 (9):1965–73.

Tsai, A. P., Y. Murakami, and A. Niikura. 2000. The Zn-Mg-Y phase diagram involving quasi-crystals. *Philosophical Magazine A-Physics of Condensed Matter Structure Defects and Mechanical Properties* 80 (5):1043–54.

Tsubakino, H., A. Yamamoto, S. Fukumoto, A. Watanabe, K. Sugahara, and H. Inoue. 2003. High-purity magnesium coating on magnesium alloys by vapor deposition technique for improving corrosion resistance. *Materials Transactions* 44 (4):504–10.

Tsujikawa, M., S.-I. Adachi, Y. Abe, S. Oki, K. Nakata, and M. Kamita. 2007. Corrosion protection of Mg-Li alloy by plasma thermal spraying of aluminum. *Plasma Processes and Polymers* 4 (S1):S593–6.

Tunold, R., H. Holtan, M.-B. H. Berge, A. Lasson, and R. Steen-Hansen. 1977. The corrosion of magnesium in aqueous solution containing chloride ions. *Corrosion Science* 17 (4):353–65.

Turhan, M. C., M. Weiser, M. S. Killian, B. Leitner, and S. Virtanen. 2011. Electrochemical polymerization and characterization of polypyrrole on Mg–Al alloy (AZ91D). *Synthetic Metals* 161 (3–4):360–4.

Turner, A.S. 2001. Animal models of osteoporosis—Necessity and limitations. *European Cells and Materials* 1:66–81.

Tziampazis, E., J. Kohn, and P. V. Moghe. 2000. PEG-variant biomaterials as selectively adhesive protein templates: Model surfaces for controlled cell adhesion and migration. *Biomaterials* 21 (5):511–20.

Uhrin, R., R. F. Belt, and V. Rosati. 1977. Preparation and crystal growth of lithium yttrium fluoride for laser applications. *Journal of Crystal Growth* 38 (1):38–44.

Ullmann, B., J. Reifenrath, D. Dziuba et al. 2011. In vivo degradation behavior of the magnesium alloy LANd442 in rabbit tibiae. *Materials* 4 (12):2197–218.

U. S. Department of Labor, Occupational Safety and Health Administration (OSHA) 2006. *Best Practices for the Safe Use of Glutaraldehyde in Health Care.* Washington, DC: OSHA. Available from https://www.osha.gov/Publications/3258-08N-2006-English.html.

van den Dolder, J., E. Farber, P. H. Spauwen, and J. A. Jansen. 2003. Bone tissue reconstruction using titanium fiber mesh combined with rat bone marrow stromal cells. *Biomaterials* 24 (10):1745–50.

Van der Stok, J., E. M. Van Lieshout, Y. El-Massoudi, G. H. Van Kralingen, and P. Patka. 2011. Bone substitutes in the Netherlands–A systematic literature review. *Acta Biomaterialia* 7 (2):739–50.

Verbrugge, J. 1933. La tolérance du tissu osseux vis-à-vis du magnésium métallique. *Presse Méd* 55:1112–14.

Verbrugge, J. 1934. *Le matériel métallique résorbable en chirurgie osseuse, par Jean Verbrugge.* Paris: Masson.

Verbrugge, J. 1937. L'utilisation du magnésium dans le traitement chirurgical des fractures. *Bull Mém Soc Nat Cir* 59 (59):813–23.

Verheyen, C. C., W. J. Dhert, L. H. Braak, and K. de Groot. 1991. Push-out test evaluated by finite element analysis. *Transactions of the Society for Biomaterials* 17:216.

Vojtech, D., J. Kubasek, J. Serak, and P. Novak. 2011. Mechanical and corrosion properties of newly developed biodegradable Zn-based alloys for bone fixation. *Acta Biomaterialia* 7 (9):3515–22.

Von Der Höh, N., D. Bormann, A. Lucas et al. 2009a. Influence of different surface machining treatments of magnesium-based resorbable implants on the degradation behavior in rabbits. *Advanced Engineering Materials* 11 (5):B47–54.

Von Der Höh, N., B. von Rechenberg, D. Bormann, A. Lucas, and A. Meyer-Lindenberg. 2009b. Influence of different surface machining treatments of resorbable magnesium alloy implants on degradation—EDX-analysis and histology results. *Materialwissenschaft und Werkstofftechnik* 40 (1–2):88–93.

Vormann, J. 2003. Magnesium: Nutrition and metabolism. *Molecular Aspects of Medicine* 24 (1):27–37.

Wada, T., F. Qin, X. Wang, M. Yoshimura, A. Inoue, N. Sugiyama, R. Ito, and N. Matsushita. 2009. Formation and bioactivation of Zr-Al-Co bulk metallic glasses. *Journal of Materials Research* 24 (9):2941–8.

Wagoner Johnson, A. J., and B. A. Herschler. 2011. A review of the mechanical behavior of CaP and CaP/polymer composites for applications in bone replacement and repair. *Acta Biomaterialia* 7 (1):16–30.

Waizy, H., J.-M. Seitz, J. Reifenrath, A. Weizbauer, F.-W. Bach, A. Meyer-Lindenberg, B. Denkena, and H. Windhagen. 2012. Biodegradable magnesium implants for orthopedic applications. *Journal of Materials Science* 48 (1):39–50.

Waizy, H., J. Diekmann, A. Weizbauer et al. 2014. In vivo study of a biodegradable orthopedic screw (MgYREZr-alloy) in a rabbit model for up to 12 months. *Journal of Biomaterials Applications* 28 (5):667–75.

Waksman, R. 2007. Promise and challenges of bioabsorbable stents. *Catheterization and Cardiovascular Interventions* 70:407–14.

Waksman, R., R. Pakala, P. K. Kuchulakanti et al. 2006. Safety and efficacy of bioabsorbable magnesium alloy stents in porcine coronary arteries. *Catheterization and Cardiovascular Interventions* 68 (4):607–17; discussion 618–19.

Waksman, R. O. N., R. Pakala, T. Okabe et al. 2007. Efficacy and safety of absorbable metallic stents with adjunct intracoronary beta radiation in porcine coronary arteries. *Journal of Interventional Cardiology* 20 (5):367–72.

Waksman, R., R. Erbel, C. Di Mario, J. Bartunek, B. de Bruyne, F. R. Eberli, P. Erne, M. Haude, M. Horrigan, and C. Ilsley. 2009. Early-and long-term intravascular ultrasound and angiographic findings after bioabsorbable magnesium stent implantation in human coronary arteries. *JACC: Cardiovascular Interventions* 2 (4):312–20.

Walker, J., S. Shadanbaz, N. T. Kirkland, E. Stace, T. Woodfield, M. P. Staiger, and G. J. Dias. 2012. Magnesium alloys: Predicting in vivo corrosion with in vitro immersion testing. *Journal of Biomedical Materials Research Part B: Applied Biomaterials* 100 (4):1134–41.

Walker, J., S. Shadanbaz, T. B. Woodfield, M. P. Staiger, and G. J. Dias. 2014. Magnesium biomaterials for orthopedic application: A review from a biological perspective. *Journal of Biomedical Materials Research Part B: Applied Biomaterials* 102 (6):1316–31.

Walter, R., and M. B. Kannan. 2011a. Influence of surface roughness on the corrosion behaviour of magnesium alloy. *Materials and Design* 32 (4):2350–4.

Walter, R., and M. B. Kannan. 2011b. In-vitro degradation behaviour of WE54 magnesium alloy in simulated body fluid. *Materials Letters* 65 (4):748–50.

Walter, R., M. B. Kannan, Y. He, and A. Sandham. 2013. Effect of surface roughness on the in vitro degradation behaviour of a biodegradable magnesium-based alloy. *Applied Surface Science* 279:343–8.

Wan, D. Q., J. C. Wang, L. Lin, Z. G. Feng, and G. C. Yang. 2008. Damping properties of Mg-Ca binary alloys. *Physica B-Condensed Matter* 403 (13–16):2438–42.

Wan, Y. Z., G. Y. Xiong, H. L. Luo, F. He, Y. Huang, and Y. L. Wang. 2008a. Influence of zinc ion implantation on surface nanomechanical performance and corrosion resistance of biomedical magnesium–calcium alloys. *Applied Surface Science* 254 (17):5514–16.

Wan, Y. Z., G. Y. Xiong, H. L. Luo, F. He, Y. Huang, and X. S. Zhou. 2008b. Preparation and characterization of a new biomedical magnesium-calcium alloy. *Materials and Design* 29 (10):2034–7.

Wan, D., J. Wang, and G. Yang. 2009. A study of the effect of Y on the mechanical properties, damping properties of high damping Mg–0.6% Zr based alloys. *Materials Science and Engineering: A* 517 (1):114–17.

Wang, J.-Y. 2009. Mechanical properties of room temperature rolled MgLiAlZn alloy. *Journal of Alloys and Compounds* 485 (1–2):241–4.

Wang, S. C., and C. P. Chou. 2008. Effect of adding Sc and Zr on grain refinement and ductility of AZ31 magnesium alloy. *Journal of Materials Processing Technology* 197 (1):116–21.

Wang, H., and Z. Shi. 2011. In vitro biodegradation behavior of magnesium and magnesium alloy. *Journal of Biomedical Materials Research Part B: Applied Biomaterials* 98 (2):203–9.

Wang, X., and C. Wen. 2014. Corrosion protection of mesoporous bioactive glass coating on biodegradable magnesium. *Applied Surface Science* 303:196–204.

Wang, Y., G. H. McArdle, S. A. Feig, S. Karasick, H. A. Koolpe, E. Mapp, V. M. Rao, R. M. Steiner, and R. J. Wechsler. 1984. Clinical applications of yttrium filters for exposure reduction. *Radiographics* 4 (3):479–505.

Wang, X., J. D. Mabrey, and C. M. Agrawal. 1998. An interspecies comparison of bone fracture properties. *Bio-medical Materials and Engineering* 8 (1):1–9.

Wang, L., Y. H. Tang, Y. Wang et al. 2002. The hydrogenation properties of Mg1.8Ag0.2Ni alloy. *Journal of Alloys and Compounds* 336 (1):297–300.

Wang, S. J., G. Q. Wu, R. H. Li, G. X. Luo, and Z. Huang. 2006. Microstructures and mechanical properties of 5 wt.% Al2Yp/Mg–Li composite. *Materials Letters* 60 (15):1863–5.

Wang, W.-Q., D.-K. Liang, D.-Z. Yang, and M. Qi. 2006. Analysis of the transient expansion behavior and design optimization of coronary stents by finite element method. *Journal of Biomechanics* 39 (1):21–32.

Wang, X. M., X. Q. Zeng, G. S. Wu, and S. S. Yao. 2006. Yttrium ion implantation on the surface properties of magnesium. *Applied Surface Science* 253 (5):2437–42.

Wang, H., Y. Estrin, H. Fu, G. Song, and Z. Zúberová. 2007. The effect of pre-processing and grain structure on the bio-corrosion and fatigue resistance of magnesium alloy AZ31. *Advanced Engineering Materials* 9 (11):967–72.

Wang, J., D. P. Zhang, D. Q. Fang, H. Y. Lu, D. X. Tang, J. H. Zhang, and J. Meng. 2008a. Effect of Y for enhanced age hardening response and mechanical properties of Mg–Ho–Y–Zr alloys. *Journal of Alloys and Compounds* 454 (1):194–200.

Wang, J.-F., X. Wu, F.-S. Pan, A.-T. Tang, P.-D. Ding, and R.-P. Liu. 2008b. Microstructure and mechanical properties of Mg-Cu-Y-Zn bulk metallic glass matrix composites prepared in low vacuum. *Transactions of Nonferrous Metals Society of China* 18:s278–82.

Wang, J., J. Yang, Y. Wu, H. Zhang, and L. Wang. 2008c. Microstructures and mechanical properties of as-cast Mg–5Al–0.4Mn–xNd (x = 0, 1, 2 and 4) alloys. *Materials Science and Engineering: A* 472 (1–2):332–7.

Wang, X. M., X. Q. Zeng, Y. Zhou, G. S. Wu, S. S. Yao, and Y. J. Lai. 2008. Early oxidation behaviors of Mg–Y alloys at high temperatures. *Journal of Alloys and Compounds* 460 (1):368–74.

Wang, Q., D. Li, J. J. Blandin, M. Suery, P. Donnadieu, and W. Ding. 2009. Microstructure and creep behavior of the extruded Mg–4Y–4Sm–0.5 Zr alloy. *Materials Science and Engineering: A* 516 (1):189–92.

Wang, Y., R. Mori, N. Ozoe, T. Nakai, and Y. Uchio. 2009. Proximal half angle of the screw thread is a critical design variable affecting the pull-out strength of cancellous bone screws. *Clinical Biomechanics* 24:781–5.

Wang, H., R. Akid, and M. Gobara. 2010a. Scratch-resistant anticorrosion sol-gel coating for the protection of AZ31 magnesium alloy via a low temperature sol-gel route. *Corrosion Science* 52 (8):2565–70.

Wang, H. X., S. K. Guan, X. Wang, C. X. Ren, and L. G. Wang. 2010b. In vitro degradation and mechanical integrity of Mg-Zn-Ca alloy coated with Ca-deficient hydroxyapatite by the pulse electrodeposition process. *Acta Biomaterialia* 6 (5):1743–8.

Wang, J., L. Wang, S. Guan, S. Zhu, C. Ren, and S. Hou. 2010. Microstructure and corrosion properties of as sub-rapid solidification Mg-Zn-Y-Nd alloy in dynamic simulated body fluid for vascular stent application. *Journal of Materials Science: Materials in Medicine* 21 (7):2001–8.

Wang, L., C. Xing, X. Hou, Y. Wu, J. Sun, and L. Wang. 2010. Microstructures and mechanical properties of as-cast Mg–5Y–3Nd–Zr–xGd (x = 0, 2 and 4 wt.%) alloys. *Materials Science and Engineering: A* 527 (7):1891–5.

Wang, Q., J. Chen, Z. Zhao, and S. He. 2010a. Microstructure and super high strength of cast Mg-8.5Gd-2.3Y-1.8Ag-0.4Zr alloy. *Materials Science and Engineering: A* 528 (1):323–8.

Wang, Q., Y. Gao, D.-D. Yin, and C.-J. Chen. 2010b. Characterization of phases in Mg-10Y-5Gd-2Zn-0.5 Zr alloy processed by heat treatment. *Transactions of Nonferrous Metals Society of China* 20 (11):2076–80.

Wang, C., J. Li, Q. Liu et al. 2011. Hydrogen-rich saline reduces oxidative stress and inflammation by inhibit of JNK and NF-κB activation in a rat model of amyloid-beta-induced Alzheimer's disease. *Neuroscience Letters* 491 (2):127–32.

Wang, D. W., Y. Cao, H. Qiu, and Z. G. Bi. 2011. Improved blood compatibility of Mg-1.0Zn-1.0Ca alloy by micro-arc oxidation. *Journal of Biomedical Materials Research Part A* 99A (2):166–72.

Wang, F., G. Yu, S.-Y. Liu et al. 2011. Hydrogen-rich saline protects against renal ischemia/reperfusion injury in rats. *Journal of Surgical Research* 167 (2):e339–44.

Wang, H., S. Guan, Y. Wang, H. Liu, H. Wang, L. Wang, C. Ren, S. Zhu, and K. Chen. 2011. In vivo degradation behavior of Ca-deficient hydroxyapatite coated Mg-Zn-Ca alloy for bone implant application. *Colloids Surf B Biointerfaces* 88 (1):254–9.

Wang, J., X. Zhu, R. Wang, Y. Xu, J. Nie, and G. Ling. 2011a. Microstructure and mechanical properties of Mg-xY-1.5 MM-0.4 Zr alloys. *Journal of Rare Earths* 29 (5):454–9.

Wang, J., L. Wang, Y. Wu, and L. Wang. 2011b. Effects of samarium on microstructures and tensile properties of Mg–5Al–0.3 Mn alloy. *Materials Science and Engineering: A* 528 (12):4115–19.

Wang, Q., L. Tan, W. Xu, B. Zhang, and K. Yang. 2011. Dynamic behaviors of a Ca–P coated AZ31B magnesium alloy during in vitro and in vivo degradations. *Materials Science and Engineering: B* 176 (20):1718–26.

Wang, Y., C. S. Lim, C. V. Lim, M. S. Yong, E. K. Teo, and L. N. Moh. 2011. In vitro degradation behavior of M1A magnesium alloy in protein-containing simulated body fluid. *Materials Science and Engineering: C* 31 (3):579–87.

Wang, B., J. H. Gao, L. G. Wang, S. J. Zhu, and S. K. Guan. 2012. Biocorrosion of coated Mg-Zn-Ca alloy under constant compressive stress close to that of human tibia. *Materials Letters* 70:174–6.

Wang, H., C. Zhao, Y. Chen, J. Li, and X. Zhang. 2012. Electrochemical property and in vitro degradation of DCPD–PCL composite coating on the biodegradable Mg–Zn alloy. *Materials Letters* 68:435–8.

Wang, J., J.-J. Nie, R. Wang, Y.-D. Xu, X.-R. Zhu, and G.-P. Ling. 2012. Effect of Y on age hardening response and mechanical properties of Mg-xY–1.5LPC –0.4Zr alloys. *Transactions of Nonferrous Metals Society of China* 22 (7):1549–55.

Wang, S., X. Guo, Y. Xie, L. Liu, H. Yang, R. Zhu, J. Gong, L. Peng, and W. Ding. 2012. Preparation of superhydrophobic silica film on Mg–Nd–Zn–Zr magnesium alloy with enhanced corrosion resistance by combining micro-arc oxidation and sol-gel method. *Surface and Coatings Technology* 213:192–201.

Wang, Y., Y. He, Z. Zhu, Y. Jiang, J. Zhang, J. Niu, L. Mao, and G. Yuan. 2012. In vitro degradation and biocompatibility of Mg-Nd-Zn-Zr alloy. *Chinese Science Bulletin* 57 (17):2163–70.

Wang, X., S. Cai, G. Xu, X. Ye, M. Ren, and K. Huang. 2013a. Surface characteristics and corrosion resistance of sol–gel derived CaO–P_2O_5–SrO–Na_2O bioglass–ceramic coated Mg alloy by different heat-treatment temperatures. *Journal of Sol-Gel Science and Technology* 67 (3):629–38.

Wang, X., L. H. Dong, X. L. Ma, and Y. F. Zheng. 2013b. Microstructure, mechanical property and corrosion behaviors of interpenetrating C/Mg-Zn-Mn composite fabricated by suction casting. *Materials Science and Engineering C-Materials for Biological Applications* 33 (2):618–25.

Wang, Y. M., J. W. Guo, Z. K. Shao, J. P. Zhuang, M. S. Jin, C. J. Wu, D. Q. Wei, and Y. Zhou. 2013a. A metasilicate-based ceramic coating formed on magnesium alloy by microarc oxidation and its corrosion in simulated body fluid. *Surface and Coatings Technology* 219:8–14.

Wang, Y., Z. Liao, C. Song, and H. Zhang. 2013b. Influence of Nd on microstructure and bio-corrosion resistance of Mg-Zn-Mn-Ca alloy. *Rare Metal Materials and Engineering* 42 (4):661–6.

Wang, Y., G. Luo, J. Zhang, Q. Shen, and L. Zhang. 2013c. Microstructure of diffusion-bonded Mg-Ag-Al multilayer composite materials. *Journal of Physics: Conference Series* 419:012023.

Wang, Y., Y. Ouyang, Y. He, D. Chen, Y. Jiang, L. Mao, J. Niu, J. Zhang, and G. Yuan. 2013d. Biocompatibility of Mg-Nd-Zn-Zr alloy with rabbit blood. *Chinese Science Bulletin* 58 (23):2903–8.

Wang, Y., X. Wang, T. Zhang, K. Wu, and F. Wang. 2013e. Role of β phase during microarc oxidation of Mg alloy AZ91D and corrosion resistance of the oxidation coating. *Journal of Materials Science and Technology* 29 (12):1129–33.

Wang, H., S. Zhu, L. Wang, Y. Feng, X. Ma, and S. Guan. 2014. Formation mechanism of Ca-deficient hydroxyapatite coating on Mg-Zn-Ca alloy for orthopaedic implant. *Applied Surface Science* 307:92–100.

Wang, J., R. Lu, D. Qin, W. Yang, and Z. Wu. 2014. Effect of arc-bending deformation on amplitude-dependent damping in pure magnesium. *Materials Science and Engineering: A* 615:296–301.

Wang, L., G. Fang, S. Leeflang, J. Duszczyk, and J. Zhou. 2014. Investigation into the hot workability of the as-extruded WE43 magnesium alloy using processing map. *Journal of the Mechanical Behavior of Biomedical Materials* 32:270–8.

Wang, S., Y. Xia, L. Liu, and N. Si. 2014. Preparation and performance of MAO coatings obtained on AZ91D Mg alloy under unipolar and bipolar modes in a novel dual electrolyte. *Ceramics International* 40 (1):93–9.

Wang, X., S. Cai, T. Liu, M. Ren, K. Huang, R. Zhang, and H. Zhao. 2014. Fabrication and corrosion resistance of calcium phosphate glass-ceramic coated Mg alloy via a PEG assisted sol–gel method. *Ceramics International* 40 (2):3389–98.

Watanabe, H., Y. Sasakura, N. Ikeo, and T. Mukai. 2015. Effect of deformation twins on damping capacity in extruded pure magnesium. *Journal of Alloys and Compounds* 626:60–4.

Waterman, J., A. Pietak, N. Birbilis, T. Woodfield, G. Dias, and M. P. Staiger. 2011. Corrosion resistance of biomimetic calcium phosphate coatings on magnesium due to varying pretreatment time. *Materials Science and Engineering: B* 176 (20):1756–60.

Watson, R. R., V. R. Preedy, and S. Zibadi. 2013. *Magnesium in Human Health and Disease.* New York: Springer.

Wearmouth, W. R., G. P. Dean, and R. N. Parkins. 1973. Role of stress in the stress corrosion cracking of a Mg-Al alloy. *Corrosion* 29 (6):251–60.

Wei, L. Y., G. L. Dunlop, and H. Westengen. 1995. The intergranular microstructure of cast Mg-Zn and Mg-Zn-rare earth alloys. *Metallurgical and Materials Transactions A-Physical Metallurgy and Materials Science* 26 (8):1947–55.

Wei, S., Y. Chen, Y. Tang, X. Zhang, M. Liu, S. Xiao, and Y. Zhao. 2009. Compressive creep behavior of Mg–Sn–La alloys. *Materials Science and Engineering: A* 508 (1):59–63.

Wei, Z., H. Du, and E. Zhang. 2011. The formation mechanism and biocorrosion property of $CaSiO_3/CaHPO_4 \cdot 2H_2O$ composite conversion coating on the extruded Mg-Zn-Ca alloy for bone implant application. *Surface and Interface Analysis* 43 (4):791–4.

Wei, S., T. Zhu, H. Hou, J. Kim, E. Kobayashi, T. Sato, M. Hodgson, and W. Gao. 2014. Effects of Pb/Sn additions on the age-hardening behaviour of Mg–4Zn alloys. *Materials Science and Engineering: A* 597:52–61.

Weishaupt, A., J. Bernhard, and B. Söderberg. 2013. Development of an atrial septal occluder using a biodegradable framework. *European Cells and Materials* 26 (S1):2.

Weisheit, A., R. Galun, and B. L. Mordike. 1998. CO_2 laser beam welding of magnesium-based alloys. *Welding Research Supplement* 74 (4):149–54.

Weizbauer, A., C. Modrejewski, S. Behrens et al. 2014. Comparative in vitro study and biomechanical testing of two different magnesium alloys. *Journal of Biomaterials Applications* 28 (8):1264–73.

Wellinghausen, N. R. L. 1998. The significance of zinc for leukocyte biology. *Journal of Leukocyte Biology* 64 (5):571–7.

Wen, C., M. Mabuchi, Y. Yamada, K. Shimojima, Y. Chino, and T. Asahina. 2001. Processing of biocompatible porous Ti and Mg. *Scripta Materialia* 45 (10):1147–53.

Wen, C., Y. Yamada, K. Shimojima, Y. Chino, H. Hosokawa, and M. Mabuchi. 2002. Novel titanium foam for bone tissue engineering. *Journal of Materials Research* 17 (10):2633–9.

Wen, C., Y. Yamada, K. Shimojima, Y. Chino, H. Hosokawa, and M. Mabuchi. 2004. Compressibility of porous magnesium foam: Dependency on porosity and pore size. *Materials Letters* 58 (3):357–60.

Wen, L., Z. Ji, and X. Li. 2008. Effect of extrusion ratio on microstructure and mechanical properties of Mg–Nd–Zn–Zr alloys prepared by a solid recycling process. *Materials Characterization* 59 (11):1655–60.

Wen, C., S. Guan, L. Peng, C. Ren, X. Wang, and Z. Hu. 2009. Characterization and degradation behavior of AZ31 alloy surface modified by bone-like hydroxyapatite for implant applications. *Applied Surface Science* 255 (13–14):6433–8.

Wen, Z., C. Wu, C. Dai, and F. Yang. 2009. Corrosion behaviors of Mg and its alloys with different Al contents in a modified simulated body fluid. *Journal of Alloys and Compounds* 488 (1):392–9.

Wessels, V., G. Le Mene, S. F. Fischerauer, T. Kraus, A. M. Weinberg, P. J. Uggowitzer, and J. F. Loffler. 2012. In vivo performance and structural relaxation of biodegradable bone implants made from Mg-Zn-Ca bulk metallic glasses. *Advanced Engineering Materials* 14 (6):B357–64.

Wexler, B. C. 1980. Pathophysiologic responses of spontaneously hypertensive rats to arterial magnesium—Aluminum wire implants. *Atherosclerosis* 36 (4):575–87.

Wieding, J., R. Souffrant, W. Mittelmeier, and R. Bader. 2013. Finite element analysis on the biomechanical stability of open porous titanium scaffolds for large segmental bone defects under physiological load conditions. *Medical Engineering and Physics* 35 (4): 422–32.

Wilflingseder, P., R. Martin, and C. Papp. 1981. Magnesium seeds in the treatment of lymph- and haemangiomata. *Chirurgia Plastica* 6 (2):105–16.

Willbold, E., A. A. Kaya, R. A. Kaya, F. Beckmann, and F. Witte. 2011. Corrosion of magnesium alloy AZ31 screws is dependent on the implantation site. *Materials Science and Engineering B-Advanced Functional Solid-State Materials* 176 (20):1835–40.

Willbold, E., K. Kalla, I. Bartsch et al. 2013. Biocompatibility of rapidly solidified magnesium alloy RS66 as a temporary biodegradable metal. *Acta Biomaterialia* 9 (10):8509–17.

Willbold, E., X. Gu, D. Albert, K. Kalla, K. Bobe, M. Brauneis, C. Janning, J. Nellesen, W. Czayka, W. Tillmann, Y. Zheng, and F. Witte. 2015. Effect of the addition of low rare earth elements (La, Nd and Ce) on the biodegradation and biocompatibility of Mg. *Acta Biomaterialia* 11:554–62.

Williams, D. 2006. New Interests in Magnesium. *Medica Device Technology* 17 (3):9–10.

Willie, B. M., X. Yang, N. H. Kelly, J. Merkow, S. Gagne, R. Ware, T. M. Wright, and M. P. Bostrom. 2010. Osseointegration into a novel titanium foam implant in the distal femur of a rabbit. *Journal of Biomedical Materials Research Part B: Applied Biomaterials* 92 (2):479–88.

Willumeit, R., J. Fischer, F. Feyerabend et al. 2011. Chemical surface alteration of biodegradable magnesium exposed to corrosion media. *Acta Biomaterialia* 7 (6):2704–15.

Wilson, R., C. J. Bettles, B. C. Muddle, and J. F. Nie. 2003. Precipitation hardening in Mg-3 wt% Nd (-Zn) casting alloys. *Materials Science Forum* 419:267–72.

Windecker, S., A. Remondino, F. R. Eberli et al. 2005. Sirolimus-eluting and paclitaxel-eluting stents for coronary revascularization. *New England Journal of Medicine* 353 (7):653–62.

Windhagen, H., K. Radtke, A. Weizbauer et al. 2013. Biodegradable magnesium-based screw clinically equivalent to titanium screw in hallux valgus surgery: Short term results of the first prospective, randomized, controlled clinical pilot study. *Biomedical Engineering Online* 12 (1):62.

Winzer, N., A. Atrens, G. Song et al. 2005. A critical review of the stress corrosion cracking (SCC) of magnesium alloys. *Advanced Engineering Materials* 7 (8):659–93.

Winzer, N., A. Atrens, W. Dietzel, G. Song, and K. U. Kainer. 2007. Stress corrosion cracking in magnesium alloys: Characterization and prevention. *JOM* 59 (8):49–53.

Wittchow, E., N. Adden, J. Riedmüller, C. Savard, R. Waksman, and M. Braune. 2013. Bioresorbable drug-eluting magnesium-alloy scaffold: Design and feasibility in a porcine coronary model. *EuroIntervention* 8 (12):1441–50.

Witte, F. 2010. The history of biodegradable magnesium implants: A review. *Acta Biomaterialia* 6 (5):1680–92.

Witte, F., V. Kaese, H. Haferkamp et al. 2005. In vivo corrosion of four magnesium alloys and the associated bone response. *Biomaterials* 26 (17):3557–63.

Witte, F., J. Fischer, J. Nellesen, and F. Beckmann. 2006a. Microtomography of magnesium implants in bone and their degradation. Paper presented at SPIE 6318, Developments in X-Ray Tomography V, September 7, 2006, San Diego, CA.

Witte, F., J. Fischer, J. Nellesen et al. 2006b. In vitro and in vivo corrosion measurements of magnesium alloys. *Biomaterials* 27 (7):1013–18.

Witte, F., F. Feyerabend, P. Maier, J. Fischer, M. Störmer, C. Blawert, W. Dietzel, and N. Hort. 2007a. Biodegradable magnesium–hydroxyapatite metal matrix composites. *Biomaterials* 28 (13):2163–74.

Witte, F., H. Ulrich, C. Palm, and E. Willbold. 2007b. Biodegradable magnesium scaffolds: Part II: Peri-implant bone remodeling. *Journal of Biomedical Materials Research Part A* 81 (3):757–65.

Witte, F., H. Ulrich, M. Rudert, and E. Willbold. 2007c. Biodegradable magnesium scaffolds: Part I: Appropriate inflammatory response. *Journal of Biomedical Materials Research Part A* 81 (3):748–56.

Witte, F., N. Hort, C. Vogt, S. Cohen, K. U. Kainer, R. Willumeit, and F. Feyerabend. 2008a. Degradable biomaterials based on magnesium corrosion. *Current Opinion in Solid State and Materials Science* 12 (5–6):63–72.

Witte, F., I. Abeln, E. Switzer, V. Kaese, A. Meyer-Lindenberg, and H. Windhagen. 2008b. Evaluation of the skin sensitizing potential of biodegradable magnesium alloys. *Journal of Biomedical Materials Research Part A* 86A (4):1041–7.

Witte, F., J. Fischer, J. Nellesen, C. Vogt, J. Vogt, T. Donath, and F. Beckmann. 2010. In vivo corrosion and corrosion protection of magnesium alloy LAE442. *Acta Biomaterialia* 6 (5):1792–9.

Wolters, L., N. Angrisani, J. Seitz et al. 2013. Applicability of degradable magnesium LAE442alloy plate-screw-systems in a rabbit model. *Biomedizinische Technik. Biomedical Engineering* 58 (S1) doi: 10.1515/bmt-2013-4059.

Wong, H. M., K. W. K. Yeung, K. O. Lam et al. 2010. A biodegradable polymer-based coating to control the performance of magnesium alloy orthopaedic implants. *Biomaterials* 31 (8):2084–96.

Wong, H. M., Y. Zhao, V. Tam et al. 2013. In vivo stimulation of bone formation by aluminum and oxygen plasma surface-modified magnesium implants. *Biomaterials* 34 (38):9863–76.

Woodard, J. R., A. J. Hilldore, S. K. Lan, C. Park, A. W. Morgan, J. A. C. Eurell, S. G. Clark, M. B. Wheeler, R. D. Jamison, and A. J. Wagoner Johnson. 2007. The mechanical properties and osteoconductivity of hydroxyapatite bone scaffolds with multi-scale porosity. *Biomaterials* 28 (1):45–54.

Wozniak, T. D., Y. Kocabey, S. Klein, J. Nyland, and D. N. M. Caborn. 2005. Influence of thread design on bioabsorbable interference screw insertion torque during retrograde fixation of a soft-tissue graft in synthetic bone. *Arthroscopy: The Journal of Arthroscopic and Related Surgery* 21 (7):815–19.

Wu, Y., and W. Hu. 2008. Comparison of the solid solution properties of Mg-RE (Gd, Dy, Y) alloys with atomistic simulation. *Physics Research International* 2008:476812.

Wu, R., and M. Zhang. 2009. Microstructure, mechanical properties and aging behavior of Mg–5Li–3Al–2Zn–xAg. *Materials Science and Engineering: A* 520 (1–2):36–19.

Wu, W., M. Qi, X.-P. Liu, D.-Z. Yang, and W.-Q. Wang. 2007a. Delivery and release of nitinol stent in carotid artery and their interactions: A finite element analysis. *Journal of Biomechanics* 40 (13):3034–40.

Wu, W., W.-Q. Wang, D.-Z. Yang, and M. Qi. 2007b. Stent expansion in curved vessel and their interactions: A finite element analysis. *Journal of Biomechanics* 40 (11):2580–5.

Wu, G. S., A. Y. Wang, K. J. Ding et al. 2008. Fabrication of Cr coating on AZ31 magnesium alloy by magnetron sputtering. *Transactions of Nonferrous Metals Society of China* 18:s329–33.

Wu, H.-Y., G.-Z. Zhou, Z.-W. Gao, and C.-H. Chiu. 2008. Mechanical properties and formability of an Mg–6%Li–1%Zn alloy thin sheet at elevated temperatures. *Journal of Materials Processing Technology* 206 (1–3):419–24.

Wu, H.-Y., Z.-W. Gao, J.-Y. Lin, and C.-H. Chiu. 2009. Effects of minor scandium addition on the properties of Mg–Li–Al–Zn alloy. *Journal of Alloys and Compounds* 474 (1–2):158–63.

Wu, R. Z., Y. S. Deng, and M. L. Zhang. 2009a. Microstructure and mechanical properties of Mg–5Li–3Al–2Zn–xRE alloys. *Journal of Materials Science* 44 (15):4132–9.

Wu, R., Z. Qu, and M. Zhang. 2009b. Effects of the addition of Y in Mg–8Li–(1,3)Al alloy. *Materials Science and Engineering: A* 516 (1–2):96–9.

Wu, B. L., Y. H. Zhao, X. H. Du, Y. D. Zhang, F. Wagner, and C. Esling. 2010. Ductility enhancement of extruded magnesium via yttrium addition. *Materials Science and Engineering: A* 527 (16):4334–40.

Wu, C., Z. Wen, C. Dai, Y. Lu, and F. Yang. 2010. Fabrication of calcium phosphate/chitosan coatings on AZ91D magnesium alloy with a novel method. *Surface and Coatings Technology* 204 (20):3336–47.

Wu, D., G. Liang, L. Li, and H. Wu. 2010. Microstructural investigation of electrochemical hydrogen storage in amorphous Mg–Ni–La alloy. *Materials Science and Engineering: B* 175 (3):248–52.

Wu, R. Z., Z. K. Qu, and M. L. Zhang. 2010. Reviews on the influences of alloying elements on the microstructure and mechanical properties of Mg–Li base alloys. *Reviews on Advanced Materials Science* 24:35–43.

Wu, W., L. Petrini, D. Gastaldi, T. Villa, M. Vedani, E. Lesma, B. Previtali, and F. Migliavacca. 2010. Finite element shape optimization for biodegradable magnesium alloy stents. *Annals of Biomedical Engineering* 38 (9):2829–40.

Wu, G., L. Gong, K. Feng, S. Wu, Y. Zhao, and P. K. Chu. 2011. Rapid degradation of biomedical magnesium induced by zinc ion implantation. *Materials Letters* 65 (4):661–3.

Wu, L., C. Cui, R. Wu, J. Li, H. Zhan, and M. Zhang. 2011. Effects of Ce-rich RE additions and heat treatment on the microstructure and tensile properties of Mg–Li–Al–Zn-based alloy. *Materials Science and Engineering: A* 528 (4):2174–9.

Wu, W., D. Gastaldi, K. Yang, L. Tan, L. Petrini, and F. Migliavacca. 2011. Finite element analyses for design evaluation of biodegradable magnesium alloy stents in arterial vessels. *Materials Science and Engineering B* 176 (20):1733–40.

Wu, X. F., Y. Kang, F. F. Wu, K. Q. Qiu, and L. K. Meng. 2011. Formation of ternary Mg-Cu-Dy bulk metallic glasses. *Bulletin of Materials Science* 34 (7):1507–10.

Wu, G., K. Feng, A. Shanaghi, Y. Zhao, R. Xu, G. Yuan, and P. K. Chu. 2012a. Effects of surface alloying on electrochemical corrosion behavior of oxygen-plasma-modified biomedical magnesium alloy. *Surface and Coatings Technology* 206 (14):3186–95.

Wu, G., R. Xu, K. Feng, S. Wu, Z. Wu, G. Sun, G. Zheng, G. Li, and P. K. Chu. 2012b. Retardation of surface corrosion of biodegradable magnesium-based materials by aluminum ion implantation. *Applied Surface Science* 258 (19):7651–7.

Wu, Q., S. Zhu, L. Wang, Q. Liu, G. Yue, J. Wang, and S. Guan. 2012. The microstructure and properties of cyclic extrusion compression treated Mg–Zn–Y–Nd alloy for vascular stent application. *Journal of the Mechanical Behavior of Biomedical Materials* 8:1–7.

Wu, W., S. Chen, D. Gastaldi, L. Petrini, D. Mantovani, K. Yang, L. Tan, and F. Migliavacca. 2012. Experimental data confirm numerical modeling of the degradation process of magnesium alloys stents. *Acta Biomaterialia* 9 (10):8730–9.

Wu, Y., L. Peng, S. Zhao et al. 2012. Ignition-proof properties of a high-strength Mg-Gd-Ag-Zr alloy. *Journal of Shanghai Jiaotong University (Science)* 17:643–7.

Wu, G., J. M. Ibrahim, and P. K. Chu. 2013. Surface design of biodegradable magnesium alloys—a review. *Surface and Coatings Technology* 233:2–12.

Wu, L., J. Dong, and W. Ke. 2013. Potentiostatic deposition process of fluoride conversion film on AZ31 magnesium alloy in 0.1M KF solution. *Electrochimica Acta* 105:554–9.

Wu, D., R. S. Chen, and W. Ke. 2014. Microstructure and mechanical properties of a sand-cast Mg–Nd–Zn alloy. *Materials and Design* 58:324–31.

Wu, G., X. Zhang, Y. Zhao, J. M. Ibrahim, G. Yuan, and P. K. Chu. 2014. Plasma modified Mg–Nd–Zn–Zr alloy with enhanced surface corrosion resistance. *Corrosion Science* 78:121–9.

Wulandari, W., G. A. Brooks, M. A. Rhamdhani, and B. J. Monaghan. 2010. Magnesium: Current and alternative production routes. *Chemeca 2010: Engineering at the Edge*; 26–29 September, Hilton Adelaide, South Australia.

Xia, Y. H., B. P. Zhang, Y. Wang, M. F. Qian, and L. Geng. 2012. In-vitro cytotoxicity and in-vivo biocompatibility of as-extruded Mg-4.0Zn-0.2Ca alloy. *Materials Science and Engineering C-Materials for Biological Applications* 32 (4):665–9.

Xia, Y. H., B. P. Zhang, C. X. Lu, and L. Geng. 2013. Improving the corrosion resistance of Mg-4.0Zn-0.2Ca alloy by micro-arc oxidation. *Materials Science and Engineering. C, Materials for Biological Applications* 33 (8):5044–50.

Xiao, W., S. Jia, J. Wang, Y. Wu, and L. Wang. 2008. Effects of cerium on the microstructure and mechanical properties of Mg–20Zn–8Al alloy. *Materials Science and Engineering: A* 474 (1):317–22.

Xiao, X., H. Yu, Q. Zhu et al. 2013. In vivo corrosion resistance of Ca-P coating on AZ60 magnesium alloy. *Journal of Bionic Engineering* 10 (2):156–61.

Xie, G. M., Z. Y. Ma, L. Geng, and R. S. Chen. 2007. Microstructural evolution and mechanical properties of friction stir welded Mg–Zn–Y–Zr alloy. *Materials Science and Engineering: A* 471 (1–2):63–8.

Xie, K., Y. Yu, Y. Pei et al. 2010a. Protective effects of hydrogen gas on murine polymicrobial sepsis via reducing oxidative stress and HMGB1 release. *Shock* 34 (1):90–7.

Xie, K., Y. Yu, Z. Zhang et al. 2010b. Hydrogen gas improves survival rate and organ damage in zymosan-induced generalized inflammation model. *Shock* 34 (5):495–501.

Xin, Y. C., and P. K. Chu. 2010. Influence of Tris in simulated body fluid on degradation behavior of pure magnesium. *Materials Chemistry and Physics* 124 (1):33–5.

Xin, Y. C., K. F. Huo, H. Tao, G. Y. Tang, and P. K. Chu. 2008. Influence of aggressive ions on the degradation behavior of biomedical magnesium alloy in physiological environment. *Acta Biomaterialia* 4 (6):2008–15.

Xin, Y., J. Jiang, K. Huo, G. Tang, X. Tian, and P. K. Chu. 2009a. Corrosion resistance and cytocompatibility of biodegradable surgical magnesium alloy coated with hydrogenated amorphous silicon. *Journal of Biomedical Materials Research. Part A* 89 (3):717–26.

Xin, Y., C. Liu, K. Huo, G. Tang, X. Tian, and P. K. Chu. 2009b. Corrosion behavior of ZrN/Zr coated biomedical AZ91 magnesium alloy. *Surface and Coatings Technology* 203 (17–18):2554–7.

Xin, Y., T. Hu, and P. K. Chu. 2010. Influence of test solutions on in vitro studies of biomedical magnesium alloys. *Journal of the Electrochemical Society* 157 (7):C238–43.

Xin, Y., K. Huo, T. Hu, G. Tang, and P. K. Chu. 2011a. Corrosion products on biomedical magnesium alloy soaked in simulated body fluids. *Journal of Materials Research* 24 (8):2711–19.

Xin, Y., T. Hu, and P. K. Chu. 2011b. In vitro studies of biomedical magnesium alloys in a simulated physiological environment: A review. *Acta Biomaterialia* 7 (4):1452–9.

Xin, Y., T. Hu, and P. K. Chu. 2011c. Degradation behaviour of pure magnesium in simulated body fluids with different concentrations of HCO-3. *Corrosion Science* 53 (4):1522–8.

Xu, B. 2006. Research on drawing process of AZ31 magnesium alloy wire, Masters Thesis, Harbin Institute of Technology: Harbin, China.

Xu, L., and A. Yamamoto. 2012a. Characteristics and cytocompatibility of biodegradable polymer film on magnesium by spin coating. *Colloids and Surfaces B: Biointerfaces* 93:67–74.

Xu, L., and A. Yamamoto. 2012b. In vitro degradation of biodegradable polymer-coated magnesium under cell culture condition. *Applied Surface Science* 258 (17):6353–8.

Xu, D. K., W. N. Tang, L. Liu, Y. B. Xu, and E. H. Han. 2007a. Effect of Y concentration on the microstructure and mechanical properties of as-cast Mg–Zn–Y–Zr alloys. *Journal of Alloys and Compounds* 432 (1):129–34.

Xu, D. K., L. Liu, Y. B. Xu, and E. H. Han. 2007b. The fatigue crack propagation behavior of the forged Mg–Zn–Y–Zr alloy. *Journal of Alloys and Compounds* 431 (1):107–11.

Xu, L. P., G. N. Yu, E. Zhang, F. Pan, and K. Yang. 2007. In vivo corrosion behavior of Mg-Mn-Zn alloy for bone implant application. *Journal of Biomedical Materials Research Part A* 83A (3):703–11.

Xu, L. P., E. L. Zhang, D. S. Yin, S. Y. Zeng, and K. Yang. 2008. In vitro corrosion behaviour of Mg alloys in a phosphate buffered solution for bone implant application. *Journal of Materials Science-Materials in Medicine* 19 (3):1017–25.

Xu, L., F. Pan, G. Yu et al. 2009a. In vitro and in vivo evaluation of the surface bioactivity of a calcium phosphate coated magnesium alloy. *Biomaterials* 30 (8):1512–23.

Xu, L., E. Zhang, and K. Yang. 2009b. Phosphating treatment and corrosion properties of Mg-Mn-Zn alloy for biomedical application. *Journal of Materials Science. Materials in Medicine* 20 (4):859–67.

Xu, Y., K. Li, Z. Yao, Z. Jiang, and M. Zhang. 2009. Micro-arc oxidation coatings on Mg-Li alloys. *Rare Metals* 28 (2):160–3.

Xu, X., P. Lu, M. Guo, and M. Fang. 2010. Cross-linked gelatin/nanoparticles composite coating on micro-arc oxidation film for corrosion and drug release. *Applied Surface Science* 256 (8):2367–71.

Xu, R., G. Wu, X. Yang, T. Hu, Q. Lu, and P. K. Chu. 2011. Controllable degradation of biomedical magnesium by chromium and oxygen dual ion implantation. *Materials Letters* 65 (14):2171–3.

Xu, Z., C. Smith, S. Chen, and J. Sankar. 2011. Development and microstructural characterizations of Mg–Zn–Ca alloys for biomedical applications. *Materials Science and Engineering: B* 176 (20):1660–5.

Xu, C., S. W. Xu, M. Y. Zheng, K. Wu, E. D. Wang, S. Kamado, G. J. Wang, and X. Y. Lv. 2012a. Microstructures and mechanical properties of high-strength Mg–Gd–Y–Zn–Zr alloy sheets processed by severe hot rolling. *Journal of Alloys and Compounds* 524:46–52.

Xu, C., M. Y. Zheng, S. W. Xu, K. Wu, E. D. Wang, S. Kamado, G. J. Wang, and X. Y. Lv. 2012b. Ultra high-strength Mg–Gd–Y–Zn–Zr alloy sheets processed by large-strain hot rolling and ageing. *Materials Science and Engineering: A* 547:93–8.

Xu, R., G. Wu, X. Yang, X. Zhang, Z. Wu, G. Sun, G. Li, and P. K. Chu. 2012a. Corrosion behavior of chromium and oxygen plasma-modified magnesium in sulfate solution and simulated body fluid. *Applied Surface Science* 258 (20):8273–8.

Xu, R., X. Yang, K. W. Suen, G. Wu, P. Li, and P. K. Chu. 2012b. Improved corrosion resistance on biodegradable magnesium by zinc and aluminum ion implantation. *Applied Surface Science* 263:608–12.

Xu, C., M. Y. Zheng, K. Wu, E. D. Wang, G. H. Fan, S. W. Xu, S. Kamado, X. D. Liu, G. J. Wang, and X. Y. Lv. 2013a. Influence of rolling temperature on the microstructure and mechanical properties of Mg–Gd–Y–Zn–Zr alloy sheets. *Materials Science and Engineering: A* 559:615–22.

Xu, C., M. Y. Zheng, S. W. Xu, K. Wu, E. D. Wang, G. H. Fan, S. Kamado, X. D. Liu, G. J. Wang, and X. Y. Lv. 2013b. Microstructure and mechanical properties of Mg–Gd–Y–Zn–Zr alloy sheets processed by combined processes of extrusion, hot rolling and ageing. *Materials Science and Engineering: A* 559:844–51.

Xu, R., X. Yang, X. Zhang, M. Wang, P. Li, Y. Zhao, G. Wu, and P. K. Chu. 2013. Effects of carbon dioxide plasma immersion ion implantation on the electrochemical properties of AZ31 magnesium alloy in physiological environment. *Applied Surface Science* 286:257–60.

Xue, D., Z. Tan, M. J. Schulz, W. J. Vanooij, J. Sankar, Y. Yun, and Z. Dong. 2012a. Corrosion studies of modified organosilane coated magnesium–yttrium alloy in different environments. *Materials Science and Engineering: C* 32 (5):1230–6.

Xue, D., Y. Yun, Z. Tan, Z. Dong, and M. J. Schulz. 2012b. In vivo and in vitro degradation behavior of magnesium alloys as biomaterials. *Journal of Materials Science and Technology* 28 (3):261–7.

Yamada, Y., K. Shimojima, Y. Sakaguchi, M. Mabuchi, M. Nakamura, T. Asahina, T. Mukai, H. Kanahashi, and K. Higashi. 1999. Compressive properties of open-cellular SG91A Al and AZ91 Mg. *Materials Science and Engineering: A* 272 (2):455–8.

Yamada, K., Y. Okubo, M. Shiono, H. Watanabe, S. Kamado, and Y. Kojima. 2006. Alloy development of high toughness Mg-Gd-Y-Zn-Zr alloys. *Materials Transactions* 47 (4):1066–70.

Yamada, K., H. Hoshikawa, S. Maki, T. Ozaki, Y. Kuroki, S. Kamado, and Y. Kojima. 2009. Enhanced age-hardening and formation of plate precipitates in Mg–Gd–Ag alloys. *Scripta Materialia* 61 (6):636–9.

Yamada, S., H. Maeda, A. Obata, U. Lohbauer, A. Yamamoto, and T. Kasuga. 2013. Cytocompatibility of siloxane-containing vaterite/poly (L-lactic acid) composite coatings on metallic magnesium. *Materials* 6 (12):5857–69.

Yamaguchi, M., H. Oishi, and Y. Suketa. 1987. Stimulatory effect of zinc on bone-formation in tissue-culture. *Biochemical Pharmacology* 36 (22):4007–12.

Yamamoto, A., and S. Hiromoto. 2009. Effect of inorganic salts, amino acids and proteins on the degradation of pure magnesium in vitro. *Materials Science and Engineering: C* 29 (5):1559–68.

Yamamoto, A., R. Honma, and M. Sumita. 1998. Cytotoxicity evaluation of 43 metal salts using murine fibroblasts and osteoblastic cells. *Journal of Biomedical Materials Research* 39 (2):331–40.

Yamasaki, M., and Y. Kawamura. 2009. Thermal diffusivity and thermal conductivity of Mg–Zn–rare earth element alloys with long-period stacking ordered phase. *Scripta Materialia* 60 (4):264–7.

Yamasaki, M., T. Anan, S. Yoshimoto, and Y. Kawamura. 2005. Mechanical properties of warm-extruded Mg–Zn–Gd alloy with coherent 14H long periodic stacking ordered structure precipitate. *Scripta Materialia* 53 (7):799–803.

Yamasaki, M., N. Hayashi, S. Izumi, and Y. Kawamura. 2007. Corrosion behavior of rapidly solidified Mg–Zn–rare earth element alloys in NaCl solution. *Corrosion Science* 49 (1):255–62.

Yamashita, A., Z. Horita, and T. G. Langdon. 2001. Improving the mechanical properties of magnesium and a magnesium alloy through severe plastic deformation. *Materials Science and Engineering: A* 300 (1):142–7.

Yamauchi, K., and S. Asakura. 2003. Galvanic dissolution behavior of magnesium-1 mass% manganese-0.5 mass% calcium alloy anode for cathodic protection in fresh water. *Materials Transactions* 44 (5):1046–8.

Yamauchi, N., N. Ueda, A. Okamoto, T. Sone, M. Tsujikawa, and S. Oki. 2007. DLC coating on Mg–Li alloy. *Surface and Coatings Technology* 201 (9–11):4913–18.

Yan, J., Y. Sun, F. Xue, S. Xue, and W. Tao. 2008. Microstructure and mechanical properties in cast magnesium–neodymium binary alloys. *Materials Science and Engineering: A* 476 (1–2):366–71.

Yan, Y. D., M. L. Zhang, Y. Xue, W. Han, D. X. Cao, and L. Y. He. 2008. Electrochemical study of the codeposition of Mg–Li–Al alloys from LiCl–KCl–MgCl2–AlCl3 melts. *Journal of Applied Electrochemistry* 39 (3):455–61.

Yan, Y. D., M. L. Zhang, Y. Xue, W. Han, D. X. Cao, X. Y. Jing, L. Y. He, and Y. Yuan. 2009a. Electrochemical formation of Mg-Li-Ca alloys by codeposition of Mg, Li and Ca from LiCl-KCl-MgCl2-CaCl2 melts. *Physical Chemistry Chemical Physics* 11 (29):6148–55.

Yan, Y. D., M. L. Zhang, Y. Xue, W. Han, D. X. Cao, and S. Q. Wei. 2009b. Study on the preparation of Mg–Li–Zn alloys by electrochemical codeposition from LiCl–KCl–MgCl2–ZnCl2 melts. *Electrochimica Acta* 54 (12):3387–93.

Yan, J., Y. Sun, F. Xue, S. Xue, Y. Xiao, and W. Tao. 2009. Creep behavior of Mg–2wt.%Nd binary alloy. *Materials Science and Engineering: A* 524 (1–2):102–7.

Yan, T., L. Tan, D. Xiong, X. Liu, B. Zhang, and K. Yang. 2010. Fluoride treatment and in vitro corrosion behavior of an AZ31B magnesium alloy. *Materials Science and Engineering: C* 30 (5):740–8.

Yan, J., Y. G. Chen, Q. L. Yuan, S. Yu, W. C. Qiu, C. G. Yang, Z. G. Wang, J. F. Gong, K. X. Ai, Q. Zheng, J. N. Li, S. X. Zhang, and X. N. Zhang. 2013. Comparison of the effects of Mg-6Zn and titanium on intestinal tract in vivo. *Journal of Materials Science-Materials in Medicine* 24 (6):1515–25.

Yan, T., L. Tan, B. Zhang, and K. Yang. 2014. Fluoride conversion coating on biodegradable AZ31B magnesium alloy. *Journal of Materials Science and Technology* 30 (7):666–74.

Yang, L., and E. Zhang. 2009. Biocorrosion behavior of magnesium alloy in different simulated fluids for biomedical application. *Materials Science and Engineering: C* 29 (5):1691–6.

Yang, S., K.-F. Leong, Z. Du, and C.-K. Chua. 2001. The design of scaffolds for use in tissue engineering. Part I. Traditional factors. *Tissue Engineering* 7 (6):679–89.

Yang, W., P. Zhang, J. Liu, and Y. Xue. 2006. Effect of long-term intake of Y3+ in drinking water on gene expression in brains of rats. *Journal of Rare Earths* 24 (3):369–73.

Yang, M. B., F. S. Pan, R. J. Cheng, and A. T. Tang. 2007. Comparison about efficiency of Al-10Sr and Mg-10Sr master alloys to grain refinement of AZ31 magnesium alloy. *Journal of Materials Science* 42 (24):10074–9.

Yang, J., J. Wang, L. Wang, Y. Wu, L. Wang, and H. Zhang. 2008. Microstructure and mechanical properties of Mg–4.5Zn–xNd (x = 0, 1 and 2, wt%) alloys. *Materials Science and Engineering: A* 479 (1):339–44.

Yang, J.-X., Y.-P. Jiao, Q.-S. Yin, Y. Zhang, and T. Zhang. 2008a. Calcium phosphate coating on magnesium alloy by biomimetic method: Investigation of morphology, composition and formation process. *Frontiers of Materials Science in China* 2 (2):149–55.

Yang, J. X., F. Z. Cui, I.-S. Lee, Y. P. Jiao, Q. S. Yin, and Y. Zhang. 2008b. Ion-beam assisted deposited C–N coating on magnesium alloys. *Surface and Coatings Technology* 202 (22–23):5737–41.

Yang, J. X., Y. P. Jiao, F. Z. Cui, I.-S. Lee, Q. S. Yin, and Y. Zhang. 2008c. Modification of degradation behavior of magnesium alloy by IBAD coating of calcium phosphate. *Surface and Coatings Technology* 202 (22–23):5733–6.

Yang, L., J. Li, X. Yu, M. Zhang, and X. Huang. 2008. Lanthanum-based conversion coating on Mg–8Li alloy. *Applied Surface Science* 255 (5):2338–41.

Yang, M. B., F. S. Pan, R. J. Cheng, and S. Jia. 2008. Comparison about effects of Sb, Sn and Sr on as-cast microstructure and mechanical properties of AZ61-0.7Si magnesium alloy. *Materials Science and Engineering A-Structural Materials Properties Microstructure and Processing* 489 (1–2):413–18.

Yang, J., F. Cui, Q. Yin, Y. Zhang, T. Zhang, and X. Wang. 2009. Characterization and degradation study of calcium phosphate coating on magnesium alloy bone implant in vitro. *Plasma Science, IEEE Transactions on* 37 (7):1161–8.

Yang, L., J. Li, Y. Zheng, W. Jiang, and M. Zhang. 2009a. Electroless Ni–P plating with molybdate pretreatment on Mg–8Li alloy. *Journal of Alloys and Compounds* 467 (1–2):562–6.

Yang, L., M. Zhang, J. Li, X. Yu, and Z. Niu. 2009b. Stannate conversion coatings on Mg–8Li alloy. *Journal of Alloys and Compounds* 471 (1–2):197–200.

Yang, M.-B., Y.-L. Ma, and F.-S. Pan. 2009. Effects of little Ce addition on as-cast microstructure and creep properties of Mg-3Sn-2Ca magnesium alloy. *Transactions of Nonferrous Metals Society of China* 19 (5):1087–92.

Yang, X., G. Wang, G. Dong, F. Gong, and M. Zhang. 2009. Rare earth conversion coating on Mg-8.5Li alloys. *Journal of Alloys and Compounds* 487 (1–2):64–8.

Yang, M. B., L. Cheng, and F. S. Pan. 2010. Comparison of as-cast microstructure, tensile and creep properties for Mg-3Sn-1Ca and Mg-3Sn-2Ca magnesium alloys. *Transactions of Nonferrous Metals Society of China* 20 (4):584–9.

Yang, H., X. Guo, G. Wu, W. Ding, and N. Birbilis. 2011. Electrodeposition of chemically and mechanically protective Al-coatings on AZ91D Mg alloy. *Corrosion Science* 53 (1):381–7.

Yang, L., F. Feyerabend, K. U. Kainer, R. Willumeit, and N. Hort. 2011a. Corrosion behavior of as-cast binary Mg-Dy alloys. *Materials Science Forum* 690:417–21.

Yang, L., Y. Huang, Q. Peng, F. Feyerabend, K. U. Kainer, R. Willumeit, and N. Hort. 2011b. Mechanical and corrosion properties of binary Mg–Dy alloys for medical applications. *Materials Science and Engineering: B* 176 (20):1827–34.

Yang, Q., B. L. Xiao, Z. Y. Ma, and R. S. Chen. 2011. Achieving high strain rate superplasticity in Mg–Zn–Y–Zr alloy produced by friction stir processing. *Scripta Materialia* 65 (4):335–8.

Yang, S., Z. Pu, D. Puleo, O. Dillon Jr., and I. Jawahir. 2011. Cryogenic processing of biomaterials for improved surface integrity and product sustainability. In *Advances in Sustainable Manufacturing*. Springer: Berlin, Heidelberg.

Yang, W., Y. Zhang, J. Yang, L. Tan, and K. Yang. 2011. Potential antiosteoporosis effect of biodegradable magnesium implanted in STZ-induced diabetic rats. *Journal of Biomedical Materials Research. Part A* 99 (3):386–94.

Yang, X., L. Guo, X. Sun, X. Chen, and X. Tong. 2011. Protective effects of hydrogen-rich saline in preeclampsia rat model. *Placenta* 32 (9):681–6.

Yang, L., Y. Huang, F. Feyerabend, R. Willumeit, K. U. Kainer, and N. Hort. 2012. Influence of ageing treatment on microstructure, mechanical and bio-corrosion properties of Mg–Dy alloys. *Journal of the Mechanical Behavior of Biomedical Materials* 13:36–44.

Yang, M. B., H. Li, W. Zhang, and T. Zhou. 2012a. Effects of holding time and temperature on Sr content of Mg-Sr master alloys produced by metallothermic reduction of SrO. *Advanced Materials Research* 403:20–3.

Yang, M. B., H. L. Li, R. J. Cheng, F. S. Pan, and H. J. Hu. 2012b. Comparison about effects of minor Zr, Sr and Ca additions on microstructure and tensile properties of Mg-5Gd-1.2Mn-0.4Sc (wt.%) magnesium alloy. *Materials Science and Engineering A-Structural Materials Properties Microstructure and Processing* 545:201–8.

Yang, L., N. Hort, D. Laipple, D. Höche, Y. Huang, K. U. Kainer, R. Willumeit, and F. Feyerabend. 2013a. Element distribution in the corrosion layer and cytotoxicity of alloy Mg–10Dy during in vitro biodegradation. *Acta Biomaterialia* 9 (10):8475–87.

Yang, L., Y. Huang, F. Feyerabend, R. Willumeit, C. Mendis, K. U. Kainer, and N. Hort. 2013b. Microstructure, mechanical and corrosion properties of Mg–Dy–Gd–Zr alloys for medical applications. *Acta Biomaterialia* 9 (10):8499–508.

Yang, M. B., T. Z. Guo, and H. L. Li. 2013a. Effects of Gd addition on as-cast microstructure, tensile and creep properties of Mg-3.8 Zn-2.2Ca (wt.%) magnesium alloy. *Materials Science and Engineering A-Structural Materials Properties Microstructure and Processing* 587:132–42.

Yang, M. B., H. L. Li, C. Y. Duan, and J. Zhang. 2013b. Effects of minor Ti addition on as-cast microstructure and mechanical properties of Mg-3Sn-2Sr (wt.%) magnesium alloy. *Journal of Alloys and Compounds* 579:92–9.

Yang, Q., B. Jiang, Y. Tian, W. Liu, and F. Pan. 2013. A tilted weak texture processed by an asymmetric extrusion for magnesium alloy sheets. *Materials Letters* 100:29–31.

Yang, W., P. Ke, Y. Fang, H. Zheng, and A. Wang. 2013. Microstructure and properties of duplex (Ti:N)-DLC/MAO coating on magnesium alloy. *Applied Surface Science* 270:519–25.

Yang, X. M., M. Li, X. Lin, L. L. Tan, G. B. Lan, L. H. Li, Q. S. Yin, H. Xia, Y. Zhang, and K. Yang. 2013. Enhanced in vitro biocompatibility/bioactivity of biodegradable Mg-Zn-Zr alloy by micro-arc oxidation coating contained Mg2SiO4. *Surface and Coatings Technology* 233:65–73.

Yang, L., X. Zhou, S.-M. Liang, R. Schmid-Fetzer, Z. Fan, G. Scamans, J. Robson, and G. Thompson. 2015. Effect of traces of silicon on the formation of Fe-rich particles in pure magnesium and the corrosion susceptibility of magnesium. *Journal of Alloys and Compounds* 619:396–400.

Yanovska, A., V. Kuznetsov, A. Stanislavov, S. Danilchenko, and L. Sukhodub. 2012. Calcium–phosphate coatings obtained biomimetically on magnesium substrates under low magnetic field. *Applied Surface Science* 258 (22):8577–84.

Yao, H. B., Y. Li, and A. T. S. Wee. 2000. An XPS investigation of the oxidation/corrosion of melt-spun Mg. *Applied Surface Science* 158 (1–2):112–19.

Yao, H. B., Y. Li, and A. T. S. Wee. 2003. Passivity behavior of melt-spun Mg-Y alloys. *Electrochimica Acta* 48 (28):4197–204.

Yao, Z., L. Li, and Z. Jiang. 2009. Adjustment of the ratio of Ca/P in the ceramic coating on Mg alloy by plasma electrolytic oxidation. *Applied Surface Science* 255 (13–14):6724–8.

Yao, C., E. B. Slamovich, J. I. Qazi, H. Rack, and T. J. Webster. 2012. Improved bone cell adhesion on ultrafine grained titanium and Ti-6A1-4V. Paper read at Ceramic Nanomaterials and Nanotechnology III: Proceedings of the 106th Annual Meeting of the American Ceramic Society, Indianapolis, IN 2004, Ceramic Transactions.

Yates, A. A., S. A. Schlicker, and C. W. Suitor. 1998. Dietary reference intakes. *Journal of the American Dietetic Association* 98 (6):699–706.

Ye, J., Y. Li, T. Hamasaki et al. 2008. Inhibitory effect of electrolyzed reduced water on tumor angiogenesis. *Biological and Pharmaceutical Bulletin* 31 (1):19–26.

Ye, X., M. Chen, M. Yang, J. Wei, and D. Liu. 2010. In vitro corrosion resistance and cytocompatibility of nano-hydroxyapatite reinforced Mg–Zn–Zr composites. *Journal of Materials Science: Materials in Medicine* 21 (4):1321–8.

Ye, X., S. Cai, Y. Dou, G. Xu, K. Huang, M. Ren, and X. Wang. 2012. Bioactive glass–ceramic coating for enhancing the in vitro corrosion resistance of biodegradable Mg alloy. *Applied Surface Science* 259:799–805.

Ye, C.-H., T.-F. Xi, Y.-F. Zheng, S.-Q. Wang, and Y.-D. Li. 2013. In vitro corrosion and biocompatibility of phosphating modified WE43 magnesium alloy. *Transactions of Nonferrous Metals Society of China* 23 (4):996–1001.

Yerby, S., C. C. Scott, N. J. Evans, K. L. Messing, and D. R. Carter. 2001. Effect of cutting flute design on cortical bone screw insertion torque and pullout strength. *Journal of Orthopaedic Trauma* 15 (3):216–21.

Yi, S. B., H. G. Brokmeier, R. E. Bolmaro, K. U. Kainer, and T. Lippmann. 2004. In situ measurements of texture variations during a tensile loading of Mg-alloy AM20 using synchrotron X-ray radiation. *Scripta Materialia* 51 (5):455–60.

Yi, J.-L., X.-M. Zhang, M.-A. Chen, R. Gu, and Y.-L. Deng. 2009. Corrosion resistance of cerium conversion film electrodeposited on Mg-Gd-Y-Zr magnesium alloy. *Journal of Central South University of Technology* 16:38–42.

Yi, J.-X., B.-Y. Tang, P. Chen, D.-L. Li, L.-M. Peng, and W.-J. Ding. 2011. Crystal structure of the mirror symmetry 10H-type long-period stacking order phase in Mg–Y–Zn alloy. *Journal of Alloys and Compounds* 509 (3):669–74.

Yin, D. S., E. L. Zhang, and S. Y. Zeng. 2008. Effect of Zn on mechanical property and corrosion property of extruded Mg-Zn-Mn alloy. *Transactions of Nonferrous Metals Society of China* 18 (4):763–8.

Yin, P., N. F. Li, T. Lei, L. Liu, and C. Ouyang. 2013. Effects of Ca on microstructure, mechanical and corrosion properties and biocompatibility of Mg-Zn-Ca alloys. *Journal of Materials Science-Materials in Medicine* 24 (6):1365–73.

Yongdong, X., H. Shengsun, L. Song, C. Donglang, and Z. Xiurong. 2011. Influence of heat treatment on the microstructure and property of Mg-Nd-Gd-Zn-Zr alloy. *Rare Metal Materials and Engineering* 40 (7):1133–7.

Yoon, K. S., X. Z. Huang, Y. S. Yoon et al. 2010. Histological study on the effect of electrolyzed reduced water-bathing on UVB radiation-induced skin injury in hairless mice. *Biological and Pharmaceutical Bulletin* 34 (11):1671–7.

Yoon, Y.-S., D.-H. Kim, S.-K. Kim et al. 2011. The melamine excretion effect of the electrolyzed reduced water in melamine-fed mice. *Food and Chemical Toxicology* 49 (8):1814–19.

You, B. S., W. W. Park, and I. S. Chung. 2000. The effect of calcium additions on the oxidation behavior in magnesium alloys. *Scripta Materialia* 42 (11):1089–94.

Yu, W., Z. Liu, H. He, N. Cheng, and X. Li. 2008. Microstructure and mechanical properties of ZK60–Yb magnesium alloys. *Materials Science and Engineering: A* 478 (1):101–7.

Yu, P., Z. Wang, X. Sun et al. 2011. Hydrogen-rich medium protects human skin fibroblasts from high glucose or mannitol induced oxidative damage. *Biochemical and Biophysical Research Communications* 409 (2):350–5.

Yu, H. J., J. Q. Wang, X. T. Shi, D. V. Louzguine-Luzgin, H. K. Wu, and J. H. Perepezko. 2013. Ductile biodegradable Mg-based metallic glasses with excellent biocompatibility. *Advanced Functional Materials* 23 (38):4793–800.

Yu, H. Y., H. G. Yan, J. H. Chen, B. Su, Y. Zheng, Y. J. Shen, and Z. J. Ma. 2014. Effects of minor Gd addition on microstructures and mechanical properties of the high strain-rate rolled Mg-Zn-Zr alloys. *Journal of Alloys and Compounds* 586:757–65.

Yuan, L., L. Yanxiang, W. Jiang, and Z. Huawei. 2005. Evaluation of porosity in lotus-type porous magnesium fabricated by metal/gas eutectic unidirectional solidification. *Materials Science and Engineering: A* 402 (1):47–54.

Yuan, J. W., K. Zhang, T. Li, X. G. Li, Y. J. Li, M. L. Maa, P. Luo, G. Q. Luo, and Y. H. Hao. 2012. Anisotropy of thermal conductivity and mechanical properties in Mg-5Zn-1Mn alloy. *Materials and Design* 40:257–61.

Yue, T. M., and K. J. Huang. 2011. Laser forming of Zr-based coatings on AZ91D magnesium alloy substrates for wear and corrosion resistance improvement. *Materials Transactions* 52 (4):810–13.

Yun, Y., Z. Dong, N. Lee, Y. Liu, D. Xue, X. Guo, J. Kuhlmann, A. Doepke, H. B. Halsall, and W. Heineman. 2009a. Revolutionizing biodegradable metals. *Materials Today* 12 (10):22–32.

Yun, Y., Z. Dong, D. Yang, M. J. Schulz, V. N. Shanov, S. Yarmolenko, Z. Xu, P. Kumta, and C. Sfeir. 2009b. Biodegradable Mg corrosion and osteoblast cell culture studies. *Materials Science and Engineering: C* 29 (6):1814–21.

Zafari, A., H. M. Ghasemi, and R. Mahmudi. 2014. An investigation on the tribological behavior of AZ91 and AZ91 + 3 wt.% RE magnesium alloys at elevated temperatures. *Materials and Design* 54:544–52.

Zagorchev, L., P. Oses, Z. W. Zhuang et al. 2010. Micro computed tomography for vascular exploration. *Journal of Angiogenesis Research* 2 (7). doi: 10.1186/2040-2384-2-7.

Zahedmanesh, H., and C. Lally. 2009. Determination of the influence of stent strut thickness using the finite element method: Implications for vascular injury and in-stent restenosis. *Medical and Biological Engineering and Computing* 47 (4):385–93.

Zakiyuddin, A., K. Yun, and K. Lee. 2014. Corrosion behavior of as-cast and hot rolled pure magnesium in simulated physiological media. *Metals and Materials International* 20 (6):1163–8.

Zartner, P., R. Cesnjevar, H. Singer, and M. Weyand. 2005. First successful implantation of a biodegradable metal stent into the left pulmonary artery of a preterm baby. *Catheterization and Cardiovascular Interventions* 66 (4):590–4.

Zartner, P., M. Buettner, H. Singer, and M. Sigler. 2007. First biodegradable metal stent in a child with congenital heart disease: Evaluation of macro and histopathology. *Catheterization and Cardiovascular Interventions* 69 (3):443–6.

Zberg, B., E. R. Arata, P. J. Uggowitzer, and J. F. Löffler. 2009a. Tensile properties of glassy MgZnCa wires and reliability analysis using Weibull statistics. *Acta Materialia* 57 (11):3223–31.

Zberg, B., P. J. Uggowitzer, and J. F. Löffler. 2009b. MgZnCa glasses without clinically observable hydrogen evolution for biodegradable implants. *Nature Materials* 8 (11):887–91.

Zeng, X. Q., Y. X. Wang, W. J. Ding, A. A. Luo, and A. K. Sachdev. 2006. Effect of strontium on the microstructure, mechanical properties, and fracture behavior of AZ31 magnesium alloy. *Metallurgical and Materials Transactions A-Physical Metallurgy and Materials Science* 37A (4):1333–41.

Zeng, R. C., W. Dietzel, J. Chen, W. J. Huang, and J. Wang. 2008a. Corrosion behavior of TiO₂ coating on magnesium alloy AM60 in Hank's solution. *Key Engineering Materials* 373–4:609–12.

Zeng, R., W. Dietzel, F. Witte, N. Hort, and C. Blawert. 2008b. Progress and challenge for magnesium alloys as biomaterials. *Advanced Engineering Materials* 10 (8):B3–14.

Zeng, X. Q., D. H. Li, J. Dong, C. Lu, and W. J. Ding. 2008. Effect of solution treatment and extrusion on evolution of microstructure in a Mg–12Dy–3Nd–0.4 Zr alloy. *Journal of Alloys and Compounds* 456 (1):419–24.

Zeng, Y., K. Fan, X. Li, B. Xu, X. Gao, and L. Meng. 2010. First-principles studies of the structures and properties of Al- and Ag-substituted Mg2Ni alloys and their hydrides. *International Journal of Hydrogen Energy* 35 (19):10349–58.

Zeng, R., L. Liu, S. Li, Y. Zou, F. Zhang, Y. Yang, H. Cui, and E.-H. Han. 2013a. Self-assembled silane film and silver nanoparticles coating on magnesium alloys for corrosion resistance and antibacterial applications. *Acta Metallurgica Sinica (English Letters)* 26 (6):681–6.

Zeng, R. C., X. X. Sun, Y. W. Song, F. Zhang, S. Q. Li, H. Z. Cui, and E. H. Han. 2013b. Influence of solution temperature on corrosion resistance of Zn-Ca phosphate conversion coating on biomedical Mg-Li-Ca alloys. *Transactions of Nonferrous Metals Society of China* 23 (11):3293–9.

Zeng, X. S., Y. Liu, Q. Y. Huang, G. Zeng, and G. H. Zhou. 2013. Effects of carbon nanotubes on the microstructure and mechanical properties of the wrought Mg-2.0Zn alloy. *Materials Science and Engineering A-Structural Materials Properties Microstructure and Processing* 571:150–4.

Zeng, R.-C., L. Sun, Y.-F. Zheng, H.-Z. Cui, and E.-H. Han. 2014. Corrosion and characterisation of dual phase Mg–Li–Ca alloy in Hank's solution: The influence of microstructural features. *Corrosion Science* 79:69–82.

Zhai, Z., X. Qu, H. Li et al. 2014. The effect of metallic magnesium degradation products on osteoclast-induced osteolysis and attenuation of NF-kappa B and NFATc1 signaling. *Biomaterials* 35 (24):6299–310.

Zhan, M.-Y., W.-W. Zhang, and D.-T. Zhang. 2011. Production of Mg-Al-Zn magnesium alloy sheets with ultrafine-grain microstructure by accumulative roll-bonding. *Transactions of Nonferrous Metals Society of China* 21 (5):991–7.

Zhang, E., L. Xu, and K. Yang. 2005. Formation by ion plating of Ti-coating on pure Mg for biomedical applications. *Scripta Materialia* 53 (5):523–7.

Zhang, E. L., and L. Yang. 2008. Microstructure, mechanical properties and bio-corrosion properties of Mg-Zn-Mn-Ca alloy for biomedical application. *Materials Science and Engineering A-Structural Materials Properties Microstructure and Processing* 497 (1–2):111–18.

Zhang, H., J. Feng, W. F. Zhu, C. Q. Liu, D. S. Wu, W. J. Yang, and J. H. Gu. 2000. Rare-earth element distribution characteristics of biological chains in rare-earth element-high background regions and their implications. *Biological Trace Element Research* 73 (1):19–27.

Zhang, Q. H., S. H. Tan, and S. M. Chou. 2004. Investigation of fixation screw pull-out strength on human spine. *Journal of Biomechanics* 37:479–85.

Zhang, Q. H., S. H. Tan, and S. M. Chou. 2006. Effects of bone materials on the screw pull-out strength in human spine. *Medical Engineering and Physics* 28 (8):795–801.

Zhang, M.-L., R.-Z. Wu, T. Wang, B. Liu, and Z.-Y. Niu. 2007a. Microstructure and mechanical properties of Mg-xLi-3Al-1Ce alloys. *Transactions of Nonferrous Metals Society of China* 17 (s1A):s381–4.

Zhang, M. L., Y. D. Yan, Z. Y. Hou, L. A. Fan, Z. Chen, and D. X. Tang. 2007b. An electrochemical method for the preparation of Mg–Li alloys at low temperature molten salt system. *Journal of Alloys and Compounds* 440 (1–2):362–6.

Zhang, X. P., Z. P. Zhao, F. M. Wu, Y. L. Wang, and J. Wu. 2007. Corrosion and wear resistance of AZ91D magnesium alloy with and without microarc oxidation coating in Hank's solution. *Journal of Materials Science* 42 (20):8523–8.

Zhang, C., X. Huang, M. Zhang, L. Gao, and R. Wu. 2008. Electrochemical characterization of the corrosion of a Mg–Li Alloy. *Materials Letters* 62 (14):2177–80.

Zhang, D. F., G. L. Shi, Q. W. Dai, W. Yuan, and H. L. Duan. 2008. Microstructures and mechanical properties of high strength Mg-Zn-Mn alloy. *Transactions of Nonferrous Metals Society of China* 18:S59–63.

Zhang, E., W. W. He, H. Du, and K. Yang. 2008. Microstructure, mechanical properties and corrosion properties of Mg-Zn-Y alloys with low Zn content. *Materials Science and Engineering A-Structural Materials Properties Microstructure and Processing* 488 (1–2):102–11.

Zhang, J., J. Wang, X. Qiu, D. Zhang, Z. Tian, X. Niu, D. Tang, and J. Meng. 2008a. Effect of Nd on the microstructure, mechanical properties and corrosion behavior of die-cast Mg–4Al-based alloy. *Journal of Alloys and Compounds* 464 (1–2):556–64.

Zhang, J., D. Zhang, Z. Tian, J. Wang, K. Liu, H. Lu, D. Tang, and J. Meng. 2008b. Microstructures, tensile properties and corrosion behavior of die-cast Mg–4Al-based alloys containing La and/or Ce. *Materials Science and Engineering: A* 489 (1):113–19.

Zhang, K., X.-G. Li, Y.-J. Li, and M.-L. Ma. 2008. Effect of Gd content on microstructure and mechanical properties of Mg-Y-RE-Zr alloys. *Transactions of Nonferrous Metals Society of China* 18:s12–16.

Zhang, E. L., L. P. Xu, G. N. Yu, F. Pan, and K. Yang. 2009a. In vivo evaluation of biodegradable magnesium alloy bone implant in the first 6 months implantation. *Journal of Biomedical Materials Research Part A* 90A (3):882–93.

Zhang, E. L., D. S. Yin, L. P. Xu, L. Yang, and K. Yang. 2009b. Microstructure, mechanical and corrosion properties and biocompatibility of Mg-Zn-Mn alloys for biomedical application. *Materials Science and Engineering C-Biomimetic and Supramolecular Systems* 29 (3):987–93.

Zhang, J. S., J. Yan, W. Liang, E. L. Du, and C. X. Xu. 2009a. Microstructures of Mg–Zn–Nd alloy including small quasicrystalline grains. *Journal of Non-Crystalline Solids* 355 (14–15):836–9.

Zhang, J., K. Liu, D. Fang, X. Qiu, P. Yu, D. Tang, and J. Meng. 2009b. Microstructures, mechanical properties and corrosion behavior of high-pressure die-cast Mg–4Al–0.4Mn–xPr (x = 1, 2, 4, 6) alloys. *Journal of Alloys and Compounds* 480 (2):810–19.

Zhang, M. L., R. Z. Wu, and T. Wang. 2009. Microstructure and mechanical properties of Mg–8Li–(0–3) Ce alloys. *Journal of Materials Science* 44 (5):1237–40.

Zhang, S. X., J. A. Li, Y. Song et al. 2009. In vitro degradation, hemolysis and MC3T3-E1 cell adhesion of biodegradable Mg-Zn alloy. *Materials Science and Engineering C-Materials for Biological Applications* 29 (6):1907–12.

Zhang, Y., G. Zhang, and M. Wei. 2009. Controlling the biodegradation rate of magnesium using biomimetic apatite coating. *Journal of Biomedical Materials Research Part B: Applied Biomaterials* 89 (2):408–14.

Zhang, C. Y., R. C. Zeng, C. L. Liu, and J. C. Gao. 2010. Comparison of calcium phosphate coatings on Mg-Al and Mg-Ca alloys and their corrosion behavior in Hank's solution. *Surface and Coatings Technology* 204 (21–22):3636–40.

Zhang, E. L., L. Yang, J. W. Xu, and H. Y. Chen. 2010. Microstructure, mechanical properties and bio-corrosion properties of Mg-Si(-Ca, Zn) alloy for biomedical application. *Acta Biomaterialia* 6 (5):1756–62.

Zhang, J., M. Zhang, J. Meng, R. Wu, and D. Tang. 2010. Microstructures and mechanical properties of heat-resistant high-pressure die-cast Mg–4Al–xLa–0.3Mn (x = 1, 2, 4, 6) alloys. *Materials Science and Engineering: A* 527 (10):2527–37.

Zhang, M., X. Meng, R. Wu, C. Cui, and L. Wu. 2010. Effect of Ce on microstructures and mechanical properties of as-cast Mg–8Li–1Al alloys. *Kovove Materialy-Metallic Materials* 48:211–16.

Zhang, S., Q. Li, X. Yang, X. Zhong, Y. Dai, and F. Luo. 2010a. Corrosion resistance of AZ91D magnesium alloy with electroless plating pretreatment and Ni–TiO2 composite coating. *Materials Characterization* 61 (3):269–76.

Zhang, S. X., X. N. Zhang, C. L. Zhao et al. 2010b. Research on an Mg-Zn alloy as a degradable biomaterial. *Acta Biomaterialia* 6 (2):626–40.

Zhang, B. P., Y. L. Hou, X. D. Wang, Y. Wang, and L. Geng. 2011a. Mechanical properties, degradation performance and cytotoxicity of Mg-Zn-Ca biomedical alloys with different compositions. *Materials Science and Engineering C-Materials for Biological Applications* 31 (8):1667–73.

Zhang, D. F., G. L. Shi, X. B. Zhao, and F. G. Qi. 2011. Microstructure evolution and mechanical properties of Mg-x%Zn-1%Mn (x = 4, 5, 6, 7, 8, 9) wrought magnesium alloys. *Transactions of Nonferrous Metals Society of China* 21 (1):15–25.

Zhang, J., Z. Leng, S. Liu, M. Zhang, J. Meng, and R. Wu. 2011a. Structure stability and mechanical properties of Mg–Al-based alloy modified with Y-rich and Ce-rich misch metals. *Journal of Alloys and Compounds* 509 (20):L187–93.

Zhang, J., Z. Leng, S. Liu, J. Li, M. Zhang, and R. Wu. 2011b. Microstructure and mechanical properties of Mg–Gd–Dy–Zn alloy with long period stacking ordered structure or stacking faults. *Journal of Alloys and Compounds* 509 (29):7717–22.

Zhang, J., Z. Leng, M. Zhang, J. Meng, and R. Wu. 2011c. Effect of Ce on microstructure, mechanical properties and corrosion behavior of high-pressure die-cast Mg–4Al-based alloy. *Journal of Alloys and Compounds* 509 (3):1069–78.

Zhang, Y., Q. Sun, B. He et al. 2011. Anti-inflammatory effect of hydrogen-rich saline in a rat model of regional myocardial ischemia and reperfusion. *International Journal of Cardiology* 148 (1):91–5.

Zhang, J., C.-S. Dai, J. Wei, and Z.-H. Wen. 2012a. Study on the bonding strength between calcium phosphate/chitosan composite coatings and a Mg alloy substrate. *Applied Surface Science* 261:276–86.

Zhang, J. S., J. D. Xu, W. L. Cheng, and J. J. Kang. 2012b. Corrosion behavior of Mg-Zn-Y alloy with long-period stacking ordered structures. *Journal of Materials Science and Technology* 28 (12):1157–62.

Zhang, W. J., M. H. Li, Q. Chen, W. Y. Hu, W. M. Zhang, and W. Xin. 2012. Effects of Sr and Sn on microstructure and corrosion resistance of Mg-Zr-Ca magnesium alloy for biomedical applications. *Materials and Design* 39:379–83.

Zhang, X., G. Chen, and T. Bauer. 2012a. Mg-based bulk metallic glass composite with high bio-corrosion resistance and excellent mechanical properties. *Intermetallics* 29:56–60.

Zhang, X., G. Yuan, and Z. Wang. 2012b. Mechanical properties and biocorrosion resistance of Mg-Nd-Zn-Zr alloy improved by cyclic extrusion and compression. *Materials Letters* 74:128–31.

Zhang, X., Z. Wang, G. Yuan, and Y. Xue. 2012c. Improvement of mechanical properties and corrosion resistance of biodegradable Mg–Nd–Zn–Zr alloys by double extrusion. *Materials Science and Engineering: B* 177 (13):1113–19.

Zhang, X. B., Y. J. Wu, Y. J. Xue, Z. Z. Wang, and L. Yang. 2012d. Biocorrosion behavior and cytotoxicity of a Mg-Gd-Zn-Zr alloy with long period stacking ordered structure. *Materials Letters* 86:42–5.

Zhang, X. B., G. Y. Yuan, L. Mao, J. L. Niu, and W. J. Ding. 2012e. Biocorrosion properties of as-extruded Mg-Nd-Zn-Zr alloy compared with commercial AZ31 and WE43 alloys. *Materials Letters* 66 (1):209–11.

Zhang, X. B., G. Y. Yuan, J. L. Niu, P. H. Fu, and W. J. Ding. 2012f. Microstructure, mechanical properties, biocorrosion behavior, and cytotoxicity of as-extruded Mg-Nd-Zn-Zr alloy with different extrusion ratios. *Journal of the Mechanical Behavior of Biomedical Materials* 9:153–62.

Zhang, X., G. Yuan, L. Mao, J. Niu, P. Fu, and W. Ding. 2012g. Effects of extrusion and heat treatment on the mechanical properties and biocorrosion behaviors of a Mg–Nd–Zn–Zr alloy. *Journal of the Mechanical Behavior of Biomedical Materials* 7:77–86.

Zhang, X., K. Zhang, X. Deng, L. Hongwei, L. Yongjun, M. Minglong, L. Ning, and Y. Wang. 2012h. Corrosion behavior of Mg–Y alloy in NaCl aqueous solution. *Progress in Natural Science: Materials International* 22 (2):169–74.

Zhang, X., K. Zhang, X. Li, X. Deng, Y. Li, M. Ma, and Y. Shi. 2012i. Effect of solid-solution treatment on corrosion and electrochemical behaviors of Mg-15Y alloy in 3.5 wt.% NaCl solution. *Journal of Rare Earths* 30 (11):1158–67.

Zhang, X. B., Y.-J. Xue, and Z.-Z. Wang. 2012j. Effect of heat treatment on microstructure, mechanical properties and in vitro degradation behavior of as-extruded Mg-2.7Nd-0.2Zn-0.4Zr alloy. *Transactions of Nonferrous Metals Society of China* 22 (10):2343–50.

Zhang, X., Q. Li, L. Li, P. Zhang, Z. Wang, and F. Chen. 2012k. Fabrication of hydroxyapatite/stearic acid composite coating and corrosion behavior of coated magnesium alloy. *Materials Letters* 88:76–8.

Zhang, Y., K. Bai, Z. Fu, C. Zhang, H. Zhou, L. Wang, S. Zhu, S. Guan, D. Li, and J. Hu. 2012a. Composite coating prepared by micro-arc oxidation followed by sol-gel process and in vitro degradation properties. *Applied Surface Science* 258 (7):2939–43.

Zhang, Y., L. Ren, M. Li, X. Lin, H. Zhao, and K. Yang. 2012b. Preliminary study on cytotoxic effect of biodegradation of magnesium on cancer cells. *Journal of Materials Science and Technology* 28 (9):769–72.

Zhang, H., G. Huang, L. Wang, H. J. Roven, and F. Pan. 2013a. Enhanced mechanical properties of AZ31 magnesium alloy sheets processed by three-directional rolling. *Journal of Alloys and Compounds* 575:408–13.

Zhang, H. J., D. F. Zhang, C. H. Ma, and S. F. Guo. 2013b. Improving mechanical properties and corrosion resistance of Mg-6Zn-Mn magnesium alloy by rapid solidification. *Materials Letters* 92:45–8.

Zhang, J., C. Dai, J. Wei, Z. Wen, S. Zhang, and C. Chen. 2013a. Degradable behavior and bioactivity of micro-arc oxidized AZ91D Mg alloy with calcium phosphate/chitosan composite coating in m-SBF. *Colloids Surf B Biointerfaces* 111C:179–87.

Zhang, J. S., C. Xin, W. L. Cheng, L. P. Bian, H. X. Wang, and C. X. Xu. 2013b. Research on long-period-stacking-ordered phase in Mg-Zn-Dy-Zr alloy. *Journal of Alloys and Compounds* 558:195–202.

Zhang, K., X. Zhang, X. Deng, X. Li, and M. Ma. 2013. Relationship between extrusion, Y and corrosion behavior of Mg–Y alloy in NaCl aqueous solution. *Journal of Magnesium and Alloys* 1 (2):134–8.

Zhang, M., Y. Yang, W. Han, M. Li, K. Ye, Y. Sun, and Y. Yan. 2013. Electrodeposition of magnesium–lithium–dysprosium ternary alloys with controlled components from dysprosium oxide assisted by magnesium chloride in molten chlorides. *Journal of Solid State Electrochemistry* 17 (10):2671–8.

Zhang, X. B., Z. X. Ba, Z. Z. Wang, X. C. He, C. Shen, and Q. Wang. 2013a. Influence of silver addition on microstructure and corrosion behavior of Mg-Nd-Zn-Zr alloys for biomedical application. *Materials Letters* 100:188–91.

Zhang, X. B., X. C. He, Y. J. Xue, Z. Z. Wang, and Q. Wang. 2013b. Effects of Sr on microstructure and corrosion resistance in simulated body fluid of as cast Mg-Nd-Zr magnesium alloys. *Corrosion Engineering, Science and Technology* 49 (5):345–51.

Zhang, X. B., G. Y. Yuan, X. X. Fang, Z. Z. Wang, and T. Zhang. 2013c. Effects of solution treatment on yield ratio and biocorrosion behaviour of as-extruded Mg-2·7Nd–0·2Zn–0·4Zr alloy for cardiovascular stent application. *Materials Science and Technology* 28 (3):155–8.

Zhang, X. B., G. Y. Yuan, and Z. Z. Wang. 2013d. Effects of extrusion ratio on microstructure, mechanical and corrosion properties of biodegradable Mg-Nd-Zn-Zr alloy. *Materials Science and Technology* 29 (1):111–16.

Zhang, X., X. Li, J. Li, and X. Sun. 2013e. Processing, microstructure and mechanical properties of biomedical magnesium with a specific two-layer structure. *Progress in Natural Science: Materials International* 23 (2):183–9.

Zhang, X., X.-W. Li, J.-G. Li, and X.-D. Sun. 2013f. Preparation and characterizations of bioglass ceramic cement/Ca–P coating on pure magnesium for biomedical applications. *ACS Applied Materials and Interfaces* 6 (1):513–25.

Zhang, L., J. H. Zhang, Z. Leng, S. J. Liu, Q. Yang, R. Z. Wu, and M. L. Zhang. 2014. Microstructure and mechanical properties of high-performance Mg-Y-Er-Zn extruded alloy. *Materials and Design* 54:256–63.

Zhang, X. B., X. C. He, Y. J. Xue, Z. Z. Wang, and Q. Wang. 2014. Microstructure and corrosion resistance of as cast Mg-Nd-Gd-Sr-Zn-Zr alloys for biomedical applications. *Materials Technology: Advanced Performance Materials* 29 (3):179–87.

Zhao, D. W., and S. B. Huang. 2013. Application of biodegradable magnesium screw in the conservative treatment in patients with avascular necrosis of the femoral head. Paper presented at Annual conference of Chinese Society for Biomaterials, Dec. 21–23, at Shenzhen, China.

Zhao, X., K. Zhang, X. Li, Y. Li, Q. He, and J. Sun. 2008. Deformation behavior and dynamic recrystallization of Mg-Y-Nd-Gd-Zr alloy. *Journal of Rare Earths* 26 (6):846–50.

Zhao, Y.-Y., E. Ma, and J. Xu. 2008. Reliability of compressive fracture strength of Mg–Zn–Ca bulk metallic glasses: Flaw sensitivity and Weibull statistics. *Scripta Materialia* 58 (6):496–9.

Zhao, L., C. Cui, Q. Wang, and S. Bu. 2010. Growth characteristics and corrosion resistance of micro-arc oxidation coating on pure magnesium for biomedical applications. *Corrosion Science* 52 (7):2228–34.

Zhao, L., C. Zhou, J. Zhang et al. 2011. Hydrogen protects mice from radiation induced thymic lymphoma in BALB/c mice. *International Journal of Biological Sciences* 7 (3):297–300.

Zhao, Y., G. Wu, H. Pan, K. W. K. Yeung, and P. K. Chu. 2012. Formation and electrochemical behavior of Al and O plasma-implanted biodegradable Mg-Y-RE alloy. *Materials Chemistry and Physics* 132 (1):187–91.

Zhao, Q., X. Guo, X. Dang, J. Hao, J. Lai, and K. Wang. 2013. Preparation and properties of composite MAO/ECD coatings on magnesium alloy. *Colloids Surf B Biointerfaces* 102:321–6.

Zhao, X., L.-L. Shi, and J. Xu. 2013a. Biodegradable Mg-Zn–Y alloys with long-period stacking ordered structure: Optimization for mechanical properties. *Journal of the Mechanical Behavior of Biomedical Materials* 18:181–90.

Zhao, X., L.-L. Shi, and J. Xu. 2013b. Mg-Zn–Y alloys with long-period stacking ordered structure: In vitro assessments of biodegradation behavior. *Materials Science and Engineering: C* 33 (7):3627–37.

Zhao, Y., G. Wu, Q. Lu, J. Wu, R. Xu, K. W. K. Yeung, and P. K. Chu. 2013. Improved surface corrosion resistance of WE43 magnesium alloy by dual titanium and oxygen ion implantation. *Thin Solid Films* 529:407–11.

Zhao, N., B. Workman, and D. Zhu. 2014. Endothelialization of novel magnesium-rare earth alloys with fluoride and collagen coating. *International Journal of Molecular Sciences* 15 (4):5263–76.

Zhao, Y., M. I. Jamesh, W. K. Li, G. Wu, C. Wang, Y. Zheng, K. W. Yeung, and P. K. Chu. 2014a. Enhanced antimicrobial properties, cytocompatibility, and corrosion resistance of plasma-modified biodegradable magnesium alloys. *Acta Biomaterialia* 10 (1):544–56.

Zhao, Y. F., J. J. Si, J. G. Song, and X. D. Hui. 2014b. High strength Mg-Zn-Ca alloys prepared by atomization and hot pressing process. *Materials Letters* 118:55–8.

Zhen, R., Y. S. Sun, F. Xue, J. J. Sun, and J. Bai. 2013. Effect of heat treatment on the microstructures and mechanical properties of the extruded Mg-11Gd-1Zn alloy. *Journal of Alloys and Compounds* 550:273–8.

Zhen, Z., T.-F. Xi, and Y.-F. Zheng. 2013. A review on in vitro corrosion performance test of biodegradable metallic materials. *Transactions of Nonferrous Metals Society of China* 23 (8):2283–93.

Zheng, Q. 2012. Mg-Cu(Ag)-Gd bulk metallic glass forming alloy: Phase selection of melt crystallization. *Science of Advanced Materials* 4 (9):969–77.

Zheng, X., X. Zheng, Y. Mao et al. 2009. Hydrogen-rich saline protects against intestinal ischemia/reperfusion injury in rats. *Free Radical Research* 43 (5):478–84.

Zheng, J., Q. Wang, Z. Jin, and T. Peng. 2010a. Effect of Sm on the microstructure, mechanical properties and creep behavior of Mg–0.5 Zn–0.4 Zr based alloys. *Materials Science and Engineering: A* 527 (7):1677–85.

Zheng, J., Q. Wang, Z. Jin, and T. Peng. 2010b. The microstructure, mechanical properties and creep behavior of Mg–3Sm–0.5 Zn–0.4 Zr (wt.%) alloy produced by different casting technologies. *Journal of Alloys and Compounds* 496 (1):351–6.

Zheng, J., K. Liu, Z. Kang et al. 2010c. Saturated hydrogen saline protects the lung against oxygen toxicity. *Undersea and Hyperbaric Medical Society* 37 (3):185–92.

Zheng, X., J. Dong, D. Yin, W. Liu, F. Wang, L. Jin, and W. Ding. 2010. Forgeability and die-forging forming of direct chill casting Mg–Nd–Zn–Zr magnesium alloy. *Materials Science and Engineering: A* 527 (16–17):3690–4.

Zheng, Y. F., X. N. Gu, Y. L. Xi, and D. L. Chai. 2010. In vitro degradation and cytotoxicity of Mg/Ca composites produced by powder metallurgy. *Acta Biomaterialia* 6 (5):1783–91.

Zheng, F. Y., Y. J. Wu, L. M. Peng, X. W. Li, P. H. Fu, and W. J. Ding. 2013. Microstructures and mechanical properties of friction stir processed Mg–2.0Nd–0.3Zn–1.0Zr magnesium alloy. *Journal of Magnesium and Alloys* 1 (2):122–7.

Zheng, Y. F., X. N. Gu, and F. Witte. 2014. Biodegradable metals. *Materials Science and Engineering Reports* 77:1–34.

Zhong, H., L. Feng, P. Liu, and T. Zhou. 2005. Design of a Mg-Li-Al-Zn alloy by means of CALPHAD approach. *Journal of Computer-Aided Materials Design* 10 (3):191–9.

Zhong, X., Q. Li, J. Hu, S. Zhang, B. Chen, S. Xu, and F. Luo. 2010. A novel approach to heal the sol-gel coating system on magnesium alloy for corrosion protection. *Electrochimica Acta* 55 (7):2424–9.

Zhou, H., and J. Lee. 2011. Nanoscale hydroxyapatite particles for bone tissue engineering. *Acta Biomaterialia* 7 (7):2769–81.

Zhou, B., T. J. Kane, G. J. Dixon, and R. L. Byer. 1985. Efficient, frequency-stable laser-diode-pumped Nd: YAG laser. *Optics Letters* 10 (2):62–4.

Zhou, X., Y. Huang, Z. Wei, Q. Chen, and F. Gan. 2006. Improvement of corrosion resistance of AZ91D magnesium alloy by holmium addition. *Corrosion Science* 48 (12):4223–33.

Zhou, Y. L., D. M. Luo, W. Y. Hu, Y. C. Li, P. D. Hodgson, and C. E. Wen. 2011. Compressive properties of hot-rolled Mg-Zr-Ca alloys for biomedical applications. *Advanced Materials Research* 197:56–9.

Zhou, H., Q.-D. Wang, J. Chen, B. Ye, and W. Guo. 2012. Microstructure and mechanical properties of extruded Mg–8.5Gd–2.3Y–1.8Ag–0.4Zr alloy. *Transactions of Nonferrous Metals Society of China* 22 (8):1891–5.

Zhou, Y. L., J. An, D. M. Luo, W. Y. Hu, Y. C. Li, P. Hodgson, and C. E. Wen. 2012. Microstructures and mechanical properties of as cast Mg-Zr-Ca alloys for biomedical applications. *Materials Technology* 27 (1):52–4.

Zhou, N., Z. Zhang, J. Dong, L. Jin, and W. Ding. 2013a. High ductility of a Mg–Y–Ca alloy via extrusion. *Materials Science and Engineering: A* 560:103–10.

Zhou, N., Z. Zhang, J. Dong, L. Jin, and W. Ding. 2013b. Selective oxidation behavior of an ignition-proof Mg-Y-Ca-Ce alloy. *Journal of Rare Earths* 31 (10):1003–8.

Zhou, W. R., Y. F. Zheng, M. A. Leeflang, and J. Zhou. 2013. Mechanical property, biocorrosion and in vitro biocompatibility evaluations of Mg-Li-(Al)-(RE) alloys for future cardiovascular stent application. *Acta Biomaterialia* 9 (10):8488–98.

Zhou, Y. L., Y. C. Li, D. M. Luo, C. E. Wen, and P. Hodgson. 2013. Microstructures, mechanical properties and in vitro corrosion behaviour of biodegradable Mg-Zr-Ca alloys. *Journal of Materials Science* 48 (4):1632–9.

Zhou, N., Z. Zhang, L. Jin, J. Dong, B. Chen, and W. Ding. 2014a. Ductility improvement by twinning and twin–slip interaction in a Mg-Y alloy. *Materials and Design* 56:966–74.

Zhou, X., L. Jiang, P. Wu, Y. Sun, Y. Yu, G. Wei, and H. Ge. 2014b. Effect of aggressive ions on degradation of WE43 magnesium alloy in physiological environment. *International Journal of Electrochemical Science* 9:304–14.

Zhu, Y. M., A. J. Morton, and J. F. Nie. 2008. Improvement in the age-hardening response of Mg–Y–Zn alloys by Ag additions. *Scripta Materialia* 58 (7):525–8.

Zhu, Y., G. Wu, Q. Zhao, Y.-H. Zhang, G. Xing, and D. Li. 2009. Anticorrosive magnesium hydroxide coating on AZ31 magnesium alloy by hydrothermal method. *Journal of Physics: Conference Series* 188:012044.

Zhu, Y. M., A. J. Morton, and J. F. Nie. 2010. Characterisation of intermetallic phases in an Mg–Y–Ag–Zn casting alloy. *Philosophical Magazine Letters* 90 (3):173–81.

Zhu, W.-J., M. Nakayama, T. Mori et al. 2011. Intake of water with high levels of dissolved hydrogen (H2) suppresses ischemia-induced cardio-renal injury in Dahl salt-sensitive rats. *Nephrology Dialysis Transplantation* 26 (7):2112–18.

Zhu, Y., G. Wu, Y.-H. Zhang, and Q. Zhao. 2011. Growth and characterization of Mg(OH)2 film on magnesium alloy AZ31. *Applied Surface Science* 257 (14):6129–37.

Zhu, Y. M., A. J. Morton, and J.-F. Nie. 2012a. Growth and transformation mechanisms of 18R and 14H in Mg–Y–Zn alloys. *Acta Materialia* 60 (19):6562–72.

Zhu, Y., Q. Zhao, Y.-H. Zhang, and G. Wu. 2012b. Hydrothermal synthesis of protective coating on magnesium alloy using de-ionized water. *Surface and Coatings Technology* 206 (11–12):2961–6.

Zhuang, H., Y. Han, and A. Feng. 2008. Preparation, mechanical properties and in vitro biodegradation of porous magnesium scaffolds. *Materials Science and Engineering: C* 28 (8):1462–6.

Zomorodian, A., F. Brusciotti, A. Fernandes, M. J. Carmezim, T. Moura e Silva, J. C. S. Fernandes, and M. F. Montemor. 2012. Anti-corrosion performance of a new silane coating for corrosion protection of AZ31 magnesium alloy in Hank's solution. *Surface and Coatings Technology* 206 (21):4368–75.

Zong, Y., G. Yuan, X. Zhang, L. Mao, J. Niu, and W. Ding. 2012. Comparison of biodegradable behaviors of AZ31 and Mg–Nd–Zn–Zr alloys in Hank's physiological solution. *Materials Science and Engineering: B* 177 (5):395–401.

Zou, Y. S., Y. F. Wu, H. Yang et al. 2011. The microstructure, mechanical and friction properties of protective diamond like carbon films on magnesium alloy. *Applied Surface Science* 258 (4):1624–9.

Zreiqat, H., C. R. Howlett, A. Zannettino, P. Evans, G. Schulze-Tanzil, C. Knabe, and M. Shakibaei. 2002. Mechanisms of magnesium-stimulated adhesion of osteoblastic cells to commonly used orthopaedic implants. *Journal of Biomedical Materials Research* 62 (2):175–84.

Zuberova, Z., L. Kunz, T. Lamark, Y. Estrin, and M. Janeček. 2007. Fatigue and tensile behavior of cast, hot-rolled, and severely plastically deformed AZ31 magnesium alloy. *Metallurgical and Materials Transactions A* 38 (9):1934–40.

Zucchi, F., V. Grassi, A. Frignani, C. Monticelli, and G. Trabanelli. 2006. Influence of a silane treatment on the corrosion resistance of a WE43 magnesium alloy. *Surface and Coatings Technology* 200 (12–13):4136–43.

Zucchi, F., A. Frignani, V. Grassi, A. Balbo, and G. Trabanelli. 2008. Organo-silane coatings for AZ31 magnesium alloy corrosion protection. *Materials Chemistry and Physics* 110 (2–3):263–8.

Zuleta, A. A., E. Correa, C. Villada, M. Sepúlveda, J. G. Castaño, and F. Echeverría. 2011. Comparative study of different environmentally friendly (Chromium-free) methods for surface modification of pure magnesium. *Surface and Coatings Technology* 205 (23):5254–9.

Index

Page numbers followed f and t indicate figures and tables, respectively.